ANNUAL REVIEW OF
ECOLOGY, EVOLUTION, AND SYSTEMATICS

EDITORIAL COMMITTEE (2004)

MICHAEL J. FOOTE
DOUGLAS J. FUTUYMA
C. DREW HARVELL
LARS O. HEDIN
MICHAEL J. RYAN
DOUGLAS W. SCHEMSKE
H. BRADLEY SHAFFER
DANIEL SIMBERLOFF

RESPONSIBLE FOR THE ORGANIZATION OF VOLUME 35 (EDITORIAL COMMITTEE, 2002)

JAMES CLARK
DOUGLAS J. FUTUYMA
C. DREW HARVELL
MICHAEL J. RYAN
DOUGLAS W. SCHEMSKE
H. BRADLEY SHAFFER
DANIEL SIMBERLOFF
SCOTT WING
DAN PAPAJ (GUEST)
J. BRUCE WALSH (GUEST)

Production Editor: ERIN WAIT
Bibliographic Quality Control: MARY A. GLASS
Electronic Content Coordinator: SUZANNE K. MOSES
Subject Indexer: CHERI WALSH

ANNUAL REVIEW OF ECOLOGY, EVOLUTION, AND SYSTEMATICS

VOLUME 35, 2004

DOUGLAS J. FUTUYMA, *Editor*
State University of New York, Stony Brook

H. BRADLEY SHAFFER, *Associate Editor*
University of California, Davis

DANIEL SIMBERLOFF, *Associate Editor*
University of Tennessee

www.annualreviews.org   science@annualreviews.org   650-493-4400

ANNUAL REVIEWS
4139 El Camino Way • P.O. Box 10139 • Palo Alto, California 94303-0139

ANNUAL REVIEWS
Palo Alto, California, USA

COPYRIGHT © 2004 BY ANNUAL REVIEWS, PALO ALTO, CALIFORNIA, USA. ALL RIGHTS RESERVED. The appearance of the code at the bottom of the first page of an article in this serial indicates the copyright owner's consent that copies of the article may be made for personal or internal use, or for the personal or internal use of specific clients. This consent is given on the condition that the copier pay the stated per-copy fee of $14.00 per article through the Copyright Clearance Center, Inc. (222 Rosewood Drive, Danvers, MA 01923) for copying beyond that permitted by Section 107 or 108 of the US Copyright Law. The per-copy fee of $14.00 per article also applies to the copying, under the stated conditions, of articles published in any *Annual Review* serial before January 1, 1978. Individual readers, and nonprofit libraries acting for them, are permitted to make a single copy of an article without charge for use in research or teaching. This consent does not extend to other kinds of copying, such as copying for general distribution, for advertising or promotional purposes, for creating new collective works, or for resale. For such uses, written permission is required. Write to Permissions Dept., Annual Reviews, 4139 El Camino Way, P.O. Box 10139, Palo Alto, CA 94303-0139 USA.

International Standard Serial Number: 1543-592X
International Standard Book Number: 0-8243-1435-2
Library of Congress Catalog Card Number: 71-135616

All Annual Reviews and publication titles are registered trademarks of Annual Reviews.

∞ The paper used in this publication meets the minimum requirements of American National Standards for Information Sciences—Permanence of Paper for Printed Library Materials. ANSI Z39.48-1992.

Annual Reviews and the Editors of its publications assume no responsibility for the statements expressed by the contributors to this *Annual Review*.

TYPESET BY TECHBOOKS, FAIRFAX, VA
PRINTED AND BOUND BY MALLOY INCORPORATED, ANN ARBOR, MI

Annual Review of Ecology, Evolution, and Systematics
Volume 35, 2004

Contents

VERTEBRATE DISPERSAL OF SEED PLANTS THROUGH TIME, *Bruce H. Tiffney*	1
ARE DISEASES INCREASING IN THE OCEAN? *Kevin D. Lafferty, James W. Porter, and Susan E. Ford*	31
BIRD SONG: THE INTERFACE OF EVOLUTION AND MECHANISM, *Jeffrey Podos, Sarah K. Huber, and Benjamin Taft*	55
APPLICATION OF ECOLOGICAL INDICATORS, *Gerald J. Niemi and Michael E. McDonald*	89
ECOLOGICAL IMPACTS OF DEER OVERABUNDANCE, *Steeve D. Côté, Thomas P. Rooney, Jean-Pierre Tremblay, Christian Dussault, and Donald M. Waller*	113
ECOLOGICAL EFFECTS OF TRANSGENIC CROPS AND THE ESCAPE OF TRANSGENES INTO WILD POPULATIONS, *Diana Pilson and Holly R. Prendeville*	149
MUTUALISMS AND AQUATIC COMMUNITY STRUCTURE: THE ENEMY OF MY ENEMY IS MY FRIEND, *Mark E. Hay, John D. Parker, Deron E. Burkepile, Christopher C. Caudill, Alan E. Wilson, Zachary P. Hallinan, and Alexander D. Chequer*	175
OPERATIONAL CRITERIA FOR DELIMITING SPECIES, *Jack W. Sites, Jr., and Jonathon C. Marshall*	199
THE NEW VIEW OF ANIMAL PHYLOGENY, *Kenneth M. Halanych*	229
LANDSCAPES AND RIVERSCAPES: THE INFLUENCE OF LAND USE ON STREAM ECOSYSTEMS, *J. David Allan*	257
LONG-TERM STASIS IN ECOLOGICAL ASSEMBLAGES: EVIDENCE FROM THE FOSSIL RECORD, *W.A. DiMichele, A.K. Behrensmeyer, T.D. Olszewski, C.C. Labandeira, J.M. Pandolfi, S.L. Wing, and R. Bobe*	285
AVIAN EXTINCTIONS FROM TROPICAL AND SUBTROPICAL FORESTS, *Navjot S. Sodhi, L.H. Liow, and F.A. Bazzaz*	323
EVOLUTIONARY BIOLOGY OF ANIMAL COGNITION, *Reuven Dukas*	347
POLLINATION SYNDROMES AND FLORAL SPECIALIZATION, *Charles B. Fenster, W. Scott Armbruster, Paul Wilson, Michele R. Dudash, and James D. Thomson*	375

ON THE ECOLOGICAL ROLES OF SALAMANDERS, *Robert D. Davic and Hartwell H. Welsh, Jr.*   405

ECOLOGICAL AND EVOLUTIONARY CONSEQUENCES OF MULTISPECIES PLANT-ANIMAL INTERACTIONS, *Sharon Y. Strauss and Rebecca E. Irwin*   435

SPATIAL SYNCHRONY IN POPULATION DYNAMICS, *Andrew Liebhold, Walter D. Koenig, and Ottar N. Bjørnstad*   467

ECOLOGICAL RESPONSES TO HABITAT EDGES: MECHANISMS, MODELS, AND VARIABILITY EXPLAINED, *Leslie Ries, Robert J. Fletcher, Jr., James Battin, and Thomas D. Sisk*   491

EVOLUTIONARY TRAJECTORIES AND BIOGEOCHEMICAL IMPACTS OF MARINE EUKARYOTIC PHYTOPLANKTON, *Miriam E. Katz, Zoe V. Finkel, Daniel Grzebyk, Andrew H. Knoll, and Paul G. Falkowski*   523

REGIME SHIFTS, RESILIENCE, AND BIODIVERSITY IN ECOSYSTEM MANAGEMENT, *Carl Folke, Steve Carpenter, Brian Walker, Marten Scheffer, Thomas Elmqvist, Lance Gunderson, and C.S. Holling*   557

ECOLOGY OF WOODLAND HERBS IN TEMPERATE DECIDUOUS FORESTS, *Dennis F. Whigham*   583

THE SOUTHWEST AUSTRALIAN FLORISTIC REGION: EVOLUTION AND CONSERVATION OF A GLOBAL HOT SPOT OF BIODIVERSITY, *Stephen D. Hopper and Paul Gioia*   623

PREDATOR-INDUCED PHENOTYPIC PLASTICITY IN ORGANISMS WITH COMPLEX LIFE HISTORIES, *Michael F. Benard*   651

THE EVOLUTIONARY ECOLOGY OF NOVEL PLANT-PATHOGEN INTERACTIONS, *Ingrid M. Parker and Gregory S. Gilbert*   675

INDEXES
  Subject Index   701
  Cumulative Index of Contributing Authors, Volumes 31–35   721
  Cumulative Index of Chapter Titles, Volumes 31–35   724

ERRATA
  An online log of corrections to *Annual Review of Ecology, Evolution, and Systematics* chapters may be found at http://ecolsys.annualreviews.org/errata.shtml

Related Articles

From the *Annual Review of Earth and Planetary Sciences*, Volume 32 (2004)

Computer Models of Early Land Plant Evolution, Karl J. Niklas

Modern Analogs in Quaternary Paleoecology: Here Today, Gone Yesterday, Gone Tomorrow? Stephen T. Jackson and John W. Williams

Genes, Diversity, and Geologic Process on the Pacific Coast, David K. Jacobs, Todd A. Haney, and Kristina D. Louie

From the *Annual Review of Entomology*, Volume 49 (2004)

Functional Ecology of Immature Parasitoids, Jacques Brodeur and Guy Boivin

Galling Aphids: Specialization, Biological Complexity, and Variation, David Wool

Population Genetics of Autocidal Control and Strain Replacement, Fred Gould and Paul Schliekelman

Phylogeny and Biology of Neotropical Orchid Bees (Euglossini), Sydney A. Cameron

Plant-Insect Interactions in Fragmented Landscapes, Teja Tscharntke and Roland Brandl

From the *Annual Review of Environment and Resources*, Volume 29 (2004)

Marine Reserves and Ocean Neighborhoods: The Spatial Scale of Marine Populations and Their Management, Stephen R. Palumbi

Plant Genetic Resources for Food and Agriculture: Assessing Global Availability, Cary Fowler and Toby Hodgkin

Grazing Systems, Ecosystem Responses, and Global Change, Gregory P. Asner, Andrew J. Elmore, Lydia P. Olander, Roberta E. Martin, and A. Thomas Harris

From the *Annual Review of Microbiology*, Volume 58 (2004)

Selection for Gene Clustering by Tandem Duplication, Andrew B. Reams and Ellen L. Neidle

The Ecology and Genetics of Microbial Diversity, Rees Kassen and Paul B. Rainey

Endangered Antarctic Environments, Don A. Cowan and Lemese Ah Tow

From the *Annual Review of Phytopathology*, Volume 42 (2004)

Evolution of Plant Parasitism Among Nematodes, J.G. Baldwin, S.A. Nadler, and B.J. Adams

Comparative Genomics Analyses of Citrus-Associated Bacteria, Leandro M. Moreira, Robson F. de Souza, Nalvo F. Almeida, Jr., João C. Setubal, Julio Cezar F. Oliveira, Luiz R. Furlan, Jesus A. Ferro, and Ana C.R. da Silva

Microbial Diversity in Soil: Selection of Microbial Populations by Plant and Soil Type and Implications for Disease Suppressiveness, P. Garbeva, J.A. van Veen, and J.D. van Elsas

Chemical Biology of Multi-Host/Pathogen Interactions: Chemical Perception and Metabolic Complementation, Andrew G. Palmer, Rong Gao, Justin Maresh, W. Kaya Erbil, and David G. Lynn

ANNUAL REVIEWS is a nonprofit scientific publisher established to promote the advancement of the sciences. Beginning in 1932 with the *Annual Review of Biochemistry*, the Company has pursued as its principal function the publication of high-quality, reasonably priced *Annual Review* volumes. The volumes are organized by Editors and Editorial Committees who invite qualified authors to contribute critical articles reviewing significant developments within each major discipline. The Editor-in-Chief invites those interested in serving as future Editorial Committee members to communicate directly with him. Annual Reviews is administered by a Board of Directors, whose members serve without compensation.

2004 Board of Directors, Annual Reviews

Richard N. Zare, *Chairman of Annual Reviews*
  Marguerite Blake Wilbur Professor of Chemistry, Stanford University
John I. Brauman, *J.G. Jackson–C.J. Wood Professor of Chemistry, Stanford University*
Peter F. Carpenter, *Founder, Mission and Values Institute, Atherton, California*
Sandra M. Faber, *Professor of Astronomy and Astronomer at Lick Observatory,*
  University of California at Santa Cruz
Susan T. Fiske, *Professor of Psychology, Princeton University*
Eugene Garfield, *Publisher*, The Scientist
Samuel Gubins, *President and Editor-in-Chief, Annual Reviews*
Steven E. Hyman, *Provost, Harvard University*
Daniel E. Koshland Jr., *Professor of Biochemistry, University of California at Berkeley*
Joshua Lederberg, *University Professor, The Rockefeller University*
Sharon R. Long, *Professor of Biological Sciences, Stanford University*
J. Boyce Nute, *Palo Alto, California*
Michael E. Peskin, *Professor of Theoretical Physics, Stanford Linear Accelerator Center*
Harriet A. Zuckerman, *Vice President, The Andrew W. Mellon Foundation*

Management of Annual Reviews

Samuel Gubins, President and Editor-in-Chief
Richard L. Burke, Director for Production
Paul J. Calvi Jr., Director of Information Technology
Steven J. Castro, Chief Financial Officer and Director of Marketing & Sales

Annual Reviews of

Anthropology	Environment and Resources	Physical Chemistry
Astronomy and Astrophysics	Fluid Mechanics	Physiology
Biochemistry	Genetics	Phytopathology
Biomedical Engineering	Genomics and Human Genetics	Plant Biology
Biophysics and Biomolecular Structure	Immunology	Political Science
	Law and Social Science	Psychology
Cell and Developmental Biology	Materials Research	Public Health
	Medicine	Sociology
Clinical Psychology	Microbiology	
Earth and Planetary Sciences	Neuroscience	
Ecology, Evolution, and Systematics	Nuclear and Particle Science	SPECIAL PUBLICATIONS
	Nutrition	Excitement and Fascination of Science, Vols. 1, 2, 3, and 4
Entomology	Pharmacology and Toxicology	

VERTEBRATE DISPERSAL OF SEED PLANTS THROUGH TIME

Bruce H. Tiffney
Department of Geological Sciences, University of California, Santa Barbara, California 93106; email: tiffney@geol.ucsb.edu

Key Words angiosperm, gymnosperm, fossil seed, fossil fruit, evolution, coevolution

■ **Abstract** Vertebrate dispersal of fruits and seeds is a common feature of many modern angiosperms and gymnosperms, yet the evolution and frequency of this feature in the fossil record remain unclear. Increasingly complex information suggests that (*a*) plants had the necessary morphological features for vertebrate dispersal by the Pennsylvanian, but possibly in the absence of clear vertebrate dispersal agents; (*b*) vertebrate herbivores first diversified in the Permian, and consistent dispersal relationships became possible; (*c*) the Mesozoic was dominated by large herbivorous dinosaurs, possible sources of diffuse, whole-plant dispersal; (*d*) simultaneously, several groups of small vertebrates, including lizards and, in the later Mesozoic, birds and mammals, could have established more specific vertebrate-plant associations, but supporting evidence is rudimentary; and (*e*) the diversification of small mammals and birds in the Tertiary established a consistent basis for organ-level interactions, allowing for the widespread occurrence of biotic dispersal in gymnosperms and angiosperms.

INTRODUCTION

The dispersal of seeds and fruits by vertebrates (Corlett 1998, Wenny 2001, Clark et al. 2001) and invertebrates (Beattie 1985, Handel & Beattie 1990) is central to individual species biology and modern ecosystem function (e.g., Howe & Smallwood 1982, Murray 1986, Sallabanks & Courtney 1992, Cain et al. 2000, Nathan & Muller-Landau 2000, Herrera 2002, Levey et al. 2002, Wang & Smith 2002). Dispersal allows escape from predators, location of favorable growth sites, and reduction of parent-offspring and sibling competition. It further influences the structure of plant communities and the distribution of individual taxa. Thirty percent of angiosperm families are biotically dispersed, and another 22% possess biotic and abiotic dispersal; 14% of species are biotically dispersed, and 42% possess both biotic and abiotic dispersal (Tiffney & Mazer 1995). This factor has been implicated in the diversification of individual clades (Eriksson & Bremer 1991, Charlesdominique 1993, Tiffney & Mazer 1995, Smith 2001), although other factors also play an important role (see Midgley & Bond 1991, Dodd et al. 1999, Magallon & Sanderson 2001). Biotic dispersal is also present in 64% of gymnosperm families (Herrera 1989a) and 46% of gymnosperm species [another 39% of species use both biotic and abiotic dispersal (B. Tiffney, unpublished data)]. Biotic dispersal also has an important effect on individual gymnosperm species success (e.g., Tomback & Linhart 1990, Vander Wall 1992, Willson et al. 1996).

Extant dispersal commonly involves endozoochory—the consumption of the disseminule, including its passage through the disperser's gut. In a variant, the disperser may consume external flesh without eating the contained disseminule—for example, monkeys consuming large-sized fruits (Lambert & Garber 1998) or ants dispersing small seeds bearing external attractant bodies (Beattie 1985, Beattie & Hughes 2002). If the flesh is the reward, the embryo and nutritive tissue are generally not damaged by the disperser. Active seed predation may also accomplish dispersal but with substantial disseminule mortality. Less frequently, biotic dispersal involves exozoochory, in which the disseminule attaches by hooks, barbs, or glue to the surface of the dispersal agent (Sorensen 1986).

Dispersal mode also has broad links to seed size. Small disseminules may be dispersed by either biotic or abiotic means, but larger ones tend to move through biotic dispersal (Hughes et al. 1994a). Smaller seeds are generally assumed to characterize plants of early successional, light-rich environments, inasmuch as their small nutritional reserve requires that they become photosynthetically self-supporting shortly after germination. Larger seeds possess greater reserves and can generate a larger leaf surface area before becoming self-sufficient. Thus, larger seeds tend to characterize closed communities with lower light intensities (Salisbury 1942, Harper et al. 1970). Grubb & Metcalfe (1996) argue that large seed size may also occur in closed communities as an adaptation for germination through dense leaf litter. In either model, larger seed size is associated with a more closed forest, which, in turn, is associated with a greater frequency of biotic dispersal.

Although contemporaneous biotic dispersal has received substantial attention (e.g., van der Pijl 1982, Herrera & Pellmyr 2002, Levey et al. 2002), its historical evolution has been less studied (e.g., Tiffney 1986a,b; Fleming & Lips 1991; Collinson & Hooker 2000; Eriksson et al. 2000a,b; Labandeira 2002). In part, this discrepancy reflects the necessary parochialism of scientific specialization, but it is also a function of a more intractable problem—taphonomy. Paleontologic evidence provides only a snapshot in time, generally based on the morphology of one organ or organism. Rarely is an investigator presented with a fossil animal with a gut full of identifiable fruits or seeds. However, dispersal is by its nature a dynamic process involving two or more players (Herrera 2002). Thus, elucidation of dispersal relations from the fossil record requires the application of "biological uniformitarianism," which is the use of morphological features of known ecological function in the present day as indicators of dynamic relations in the past. This approach is inherently antievolutionary because it assumes little change over time. However, the frequency with which convergence (the acquisition of similar morphological features by unrelated organisms facing similar ecological challenges) occurs and can be identified in the modern day strongly suggests that certain morphological features are reasonable—but not failsafe—guides to the interpretation of past plant-animal interactions.

Classically, it was assumed that dispersal involved "coevolution," the reciprocal selective influence of the dispersing animal and dispersed disseminule (Tewksbury 2002). Furthermore, it was assumed that this ongoing evolutionary interaction tightly bound the two organisms together, to the point that the extinction of one member would greatly reduce the fitness or cause the extinction of the other (Howe 1977). Recent experiments and observations, together with evidence from the fossil record, clearly indicate that such "tight" coevolution is rare if it exists at all (Schemske 1983, Witmer & Cheke 1991, Herrera 1995, Jordano 1995, Eby 1998, Wenny 2001; but see Tralau 1968, Janzen & Martin 1982, Tiffney 1984). Rather, the norm is "diffuse" coevolution, in which a distinct interaction exists between classes of dispersers and plants (e.g., between birds or mammals and fleshy fruits or seeds of a certain size). The dispersal agent may influence certain aspects of the disseminule (e.g., seed shape in some bird-dispersed plants; Mazer & Wheelwright 1993), but the disperser and disseminule are not tightly coadapted.

Finally, a word to the nonpaleontologist. The fossil record is wonderfully informative, but its data are not without biases. Particular to this paper, I must note three. First, paleontology is dominated by data from the modern temperate latitudes. These zones spawned the industrial revolution and the commensurate growth of an academic infrastructure that can spend the time to study the natural world. Furthermore, the modern tropics present difficulties in finding rocks buried beneath luxuriant greenery. Additionally, rocks that are found in such climates are often deeply weathered, and fossil evidence has been destroyed. Thus, with some very useful exceptions, most of the data come from the present-day latitudes 35° N to 50° N. Second, continents have moved over time. At the time of the probable initiation of vertebrate dispersal of plants, most modern Eurasian and North American localities were at or possibly slightly south of the paleoequator (Smith et al. 1981). Subsequent plate motion has influenced terrestrial environments at any one geographic point. Third, global climate has changed over time (Crowley & North 1991). If the incidence of vertebrate dispersal syndromes is related to climate, as the data suggest, the fossil record will not provide us with a global picture, but only with regional snapshots from successive time periods and paleoclimates.

With those caveats, I briefly review various lines of evidence for vertebrate dispersal of plant disseminules through time and scenarios for its origin, followed by a summation of this evidence from the Paleozoic, Mesozoic, and Cenozoic. I conclude with a consideration of outstanding patterns and future possible directions of inquiry. I intentionally exclude ant dispersal because of a lack of fossil evidence for ant-dispersed disseminules until the later Tertiary, and then only through inference from related living genera and species. It is potentially significant because ants are present in the Late Cretaceous and become important members of the terrestrial ecosystem in the early Tertiary (Grimaldi & Agosti 2000).

THE NATURE OF THE EVIDENCE

The evidence available for scrutiny using biological uniformitarianism comes from two sources: the fossil record of animals and that of plants.

From the Disperser's Viewpoint

COPROLITES AND GUT CAVITY FOSSILS Coprolites (fossil feces) give an insight into diet but are often difficult to attribute to a source (Thulborn 1991, Chin 1997), whereas gut contents are animal specific but open to varied interpretation (Barrett & Willis 2001). The oldest evidence of either form is from the Permian. Weigelt (1930) reported nearly 60 seeds in the gut cavity of a Late Permian rhynchocephalian *Protorosaurus* from Germany. He initially allied the seeds with the Podocarpaceae, but Schweitzer (1968) ascribed them to the primitive conifer *Pseudovoltzia*, which is wind dispersed (Taylor & Taylor 1993). This evidence leads one to suspect that the association was spurious or that the ingestion was accidental. However, Munk & Sues (1993) report the occurrence of similar seeds in the gut cavity of a second specimen of *Protorosaurus*. This finding suggests that the seeds were consumed in the absence of any morphological adaptation to vertebrate dispersal. Post-Permian coprolites and gut contents confirm continuing consumption of gymnosperm and angiosperm disseminules (Harris 1945; Hill 1976; Richter 1987, 1988; Nambudiri & Binda 1989; Rodriguez-de la Rosa et al. 1998; Collinson & Hooker 2000).

DENTITION Tooth structure and the arrangement of teeth within the mouth can be used to distinguish vertebrate herbivores from carnivores in a general sense (Sues 2000). Mammals possess a particularly complex tooth structure, which allows identification of specialized herbivory (i.e., granivory, frugivory, folivory, etc.) (Janis 2000, Rensberger 2000). In one case, this specialization has allowed direct association of disseminule and potential disperser (Collinson & Hooker 2000). Tooth marks on an Eocene *Stratiotes* (Hydrocharitaceae) seed can be attributed to a coeval rodent. However, this report is unusual, and the inference of diet from dentition has two substantial pitfalls.

The first pitfall is Janzen's (1984) hypotheses that "the foliage is the fruit," wherein animals with grazing or browsing dentition disperse fruits or seeds inadvertently while consuming foliage [for example, *Trillium* (Liliaceae), which is normally ant-dispersed, may also be moved by deer (Vellend et al. 2003)]. The second pitfall is that fruits and seeds may be effectively distributed by organisms with carnivorous dentition (Herrera 1989b), including bears (Rogers & Applegate 1983), coyotes and raccoons (Cypher 1999), foxes (Milton & Dean 2001), and even wolves (Motta-Junior & Martins 2002). The Permian *Protorosaurus* noted above (Munk & Sues 1993) has carnivorous dentition. Similarly, the absence of teeth in early birds could be associated with some form of herbivory, but living toothless birds range from frugivorous (e.g., parrots) to omnivorous (e.g., crows) to carnivorous (e.g., owls). Thus, although we have an instinctive tendency to associate biotic fruit and seed dispersal with herbivores, reference to living forms indicates a far greater fluidity of resource use. Reliance on dentition may lead us astray in the fossil record, although not wildly so. Rather, the reliability of dentition is a matter of degree. In the absence of herbivorous dentition, dispersal may still be occurring, but in the presence of herbivorous dentition, or of specialized frugivorous or granivorous characters, dispersal is certainly more common. With rare exceptions (e.g., gnaw marks; Collinson & Hooker 2000), distinguishing simple dispersal from that associated with seed/fruit predation in the fossil record will be impossible.

From the Plant's Viewpoint

The structure of fruits and seeds provides (sometimes ambiguous) clues to their mode of dispersal (van der Pijl 1982, Tiffney 1986a). Very small disseminules (Hughes et al. 1994a), or those with well-developed wings, suggest dispersal by wind; those with well-developed corky tissue suggest dispersal by water. Disseminules with surficial hooks suggest exozoochorous dispersal.

The most common indication of biotic dispersal is external flesh. Occasionally, preservational circumstances (permineralization; less commonly, compression or impression fossils) allow retention of the fleshy layer, but in most cases, it is degraded before final preservation. Even when flesh is present in fossil compressions and impressions, distortion can limit one's ability to estimate whether the surface was attractive or simply tough and leathery. Often, the interpretation of endozoochory is made by reference to the nearest living relatives. The preserved sclerotestas of Mesozoic cycads and *Ginkgo* generally lack evidence of flesh, but we assume its presence from their modern descendants. Similarly, in the absence of preserved flesh, many Tertiary fruits and seeds are interpreted as having been fleshy [e.g., *Vitis* (Vitaceae), Annonaceae, etc.] on the basis of nearest living relatives.

In some disseminules, largely nuts, the reward to the disperser is the contents of the disseminule. The "adaptation" to biotic dispersal is relatively large size (e.g., Juglandaceae, Fagaceae; Vander Wall 2001), although large size may also favor barochory or hydrochory. Occasionally these disseminules may show direct evidence of association with animals (e.g., gnaw marks; Schmidt et al. 1958) or scatter hoarding [e.g., cached hickory nuts from Miocene sediments of Washington (see figure 34a in Manchester 1987) or cached *Castanopsis* from the Miocene of Germany (see Gee et al. 2003)].

ORIGINS OF VERTEBRATE DISPERSAL

How did vertebrate dispersal arise, and did it arise independently in separate lineages, as seems likely within the rubric of "diffuse coevolution"? The fossil record is not likely to answer these questions, but it can inform the inquiry. Tiffney (1986a) suggested the dispersal of fleshy seeds of seed ferns by early herbivorous synapsids (*Edaphosaurus*) and the dispersal of cordaitalean seeds by fish. This implies the direct evolution of flesh for purpose of dispersal.

An alternative and perhaps more likely possibility is that biotic dispersal arose through a transfer of function; flesh evolved for one purpose and was subsequently co-opted for biotic dispersal. Could oil-rich flesh assist in water dispersal (Tiffney 1986a)? Although precise modern analogs are lacking, abiotic dispersal can be common in modern riverine communities (Gordon 1998), but this hypothesis could be difficult to test in the fossil record.

Building on observations of C. Herrera (1982), J. Herrera (1987), Cipollini & Stiles (1992), and Cipollini & Levey (1997) (see also Willson & Whelan 1990, Tewksbury 2002), Mack (2000) suggested that fleshy coverings on fruits or seeds could have originated to retain chemicals to defend the fruits or seeds against invertebrate predation or fungal attack, as well as to serve as a mechanical barrier in its own right. Cipollini & Levey (1997) additionally note that the same chemicals could help define the time of germination through inhibition. Perhaps such chemicals could have been first degraded by insects or bacteria, a role taken over by passage through the gut of the herbivore.

The interaction of disseminules, vertebrate dispersers, and insects is complex in the present day (Sallabanks & Courtney 1992), and the evolution of such interaction in the fossil record is unstudied. Many authors have surmised that glandular structures commonly found on Paleozoic seed plants and ovules/seeds may represent an insect deterrent (e.g., Scott & Taylor 1983), and insects certainly attacked Pennsylvanian seeds (Baxendale 1979, Scott & Taylor 1983, Hilton et al. 2002, Labandeira 2002). Further, insect herbivory was more pronounced in Permian communities than in the present day (Beck & Labandeira 1998). Thus, fleshy fruits or seeds might not be an accurate indicator of biotic dispersal within a group but rather of an initial defense against insects and, possibly, pathogens. If surficial flesh were established for such a purpose, it could have then imitated carrion and attracted opportunistic tetrapod carnivores. The elaiosomes of ant-dispersed seeds have a chemical makeup similar to insect prey and thus attract predatory ants (Hughes et al. 1994b).

Mack (2000) argues that such a transfer of function may have occurred independently in several lineages and most recently within angiosperms. Certainly, fruit structure and dispersal mode is flexible within angiosperms (Tiffney 1984, Knapp 2002) and warns against using phylogenetic affinity to predict a particular dispersal solution.

FOSSIL EVIDENCE

The fossil record of terrestrial animals and plants spans three geologic eras; although these geological units are defined by biological changes, in some cases patterns transgress era boundaries.

Paleozoic Era

PLANT DATA   Seeds first appeared in the late–Middle or Late Devonian (Bateman & DiMichele 1994, Marshall & Hemsley 2003). The initial seeds were small, but their size and morphological diversity rapidly expanded through the Mississippian and Pennsylvanian to encompass a wide range of winged, spinose, plumose, and fleshy forms (Sims 1997, 2000). Thus, many of the morphologies associated with major dispersal modes were established early in the history of the seed. This rise in morphological diversity was paralleled by a rise in seed plant diversity (Sims 1999).

Early seeds were likely dispersed by wind or water (Gensel & Skog 1977, Tiffney 1986a). Although the seeds of some seed ferns bore flattened wings suggestive of wind dispersal (Vega & Archangelsky 2000), similar to coeval conifers (Taylor & Taylor 1993), by the Pennsylvanian, large, flesh-covered seeds had evolved in both seed ferns and cordaitaleans (conifer relatives) (Tiffney 1986a, Taylor & Taylor 1993). Among medullosan seed ferns, the average seed ranged from 20,000 mm^3 to nearly 50,000 mm^3, and one species exceeded 300,000 mm^3 (sizes range from a pecan to a mango; see Taylor 1965, Tiffney 1986a, Taylor & Taylor 1993). These disseminules are some of the largest reported in the fossil record until the radiation of large angiosperm fruits in the early Tertiary. Smaller, fleshy seed fern seeds also existed (Tiffney 1986a, Taylor & Taylor 1993) and could have been dispersed by vertebrates. Thus, Klavins et al. 2001 observe that fleshy appendages on the apex of a Permian seed could have attracted an arthropod or vertebrate. However, early seeds often attained large size prior to pollination, and these appendages could also have served an earlier role in pollination (see Krasilov 1999). Cordaitalean seeds were smaller but still large, ranging from 122 mm^3 to over 12,000 mm^3 in volume.

ANIMAL DATA   The initial terrestrial herbivores were insects (Scott & Taylor 1983, Taylor & Scott 1983, Milner et al. 1986, Labandeira 1998). Hotton et al. (1997) have advanced evidence for the presence of low-fiber herbivory and omnivory evolving in later Mississippian and Early to Middle Pennsylvanian tetrapods. The supporting evidence is largely inferential and involves (*a*) analogy to living carnivorous lizards known to eat vegetation and fruit, (*b*) predicted jaw mechanics, (*c*) tooth-wear patterns, and (*d*) head-body ratios. Although the individual points are debatable, the resulting conclusion is not refutable in light of the often generalized diet of many living taxa.

Dedicated generalized herbivorous tetrapods radiated in the Early Permian (Carroll 1988; Modesto 1992, 1995; Modesto & Reisz 1992). The edaphosaurs were the earliest diverse herbivorous tetrapods, appearing in the Late Pennsylvanian and diversifying in the Early Permian before giving way to caseids and diadectids. Caseids and diadectids were succeeded in the Middle and Late Permian by the radiation of herbivorous therapsids, the dicynodonts. The latter group included small to large herbivores, many of which showed adaptations to digging, either for food or shelter (King 1990, Cox 1998, Ray & Chinsamy 2003). Whereas earlier studies suggested a great diversity of dicynodonts, including some portrayed as frugivores (see illustration on p. 57 of Czerkas & Czerkas 1990), recent studies suggest that dicynodonts occurred with much lower diversity and in smaller numbers than originally thought and were generalist herbivores (Cox 1998) that possessed slicing teeth and fed close to the ground, perhaps on herbaceous sphenopsids (Rayner 1992). Dicynodonts were accompanied by other less diverse herbivorous reptiles, including procolophonids and pareiasaurs, a few of which also crossed into the early Mesozoic era (King 1996).

SUMMARY A wide range of seed morphologies had evolved by the end of the Pennsylvanian. Van der Pijl (1982) was the first to suggest that reptiles dispersed the large, fleshy forms. Hotton et al. (1997) elaborated that some amphibia, reptiles, and early synapsids could have provided dispersal sources in the latest Mississippian through later Pennsylvanian but that dedicated "high-fiber" vertebrate herbivores did not appear until the latest Pennsylvanian. Seed dispersal may also have been accomplished by fish (W. Chaloner, personal communication, 1981) because many plant taxa clearly lived adjacent to river and estuarine channels, and fish dispersal is an important feature in some situations in the present day (Goulding 1980).

These hypotheses lead to several intertwined possibilities. If we assume that only obligate herbivores are potential dispersal agents, then flesh clearly evolved for some purpose other than dispersal and assumed that role secondarily. However, large amphibian and reptilian carnivores and omnivores coexisted with these early fleshy seeds, and the possibility remains that these animals were the initial tetrapod dispersal agents, perhaps mistaking seeds for carrion. When synapsid herbivores radiated in the Permian, they supplanted the earlier forms, which were in decline because of drying climates—perhaps an early instance of diffuse coevolution.

The fleshy seeds of seed ferns and cordaitaleans appear to lack dormancy or germination deferment (fossils are found either empty of contents or germinated, but not ungerminated and containing an embryo). If dormancy was absent or poorly developed in these early forms, a trip through a gut may have spelled death. Furthermore, most reconstructions suggest that the early tetrapods were equatorial (Berman et al. 1997) and, thus, would have limited effects upon plant ranges. However, Permian synapsid and reptilian herbivores, although lacking apparent specific adaptations to seed dispersal, were both widespread and increasingly likely both to consume and to move seeds.

Mesozoic Era

PLANT DATA Tiffney (1986a) summarized diaspore size and inferred dispersal modes for the Triassic, Jurassic, and Cretaceous. With the exception of the data for angiosperms, little has changed in this picture. Mesozoic gymnosperms involved a wide range of groups, some extant (conifers and cycads) and others extinct (various seed ferns and lesser known clades). The seeds exhibit a wide range of sizes and dispersal modes (Tiffney 1986a) (Figure 1), including many large forms with fleshy exteriors suggestive of biotic dispersal such as cycads, *Ginkgo*, *Podocarpus* (Coniferae), and lesser known groups such as *Pentoxylon*, a Jurassic and Cretaceous plant of the Southern Hemisphere that possessed a head of fleshy seeds (Taylor & Taylor 1993, Howe & Cantrill 2001), *Caytonia*, which has fleshy "fruit" surrounding several seeds (Taylor & Taylor, 1993; but see also Reymanowna 1973), and the *"Brenneria"* plant, which had small, fleshy seeds (Pedersen et al. 1993). This evidence supports the inference of biotic dispersal in many Mesozoic gymnosperms (Weishampel 1984, Tiffney 1986a).

Angiosperms appeared and diversified in the later portions of the Early Cretaceous through the Late Cretaceous. The initial radiation involved shrubby or perhaps herbaceous plants growing in disturbed sites, particularly along rivers (Wing & Boucher 1998, Friis et al. 1999). Recent studies of floral material indicate the presence of many living families and orders (Crane & Herendeen 1996, Friis et al. 2000, Takahashi et al. 2002). However, a distinctive feature of these early angiosperms is the dominance (relative to the modern flora) of very small flowers, fruits, and seeds (Tiffney 1984; Crane & Herendeen 1996; Eriksson et al. 2000a,b). These seeds tend to range from less than 0.1 mm^3 to about 10 mm^3 (Eriksson et al. 2000a). Although many temperate floras of the present day have a similar mean seed mass (e.g., Leishman et al. 2000), the range includes seeds much larger than those observed in the Cretaceous. In a separate contrast, the mean seed volume of Mesozoic gymnosperm–dominated floras is generally two to three orders of magnitude greater than that of coeval angiosperm floras (Tiffney 1986a) (Figure 1). These distinctions suggest that Mesozoic and Cenozoic angiosperm communities are distinct and that strong differences exist in the ecology of dispersal and seedling establishment between Mesozoic angiosperms and many coexisting gymnosperms. On the basis of the small size and the structure of these seeds, Tiffney (1984) posited that Cretaceous angiosperms were dominated by abiotic dispersal. More recently, Eriksson et al. (2000b) described a diverse flora (106 angiosperm fruit and seed taxa) in which nearly 25% of the species were fleshy as interpreted by the presence of a nonmechanical external layer or a textured exocarp.

Larger Cretaceous disseminules exist (Monteillet & Lappartient 1981, Krasilov & Martinson 1982, Lamb 2001). In two cases, these disseminules are littoral (Monteillet & Lappartient 1981, Lamb 2001), suggesting possible association with water transport. However, a bias is introduced by paleolatitude. Save for Chesters (1955) and Monteillet & Lappartient (1981), all of the reports of Cretaceous fruits and seeds come from Europe and North America, which lay at roughly 35° N to 55° N paleolatitude in the Cretaceous. We know almost nothing of Cretaceous equatorial floras.

ANIMAL DATA The primary herbivores in the Mesozoic era are reptiles. However, many of the smaller reptiles tend to be facultative rather than obligate herbivores (King 1996), making assured attribution of dispersal to a particular group difficult. As a result, much evidence is anecdotal or conjectural.

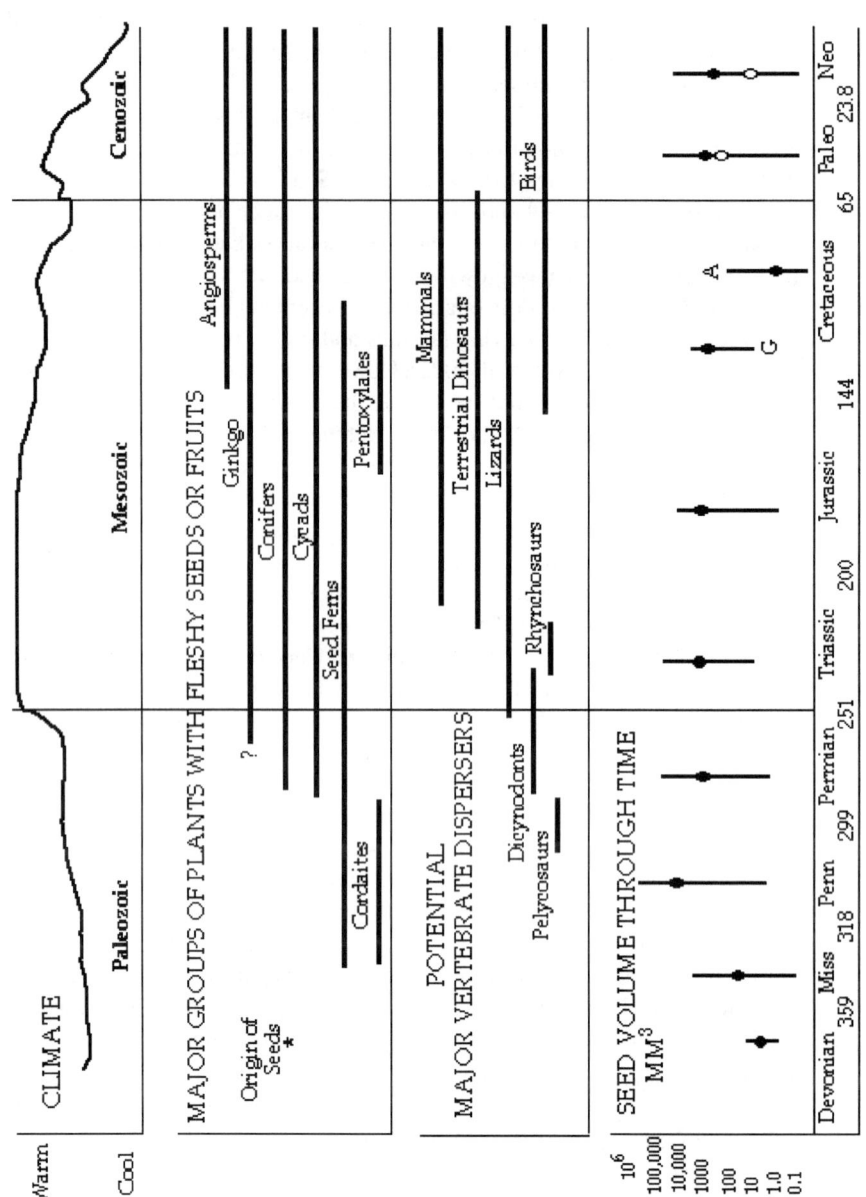

Early to Middle Triassic The dicynodonts carried from the later Permian into the Early Triassic and became extinct. These were followed by a Middle and Late Triassic radiation of generalist herbivorous archosaurs, the rhynchosaurs (Carroll 1988), accompanied by aetosaurs and lesser groups of reptilian herbivores (King 1996). In turn, this grouping became extinct, and their role was assumed by the herbivorous dinosaurs.

Dinosaurs Although dinosaurian tooth structure allows the separation of herbivores and carnivores, it is generalized enough to preclude identification of particular specializations (e.g., frugivory). Furthermore, most dinosaurs are sufficiently massive to suggest that they were generalist, rather than specialist, herbivores. Clear arguments have been made for their probable importance in dispersal (Weishampel 1984), but direct evidence is difficult to come by. Gut contents have been reported for herbivorous sauropods (Stokes 1964, Ash et al. 1992, Mohabey 2001) and ornithischians (Sternberg 1909, Kräusel 1922, Weigelt 1927), but the authenticity of some of these findings has been questioned (see Currie et al. 1995, Barrett & Willis 2001). The most interesting report is of an Early Cretaceous ankylosaur (Molnar & Clifford 2000, 2001) whose gut contents included three 4.5 mm long fruits and several 0.3 mm diameter seeds. From this and other data, the authors interpret this small (about 1.4 m long) ornithischian as engaging in the selective nipping of particular plant parts, although the presence of fern sporangia and much material interpreted as leaf remains suggests a facultative, not obligate, dispersal agent.

Birds The fossil record of birds commences in the Jurassic (Padian & Chiappe 1998) and involves two periods of radiation, one in the Cretaceous and a second, explosive one in the Tertiary (Chiappe & Dyke 2002, Feduccia 2003). The latter

---

Figure 1 Summation of major features and clades involved in the evolution of vertebrate dispersal from the Devonian through Tertiary periods. Climate: a relative track (cooler, warmer) of global climate (from Crowley & North 1991, Stanley 1999). Plant groups: geologic occurrence of major groups of seed plants that may possess fleshy disseminules (ranges from Taylor & Taylor 1993). Vertebrate dispersers: geologic occurrence of major groups of potential vertebrate dispersers (ranges from Carroll 1988). Seed volume: average seed volume in cubic mm by period from Tiffney (1986a) modified after Eriksson (2000a). Bar represents the full range; black dot represents the mean value. In the Cretaceous period, G = gymnosperms only, and A = angiosperms only. The Tertiary is broken into the Paleogene and Neogene; the white dots are data from Eriksson (2000a) that have slightly lower Tertiary values than data from Tiffney (1986a) because of different floras sampled. This illustrates the average decline in seed volume in the Northern Hemisphere as climates cooled. The horizontal axis is geologic time, with the boundaries between the periods noted in millions of years. Miss = Mississippian, Penn = Pennsylvanian, Paleo = Paleogene, Neo = Neogene.

diversification clearly involved many frugivorous and seed-eating birds, particularly among the passerines (Ericson et al. 2003). However, the role of birds in seed dispersal in the Cretaceous is less clear. Early birds were classically seen as toothed carnivores, particularly those associated with aquatic habitats. However, research in the past 20 years has demonstrated a greater diversity of Mesozoic birds and the existence of a whole clade (Enantiornithines) that became extinct at the end of the Cretaceous (Chiappe & Dyke 2002, Feduccia 2003). We have no living analogs by which to judge the habits of these latter organisms. Although the role of Cretaceous birds as dispersal agents is largely conjectural, two examples provide tantalizing evidence. Zhou & Zhang (2002) report a middle–Early Cretaceous bird of about 60 cm in length with remains of *Carpolithus* sp. (a form genus for seeds of unidentifiable affinity) in the area of its stomach. Because of the size of the plant objects and the associated vegetation (Zhou et al. 2003), the most parsimonious suggestion is that this bird was feeding on a gymnosperm. An associated genus from the same area, *Sapeornis*, possessed gastroliths and reduced teeth, suggesting it, too, fed on plants; the associated foot structure indicated that it could climb (Zhou & Zhang 2003).

Smaller terrestrial reptiles Many modern lizards are omnivorous (Barrett 2000) and are known to disperse fruits and seeds (e.g., Lord & Marshall 2001, Valido & Nogales 2003). Lizards were widespread in the Mesozoic era and could have played an important role in moving smaller disseminules. Similarly, turtles can be effective dispersal agents in the present day (Moll & Jansen 1995) and are likely the source of a Campanian age coprolite containing nearly 200 achenes of early Ranunculaceae (Rodriguez-de la Rosa et al. 1998).

Pterosaurs Fleming & Lips (1991) argued that pterosaurs were potentially important seed dispersers, particularly of early angiosperms. The majority of pterosaurs display a dentition suited to piscivory or insectivory; the few forms that are edentulous may be demonstrated either by stomach contents to be fish eaters (*Pteranodon*) or by their sheer size (*Pteranodon, Nyctosaurus*) unlikely to be effective seed or fruit dispersers because they would have had great difficulty perching on vegetation. The stratigraphic distribution of the major clades suggests that only the Pteranodontidae and Nyctosauridae radiated in the middle or later Cretaceous, in parallel with the appearance of angiosperms. One very odd group, the Barremian to Albian Tapjaridae, has an edentulous bill, relatively small size (although still large with an estimated 1.5 to 2.0 m wingspan), and an odd head shape reminiscent of a moa or an emu; it could have been involved in dispersal of larger seeds. Fleming & Lips (1991) and Wellnhofer (1991) argued that fruit-eating pterosaurs likely lived in more terrestrial settings with lower potential for fossilization. Wellnhofer concluded, "If there were fruit-eaters—which is highly probable—we are unlikely to ever find their fossil remains" (Wellnofer 1991, p. 161). If this argument were valid, the fossil record of birds in the Cretaceous and Tertiary would be entirely of aquatic forms, which it decidedly is not. Although evidence suggests pterosaurs were at best a minor source of biotic dispersal, caution must be exercised in light of evidence for "carnivores" consuming fleshy fruits or seeds in other groups.

Synapsids Early Triassic herbivore communities were dominated by dicynodont synapsids, carrying on the tradition of the Permian (Carroll 1988, Schwanke 1998). The dicynodonts became extinct in the Middle Triassic, and synapsids were represented for most of the rest of the Mesozoic era by insectivores, including mammals. A timely summary (Kielan-Jaworowska et al. 2004) allows access to the most recent thinking on Mesozoic mammalian diets. The multituberculates were classically identified as the most diverse and important mammalian herbivores of the middle and late Mesozoic based on their dentition. Their demise in the early Paleocene in the face of the explosive diversification of rodents was seen as further evidence of their occupying a rodent-like niche, although differences in reproductive features may better explain this extinction (Kielan-Jaworowska et al. 2004). More recent examination of the jaw mechanics and biology of modern analogs now indicates that multituberculates were omnivores (Clemens & Kielan-Jaworowska 1979, Kielan-Jaworowska et al. 2004). Del Tredici (1989) invoked them as dispersal agents for *Ginkgo*, but Wall & Krause (1992) demonstrated that at least *Ptilodotus*, an advanced multituberculate, would have been unable to crack the seed of *Ginkgo*. Similarly, some Cretaceous marsupials (e.g., *Turgidodon* of the later Cretaceous) were likely omnivorous and could have dispersed fruits or seeds. Only one species, the Late Cretaceous marsupial *Glasbius intricatus* of North America, is specifically identified as a possible early frugivore (Kielan-Jaworowska et al. 2004). A third group of generalist herbivores were the "zhelestid" eutherians, also referred to as archaic hoofed mammals (Nessov et al. 1998). These small herbivores of Central Asia possessed low crowned teeth that were appropriate for browsing.

SUMMARY Sparse evidence from gut contents and coprolites indicates seed and fruit dispersal occurred (Nambudiri & Binda 1989, Molnar & Clifford 2000, Zhou & Zhang 2002) and affected angiosperms (Rodriguez-de la Rosa et al. 1998). However, the relative frequency of the dispersal is unclear. Potential dispersal agents had been available to gymnosperms throughout the Mesozoic era, and the large size of many saurischian herbivores suggests that dispersal could occur in the absence of specific animal or plant adaptations. In the middle to Late Cretaceous the rise of angiosperms was paralleled by the diversification of ornithischian herbivores, at least in North America (Weishampel & Norman 1989; Tiffney 1992, 1997). This development suggests, but does not demonstrate, that the potential existed for generalized angiosperm dispersal by ornithischians, much as ungulates serve in this role in the present day (Carrano et al. 1999).

Fleshy fruits formed an important component of one Early Cretaceous flora (Eriksson et al. 2000b). This finding could indicate dispersal by smaller dispersal agents, or it could be related to aspects of germination ecology or plant pathogen deterrence (Mack 2000). If dispersers were involved, they might include lizards, birds, and, in the Late Cretaceous period, multituberculates, "zhelestids," and the occasional marsupial. However, no compelling evidence presently exists for obligate fruit or seed dispersal by mammals, although only the North American mammal record is known in detail (Clemens 2001); future discoveries may alter this conclusion.

Cenozoic Era

PLANT DATA The angiosperms underwent a major burst of taxonomic and morphological diversification in the Tertiary period that involved the appearance of many modern genera (Niklas et al. 1985), often with fruit and seed sizes and morphologies close to those of their living descendants (Tiffney 1990, Dilcher 2000). This development is displayed in several Eocene fruit and seed floras (e.g., Chandler 1964, Collinson 1983, Manchester 1994). These floras are frequently associated with diverse mammalian faunas that contain potential dispersal agents; for example, the Paleocene-Eocene Dormaal locality of Belgium (Fairon-Demaret & Smith 2002) and the coeval faunas and floras of southern England (Collinson & Hooker 1987).

The initial Tertiary radiation generally involved plants of closed communities, often with large disseminule sizes (Tiffney 1986a, Eriksson et al. 2000a). Cooling and drying climates of the later Paleogene led to the evolution of more open communities in middle latitudes, and by the middle Tertiary another angiosperm radiation involving annual and biennial herbs and grasses was underway (Friis 1975; Tiffney 1984, 1986a; Jacobs et al. 1999). Although large, animal-dispersed disseminules were present in virtually all of these communities, the overall trend was toward a reduction in fruit and seed size (Tiffney 1986a, Eriksson et al. 2000a). Many of these smaller disseminules (e.g., Compositae) could be as easily spread by wind as by granivorous birds or rodents. Indeed, the seeds of most herbs do not exhibit clear morphological adaptations to vertebrate dispersal.

This second radiation is well encompassed by the north temperate fossil record, but we have no evidence from the tropics to tell whether the climatic deterioration of the mid-Tertiary had an influence on the structure of tropical forests or their associated dispersal syndromes. Thus, the evidence from the fossil record for the rise of abiotic dispersal mechanisms in the later Tertiary is a regional, not global, feature.

ANIMAL DATA The radiation of birds in the Tertiary (Feduccia 1995, 2003) suggests a strong escalation of bird-plant interaction. Initially, this interaction involved several lineages focused on larger, fleshy fruits, but in the middle Tertiary, the radiation of the passerines layered on a major group of frugivores and granivores.

Recent phylogenetic studies may place a twist on this story. Passerines originated in Australia (Ericson et al. 2003) and spread to the rest of the world as Australia drifted north, starting in the Eocene. As a result, passerine-plant interactions could have early Tertiary or Late Cretaceous roots in Australia.

Similarly, many important mammal groups radiated in the early Tertiary. Several of these groups displayed clear adaptations to frugivory and granivory (e.g.,

Collinson & Hooker 1991). Primates in particular showed early adaptations to frugivory (Szalay 1968), radiated in the early Tertiary (Sussman 1991, 1995), and serve an important role in dispersing modern tropical angiosperms (Sussman 1995, Lambert & Garber 1998, Regan et al. 2001). Organ-specific biotic dispersal agents also became important in more temperate lineages as families with small Cretaceous fruits (Juglandaceae, Fagaceae) evolved larger nuts in the early Tertiary (Tiffney 1986b, Vander Wall 2001). Bats first appeared in the Eocene (Jepson 1970, Benton 1993) and diversified through the Tertiary. The earliest forms were clearly insectivorous. The oldest frugivorous bat is a middle Oligocene megachiropteran, and the oldest microchiropteran frugivore is from the Miocene (University of Michigan Museum of Zoology 2004). Because frugivorous bats are essentially tropical and because of the relative paucity of fossil material from the equatorial portions of Earth, this adaptation possibly evolved earlier in the history of the group.

Subsequent middle Tertiary radiations of "whole-plant" grazers (even-toed and odd-toed ungulates and others) added a further layer of diffuse dispersal relationships (Janis 1993, Novacek 1999, Janis et al. 2002).

SUMMARY   A clear upsurge occurred in the importance of biotic dispersal in the Tertiary. Central to this are birds and mammals that, because of their small size, have organ-level interactions with plants (Wing & Tiffney 1987a,b; Tiffney 1992), particularly in closed communities (Clark et al. 2001). The question of what the forcing factor is remains to be answered. Wing & Tiffney (1987a,b) and Tiffney (1992) postulated that the radiation of small vertebrates led to a rise in dispersal and the subsequent establishment of closed-canopy forests. Eriksson et al. (2000a) note that increased precipitation and warmer climates would favor closed forests, which, in turn, would favor larger disseminules and would thus drive the radiation of smaller vertebrates scaled to fit within this community. This hypothesis is considered in more detail below.

Fossil evidence suggests that subsequent climatic cooling led to the spread of herbaceous angiosperms, many of which possessed smaller seeds and fruits, amenable to biotic and abiotic dispersal. By the Tertiary the continents were near enough to their present positions that the fossil record, restricted to the temperate latitudes, provides evidence for only a portion of the evolution of Tertiary dispersal spectra.

DISCUSSION

I have summarized the basic patterns of vertebrate dispersal within each of the geologic eras (Figure 1); here my focus is on common threads.

Data Quality

Neontological studies delineate granivores, frugivores, and herbivores in contrast to carnivores, leaving the impression that feeding mode may be clearly recognized from morphological characters. In fact, the relationship between diet and dentition appears as diffuse as coevolution—so diffuse as to greatly limit our ability to infer the absolute presence or absence of biotic dispersal from plant or animal morphology in the fossil record. Flesh could possibly serve purposes other than dispersal, either at its evolutionary inception or while dispersal is actively under way. Carnivorous dentition is no guarantor of diet because fossil gut contents of reptiles that would otherwise be considered carnivores indicate that they could serve as dispersal agents. In short, it is difficult to conclusively demonstrate the presence or absence of dispersal except through gut contents or coprolites. However, the relative frequency of herbivorous dentition, biotic dispersal–related fruit and seed characters, or both within a fossil assemblage allows us to establish probabilities. I suggest from present data that biotic dispersal has been with us from at least the Permian and possibly the later Pennsylvanian, but that it became more widespread and important in the Tertiary than in previous periods.

Dispersal is Very Diffuse Coevolution

The foregoing observations, particularly the ability of carnivores to act as dispersal agents, strengthens Herrera's (1995) contention that coevolutionary dispersal relationships are diffuse. A broad perspective suggests that dispersal can be achieved by any organism with the ability to reach and manipulate the dispersal unit. That certain groups become important dispersal agents in particular environments (e.g., birds and primates in tropical forests) might primarily be a function of their ability to move within the canopy and secondarily of other morphological features related to frugivory.

BIOLOGICAL UNIFORMITARIANISM TAKES US JUST SO FAR   We can use the morphology and associated ecological functions of living plants to infer function from the morphological remains of the past. I believe such inferences about dispersal are viable back to the beginning of the Tertiary. Beyond that, this approach may encounter a "stem group/crown group" problem.

Molecular and morphological systematists have battled in the past decade over the disconnect between molecular evidence that suggests a far earlier cladogenesis than is observed in the morphological fossil record. With time, most have agreed that the solution lies in disconnecting phylogenetic radiation from ecological radiation. The two may occur simultaneously, but more often (early metazoa, mammalian orders), phylogenetic diversification occurs within a stem group that is ecologically homogeneous (e.g., Knoll & Carroll 1999). Subsequent environmental change releases different taxa to pursue different ecological solutions, resulting in a later burst of crown group morphological diversification. These two alternatives are summed in the terms "pattern congruency" (simultaneous phylogenetic and ecological diversification) and "pattern lag" (ecological diversification follows phylogenetic diversification), coined by Wing & Boucher (1998) in their summary of Cretaceous angiosperm evolution. Although Cretaceous floras harbor many modern angiosperm families, structural features suggest the group occupied a far narrower ecological niche than in the Tertiary or in the present. Barring future data to the contrary, inference of Cretaceous ecological function from living representatives of these families is thus fraught with uncertainty.

What if no living clade is available for reference? We then assume that morphological characteristics that cross clades reflect shared function; that is, flesh promotes biotic dispersal in the seeds of widely divergent clades. However, in examining the fossil record, we should remain alert to the possibility that some characteristics have changed function over time. I am increasingly dubious that the fleshy seeds of the Mississippian and Pennsylvanian evolved for the purpose of biotic dispersal, although the function of that tissue remains conjectural. At the very least, we should be careful in using the present as a guide to the past, lest we turn the past into a precise parallel to the present, just with different plants and animals plugged into their respective roles.

FUTURE ENDEAVORS

Early Dispersal

The time at which significant vertebrate dispersal of seeds first appeared is conjectural. Vertebrate dispersal was certainly in place by the Late Permian (Weigelt 1930, Munk & Sues 1993) and likely by the Early Permian, as indicated by the presence of obligate herbivores (Hotton et al. 1997). It could have occurred earlier, in the Pennsylvanian, accomplished by "soft tissue" herbivores and omnivores (Hotton et al. 1997) or even earlier by apparent carnivores. Resolution of this problem using present morphological data is not possible, but paleobiogeographic distributions may provide some insight.

Pennsylvanian and Early Permian herbivores occur within 10° N and 10° S of the paleoequator, and spread beyond this belt only in the Late Permian (Milner 1993, Berman et al. 1997). This development parallels the distribution of many groups with fleshy-seeded plants. However, if fleshy-seeded gymnosperms occurred beyond the paleoequatorial range of early tetrapods, then flesh evolved for a purpose other than dispersal, even if it was involved in facultative dispersal relationships where tetrapods existed.

Seed Coat Thickness

Vertebrate dispersal required a seed coat that is light enough to be moved by a dispersal agent but resistant enough to protect the embryo from being consumed. An examination of trends in seed coat thickness (normalized for seed size to account for scaling effects) through time, particularly with reference to the evolution and demise of various potential disperser lineages, would be informative.

Tropical Data: A Key Missing Element

A core requirement for a better understanding of the evolution of dispersal relationships is better fossil data from the tropics, particularly because modern vertebrate dispersal is so important in these communities (Herrera 2002). Tertiary communities of Europe and North America began in a greenhouse world of subtropical environments that gave way to increasingly temperate ones. This circumstance enforces a bias in that we cannot track dispersal-mode changes within a single community type in one climate regime over time. Optimally, we would like to observe Eocene to present patterns within plant communities living under similar climates.

In the Cretaceous, we lack clear evidence of what is occurring at the tropics. Perhaps lower-middle latitude communities were the tropics. If they were not, however, we might be missing an important part of the picture. The phylogenetic and ecological radiation of angiosperms might have occurred concurrently in a tropical setting we have yet to sample, but the results leaked stepwise to more temperate latitudes, simulating pattern lag.

Palms

Palms appear in the Santonian stage of the Cretaceous, represented by wood and leaves (Uhl & Dransfield 1987). No unequivocal Cretaceous palm fruits have been found (excluding the dubious Aptian *Hyphaeneocarpon*; Vaudois-Miéja & Lejal-Nicol 1987), although fruits are widely known from the Tertiary (Mai 2000). Modern palms are almost entirely animal dispersed (Uhl & Dransfield 1987). Although excellent evidence suggests that dispersal modes changed from abiotic in the Cretaceous to biotic in the Tertiary in some angiosperm lineages (Tiffney 1986b), vestiges of the Cretaceous dispersal mode persisted in these groups in the Tertiary and in the present. Not so in palms. Either we are missing abiotically dispersed palm fruits in the Cretaceous and Tertiary, or we are missing biotically dispersed palm fruits in the Cretaceous. One possible exception is the enigmatic *"Ficus" ceratops* of the Campanian age of western North America (Shoemaker 1977), whose morphology is suggestive of a cast of an eroded palm fruit.

Latitudinal Gradients

Climate influences the modes of dispersal within communities through its effect on vegetation. The cooling Tertiary climates in Europe and North America led to a radiation of smaller seeds and fruits with a higher incidence of abiotic dispersal (Tiffney 1984, Eriksson et al. 2000a). This same effect is observed in the latitudinal distribution of modern seed size; Moles & Westoby (2003) record that, for every roughly 23° distance from the equator, average seed mass decreases 10-fold, although some intraspecific variation exists (Hampe 2003).

If the ecological relationships between Cretaceous angiosperms, dispersal agents, and climates were similar to those of the Tertiary and the present, we would predict that a similar pattern would emerge from a time-limited, pole-to-equator summation of Cretaceous seed size. Available data do not allow such a summation, and this becomes another goal for future exploration. If a pattern emerges similar to that observed in the Tertiary and the present, then one might infer similar ecological relationships between seed size, dispersal, and community ecology. If the pattern (once corrected for the lower equator-to-pole temperature gradients of the Cretaceous) is not comparable, it may suggest that Cretaceous and Tertiary angiosperm communities functioned differently.

This pattern should also be separately checked for gymnosperms. In the present day, the intuitive response is that the pattern is parallel to that of angiosperms (Pinaceae dominating the far northern latitudes, whereas larger-seeded *Ginkgo* is more temperate, and cycads are subtropical). However, will this pattern hold in the Cretaceous?

Do Dispersers Structure Vegetation or Does Vegetation Drive Dispersers?

One of the most interesting historical questions raised with respect to biotic dispersal involves the Cretaceous-Tertiary transition. Disseminule size and the occurrence of biotic dispersal both markedly increase across this boundary; however, the cause is open to debate. Tiffney (1984) and Wing & Tiffney (1987a,b) suggested that the radiation of birds and mammals established dispersal agents scaled to move organs, which, in turn, allowed larger disseminule size and ultimately the establishment of closed forests with low-light germination regimes. Eriksson et al. (2000a) countered that these relations were inverted and that a warmer climate and increased precipitation allowed the establishment of closed forest, which shifted the ecological constraints on seed recruitment (Eriksson & Jakobsson 1999) and possibly seedling ecology (Ibarra-Manriquez et al. 2001) and led to larger seeds. The dispersal agents therefore tracked, not led, the increase in seed size.

Plant stature and community density have relationships both with prevailing climate and with mode of dispersal; disentangling the influence of these two forcing factors may prove difficult. However, I think that the basis of distinguishing the two will be found in examining two features of Cretaceous community function.

First, one must examine whether the rise in precipitation and warmth at the Cretaceous-Tertiary boundary is a valid trigger of the radiation of closed angiosperm communities. Certainly such a climatic change occurred at the boundary (Wolfe & Upchurch 1986, 1987). However, warm and wet climates existed in areas occupied by angiosperms earlier in the Cretaceous. White et al. (2001) and Ufnar et al. (2002) advance isotopic evidence that the western margin of eastern North America may have had as much as 2500 to 4100 mm of rain per year in the Albian through Cenomanian ages, distributed evenly throughout the year. Contemporaneous gravel-choked, low-gradient stream channels support this interpretation (Brenner et al. 2003). Although these data come from the eastern shore of the Mid-Continental Seaway, prevailing wind currents (White et al. 2001, Ufnar et al. 2002) would carry airmasses northeast, providing at least adequate rain to the east coast of North America, where moisture-indicating coals are recognized in the earlier Cretaceous (Ziegler et al. 1987). If the hypothesis of Eriksson et al. (2000a) is correct, floras from these areas (and others shown to have high rainfall; see Parrish et al. 1982) should display a tendency toward larger seed sizes. The data are presently limited, but the Santonian-Campanian floras of New Jersey and New England appear to exhibit small fruit and seed size (Tiffney 1977, Crepet et al. 2001). If further exploration supports higher moisture being associated with small fruit and seed size, then moisture is unlikely to be the feature that triggered the evolution of larger fruit and seed size at the Cretaceous-Tertiary boundary.

Second, one must examine whether the ecosystem dynamic of the early Tertiary is homologous to that of the middle Cretaceous. Whereas Cretaceous angiosperms were dominated by small seed size, Cretaceous gymnosperm groups included taxa with both small (e.g., Bennititales) and large (cycads, *Ginkgo*, several conifers) (Figure 1) seeds. Unless a different ecologic metric exists in seed recruitment and response to climate in gymnosperms and angiosperms, both groups should roughly parallel each other in seed size and dispersal mode when growing in similar climates. This disparity may provide clues to the Cretaceous-Tertiary transition and involve coevolution on a very broad scale. Reptiles and gymnosperms both radiated in the Triassic and held sway through the Mesozoic. Terrestrial dinosaurs and enantiornithine birds became extinct at the boundary, ending approximately 150 million years of association. Mammals, modern birds, and angiosperms then radiated, establishing new and broad coevolutionary relationships in the Tertiary (Wing & Tiffney 1987a,b; Tiffney 1992, 1997). I suggest that, although the Mesozoic and Cenozoic eras both involved plant-animal interactions, the Mesozoic dynamic was dominated by reptile-gymnosperm interactions to which the angiosperms were newcomers. Only with the demise of terrestrial dinosaurs at the Creatceous-Tertiary boundary did an angiosperm-bird-mammal dynamic become fully established. In short, the terrestrial Mesozoic and Cenozoic had different ecological dynamics, much as Bambach et al. (2002) have demonstrated for the marine realm. Further research can test both of these contentions.

ACKNOWLEDGMENTS

I express my deep appreciation to Zofia Kielan-Jaworowska, Richard Cifelli, and Zhe-Xi Luo for advance access to the text of their forthcoming book, *Mammals from the Age of Dinosaurs: Origins, Evolution and Structure*. I also thank Kevin Padian and Chris Bennett for their insights on pterosaurs, and Angela Moles for a critical reading of the manuscript and for suggesting the figure.

The *Annual Review of Ecology, Evolution, and Systematics* is online at
http://ecolsys.annualreviews.org

LITERATURE CITED

Ash S, Tidwell WD, Madsen JA Jr, Stokes LL. 1992. The last supper of a Jurassic dinosaur. *Prog. Abstr. Geol. Soc. Am. Rocky Mountain Sect.* 9:1 (Abstr.)

Bambach RK, Knoll AN, Sepkoski JJ. 2002. Anatomical and ecological constraints on Phanerozoic animal diversity in the marine realm. *Proc. Natl. Acad. Sci. USA* 99:6854–59

Barrett PM. 2000. Prosauropod dinosaurs and iguanas: speculations on the diets of extinct reptiles. See Sues 2000, pp. 42–78

Barrett PM, Willis KJ. 2001. Did dinosaurs invent flowers? Dinosaur-angiosperm coevolution revisited. *Biol. Rev. Camb. Philos. Soc.* 76:411–47

Bateman RM, DiMichele WA. 1994. Heterospory: the most iterative key innovation in the evolutionary history of the plant kingdom. *Biol. Rev. Camb. Philos. Soc.* 69:345–417

Baxendale RW. 1979. Plant-bearing coprolites from North American Pennsylvanian coal balls. *Palaeontology* 22:537–48

Beattie AJ. 1985. *The Evolutionary Ecology of Ant-Plant Mutualisms.* Cambridge, UK: Cambridge Univ. Press. 182 pp.

Beattie AJ, Hughes L. 2002. Ant-plant interactions. See Herrera & Pellmyr 2002, pp. 211–35

Beck A, Labandeira CC. 1998. Early Permian insect folivory on a gigantopterid-dominated riparian flora from north-central Texas. *Palaeogeogr. Palaeoclimatol. Palaeoecol.* 142:139–73

Benton MJ, ed. 1993. *The Fossil Record 2.* London: Chapman & Hall. 845 pp.

Berman DS, Sumida SS, Lombard RE. 1997. Biogeography of primitive amniotes. See Sumida & Martin 1997, pp. 85–139

Brenner RL, Ludvigson GA, Witzke BL, Phillips PL, White TS, et al. 2003. Aggradation of gravels in tidally influenced fluvial systems: upper Albian (Lower Cretaceous) on the cratonic margin of the North American western interior foreland basin. *Cretac. Res.* 24:439–48

Cain ML, Milligan BG, Strand AE. 2000. Long-distance seed dispersal in plant populations. *Am. J. Bot.* 87:1217–27

Carrano MT, Janis CM, Sepkoski JJ. 1999. Hadrosaurs as ungulate parallels: Lost lifestyles and deficient data. *Acta Palaeontol. Pol.* 44:237–61

Carroll RL. 1988. *Vertebrate Paleontology and Evolution.* New York: Freeman. 698 pp.

Chandler MEJ. 1964. *The Lower Tertiary Floras of Southern England, IV. A Summary and Survey of Findings in the Light of Recent Botanical Observations.* London: Br. Mus. Nat. Hist. 151 pp.

Charlesdominique P. 1993. Speciation and coevolution—An interpretation of frugivory phenomena. *Vegetatio* 108:75–84

Chesters KIM. 1955. Some plant remains from the Upper Cretaceous and Tertiary of West Africa. *Ann. Mag. Nat. Hist. Ser. 12* 8:498–504

Chiappe LM, Dyke GJ. 2002. The Mesozoic radiation of birds. *Annu. Rev. Ecol. Syst.* 33:91–124

Chin K. 1997. What did dinosaurs eat? Coprolites and other direct evidence of dinosaur diets. In *The Complete Dinosaur*, ed. JO Farlow, MK Brett-Surman, pp. 371–82. Bloomington: Indiana Univ. Press. 752 pp.

Cipollini ML, Levey DJ. 1997. Secondary metabolites of fleshy vertebrate-dispersed fruits: adaptive hypotheses and implications for seed dispersal. *Am. Nat.* 150:346–72

Cipollini ML, Stiles EW. 1992. Antifungal activity of ripe ericaceous fruits: phenolic-acid interactions and palatability for dispersers. *Biochem. Syst. Ecol.* 20:501–14

Clark CJ, Poulsen JR, Parker VT. 2001. The role of arboreal seed dispersal groups on the seed rain of a lowland tropical forest. *Biotropica* 33:606–20

Clemens WA. 2001. Patterns of mammalian evolution across the Cretaceous-Tertiary boundary. *Mitt. Mus. Naturkd. Berl. Zool. Reihe* 77:175–91

Clemens WA, Kielan-Jaworowska Z. 1979. Multituberculata. In *Mesozoic Mammals. The First Two-Thirds of Mammalian History*, ed. JA Lillegraven, Z Kielan-Jaworowska, WA Clemens, pp. 99–149. Berkeley: Univ. Calif. Press

Collinson ME. 1983. Fossil plants of the London Clay. *The Palaeont. Assoc. Field Guide Fossils* 1:1–121

Collinson ME, Hooker JJ. 1987. Vegetational and mammalian faunal changes in the early Tertiary of southern England. See Friis et al. 1987, pp. 259–304

Collinson ME, Hooker JJ. 1991. Fossil evidence of interactions between plants and plant-eating mammals. *Philos. Trans. R. Soc. London Ser. B* 33:197–208

Collinson ME, Hooker JJ. 2000. Gnaw marks in Eocene seeds: evidence for early rodent behaviour. *Palaeogeogr. Palaeoclimatol. Palaeoecol.* 157:127–49

Corlett RT. 1998. Frugivory and seed dispersal by vertebrates in the Oriental (Indomalaysian) region. *Biol. Rev. Camb. Philos. Soc.* 73:413–48

Cox CB. 1998. The jaw function and adaptive radiation of the dicynodont mammal-like reptiles of the Karoo Basin of South Africa. *Zool. J. Linn. Soc.* 122:349–84

Crane PR, Herendeen PS. 1996. Cretaceous floras containing angiosperm flowers and fruits from eastern North America. *Rev. Palaeobot. Palynol.* 90:319–37

Crepet WL, Nixon KC, Gandolfo MA. 2001. Turonian flora of New Jersey, USA. In *7th Int. Symp. Mesozoic Terrestrial Ecosystems*, ed. HA Leanza. *Publ. Especial—Asoc. Paleontol. Argent.* 7:61–69

Crowley TJ, North GR. 1991. *Paleoclimatology*. Oxford: Oxford Univ. Press. 349 pp.

Currie PJ, Koppelhus EB, Muhammad AF. 1995. "Stomach" contents of a hadrosaur from the Dinosaur Park Formation (Campanian, Upper Cretaceous) of Alberta, Canada. In *6th Symp. Mesozoic Terrestrial Ecosystems and Biota, Short Papers*, ed. A Sun, Y Wang, pp. 111–14. Beijing: China Ocean Press. 250 pp.

Cypher BL. 1999. Germination rates of tree seeds ingested by coyotes and raccoons. *Am. Midl. Nat.* 142:71–76

Czerkas SJ, Czerkas SA. 1990. *Dinosaurs: A Global View*. Surrey, UK: Dragon's World. 247 pp.

Del Tredici P. 1989. Ginkgos and multituberculates: evolutionary interactions in the Tertiary. *Biosystems* 22:327–39

Dilcher D. 2000. Toward a new synthesis: major trends in the angiosperm fossil record. *Proc. Natl. Acad. Sci. USA* 94:7030–36

Dodd ME, Silvertown J, Chase MW. 1999. Phylogenetic analysis of trait evolution and species diversity variation among angiosperm families. *Evolution* 53:732–44

Eby P. 1998. An analysis of diet specialization in frugivorous *Pteropus poliocephalus* (Megachiroptera) in Australian subtropical rainforests. *Aust. J. Ecol.* 23:443–56

Ericson GP, Irestedt M, Johansson US. 2003. Evolution, biogeography and patterns of diversification of passerine birds. *J. Avian Biol.* 34:3–15

Eriksson O, Bremer B. 1991. Fruit characteristics, life forms and species richness in the plant family Rubiaceae. *Am. Nat.* 138:751–61

Eriksson O, Friis EM, Lofgren P. 2000a. Seed size, fruit size and dispersal systems in angiosperms from the Early Cretaceous to the late Tertiary. *Am. Nat.* 156:47–58

Eriksson O, Friis EM, Pedersen KR, Crane PR. 2000b. Seed size and dispersal systems of Early Cretaceous angiosperms from Famalicão, Portugal. *Int. J. Plant Sci.* 161:319–29

Eriksson O, Jakobsson A. 1999. Recruitment trade-offs and the evolution of dispersal mechanisms in plants. *Evol. Ecol.* 13:411–23

Fairon-Demaret M, Smith T. 2002. Fruits and seeds of the Tienen Formation at Dormaal, Paleocene-Eocene transition in eastern Belgium. *Rev. Palaeobot. Palynol.* 122:47–62

Feduccia A. 1995. Explosive evolution in Tertiary birds and mammals. *Science* 267:637–38

Feduccia A. 2003. 'Big bang' for Tertiary birds? *Trends Ecol. Evol.* 18:172–76

Fleming TH, Lips KR. 1991. Angiosperm endozoochory: were pterosaurs Cretaceous seed dispersers? *Am. Nat.* 138:1058–65

Friis EM. 1975. Climatic implications of microcarpological analyses of the Miocene Fasterholt flora, Denmark. *Bull. Geol. Soc. Den.* 24:179–91

Friis EM, Chaloner WG, Crane PR, eds. 1987. *The Origins of Angiosperms and Their Biological Consequences*. Cambridge, UK: Cambridge Univ. Press. 358 pp.

Friis EM, Pedersen KR, Crane PR. 1999. Early

angiosperm diversification: the diversity of pollen associated with angiosperm reproductive structures in Early Cretaceous floras from Portugal. *Ann. Mo. Bot. Gard.* 86:259–96

Friis EM, Pedersen KR, Crane PR. 2000. Reproductive structure and organization of basal angiosperms from the Early Cretaceous (Barremian or Aptian) of Western Portugal. *Int. J. Plant Sci.* 161(Suppl. 6):S169–82

Gee CT, Sander PM, Petzelberger BEM. 2003. A Miocene rodent nut cache in coastal dunes of the Lower Rhine Embayment, Germany. *Palaeontology* 46:1133–49

Gensel PG, Skog JE. 1977. Two Early Mississippian seeds from the Price Formation of southwestern Virginia. *Brittonia* 29:332–51

Gordon E. 1998. Seed characteristics of plant species from riverine wetlands in Venezuela. *Aquat. Bot.* 60:417–31

Goulding M. 1980. *The Fishes and the Forest.* Berkeley: Univ. Calif. Press. 280 pp.

Grimaldi D, Agosti D. 2000. A formicine in New Jersey Cretaceous amber (Hymenoptera: Formicidae) and early evolution of the ants. *Proc. Natl. Acad. Sci. USA* 97:13678–83

Grubb PJ, Metcalfe DJ. 1996. Adaptation and inertia in the Australian tropical lowland rain-forest flora: contradictory trends in intergeneric and intrageneric comparisons of seed size in relation to light demand. *Funct. Ecol.* 10:512–20

Hampe A. 2003. Large-scale geographical trends in fruit traits of vertebrate-dispersed temperate plants. *J. Biogeogr.* 30:487–96

Handel SN, Beattie AJ. 1990. Seed dispersal by ants. *Sci. Am.* 263:76–83

Harper JL, Lovell PH, Moore KG. 1970. The shapes and sizes of seeds. *Annu. Rev. Ecol. Syst.* 1:327–56

Harris TM. 1945. Notes on the Jurassic flora of Yorkshire. 21. A coprolite of *Caytonia* pollen. *Ann. Mag. Nat. Hist. Ser. II* 12:373–78

Herrera CM. 1982. Defense of ripe fruit from pests: its significance in relation to plant-disperser interactions. *Am. Nat.* 120:218–41

Herrera CM. 1989a. Seed dispersal by animals: A role in angiosperm diversification? *Am. Nat.* 133:309–22

Herrera CM. 1989b. Frugivory and seed dispersal by carnivorous mammals, and associated fruit characteristics, in undisturbed Mediterranean habitats. *Oikos* 55:250–62

Herrera CM. 1995. Plant-vertebrate seed dispersal systems in the Mediterranean: ecological, evolutionary and historical determinants. *Annu. Rev. Ecol. Syst.* 26:705–27

Herrera CM. 2002. Seed dispersal by vertebrates. See Herrera & Pellmyr 2002, pp. 185–208

Herrera CM, Pellmyr O, eds. 2002. *Plant-Animal Interactions: An Evolutionary Approach.* Oxford: Blackwell. 313 pp.

Herrera J. 1987. Flower and fruit biology in southern Spanish Mediterranean shrublands. *Ann. Mo. Bot. Gard.* 74:69–78

Hill CR. 1976. Coprolites of *Ptilophyllum* cuticles from the Middle Jurassic of North Yorkshire. *Bull. Br. Mus. Nat. Hist. Geol.* 27:289–94

Hilton J, Wang SJ, Zhu WQ, Tian BL, Galtier J, Wei AH. 2002. *Callospermarion* ovules from the Early Permian of northern China: palaeofloristic and palaeogeographic significance of callistophytalean seed-ferns in the Cathaysian flora. *Rev. Palaeobot. Palynol.* 120:301–14

Hotton N III, Olson EC, Beerbower R. 1997. The amniote transition and the discovery of herbivory. See Sumida & Martin 1997, pp. 207–64

Howe HF. 1977. Bird activity and seed dispersal of a tropical wet forest tree. *Ecology* 58:539–50

Howe HF, Smallwood J. 1982. Ecology of seed dispersal. *Annu. Rev. Ecol. Syst.* 13:201–28

Howe J, Cantrill DJ. 2001. Palaeoecology and taxonomy of Pentoxylales from the Albian of Antarctica. *Cretac. Res.* 22:779–93

Hughes L, Dunlop M, French K, Leishman WR, Rice B, et al. 1994a. Predicting dispersal spectra: a minimal set of hypotheses based on plant attributes. *J. Ecol.* 82:933–50

Hughes L, Westoby M, Jurado E. 1994b.

Convergence of elaiosomes and insect prey: evidence from ant foraging behaviour and fatty acid composition. *Funct. Ecol.* 8:358–65

Ibarra-Manriquez G, Ramos MM, Oyama K. 2001. Seedling functional types in a lowland rain forest in Mexico. *Am. J. Bot.* 88:1801–12

Jacobs BF, Kingston JD, Jacobs LL. 1999. The origin of grass-dominated ecosystems *Ann. Mo. Bot. Gard.* 86:590–643

Janis CM. 1993. Tertiary mammal evolution in the context of changing climates, vegetation and tectonic events. *Annu. Rev. Ecol. Syst.* 24:467–500

Janis CM. 2000. Patterns in the evolution of herbivory in large terrestrial mammals. See Sues 2000, pp. 168–222

Janis CM, Damuth J, Theodora JM. 2002. The origins and evolution of the North American grassland biome: the story from the hoofed mammals. *Palaeogeogr. Palaeoclimatol. Palaeoecol.* 177:183–98

Janzen DH. 1984. Dispersal of small seeds by big herbivores: foliage is the fruit. *Am. Nat.* 123:338–53

Janzen DH, Martin PS. 1982. Neotropical anachronisms: the fruits the gomphotheres ate. *Science* 215:19–27

Jepson GL. 1970. Bat origins and evolution. In *Biology of Bats*, ed. WA Wimsatt. 1:1–64. New York: Academic. 406 pp.

Jordano P. 1995. Angiosperm fleshy fruits and seed dispersers—a comparative analysis of adaptation and constraints in plant-animal interactions. *Am. Nat.* 145:163–91

Kielan-Jaworowska Z, Cifelli RL, Luo Z-X. 2004. *Mammals from the Age of Dinosaurs: Origins, Evolution and Structure.* New York: Columbia Univ. Press. In press

King GM. 1990. The *Dicynodonts: A Study in Palaeobiology.* London: Chapman & Hall. 233 pp.

King GM. 1996. *Reptiles and Herbivory.* London: Chapman & Hall. 160 pp.

Klavins SD, Taylor EL, Krings M, Taylor TN. 2001. An unusual, structurally preserved ovule from the Permian of Antarctica. *Rev. Palaeobot. Palynol.* 115:107–17

Knapp S. 2002. Tobacco to tomatoes; a phylogenetic perspective on fruit diversity in the Solanaceae. *J. Exp. Bot.* 53:2001–22

Knoll AH, Carroll SB. 1999. Early animal evolution: emerging views from comparative biology and geology. *Science* 284:2129–37

Krasilov VA. 1999. A reappearance of an archaic structure in the latest Permian seeds. *Paleontol. Zh.* 33:330–33

Krasilov VA, Martinson GG. 1982. Fossil fruit from the Upper Cretaceous deposits of Mongolia. *Paleontol. J.* 1982(1):113–21

Kräusel R. 1922. Die Nahrung von Trachodon. *Paläontol. Zeit.* 4:80

Labandeira CC. 1998. Plant-insect associations from the fossil record. *Geotimes* 43:18–24

Labandeira CC. 2002. The history of associations between plants and animals. See Herrera & Pellmyr 2002, pp. 26–74

Lamb JP. 2001. Late Cretaceous (Santonian) vegetation from the Gulf Coast. *Geol. Soc. Am. Abstr. Prog.* 33(2):16

Lambert JE, Garber PA. 1998. Evolutionary and ecological implications of primate seed dispersal. *Am. J. Primatol.* 45:9–28

Leishman MR, Wright IJ, Moles AT, Westoby M. 2000. The evolutionary ecology of seed size. In *Seeds—The Ecology of Regeneration in Plant Communities*, ed. M Fenner, pp. 31–57. Wallingford: CAB Int. 373 pp.

Levey DJ, Silva WR, Galetti M, eds. 2002. *Seed Dispersal and Frugivory: Ecology, Evolution and Conservation.* Wallingford: CAB Int. 511 pp.

Lord JM, Marshall J. 2001. Correlations between growth form, habitat, and fruit colour in the New Zealand flora, with references to frugivory by lizards. *N. Z. J. Bot.* 39:567–76

Mack AL. 2000. Did fleshy fruit pulp evolve as a defense against seed loss rather than as a dispersal mechanism? *J. Biosci.* 25:93–97

Magallon S, Sanderson MJ. 2001. Absolute diversification rates in angiosperm clades. *Evolution* 55:1762–80

Mai DH. 2000. Palm trees in the past—paleoclimatological and paleoecological indicators. *GFF* 122:97–98

Manchester SR. 1987. The fossil history of the

Juglandaceae. *Monogr. Syst. Bot. Mo. Bot. Gard.* 21:1–137

Manchester SR. 1994. Fruits and seeds of the Middle Eocene Nut Beds Flora, Clarno Formation, Oregon. *Palaeontogr. Am.* 58:1–205

Marshall JE, Hemsley AR. 2003. A Mid Devonian seed-megaspore from East Greenland and the origin of seed plants. *Palaeontology* 46:647–70

Mazer SJ, Wheelwright NT. 1993. Fruit size and shape: allometry at different taxonomic levels in bird-dispersed plants. *Evol. Ecol.* 7:556–75

Midgley JJ, Bond WJ. 1991. How important is biotic pollination and dispersal to the success of the angiosperms? *Philos. Trans. R. Soc. London Ser. B* 333:209–15

Milner AR. 1993. Biogeography of Paleozoic tetrapods. In *Palaeozoic Vertebrate Biostratigraphy and Biogeography*, ed. JA Long, pp. 324–53. London: Belhaven. 369 pp.

Milner AR, Smithson TR, Milner AC, Coates MI, Rolfe WDI. 1986. The search for early tetrapods. *Mod. Geol.* 10:1–28

Milton SJ, Dean WRJ. 2001. Seeds dispersed in dung of insectivores and herbivores in semi-arid southern Africa. *J. Arid. Environ.* 47:465–83

Modesto SP. 1992. Did herbivory foster early amniote diversification? *J. Vertebr. Paleontol.* 12(Suppl.):44a (Abstr.)

Modesto SP. 1995. The skull of the herbivorous synapsid *Edaphosaurus boanerges* from the Lower Permian of Texas. *Palaeontology* 38:213–39

Modesto SP, Reisz RR. 1992. Restudy of Permo-Carboniferous synapsid *Edaphosaurus novomexicanus* Williston and Case, the oldest known herbivorous amniote. *Can. J. Earth Sci.* 29:2653–62

Mohabey DM. 2001. Dinosaur eggs and dung (fecal mass) from Late Cretaceous of central India; dietary implications. *Geol. Soc. India Spec. Publ.* 64:605–15

Moles AT, Westoby M. 2003. Latitude, seed predation and seed mass. *J. Biogeogr.* 30:105–28

Moll D, Jansen KP. 1995. Evidence for a role in seed dispersal by two tropical herbivorous turtles. *Biotropica* 27:121–27

Molnar RE, Clifford HT. 2000. Gut contents of a small ankylosaur. *J. Vertebr. Paleontol.* 20:194–96

Molnar RE, Clifford HT. 2001. An ankylosaurian cololite from the Lower Cretaceous of Queensland, Australia. In *The Armored Dinosaurs*, ed. K Carpenter, pp. 399–412. Bloomington: Indiana Univ. Press. 526 pp.

Monteillet J, Lappartient J-R. 1981. Fossil fruits and seeds of the Upper Cretaceous from the Paki Quarry, Senegal. *Rev. Palaeobot. Palynol.* 34:331–44

Motta-Junior JC, Martins K. 2002. The frugivorous diet of the Maned Wolf, *Chrysocyon brachyurus*, in Brazil: ecology and conservation. See Levey et al. 2002, pp. 291–303

Munk W, Sues H-D. 1993. Gut contents of *Parasaurus* (Pareiasauria) and *Protorosaurus* (Archosauromorpha) from the Kupferschiefer (Upper Permian) of Hessen, Germany. *Paläeontol. Z.* 67:169–76

Murray DR, ed. 1986. *Seed Dispersal.* Sydney: Academic. 322 pp.

Nambudiri EM, Binda PL. 1989. Dicotyledonous fruits associated with coprolites from the Upper Cretaceous (Maastrichtian) Whitemud Formation, southern Saskatchewan, Canada. *Rev. Palaeobot. Palynol.* 59:57–66

Nathan R, Muller-Landau HC. 2000. Spatial patterns of seed dispersal, their determinants and consequences for recruitment. *Trends Ecol. Evol.* 15:278–85

Nessov LA, Archibald JD, Kielan-Jaworowska Z. 1998. Ungulate-like mammals from the Late Cretaceous of Uzbekistan and a phylogenetic analysis of Ungulatomorpha. In *Dawn of the Age of Mammals in Asia*, ed. KC Beard, MR Dawson. *Bull. Carnegie Mus. Nat. Hist.* 34:40–88

Niklas KJ, Tiffney BH, Knoll AH. 1985. Patterns in vascular land plant diversification: an analysis at the species level. In *Phanerozoic Diversity Patterns: Profiles in Macroevolution*, ed. JW Valentine, pp. 97–128. Princeton, NJ: Princeton Univ. Press. 441 pp.

Novacek MJ. 1999. 100 million years of land

vertebrate evolution: the Cretaceous–Early Tertiary transition. *Ann. Mo. Bot. Gard.* 86: 230–58

Padian K, Chiappe LM. 1998. The origin and early evolution of birds. *Biol. Rev. Camb. Philos. Soc.* 73:1–42

Parrish JT, Ziegler AM, Scotese CR. 1982. Rainfall patterns and the distribution of coals and evaporites in the Mesozoic and Cenozoic. *Palaeogeogr. Palaeoclimatol. Palaeoecol.* 40:67–101

Pedersen KR, Friis EM, Crane PR. 1993. Pollen organs and seeds with *Decussosporites* Brenner from Lower Cretaceous Potomac Group sediments in eastern USA. *Grana Palynol.* 32:273–89

Ray S, Chinsamy A. 2003. Functional aspects of the postcranial anatomy of the Permian dicynodont *Diictodon* and their ecological implications. *Palaeontology.* 46:151–83

Rayner RJ. 1992. *Phyllotheca*—the pastures of the Late Permian. *Palaeogeogr. Palaeoclimatol. Palaeoecol.* 92:31–40

Regan BC, Julliot C, Simmen B, Vienot F, Charlesdominique P, Mollon JD. 2001. Fruits, foliage and the evolution of primate colour vision. *Philos. Trans. R. Soc. London Ser. B Biol. Sci.* 356:229–83

Rensberger JM. 2000. Dental constraints in the early evolution of mammalian herbivory. See Sues 2000, pp. 144–67

Reymanowna M. 1973. The Jurassic flora from Grojec near Krakow in Poland. Part II. Caytoniales and anatomy of *Caytonia. Acta Palaeobot.* 14:45–87

Richter G. 1987. Untersuchung zur Ernährung eozäner Säuger aus der Fossilfundstätte Messel bei Darmstadt. *Cour. Forsch.-Inst. Senckenb.* 99:1–33

Richter G. 1988. Problems in the analysis of stomach contents of Eocene mammals from the Messel oil shale layers. *Cour. Forsch.-Inst. Senckenb.* 107:121–27

Rodriguez-de la Rosa RA, Cevallos-Ferriz SRS, Silva-Pineda A. 1998. Paleobiological implications of Campanian coprolites. *Palaeogeogr. Palaeoclimatol. Palaeoecol.* 142:231–54

Rogers LL, Applegate RD. 1983. Dispersal of fruit and seeds by black bears. *J. Mamm.* 64:310–11

Salisbury EJ. 1942. *The Reproductive Capacity of Plants: Studies in Quantitative Biology.* London: Bell. 244 pp.

Sallabanks R, Courtney SP. 1992. Frugivory, seed predation and insect-vertebrate interactions. *Annu. Rev. Entomol.* 37:377–400

Schemske DW. 1983. Limits to specialization and co-evolution in plant-animal mutualisms. In *Coevolution*, ed. MN Nitecki, pp. 67–109. Chicago: Univ. Chicago Press. 392 pp.

Schmidt W, Schürmann M, Teichmuller M. 1958. Biss-Spuren an Früchten des Miozän-Waldes der neiderrheinischen Braunkohlen-Formation. *Fortschr. Geol. Rheinl. Westfalen* 2:563–72

Schwanke C. 1998. Herbivory in dicynodonts and the evidence of coevolutionary patterns. *J. Vertebr. Paleontol.* 18(Suppl.):A76 (Abstr.)

Schweitzer H-J. 1968. Die Flora des Oberen Perms in Mitteleuropa. *Naturwissens. Rund.* 21:93–102

Scott AC, Taylor TN. 1983. Plant/animal interactions during the Upper Carboniferous. *Bot. Rev.* 49:259–307

Shoemaker RE. 1977. Fossil fig-like objects from the Upper Cretaceous sediments of the western interior of North America. *Palaeontogr. Abt. B* 161:165–75

Sims HJ. 1997. Evolutionary trends in seed size during the late Paleozoic. *Geol. Soc. Am. Abstr. Prog.* 29(6):461

Sims HJ. 1999. The latest Paleozoic radiation of seed plants; patterns in taxonomic diversity. *Geol. Soc. Am. Abstr. Prog.* 31:364

Sims HJ. 2000. The accretion of morphological and ecological diversity in the late Paleozoic radiation of seeds and seed plants. *Geol. Soc. Am. Abstr. Prog.* 32:(7):194–95

Smith AG, Hurley AM, Briden JC. 1981. *Phanerozoic Paleocontinental World Maps.* Cambridge, UK: Cambridge Univ. Press. 102 pp.

Smith JF. 2001. High species diversity in fleshy-fruited tropical understory plants. *Am. Nat.* 157:646–53

Sorensen AE. 1986. Seed dispersal by adhesion. *Annu. Rev. Ecol. Syst.* 17:443–63

Stanley SM. 1999. *Earth System History.* New York: Freeman. 615 pp.

Sternberg CH. 1909. A new *Trachodon. Science* 29:753–54

Stokes WL. 1964. Fossilized stomach contents of a sauropod dinosaur. *Science* 143:576–77

Sues H-D, ed. 2000. *Evolution of Herbivory in Terrestrial Vertebrates. Perspectives from the Fossil Record.* Cambridge, NY: Cambridge Univ. Press. 256 pp.

Sumida SS, Martin KLM, eds. 1997. *Amniote Origins: Completing the Transition to Land.* San Diego: Academic. 510 pp.

Sussman RW. 1991. Primate origins and the evolution of angiosperms. *Am. J. Primatol.* 23:209–23

Sussman RW. 1995. How primates invented the rainforest and vice versa. In *Creatures of the Dark: The Nocturnal Prosimians*, ed. L Alterma, G Doyle, M Izard, pp. 1–10. New York: Plenum. 571 pp.

Szalay FS. 1968. The beginnings of primates. *Evolution* 22:19–36

Takahashi M, Crane PR, Manchester SR. 2002. *Hironoia fusiformis* gen. et sp. nov.; a cornalean fruit from the Kamikitaba locality (Upper Cretaceous, Lower Coniacian) in northeastern Japan. *J. Plant Res.* 115:463–73

Taylor TN. 1965. Paleozoic seed studies: a monograph of the American species of *Pachytesta. Palaeontogr. Abt. B* 117:1–46

Taylor TN, Scott AC. 1983. Interactions of plants and animals during the Carboniferous. *BioScience* 33:488–93

Taylor TN, Taylor EL. 1993. *The Biology and Evolution of Fossil Plants.* Englewood Cliffs: Prentice Hall. 982 pp.

Tewksbury JJ. 2002. Fruits, frugivores and the evolutionary arms race. *New Phytol.* 156:137–39

Thulborn RA. 1991. Morphology, preservation and paleobiological significance of dinosaur coprolites. *Palaeogeogr. Palaeoclimatol. Palaeoecol.* 83:341–66

Tiffney BH. 1977. Dicotyledonous angiosperm flower from the Upper Cretaceous of Martha's Vineyard, Massachusetts. *Nature* 265:136–37

Tiffney BH. 1984. Seed size, dispersal syndromes, and the rise of the angiosperms: evidence and hypothesis. *Ann. Mo. Bot. Gard.* 71:551–76

Tiffney BH. 1986a. Evolution of seed dispersal syndromes according to the fossil record. In *Seed Dispersal*, ed. DR Murray, pp. 274–305. Sydney: Academic. 322 pp.

Tiffney BH. 1986b. Fruit and seed dispersal and the evolution of the Hamamelidae. *Ann. Mo. Bot. Gard.* 73:394–416

Tiffney BH. 1990. The collection and study of dispersed angiosperm fruits and seeds. *Palaios* 5:499–519

Tiffney BH. 1992. The role of vertebrate herbivory in the evolution of land plants. *Palaeobotanist* 41:87–97

Tiffney BH. 1997. Land plants as food and habitat in the age of dinosaurs. In *The Complete Dinosaur*, ed. JO Farlow, MK Brett-Surman, pp. 352–70. Bloomington: Indiana Univ. Press. 752 pp.

Tiffney BH, Mazer SJ. 1995. Angiosperm growth habit, dispersal and diversification reconsidered. *Evol. Ecol.* 9:93–117

Tomback DF, Linhart YB. 1990. The evolution of bird-dispersed pines. *Evol. Ecol.* 4:185–219

Tralau H. 1968. Evolutionary trends in the genus *Ginkgo. Lethaia* 1:63–101

Ufnar DF, Gonzalez LA, Ludvigson GA, Brenner RL, Witzke BJ. 2002. The mid-Cretaceous water bearer: isotope mass balance quantification of the Albian hydrologic cycle. *Palaeogeogr. Palaeoclimatol. Palaeoecol.* 188:51–71

Uhl NW, Dransfield J. 1987. *Genera Palmarum. A Classification of Palms Based on the Work of Harold E. Moore, Jr.* Lawrence, KS: Allan Press. 610 pp.

Univ. Mich. Mus. Zool. 2004. *Animal diversity web.* http://animaldiversity.ummz.umich.

edu/chordata/mammalia/chiroptera/pteropodidae.html; http://animaldiversity.ummz.umich.edu/chordata/mammalia/chiroptera/phyllostomidae.html

Valido A, Nogales M. 2003. Digestive ecology of two omnivorous Canarian lizard species (Gallotia, Lacertidae). *Amphibia-Reptilia.* 24:331–44

van der Pijl L. 1982. *Principles of Dispersal in Higher Plants.* Berlin: Springer-Verlag. 214 pp. 3rd ed.

Vander Wall SB. 1992. The role of animals in dispersing a wind-dispersed pine. *Ecology* 73:614–21

Vander Wall SB. 2001. The evolutionary ecology of nut dispersal. *Bot. Rev.* 67:74–117

Vaudois-Miéja N, Lejal-Nicol A. 1987. Paléocarpologie africaine: apparition dès l'Aptien en Égypte d'un palmier (*Hyphaeneocarpon aegyptiacum* n. sp.). *C. R. Acad. Sci. Paris Sér. II* 304:233–38

Vega JC, Archangelsky S. 2000. *Jejenia* gen. nov., a new Carboniferous disseminule from San Juan, Argentina. *Bol. Acad. Nac. Cienc.* 64:61–69

Vellend M, Myers JA, Gardescu S, Marks PL. 2003. Dispersal of *Trillium* seeds by deer: implications for long-distance migration of forest herbs. *Ecology* 84:1067–72

Wall CE, Krause DW. 1992. A biomechanical analysis of the masticatory apparatus of *Ptilodus* (Multituberculata). *J. Vertebr. Paleontol.* 12:172–87

Wang BC, Smith TB. 2002. Closing the seed dispersal loop. *Trends Ecol. Evol.* 17:379–85

Weigelt J. 1927. *Rezente Wirbeltierleichen und ihre paläobiologische Bedeutung.* Leipzig: Verlag von Max Weg. 227 pp.

Weigelt J. 1930. Über die vermutliche Nahrung von *Protorosaurus* und über einen körperlich erhaltenen Fruchtstand von *Archaeopodocarpus germanica* aut. *Leopoldina* 6:269–80

Weishampel DB. 1984. Interactions between Mesozoic plants and vertebrates: fructifications and seed predation. *Neues Jahrb. Geol. Palaeontol. Abh.* 167:224–50

Weishampel DB, Norman DB. 1989. Vertebrate herbivory in the Mesozoic: jaws, plants and evolutionary metrics. *Geol. Soc. Am. Spec. Pap.* 238:87–100

Wellnhofer P. 1991. *The Illustrated Encyclopedia of Pterosaurs.* New York: Crescent Books. 192 pp.

Wenny DG. 2001. Advantages of seed dispersal: a re-evaluation of directed dispersal. *Evol. Ecol. Res.* 3:51–74

White T, Gonzalez L, Ludvigson G, Poulsen C. 2001. Middle Cretaceous greenhouse hydrologic cycle of North America. *Geology* 29:363–66

Willson MF, Sabag C, Figueroa J, Armesto JJ. 1996. Frugivory and seed dispersal of *Podocarpus nubigena* in Chiloe, Chile. *Rev. Chil. Hist. Nat.* 69:343–49

Willson MF, Whelan CJ. 1990. The evolution of fruit-color in fleshy-fruited plants. *Am. Nat.* 136:790–809

Wing SL, Boucher LD. 1998. Ecological aspects of the Cretaceous flowering plant radiation. *Annu. Rev. Earth Planet. Sci.* 26:379–421

Wing SL, Tiffney BH. 1987a. Interactions of angiosperms and herbivorous tetrapods through time. See Friis et al. 1987, pp. 203–24

Wing SL, Tiffney BH. 1987b. The reciprocal interaction of angiosperm evolution and tetrapod herbivory. *Rev. Palaeobot. Palynol.* 50:179–210

Witmer MC, Cheke AS. 1991. The dodo and the tambalacoque tree—an obligate mutualism reconsidered. *Oikos* 61:133–37

Wolfe JA, Upchurch GR Jr. 1986. Vegetation, climatic and floral changes at the Cretaceous-Tertiary boundary. *Nature* 324:148–52

Wolfe JA, Upchurch GR Jr. 1987. North American nonmarine climates and vegetation during the Late Cretaceous. *Palaeogeogr. Palaeoclimatol. Palaeoecol.* 61:33–77

Zhou Z, Barrett PM, Hilton J. 2003. An exceptionally preserved lower Cretaceous ecosystem. *Nature* 421:807–14

Zhou Z, Zhang F. 2002. A long-tailed, seed-eating bird from the Early Cretaceous of China. *Nature* 418:405–9

Zhou Z, Zhang F. 2003. Anatomy of the primitive bird *Sapeornis chaoyangensis* from the Early Cretaceous of Liaoning, China. *Can. J. Earth Sci.* 40:731–47

Ziegler AM, Raymond AL, Gierlowski TC, Horrell MA, Rowley DB, Lottes AL. 1987. Coal, climate and terrestrial productivity: the present and Early Cretaceous compared. In *Coal and Coal-Bearing Strata: Recent Advances*, ed. AC Scott. *Geol. Soc. Spec. Publ.* 32:25–49

ARE DISEASES INCREASING IN THE OCEAN?*

Kevin D. Lafferty,[1] James W. Porter,[2] and Susan E. Ford[3]

[1]U.S. Geological Survey, Western Ecological Research Center, c/o Marine Science Institute, University of California, Santa Barbara, California 93106; email: Lafferty@lifesci.ucsb.edu
[2]Institute of Ecology, University of Georgia, Athens, Georgia 30602; email: jporter@uga.edu
[3]Haskin Shellfish Research Laboratory, Rutgers University, Port Norris, New Jersey 08349; email: susan@hsrl.rutgers.edu

Key Words mass mortality, bleaching, global warming, disease, marine

■ **Abstract** Many factors (climate warming, pollution, harvesting, introduced species) can contribute to disease outbreaks in marine life. Concomitant increases in each of these makes it difficult to attribute recent changes in disease occurrence or severity to any one factor. For example, the increase in disease of Caribbean coral is postulated to be a result of climate change and introduction of terrestrial pathogens. Indirect evidence exists that (*a*) warming increased disease in turtles; (*b*) protection, pollution, and terrestrial pathogens increased mammal disease; (*c*) aquaculture increased disease in mollusks; and (*d*) release from overfished predators increased sea urchin disease. In contrast, fishing and pollution may have reduced disease in fishes. In other taxa (e.g., sea grasses, crustaceans, sharks), there is little evidence that disease has changed over time. The diversity of patterns suggests there are many ways that environmental change can interact with disease in the ocean.

INTRODUCTION

The perception of an ecological crisis in the oceans has led to research on signs of deterioration in ocean health. Recent mass mortalities in marine systems, including Caribbean sea urchins (Lessios 1988), phocine distemper virus (Heide-Jorgensen et al. 1992), pilchard mortalities (Jones et al. 1997), and especially coral bleaching (Hoegh-Guldberg 1999), have consequently received the attention and concern of marine ecologists (e.g., Goreau et al. 1998, Greenstein et al. 1998, Hayes & Goreau 1998, Hoegh-Guldberg 1999, Porter et al. 2001, Wilkinson 2002). The scale of these events and their impact on populations and associated communities have

---

*The U.S. Government has the right to retain a nonexclusive, royalty-free license in and to any copyright covering this paper.

been dramatic in some cases. Some species, such as black abalone in California (Lafferty & Kuris 1993), sun stars in the Sea of Cortez (Dungan et al. 1982), staghorn coral (Diaz-Soltero 1999), and long-spined sea urchin in the Caribbean (Lessios 1988), were reduced to such low population densities by infectious disease that recovery is in question. That these events seem unprecedented has led to the hypothesis that disease outbreaks in marine organisms have increased in recent years (Epstein 1996, Harvell et al. 1999, Hayes et al. 2001, Williams & Bunkley-Williams 1990). Here, we review the little systematic evidence available on changes in marine disease and outline factors that can increase (or decrease) disease in the ocean.

Coral diseases have received increasing attention as an explanation for the recent decline in coral reefs and this notable situation provides a fitting start for our review. Wilkinson (2002) estimates that 27% of reefs worldwide have already been lost, with an additional 16% at serious risk. Between 1996 and 2001, coral reefs of the Florida Keys lost 37% of their living coral (Porter et al. 2002), and some species, such as the massive, branching elkhorn coral (*Acropora palmata*), have declined by 91% (Patterson et al. 2002). Surveys (Connell 1997, Dustan 1999, Porter et al. 2002, Porter & Meier 1992) reveal that disease is a primary cause of this coral reef decline (Aronson & Precht 1998, Gladfelter 1982, Kim & Harvell 2004). Elevated sea surface temperature from El Niño events is a common explanation for coral bleaching (Hoegh-Guldberg 1999, Williams & Bunkley-Williams 1990), and such a rise in temperature could increase coral susceptibility to infectious diseases (Harvell et al. 2001). Diseases have had a major impact on the biodiversity of coral reefs (Chadwick-Furman 1996), eliminating rare species and severely reducing the abundance of common ones (Porter et al. 2001). For example, species within the branching acroporidae reproduce almost exclusively by fragmentation rather than by sexual reproduction (Aronson & Precht 2001, Knowlton 1992). Although this reproductive strategy works well after storms, it is ineffective after epizootics that destroy all the living tissue (Patterson et al. 2002). *Acropora palmata*, the most common coral in the Caribbean, experienced such rapid and widespread losses that it has been proposed for inclusion on the Endangered Species List (Diaz-Soltero 1999). The species susceptible to the greatest number of discases are the following massive reef building corals: *Montastraea annularis* (nine diseases), *Colpophyllia natans* (eight diseases), *M. faveolata* (six diseases), *Acropora palmata* (six diseases), *M. frankski* (five diseases), *M. cavernosa* (five diseases), and *Diploria labyrinthiformis* (five diseases). Loss of massive reef builders has obvious large-scale effects on coral reef ecosystems.

In this paper, we assess whether there is direct or indirect evidence for a change in disease over time. We then consider difficulties in making simple predictions about stress and disease. We discuss how increasing anthropogenic factors, namely climate warming, pollution, changes in density (through fishing and conservation), and pathogen pollution, could affect disease in the ocean. We conclude with a case study of increasing disease in oysters.

EVIDENCE FOR CHANGE

Determining whether disease is changing over time is problematic because the lack of baseline data for most marine communities precludes a direct test (Ward & Lafferty 2004). In this section, we consider some of the empirical evidence as well as suggestive patterns from reviews of the literature.

Direct Observations of Disease

Systematically collected data for changes in disease prevalence or severity are rare compared with lists of recent, dramatic events. Proof that change has occurred in any complex system requires baseline data from before a disease event that can be compared with data collected during and postepizootic. If monitoring programs track disease, they can reveal trends in disease over time.

We reanalyzed data on sea urchins (from the Channel Islands in southern California) from Lafferty (2004) to track changes over 20 years in the percent of stations where bacterial disease was observed (Figure 1). Over the course of the study, urchin disease significantly increased with time ($R = 0.87, N = 20, P < 0.0001$). This is because disease was absent for the first decade of the study and,

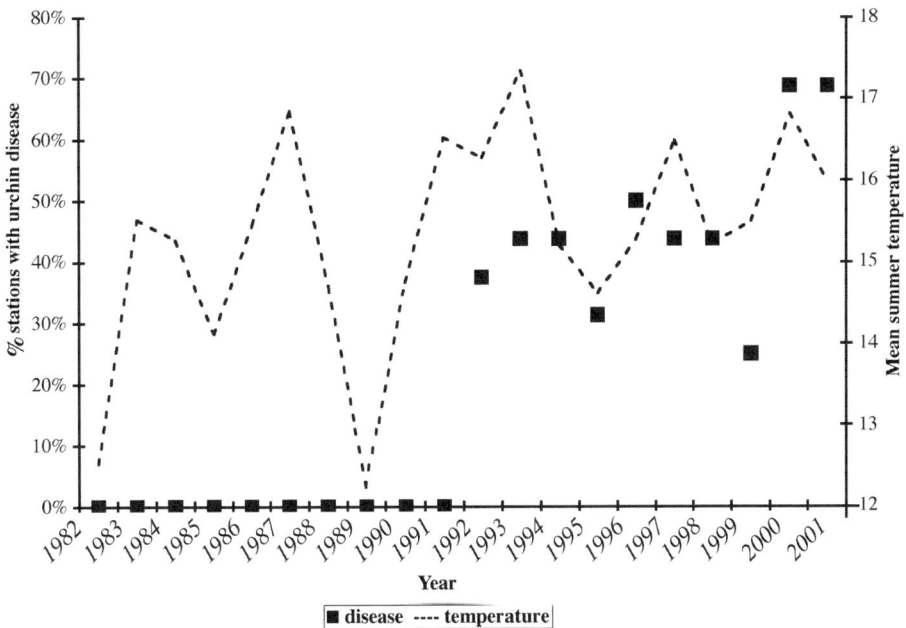

Figure 1 The percentage of 16 Channel Islands Kelp Forest monitoring stations with disease in sea urchins. Disease increased significantly over 20 years, but not evidently because of warming (Lafferty 2004).

after appearing dramatically in 1992, had a slight, but insignificant, increase (R = 0.5, N = 10, P = 0.14). Variation in the percent of stations with disease (following the arrival of the disease) was not significantly correlated with summer temperatures (R = 0.4, N = 10, P = 0.27). These results support an emerging pathogen scenario, but not one of increasing disease because of warming. As we describe below, fishing lobsters appears to be the main factor explaining disease in this system.

The Coral Reef Monitoring Project was established in the Florida Keys National Marine Sanctuary in 1996 (Aronson & Precht 1998, Gladfelter 1982, Porter et al. 2001). Between 1996 and 2003, there were significant increases in the number of stations containing diseased coral (Figure 2), the number of species with disease, and the number of different types of disease. These data show definitively that (a) coral disease increased throughout the Florida Keys over a six-year period and (b) these increases were not due to observational inadequacies. Although the short time series does not necessarily allow the detection of a long-term trend, the most dramatic increases in disease fell on the heels of the 1997–1998 El Niño event, suggesting a link between disease and climate. Between 1996 and 1998, the number of coral species with disease in the Florida Keys survey increased from 11 to 36 (85% of all species are now affected by disease, a 218% increase over 1996 values) (Porter et al. 2001).

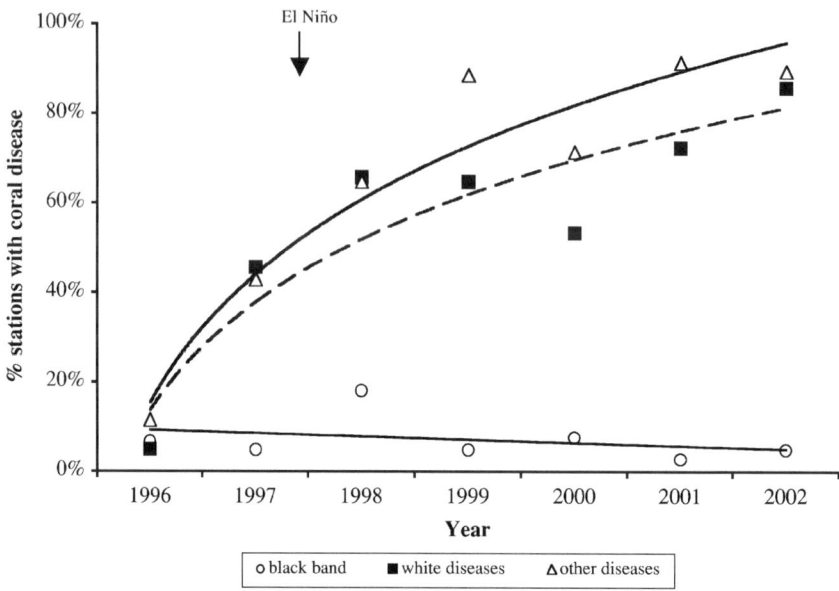

Figure 2 The percentage of the Florida Keys Coral Reef Monitoring Project stations with disease increased significantly between 1996 and 2002, with 1997–1998 El Niño noted. These data come from 105 randomly chosen stations that were censused annually by uniform survey techniques.

Patterns in the Literature

Because few marine studies systematically track disease over time, an indirect approach has been to determine if reports of disease in the scientific literature have increased over time. In one of the first efforts, Epstein (1996) and Epstein et al. (1998) plotted reports of various marine events over time. Many of these events were disease related, but they also included other mass mortalities (e.g., harmful algal blooms). Most increased in frequency from 1970 to the late 1990s. The authors suggest this corresponds to climatic changes, most notably higher frequencies of high sea-surface temperature. Subsequently, Harvell et al. (1999) published a table of examples of marine diseases over time as evidence consistent with the hypothesis that the rate of disease events had increased, but noted that the apparent increase could be an artifact of increased detection ability. For example, the advent of molecular techniques has improved diagnostics for viruses and other pathogens difficult to assay by traditional means. Harvell et al. (1999) also listed environmental correlates associated with each event, noting that high temperature was often a factor in marine disease, and suggested a role for pollution in diseases of marine mammals. They further recognized that many of the diseases appeared novel, resulting from host shifts, range shifts, or species introductions. Hayes et al. (2001) statistically analyzed the selected examples from the Harvell et al. (1999) and Epstein (1996) studies and found a significant increase over time in both the variety of disease organisms and the absolute number of episodes of disease (though it is questionable whether statistical analysis of selected examples is appropriate). Hayes et al. (2001) also noted that the timing of a climate regime shift in the mid-1970s corresponded with the increase in reports of disease. They further postulated that changes in atmospheric circulation resulted in an increased deposition of iron-rich dust that they hypothesized could favor the growth of opportunistic pathogens. Predictions of future increases in disease in the ocean owing to climate change lend new urgency to the desire to understand causes and trends of marine disease outbreaks (Harvell et al. 2002).

Combining trends among host taxa may obscure trends specific to a particular group. In an effort to determine if coral diseases have increased over time, we show reports for the number of coral disease agents described over time. Because this is a cumulative list, it will increase or stay the same from year to year. It is the shape of the increase that informs whether disease is increasing or decreasing. Figure 3 shows an exponential increase in the number of described coral diseases from 1965 to 2003 (Sutherland et al. 2004). Although it is impossible to determine if the exponential increase is due to an increase in observational intensity or to real increases in the number of diseases, many of the most recently described diseases, such as white pox (Patterson et al. 2002), have such clear manifestations that it is unlikely they were historically common. Fortunately, corals leave an excellent fossil record. This has allowed researchers to determine that the recent shift of dominant corals (from *Acropora* to *Agaricia*) on reefs owing to white band disease was unprecedented in the last 3000 years (Aronson et al. 2002).

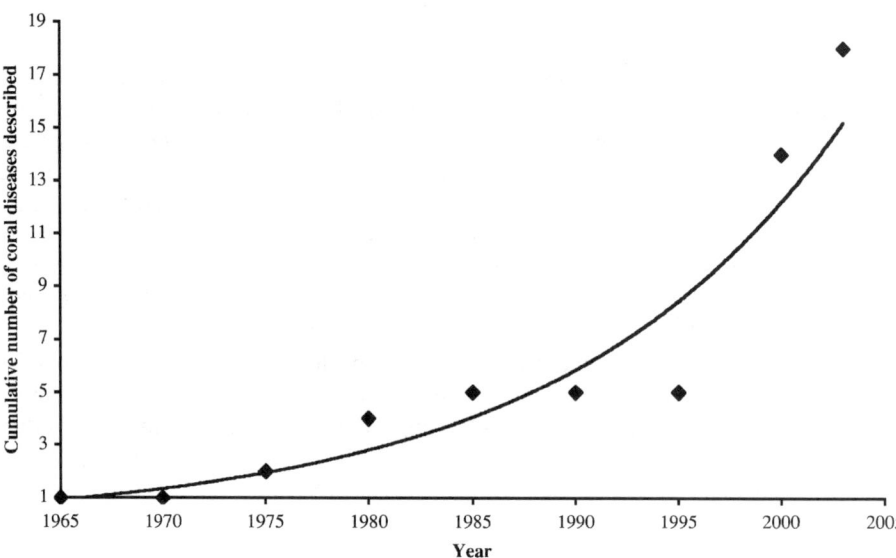

Figure 3 The cumulative increase in described coral diseases shows an exponential rise since 1965 (Sutherland et al. 2004).

Ward & Lafferty (2004) directly addressed the possibility that the increase in reports of disease could be a simple artifact of burgeoning publication rates. By systematically searching a literature database, they acquired a data set on disease reports over time that was more representative and comprehensive than previous studies. They found that the absolute number of studies of disease had increased in all taxa. However, they determined that this increase was, at least in part, a consequence of the overall growth of the scientific literature and was driven by the vast number of studies on fishes. To account for the underlying bias in publication rates, they normalized reports of disease as a proportion of all reports (i.e., reports on disease divided by all reports). To distinguish trends among taxa, they did this separately for nine marine taxonomic groups. Several patterns emerged when these normalized reports were analyzed for temporal trends between 1970 and 2001. The proportion of studies on disease increased significantly with time in turtles, corals, mammals, sea urchins, and mollusks. A closer inspection of the coral data indicated that reports of noninfectious coral bleaching increased significantly over the past three decades, but reports of infectious coral disease have not. However, reports of coral disease were high following the two large El Niño events (Figure 4). No significant trends were detected for sea grasses, decapods, or sharks/rays (though disease occurred in these groups). Surprisingly, the normalized reports of disease decreased in fishes. The finding that disease did not increase in all taxa indicates that increases were not exclusively the result of increased study or detection ability of disease by marine biologists. The finding that disease reports increased in most taxa is strong support for the hypothesis that there has been a real increase in some

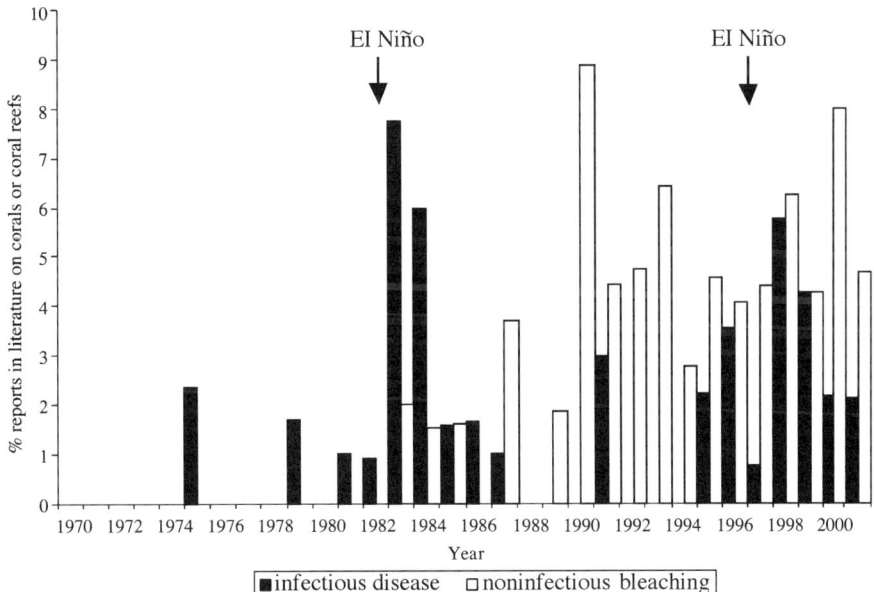

Figure 4 The percentage of the scientific literature on corals or coral reefs that was related to disease. Data are from Ward & Lafferty (2004). Over time, there is a significant overall increase in disease reports as well as a significant increase for bleaching. Infectious coral disease does not show a significant trend over time but instead shows a bimodal pattern coincident with the two large El Niño events of the last three decades.

diseases. Below, we consider what factors might be associated with increases or decreases in disease.

PROBLEMS WITH SIMPLE PREDICTIONS

Several anthropogenic stressors are changing in concert: Climate warming may lead to thermal stress; coastal development leads to habitat destruction and degradation; introduced species bring disease; fishing reduces once common species and favors others; runoff and discharge contain nutrients and pathogens along with fertilizers, pesticides, antibiotics, anthelminthics, and herbicides. This list of potential stressors is matched by a diverse set of infectious and noninfectious diseases, most of which are poorly understood. In this review, we focus primarily on infectious diseases, and such infectious diseases clearly have always been common in natural communities (Dobson et al. 1992, Kennedy et al. 1986). Moreover, although infectious diseases are usually perceived as negative, some play positive roles in promoting biodiversity and maintaining ecosystems (Lafferty 2003). A vast diversity of parasitic strategies naturally occur in the ocean, including typical parasites (e.g., intestinal tapeworms), pathogens (e.g., many viruses, fungi, and protozoa), parasitoids (e.g., hyperiid amphipods), parasitic castrators (e.g., rhizocephalan barnacles), trophically transmitted parasites (e.g., larval acanthocephalans), and, by some considerations, micropredators (e.g., leeches) (Lafferty & Kuris 2002). Some (perhaps most) of these infectious agents have no detectable effect on host populations, whereas others cause mortality, reduce growth, lower fecundity, alter behavior, or lower social status. Furthermore, different parasites and pathogens might interact differently with environmental change. Given the diversity of interactions between environmental disturbance and infectious disease, we cannot overly generalize about whether diseases should increase in association with environmental degradation. Disease may increase or decrease, depending on the infectious agent and the factors affecting its host population (Lafferty 1997).

EFFECTS OF ANTHROPOGENIC CHANGE ON DISEASE

The foremost assumption related to the hypothesis that disease is increasing in marine environments is that increasing stressors, such as toxic chemicals (Khan 1990), pollutants (Harvell et al. 1999), and warming (Harvell et al. 2002), increase individual susceptibility to infectious diseases. Under this view, Porter & Tougas (2001) proposed a general coral disease model that integrates stress factors from climate change predictions and pollution. Global factors include predicted rise in ocean temperature (Hoegh-Guldberg 1999), storm frequency (Birkeland 1997), and oceanic carbon dioxide concentrations (Kleypas et al. 1999). Pollution factors include elevated nutrients (Lapointe et al. 2002), elevated toxicants (Glynn et al. 1984), and reduced water clarity (Cook et al. 2002). This model predicts that deteriorating environmental conditions will be expressed by increases in the number of places and number of species simultaneously experiencing epizootics. These predictions match current patterns of disease outbreaks in the Florida Keys (Porter et al. 2001).

Predictions become complicated, however, because some stressors can have a more negative impact on parasites than on hosts (Lafferty 1997). This will increase recovery rates of infected individuals and mitigate the population-level impacts of disease. Infected hosts might also experience differentially high mortality when under stress. This will remove parasites more rapidly from the host population than would occur without the stressor. Although this increases the impact of disease on infected individuals, it simultaneously decreases the spread of an epizootic through the host population. Such a relationship underscores the point that effects of stress and infectious disease at the population level cannot necessarily be predicted from their effects on individuals (Lafferty & Holt 2003).

Other predictions emerge if one considers how contact between individuals affects the dynamics of infectious disease. High host density increases contact rates between infected and uninfected individuals (Anderson & May 1986, Stiven 1964). For this reason, dense populations tend to have more parasites (Arneberg et al. 1998), meaning that some epizootics could be due to increasing host density, not to outside stressors. In contrast, if stress impairs host vital rates (birth, death, growth), this will depress host population density, thus reducing the chance of an epizootic. Again, this is the opposite of the prediction that stress increases disease.

Perhaps the main benefit of explicitly listing these hypotheses is that it acknowledges how anthropogenic effects can decrease disease as well as increase disease. In addition, none of the above hypotheses are exclusive. Thus, a particular stress can have multiple effects on hosts and parasites, such as increasing host susceptibility to disease while impairing host vital rates. This can make it difficult to predict how a particular stressor will affect disease in a host population. For example, although stressed individuals should be more susceptible to infection if exposed, the stressor could also reduce contact rates between infected and uninfected individuals. Simulation models help resolve the opposing predictions stemming from these alternative effects (Lafferty & Holt 2003). Although many interactions between stress and disease are theoretically possible, stress seems most likely to reduce the impact of infectious diseases as long as transmission is closed (infections tend to be transmitted between members of the same population) and host specificity is high. Stress will generally increase the impact of noninfectious diseases or infectious diseases that are not host specific or that maintain themselves in large (often relatively resistant) reservoir populations (Lafferty & Holt 2003).

Climate Warming

Predictions for coastal marine areas include changes in air and sea temperatures, ocean currents, atmospheric storms, freshwater inputs from land, and sea levels. Of these, the most notable prediction is for widespread increases in average sea surface temperatures driven by elevated greenhouse gases (Hoegh-Guldberg 1999, Houghton et al. 1996, Kleypas et al. 1999). The ocean has already increased in temperature, on average, a third of a degree in the last half century (Levitus et al. 2000). This aspect of climate change should alter the geographic distribution of infectious disease by shifting host and parasite latitudinal ranges poleward (Marcogliese 2001a).

Higher temperatures may stress organisms, increasing their susceptibility to disease (Holmes 1996, Kim et al. 2000, Scott 1988), but an increase in susceptibility will depend on the relative sensitivity of hosts and parasites to temperature. Hosts and their parasites exhibit a peak performance at a thermal optimum (Harvell et al. 2002). If the optimum differs between host and parasite, the resulting gene-by-gene-by-environment interaction will either increase or decrease susceptibility to disease at a given temperature (Elliot et al. 2002). Tropical marine organisms are naturally much closer to their upper lethal temperature than to their lower lethal temperature, and this may put them at a disadvantage in fighting disease if temperature increases (Porter et al. 1989). For example, the optimal temperature of a fungal pathogen is higher than the optimal temperature of its sea fan host, placing the sea fan at risk to climate warming (Alker et al. 2001). One aspect of warming is universally stressful for marine invertebrates and fishes. As temperature rises, oxygen levels decrease and metabolic rates increase, suggesting that warming could lead to respiratory stress.

Some parasites have higher growth rates (Chubb 1980) (and, subsequently, higher pathology), decreased generation times (Pojmanska et al. 1980), or higher reproductive output (Chubb 1979) at warmer temperatures, suggesting that the severity of disease could increase with temperature. This may be why withering syndrome in black abalone results in faster die-offs in warm water (Lafferty & Kuris 1993). Climate warming may also favor disease in sea turtles. Green turtle fibropapilloma tumors are hypothesized to grow more rapidly in warm water and the prevalence of this disease has increased since the 1980s (Herbst 1994). Reports of diseases of sea turtles have increased greatly over the last three decades (Ward & Lafferty 2004), consistent with the evidence that warming favors this disease.

Although several stressors can produce bleaching in corals, the vast majority of bleaching is caused by elevated water temperature (Brown 1997). This may explain why reports of noninfectious coral bleaching have significantly increased in the last three decades (Ward & Lafferty 2004). Several other coral diseases, such as aspergillosis (Alker et al. 2001), black band (Carlton & Richardson 1995, Edmunds 1991, Rutzler et al. 1983), white pox (Patterson et al. 2002), and dark spot disease (Gil-Agudelo & Garzon-Ferreira 2001), also grow more rapidly under elevated temperature. Because the growth rate and spread of many coral diseases increase during the late summer period, one of the likeliest outcomes of global warming is for the summer disease season to become (a) more severe, as summer temperature maxima increase; and (b) longer, as these elevated thermal regimes start earlier and persist later in the season. Also, winter warming may release some infectious diseases from the low-temperature control that provides hosts with a seasonal escape from disease (Harvell et al. 2002). Nevertheless, disease will not always increase with warming. For example, aptly named coldwater disease in salmonids disappears as water temperature increases (Holt et al. 1989).

Pollution

Toxic pollution affects both parasites and hosts, making it difficult to generalize broadly about its effects on disease (Lafferty 1997). In general, toxicants increase an individual's susceptibility to disease by impairing defenses. For example, the intensity and prevalence of parasitic gill ciliates and monogenes of fishes increase with a wide range of pollutants (Khan & Thulin 1991) because toxicants impair mucus production, a fish's main defense against gill parasites (Khan 1990). Similarly, marine mammals bioaccumulate lipophillic toxicants (O'Shea 1999) that can affect the mammalian immune system (Swart et al. 1994) by lowering killer cell activity, responses to T and B cell mitogens, and antibody responses (DeStewart et al. 1996). Such immunosuppression may increase pathological conditions in seals exposed to morbillivirus (Van Loveren et al. 2000), Phocine Distemper (Harder et al. 1992), leptospirosis, and calicivirus (Gilmartin et al. 1976).

This suggests that individuals exposed to infectious disease are more likely to become infected or to suffer more from disease when in toxic environments.

Most helminth parasites, however, decline in polluted areas (Lafferty 1997). This is because, in addition to impacting host defenses, toxicants increase parasite mortality rates, thereby reducing parasitic disease (Lafferty 1997). For example, selenium is more toxic to tapeworms than to fish hosts (Riggs et al. 1987). Also, free-swimming trematode larvae are sensitive to heavy metals (Siddall & Clers 1994), as are their mollusk first intermediate hosts. This is probably why reduced trematode diversity occurs at sites contaminated with heavy metals (Lefcort et al. 2002). Toxic pollution, therefore, should impede trematode transmission to vertebrates that serve as final hosts. Toxic pollution can also negatively affect parasites if infected hosts are differentially killed by pollution (Guth et al. 1977, Stadnichenko et al. 1995). For instance, cadmium kills amphipods infected with larval acanthocephalans more readily than it kills uninfected amphipods (Brown & Pascoe 1989).

Changes in Population Density

Anthropogenic change can affect the density of marine organisms in several ways, and such changes in host density affect infectious disease transmission. Toxicants can reduce host density by reducing host vital rates (Johnson 2001). Therefore, even if toxicants make uninfected individuals more susceptible to infectious disease, the contact between uninfected and infected individuals may be sufficiently low in a polluted environment that infectious disease cannot persist in the host population.

Habitat degradation and human disturbance may drive some species, particularly top predators, from coastal areas. Take, for instance, the assumption that a diverse and abundant trematode community in snails is impossible without a diverse and abundant final host community. This is because final hosts are the source of trematode stages infectious to snails, and final hosts vary in the type of adult trematode communities they harbor. Trematode infections in snails should be higher at locations where final host birds are abundant (Bustnes et al. 2000, Hoff 1941, Marcogliese 2001b, Robson & Williams 1970, Smith 2001). A salt marsh restoration project provided an opportunity to test the hypothesis that habitat degradation decreases parasitism of snails by trematodes (Huspeni & Lafferty 2004). Before restoration, snails at impacted sites had significantly fewer larval trematodes (lower prevalence and species richness) than did snails at control sites located in intact salt marsh habitat. After restoration, the impacted sites increased in trematode prevalence and species richness, eventually surpassing controls. This is because birds prefer areas with diverse and abundant prey populations. This example indicates how some parasites may actually be positive indicators of healthy ecosystems.

Fishing has dramatically reduced the abundance of many species (Jackson et al. 2001, Myers & Worm 2003). If a stock is fished to a density below the host density threshold for transmission, a fishery can fish out parasites (Dobson & May 1987). For instance, fishing substantially reduced the prevalence of a tapeworm in whitefish (Amundsen & Kristoffersen 1990), apparently extirpated a swim bladder nematode from native trout in the Great Lakes (Black 1983), and dramatically reduced the prevalence of bucephalid trematodes in scallops (Sanders 1966). Fishing out a parasite is most likely when the parasite has a recruitment system that is relatively closed compared with the recruitment of its host (Kuris & Lafferty 1992). The significant decline in the normalized reports of disease in marine fish (Ward & Lafferty 2004) was based almost entirely on data from commercially fished species. This suggests the hypothesis that exploitation has reduced diseases in fishes by making transmission more difficult. In contrast, aquaculture intentionally increases species densities, and this should favor diseases (as detailed in our case study on oysters below).

Fishing and hunting can indirectly increase diseases in species at lower levels of the food web (Hochachka & Dhondt 2000, Jackson et al. 2001). At the California Channel Islands, extirpation of sea otters (and removal of Native American hunter-gatherers) facilitated an increase in black abalone populations to great abundance. Under these conditions, a previously unknown rickettsial disease caused a catastrophic collapse of black abalone populations (Lafferty & Kuris 1993). In this same area, lobsters and sea otters historically kept sea urchin populations at low levels, and kelp forests developed in a community-level trophic cascade (Tegner & Levin 1983). Where lobsters were fished (nearly everywhere but in marine reserves), urchin populations increased and often overgrazed kelps. An urchin-specific bacterial disease recently entered the area where urchin densities well exceeded the host-threshold density for transmission; epizootics followed, except in a marine reserve where lobsters were protected (Lafferty 2004). Similarly, in the 1980s, an unknown pathogen led to a 98% die-off of the long-spined sea urchin, *Diadema antillarum*, throughout the Caribbean (Lessios 1988). Fishing of sea urchin predators and the subsequent increase in sea urchin populations is common (Babcock et al. 1999, Pinnegar et al. 2000, Sala et al. 1998, Shears & Babcock 2002). This suggests that increased reports of disease in echinoderms (Ward & Lafferty 2004) may be more an indirect result of fishing top predators than of climate stress. Indeed, fishing, rather than climate change, may be the major cause of increased diseases in marine organisms at lower trophic levels (Jackson et al. 2001). Protection from fishing/hunting can also increase the density of exploited species and facilitate disease transmission. For example, protected pinniped populations have soared in some areas (Stewart & Yochem 2000). The combination of increased susceptibility owing to stressors and increased population density owing to successful marine mammal protection regulations may explain why reports of disease in marine mammals have increased (Ward & Lafferty 2004).

Eutrophication is a common consequence of coastal development. However, unlike toxic pollution (with which it is often associated), eutrophication may raise rates of parasitism because the associated increased productivity can increase the abundance of hosts and, subsequently, parasites. In addition, some coral pathogens can directly use nutrients to increase their growth and pathogenicity (Bruno et al.

2003). Parasites that increase under eutrophic conditions tend to be host generalists and have local recruitment, such as cestodes with short life cycles, or, like trematodes, have intermediate hosts, such as snails, that benefit from enrichment (Marcogliese 2001a).

Pathogen Pollution

Pathogen pollution, or the introduction of new disease-causing agents, is unequivocally increasing disease in the ocean and elsewhere. Global trade and travel are increasingly introducing species, particularly to estuarine habitats, where introduced species can make up the bulk of the fauna (Ruiz et al. 2000). Averaging across several taxa, introduced animals bring a mean of 16% of their parasite species when they invade (Torchin et al. 2003). Although most parasites of introduced species are left behind, a fraction of those that do invade can have severe consequences. This is best illustrated by the fact that when epizootics decimate formerly common species, the source of the disease is usually a new pathogen (Lafferty & Gerber 2002). Many successful introduced pathogens have broad host specificity and are more pathogenic in naive hosts than in their original, abundant (exotic) hosts (Gog et al. 2002, McCallum & Dobson 1995, Woodroffe 1999). When the Caspian stellate sturgeon invaded the Aral Sea, it brought a monogene that infected gills of the native spiny sturgeon, leading to mass mortalities of this naive host (Dogiel & Lutta 1937). Similarly, when European trout were introduced to North America, whirling disease spread from stocked trout to native trout, with severe consequences for the natives (Bergersen & Anderson 1997, Gilbert & Granath 2003). A similar type of pathogen pollution occurs where domestic animals create a large source of disease. Such disease reservoirs can cause rare species to decline (Lafferty & Gerber 2002). This can theoretically occur in marine systems when mariculture operations maintain a continual source of disease transmission to closely related native species. One possible example is in the rapidly expanding shrimp farming industry in which several viruses of marine penaeid shrimp have spread through farms and wild populations (Lightner 1996). Other, better-documented cases are sea lice, *Lepeophtheirus salmonis*, and the bacterial agent of furunculosis, *Aeromonas salmonicida*, which can escape from salmon farms into wild stocks.

Unusual contact events between typically segregated species have the potential to spread disease to new host species. Canine distemper virus (CDV) and related morbilliviruses, i.e., phocine distemper virus (PDV), have recently emerged in a number of marine hosts. This may be a viral host shift from domestic dogs to marine mammals. For example, a 1955 mass mortality of Antarctic crab-eater seals was attributed to CDV introduced by sled dogs (Bengston & Boveng 1991). In 1987–1988, PDV emerged in European harbor seals (Osterhaus & Vedder 1988), perhaps from contact with infected harp seals or terrestrial canids (Heide-Jorgensen & Harkonen 1992), causing ~60% mortality in most regions (Heide-Jorgensen & Harkonen 1992, Osterhaus & Vedder 1988).

Another aspect of pathogen pollution involves infective agents, "pollutogens," that have a source outside the ecosystem. For instance, *Aspergillus sydowii* is a common terrestrial fungus that may enter the marine environment via local sediment runoff (Smith et al. 1996) or by long-distance transport in African dust (Shinn et al. 2000). In the marine environment, this is a widespread pathogen of sea fans throughout the Caribbean (Kim & Harvell 2002, Kim et al. 2000). Pollutogens may reproduce within a host and elicit defensive responses but have little or no infectious dynamic within the host population (Lafferty & Kuris 2004). Because they do not require contact between hosts for transmission, pollutogens should increase with host stress. Two diseases of California sea otters are good examples of pollutogens; valley fever is caused by a fungus that enters the marine environment from eroded soil and toxoplasmosis is caused by a protozoan that enters the ocean along with feces from domestic cats (Kreuder et al. 2003).

Gathering evidence suggests that pollutogens may be related to coral decline (Richmond 1993). Within the Caribbean, populations of the most common reef-building coral, *Acropora palmata*, are being decimated by white pox disease and other causes, with losses of living cover in the Florida Keys averaging 85% or greater (Miller et al. 2002, Patterson et al. 2002). *Serratia marcescens* is the cause of white pox disease (Patterson et al. 2002). This is a common gram-negative bacterium classified as a fecal coliform of humans and other animals. Concurrent studies also show that human sewage markers (e.g., human enteric viruses) are prevalent among near-shore corals and environments of the Florida Keys (Griffin et al. 2000, X). Santavy et al. (2001) have further demonstrated that the incidence of white pox disease on *A. palmata* is significantly greater on coral reefs near Key West than on reefs in the Dry Tortugas. They speculate that this correlation is due to the superior water quality in the Dry Tortugas (Boyer & Jones 2002). These studies strongly suggest that pollutogens affect coral health.

Pathogen pollution is likely to increase with expanding coastal human populations and the increased precipitation predicted from climate change models. This is also an impact for which management can provide some relief, such as by tighter controls on species introductions and growing pressure by coastal communities to improve sewage run off. Concerns for the effects of pollutogens on sea otters and coral reefs have led to public pressure for better sewage management in the United States.

OYSTERS: A CASE STUDY

Oysters, in particular, offer an attractive model for gauging whether diseases are increasing in the marine environment. They are sessile, and mortalities are relatively easy to observe because shells remain in situ after death. They are probably the most valuable commercial molluscan species worldwide and consequently have drawn the attention of local and national governments for centuries. Production figures date to the 1800s in both Europe and the United States (MacKenzie et al.

1997), at which time oysters also became the object of numerous descriptive and scientific studies (Brooks 1880, Ingersoll 1881, Nelson 1889).

Some of the earliest accounts of oyster fisheries mention mortalities large enough to have reduced harvestable supplies (Hoek 1879, cited in Dijkema 1997; Ingersoll 1881; LaFont 1874). In addition to predation, siltation, and freshwater runoff (Goulletquer & Heral 1997, Nelson 1889), disease was sometimes listed as a cause of the deaths, but without being attributed to a specific agent (Strand & Vølstad 1997). Shell disease, first described in 1878 (Alderman & Jones 1971), was probably the first scientifically recognized oyster disease caused by an infectious agent (Table 1). Pathogens were suspected in epizootics in Canada (1915–1916) and Europe (1920–1921) but never identified. Since the mid-1900s, a series of oyster diseases and disease-causing organisms has been described, many causing epizootic mortalities and ruinous declines in oyster production in the United States and Western Europe (Table 1).

Several factors underlie the recent increase in reported oyster (and other molluscan) disease outbreaks. Climate change has been implicated in the northward expansion of dermo and possibly MSX diseases of oysters in the United States (Cook et al. 1998, Hofmann et al. 2001). Parasites have been introduced into new areas through increased shipment of host oysters for fisheries and aquaculture, and increased shipping may have introduced parasites in ballast water or in oysters attached to ships' hulls (Bustnes et al. 2000, Elston et al. 1986, Farley 1992). Newly introduced animals may be susceptible to local pathogens (Ford et al. 2002, Maes & Paillard 1992). Sometimes, diseases appear to spread rapidly around the globe (Table 1), either because of transport of oysters or the spread of recognition among growers. Molluscan aquaculture, which has increased markedly over the past few decades, is also the source of disease outbreaks caused by culture conditions themselves. The high densities under which animals are grown and the high temperatures sustained in hatcheries favor the proliferation and transmission of opportunistic pathogens (Elston & Wilkinson 1985, LeDeuff et al. 1996).

Although normalized reports of disease in mollusks have increased over time (Ward & Lafferty 2004), perceived increases in disease could be an artifact of improvements in detection (Harvell et al. 1999). The level of scientific observation has grown significantly since the oyster epizootics of the 1950s–1970s. Many governments now require health inspections of mollusks before importation. They also conduct pathological surveys of wild and cultured populations. Both efforts uncover previously undescribed parasites and diseases, although anecdotal and documentary evidence of earlier unexplained mortalities suggest that many of the diseases were not new when first described. Finally, increased funding for shellfish disease research in many countries has resulted in (*a*) the creation by universities and other research agencies of positions in shellfish pathology and (*b*) the training of students to fill those positions—where they expand the potential for discovering, documenting, and investigating diseases and disease agents. Although some oyster disease outbreaks are clearly new, others may be long-standing conditions, noted and accepted for years by shellfish harvesters, but they become newly discovered

TABLE 1 Diseases of oysters over time, their causative agents (where known), host species, areas reported, and year first reported[a]

Disease/Agent	Host(s)	Region	Reported
Shell disease/*Ostracoblabe implexa* (fungus)	*Ostrea edulis, Crassostrea angulata*	Europe	1878[b]
Malpeque disease/Unknown	*C. virginica*	Eastern Canada	1915
Australian Winter disease/ *Mikrocytos mackini* (protozoan)	*Saccostrea commercialis*	Australia	1924[b]
Dermo disease/*Perkinsus marinus* (protozoan)	*C. virginica, C. gigas, C. ariakensis*	Gulf Coast and eastern United States	1940s[b] (1900s)
Summer mortality/Bacteria and viruses?	*C. gigas*[c]	Japan, western United States, France	1940s to 1980s
MSX disease/*Haplosporidium nelsoni* (protozoan)	*C. virginica, C. gigas*	Eastern and western United States, Asia[d]	1957
SSO disease/*H. costale* (protozoan)	*C. virginica*	Eastern United States	1959[b]
Marteiliosis/*Marteilia refringens* (protozoan)	*O. edulis, O. angasi, Tiostrea chilensis*	Western Europe	1967[b]
QX disease/*Marteilia sydneyi* (protozoan)	*Saccostrea commercialis*	Australia	1969[b]
Denman Island disease/ *Mikrocytos mackini* (protozoan)	*C. gigas, C. virginica, O. edulis, Ostreola conchaphila*	Western Canada	1960[b]
Herpes disease/Herpes-like virus(es)	Numerous bivalves[c]	Worldwide	1970s to 1990s
Juvenile Oyster disease/Bacteria?	*Crassostrea virginica*[c]	Eastern United States	1988
Bonamiosis/*Bonamia ostrea* (protozoan)	Five *Ostrea* spp., *O. conchaphila, Tiostrea chilensis, C. ariakensis*?	Western Europe, northeastern and northwestern United States	1979[b] (eastern United States, 1960s)
Bonamiosis/*Bonamia exositosa*	*Tiostrea chilensis*	New Zealand	1986

[a]For Supplemental References to Table 1: Follow the Supplemental Material link from the Annual Reviews home page at http://www.annualreviews.org.
[b]Probably present earlier.
[c]In culture/hatcheries.
[d]Introduced from Asia to United States.

diseases when investigated. Despite this potential for bias, the evidence suggests that diseases have increased in oysters over time.

CONCLUSIONS

A high economic and social value is placed on the abundance of marine species, making increases in disease a concern for society. Unfortunately, disease appears to be increasing in several marine taxa. Climate warming may be responsible for some of the more notable examples. Increasing temperature facilitates the spread of warm-water parasites or weakens the defenses of those marine organisms that are already near the upper end of their thermal tolerance. The resulting stress should especially increase noninfectious and generalist diseases and parasites. This appears to be leading to unprecedented bleaching events in corals. Infectious coral diseases may be related to El Niño events, which seem likely to increase in frequency with climate warming. Because corals create habitat for whole communities, an increase in coral disease could lead to dramatic changes in tropical near-shore communities. These effects are alarming, and the effects of climate warming on coral disease should generate considerable concern. Other aspects of global change can increase disease in the ocean. Increasing populations, such as seen in many protected marine mammals, or as an indirect result of fishing top predators (such as for the urchin disease example mentioned above), or of eutrophication, provide increased opportunities for disease transmission. Increases in introduced species, aquaculture, and contacts between terrestrial and marine species also leads to the emergence of new pathogens. Terrestrial runoff is increasingly polluted, raising the chance that terrestrial pathogens may affect marine species, even if marine hosts are dead ends for the "pollutogen." Several of these factors that increase disease are indications of how much humans stress marine ecosystems.

Contrary to most views on disease in the ocean, some diseases will decline with environmental degradation. Some parasites are more sensitive to toxic pollution than are their hosts. Perhaps more importantly, fishing may result in widescale losses of infectious disease. Given the vast scale of fisheries, parasite loss may be the hidden, but dominant, effect of anthropogenic change on disease in the ocean. Although few will mourn the loss of fish diseases, their disappearance should be alarming if they indicate broadscale overfishing.

Scientists increasingly realize that disease is an important aspect of ecology and that environmental degradation and disease interact in a complex manner. Roughly half of the studies of parasite ecology now concern how changes in the environment affect parasites (Lafferty 2003). Two thirds of these environmental studies focus on emerging diseases or increases in infectious disease associated with environmental change, whereas the other third find that changes in the environment reduce infectious disease. However, few studies acknowledge that environmental change can either increase or decrease infectious disease, depending on the disease and the type of change. Progress in understanding and managing diseases in the ocean will require an acknowledgment of the complex interactions between disease and the environment. It will also require a dedication to systematic, long-term studies that track changes in disease over time across several taxa and habitats.

ACKNOWLEDGMENTS

This work was conducted as part of the Diseases in the Ocean Working Group supported by the National Center for Ecological Analysis and Synthesis, and the University of California, Santa Barbara. We thank D. Harvell, K. Kim, J. Ward, M. Torchin, K. Smith, H. McCallum, and A. Dobson for discussion and comments. We also thank K. Patterson, J. Ward, and the Channel Islands National Park for providing data. D. Harvell and J. Ward provided useful comments on the final draft of the manuscript. This research was partially supported by NSF through the NIH/NSF Ecology of Infectious Disease Program (DEB-02,24565) awarded to KDL.

The *Annual Review of Ecology, Evolution, and Systematics* is online at
http://ecolsys.annualreviews.org

LITERATURE CITED

Alderman DJ, Jones EBG. 1971. Shell disease of oysters. *Fish. Invest. London Ser. I* 26:1–19

Alker AP, Smith GW, Kim K. 2001. Characterization of *Aspergillus sydowii* (Thom et Church), a fungal pathogen of Caribbean sea fan corals. *Hydrobiologia* 460:105–11

Amundsen PA, Kristoffersen R. 1990. Infection of whitefish (*Coregonus lavaretus* L. s.l.) by *Triaenophorus crassus* Forel (Cestoda: Pseudophyllidea): a case study in parasite control. *Can. J. Zool.* 68:1187–92

Anderson RM, May RM. 1986. The invasion, persistence and spread of infectious diseases within animal and plant communities. *Philos. Trans. R. Ser. B* 314:533–70

Arneberg P, Skorping A, Grenfell B, Read A. 1998. Host densities as determinants of abundance in parasite communities. *Philos. R. Soc. London Ser. B* 265:1283–89

Aronson RB, Macintyre IG, Precht WF, Murdoch TJT, Wapnick CM. 2002. The expanding scale of species turnover events on coral reefs in Belize. *Ecol. Monogr.* 72:233–49

Aronson RB, Precht WF. 1998. Extrinsic control of species replacement on a Holocene reef in Belize: the role of coral disease. *Coral Reefs* 17:223–30

Aronson RB, Precht WF. 2001. White-band diseases and the changing face of Caribbean coral reef. *Hydrobiologia* 460:25–38

Babcock RC, Kelly S, Shears NT, Walker JW, Willis TJ. 1999. Changes in community structure in temperate marine reserves. *Mar. Ecol. Prog. Ser.* 189:125–34

Bengston JL, Boveng P. 1991. Antibodies to canine distemper virus in Antarctic seals. *Mar. Mammal Sci.* 7:85–87

Bergersen EP, Anderson DE. 1997. The distribution and spread of *Myxobolus cerebralis* in the United States. *Fisheries* 22:6–7

Birkeland C. 1997. *Life and Death of Coral Reefs*. New York: Chapman & Hall

Black GA. 1983. Taxonomy of a swimbladder nematode, *Cystidicola stigmatura* (Leidy), and evidence of its decline in the Great Lakes. *Can. J. Fish. Aquat. Sci.* 40:643–47

Boyer JN, Jones RD. 2002. A view from the bridge: external and internal forces affecting the ambient water quality of the Florida Keys National Marine Sanctuary. See Porter & Porter 2002, pp. 609–28

Brooks WK. 1880. Development of the American oyster (*Ostrea virginica* L.). *Stud. Biol. Lab., Johns Hopkins Univ.* 4:1–81

Brown AF, Pascoe D. 1989. Parasitism and host sensitivity to cadmium: an acanthocephalan infection of the freshwater amphipod *Gammarus pulex*. *J. Appl. Ecol.* 26:473–87

Brown BE. 1997. Coral bleaching: causes and consequences. *Coral Reefs* 16:129–38

Bruno JF, Petes LE, Harvell CD, Hettinger A. 2003. Nutrient enrichment can increase the severity of coral diseases. 6:1056–61

Bustnes JO, Galaktionov KV, Irwin SWB. 2000. Potential threats to littoral biodiversity: Is increased parasitism a consequence of human activity? *Oikos* 90:189–90

Carlton RG, Richardson LL. 1995. Oxygen and sulfide dynamics in a horizontally migrating cyanobacterial mat: black band disease of corals. *FEMS Microbiol. Ecol.* 18:155–62

Chadwick-Furman NE. 1996. Reef coral diversity and global change. *Global Change Biol.* 2:559–68

Chubb JC. 1979. Seasonal occurrences of helminths in freshwater fishes. Part II. Trematoda. *Adv. Parasitol.* 17:141–313

Chubb JC. 1980. Seasonal occurrence of helminths in freshwater fishes. Part III. Larval Cestoda and Nematoda. *Adv. Parasitol.* 18:1–120

Connell JH. 1997. Disturbance and recovery of coral assemblages. *Coral Reefs* 16:101–13

Cook CB, Mueller EM, Ferrier MD, Annis E. 2002. The influence of nearshore waters on corals of the Florida Reef Tract. See Porter & Porter 2002, pp. 770–83

Cook T, Folli M, Klinck J, Ford S, Miller J. 1998. The relationship between increasing sea surface temperature and the northward spread of *Perkinsus marinus* (Dermo) disease epizootics in oysters. *Estuar. Coast. Shelf Sci.* 40:587–97

DeStewart RL, Ross PS, Voss JG, Osterhaus ADME. 1996. Impaired immunity in harbour seals (*Phoca vitulina*) fed environmentally contaminated herring. *Vet. Q.* 18:S127–28

Diaz-Soltero H. 1999. Endangered and threatened species: a revision of candidate species list under the Endangered Species Act. *Fed. Regist.* 64:33466–68

Dijkema R. 1997. Molluscan fisheries and culture in the Netherlands. See MacKenzie et al. 1997, pp. 115–36

Dobson AP, Hudson PJ, Lyles AM. 1992. Macroparasites: worms and others. In *Natural Enemies. The Population Biology of Predators, Parasites and Diseases*, ed. MJ Crawley, pp. 329–48. Oxford, UK: Blackwell

Dobson AP, May RM. 1987. The effects of parasites on fish populations—theoretical aspects. *Int. J. Parasitol.* 17:363–70

Dogiel VA, Lutta A. 1937. Mortality among spiny sturgeon of the Aral Sea in 1936. *Rybn Khoz* 12:26–27

Dungan M, Miller T, Thomson D. 1982. Catastrophic decline of a top carnivore in the Gulf of California rocky intertidal zone. *Science* 216:989–91

Dustan P. 1999. Coral Reefs under stress: sources of mortality in the Florida Keys. *Nat. Resour. Forum* 23:147–55

Edmunds PJ. 1991. Extent and effect of black band disease on a Caribbean reef. *Coral Reefs* 10:161–65

Elliot SL, Blanford S, Thomas MB. 2002. Host-pathogen interactions in a varying environment: temperature, behavioural fever and fitness. *Philos. R. Soc. London Ser. B* 269:1599–607

Elston RA, Farley CA, Kent ML. 1986. Occurrence and significance of bonamiasis in European flat oysters *Ostrea edulis* in North America. *Dis. Aquat. Organ.* 2:49–54

Elston RA, Wilkinson MT. 1985. Pathology, management and diagnosis of oyster velar virus disease (OVVD). *Aquaculture* 48:189–210

Epstein P. 1996. Emergent stressors and public health implications in large marine ecosystems: an overview. In *The Northeast Shelf Ecosystem: Assessment, Sustainability, and Management*, ed. T Smayda, pp. 417–38. Cambridge, MA: Blackwell

Epstein PR, Sherman B, Spanger-Siegfried E, Langston A, Prasad S, McKay B. 1998. *Marine Ecosystems: Emerging Diseases as*

Indicators of Change. Boston, MA: Cent. Health Glob. Environ., Harvard Med. Sch. 85 pp.

Farley CA. 1992. Dispersal of pathogens, parasites, pests, predators and competitors. In *Dispersal of Living Organisms into Aquatic Ecosystems*, ed. A Rosenfield, R Mann, pp. 139–54. College Park, MD: Maryland Sea Grant

Ford SE, Kraeuter JN, Barber RD, Mathis G. 2002. Aquaculture associated factors in QPX disease of hard clams: density and seed source. *Aquaculture* 208:23–38

Gil-Agudelo DL, Garzon-Ferreira J. 2001. Spatial and seasonal variation of Dark Spots Disease in coral communities of the Santa Marta area (Colombian Caribbean). *Bull. Mar. Sci.* 69:619–29

Gilbert MA, Granath W. 2003. Whirling Disease of salmonid fish: life cycle, biology and disease. *J. Parasitol.* 89:658–67

Gilmartin WG, DeLong RL, Smith AW, Sweeney JC, Lappe BWD, et al. 1976. Premature parturition of the California sea lion. *J. Wildlife Dis.* 12:104–15

Gladfelter WB. 1982. White-Band Disease in *Acropora palmata*: implications for the structure and growth of shallow reefs. *Bull. Mar. Sci.* 32:639–43

Glynn PW, Howard LS, Corcoran E, Freay AD. 1984. The occurrence and toxicity of herbicides in reef building corals. *Mar. Pollut. Bull.* 15:370–74

Gog J, Woodroffe R, Swinton J. 2002. Disease in endangered metapopulations: the importance of alternative hosts. *Philos. R. Soc. London Ser. B* 269:671–76

Goreau TJ, Cervino J, Goreau M, Hayes R, Hayes M, et al. 1998. Rapid spread of diseases in Caribbean coral reefs. *Rev. Biol. Trop.* 46:157–71

Goulletquer P, Heral M. 1997. Marine molluscan production trends in France: from fisheries to aquaculture. See MacKenzie et al. 1997, pp. 137–64

Greenstein BJ, Curran HA, Pandolfi JM. 1998. Shifting ecological baselines and the demise of *Acropora cervicornis* in the western North Atlantic and Caribbean Province: a Pleistocene perspective. *Coral Reefs* 17:249–61

Griffin DW, Gibson CJ, Lipp EK, Riley K, Paul JH, Rose JB. 2000. Detection of viral pathogens by reverse transcriptase PCR and of microbial indicators by standard methods in the canals of the Florida Keys. *Appl. Environ. Microbiol.* 66:876

Guth DJ, Blankespoor HD, Cairns J. 1977. Potentiation of zinc stress caused by parasitic infection of snails. *Hydrobiologia* 55:225–29

Harder TC, Willhus T, Leibold W, Liess B. 1992. Investigations on the course and outcome of phocine distemper virus infection in harbor seals (*Phoca vitulina*) exposed to polychlorinated biphenyls. *J. Vet. Med. Ser. B* 39:19–31

Harvell CD, Kim K, Burkholder JM, Colwell RR, Epstein PR, et al. 1999. Emerging marine diseases—climate links and anthropogenic factors. *Science* 285:1505–10

Harvell D, Kim K, Quirolo C, Weir J, Smith G. 2001. Coral bleaching and disease: contributors to 1998 mass mortality in *Briareum asbestinum* (Octocorallia, Gorgonacea). *Hydrobiologia* 460:97–104

Harvell D, Mitchell CE, Ward JR, Altizer S, Dobson A, et al. 2002. Climate warming and disease risks for terrestrial and marine biota. *Science* 296:2158–62

Hayes ML, Bonaventura J, Mitchell TP, Prospero JM, Shinn EA, et al. 2001. How are climate and marine biological outbreaks functionally linked? *Hydrobiolgia* 460:213–20

Hayes RL, Goreau NI. 1998. The significance of emerging diseases in the tropical coral reef ecosystem. *Rev. Biol. Trop.* 46:173–85

Heide-Jorgensen M-P, Harkonen T, Dietz R, Thompson PM. 1992. Retrospective of the 1988 European seal epizootic. *Dis. Aquat. Organ.* 13:37–62

Heide-Jorgensen MP, Harkonen T. 1992. Epizootiology of the seal disease in the Eastern North Sea. *J. Appl. Ecol.* 29:99–107

Herbst LH. 1994. Fibropapillomatosis of marine turtles. *Annu. Rev. Fish Dis.* 4:389–425

Hochachka WM, Dhondt A. 2000. Density-dependent decline of host abundance resulting from a new infectious disease. *Proc. Natl. Acad. Sci. USA* 97:5303–6

Hoegh-Guldberg O. 1999. Climate change, coral bleaching and the future of the world's coral reefs. *Mar. Freshw. Res.* 50:839–66

Hoff CC. 1941. A case of correlation between infection of snail hosts with *Cryptocotyle lingua* and the habits of gulls. *J. Parasitol.* 27:539

Hofmann E, Ford S, Powell E, Klinck J. 2001. Modeling studies of the effect of climate variability on MSX disease in eastern oyster (*Crassostrea virginica*) populations. *Hydrobiologia* 460:195–212

Holmes JC. 1996. Parasites as threats to biodiversity in shrinking ecosystems. *Biodivers. Conserv.* 5:975–83

Holt RA, Amandi A, Rohovec JS, Fryer JL. 1989. Relation of water temperature to bacterial coldwater disease in coho salmon, chinook salmon, and rainbow trout. *J. Aquat. Anim. Health* 1:94–101

Houghton JT, Filho LGM, Callandar BA, Harris N, Kattenberg A, Maskell K, eds. 1996. *Climate Models—Projections of Future Climate*. New York: Cambridge Univ. Press

Huspeni TC, Lafferty KD. 2004. Using larval trematodes that parasitize snails to evaluate a salt-marsh restoration project. *Ecol. Appl.* 14(3):795–804

Ingersoll E. 1881. *The Oyster Industry*. Washington, DC: US Dep. Interior. 251 pp.

Jackson JBC, Kirby MX, Berger WH, Bjorndal KA, Botsford LW, et al. 2001. Historical overfishing and the recent collapse of coastal ecosystems. *Science* 293:629–38

Johnson LL. 2001. *An analysis in support of sediment quality thresholds for polycyclic aromatic hydrocarbons to protect estuarine fish*. Rep. NMFS-NWFSC-47, US Dep. Commer., NOAA

Jones JB, Hyatt AD, Hine PM, Whittington RJ, Griffin DA, Bax NJ. 1997. Special topic review: Australasian pilchard mortalities. *World J. Microb. Biot.* 13:383–92

Kennedy CR, Bush AO, Aho JM. 1986. Patterns in helminth communities: Why are birds and fish different? *Parasitology* 93:205–15

Khan RA. 1990. Parasitism in marine fish after chronic exposure to petroleum hydrocarbons in the laboratory and to the Exxon Valdez oil spill. *Bull. Environ. Contam. Toxicol.* 44:759–63

Khan RA, Thulin J. 1991. Influence of pollution on parasites of aquatic animals. *Adv. Parasitol.* 30:201–38

Kim K, Harvell CD. 2002. Aspergillosis of sea fan corals: disease dynamics in the Florida Keys. In *The Everglades, Florida Bay, and the Coral Reefs of the Florida Keys: An Ecosystem Sourcebook*, ed. JW Porter, KG Porter, pp. 824–913. Boca Raton, FL: CRC Press

Kim K, Harvell CD. 2004. The rise and fall of a 6 year coral-fungal epizootic. *Am. Nat.* In press

Kim K, Harvell CD, Kim PD, Smith GW, Merkel SM. 2000. Fungal disease resistance of Caribbean sea fan corals (*Gorgonia* spp.). *Mar. Biol.* 136:259–67

Kleypas JA, Buddemeier RW, Archer D, Gattuso JP, Langdon C, Opdyke BN. 1999. Geochemical consequences of increased atmospheric carbon dioxide on coral reefs. *Science* 284:118–20

Knowlton N. 1992. Thresholds and multiple stable states in coral reef community dynamics. *Am. Zool.* 32:674–82

Kreuder C, Miller MA, Jessup DA, Lowenstein LJ, Harris MD, et al. 2003. Patterns of mortality in southern sea otters (*Enhydra lutris nereis*) from 1998–2001. *J. Wildlife Dis.* 39:495–509

Kuris AM, Lafferty KD. 1992. Modelling crustacean fisheries: effects of parasites on management strategies. *Can. J. Fish. Aquat. Sci.* 49:327–36

Lafferty KD. 1997. Environmental parasitology: What can parasites tell us about human impacts on the environment? *Parasitol. Today* 13:251–55

Lafferty KD. 2003. Is disease increasing or decreasing, and does it impact or maintain biodiversity? *J. Parasitol.* 89:S101–5

Lafferty KD. 2004. Fishing for lobsters indirectly increases epidemics in sea urchins. *Ecol. Appl.* In press

Lafferty KD, Gerber L. 2002. Good medicine for conservation biology: the intersection of epidemiology and conservation theory. *Conserv. Biol.* 16:593–604

Lafferty KD, Holt RD. 2003. How should environmental stress affect the population dynamics of disease? *Ecol. Lett.* 6:654–64

Lafferty KD, Kuris AM. 1993. Mass mortality of abalone *Haliotis cracherodii* on the California Channel Islands: tests of epidemiological hypotheses. *Mar. Ecol. Prog. Ser.* 96:239–48

Lafferty KD, Kuris AM. 2002. Trophic strategies, animal diversity and body size. *Trends. Ecol. Evol.* 17:507–13

Lafferty KD, Kuris AM. 2004. Parasitism and environmental disturbances. In *Parasites and Ecosystems*, ed. F Thomas, JF Guégan, F Renaud. Oxford: Oxford Univ. Press. In press

LaFont A. 1874. *Note sur les Huitrières du Bassin d'Archacon.* Bordeaux: Chez Feret, Libraire. 52 pp.

Lapointe BE, Matzie WR, Barille PJ. 2002. Biotic phase shifts in Florida Bay and back reef communities in the Florida Keys: linkages with historical freshwater flows and nitrogen loading from Everglades runoff. In *The Everglades, Florida Bay, and the Coral Reefs of the Florida Keys*, ed. JW Porter, KG Porter, pp. 629–53. Boca Raton, FL: CRC Press

LeDeuff RM, Renault T, Gerard A. 1996. Effects of temperature on herpes-like virus detection among hatchery-reared larval Pacific oyster *Crassostrea gigas. Dis. Aquat. Organ.* 24:149–57

Lefcort H, Aguon MQ, Bond KA, Chapman KR, Chaquette R, et al. 2002. Indirect effects of heavy metals on parasites may cause shifts in snail species compositions. *Arch. Environ. Contam. Toxicol.* 43:34–41

Lessios HA. 1988. Mass mortality of *Diadema antillarum* in the Caribbean: What have we learned? *Annu. Rev. Ecol. Syst.* 19:371–93

Levitus S, Antonov JI, Boyer TP, Stephans C. 2000. Warming of the world ocean. *Science* 287:2225–29

Lightner DV. 1996. Epizootiology, distribution and the impact on international trade of two penaeid shrimp viruses in the Americas. *Rev. Sci. Tech. Off. Int. Epizoot.* 15:579–601

Lipp EK, Jarrell JL, Griffin DW, Lukasik J, Jacukiewicz J, Rose JB. 2002. Preliminary evidence for human fecal contamination in corals of the Florida Keys, USA. *Mar. Pollut. Bull.* 44:666–70

MacKenzie CLJ, Burrell VGJ, Rosenfield A, Hobart WL, eds. 1997. *The History, Present Condition, and Future of the Molluscan Fisheries of North and Central America and Europe.* Vol. 129. Washington, DC: US Dep. Commer. 240 pp.

Maes P, Paillard C. 1992. Effect du *Vibrio* P1, pathogene de *Ruditapes philippinarum*, sur d'autres espèces de bivalves. *Les Molluscques Marins, Biologies et Aquaculture. IFREMER, Actes de Colloques* 14:141–48

Marcogliese DJ. 2001a. Implications of climate change for parasitism of animals in the aquatic environment. *Can. J. Zool.* 79:1331–52

Marcogliese DJ. 2001b. Pursuing parasites up the food chain: implications of food web structure and function on parasite communities in aquatic systems. *Acta Parasitol.* 46:82–93

McCallum HI, Dobson AP. 1995. Detecting disease and parasite threats to endangered species and ecosystems. *Trends Ecol. Evol.* 10:190–94

Miller MW, Bourque AS, Bohnsack JA. 2002. An analysis of the loss of acroporid corals at Looe Key, Florida, USA: 1983–2000. *Coral Reefs* 21:179–82

Myers RA, Worm B. 2003. Rapid worldwide depletion of predatory fish communities. *Nature* 423:280–83

Nelson J. 1889. *Oyster interests of New Jersey. Rep. Spec. Bull. E*, NJ Agric. Coll. Exp. Stn., New Brunswick

O'Shea T. 1999. Environmental contaminants and marine mammals. In *Biology of Marine*

Mammals, ed. SA Rommel, pp. 485–564. Washington, DC: Smithsonian Inst. Press

Osterhaus ADME, Vedder EJ. 1988. Identification of virus causing recent seal deaths. *Nature* 335:20

Patterson KL, Porter JW, Ritchie KE, Polson SW, Mueller E, et al. 2002. The etiology of white pox, a lethal disease of the Caribbean elkhorn coral, *Acropora palmata*. *Proc. Natl. Acad. Sci. USA* 99:8725–30

Pinnegar JK, Polunin NVC, Francour P, Badalamenti F, Chemello R, et al. 2000. Trophic cascades in benthic marine ecosystems: lessons for fisheries and protected-area management. *Environ. Conserv.* 27:179–200

Pojmanska T, Grabda-Kazubska B, Kazubski SL, Machalska J, Niewiadomska K. 1980. Parasite fauna of five fish species from the Konin lakes complex, artificially heated with thermal effluents, and from Goplo Lake. *Acta Parasitol.* 27:319–57

Porter JW, Dustan P, Jaap WC, Patterson KL, Kosmynin V, et al. 2001. Patterns of spread of coral disease in the Florida Keys. *Hydrobiologia* 460:1–24

Porter JW, Fitt WK, Spero HJ, Rogers CS, White MW. 1989. Bleaching in reef corals—physiological and stable isotopic responses. *Proc. Natl. Acad. Sci. USA* 86:9342–46

Porter JW, Kosmynin V, Patterson K, Porter KG, Jaap WC, et al. 2002. Detection of coral reef change by the Florida Keys Coral Reef Monitoring Project. See Porter & Porter 2002, pp. 749–69

Porter JW, Meier OW. 1992. Quantification of loss and change in Floridian reef coral populations. *Am. Zool.* 32:625–40

Porter JW, Porter KG, eds. 2002. *The Everglades, Florida Bay, and Coral Reefs of the Florida Keys*. Boca Raton, FL: CRC Press

Porter JW, Tougas JI. 2001. Reef ecosystems: threats to their biodiversity. *Encyclop. Biodiv.* 5:73–95

Richmond RH. 1993. Coral reefs—present problems and future concerns resulting from anthropogenic disturbance. *Am. Zool.* 33:524–36

Riggs MR, Lemly AD, Esch GW. 1987. The growth, biomass, and fecundity of *Bothriocephalus acheilognathi* in a North Carolina cooling reservoir. *J. Parasitol.* 73:893–900

Robson EM, Williams IC. 1970. Relationships of some species of digenea with the marine prosobranch *Littorina littorea*. Part 1. The occurrence of larval digenea in *Littorina littorea* on the North Yorkshire coast. *J. Helminthol.* 44:153–68

Ruiz GM, Fofonoff P, Carlton JT, Wonham MJ, Hines AH. 2000. Invasions of coastal marine communities in North America: apparent patterns, processes, and biases. *Annu. Rev. Ecol. Syst.* 31:481–531

Rutzler K, Santavy DL, Antonius A. 1983. The Black Band Disease of Atlantic reef corals. III. Distribution, ecology and development. *PSZNI Mar. Ecol.* 4:329–58

Sala E, Boudouresque CF, Harmelin-Vivien M. 1998. Fishing, trophic cascades, and the structure of algal assemblages: evaluation of an old but untested paradigm. *Oikos* 82:425–39

Sanders MJ. 1966. Parasitic castration of scallop *Pecten alba* (Tate) by a bucephalid trematode. *Nature* 212:307

Santavy DL, Mueller E, Peters EC, MacLaughlin L, Porter JW, et al. 2001. Quantitative assessment of coral diseases in the Florida Keys: strategy and methodology. *Hydrobiologia* 460:39–52

Scott ME. 1988. The impact of infection and disease on animal populations: implications for conservation biology. *Conserv. Biol.* 2:40–56

Shears NT, Babcock RC. 2002. Marine reserves demonstrate top-down control of community structure on temperate reefs. *Oecologia* 132:131–42

Shinn EA, Smith GW, Prospero JM, Betzer P, Hayes ML, et al. 2000. African dust and the demise of Caribbean coral reefs. *Geophys. Res. Lett.* 27:3029–32

Siddall R, Clers SD. 1994. Effect of sewage sludge on the miracidium and cercaria of *Zoogonoides viviparus* (Trematoda: Digenea). *Helminthologia* 31:143–53

Smith GW, Ives LD, Nagelkerken IA, Ritchie

KB. 1996. Caribbean sea-fan mortalities. *Nature* 383:487

Smith NF. 2001. Spatial heterogeneity in recruitment of larval trematodes to snail intermediate hosts. *Oecologia* 127:115–22

Stadnichenko AP, Ivanenko LD, Gorchenko IS, Grabinskaya OV, Osadchuk LA, Sergeichuk SA. 1995. The effect of different concentrations of nickel sulphate on the horn snail (Mollusca: Bulinidae) infected with the trematode *Cotylurus cornutus* (Strigeidae). *Parazitologiya* 29:112–16

Stewart BS, Yochem PK. 2000. Community ecology of California Channel Islands pinnipeds. In *Fifth California Islands Symposium*, ed. DR Brown, KL Mitchell, HW Chang, pp. 413–20. Santa Barbara, CA: Miner. Manag. Serv.

Stiven AE. 1964. Experimental studies on the host parasite system hydra and *Hydramoeba hydroxena* (Entz.). II. The components of a single epidemic. *Ecol. Monogr.* 34:119–42

Strand O, Vølstad JH. 1997. Molluscan fisheries and culture of Norway. See MacKenzie et al. 1997, pp. 7–24

Sutherland KP, Porter JW, Torres C. 2004. Disease and immunity in Caribbean and Indo-Pacific zooxanthellate corals. *Mar. Ecol. Prog. Ser.* 266:273–302

Swart RL, Ross PS, Vedder LJ, Timmerman HH, Heisterkamp S, et al. 1994. Impairment of immune function in harbor seals (*Phoca vitulina*) feeding on fish from polluted waters. *Ambio* 23:155–59

Tegner MJ, Levin LA. 1983. Spiny lobsters and sea urchins: analysis of a predator-prey interaction. *J. Exp. Mar. Biol. Ecol.* 73:125–50

Torchin ME, Lafferty KD, Dobson AP, McKenzie VJ, Kuris AM. 2003. Introduced species and their missing parasites. *Nature* 421:628–30

Van Loveren H, Ross PS, Osterhaus A, Vos JG. 2000. Contaminant induced immunosuppression and mass mortalities among harbor seals. *Toxicol. Lett.* 112:319–24

Ward JR, Lafferty KD. 2004. The elusive baseline of marine disease: Are diseases in ocean ecosystems increasing? *PLoS Biol.* 2:542–47

Wilkinson C, ed. 2002. *Status of Coral Reefs of the World: 2002*. Townsville: Aust. Inst. Mar. Sci.

Williams J, Ernest H, Bunkley-Williams L. 1990. The world-wide coral reef bleaching cycle and related sources of coral mortality. *Atoll Res. Bull.* 335:1–71

Woodroffe R. 1999. Managing threats to wild mammals. *Anim. Conserv.* 2:185–93

BIRD SONG: The Interface of Evolution and Mechanism

Jeffrey Podos, Sarah K. Huber, and Benjamin Taft

Department of Biology and Graduate Program in Organismic and Evolutionary Biology, University of Massachusetts, Amherst, Massachusetts 01003; email: jpodos@bio.umass.edu, shuber@bio.umass.edu, btaft@bio.umass.edu

Key Words sexual selection, cultural transmission, neurobiology, biomechanics

■ **Abstract** Bird song provides an unusually impressive illustration of vertebrate behavioral diversification. Research on bird song evolution traditionally focuses on factors that enhance song diversity, such as cultural transmission and sexual selection. Recent advances in the study of proximate mechanisms of vocal behavior, however, provide opportunities for studying mechanistic constraints on song evolution. The main goal of this review is to examine, from both conceptual and empirical perspectives, how proximate mechanisms might temper patterns of song evolution. We provide an overview of the two "substrates" of song evolution, memes and vocal mechanisms. We argue that properties of vocal mechanisms (control, production, and ontogeny) constrain vocal potential and may thus limit pathways of meme evolution. We then consider how vocal mechanisms may constrain song evolution under five scenarios of drift and selection and examine four specific song traits for which mechanistic constraints appear to counter the diversifying effects of sexual selection. These examples illustrate the interplay between meme evolution as a diversifying influence and proximate limitations as a barrier to song divergence. We conclude by suggesting that vocal mechanisms not only constrain song evolution but also can facilitate the evolution of novel vocal features.

INTRODUCTION

The songs of songbirds are among the most celebrated examples of behavioral diversification in the animal kingdom (Catchpole & Slater 1995, Baptista & Kroodsma 2001). Most species' songs are distinct, a fact long appreciated by naturalists. Studies during recent decades have documented extensive song variation at additional levels of organization including within species, populations, and individuals. We have learned, for example, that the songs of rufous-collared sparrows (*Zonotrichia capensis*) vary broadly in the structure of their "terminal trills" (Handford & Lougheed 1991), that individual brown thrashers (*Toxostoma rufum*) can produce thousands of distinct song types (Kroodsma & Parker 1977), and that Lincoln's sparrows (*Melospiza lincolnii*) introduce subtle variations into song types nearly each time they are sung (Cicero & Benowitz-Fredericks 2000). Explaining the evolutionary origins and maintenance of phenotypic diversity is a principal challenge in the study of bird song.

Current research on bird song tends to emphasize the lability of song structure in evolution. As is true for many mating signals, sexual selection by female choice is thought to facilitate the evolution of elaborate and distinct song traits (reviewed by Searcy & Yasukawa 1996). Intrasexual selection may also contribute to song diversification by favoring individually distinctive song patterns. In species that learn to sing by imitation, songs can diversify rapidly through the accumulation of copying inaccuracies during cultural transmission (Payne 1996). Vocal diversity also appears to be enhanced through the correlated evolution of song structure and life history or ecological variables such as migratory habits (Nelson et al. 1996, Peters et al. 2000, Kroodsma et al. 2002) and the density of breeding territories (Kroodsma 1977, Catchpole 1980).

Although bird song may be unusually susceptible to change over evolutionary time, it is still not free to evolve in any direction at any time. All phenotypic traits are tempered in their evolution by intrinsic constraints on genetic variation and mechanical design (Gould 1980, Wake 1991, Arnold 1992). The strong genetic covariance of traits, for example, can bias phenotype evolution in directions that do not necessarily match those specified by selection pressures (Schluter 1996). Behavioral traits such as song can be further tempered in their evolution by limits on neuromuscular capacity and performance abilities. Rapidly modulated communication displays, such as electric organ discharges in fishes or rattle signals in snakes, are circumscribed in their structure by maximum rates of neural activity and muscular contraction (Kramer 1990, Rome et al. 1996). Studies of intrinsic constraints are of broad value because they help to specify the range of phenotypes that can be expressed and on which selection can act (Perrin & Travis 1992).

In the study of bird songs, the idea of constraints is often raised in conjunction with evidence for stasis during song learning and evolution. Experimental studies of song learning have revealed innate predispositions in some species for copying conspecific song models in preference to heterospecific song models (e.g., Marler & Peters 1977, Nelson & Marler 1993). Studies of isolate-reared, cross-species-reared, and hybrid songbirds have documented high genetic heritability in numerous song features such as note and song duration, rhythm, and repertoire size (Baptista 1996). Furthermore, comparative surveys both within and across species indicate that some song traits are strongly conserved over evolutionary time (Marler & Pickert 1984, Payne 1986, Price & Lanyon 2002). These lines of evidence suggest that genetic biases guide young birds to develop and produce specific kinds of sounds, even among species that learn by imitation. If sustained over a sufficient evolutionary time frame, genetic biases presumably could limit the diversification of song phenotypes. Evidence for stasis during song development and evolution, however, might also be explained by stabilizing selection and, thus, provides only weak insight into constraint-based hypotheses.

A stronger argument for evolutionary constraints on behavior can be made through functional analysis of proximate mechanisms such as neural organization and biomechanical performance (e.g., Garland & Losos 1994, Herrel et al. 2002). In recent decades, bird song has become a particularly valuable model system for studies of the neural bases of complex, learned behavior (Konishi 1985, Brainard & Doupe 2002), and great strides have also been made in understanding how vocal morphology and physiology contribute to song production (Suthers et al. 1999). Studies on both topics have been motivated largely by curiosity about behavioral mechanisms. Yet, advances in our understanding of vocal mechanisms might also point to biases and opportunities in the evolution of vocal traits, in ways that are just beginning to be explored (Nowicki et al. 1992, Lambrechts 1996, Suthers & Goller 1997, Gil & Gahr 2002, Ten Cate et al. 2002, Podos & Nowicki 2004).

Our goal here is to examine, from both conceptual and empirical perspectives, how vocal mechanisms might impose constraints on song evolution. We begin with an overview of the two "substrates" of song evolution, memes and vocal mechanisms. Next, we consider how vocal mechanisms may impose constraints under different scenarios of drift and selection. We examine in detail four song traits for which recent evidence suggests that mechanistic features constrain song elaboration in opposition to sexual selection. Finally, we argue that vocal mechanisms not only constrain song evolution but also can facilitate the evolution of novel song features.

SUBSTRATES OF SONG EVOLUTION

A first step in any discussion of song evolution is to identify the units that are transmitted across generations. Like other complex behavioral phenotypes, bird vocalizations can be regarded as composites of discrete traits, arranged at different levels of organization and varying across time (Tinbergen 1951). Ornithologists typically label the most basic units of organization in bird songs as notes, identified on sound spectrographs as continuous traces separated by silent gaps (Figure 1). Individual notes are defined by their time-varying frequency structure ("phonology"), amplitude profiles, and timbre, which is the distribution of acoustic energy across fundamental frequencies and harmonic overtones (e.g., Marler 1969, Clark et al. 1987). Notes typically are arranged in higher-order groupings such as "note complexes" or "syllables" (Figure 1), which are produced by birds in sequence to form songs. The structure of songs across species is immensely variable (Figure 1). Additional traits occur at higher levels of song organization, such as in patterns of song delivery and in the occurrence of song repertoires.

Because songs are composite entities, vocal evolution is perhaps best viewed as the product of evolutionary change (or stasis) in multiple, potentially independent traits. This view is consistent with the traditional ethological view for which multitrait analyses have revealed independent evolutionary trajectories in some behavioral display components (e.g., Lorenz 1950, Prum 1990). In the study of

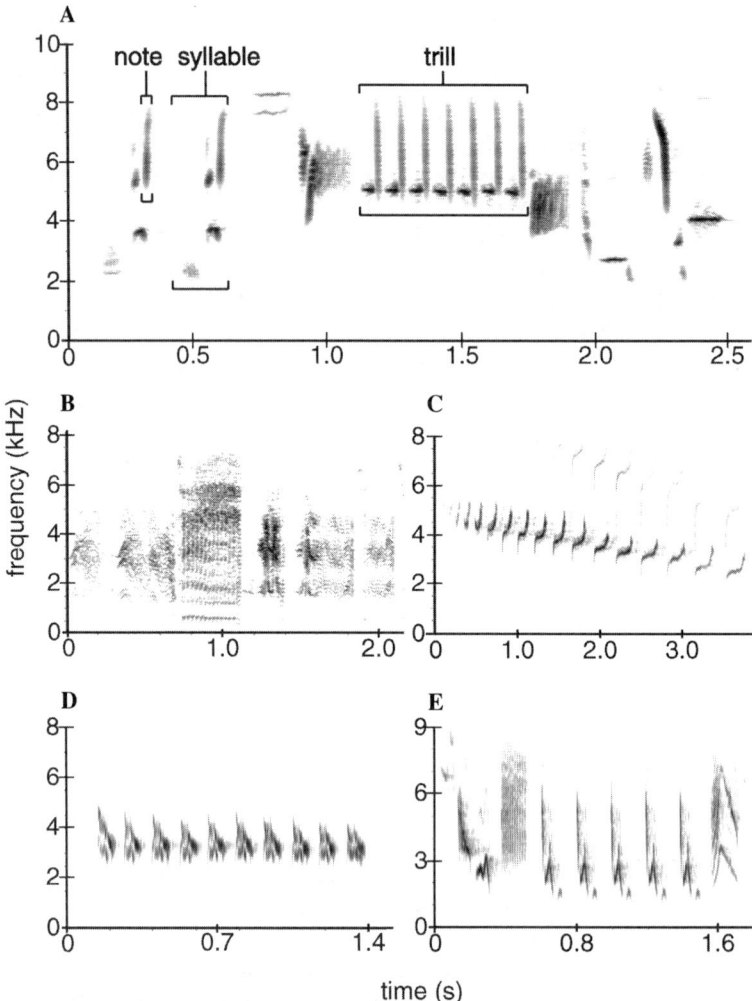

Figure 1 Spectrograms of five representative passerine songs. Songs exhibit different levels of organization and significant structural variation among species. (*A*) Song sparrow, *Melospiza melodia* (Emberizidae). A representative note, syllable, and trill are marked with brackets. (*B*) Satin bowerbird, *Ptilonorhynchus violaceus* (Ptilonorhynchidae). (*C*) Canyon wren, *Catherpes mexicanus* (Troglodytidae). (*D*) Small tree finch, *Camarhynchus parvulus* (Emberizidae). (*E*) Thrush nightingale, *Luscinia megarhynchos* (Turdidae).

bird songs, phylogenetic analyses have revealed independent trajectories for the evolution of different song traits within lineages, such as in the oropendolas and caciques (Price & Lanyon 2002, 2004; see also Payne 1996, Slabbekoorn & Smith 2002b).

How, then, are song traits transmitted across generations? In this review, we focus our discussion on songbirds that learn their songs by imitation. We argue that patterns of song evolution can be attributed to changes in two distinct substrates: (*a*) memes and (*b*) mechanisms of vocal control, production, and ontogeny.

Memes

Memes refer to cultural traditions that are transmitted between individuals through imitation (Dawkins 1976, Bonner 1980, Boyd & Richerson 1985). The best-known examples of memes in nonhuman animals are social feeding strategies and vocal traditions (e.g., Heyes 1994, Noad et al. 2000). Memes may evolve through population-level sorting among transmitted units, as illustrated in changing patterns of word use or loss within human languages (Cavalli-Sforza & Feldman 1981). Alternatively, memes may evolve through the progressive transformation of transmitted units themselves, as has been demonstrated within the songs of humpback whales (Payne et al. 1983). Our discussion of bird song evolution focuses on the latter process; detailed reviews of bird song "population memetics" are available in Lynch (1996) and Payne (1996).

Observations of avian song learning before the 1950s emphasized the temporal continuity of song memes. Bird breeders long recognized that "student" birds often faithfully reproduce song patterns of their vocal "tutors" (Thorpe 1961, Konishi 1985). The advent of spectrographic analyses made possible the examination of patterns of song change as well. Some of the first studies of song evolution focused on geographic variation in song traits, following the assumption that variation in space provides a window (albeit indirect) into causes of change over time. Marler (1960) documented song variations within several songbird species and hypothesized that these variations had emerged as a result of copying imprecision. Subsequent descriptive studies of geographic variation in song have supported and refined this hypothesis (e.g., Lemon 1975, Bitterbaum & Baptista 1979). In some species, song types tend to vary gradually across locality, with neighboring population pairs producing the most similar songs (Morton 1987, Irwin 2000). Such "clinal" variation suggests a free interchange of song memes between neighboring populations. Other species exhibit more pronounced geographic discontinuities in song structure, resulting in song dialects (Marler & Tamura 1962, Baptista 1975). Dialects generally occur in species with specific life-history and behavioral ecological traits such as limited dispersal, early song acquisition, and comparatively simple repertoires. This pattern is consistent with the hypothesis that copy inaccuracies contribute significantly to song evolution (Slater 1986, Nelson 2000).

Insight into the process by which song memes evolve has also been provided by laboratory studies of song learning. Marler & Tamura (1964) showed that laboratory-reared white-crowned sparrows (*Zonotrichia leucophrys*) are able to produce remarkably precise copies of training models from both native and foreign dialects. Accurate song model reproduction has since been demonstrated in numerous additional species (e.g., Marler & Peters 1977, Kroodsma & Pickert 1980, Slater et al. 1988). Even the most accurate copies of song are not perfectly precise, however, and imprecisions can accumulate through time and result in song evolution. Thorpe (1961) identified two additional causes of song evolution: the rearrangement of accurately copied material and the invention of new material. Both possibilities have been documented in laboratory studies. For example, song sparrows (*Melospiza melodia*) were shown to combine elements from multiple tutor songs into common "hybrid" songs (Marler & Peters 1987, 1988; Beecher 1996), and sedge wrens (*Cistothorus platensis*) were shown to improvise novel songs as well as imitate members of their cohort during song development (Kroodsma 1996).

Laboratory results are mirrored by patterns described in longitudinal field studies. Studies tracking song changes in island populations have documented modifications in the structure of individual notes that were retained in subsequent generations (Jenkins 1978, Grant & Grant 1996). Other studies documented instances in which notes were deleted or reordered during cultural transmission (Ince et al. 1980, Payne 1996) or in which notes from multiple models were blended together within copied song types (Slater 1989, Payne 1996; see also Slabbekoorn et al. 2003). Island birds have also been reported to evolve song types with novel syntax (e.g., Baker et al. 2003). Results from field and laboratory studies together are consistent with broader definitions of meme evolution (Dawkins 1976, Boyd & Richerson 1985). Like other memes, songs are encoded as neural representations, are transmitted between individuals through environmental media, and incur inaccuracies during transmission that can be retained over time.

Mechanisms

Patterns of song meme evolution—whether identified in the laboratory or field and whether arising through copy error, innovation, or improvisation—are normally, if implicitly, attributed to properties of the neural mechanisms that underlie song perception and memorization. For example, gradual shifts in song structure during evolution have been attributed to limits on how well birds are able to perceive and memorize song models during the "sensory phase" of song learning (Marler 1976, 1984; Slater 1989). Perceptual and memory-based hypotheses have also been proposed to explain more rapid meme diversification, such as when birds incorporate elements of different models into individual copies. To illustrate, Marler & Peters (1987, p. 99) speculated that birds "hybridize" acoustic elements from multiple models through a neural process in which stored songs are "consolidated and reorganized" before the onset of vocal motor ontogeny. The other side of the coin—meme stability—is also normally attributed to mechanisms of perception and memorization. Kroodsma et al. (1999), for example, proposed that the remarkable structural uniformity of black-capped chickadee songs across mainland North America reflects a tendency of these birds to memorize and "average" multiple song models. The neural mechanisms that underlie these kinds of processes (e.g., consolidation, reorganization, and averaging) are presently unspecified. In general, much more work will be needed to link species' perceptual and memory-based predispositions and capabilities to properties of their neural mechanisms.

In contrast to our lack of understanding of mechanisms of perception and memorization, we have made great strides in characterizing the proximate mechanisms that underlie song control, production, and ontogeny. In the remainder of this section, we provide a brief description of these mechanisms and consider how modifications to them may shape patterns of meme evolution.

The source of vocalizations in birds is the syrinx, an organ unique to the class Aves (Greenewalt 1968). The syrinx is located within the thoracic cavity at or near the base of the trachea. Sound is produced when air flow from the lungs causes syringeal tissues to vibrate (Goller & Larsen 1997, Larsen & Goller 1999). Vocal production depends also on a suite of complementary motor and neural systems. Simultaneous recording of bronchial airflow and thoracic air-sac pressure has revealed that respiratory and syringeal motor patterns are tightly coupled (Suthers et al. 1994, Suthers 1997), insofar as vocal output normally occurs only during expirations (cf. Goller & Daley 2001). Anatomical components of the vocal tract anterior to the syrinx (the trachea, larynx, and beak) modify the spectral structure of vocalizations produced by the syrinx (Nowicki 1987, Westneat et al. 1993, Beckers et al. 2003). Sound generated by the syrinx contains acoustic energy at wide ranges of frequencies. As sound passes through the vocal tract, however, harmonic overtones tend to be selectively attenuated, whereas fundamental frequencies tend to pass unimpeded. The vocal tract is, thus, responsible for the pure-tonal, musical quality of many bird songs (Nowicki & Marler 1988).

Nearly three decades of work has shown that the activity of the vocal apparatus is mediated by a series of interconnected brain nuclei that function in a hierarchical manner (Nottebohm et al. 1976, Yu & Margoliash 1996, Hahnloser et al. 2002). The "motor pathway" of song production features prominent song nuclei that project to syringeal and respiratory musculature. Lesion studies by Nottebohm et al. (1976) showed that the motor pathway is required for successful song production in adult birds (see also McCasland 1987, Simpson & Vicario 1990). Two of the most prominent brain nuclei, HVc and RA, undergo annual changes in size consistent with seasonal variation in vocal output (Nottebohm et al. 1986). Furthermore, these nuclei and their constituent cells tend to be larger in males than in females in species in which only males sing (Nottebohm & Arnold 1976, Arnold 1990), whereas no such difference is found in a duetting species in which both males and females sing (Brenowitz et al. 1985). Information about the organization of song nuclei has enabled direct exploration of the cellular bases of song development and production (Brainard & Doupe 2002).

Vocal ontogeny refers to the translation of memorized song models into stereotyped vocal motor patterns during the "sensorimotor" phase of song learning (Slater 1989). During this phase, the songs of young birds progress from a babblelike rudimentary form (subsong) to a highly stereotyped adult form (crystallized song; Thorpe 1961, Marler & Peters 1982). This progression appears to reflect a need for young birds to practice and refine their ability to reproduce previously memorized models (Marler 1984, Podos et al. 1995). Studies of deafened birds established that successful vocal learning requires auditory feedback in which birds compare the auditory experience of their own vocalizations to neural "templates" of previously memorized models (Konishi 1965, 1985). Neurobiologists have made substantial progress in identifying the neural control pathways that mediate auditory feedback (Bottjer et al. 1984, Williams & Mehta 1999) and in characterizing cellular changes that occur in these pathways during ontogeny (Livingston et al. 2000).

Advances in our understanding of the proximate mechanisms of vocal control, production, and ontogeny provide new opportunities for studying constraints on song evolution. The central premise we adopt here is that song features may be limited in their expression by properties of vocal mechanisms, and these properties may persist over evolutionary time. On the smallest of evolutionary time scales, from one generation to the next, evidence for mechanistic constraints on song evolution has been limited. On the contrary, some species have been shown to be able to reproduce the song models of heterospecific tutors accurately (e.g., Baptista & Petrinovich 1984), which suggests that song development is normally not restricted by limits on how birds are able to produce sound (Marler 1976, 1984; Konishi 1985). Over broader evolutionary time scales, however, we expect meme diversification to be tempered by limits of mechanical possibility or at least to be biased to evolve in specific directions because of mechanistic features.

One way to articulate this point is to consider the distinction between potential and actual (or realized) song phenotypes, as conceptualized in Figure 2 (see also Podos 1996, 1997; Slabbekoorn & Smith 2002a). Mechanistic constraints circumscribe a lineage's range of potential song phenotypes, within which memes are free to evolve (Figure 2*B*). However, memes and mechanistic boundaries may sometimes intersect because of evolutionary changes in either or both (Figure 2*D,F*), and mechanisms may, thus, limit or at least bias subsequent patterns of song evolution. It is important to note that constraints need not be absolute but rather can be modified over the course of evolution (Figure 2*C,E,F*). To illustrate, song nucleus volume is thought to exert mechanistic limits on repertoire size (see section below, Song Trait Evolution). If selection pressure on song repertoire size is sufficiently strong, we might, therefore, expect to observe increases in song nucleus volume. Yet, such an adjustment would not come without a cost. In our example, increases in brain volume would presumably result in reduced allocation of developmental or metabolic resources to other traits. Such trade-offs have also been regarded as evidence for constraints (Maynard Smith et al. 1985, Perrin & Travis 1992).

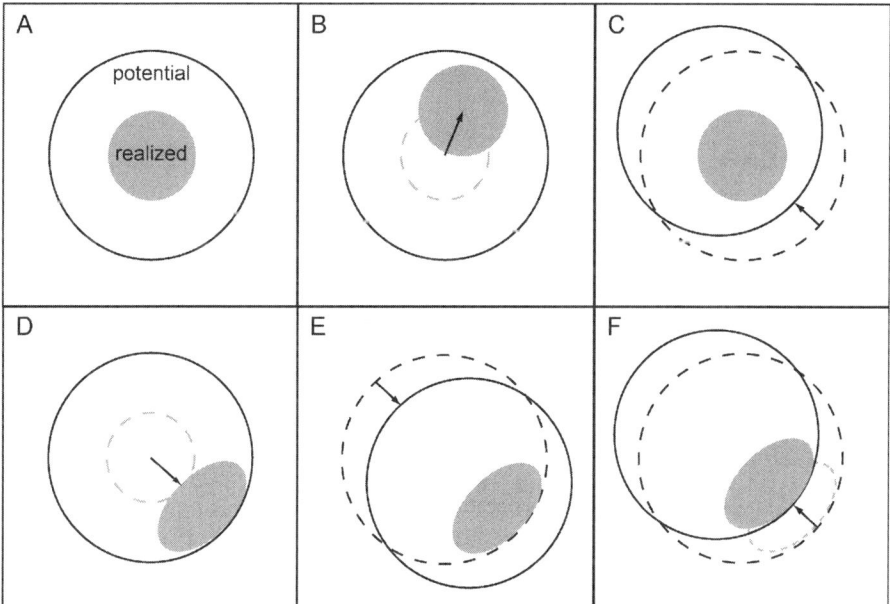

Figure 2 Schematic overview of some possible interactions between memes and mechanisms during song evolution. This illustration expands upon figure 3 of Slabbekoorn & Smith (2002). Gray shaded areas indicate realized song phenotypes, i.e., all memes within a population, and black lines (*circles*) encompass the range of potential song phenotypes, as delimited by mechanistic constraints. Arrows indicate pathways of evolution caused by selection, drift, or both. (*A*) Realized song variation may not intersect with potential song variation. In this scenario, initial evolutionary changes in song (*B*) will occur unimpeded, and song evolution will be tantamount to meme evolution. (*C*) When realized and potential song phenotypes initially do not intersect, minor evolutionary changes in vocal potential should not affect song evolution. (*D*) Evolutionary pressures, especially sexual selection, should eventually push memes to the boundary of physical possibility. Further song evolution in the direction indicated will be constrained at least in the absence of any changes in vocal mechanisms. (*E*) Subsequent evolutionary changes in vocal mechanisms may occur independently of meme evolution. (*F*) Alternatively, evolutionary changes in vocal mechanisms may set further constraints on song evolution.

CAUSES OF SONG EVOLUTION

Like other phenotypic traits, song evolves through a combination of drift and selection. We now consider if and how mechanistic constraints might influence song evolution under five different scenarios of drift and selection.

Cultural Drift

Cultural drift refers to changes in song memes driven by chance (nonfunctional) variation in their propagation across generations. Cultural drift in song evolution has been discussed primarily in two specific contexts: during initial stages of dialect formation and during the colonization of islands. As discussed earlier, dialects appear to form through a process of cultural diversification in which copy errors generate seemingly random song "mutations" (Lemon 1975). Vocal forms in island populations often represent restricted and simplified subsets of vocal forms found in mainland populations. This reduction in diversity has been attributed to the susceptibility of small colonizing populations to founder effects and rapid cultural drift (Thielcke 1972, Baker et al. 2003). No empirical evidence to date supports a role for mechanistic constraints on meme drift (Figure 2*B*).

Genetic Drift

Random drift in the genetic loci that underlie vocal mechanisms presumably could also influence song evolution by modifying a lineage's vocal potential, although we are not aware of any such demonstration in the literature. We expect that genetic drift would influence song evolution only when memes are initially produced at or near some mechanistic limit, and when drift occurs in directions that cause the proximate limits of vocal mechanisms to intersect with meme properties (Figure 2*F*). We illustrate this argument using a hypothetical example. The acoustic frequencies of bird songs generally correlate negatively with syrinx mass (Cutler 1970, Wallschläger 1980). This is because the minimum vocal frequencies that a bird can produce depend on syrinx mass, with smaller birds (with lighter syringes) constrained to sing at higher frequencies. What would happen, then, to the acoustic frequencies of song in the event of genetic drift in syrinx mass? We envision three possibilities: (*a*) Song frequencies may not initially have challenged the frequency limitations of the syrinx, for instance by being produced well within the range of possibilities determined by syrinx mass. Minor changes in syrinx mass and corresponding vocal potential would thus have little or no bearing on the evolution of vocal frequencies, at least initially (Figure 2*C*). (*b*) Song frequencies might initially have been produced at mechanistic limits, for instance at the lowest frequencies that a syrinx of given mass could support. Subsequent drift in syrinx mass may occur but in a direction that does not influence song evolution (Figure 2*E*). For instance, increases in syrinx mass presumably would not limit the accurate reproduction of low-frequency song components during meme transmission. (*c*) Frequencies may have initially been produced at mechanistic limits, and changes in vocal potential occur in specific directions that constrain song frequency evolution (Figure 2*F*). If our hypothetical population initially had produced song components at their lowest possible frequencies, then subsequent reductions in syrinx mass would presumably limit copying accuracy and thus bias songs to evolve toward higher frequencies. As this example is intended to illustrate, evolutionary changes in vocal mechanisms need to be directional and of significant magnitude if they are to influence song evolution.

Cultural Selection

Cultural selection refers to the differential propagation of memes across generations based on variation in their effectiveness as communication signals. Moreover, unlike natural and sexual selection (see below), cultural selection on song memes is not powered by variation in the relative fitness of birds in a population. Perhaps the best illustration of this process involves selection for optimal sound transmission. Songs vary in how well they transmit in different acoustic environments. In forested habitats, for example, songs with slow repetitions and low frequencies suffer less degradation than do songs with fast repetitions and high frequencies, whereas the opposite trend is observed in open habitats (Marten & Marler 1977, Richards & Wiley 1980, Brown & Handford 2000). Thus, for populations with multiple song types, the types that transmit best in a specific acoustic environment should be more effective and thus favored as communication signals. Hansen (1979) hypothesized that song types that transmit best in a given habitat will be heard more often by young birds as training models and will thus be inherited at disproportionately high rates across generations. Documented correlations between acoustic habitats and song structure are consistent with these hypotheses (e.g., Wiley 1991), although, as noted by Morton (1975, p. 33), such correlations may also arise through natural or sexual selection.

Cultural selection seems most relevant as an explanation for song differences among closely related populations (e.g., Hunter & Krebs 1979, Handford & Lougheed 1991, Doutrelant et al. 1999). At this level, songs appear to adapt readily to varying environmental conditions and to evolve relatively independently of mechanistic constraints (Figure 2B). As temporal and spatial scales broaden, however, gene flow among populations diminishes (MacDougall-Shackleton & MacDougall-Shackleton 2001), and mechanistic constraints should thus take on greater relevance. Broad comparative analyses by Ryan & Brenowitz (1985) and Bertelli & Tubaro (2002), for example, showed that song evolution can be influenced not only by cultural selection but also by genetic (body mass) constraints.

Natural Selection

Natural selection on song evolution may be direct, acting on memes, or indirect, acting on vocal mechanisms. The most widely discussed hypotheses of natural selection on bird song focus on direct selection against song meme overlap, through a process of "reinforcement" and reproductive character displacement (Butlin & Ritchie 1994). Many songbird species retain the ability to produce hybrids long after their divergence from a common ancestor (Grant & Grant 1992). Hybrid production can be disadvantageous, however, because hybrid offspring often suffer from genetic deficiencies (for example, leading to reduced fertility) or poor fit to available ecological niches. Natural selection is, therefore, expected to favor individuals that select conspecific rather than heterospecific mates. Songs are often the principal source of information about singer identity and are thus expected to evolve toward increasing species-level distinctiveness (Marler 1957, 1960; Ptacek 2000).

Reinforcement hypotheses have received support from descriptive and experimental sources. Nelson & Marler (1990) showed that the songs of 13 sympatric species were distributed widely across "acoustic space," a pattern consistent with a history of acoustic competition and divergence. Songs of birds in island populations, which face relatively few acoustic competitors, are often more variable than the songs of continental conspecifics (Thielcke 1972, Naugler & Ratcliffe 1994; but see Espmark 1999), a pattern that suggests a history of relaxed selection for species distinctiveness on islands. Playback studies in many species have shown that birds respond most strongly to songs with conspecific parameters (e.g., Brémond 1976, Becker 1982, Baker 1996). More relevant are playback studies in species pairs with allopatric and sympatric populations. In these studies, sympatric populations tend to discriminate against each other more strongly than do allopatric populations (Irwin & Price 1999; see also De Kort & Ten Cate 2001).

Natural selection through reinforcement appears to shape meme evolution primarily by modifying birds' perceptual and learning preferences, which are expected to evolve together with evolutionary changes and divergence in song structure (Grant & Grant 1997, Irwin & Price 1999). Other forms of direct natural selection are possible. Evolutionary trends toward increased song output (and thus conspicuousness) might be countered by increased susceptibility to predation or parasitism (Zuk & Kolluru 1998), although we are not aware of any such demonstration in birds. In all the circumstances described above, natural selection acts directly on meme function and may occur independently of vocal potential (Figure 2B).

A contrasting class of natural selection pressures alters song evolution indirectly by adjusting the range of potential phenotypes a population can produce. The effects of indirect natural selection on song evolution depend on the relationship between vocal potential and the memes produced in a population, as outlined earlier in our discussion of genetic drift. Darwin's finches provide a good illustration of how indirect evolution of song mechanisms may constrain meme evolution. These birds are well known for their diversity in beak form and function. This diversity is the result of natural selection for varying ecological conditions, such as food availability and interspecific competition (PR Grant & BR Grant 2002). Like other songbirds, Darwin's finches also use their beaks in sound production to rapidly modify the resonance properties of the vocal tract as they shift vocal frequencies (Podos et al. 2004b). Analyses of song features in Darwin's finches in relation to the songs of other species (Podos 1997) and in relation to beak morphology (Podos 2001) suggest that large-beaked birds have faced persistent constraints on the evolution of vocal features that require rapid vocal tract dynamics. Thus, the evolutionary diversification of feeding morphology in these birds appears to have had indirect effects on vocal performance and song evolution (Figure 2*F*) (Podos & Nowicki 2004). As a caveat, beak evolution by natural selection need not influence song evolution, as indicated in studies of sharp-beaked ground finches (*Geospiza difficilis*; BR Grant & PR Grant 2002) and black-bellied seed-crackers (*Pyrenestes ostrinus*; Slabbekoorn & Smith 2000). For these species, memes may not initially have been produced at mechanistic limits, so that subsequent beak adaptations imposed no detectable constraints on song phenotypes (Figure 2*C,E*) (Podos & Nowicki 2004).

Sexual Selection

Bird songs are produced most often by males and appear to function primarily in male-male competition and mate attraction (Searcy & Andersson 1986, Catchpole & Slater 1995). Research on song evolution and diversification has, therefore, emphasized sexual selection as a mechanism of evolutionary change. The theory of sexual selection posits that different display variants within a population vary in their effectiveness with regard to mate competition, that more effective song variants are retained over evolutionary time, and that female preferences evolve together with male display traits (Darwin 1871, Andersson 1994). The evolution of female preferences may be initiated by a number of factors, including natural selection or genetic correlations between female preferences and male traits (reviewed by Andersson 1994).

Through the early 1990s, studies of sexual selection on bird songs and other vertebrate vocal displays focused on the factors responsible for the evolution of vocal complexity (Catchpole 1980, Searcy & Andersson 1986, Ryan & Keddy-Hector 1992). This focus was consistent with Darwin's (1871) emphasis on sexual selection as a cause for the evolutionary elaboration of sexual ornaments and displays. Female preferences for complex male displays are thought to evolve under a variety of nonexclusive scenarios (reviewed by Kirkpatrick & Ryan 1991). Fisher (1930) hypothesized that female preferences and male trait complexity coevolve in a process of "runaway" sexual selection, in which genetic correlations between trait and preference drive their mutual escalation. Females gain indirect benefits by mating with elaborate males; their male offspring are favored in subsequent generations as "sexy sons." Holland & Rice (1998) offered an interesting twist on this idea by suggesting that females evolve increasingly strong resistance to (rather than preferences for) male traits and that trait complexity evolves in response to this resistance. "Sensory exploitation" hypotheses suggest that female preferences are derived from preexisting sensory biases that are exploited by signaling males (Basolo 1990, Ryan 1990, Proctor 1991). Few empirical data support any of these hypotheses for bird songs (but see Searcy 1992, Gray & Hagelin 1996) for at least two practical reasons: (*a*) The influence of song learning makes identifying vocal genetic traits that might correlate with female preferences difficult, and (*b*) a paucity of strong cladistic hypotheses of avian relationships has made conducting comparative tests difficult.

Another set of hypotheses that has drawn considerable recent attention views songs as reliable indicators of male quality. The idea that signals provide honest indicators of male quality traces to Zahavi's (1975) handicap model of signal evolution, which posits that males incur costs in signal production and that females choose mates on the basis of signals that best reflect male quality (Hamilton & Zuk 1982, Kodric-Brown & Brown 1984). The benefits females gain through choice of high-quality males can be direct (material) or indirect (genetic). In territorial birds, direct benefits are often determined by territory quality. Males in high-quality territories can provide females and offspring with superior food supplies and nesting locations and, thus, enhance a female's survival and fecundity.

Numerous lines of empirical evidence support a role for female choice in the evolution of bird song complexity (Searcy & Yasukawa 1996). In laboratory experiments, researchers have isolated the influence of song on mate choice, through playback of songs to females in the absence of males. Females have been shown to respond more vigorously to complex vocal patterns than to simple vocal patterns (Catchpole et al. 1984, Searcy 1984). In the field, researchers have identified correlations between vocal complexity and male mating success. Perhaps the clearest demonstration of this finding comes from Hasselquist et al. (1996), who showed that female great reed warblers (*Acrocephalus arundinaceus*) tend to seek extra-pair copulations with males with song repertoires larger than those of their social mates.

Researchers have also explored how intrasexual selection may have influenced the evolution of vocal complexity, and results have been mixed. In some species, such as the *Acrocephalus* warblers, birds normally produce only simple song types in the context of territorial defense and reserve more complex song types for courtship (Catchpole 2000). Simpler song types apparently suffice in territorial defense for these birds. In other species, however, recent evidence points to a functional advantage in male-male contests for diverse song repertoires or variable song delivery patterns (Spector 1992, Byers 1996). An intriguing recent example was shown in song sparrows (*Melospiza melodia*) of western North America. In these birds, a repertoire of distinct song types appears to allow males to send "graded" signals for use in different territorial interactions. Direct song-type matching signals a high level of threat to territorial boundaries, whereas "repertoire matching," in which birds respond to the song of a neighbor with a distinct yet shared song type, signals a less-pronounced threat (Beecher et al. 1996). Beecher et al. (2000) suggest that large repertoires in song sparrows are less a product of female choice than of selection for male-male song sharing and the opportunities provided therein for graded territorial signaling. In general, the ability to adjust information content during signaling interactions may favor large and diverse song repertoires (Naguib & Todt 1998, Vehrencamp 2001).

Sustained directional sexual selection in any of the above scenarios should eventually drive song memes up against mechanistic boundaries (Figure 2D). This expectation is consistent with the argument that mating displays provide females with honest indicators of male status, which can only happen if signal expression involves significant costs (Grafen 1990). The costs of bird song might be incurred in a diversity of ways (see reviews by Gil & Gahr 2002, Ten Cate et al. 2002). Nowicki et al. (1998, 2002) propose that learned songs may serve as honest indicators of male quality because their accurate reproduction requires successful brain development in the face of nutritional stresses experienced early in life. According to this hypothesis, males express heritable variation in how much stress they experience during development, as well as in how well they respond to these stresses with respect to brain development. Males that successfully reproduce complex songs effectively advertise the high quality of their genes and developmental histories. Under such circumstances, females should be selected to evaluate male quality by assessing their songs (Nowicki et al. 1998, 2002). Particularly strong sexual selection may drive not only meme evolution but also evolutionary changes in vocal mechanisms to accommodate further meme evolution (Figure 2E).

SONG TRAIT EVOLUTION

We now focus our discussion on specific examples of song traits that appear to be limited in their evolution by mechanistic constraints (Figure 2D,F). We examine four traits that meet two criteria: (a) available evidence suggests they evolve by sexual selection and (b) tangible progress has been made in understanding their mechanistic bases (Table 1).

Repertoire Size

Song repertoires vary greatly among species in their organization and size, even within closely related groups (Kroodsma 1982). Within the Sylviid warblers, for example, the songs of *Locustella* species contain two to four distinct syllable types, whereas the songs of *Acrocephalus* species contain between 25 and 100 syllable types (Székely et al. 1996). Numerous studies suggest that large repertoires are favored by sexual selection, although some concerns have been raised about the design and interpretation of these studies (Kroodsma 2004). In the laboratory, females tend to solicit copulations more often when they hear playback of multiple song types than when they hear playback of a single song type (Searcy 1984, 1992; Catchpole et al. 1986; Baker et al. 1987). Eens et al. (1991) showed that female starlings in aviaries choose males with larger song repertoires as their social mates. Field studies, some of which have controlled for potentially confounding factors such as age and territory quality, have identified correlations between male repertoire size and mating success (Searcy & Yasukawa 1990, Horn et al. 1993, Mountjoy & Lemon 1996, Buchanan & Catchpole 1997, Lampe & Espmark 2003). Krebs et al. (1978) showed that song repertoires are more effective than single song types as territorial "keep-out" signals. Subsequent research has identified a relationship in some species between repertoire size and successful territory

TABLE 1 Evidence of sexual selection and mechanistic constraints for four song traits

Song trait	Evidence for sexual selection	References	Mechanistic constraint	References
Repertoire size	Repertoire size correlates with: early pairing date	(Mountjoy & Lemon 1996) (Buchanan & Catchpole 1997) (Hasselquist et al. 1996)	Size of brain nuclei	(Nottebohm et al. 1981) (Kroodsma & Canady 1985) (Brenowitz et al. 1995) (Székely et al. 1996) (MacDougall-Shackleton et al. 1998) (Airey et al. 2000a)
	extrapair fertilizations			
	Playback of larger repertoires induces more female display	(Catchpole et al. 1984) (Searcy 1984)	Developmental stress	(Nowicki et al. 2002)
	Greater utility of large repertoires in mediating territorial interactions	(Krebs et al. 1978) (Beecher et al. 2000)	Endocrine state	(Johnson & Bottjer 1993)
Song rate	Higher rate is correlated with: early arrival	(Arvidsson & Neergaard 1991) (Nystrom 1997)	Metabolic condition: cost of singing	(Oberweger & Goller 2001) (Ward et al. 2003)
	territory quality	(Alatalo et al. 1990)		(Thomas 1999)
	early pairing date	(Hofstad et al. 2002)	foraging success	(Lucas et al. 1999)
	female fertility	(Mace 1987)	temperature	(Thomas & Cuthill 2002)
		(Pinxten & Eens 1998)	immunocompetence	(Moller 1991a)
	increase in paternity	(Moller et al. 1998)		(Duffy & Ball 2002)
			Endocrine state	(Nowicki & Ball 1989) (Wingfield & Hahn 1994) (Hunt et al. 1997)
	Playback of high song rate induces more female display	(Wasserman & Cigliano 1991) (Collins et al. 1994) (Balzer & Williams 1998)	Developmental stress	(Buchanan et al. 2003)
Trill performance	Trills with more rapid rates induce more female display	(Vallet & Kreutzer 1995) (Vallet et al. 1998) (Draganoiu et al. 2002)	Performance capacities	(Hartley & Suthers 1989) (Podos 1996, 1997, 2001) (Podos et al. 1999, 2004a)
	Wider frequency bandwidths induce more female display	(Ballentine et al. 2004)		(Westneat et al. 1993)
			Trade-offs with stereotypy	(Lambrechts 1997)
Pure-tonal structure	Pure-tonal songs elicit stronger male territorial responses	(Strote & Nowicki 1996)	Vocal tract function	(Nowicki 1987) (Hoese et al. 2000) (Beckers et al. 2003)

defense (Hiebert et al. 1989, Balsby & Dabelsteen 2001; but see Beecher et al. 2000).

Recent studies have focused on the hypothesis that repertoire size provides an honest indicator of male quality. In sedge warblers (*Acrocephalus schoenobaenus*), repertoire size was found to correlate positively with parental effort (Buchanan & Catchpole 2000). In a comparative analysis, immune responses were found to be more robust in species with large repertoire sizes, which implies an "arms race" between virulent pathogens and signaling males (Garamszegi et al. 2003). Moreover, studies of several species have found positive correlations between repertoire size and both offspring viability (Gil & Slater 2000) and lifetime reproductive success (McGregor et al. 1981, Hasselquist 1998).

A male's ability to invest in brain tissue appears to be the primary physiological mechanism that links repertoire size to male quality. Neurobiological studies have shown consistent correlations between repertoire size and the structure and size of two telencephalic nuclei, RA and HVc (Brenowitz 1997). RA is part of the song production pathway, and HVc is part of both the song production and the song memorization pathway. Comparative studies between individuals (Nottebohm et al. 1981), populations (Kroodsma & Canady 1985), species (Devoogd et al. 1993), and genera (Székely et al. 1996) have documented positive correlations between brain nucleus size and repertoire size. HVc has recently been shown to have a strong heritable component (Ward et al. 1998, Airey et al. 2000b, Williams et al. 2003), a finding which supports a possible role for sexual selection on repertoire size.

The size of brain nuclei can be regarded as a primary constraint on repertoire size. Various factors may secondarily limit repertoire size through their effects on brain nuclei. One such constraint is developmental stress (Nowicki et al. 2002). Brain tissue is costly to produce and maintain, and perhaps only high-quality individuals can meet the neurological demands imposed by a large song repertoire. These costs may be highest during development, when the song nuclei are growing rapidly in size and synaptic complexity. To illustrate, starlings subjected to nutritional stress during development sang shorter songs and fewer types of songs than their unstressed counterparts (Buchanan et al. 2003). The most direct evidence on this point to date is provided by Nowicki et al. (2002), who demonstrated that swamp sparrows reared under conditions of nutritional stress developed HVc nuclei of comparatively low volume. Endocrine state also exerts an important influence on song nucleus development. Injections of the stress hormone corticosterone led to reduced motif repertoire sizes in zebra finches (Spencer et al. 2003). In canaries, testosterone regulates HVc size in males (Johnson & Bottjer 1993) and can promote HVc growth in females (Nottebohm 1980, Rasika et al. 1994). Tramontin et al. (2003) showed that HVc volume increased in birds implanted with testosterone, 5α-dihydrotestosterone, or estradiol, although the effect of the hormone treatments on singing behavior was less clear. These lines of evidence together support the hypothesis that the development and expression of repertoire size reflects birds' nutritional history and hormonal state (Nowicki et al. 1998).

High heritability in how well birds respond to developmental stress would enable repertoire size to serve as an accurate indicator of male genetic quality (Andersson 1994).

Song Rate

Song rate refers to the amount of song a bird produces per unit of time. Individual males often vary in their song rate, and females may use this variation as a basis for mate choice. Males that arrive earlier on breeding territories, that have higher quality territories, and that pair earlier tend to sing at relatively high rates (Alatalo et al. 1990, Arvidsson & Neergaard 1991, Nystrom 1997). Nystrom (1997) notes that earlier pairing may be the result of female preferences for high-quality territory rather than high song rate per se, and, thus, these two variables may be difficult to separate. In laboratory studies, females performed more copulation solicitation displays to song stimuli presented at high rates compared with song stimuli presented at low rates (Wasserman & Cigliano 1991, Collins et al. 1994, Balzer & Williams 1998). Field studies indicate that males sometimes adjust their song rate during the breeding season to match periods of female fertility (Moller 1991b; but see Gil et al. 1999). In some species, song rate peaks during maximum female fertility, just before or during egg laying (Mace 1987, Pinxten & Eens 1998). Male barn swallows (*Hirundo rustica*) with high song rates were shown to have higher rates of within-brood paternity (Moller et al. 1998).

Several hypotheses address the mechanistic bases of song rate. Song rate is most commonly attributed to metabolic condition. Prestwich (1994) reports that the metabolic rate of signaling in insects and anurans can be up to 21 times resting metabolic rate. Recent studies of oxygen consumption in singing zebra finches, canaries, and starlings, however, suggest that bird song production imposes comparatively modest costs [maximum 2.4 times increase in metabolic rate (Oberweger & Goller 2001, Ward et al. 2003; but see Eberhardt 1994)]. Yet, several indirect lines of evidence point to high metabolic costs in bird song production. Thomas (2002) found that fat stores in nightingales dropped considerably during nighttime singing bouts. For daytime singers, low nighttime temperatures (and correspondingly high overnight metabolic expenditures) were correlated with low song rates at dawn (Thomas & Cuthill 2002). When the diets of birds were supplemented both in the field and in the laboratory, the song rate of males increased, although this increase may be more a result of reduced time required for foraging than reduced energetic constraints (Nystrom 1997, Lucas et al. 1999, Thomas 1999). Song rate may serve as an honest indicator of male immunocompetence. Duffy & Ball (2002) found that starlings with high song rates had enhanced immune function (but see Birkhead et al. 1998, Buchanan et al. 1999).

Two additional mechanistic influences on song rate have received recent support: endocrine state and developmental stress. Testosterone plays an important role mediating seasonal changes in behavior, and during the breeding season, testosterone production and song rate increase together. Correlations between testosterone levels and song rate were also found in laboratory studies in which males were implanted with testosterone (Nowicki & Ball 1989, Wingfield & Hahn 1994, Hunt et al. 1997). However, song rate can be high even when testosterone levels are low. In the song sparrow, song rates tend to increase in the fall while testosterone levels remain low (Logan & Wingfield 1990; see also Hau et al. 2000). Both testosterone and corticosterone influence the development of brain song nuclei (Ball et al. 2002), and corticosterone is produced as a response to stress (Buchanan 2000). Stress during development may have long-term effects on song rate. Buchanan et al. (2003) found that young starlings that were subject to nutritional stresses during development showed decreased song rates. Their study also provided evidence for a positive relationship between all three mechanistic factors discussed here; metabolic condition, immune function, and song rate were all impacted by developmental stress.

Trill Performance

Many bird songs include sequences in which syllables are repeated in trills (Figure 1). Fast trills are expected to be comparatively difficult to produce because they demand higher levels of vocal performance that include more rapid respiration and vocal tract movements (Hartley & Suthers 1989, Westneat et al. 1993, Podos & Nowicki 2004, Zollinger & Suthers 2004). Direct evidence that trill rates are limited by vocal performance is provided by two studies in which young swamp sparrows were reared with training models in which trill rates were artificially increased (Podos 1996, Podos et al. 2004a). Young birds in these studies were unable to produce accurate copies of rapid models and instead reproduced these models with reduced trill rates, note omissions, or pauses between syllables. Thus, selection for rapid trill rates in this species should be countered by motor limits on trill rate production (Podos 1996, see also Zollinger & Suthers 2004).

The occurrence of trade-offs between trill rate and other vocal features provides additional evidence of performance limits on trill production (Lambrechts 1996, Podos & Nowicki 2004). In great tits (*Parus major*), note frequency shows lower levels of stereotypy in faster trills (Lambrechts 1997). Another vocal feature that expresses a trade-off with trill rate is frequency bandwidth, defined as the range of frequencies expressed in songs (Figure 3*A*) (Podos 1997, Draganoiu et al. 2002, Ballentine et al. 2004). This trade-off is consistent with a hypothesis of performance constraints (Figure 3*B*) (Podos & Nowicki 2004). Activity of the peripheral vocal apparatus has been observed to correspond to frequency bandwidth values. In swamp sparrows, high bandwidth trills are accompanied by relatively broad changes in beak gape (Westneat et al. 1993). These changes allow birds to adjust vocal tract resonances and maintain pure-tonal song structure across shifting song frequencies (Nowicki 1987, Hoese et al. 2000). As trilled song sequences evolve, increases in frequency bandwidth should constrain the evolution of trill rate because of a trade-off between rates and magnitudes of possible vocal tract modulations (Figure 3*B*) (Podos 1997). This hypothesis is further supported by a study of Darwin's finches in which birds with relatively large beaks were shown to

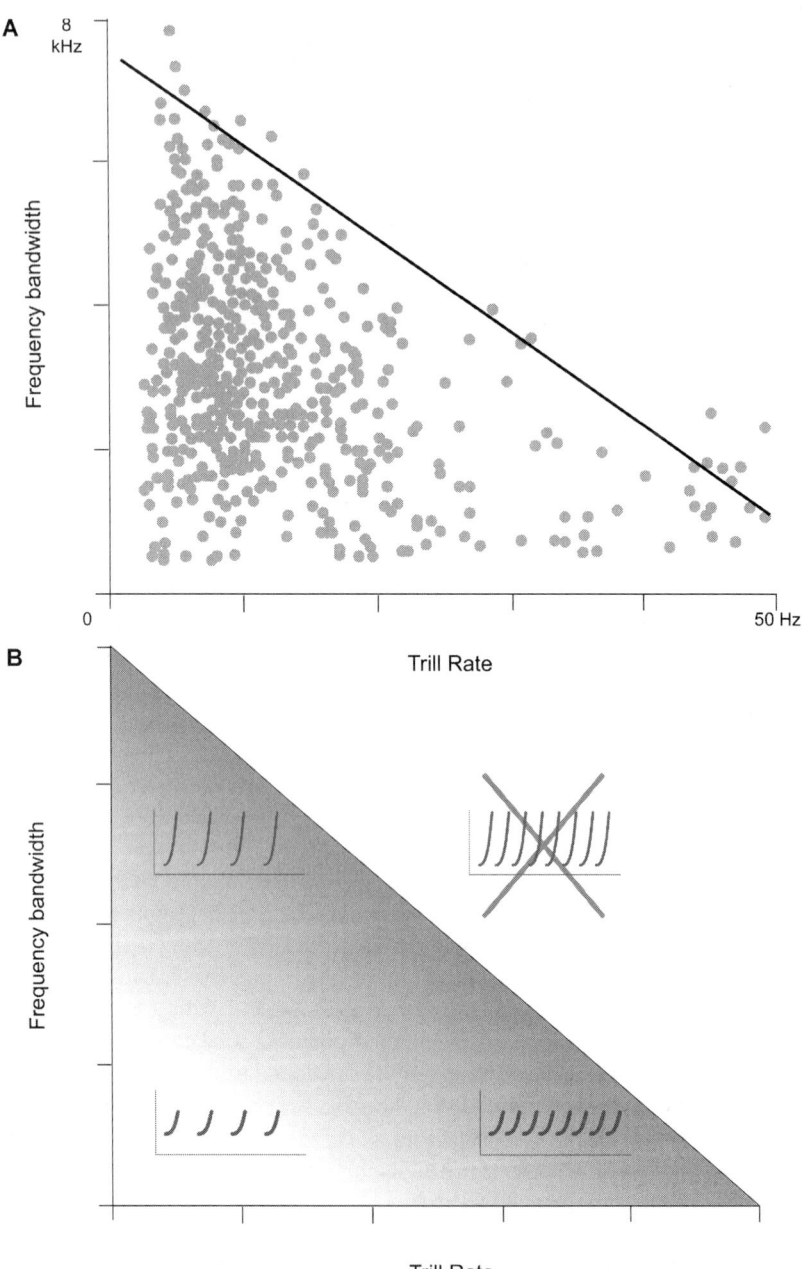

produce songs with comparatively low trill rates and frequency bandwidths (Podos 2001).

In three laboratory studies, playback of songs with more rapid trill rates elicited more elaborate or frequent copulation solicitation displays from females (Vallet & Kreutzer 1995, Vallet et al. 1998, Draganoiu et al. 2002). A fourth study failed to detect a difference in female responses to playback of control versus rapid trills (Nowicki et al. 2001). Ballentine et al. (2004) provide evidence that both frequency bandwidth and trill rate are subject to sexual selection. They presented female swamp sparrows with pairs of songs that differed only in their deviation from a "performance boundary" defined by natural variation in trill structure. Birds displayed more often to songs with faster trill rates and wider bandwidths, that is, songs that reflected greater levels of production difficulty.

Pure-Tonal Structure

Songbird vocalizations are notable for their pure-tonal quality, which means that acoustic frequencies at any given moment tend to be concentrated within a narrow range (Marler 1969). The production of pure-tonal songs appears to be enabled by the function of a vocal tract resonance filter that attenuates acoustic overtones (see section above, Mechanisms). Many bird songs include dynamic changes in fundamental frequencies. Birds maintain the pure-tonal structure of songs that vary in frequency by means of precise and often rapid adjustments to vocal tract volume such as through changes in beak gape (Westneat et al. 1993, Podos et al. 2004b). A recent study demonstrated that a loss of precision in beak gape adjustments leads to reductions in songs' pure-tonal quality (Hoese et al. 2000).

Strote & Nowicki (1996) tested the influence of pure-tonal quality on the responses of males to simulated territorial intrusion. They presented birds with two

←

Figure 3 Trill rates and frequency bandwidths can be used to depict the "acoustic space" that vocal trills occupy. (*A*) Plot of trill rate and frequency bandwidth for trilled components of 740 song types of 34 species from the family Emberizidae. Triangular distributions are also evident for individual species and genera and support the vocal tract constraint hypothesis (Podos 1997). (*B*) Songs produced with low trill rates and limited frequency bandwidth (*lower left spectrogram*) require minimal vocal tract movements and, therefore, should be easily produced. As songs increase in either frequency bandwidth (*upper left spectrogram*) or trill rate (*lower right spectrogram*), the demands required of vocal performance are elevated. Increases in bandwidth require wider gapes, and increases in trill rates require greater rates of beak opening and closing. The physical challenge of producing songs with faster trill rates or broader frequency bandwidths should increase steadily, as indicated here by the light-dark gradient, until a performance limit is reached. Songs beyond this boundary (*upper right spectrogram*) cannot be produced, because they would require vocal tract movements exceeding birds' performance capabilities (Podos & Nowicki 2004).

categories of song: control songs and songs recorded from birds singing in a helium-oxygen atmosphere in which harmonic overtones were expressed. Reduced responses to experimental songs suggest that sexual selection favors males who are able successfully to adjust vocal tract resonance properties in coordination with syrinx activity.

Other Traits?

Numerous other traits met only one of our two criteria and were, therefore, not included in Table 1. Song frequency has a clear mechanistic link to syrinx size, although we know of no evidence demonstrating that frequency is influenced by sexual selection. Measures of song versatility, such as rates of song-type switching, have been identified as correlates of mating success and thus may be subject to sexual selection (e.g., Lampe & Espmark 1994). Lambrechts & Dhondt (1988) hypothesized that rapid song-type switching allows birds to minimize exhaustion and, thereby, presumably maintain a higher song rate, although we know of no empirical evidence that supports this hypothesis. A study by Forstmeier et al. (2002) suggests that sexual selection may drive increases in song amplitude. Although the mechanistic basis of song amplitude seems clear—louder sounds should require more metabolic output—we know of no data that directly support this link.

All song traits discussed so far can be measured directly from birds' repertoires. Another relevant trait, copying accuracy, requires information about model songs and is, therefore, operationally more difficult to characterize. Copying accuracy is often discussed as being favored by natural selection, with local song dialects pointing to genetically "coadapted" populations (Baker & Cunningham 1985). Local dialect production may also serve as an honest indicator of male quality and, thus, may be subject to sexual selection (Rothstein & Fleischer 1987). Searcy et al. (2002) hypothesize that the accurate reproduction of local models honestly reflects a male's success in developing the neural mechanisms required for song learning.

CONCLUSIONS

Bird song has been a popular model system in animal behavior in part because it well illustrates the complementary influences of genetics and learning in phenotype development. Drawing a distinction between genes and environment in vocal ontogeny has been criticized on conceptual grounds (e.g., Johnston 1988) but is supported by empirical evidence (Marler & Sherman 1985, Searcy 1988, Baptista 1996). For evolutionary analyses, we suggest that focusing on another distinction, between memes and vocal mechanisms as substrates of song evolution, is more useful. Memes can evolve as a result of genetic changes (e.g., innate perceptual predispositions) and as a result of learning (e.g., song copying), whereas mechanistic evolution only occurs through genetic modifications (e.g., in syrinx mass).[1] The idea of mechanistic limits on song evolution concords with broader concepts of performance as a determinant of the relationship between patterns of morphological and behavioral evolution (Garland & Losos 1994).

Throughout this essay, we have regarded vocal mechanisms as a constraining factor in song evolution. A number of recent studies suggest that vocal mechanisms can also provide creative opportunities in song evolution, especially under pressure from meme evolution (e.g., Figure 2D). We end with three examples:

1. Fitch (1999) provides evidence that the evolution of exaggerated trachea length in numerous bird groups, including cranes, swans, and geese, has enabled these birds to reduce the "formant frequency dispersion" of their calls, a vocal trait that normally provides an honest cue of body size. Morphological plasticity under the pressure of sexual selection allowed these birds to evolve elongated trachea and novel call properties.

2. White-crowned sparrows normally only copy acoustic models that begin with whistled components, and these birds are, thus, normally constrained to copy species-specific models. Soha & Marler (2000) demonstrated, however, that birds will freely copy heterospecific vocal material (even ground squirrel alarm calls!) if positioned after an introductory whistle. Therefore, perceptual rules that normally constrain song learning can provide a means for integrating novel vocal components into birds' song repertoires.

3. Mechanistic constraints expressed during song ontogeny may physically drive the expression of novel song features. Young swamp sparrows trained with rapid models were observed to reconfigure some of these models with a novel species-atypical syntax, presumably to ensure the accurate reproduction of other song model features (Podos 1996; Podos et al. 1999, 2004a). Given these and similar studies, we conclude by suggesting that mechanistic constraints examined in this review not only act as a barrier to song evolution but also may work in tandem with meme evolution as a creative force in the evolution of bird songs.

ACKNOWLEDGMENTS

We gratefully acknowledge Barbara Ballentine, Bruce Byers, Don Kroodsma, Stephen Nowicki, Jordan Price, Michael Ryan, William Searcy, and Heather Williams for critical and thought-provoking feedback on earlier versions of this manuscript. Funding was provided by National Science Foundation grant IBN 0347291 to J.P.

---

[1]We do not, of course, mean to imply that "purely genetic" traits evolve in the absence of environmental information. All traits are subject to such information during development, for example, as encompassed in the concept of the "norm of reaction." Rather, we mean that the intergenerational transmission of mechanisms of vocal production is not influenced by the experiences of individual birds.

The *Annual Review of Ecology, Evolution, and Systematics* is online at
http://ecolsys.annualreviews.org

LITERATURE CITED

Airey DC, Buchanan KL, Székely T, Catchpole CK, DeVoogd TJ. 2000a. Song, sexual selection, and a song control nucleus (HVc) in the brains of European sedge warblers. *J. Neurobiol.* 44:1–6

Airey DC, Castillo-Juarez H, Casella G, Pollak EJ, DeVoogd TJ. 2000b. Variation in the volume of zebra finch song control nuclei is heritable: developmental and evolutionary implications. *Proc. R. Soc. London Ser. B* 267:2099–104

Alatalo RV, Glynn C, Lundberg A. 1990. Singing rate and female attraction in the pied flycatcher: an experiment. *Anim. Behav.* 39:601–3

Andersson M. 1994. *Sexual Selection*. Princeton, NJ: Princeton Univ. Press. 599 pp.

Arnold AP. 1990. The passerine bird song system as a model in neuroendocrine research. *J. Exp. Zool.* S4:22–30

Arnold SJ. 1992. Constraints on phenotypic evolution. *Am. Nat.* 140:S85–107

Arvidsson BL, Neergaard R. 1991. Mate choice in the willow warbler: a field experiment. *Behav. Ecol. Sociobiol.* 29:225–29

Baker MC. 1996. Female buntings from hybridizing populations prefer conspecific males. *Wilson Bull.* 108:771–75

Baker MC, Baker MSA, Baker EM. 2003. Rapid evolution of a novel song and an increase in repertoire size in an island population of an Australian songbird. *Ibis* 145:465–71

Baker MC, Bjerke TK, Lampe HM, Espmark YO. 1987. Sexual response of female yellowhammers to differences in regional song dialects and repertoire sizes. *Anim. Behav.* 35:395–401

Baker MC, Cunningham MA. 1985. The biology of bird-song dialects. *Behav. Brain Sci.* 8:85–133

Ball GF, Riters LV, Balthazart J. 2002. Neuroendocrinology of song behavior and avian brain plasticity: multiple sites of action of sex steroid hormones. *Front. Neuroendocrinol.* 23:137–78

Ballentine B, Hyman J, Nowicki S. 2004. Vocal performance influences female response to male bird song: an experimental test. *Behav. Ecol.* 15:163–68

Balsby TJS, Dabelsteen T. 2001. The meaning of song repertoire size and song length to male whitethroats *Sylvia communis*. *Behav. Process.* 56:75–84

Balzer AL, Williams TD. 1998. Do female zebra finches vary primary reproductive effort in relation to mate attractiveness? *Behaviour* 135:297–309

Baptista LF. 1975. Song dialects and demes in sedentary populations of the white-crowned sparrow (*Zonotrichia leucophrys nuttalli*). *Univ. Calif. Publ. Zool.* 105:1–52

Baptista LF. 1996. Nature and its nurturing in avian vocal development. See Kroodsma & Miller 1996, pp. 39–60

Baptista LF, Kroodsma DE. 2001. Avian bioacoustics: a tribute to Luis Baptista. In *Handbook of the Birds of the World, Vol. 6, Mousebirds to Hornbills*. ed. J del Hoyo, A Elliott, J Sargatal, pp. 10–52. Barcelona: Lynx Ed.

Baptista LF, Petrinovich L. 1984. Social interaction, sensitive phases and the song template hypothesis in the white-crowned sparrow. *Anim. Behav.* 32:172–81

Basolo AL. 1990. Female preference predates the evolution of the sword in swordtail fish. *Science* 250:808–10

Becker PH. 1982. The coding of species-specific characteristics in bird sounds. See Kroodsma & Miller 1982, pp. 213–52

Beckers GJL, Suthers RA, Ten Cate C. 2003. Pure-tone birdsong by resonance filtering of harmonic overtones. *Proc. Natl. Acad. Sci. USA* 100:7372–76

Beecher MD. 1996. Birdsong learning in the laboratory and field. See Kroodsma & Miller 1996, pp. 61–78

Beecher MD, Campbell SE, Nordby JC. 2000. Territory tenure in song sparrows is related to song sharing with neighbours, but not to repertoire size. *Anim. Behav.* 59:29–37

Beecher MD, Stoddard PK, Campbell SE, Horning CL. 1996. Repertoire matching between neighbouring song sparrows. *Anim. Behav.* 51:917–23

Bertelli S, Tubaro PL. 2002. Body mass and habitat correlates of song structure in a primitive group of birds. *Biol. J. Linn. Soc.* 77:423–30

Birkhead TR, Fletcher F, Pellatt EJ. 1998. Sexual selection in the zebra finch *Taeniopygia guttata*: condition, sex traits and immune capacity. *Behav. Ecol. Sociobiol.* 44:179–91

Bitterbaum E, Baptista LF. 1979. Geographical variation in songs of California house finches (*Carpodacus mexicanus*). *Auk* 96:462–74

Bonner JT. 1980. *The Evolution of Culture in Animals*. Princeton, NJ: Princeton Univ. Press. 232 pp.

Bottjer SW, Miesner EA, Arnold AP. 1984. Forebrain lesions disrupt development but not maintenance of song in passerine birds. *Science* 224:901–3

Boyd R, Richerson PJ. 1985. *Culture and the Evolutionary Process*. Chicago: Chicago Univ. Press. 332 pp.

Brainard MS, Doupe AJ. 2002. What songbirds teach us about learning. *Nature* 417:351–58

Brémond JC. 1976. Specific recognition in the song of Bonelli's warbler (*Phylloscopus bonelli*). *Behaviour* 58:99–116

Brenowitz EA. 1997. Comparative approaches to the avian song system. *J. Neurobiol.* 33:517–31

Brenowitz EA, Arnold AP, Levin RN. 1985. Neural correlates of female song in tropical duetting birds. *Brain Res.* 343:104–12

Brenowitz EA, Lent K, Kroodsma DE. 1995. Brain space for learned song in birds develops independently of song learning. *J. Neurosci.* 15:6281–86

Brown TJ, Handford P. 2000. Sound design for vocalizations: quality in the woods, consistency in the fields. *Condor* 102:81–92

Buchanan KL. 2000. Stress and the evolution of condition-dependent signals. *Trends Ecol. Evol.* 15:156–60

Buchanan KL, Catchpole CK. 1997. Female choice in the sedge warbler, *Acrocephalus schoenobaenus*: multiple cues from song and territory quality. *Proc. R. Soc. London Ser. B* 264:521–26

Buchanan KL, Catchpole CK. 2000. Song as an indicator of male parental effort in the sedge warbler. *Proc. R. Soc. London Ser. B.* 267:321–26

Buchanan KL, Catchpole CK, Lewis JW, Lodge A. 1999. Song as an indicator of parasitism in the sedge warbler. *Anim. Behav.* 57:307–14

Buchanan KL, Spencer KA, Goldsmith AR, Catchpole CK. 2003. Song as an honest signal of past developmental stress in the European starling (*Sturnus vulgaris*). *Proc. R. Soc. London Ser. B* 270:1149–56

Butlin RK, Ritchie MG. 1994. Behavior and speciation. In *Behavior and Evolution*, ed. PJB Slater, TR Halliday, pp. 43–79. Cambridge, UK: Cambridge Univ. Press

Byers BE. 1996. Messages encoded in the songs of chestnut-sided warblers. *Anim. Behav.* 52:691–705

Catchpole CK. 1980. Sexual selection and the evolution of complex songs among European warblers of the genus *Acrocephalus*. *Behaviour* 74:149–66

Catchpole CK. 2000. Sexual selection and the evolution of song and brain structure in *Acrocephalus* warblers. In *Advances in the Study of Behavior*, ed. PJB Slater, JS Rosenblatt, CT Snowden, TJ Roper, 29:45–97. San Diego: Academic

Catchpole CK, Dittami J, Leisler B. 1984. Differential responses to male song repertoires in female songbirds implanted with oestradiol. *Nature* 312:563–64

Catchpole CK, Leisler B, Dittami J. 1986. Sexual differences in the responses of captive great reed warblers (*Acrocephalus arundinaceus*) to variation in song structure and repertoire size. *Ethology* 73:69–77

Catchpole CK, Slater PJB. 1995. *Bird Song: Biological Themes and Variations.* Cambridge, UK: Cambridge Univ. Press. 248 pp.

Cavalli-Sforza LL, Feldman MW. 1981. *Cultural Transmission and Evolution: A Quantitative Approach.* Princeton, NJ: Princeton Univ. Press. 388 pp.

Cicero C, Benowitz-Fredericks ZM. 2000. Song types and variation in insular populations of Lincoln's sparrow (*Melospiza lincolnii*), and comparisons with other *Melospiza. Auk* 117:52–64

Clark CW, Marler P, Beeman K. 1987. Quantitative analysis of animal vocal phonology: an application to swamp sparrow song. *Ethology* 76:101–15

Collins SA, Hubbard C, Houtman AM. 1994. Female mate choice in the zebra finch: the effect of male beak color and male song. *Behav. Ecol. Sociobiol.* 35:21–25

Cutler B. 1970. *Anatomical Studies of the Syrinx of Darwin's Finches.* San Francisco, CA: San Francisco State Univ. 272 pp.

Darwin C. 1871. *The Descent of Man, and Selection in Relation to Sex.* London: John Murray

Dawkins R. 1976. *The Selfish Gene.* Oxford: Oxford Univ. Press. 224 pp.

De Kort SR, Ten Cate C. 2001. Response to interspecific vocalizations is affected by degree of phylogenetic relatedness in *Streptopelia* doves. *Anim. Behav.* 61:239–47

Devoogd TJ, Krebs JR, Healy SD, Purvis A. 1993. Relations between song repertoire size and the volume of brain nuclei related to song: comparative evolutionary analyses amongst oscine birds. *Proc. R. Soc. London Ser. B* 254:75–82

Doutrelant C, Leitao A, Giorgi H, Lambrechts MM. 1999. Geographical variation in blue tit song, the result of an adjustment to vegetation type? *Behaviour* 136:481–93

Draganoiu TI, Nagle L, Kreutzer M. 2002. Directional female preference for an exaggerated male trait in canary (*Serinus canaria*) song. *Proc. R. Soc. London Ser. B* 269:2525–31

Duffy DL, Ball GF. 2002. Song predicts immunocompetence in male European starlings (*Sturnus vulgaris*). *Proc. R. Soc. London Ser. B* 269:847–52

Eberhardt LS. 1994. Oxygen consumption during singing by male Carolina wrens (*Thryothorus ludovicianus*). *Auk* 111:124–30

Eens M, Pinxten R, Verheyen RF. 1991. Male song as a cue for mate choice in the European starling. *Behaviour* 116:210–38

Espmark Y. 1999. Song of the snow bunting (*Plectrophenax nivalis*) in areas with and without sympatric passerines. *Can. J. Zool. Rev. Can. Zool.* 77:1385–92

Fisher RA. 1930. *The Genetical Theory of Natural Selection.* Oxford: Clarendon. 272 pp.

Fitch WT. 1999. Acoustic exaggeration of size in birds via tracheal elongation: comparative and theoretical analyses. *J. Zool.* 248:31–48

Forstmeier W, Kempenaers B, Meyer A, Leisler B. 2002. A novel song parameter correlates with extra-pair paternity and reflects male longevity. *Proc. R. Soc. London Ser. B* 269:1479–85

Garamszegi LZ, Moller AP, Erritzoe J. 2003. The evolution of immune defense and song complexity in birds. *Evolution* 57:905–12

Garland T Jr, Losos JB. 1994. Ecological morphology of locomotor performance in squamate reptiles. In *Ecological Morphology: Integrative Organismal Biology*, ed. PC Wainwright, SM Reilly, pp. 240–302. Chicago: Univ. Chicago Press

Gil D, Gahr M. 2002. The honesty of bird song: multiple constraints for multiple traits. *Trends Ecol. Evol.* 17:133–41

Gil D, Graves JA, Slater PJB. 1999. Seasonal patterns of singing in the willow warbler: evidence against the fertility announcement hypothesis. *Anim. Behav.* 58:995–1000

Gil D, Slater PJB. 2000. Song organisation and singing patterns of the willow warbler, *Phylloscopus trochilus. Behaviour* 137:759–82

Goller F, Daley MA. 2001. Novel motor gestures for phonation during inspiration enhance the acoustic complexity of birdsong. *Proc. R. Soc. London Ser. B* 268:2301–5

Goller F, Larsen ON. 1997. A new mechanism

of sound generation in songbirds. *Proc. Natl. Acad. Sci. USA* 94:14787–91

Gould SJ. 1980. The evolutionary biology of constraint. *Daedalus* 109:39–52

Grafen A. 1990. Biological signals as handicaps. *J. Theor. Biol.* 144:517–546

Grant BR, Grant PR. 1996. Cultural inheritance of song and its role in the evolution of Darwin's finches. *Evolution* 50:2471–87

Grant BR, Grant PR. 2002. Simulating secondary contact in allopatric speciation: an empirical test of premating isolation. *Biol. J. Linn. Soc.* 76:545–56

Grant PR, Grant BR. 1992. Hybridization of bird species. *Science* 256:193–97

Grant PR, Grant BR. 1997. Hybridization, sexual imprinting, and mate choice. *Am. Nat.* 149:1–28

Grant PR, Grant BR. 2002. Adaptive radiation of Darwin's finches. *Am. Sci.* 90:130–39

Gray DA, Hagelin JC. 1996. Song repertoires and sensory exploitation: reconsidering the case of the common grackle. *Anim. Behav.* 52:795–800

Greenewalt CH. 1968. *Bird Song: Acoustics and Physiology*. Washington, DC: Smithsonian Inst. Press. 194 pp.

Hahnloser RHR, Kozhevnikov AA, Fee MS. 2002. An ultra-sparse code underlies the generation of neural sequences in a songbird. *Nature* 419:65–70

Hamilton WD, Zuk M. 1982. Heritable true fitness and bright birds: a role for parasites? *Science* 218:384–87

Handford P, Lougheed SC. 1991. Variation in duration and frequency characters in the song of the Rufous-collared sparrow, *Zonotrichia capensis*, with respect to habitat, trill dialects and body size. *Condor* 93:644–58

Hansen P. 1979. Vocal learning: its role in adapting sound structures to long-distance propagation and a hypothesis on its evolution. *Anim. Behav.* 27:1270–71

Hartley RS, Suthers RA. 1989. Airflow and pressure during canary song: direct evidence for mini-breaths. *J. Comp. Physiol. A* 165:15–26

Hasselquist D. 1998. Polygyny in great reed warblers: a long-term study of factors contributing to male fitness. *Ecology* 79:2376–90

Hasselquist D, Bensch S, von Schantz T. 1996. Correlation between male song repertoire, extra-pair paternity and offspring survival in the great reed warbler. *Nature* 381:229–32

Hau M, Wikelski M, Soma KK, Wingfield JC. 2000. Testosterone and year-round territorial aggression in a tropical bird. *Gen. Comp. Endocrinol.* 117:20–33

Herrel A, O'Reilly JC, Richmond AM. 2002. Evolution of bite performance in turtles. *J. Evol. Biol.* 15:1083–94

Heyes CM. 1994. Social learning in animals: categories and mechanisms. *Biol. Rev. Camb. Philos. Soc.* 69:207–31

Hiebert SM, Stoddard PK, Arcese P. 1989. Repertoire size, territory acquisition and reproductive success in the song sparrow. *Anim. Behav.* 37:266–73

Hoese WJ, Podos J, Boetticher NC, Nowicki S. 2000. Vocal tract function in birdsong production: experimental manipulation of beak movements. *J. Exp. Biol.* 203:1845–55

Hofstad E, Espmark Y, Moksnes A, Haugan T, Ingebrigtsen M. 2002. The relationship between song performance and male quality in snow buntings (*Plectrophenax nivalis*). *Can. J. Zool. Rev. Can. Zool.* 80:524–31

Holland B, Rice WR. 1998. Perspective: chase-away sexual selection: antagonistic seduction versus resistance. *Evolution* 52:1–7

Horn AG, Dickinson TE, Falls JB. 1993. Male quality and song repertoires in western meadowlarks (*Sturnella neglecta*). *Can. J. Zool. Rev. Can. Zool.* 71:1059–61

Hunt KE, Hahn TP, Wingfield JC. 1997. Testosterone implants increase song but not aggression in male Lapland longspurs. *Anim. Behav.* 54:1177–92

Hunter ML, Krebs JR. 1979. Geographical variation in the song of the great tit (*Parus major*) in relation to ecological factors. *J. Anim. Ecol.* 48:759–85

Ince SA, Slater PJB, Weismann C. 1980. Changes with time in the songs of a population of chaffinches. *Condor* 82:285–90

Irwin DE. 2000. Song variation in an avian ring species. *Evolution* 54:998–1010

Irwin DE, Price T. 1999. Sexual imprinting, learning and speciation. *Heredity* 82:347–54

Jenkins PF. 1978. Cultural transmission of song patterns and dialect development in a free-living bird population. *Anim. Behav.* 26:50–78

Johnson F, Bottjer SW. 1993. Hormone-induced changes in identified cell populations of the higher vocal center in male canaries. *J. Neurobiol.* 24:400–18

Johnston TD. 1988. Developmental explanation and the ontogeny of birdsong: nature/nurture redux. *Behav. Brain Sci.* 11:617–29

Kirkpatrick M, Ryan MJ. 1991. The evolution of mating preferences and the paradox of the lek. *Nature* 350:33–38

Kodric-Brown A, Brown JH. 1984. Truth in advertising: the kinds of traits favored by sexual selection. *Am. Nat.* 124:309–23

Konishi M. 1965. The role of auditory feedback in the control of vocalization in the white-crowned sparrow. *Z. Tierpsychol.* 22:770–83

Konishi M. 1985. Birdsong: from behavior to neuron. *Annu. Rev. Neurosci.* 8:125–70

Kramer B. 1990. *Electrocommunication in Teleost Fishes: Behavior and Experiments*. Berlin: Springer-Verlag. 240 pp.

Krebs JR, Ashcroft R, Webber M. 1978. Song repertoires and territory defence in the great tit. *Nature* 271:539–42

Kroodsma DE. 1977. Correlates of song organization among North American wrens. *Am. Nat.* 111:995–1008

Kroodsma DE. 1982. Song repertoires: problems in their definition and use. See Kroodsma & Miller 1982, pp. 125–46

Kroodsma DE. 1996. Ecology of passerine song development. See Kroodsma & Miller 1996, pp. 3–19

Kroodsma DE. 2004. Diversity and plasticity of bird song. In *Nature's Music: The Science of Bird Song*, ed. P Marler, H Slabbekoorn. New York: Academic. In press

Kroodsma DE, Byers BE, Halkin SL, Hill C, Minis D, et al. 1999. Geographic variation in black-capped chickadee songs and singing behavior. *Auk* 116:387–402

Kroodsma DE, Canady RA. 1985. Differences in repertoire size, singing behavior, and associated neuroanatomy among marsh wren populations have a genetic basis. *Auk* 102:439–46

Kroodsma DE, Miller EH, eds. 1982. *Acoustic Communication in Birds*. New York: Academic

Kroodsma DE, Miller EH, eds. 1996. *Ecology and Evolution of Acoustic Communication in Birds*. Ithaca, NY: Cornell Univ. Press

Kroodsma DE, Parker LD. 1977. Vocal virtuosity in the brown thrasher. *Auk* 94:783–85

Kroodsma DE, Pickert R. 1980. Environmentally dependent sensitive periods for avian vocal learning. *Nature* 288:477–79

Kroodsma DE, Woods RW, Goodwin EA. 2002. Falkland Island sedge wrens (*Cistothorus platensis*) imitate rather than improvise large song repertoires. *Auk* 119:523–28

Lambrechts MM. 1996. Organization of birdsong and constraints on performance. See Kroodsma & Miller 1996, pp. 305–20

Lambrechts MM. 1997. Song frequency plasticity and composition of phrase versions in great tits *Parus major*. *Ardea* 85:99–109

Lambrechts MM, Dhondt AA. 1988. The antiexhaustion hypothesis: a new hypothesis to explain song performance and song switching in the great tit. *Anim. Behav.* 36:327–34

Lampe HM, Espmark YO. 1994. Song structure reflects male quality in pied flycatchers, *Ficedula hypoleuca*. *Anim. Behav.* 47:869–76

Lampe HM, Espmark YO. 2003. Mate choice in pied flycatchers *Ficedula hypoleuca*: Can females use song to find high-quality males and territories? *Ibis* 145:E24–33

Larsen ON, Goller F. 1999. Role of syringeal vibrations in bird vocalizations. *Proc. R. Soc. London Ser. B* 266:1609–15

Lemon RE. 1975. How birds develop song dialects. *Condor* 77:385–406

Livingston FS, White SA, Mooney R. 2000. Slow NMDA-EPSCs at synapses critical for song development are not required for song learning in zebra finches. *Nat. Neurosci.* 3: 482–88

Logan CA, Wingfield JC. 1990. Autumnal territorial aggression is independent of plasma testosterone in mockingbirds. *Horm. Behav.* 24:568–81

Lorenz KZ. 1950. The comparative method in studying innate behaviour patterns. *Symp. Soc. Exp. Biol.* 4:221–68

Lucas JR, Schraeder A, Jackson C. 1999. Carolina chickadee (Aves, Paridae, *Poecile carolinensis*) vocalization rates: effects of body mass and food availability under aviary conditions. *Ethology* 105:503–20

Lynch A. 1996. The population mimetics of birdsong. See Kroodsma & Miller 1996, pp. 181–97

MacDougall-Shackleton EA, MacDougall-Shackleton SA. 2001. Cultural and genetic evolution in mountain white-crowned sparrows: song dialects are associated with population structure. *Evolution* 55:2568–75

MacDougall-Shackleton SA, Hulse SH, Ball GF. 1998. Neural correlates of singing behavior in male zebra finches (*Taeniopygia guttata*). *J. Neurobiol.* 36:421–30

Mace R. 1987. The dawn chorus in the great tit *Parus major* is directly related to female fertility. *Nature* 330:745–46

Marler P. 1957. Specific distinctiveness in the communication signals of birds. *Behaviour* 11:13–39

Marler P. 1960. Bird songs and mate selection. In *Animal Sounds and Communication*, ed. WE Lanyon, WN Tavolga, pp. 348–67. Washington, DC: Am. Inst. Biol. Sci.

Marler P. 1969. Tonal quality of bird sounds. In *Bird Vocalizations*, ed. RA Hinde, pp. 5–18. Cambridge, UK: Cambridge Univ. Press

Marler P. 1976. Sensory templates in species-specific behavior. In *Simpler Networks and Behavior*, ed. J Fentress, pp. 314–29. Sunderland, MA: Sinauer

Marler P. 1984. Song learning: innate species differences in the learning process. In *The Biology of Learning*, ed. P Marler, HS Terrace, pp. 289–309. Berlin: Springer-Verlag

Marler P, Peters S. 1977. Selective vocal learning in a sparrow. *Science* 198:519–21

Marler P, Peters S. 1982. Subsong and plastic song: their role in the vocal learning process. See Kroodsma & Miller 1982, pp. 25–50

Marler P, Peters S. 1987. A sensitive period for song acquisition in the song sparrow, *Melospiza melodia*: a case of age-limited learning. *Ethology* 76:89–100

Marler P, Peters S. 1988. The role of song phonology and syntax in vocal learning preferences in the song sparrow, *Melospiza melodia*. *Ethology* 77:125–49

Marler P, Pickert R. 1984. Species-universal microstructure in the learned song of the swamp sparrow, *Melospiza georgiana*. *Anim. Behav.* 32:673–89

Marler P, Sherman V. 1985. Innate differences in singing behavior of sparrows reared in isolation from adult conspecific song. *Anim. Behav.* 33:57–71

Marler P, Tamura M. 1962. Song "dialects" in three populations of white-crowned sparrows. *Condor* 64:368–77

Marler P, Tamura M. 1964. Culturally transmitted patterns of vocal behavior in sparrows. *Science* 146:1483–86

Marten K, Marler P. 1977. Sound transmission and its significance for animal vocalization. I. Temperate habitats. *Behav. Ecol. Sociobiol.* 2:271–90

Maynard Smith J, Burian R, Kauffman S, Alberch P, Campbell J, et al. 1985. Developmental constraints and evolution. *Q. Rev. Biol.* 60:265–87

McCasland JS. 1987. Neuronal control of bird song production. *J. Neurosci.* 7:23–39

McGregor PK, Krebs JR, Perrins CM. 1981. Song repertoires and lifetime reproductive success in the great tit (*Parus major*). *Am. Nat.* 118:149–59

Moller AP. 1991a. Parasite load reduces song output in a passerine bird. *Anim. Behav.* 41:723–30

Moller AP. 1991b. Why mated songbirds sing so much: mate guarding and male announcement of mate fertility status. *Am. Nat.* 138:994–1014

Moller AP, Saino N, Taramino G, Galeotti P, Ferrario S. 1998. Paternity and multiple signaling: effects of a secondary sexual character and song on paternity in the barn swallow. *Am. Nat.* 151:236–242

Morton ES. 1975. Ecological sources of selection on avian sounds. *Am. Nat.* 109:17–34

Morton ES. 1987. The effects of distance and isolation on song-type sharing in the Carolina wren. *Wilson Bull.* 99:601–10

Mountjoy DJ, Lemon RE. 1996. Female choice for complex song in the European starling: a field experiment. *Behav. Ecol. Sociobiol.* 38:65–71

Naguib M, Todt D. 1998. Recognition of neighbors' song in a species with large and complex song repertoires: the thrush nightingale. *J. Avian Biol.* 29:155–60

Naugler CT, Ratcliffe L. 1994. Character release in bird song: a test of the acoustic competition hypothesis using American tree sparrows *Spizella arborea. J. Avian Biol.* 25:142–48

Nelson DA. 2000. Song overproduction, selective attrition and song dialects in the white-crowned sparrow. *Anim. Behav.* 60:887–98

Nelson DA, Marler P. 1990. The perception of birdsong and an ecological concept of signal space. In *Comparative Perception, Vol. 2, Complex Signals*, ed. WC Stebbins, MA Berkley, pp. 443–78. New York: Wiley

Nelson DA, Marler P. 1993. Innate recognition of song in white-crowned sparrows: a role in selective vocal learning? *Anim. Behav.* 46:806–8

Nelson DA, Marler P, Morton ML. 1996. Overproduction in song development: an evolutionary correlate with migration. *Anim. Behav.* 51:1127–40

Noad MJ, Cato DH, Bryden MM, Jenner MN, Jenner KCS. 2000. Cultural revolution in whale songs. *Nature* 408:537

Nottebohm F. 1980. Testosterone triggers growth of brain vocal control nuclei in adult female canaries. *Brain Res.* 189:429–36

Nottebohm F, Arnold AP. 1976. Sexual dimorphism in vocal control areas of the songbird brain. *Science* 194:211–13

Nottebohm F, Kasparian S, Pandazis C. 1981. Brain space for a learned task. *Brain Res.* 213:99–109

Nottebohm F, Nottebohm ME, Crane L. 1986. Developmental and seasonal changes in canary song and their relation to changes in the anatomy of song control nuclei. *Behav. Neural Biol.* 46:445–71

Nottebohm F, Stokes TM, Leonard CM. 1976. Central control of song in the canary, *Serinus canarius. J. Comp. Neurol.* 165:457–86

Nowicki S. 1987. Vocal-tract resonances in oscine bird sound production: evidence from birdsongs in a helium atmosphere. *Nature* 325:53–55

Nowicki S, Ball GF. 1989. Testosterone induction of song in photosensitive and photorefractory male sparrows. *Horm. Behav.* 23:514–25

Nowicki S, Marler P. 1988. How do birds sing? *Music Perception* 5:391–426

Nowicki S, Peters S, Podos J. 1998. Song learning, early nutrition and sexual selection in songbirds. *Am. Zool.* 38:179–90

Nowicki S, Searcy WA, Hughes M, Podos J. 2001. The evolution of bird song: male and female response to song innovation in swamp sparrows. *Anim. Behav.* 62:1189–95

Nowicki S, Searcy WA, Peters S. 2002. Brain development, song learning and mate choice in birds: a review and experimental test of the "nutritional stress hypothesis." *J. Comp. Physiol. A* 188:1003–14

Nowicki S, Westneat MW, Hoese WJ. 1992. Birdsong: motor function and the evolution of communication. *Semin. Neurosci.* 4:385–90

Nystrom KGK. 1997. Food density, song rate, and body condition in territory-establishing willow warblers (*Phylloscopus trochilus*). *Can. J. Zool. Rev. Canad. Zool.* 75:47–58

Oberweger K, Goller F. 2001. The metabolic cost of birdsong production. *J. Exp. Biol.* 204:3379–88

Payne K, Tyack P, Payne R. 1983. Progressive changes in the songs of humpback whales (*Megaptera novaeangliae*): a detailed analysis of two seasons in Hawaii. In *Communication and Behavior of Whales*, ed. R Payne, pp. 9–57. Boulder, CO: Westview

Payne RB. 1986. Bird songs and avian systematics. *Curr. Ornithol.* 3:87–126

Payne RB. 1996. Song traditions in indigo buntings: origin, improvisation, dispersal, and extinction in cultural evolution. See Kroodsma & Miller 1996, pp. 198–220

Perrin N, Travis J. 1992. On the use of constraints in evolutionary biology and some allergic reactions to them. *Funct. Ecol.* 6:361–63

Peters S, Searcy WA, Beecher MD, Nowicki S. 2000. Geographic variation in the organization of song sparrow repertoires. *Auk* 117:936–42

Pinxten R, Eens M. 1998. Male starlings sing most in the late morning, following egg-laying: A strategy to protect their paternity? *Behaviour* 135:1197–211

Podos J. 1996. Motor constraints on vocal development in a songbird. *Anim. Behav.* 51:1061–70

Podos J. 1997. A performance constraint on the evolution of trilled vocalizations in a songbird family (Passeriformes:Emberizidae). *Evolution* 51:537–51

Podos J. 2001. Correlated evolution of morphology and vocal signal structure in Darwin's finches. *Nature* 409:185–88

Podos J, Nowicki S. 2004. Performance limits on birdsong. In *Nature's Music: The Science of Bird Song*, ed. P Marler, H Slabbekoorn. New York: Academic. In press

Podos J, Nowicki S, Peters S. 1999. Permissiveness in the learning and development of song syntax in swamp sparrows. *Anim. Behav.* 58:93–103

Podos J, Peters S, Nowicki S. 2004a. Calibration of song learning targets during vocal ontogeny in swamp sparrows (*Melospiza georgiana*). *Anim. Behav.* In press

Podos J, Sherer JK, Peters S, Nowicki S. 1995. Ontogeny of vocal tract movements during song production in song sparrows. *Anim. Behav.* 50:1287–96

Podos J, Southall JA, Rossi-Santos MR. 2004b.

Vocal mechanics in Darwin's finches: correlation of beak gape and song frequency. *J. Exp. Biol.* 207:607–19

Prestwich KN. 1994. The energetics of acoustic signaling in anurans and insects. *Am. Zool.* 34:625–43

Price JJ, Lanyon SM. 2002. Reconstructing the evolution of complex bird song in the oropendolas. *Evolution* 56.1514–29

Price JJ, Lanyon SM. 2004. Patterns of song evolution and sexual selection in the oropendolas and caciques. *Behav. Ecol.* 15:485–97

Proctor HC. 1991. Courtship in the water mite *Neumania papillator*: males capitalize on female adaptations for predation. *Anim. Behav.* 42:589–98

Prum RO. 1990. Phylogenetic analysis of the evolution of display behavior in the Neotropical manakins (Aves, Pipridae). *Ethology* 84:202–31

Ptacek MB. 2000. The role of mating preferences in shaping interspecific divergence in mating signals in vertebrates. *Behav. Process.* 51:111–34

Rasika S, Nottebohm F, Alvarez-Buylla A. 1994. Testosterone increases the recruitment and/or survival of new high vocal center neurons in adult female canaries. *Proc. Natl. Acad. Sci. USA* 91:7854–58

Richards DG, Wiley RH. 1980. Reverberations and amplitude fluctuations in the propagation of sound in a forest: implications for animal communication. *Am. Nat.* 115:381–99

Rome LC, Syme DA, Hollingworth S, Lindstedt SL, Baylor SM. 1996. The whistle and the rattle: the design of sound producing muscles. *Proc. Natl. Acad. Sci. USA* 93:8095–100

Rothstein SI, Fleischer RC. 1987. Vocal dialects and their possible relation to honest signalling in the brown-headed cowbird. *Condor* 89:1–23

Ryan MJ. 1990. Sexual selection, sensory systems, and sensory exploitation. *Oxford Surv. Evol. Biol.* 7:156–95

Ryan MJ, Brenowitz EA. 1985. The role of body size, phylogeny, and ambient noise in the evolution of bird song. *Am. Nat.* 126:87–100

Ryan MJ, Keddy-Hector A. 1992. Directional patterns of female mate choice and the role of sensory biases. *Am. Nat.* 139:S4–35

Schluter D. 1996. Adaptive radiation along genetic lines of least resistance. *Evolution* 50:1766–74

Searcy WA. 1992. Song repertoire and mate choice in birds. *Am. Zool.* 32:71–80

Searcy WA. 1984. Song repertoire size and female preferences in song sparrows. *Behav. Ecol. Sociobiol.* 14:281–86

Searcy WA. 1988. Song development from evolutionary and ecological perspectives. *Behav. Brain Sci.* 11:647–48

Searcy WA, Andersson M. 1986. Sexual selection and the evolution of song. *Annu. Rev. Ecol. Syst.* 17:507–33

Searcy WA, Nowicki S, Hughes M, Peters S. 2002. Geographic song discrimination in relation to dispersal distances in song sparrows. *Am. Nat.* 159:221–30

Searcy WA, Yasukawa K. 1990. Use of the song repertoire in intersexual and intrasexual contexts by male red-winged blackbirds. *Behav. Ecol. Sociobiol.* 27:123–28

Searcy WA, Yasukawa K. 1996. Song and female choice. See Kroodsma & Miller 1996, pp. 454–73

Simpson HB, Vicario DS. 1990. Brain pathways for learned and unlearned vocalizations differ in zebra finches. *J. Neurosci.* 10:1541–56

Slabbekoorn H, Jesse A, Bell DA. 2003. Microgeographic song variation in island populations of the white-crowned sparrow (*Zonotrichia leucophrys nutalli*): innovation through recombination. *Behaviour* 140:947–63

Slabbekoorn H, Smith TB. 2000. Does bill size polymorphism affect courtship song characteristics in the African finch *Pyrenestes ostrinus*? *Biol. J. Linn. Soc.* 71:737–53

Slabbekoorn H, Smith TB. 2002a. Bird song, ecology and speciation. *Philos. Trans. R. Soc. London Ser. B* 357:493–503

Slabbekoorn H, Smith TB. 2002b. Habitat-dependent song divergence in the little greenbul: an analysis of environmental selection pressures on acoustic signals. *Evolution* 56:1849–58

Slater PJB. 1986. The cultural transmission of bird song. *Trends Ecol. Evol.* 1:94–97

Slater PJB. 1989. Bird song learning: causes and consequences. *Ethol. Ecol. Evol.* 1:19–46

Slater PJB, Eales LA, Clayton NS. 1988. Song learning in zebra finches (*Taeniopygia guttata*): progress and prospects. *Adv. Stud. Behav.* 18:1–34

Soha JA, Marler P. 2000. A species-specific acoustic cue for selective song learning in the white-crowned sparrow. *Anim. Behav.* 60:297–306

Spector DA. 1992. Wood-warbler song systems: a review of paruline singing behaviors. In *Current Ornithology*, ed. DM Power, 9:199–238. New York: Plenum

Spencer KA, Buchanan KL, Goldsmith AR, Catchpole CK. 2003. Song as an honest signal of developmental stress in the zebra finch (*Taeniopygia guttata*). *Horm. Behav.* 44:132–39

Strote J, Nowicki S. 1996. Responses to songs with altered tonal quality by adult song sparrows (*Melospiza melodia*). *Behaviour* 133:161–72

Suthers RA. 1997. Peripheral control and lateralization of birdsong. *J. Neurobiol.* 33:632–52

Suthers RA, Goller F. 1997. Motor correlates of vocal diversity in songbirds. In *Current Ornithology*, ed. V Nolan Jr, ED Ketterson, CF Thompson, pp. 235–88. New York: Plenum

Suthers RA, Goller F, Hartley RS. 1994. Motor dynamics of song production by mimic thrushes. *J. Neurobiol.* 25:917–36

Suthers RA, Goller F, Pytte C. 1999. The neuromuscular control of birdsong. *Philos. Trans. R. Soc. London Ser. B* 354:927–39

Székely T, Catchpole CK, DeVoogd A, Marchl Z, DeVoogd TJ. 1996. Evolutionary changes in a song control area of the brain (HVC) are associated with evolutionary changes in song repertoire among European warblers (Sylviidae). *Proc. R. Soc. London Ser. B* 263:607–10

Ten Cate C, Slabbekoorn H, Ballintijn MR. 2002. Bird song and male-male competition: causes and consequences of vocal variability in the collared dove (*Streptopelia decaocto*). *Adv. Stud. Behav.* 31:31–75

Thielcke G. 1972. On the origin of divergence of learned signals (songs) in isolated populations. *Ibis* 115:511–16

Thomas RJ. 1999. The effect of variability in the food supply on the daily singing routines of European robins: a test of a stochastic dynamic programming model. *Anim. Behav.* 57:365–69

Thomas RJ. 2002. The costs of singing in nightingales. *Anim. Behav.* 63:959–66

Thomas RJ, Cuthill IC. 2002. Body mass regulation and the daily singing routines of European robins. *Anim. Behav.* 63:285–92

Thorpe WH. 1961. *Bird Song*. Cambridge, UK: Cambridge Univ. Press. 143 pp.

Tinbergen N. 1951. *The Study of Instinct*. Oxford: Clarendon. 228 pp.

Tramontin AD, Wingfield JC, Brenowitz EA. 2003. Androgens and estrogens induce seasonal-like growth of song nuclei in the adult songbird brain. *J. Neurobiol.* 57:130–40

Vallet E, Beme I, Kreutzer M. 1998. Two-note syllables in canary songs elicit high levels of sexual display. *Anim. Behav.* 55:291–97

Vallet E, Kreutzer M. 1995. Female canaries are sexually responsive to special song phrases. *Anim. Behav.* 49:1603–10

Vehrencamp SL. 2001. Is song-type matching a conventional signal of aggressive intentions? *Proc. R. Soc. London Ser. B* 268:1637–42

Wake DB. 1991. Homoplasy: the result of natural selection, or evidence of design limitations? *Am. Nat.* 138:543–67

Wallschläger D. 1980. Correlation of song frequency and body weight in passerine birds. *Experientia* 36:412

Ward BC, Nordeen EJ, Nordeen KW. 1998. Individual variation in neuron number predicts differences in the propensity for avian vocal imitation. *Proc. Natl. Acad. Sci. USA* 95:1277–82

Ward S, Speakman JR, Slater PJB. 2003. The energy cost of song in the canary, *Serinus canaria*. *Anim. Behav.* 66:893–902

Wasserman FE, Cigliano JA. 1991. Song output and stimulation of the female in white-throated sparrows. *Behav. Ecol. Sociobiol.* 29:55–59

Westneat MW, Long JH Jr, Hoese W, Nowicki S. 1993. Kinematics of birdsong: functional correlation of cranial movements and acoustic features in sparrows. *J. Exp. Biol.* 182:147–71

Wiley RH. 1991. Associations of song properties with habitats for territorial oscine birds of eastern North America. *Am. Nat.* 138:973–93

Williams H, Connor DM, Hill JW. 2003. Testosterone decreases the potential for song plasticity in adult male zebra finches. *Horm. Behav.* 44:402–12

Williams H, Mehta N. 1999. Changes in adult zebra finch song require a forebrain nucleus that is not necessary for song production. *J. Neurobiol.* 39:14–28

Wingfield JC, Hahn TP. 1994. Testosterone and territorial behavior in sedentary and migratory sparrows. *Anim. Behav.* 47:77–89

Yu AC, Margoliash D. 1996. Temporal hierarchical control of singing in birds. *Science* 273:1871–75

Zahavi A. 1975. Mate selection: a selection for a handicap. *J. Theor. Biol.* 53:205–14

Zollinger SA, Suthers RA. 2004. Motor mechanisms of a vocal mimic: implications for birdsong production. *Proc. R. Soc. London Ser. B* 271:483–91

Zuk M, Kolluru GR. 1998. Exploitation of sexual signals by predators and parasitoids. *Q. Rev. Biol.* 73:415–38

APPLICATION OF ECOLOGICAL INDICATORS*

Gerald J. Niemi[1] and Michael E. McDonald[2]

[1]*Natural Resources Research Institute and Department of Biology, University of Minnesota, Duluth, Minnesota 55811; email: gniemi@d.umn.edu*
[2]*U.S. Environmental Protection Agency, Environmental Monitoring and Assessment Program, Reston, Virginia 20191; email: McDonald.Michael@epa.gov*

Key Words assessment, condition, monitoring, responses, stressors

■ **Abstract** Ecological indicators have widespread appeal to scientists, environmental managers, and the general public. Indicators have long been used to detect changes in nature, but the scientific maturation in indicator development primarily has occurred in the past 40 years. Currently, indicators are mainly used to assess the condition of the environment, as early-warning signals of ecological problems, and as barometers for trends in ecological resources. Use of ecological indicators requires clearly stated objectives; the recognition of spatial and temporal scales; assessments of statistical variability, precision, and accuracy; linkages with specific stressors; and coupling with economic and social indicators. Legislatively mandated use of ecological indicators occurs in many countries worldwide and is included in international accords. As scientific advancements and innovation in the development and use of ecological indicators continue through applications of molecular biology, computer technology such as geographic information systems, data management such as bioinformatics, and remote sensing, our ability to apply ecological indicators to detect signals of environmental change will be substantially enhanced.

INTRODUCTION

Humans trying to understand the current condition or predict the future condition of ecosystems have often resorted to simple, easily interpreted surrogates. Often these surrogates have been indicators that allow humans to isolate key aspects of the environment from an overwhelming array of signals [National Research Council (NRC) 2000]. Early humans used indicators like seasonal migratory movements of animals or flowering by spring flora to provide insight into changing environmental conditions. The first reference to environmental indicators is attributed to Plato, who cited the impacts of human activity on fruit tree harvest (Rapport 1992). Morrison (1986) reviewed the work of Clements (1920) and noted that the

---

*The U.S. Government has the right to retain a nonexclusive, royalty-free license in and to any copyright covering this paper.

concept of indicators for plant and animal communities can be traced to the 1600s. Clements's (1920) work set the scientific stage for using plants as indicators of physical processes, changes to soil conditions, and other factors. In the 1920s, indicators were also being successfully used to determine changing environmental conditions, such as water clarity (Rapport 1992) or air quality with "the canary in the mine" (Burrell & Siebert 1916), which we continue to use (Van Biema 1995). One of the more elaborate early environmental indicators was the saprobian system (Kolkwitz & Marsson 1908), which used benthic and planktonic plants as indicator species for classifying stream decomposition zones.

The past 40 years have seen a rapid acceleration of scientific interest in the development and application of ecological indicators. This focus on indicators stems from the need to assess ecological condition in making regulatory, stewardship, sustainability, or biodiversity decisions. For example, the Clean Water Act of 1972 requires that each state produce a report every two years on the condition of all its waters to the U.S. Environmental Protection Agency (US EPA) for Congress. Decisions regarding sustainability and biodiversity involve both research and policy issues (e.g., Mann & Plummer 1999, Ostrom et al. 1999, Tilman 1999). In the United States, this research has been legislatively mandated to various federal agencies, in particular to the U.S. Department of Agriculture through the National Forest Management Act of 1976, to the U.S. Department of Interior (1980, Parsons 2004), and to the US EPA (2002b). These mandates have resulted from the increasing concern for the loss of species, deteriorating water quality and quantity, sustainability of resource use, climate change, and overall condition of the environment. This interest has generated many new books, articles, and reviews on ecological indicators (e.g., McKenzie et al. 1992; US EPA 2002b,c), as well as a new journal (*Ecological Indicators* in 2001).

The public has increasingly demanded a better accounting of the condition or health of the environment and whether it is improving or getting worse (Heinz Center 1999; www.heinzctr.org/ecosystems). Developing scientifically defensible indicators to establish environmental baselines and trends is a universal need at a variety of levels. For instance, federal governments in the United States and Canada (Environment Canada and US EPA 2003), Europe (www.eionet.eu.int), and Australia (www.csiro.au/csiro/envind/index.htm) have developed or are developing programs for routine reporting on ecological indicators. Recent international accords (e.g., RIO Accord) have demanded an accounting and reporting of indicators on the state of the environment. The Montréal Process (www.mpci.org) representing 12 countries was established in 1994 to develop and implement internationally agreed upon criteria and indicators for the conservation and sustainable management of temperate and boreal forests. In 2003, US EPA (2003a) released its first state of the environment report (www.epa.gov/indicators/roe/index.htm). As the world human population continues to increase exponentially (Cohen 2003), and with consequent environmental demands, the applications of indicators to determine status and trends in environmental condition will continue to grow.

DEFINITIONS

Early uses of indicators primarily reflected environmental conditions, and the terms environmental and ecological indicators have often been used interchangeably. Environmental indicators should reflect all the elements of the causal chain that links human activities to their ultimate environmental impacts and the societal responses to these impacts (Smeets & Weterings 1999). Ecological indicators are then a subset of environmental indicators that apply to ecological processes (NRC 2000). For policy makers, the amount of ecological data is often overwhelming. Environmental indicators are an attempt to reduce the information overload, isolate key aspects of the environmental condition, document large-scale patterns, and help determine appropriate actions (Niemeijer 2002). An example of a large-scale, policy-relevant environmental indicator is the environmental sustainability index (ESI). The ESI was developed to allow quantitative international comparisons of environmental conditions (World Economic Forum 2002). The ESI has five major categories: environmental systems, reducing environmental stresses, reducing human vulnerability, social and institutional capacity, and global stewardship. In 2001 the ESI included information on 68 indicators within these categories from 142 countries (World Economic Forum 2002).

Ecological indicators embody various definitions of ecology, such as the "interactions that determine the distribution and abundance of organisms" (Krebs 1978), or more broadly the "structure and function of nature" (Odum 1963). Thus, they are often primarily biological and respond to chemical, physical, and other biological (e.g., introduced species) phenomena. We have chosen to combine the definitions of the US EPA (2002b) and the hierarchy of Noss (1990), and we define ecological indicators as: measurable characteristics of the structure (e.g., genetic, population, habitat, and landscape pattern), composition (e.g., genes, species, populations, communities, and landscape types), or function (e.g., genetic, demographic/life history, ecosystem, and landscape disturbance processes) of ecological systems.

Ecological indicators are derived from measurements of the current condition of ecological systems in the field and are either used directly or combined into one or more summary values (US EPA 2002b). These ecological indicators can be aggregated into ecological attributes with reporting categories, such as biotic condition, chemical and physical characteristics, ecological processes, and disturbance (Harwell et al. 1999, US EPA 2002b). Ecosystem disturbance can be natural (e.g., fire, wind, and drought) and part of the functional attributes of ecosystems (Noss 1990), or it can be anthropogenic. Ecological indicators have been applied in many ways in the context of both natural disturbances and anthropogenic stress. However, the primary role of ecological indicators is to measure the response of the ecosystem to anthropogenic disturbances, but not necessarily to identify specific anthropogenic stress(es) causing impairment (US EPA 2002b). These indicators have been referred to as "state indicators" in the State of the Lakes Ecosystem Conference (SOLEC), which is a joint effort of Canada and the United States to develop indicators for the Great Lakes (Environment Canada and US EPA 2003).

SOLEC defines state indicators as response variables (e.g., fish, bird, amphibian populations) and pressure indicators as the stressors (e.g., phosphorus concentrations, atmospheric deposition of toxic chemicals, or water level fluctuations).

In this review we focus on ecological indicators, but clearly they can be integrated with the broader issues of ecosystem health (Rapport et al. 1998) and ultimately with economic indicators (Milon & Shogren 1995) to be even more useful for making policy decisions. There is a continuing debate on how to accomplish this integration. A common goal of linking economic and environmental indicators is often based on the concept of sustainability. For example, Ekins et al. (2003) provided a framework for linking economic, social, and environmental sustainability. Their approach identified how economic and social options were constrained if critical environmental functions were sustained. Lawn (2003) explored the theoretical foundation of several indexes of sustainability, including the Index of Sustainable Economic Welfare and the Genuine Progress Indicator. He found that these indexes were theoretically sound, but more robust valuation methods were necessary. Although progress is being made, there are no indicators that link economic, social, or environmental trends in a way that is meaningful to the public.

USE OF ECOLOGICAL INDICATORS

Ecological indicators are primarily used either to assess the condition of the environment (e.g., as an early-warning system) or to diagnose the cause of environmental change (Dale & Beyeler 2001). The widespread decline of the peregrine falcon (*Falco peregrinus*) in the 1950s is an excellent example of both uses. The catastrophic decline of the species served as an early-warning signal of problems in the environment, and research on the cause of the decline led to the diagnosis of widespread contamination by chlorinated hydrocarbons such as DDT (Ratcliffe 1980). The widespread decline of amphibians has also been viewed as an early-warning signal of problems in the environment, yet further research has failed to identify a specific cause for the decline. Amphibian declines are likely due to a variety of factors, including habitat change, global climate change, chemical contamination, disease and pathogens, invasive species, and commercial exploitation (Blaustein & Wake 1995, Semlitsch 2003).

The information gathered by ecological indicators can also be used to forecast future changes in the environment, to identify actions for remediation, or, if monitored over time, to identify changes or trends in indicators (Figure 1). As the complexity of the system being monitored increases (e.g., greater spatial scales and levels of biological organization) or as the temporal scale increases, the cost of gathering, analyzing, and reporting on indicators increases. Complexity also arises from the need to quantify linkages between specific stressors and ecological indicators (Table 1). In the few cases in which such relationships have been determined, these ecological indicators are often considered diagnostic; however, these linkages have seldom been made (Suter et al. 2002). A major challenge

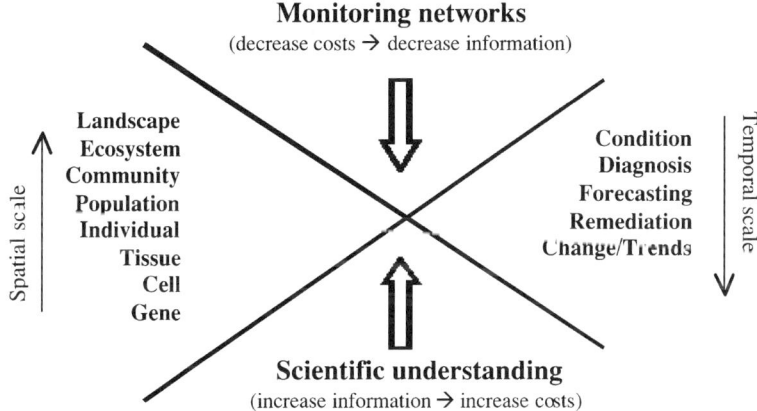

Figure 1 Illustration of the suite of ecological indicators (*left*) for which a suite of assessment capabilities (*right*) are desired. Constraints on the development of ecological indicators at all levels for all assessment endpoints are due to a lack of scientific understanding and the predominance of policies requiring low cost monitoring. Goals in applications generally include a compromise between cost-effectiveness and the ability to defend the ecological indicator scientifically at the spatial and temporal scale appropriate to answer the desired management objectives.

continues to be the difficulty of discerning specific stressor-response relationships in a multiple stressor environment and the difficulty of separating anthropogenic from natural sources of variation (Niemi et al. 2004).

Ecological indicators are usually developed by scientists and focused on aspects of ecosystems they believe are important for the assessment of condition. However, environmental managers and policy makers require indicators that are understood by the public (Schiller et al. 2001). Ideally, policy-relevant indicators would allow: (*a*) assessment of both existing and emerging problems; (*b*) diagnosis of the anthropogenic stressors leading to impairments; (*c*) establishment of trends in condition for measuring environmental policy and program performance; and (*d*) ease of communication to the public. Besides capturing the complexities of an ecosystem and being easy to communicate, an indicator should also be easily and routinely measurable (Dale & Beyeler 2001). Moreover, the cost of monitoring and subsequent analyses is also a consideration for state and federal agencies. Classifications of indicators that include scientific performance, policy relevance, and public acceptance have been proposed (Noss 1990, Cairns et al. 1993). However, the final choice of indicators should depend on the questions being asked and the quality of the science supporting the indicator.

Frost et al. (1992) suggest that ecological indicators should trade off two potentially contradictory endpoints. They should be sensitive enough to react in a detectable way when a system is affected by anthropogenic stress, and they should also remain reasonably predictable in unperturbed ecosystems. McGeoch (1998)

TABLE 1 Examples of ecological responses to natural and anthropogenic stress

Stress	Ecological response	Reference
Natural disturbance		
Forest fire	Landscape pattern	Turner et al. 1994
Drought	Bird populations	Blake et al. 1994
Herbivory	Vegetation and litter	McInnes et al. 1992
Wind	Forest stands	Foster 1988
Anthropogenic stress		
Acid rain	Feather moss	Hutchinson & Scott 1988
Eutrophication	Aquatic macrophytes	Kangas et al. 1982
Introduced species	Bird populations	Savidge 1984
Sedimentation	Shrubs	Johnston 2003
Logging	Landscape pattern	Franklin & Forman 1987
Heavy metals	Mosquito gene frequencies	Guttman 1994
Urbanization	Bird guilds	O'Connell et al. 2000
Air pollution	Plant species	Stolte & Mangis 1992
Air quality	Lichens	Kinnunen et al. 2003

provided an extensive list of suggested criteria to consider in the selection of bioindicators that included cost, species abundance, baseline data on species biology, and sensitivity to stress. Indicator selection will always be a compromise among many factors and must be optimized for the intended purpose.

There has been a strong recent interest in reporting on the ecological condition of the environment (Heinz Center 2002, US EPA 2003a) for the purposes of planning, management, and public reporting (US EPA 2003a). To accomplish these goals, the ecological indicators must be able to detect anthropogenic change against a background of natural variability. At issue is that ecological indicators at the population or community levels are not tightly coupled to the primary biological effects of stressors, which results in a slower response time, high natural variability, and low sensitivity (Jenkins & Sanders 1992). Confounding this further, communities and populations are responding to many other factors, some of which are not necessarily stressor related. In some cases it may be our attempts at aggregating the data that affect change detection. Frost et al. (1992) found that the level of taxonomic aggregation of the zooplankton populations affected the detection of change in an acid sensitive lake. Natural variability decreased with increasing taxonomic aggregation, but sensitivity had a less straightforward response. At intermediate levels of taxonomic aggregation, zooplankton populations exhibited the highest sensitivity (Frost et al. 1992). Thus, some intermediate level of taxonomic aggregation may be required to optimize the trade-off between sensitivity and variability in producing a useful ecological indicator.

At the organism level, important physiological processes can change in response to anthropogenic stress (e.g., growth, fecundity, developmental rates). However, the initial biological response often occurs at the cellular or subcellular level. This fact has advanced the use of biomarkers to detect physiologic condition and exposure to stressors (Jenkins & Sanders 1992). These biochemical and cellular indicators tend to be more sensitive to contaminants and more responsive than higher level indicators. Metallothionein induction is an example. Metallothioneins are low molecular weight, metal binding proteins that are involved in homeostasis regulation and compartmentalization of essential metals (e.g., Brouwer et al. 1989). In the presence of excess metals, induction of metallothioneins is enhanced for detoxification (Hamer 1986). The drawback to this strong indicator of anthropogenic insult is in its interpretation in the broader ecological context. For instance, it is unclear whether the biomarker is related to the condition or population of the species.

The questions, goals, and objectives of a monitoring program determine which ecological indicators are used (Dixon et al. 1998). Ecological indicators have been applied from the level of the gene (e.g., Rublee et al. 2001) to the landscape (e.g., Lausch & Herzog 2002) (Table 2). Researchers need to recognize which part of the ecological indicator spectrum is relevant to the objectives of their investigation. For example, is the indicator an early-warning system that may be related to a specific stress? Is the indicator a measurement of the condition of the ecological system? Is it important to know the cause of any change in the indicator? Clearly stated goals and objectives are essential (Yoccoz et al. 2001).

Most ecological monitoring programs are based on an aggregation of selected sites (Olsen et al. 1999), and researchers often infer regional trends from the accumulation of these site-specific trends (Urquhart et al. 1998). Many of these studies are useful for establishing temporal variability and mechanistic relationships, but larger regional trends cannot be compiled from these site-specific data unless the sites were selected in a representative and unbiased manner within the region of interest (Urquhart et al. 1998). Thus, researchers must integrate the selection of ecological indicators for examining large-scale spatial trends in an ecosystem within an appropriate statistical design (McDonald et al. 2004).

INDICATOR SPECIES

Most applications of ecological indicators have focused at the species level owing to concerns arising from endangered species and species conservation issues (Fleishman et al. 2001). For instance, Noss (1990) stated that "the use of indicator *species* to monitor or assess environmental conditions is a firmly established tradition in ecology, environmental toxicology, pollution control, agriculture, forestry, and wildlife and range management." The measurement of an indicator species assumes that a single species represents many species with similar ecological requirements (Landres et al. 1988). Typically, ecological indicator species tend to be from the macroflora and macrofauna, especially aquatic macroinvertebrates, fish, birds, and vascular plants. The primary reasons for their use are: (*a*) relative

TABLE 2 Examples of indicators that have been applied within different ecological levels of organization (modified from Noss 1990 and US EPA 2003)

Type		Example	References
Compositional	Genes	Species differentiation	Rudi et al. 2000
	Cell and subcellular	Immune response	Anderson et al. 1989
	Tissue	Metal concentration	Pérez-Lopéz et al. 2003
	Species	Butterflies	MacNally & Fleischman 2002
	Populations	Birds	Browder et al. 2002
	Communities	Floristic quality	Lopez & Fennessy 2002
	Ecosystems	Lakes	Whittier et al. 2002
	Landscape types	Land use/cover	Lausch & Herzog 2002
Functional	Genetic processes	Mutation rates	Ames et al. 1973
	Behavior	Feeding rate	Sierszen & Frost 1990
	Life history	Species traits	Hausner et al. 2003
	Demographic processes	Productivity	Underwood & Roth 2002
	Ecosystem processes	Growth	Marwood et al. 2001
	Landscape processes	Diatoms	Dixit et al. 1992
Structural	Genetic structure	Zooplankton genotypic differentiation	Baird et al. 1990
	Population structure	Bird guilds	Croonquist & Brooks 1991
	Habitat physiognomy	Forest structure	Lindenmayer et al. 2000
	Landscape patterns	Fragmentation	O'Neill et al. 1988
Integrative	Index of biotic integrity	Fish	Karr 1981
	AMOEBA	Multiple taxa	ten Brink et al. 1991
	Multivariate	Biomarkers	Cormier & Racine 1992
	Species assemblages	Beetles	Dufréne & Legendre 1997
	Index of environmental integrity	Multiple indices	Paul 2003

ease of identification, (b) interest to the public, (c) relative ease of measurement, (d) relatively large number of species with known responses to disturbance, and (e) relatively low cost.

We use indicator species as a general term to refer to approaches that use one or a few species to "indicate" condition or a response to stress that may apply to other species with similar ecological requirements. Lawton & Gaston (2001) suggest that indicator species are used in three distinct ways: (a) to reflect the biotic or abiotic state of the environment; (b) to reveal evidence for the impacts of environmental change; or (c) to indicate the diversity of other species, taxa, or communities within an area. The first two reflect the common uses of indicators as measures of condition and the diagnosis of potential cause(s) of environmental change. The third expands the concept of indicators to incorporate the idea of a single species serving as a surrogate for many other species. This idea has been largely untested and has been the focus of much debate and criticism in applications (see below). Because of the criticism, many new approaches and terms have been developed to refine the indicator species concept. These include focal species, umbrella species, flagship species, or guilds as indicators (Verner 1984, Landres et al. 1988, Lambeck 1997, Simberloff 1998, Noss 1999).

The term focal species has been used in many ways in the literature. For example, Cox et al. (1994) identified 44 focal species to serve as umbrella or indicator species of biological diversity in Florida. Lambeck (1997) identified focal species as a subset of the total pool of species in a landscape. Carroll et al. (2001) used carnivores as focal species in regional conservation planning because their distributional patterns reflected regional-scale population processes. There is not a consistent definition of a focal species, except when they are selected by various means as the "focus" of study. Focal species tend to differ from indicator species in that they do not necessarily serve to measure ecological condition nor do they convey a stress-response relationship. The focal species concept has generally been used in conservation planning, landscape ecology, and protection of biological diversity.

Fleischman et al. (2001) define umbrella species as those "whose conservation confers a protective umbrella to numerous co-occurring species." They also point out that "blurred discriminations between umbrella and indicator species have led to misunderstandings over how umbrella species should be selected." Flagship species are those that have large public appeal, such as charismatic megafauna like bears and tigers. The guild concept has been explored as an alternative to indicator species both in wildlife management and in determining regional condition (e.g., Verner 1984, O'Connell et al. 2000). Guilds were originally defined by Root (1967) as a "group of species that exploit the same class of environmental resources in a similar way"; Verner (1984) concluded that the guild concept held promise but required more testing. O'Connell et al. (2000) distinguished 16 behavioral and physiological response guilds of birds and were able to combine their bird community data into a bird community index (BCI). They related the BCI to landscape condition and change from forested to nonforested areas.

Lambeck (1997) expanded on the concepts of umbrella and focal species to incorporate more specific responses to landscape and management regimes. His analysis focused on identifying focal species with the most demanding survival requirements for several parameters threatened by anthropogenic stressors. Noss (1999) further extended and combined these concepts for indicator, focal, and umbrella species in a forest management context. In his approach, indicator species were represented by a suite of focal species, each of which was defined by different attributes that had to be present in the landscape to retain the biota. Landscape attributes included: (a) area-limited species, (b) dispersal-limited species, (c) resource-limited species, (d) process-limited species, (e) keystone species, (f) narrow endemic species, and (g) special cases such as flagship species that are of public concern in the region. Noss (1999) suggested that at least for the first four categories umbrella species could be defined that are the most sensitive to the landscape attribute.

Birds have been the primary focus for most terrestrial applications of indicator species, but insects hold great promise because of their species richness, biomass, and role in ecosystem functioning. McGeoch (1998) recognized the potential applications of insects as indicator species and defined their use as environmental, ecological, or biodiversity indicators. Researchers have attempted to examine vertebrates as possible umbrella species for insects, especially for butterflies. For example, Rubinoff (2001) analyzed the use of a bird species, the California gnatcatcher (*Poliotila californica*), as a potential umbrella species for three species of butterflies. However, the gnatcatcher was a poor indicator primarily because of its ubiquity in the landscape studied. Insects and other microfauna offer excellent potential as indicator species. They are of limited use in terrestrial systems because of the cost of sampling and processing and because there is limited acceptance by resource managers, politicians, and the general public.

Researchers have developed other indexes to provide more holistic approaches to ecological condition. These indexes range from simple diversity indexes, such as the Shannon and Wiener Index (Shannon & Weaver 1949), to multimetric indexes (e.g., Karr 1981, Kerans & Karr 1994, Karr 2000, Simon 2003). Multimetric ecological indicators are sets of mathematically aggregated or weighted indicators (US EPA 2000a, Kurtz et al. 2001) that combine attributes of entire biotic communities into a useful measure of condition (US EPA 2002a). The US EPA has recently used an index for biotic integrity (IBI) for estuarine invertebrates as one of the indicators in the assessment of the condition of the nation's estuaries (US EPA 2004). Because of the increasing use of multimetric and other indicators, researchers have developed specific guidelines for evaluating their performance (US EPA 2000a).

Another aspect of ecological indicators is whether to use them as relative or absolute measures. As a relative measure, the initial measurement becomes the baseline for comparison of future measures. Most monitoring agencies prefer or require a more quantitative benchmark for measuring and regulating changes in ecosystems. These benchmark or reference conditions can be defined as the conditions of ecological resources under minimal contemporary human disturbance (McDonald et al. 2004). As these conditions are often not available for direct measurement, models and historic information are often invoked as best approximations. However, the selection of reference conditions remains problematic (NRC 1990).

In summary, focal species represent those selected as a focus for a specific investigation. There is no consistent definition of focal species, but the concept has been expanded for use in conservation and management. Focal species have been used to identify potential indicator species when there is a desire to describe ecological condition or measure the response to a disturbance. Either a focal species or an indicator species may serve as an umbrella species if the goals are to monitor or manage one species as a surrogate for other species or to identify conservation areas for preservation. Focal, indicator, or umbrella species could be flagship species if they have a high profile or interest to the public. Moreover, any of them could be keystone species (sensu Paine 1969a,b) if they are particularly important in establishing or maintaining key ecological processes or structure for other species within an ecosystem (Simberloff 1998). Before any investigation, researchers must clearly define these terms and rigorously test whether the species can fulfill its purpose as an indicator, umbrella, or keystone species.

COMPLEXITIES IN APPLICATIONS OF ECOLOGICAL INDICATORS

Monitoring for ecosystem or resource management often requires data about a specific site or sites, whereas public policy decisions typically require information across broader geographical regions (Olsen et al. 1999). Many of the existing ecological monitoring programs are periodically or continually used at certain sites, which may lead to a better understanding of the temporal variability at the site but may not be representative of a larger area (Urquhart et al. 1999). Thus, ecological indicators are needed to assess status and trends in ecological systems and to diagnose cause(s) of declining condition across a range of spatial and temporal scales (Kratz et al. 1995, NRC 2000, Dale & Beyeler 2001, Niemi et al. 2004).

Each ecological indicator responds over different spatial and temporal scales; thus, the context of these scales must be explicitly stated for each ecological indicator. Furthermore, understanding the response variability in ecological indicators is essential for their effective use (US EPA 2002c). Without such an understanding, it is impossible to differentiate measurement error from changing condition, or an anthropogenic signal from background variation. Work has begun on understanding how natural and anthropogenic variability of indicators can affect status and trend detection, but it is closely tied to different statistical design considerations (Larsen et al. 1995). Specific monitoring designs and indicators can be implemented to detect changes across temporal and spatial scales (US EPA 1997, 2002d).

In general, as one moves up levels of organization from cellular phenomena to landscape processes, the spatial and temporal scales of application increase immensely. Similarly, as larger spatial and temporal scales are considered, the linkage to specific stressors can be either obscured or refined depending on the stressor. For example, one of the largest and most successful monitoring programs is the U.S. Geological Survey Breeding Bird Survey (BBS), which has gathered data over a 38-year period (1966 to present) (Sauer et al. 2003). The BBS is intended to indicate breeding bird species trends over relatively large regional and national spatial scales. Thus, researchers must exercise caution in interpreting results for specific regions or combining results from different regions (James et al. 1996). In contrast, nesting tree swallows (*Tachycineta bicolor*) are an effective wildlife indicator species of sediment chemical contaminant problems. Because nestlings are fed flying adult insects, which typically have aquatic early life histories, the uptake of chemicals by nestlings can be related to sediment chemical levels near the nesting site (Nichols et al. 1995, Jones 2003). Changes in bird trends from the BBS over large areas are powerful because of the large number of sample routes and the a priori experimental design, but the causes of changes in species trends are speculative. In contrast, contaminant uptake in nestling tree swallows and potential risk to wildlife can clearly be connected to food supplies derived from sediment. Many of the same problems exist for multimetric indexes commonly used to assess condition of surface waters across large regions. These indexes can distinguish degraded sites from sites with little or no human impact, but they do not diagnose the causes of impairment by themselves (US EPA 2003b).

ADVANCEMENTS IN ECOLOGICAL INDICATORS OF BIOLOGICAL COMMUNITIES

Historically, ecological indicators were primarily based on parameters associated with individual species (e.g., presence) or simple community metrics (e.g., species richness or diversity). However, many of these indicators did not fully represent the entire biological community of organisms present. Hence, Karr (1981) introduced the IBI using stream fish communities. This index was a numerical summation of subsets of the fish community from one area compared with a suitable reference area. These reference areas ideally represented areas that were natural or undisturbed from the same geographic area and with the same general ecological condition. Karr (1981) calculated the IBI using fish community data for a specific area and subdivided these data into 12 metrics, including the number of individuals and species found in the sample, the relative abundance of guilds (e.g., carnivores), specific species in the sample, and other categories (e.g., sunfish species). The IBI was expressed as deviations from the suitable reference area such that larger values represented communities similar to the reference area, whereas lower values represented areas that deviate from the reference, potentially because of stress. The IBI has received considerable attention and application over the past 20 years, including applications to fish (Fausch et al. 1984, Angermeier & Karr 1986, Karr et al. 1986, Simon & Emery 1995), macroinvertebrates (Kerans & Karr 1994, Klemm et al. 2003), plant communities (Simon et al. 2001, DeKeyser et al. 2003), aquatic communities (Simon et al. 2000), and birds (O'Connell et al. 2000).

Many other multimetric indexes have evolved over the past 20 years, such as the Hilsenhoff biotic index (Hilsenhoff 1982) and biological response signatures (Simon 2003). In contrast to multimetric indexes, multivariate indexes (Reynoldson et al. 1997, Karr 2000) are statistical analyses of the biological community using a host of multivariate techniques, such as principal components analysis (O'Connor et al. 2000), canonical correspondence analysis (Kingston et al.

1992), and combinations of multivariate analyses (Dufréne & Legendre 1997). For example, O'Connor et al. (2000) integrated information from five different taxonomic groups (diatoms, benthos, zooplankton, fish, and birds) to provide an index of the ecological condition of lakes. Their approach was effective in relating the gross condition of the lakes across taxa, but it was also effective in identifying a differential response by fish to nearshore conditions. Dufréne & Legendre (1997) used a combination of multivariate analyses of carabid beetle community data to determine indicator species and species assemblages for groups of sites. Their approach also includes a randomization procedure to test the significance of the indicator values.

Many analytical approaches of biological community data are currently being developed, tested, and used for ecological indicators. For instance, Andreasen et al. (2001) and Paul (2003) have recently introduced indexes of ecological and environmental integrity, respectively. These indexes combine information from several levels (e.g., biological communities, habitat, expert opinion) into an overall measure of integrity. The exploration and debate of these approaches will likely continue in the future.

CRITICISMS OF INDICATORS

Virtually all attempts to use ecological indicators have been heavily criticized, and many criticisms are well deserved. For instance, many existing monitoring programs of indicators suffer from two deficiencies: lack of well-articulated objectives and neglect of different sources of error (Yoccoz et al. 2001). Indicator species have been especially criticized in the context of forest management–related issues (Landres et al. 1988, Landres 1992, Niemi et al. 1997, Rolstad et al. 2002, Failing & Gregory 2003). Many of these criticisms have focused on the lack of: (*a*) identification of the appropriate context (spatial and temporal) for the indicator, (*b*) a conceptual framework for what the indicator is indicating, (*c*) integration of science and values, and (*d*) validation of the indicator.

Many of these criticisms have led to more focused efforts on individual species and to the development of additional concepts such as focal species or umbrella species (Lambeck 1997, Fleischman et al. 2001). Roberge & Angelstam (2004) recently reviewed the umbrella species concept and concluded that multispecies strategies were more compelling. Lawler et al. (2003) evaluated several indicator groups (e.g., birds, fish, mammals, and mussels) to test whether one group could provide habitat protection for other taxa in a large area of the Mid-Atlantic region of the United States. No single taxonomic indicator group could provide adequate protection for another group, especially for at-risk species within each of the groups. The failure was likely attributable to the narrow geographic ranges and restricted habitat distribution of rare species. Hence, information on rare species and those that are at risk was essential, yet gathering data on rare species is generally difficult, time-consuming, and expensive. In contrast to the indicator species approach, Manley et al. (2004) evaluated an innovative, multispecies monitoring approach that included all terrestrial vertebrate species over a large ecoregional scale (7 million ha). The design of this comprehensive approach reduced the emphasis on indicator species because the spatial coverage allowed many species to be adequately monitored. A fundamental problem with these approaches continues to be the inability to link species presence or relative abundance with relevant aspects of habitat quality (Van Horne 1983), such as productivity.

Many of these same criticisms apply to indexes (Suter 1993). Indexes have been viewed as oversimplifications and generalizations of biological processes, in which important data can be lost (May 1985). There are also concerns about how these indexes are calibrated (Seegert 2001) and whether or how they are evaluated across gradients (US EPA 2000a). Despite such criticisms, these indexes can play an important management role by helping characterize ecological condition (Rakocinski et al. 1997).

Ecological indicators span broad levels of biological, spatial, and temporal organization within ecosystems. When establishing a monitoring program and selecting indicators, it is imperative that researchers articulate a clear statement of goals. Once the goals are unambiguously stated, the scientific soundness and objectivity of the indicator become a central issue (Niemeijer 2002). Researchers must recognize these complexities and limitations in the application and use of ecological indicators (Dale & Beyeler 2001). However, having effective indicators is only one component of the problem. Sound program design and effective data management, analysis, synthesis, and interpretation are also needed to implement monitoring and assessment programs successfully (NRC 1990). In the past five years, many publications have provided excellent guidance on how ecological indicators can be improved, including documents by the NRC (2000) and US EPA (2000a, 2002b).

FUTURE OF ECOLOGICAL INDICATORS AND CONCLUSIONS

Advances in science and technology at all levels of biological organization will greatly improve our ability to apply ecological indicators in the future. Recently developed techniques in molecular biology such as biomarkers have proven useful in rapid identification of problems in ecological systems caused by pollution stress (e.g., Cormier & Racine 1992, Huggett et al. 1992). For example, Arcand-Hoy & Metcalfe (1999) found that both fluorescent aromatic compounds in bile and hepatic ethoxyresorufin-O-deethylase in fish could be used to detect exposure to polynuclear aromatic hydrocarbons in the lower Great Lakes. Evendon & Depledge (1997) identified the potential usefulness of genetically susceptible populations to environmental contaminants. Paerl et al. (2003) have recently used diagnostic photopigments of various phytoplankton groups as ecological indicators to detect changes in nutrients, noxious algal blooms, and overall water quality. Investigators are optimistic about applying molecular techniques to address specific ecological problems and to act as early-warning signals of potential problems. However, research is necessary to illustrate how these techniques can be scaled up to address environmental problems over large regions. There is tremendous potential for application of these new techniques to provide real-time, remotely sensed condition assessments of environmental problems (Kerr & Ostrovsky 2003).

Global positioning systems (GPS), geographic information systems (GIS), remote-sensing technology, and computer hardware and software hold great potential for advancing the science of ecological indicators. GPS allows precise location of repeated field measurements, thus reducing errors associated with spatial variation. GIS provides unprecedented abilities to organize, analyze, synthesize, and display information gathered in the field over both space and time. Remote-sensing technology has also advanced substantially in resolution from 30 m to <4 m resolution (e.g., Kerr & Ostrovsky 2003, Clark et al. 2004). Database storage and software to manipulate these data have increased exponentially in the past ten years and have resulted in the new field of bioinformatics. These techniques in combination with data gathered in the field or combined with existing databases have proven effective in a myriad of applications, such as change detection in forest systems (Wolter & White 2002), mapping biodiversity patterns (Stockwell & Peterson 2003), and forecasting animal distributions and abundance over large regions (Venier et al. 2004).

Researchers have addressed many of the criticisms and failures that have plagued the applications of ecological indicators, resulting in substantial improvements in assessing condition in many areas (e.g., US EPA 1998, 2000b, 2003c). Guidelines for ecological indicator development need to be heeded (Kurtz et al. 2001). Depending on the indicator's use and the spatial scales of application, experimental design considerations are crucial for appropriate inferences once the data are gathered (Urquhart et al. 1998, Olsen et al. 1999, Danz et al. 2004).

Of increasing interest is the integration of environmental indicators with other well-known economic and social indicators like the gross national product or the consumer price index (Milon & Shogren 1995, NRC 1997). The International Society for Ecological Economics, which recently began publishing the journal *Ecological Economics*, was formed partially to integrate this thinking into a "transdiscipline" aimed at developing a sustainable world (www.ecologicaleconomics.org/about/index.htm). Moreover, a variety of authors have emphasized the need to consider human health and link it to environmental health (Pimentel et al. 2000, Karr 2002), as well as to establish an economic valuation system for ecological resources (Costanza 1997, Daily 1997) or for ecological sustainability (Armsworth & Roughgarden 2003). The motivation for this integration stems largely from managers' need to better quantify ecological changes resulting from such issues as global climate change; species extinction rates; contaminated air, water, and soil; declining fish populations; human conflicts over resources such as water; and the emergence of new diseases (e.g., Pimentel et al. 2000, Brown 2003, Karr 2002) in relevant human social and economic terms. Clearly, the general public currently has a paucity of information on which to judge the ecological condition of the environment or how the condition might relate to human health or to the economy. Yet, with such information, individuals make daily and long-term decisions on the basis of health indicators (e.g., blood pressure), economic indicators (e.g., NASDAQ, Dow Jones Industrial Average), and environmental indicators (e.g., weather forecasts). Despite three decades of discussion of the integration of economic and ecological indicators, there are limited applications of integrated analysis (Milon & Shogren 1995). US EPA's (2003a) state of the environment report in 2003 is one of the first steps in informing the public of the ecological condition of the nation's resources. Future reports on integrated and understandable measures will be welcome additions as indicators of environmental sustainability, but their acceptance and impacts on policy and public opinion will have to be determined.

Strong public interest and legislative mandates exist at local, state, federal, and international levels to understand the condition, trends, and cause for change in our ecosystems. A large array of ecological indicators are available for application to environmental problems; moreover, the number of tools and techniques that are available is rapidly increasing. Therein lies the challenge for the future: to select appropriate monitoring designs and ecological indicators that will provide convincing scientific underpinnings for management and policy decisions on real-world problems.

ACKNOWLEDGMENTS

Although the research described in this article has been funded wholly or in part by the United States Environmental Protection Agency's Science to Achieve Results (STAR) program through cooperative agreement R828675-00 to the University of Minnesota, it has not been subjected to the agency's required peer and policy review and therefore does not necessarily reflect the views of the agency, and no official endorsement should be inferred. We thank James Cox, Robert Howe, and Lucinda Johnson for comments on an earlier version of this manuscript. This is publication number 359 of the Center for Water and the Environment, Natural Resources Research Institute, University of Minnesota, Duluth.

The *Annual Review of Ecology, Evolution, and Systematics* is online at
http://ecolsys.annualreviews.org

LITERATURE CITED

Ames BN, Lee F, Durston W. 1973. An improved bacterial test system for the detection and classification of mutagens and carcinogens. *Proc. Natl. Acad. Sci. USA* 70:782–86

Anderson DP, Dixon OW, Bodammer JE, Lizio EF. 1989. Suppression of antibody-producing cells in rainbow trout spleen sections exposed to copper in vitro. *J. Aquat. Anim. Health* 1:57–61

Andreasen JK, O'Neill RV, Noss R, Slosser NC. 2001. Considerations for the development of a terrestrial index of ecological integrity. *Ecol. Indic.* 1:21–35

Angermeier PL, Karr JR. 1986. Applying an

index of biotic integrity based on stream fish communities: considerations in sampling and interpretation. *North Am. J. Fish. Manage.* 6:418–29

Arcand-Hoy LD, Metcalfe CD. 1999. Biomarkers of exposure of brown bullheads (*Ameiurus nebulosus*) to contaminants in the lower Great Lakes, North America. *Environ. Toxicol. Chem.* 18:740–49

Armsworth PR, Roughgarden JE. 2003. The economic value of ecological stability. *Proc. Natl. Acad. Sci. USA* 100(12):7147–51

Baird DJ, Barber I, Calow P. 1990. Clonal variation in general responses of *Daphnia magna* Straus to toxic stress. I: Chronic life-history effects. *Funct. Ecol.* 4:399–408

Blake JG, Hanowski JM, Niemi GJ, Collins PT. 1994. Annual variation in bird populations of mixed conifer–northern hardwood forests. *Condor* 96:381–99

Blaustein AR, Wake DB. 1995. The puzzle of declining amphibian populations. *Sci. Am.* 272:52–57

Brouwer M, Winge DR, Gray WR. 1989. Structural and functional diversity of copper-metallothionein from the American lobster, *Homarus americanus*. *J. Inorg. Biochem.* 35:289–303

Browder SF, Johnson DH, Ball IJ. 2002. Assemblages of breeding birds as indicators of grassland condition. *Ecol. Indic.* 2:257–70

Brown LR. 2003. *Plan B: Rescuing a Planet Under Stress and a Civilization in Trouble.* New York/London: WW Norton

Burrell GA, Siebert FM. 1916. Gases found in coal mines. *Miner's Circular 14.* Bureau Mines, US Dep. Inter., Washington, DC

Cairns J Jr, McCormick PV, Niederlehner BR. 1993. A proposed framework for developing indicators of ecosystem health. *Hydrobiologia* 263:1–44

Carroll C, Noss RF, Paquet PC. 2001. Carnivores as focal species for conservation planning in the Rocky Mountain region. *Ecol. Appl.* 11:961–80

Clark DB, Read JM, Clark ML, Cruz AM, Dotti MF, et al. 2004. Application of 1-m and 4-m resolution satellite data to ecological studies of tropical rain forests. *Ecol. Appl.* 14:61–74

Clements FC. 1920. *Plant Indicators.* Washington, DC: Carnegie Inst.

Cohen JE. 2003. Human population: the next half century. *Science* 302:1172–75

Cormier SM, Racine RN. 1992. Biomarkers of environmental exposure and multivariate approaches for assessment and monitoring. See McKenzie et al. 1992, 1:229–42

Costanza R, d'Arge R, deGroot R, Farber S, Grasso M, et al. 1997. The value of the world's ecosystem services and natural capital. *Nature* 387:253–60

Cox J, Kautz R, MacLaughlin M, Gilbert T. 1994. *Closing the gaps in Florida's wildlife habitat conservation system.* Rep., Off. Environ. Serv., Fla. Game Fresh Water Fish Comm., Tallahassee

Croonquist MJ, Brooks RP. 1991. Use of avian and mammalian guilds as indicators of cumulative impacts in riparian-wetland areas. *J. Environ. Manage.* 15:701–14

Daily GC, ed. 1997. *Nature's Services: Societal Dependence on Natural Ecosystems.* Washington, DC: Island

Dale VH, Beyeler SC. 2001. Challenges in the development and use of ecological indicators. *Ecol. Indic.* 1:3–10

Danz NP, Regal RR, Niemi GJ, Brady VJ, Hollenhorst T, et al. 2004. Environmentally stratified sampling design for the development of Great Lakes environmental indicators. *J. Environ. Monitor. Assess.* In press

DeKeyser ES, Kirby DR, Ell MJ. 2003. An index of plant community integrity: development of methodology for assessing prairie wetland plant communities. *Ecol. Indic.* 3:119–33

Dixit SS, Smol JP, Kingston JC, Charles DF. 1992. Diatoms: powerful indicators of environmental change. *Environ. Sci. Tech.* 26:22–33

Dixon PM, Olsen AR, Kahn BM. 1998. Measuring trends in ecological resources. *Ecol. Appl.* 8:225–27

Dufréne M, Legendre P. 1997. Species assemblages and indicator species: the need for a flexible asymmetrical approach. *Ecol. Monogr.* 67:345–66

Ekins P, Simon S, Deutsch L, Folke C, DeGroot R. 2003. A framework for the practical application of the concepts of critical natural capital and strong sustainability. *Ecol. Econ.* 44:165–85

Environment Canada and US EPA. 2003. State of the Great Lakes 2003. *EPA 905-R-03-004*, Governments of Canada and the US

Evenden AJ, Depledge MH. 1997. Genetic susceptibility in ecosystems: the challenge for ecotoxicology. *Environ. Health Perspect.* 105(Suppl. 4):849–54

Failing L, Gregory RS. 2003. Ten common mistakes in designing biodiversity indicators for forest policy. *J. Environ. Manage.* 68:121–32

Fausch KD, Karr JR, Yant PR. 1984. Regional application of an index of biotic integrity based on stream-fish communities. *Trans. Am. Fish. Soc.* 113:39–55

Fleischman E, Blair RB, Murphy DD. 2001. Empirical validation of a method for umbrella species selection. *Ecol. Appl.* 11:1489–501

Foster DR. 1988. Species and stand responses to catastrophic wind in central New England, USA. *J. Ecol.* 76:135–51

Franklin JF, Forman RTT. 1987. Creating landscape patterns by forest cutting: ecological consequences and principles. *Landsc. Ecol.* 1:5–18

Frost TM, Carpenter SR, Kratz TK. 1992. Choosing ecological indicators: effects of taxonomic aggregation on sensitivity to stress and natural variability. See McKenzie et al. 1992, 1:215–27

Guttman SI. 1994. Population genetic structure and ecotoxicology. *Environ. Health Perspect.* 102(Suppl. 12):97–100

Hamer DH. 1986. Metallothionein. *Annu. Rev. Biochem.* 55:913–51

Harwell MA, Myers V, Young T, Bartuska A, Gassman N, et al. 1999. A framework for an ecosystem integrity report card. *BioScience* 49:543–56

Hausner VH, Yoccoz NG, Ims RA. 2003. Selecting indicator traits for monitoring land use impacts: birds in northern coastal birch forests. *Ecol. Appl.* 13:999–1012

Heinz Center. 1999. *Designing a report on the state of the nation's ecosystems: selected measurements for croplands, forests, and coasts and oceans*. Rep., H John Heinz III Center, Washington, DC

Heinz Center. 2002. *The State of the Nation's Ecosystems: Measuring the Lands, Waters, and Living Resources of the United States.* New York: Cambridge Univ. Press

Hilsenhoff WL. 1982. *Using a biotic index to evaluate water quality of streams*. Tech. Bull. No. 132, Dep. Natural Resources, Madison, WI

Huggett RJ, Kimerle RA, Mehrle PM Jr, Bergman HL, eds. 1992. *Biomarkers: Biochemical, Physiological, and Histological Markers of Anthropogenic Stress*. Boca Raton, FL: Lewis

Hutchinson TC, Scott MG. 1988. The response of feather moss, *Pleurozium schreberi*, to 5 years of simulated acid precipitation in the Canadian boreal forest. *Can. J. Bot.* 66:82–88

James FC, McCulloch CE, Wiedenfeld DA. 1996. New approaches to the analysis of population trends in land birds. *Ecology* 77:13–27

Jenkins KD, Sanders BM. 1992. Monitoring with biomarkers: a multi-tiered framework for evaluating the ecological impacts of contaminants. See McKenzie et al. 1992, 2:1279–93

Johnston CA. 2003. Shrub species as indicators of wetland sedimentation. *Wetlands* 23:911–20

Jones J. 2003. Tree swallows (*Tachycineta bicolor*): A new model organism? *Auk* 120:591–99

Kangas P, Autio H, Hallifors G, Luther H, Niemi A, et al. 1982. A general model of the decline of *Fucus vesiculosus* at Tvarminne, south coast of Finland in 1977–1981. *Acta Bot. Fenn.* 118:1–27

Karr JR. 1981. Assessment of biological integrity using fish communities. *Fisheries* 6:21–27

Karr JR. 2000. Health, integrity, and biological assessment: the importance of measuring whole things. In *Ecological Integrity: Integrating Environment, Conservation, and Health*, ed. D Pimentel, L Westra, RF Noss, pp. 209–26. Washington, DC: Island

Karr JR. 2002. Understanding the consequences of human actions: indicators from GNP to IBI. In *Just Ecological Integrity: The Ethics of Maintaining Planetary Life*, ed. P Miller, L Westra, pp. 98–110. Lanham, MD: Rowman & Littlefield

Karr JR, Fausch KD, Angermeier PL, Yant PR, Schlosser IJ. 1986. Assessing biological integrity in running waters: a method and its rationale. *Spec. Publ. 5*, Ill. Nat. Hist. Surv., Champaign

Kerans BL, Karr JR. 1994. A benthic index of biotic integrity (B-IBI) for rivers of the Tennessee Valley. *Ecol. Appl.* 4:768–85

Kerr JT, Ostrovsky M. 2003. From space to species: ecological applications for remote sensing. *Trends Ecol. Evol.* 18:299–305

Kingston JC, Birks HJB, Uutala AJ, Cumming BF, Smol JP. 1992. Assessing trends in fishery resources and lake water aluminum from paleolimnological analyses of siliceous algae. *Can. J. Fish. Aquat. Sci.* 49:116–27

Kinnunen H, Holopainen T, Kärenlampi L. 2003. Sources of error in epiphytic lichen variables mapped as bioindicators: needs to modify the Finnish standard. *Ecol. Indic.* 3:1–11

Klemm DJ, Blocksom KA, Fulk FA, Herlihy AT, Hughes R, et al. 2003. Development and evaluation of a macroinvertebrate biotic integrity index (MBII) for regionally assessing Mid-Atlantic Highlands streams. *Environ. Manage.* 31:656–69

Kolkwitz R, Marsson M. 1908. Ökologie der pflanzlichen Saprobien. *Ber. Dt. Bot. Ges.* 26:505–19

Kratz TK, Magnuson JJ, Bayley P, Benson BJ, Barish CW, et al. 1995. Temporal and spatial variability as neglected ecosystem properties: lessons learned from 12 American ecosystems. In *Evaluating and Monitoring the Health of Large-Scale Ecosystems*, ed. D

Rapport, P Calow, pp. 359–83. New York: Springer-Verlag

Krebs CJ. 1978. *Ecology: The Experimental Analysis of Distribution and Abundance*. New York: Harper & Row. 2nd ed.

Kurtz JC, Jackson LE, Fisher WS. 2001. Strategies for evaluating indicators based on guidelines from the Environmental Protection Agency's Office of Research and Development. *Ecol. Indic.* 1:49–60

Lambeck RJ. 1997. Focal species: a multispecies umbrella for nature conservation. *Conserv. Biol.* 11:849–56

Landres PB, Verner J, Thomas JW. 1988. Ecological uses of vertebrate indicator species: a critique. *Conserv. Biol.* 2:1–13

Landres PB. 1992. Ecological indicators: panacea or liability? See McKenzie et al. 1992, 2:1295–318

Larsen DP, Urquhart NS, Kugler DL. 1995. Regional scale trend monitoring of indicators of trophic condition of lakes. *Water Resour. Bull.* 31:117–40

Lausch A, Herzog F. 2002. Applicability of landscape metrics for the monitoring of landscape change: issues of scale, resolution and interpretability. *Ecol. Indic.* 2:3–15

Lawler JJ, White D, Sifneos JC, Master LL. 2003. Rare species and the use of indicator groups for conservation planning. *Conserv. Biol.* 17:875–82

Lawn PA. 2003. A theoretical foundation to support the Index of Sustainable Economic Welfare (ISEW), Genuine Progress Indicator (GPI), and other related indexes. *Ecol. Econ.* 44:105–18

Lawton JH, Gaston KJ. 2001. Indicator species. In *Encyclopedia of Biodiversity*, ed. S Levin, 3:437–50. San Diego: Academic

Lindenmayer DB, Margules CR, Botkin DB. 2000. Indicators of biodiversity for ecologically sustainable forest management. *Conserv. Biol.* 14:941–50

Lopez RD, Fennessy MS. 2002. Testing the floristic quality assessment index as an indicator of wetland condition. *Ecol. Appl.* 12:487–97

MacNally R, Fleishman E. 2002. Using

'indicator' species to model species richness: model development and predictions. *Ecol. Appl.* 12:79–92

Manley PN, Zielinski WJ, Schlesinger MD, Mori SR. 2004. Evaluation of a multiple-species approach to monitoring species at the ecoregional scale. *Ecol. Appl.* 14:296–310

Mann CC, Plummer ML. 1999. Call for "sustainability" in forests sparks a fire. *Science* 283:1996–98

Marwood CA, Solomon KR, Greenberg BM. 2001. Chlorophyll fluorescence as a bioindicator of effects on growth in aquatic macrophytes from mixtures of polycyclic aromatic hydrocarbons. *Environ. Toxicol. Chem.* 20:890–98

May RM. 1985. Evolution of pesticide resistance. *Nature* 315:12–13

McDonald M, Blair R, Bolgrien D, Brown B, Dlugosz J, et al. 2004. The Environmental Protection Agency's Environmental Monitoring and Assessment Program. In *Environmental Monitoring*, ed. GB Wiersma, pp. 649–68. New York: CRC

McGeoch MA. 1998. The selection, testing and application of terrestrial insects as bioindicators. *Biol. Rev.* 73:181–201

McInnes PF, Naiman RJ, Pastor J, Cohen Y. 1992. Effects of moose browsing on vegetation and litter of the boreal forest, Isle Royale, Michigan, USA. *Ecology* 73:2059–75

McKenzie DH, Hyatt DE, McDonald VJ, eds. 1992. *Ecological Indicators*, Vols. 1, 2. London/New York: Elsevier Applied Science

Milon JW, Shogren JF, eds. 1995. *Integrating Economic and Ecological Indicators*. Westport, CT: Praeger. 214 pp.

Morrison ML. 1986. Bird populations as indicators of environmental change. In *Current Ornithology*, ed. RF Johnston, 3:429–51. New York: Plenum

National Research Council. 1990. *Managing Troubled Waters*. Washington, DC: Natl. Acad.

National Research Council. 1997. *Building a Foundation for Sound Environmental Decisions*. Washington, DC: Natl. Acad.

National Research Council. 2000. *Ecological Indicators for the Nation*. Washington, DC: Natl. Acad.

Nichols JW, Larsen CP, McDonald ME, Niemi GJ, Ankley GT. 1995. Bioenergetics-based model for accumulation of polychlorinated biphenyls by nestling tree swallows, *Tachycineta bicolor*. *Environ. Sci. Technol.* 29:604–12

Niemeijer D. 2002. Developing indicators for environmental policy: data-driven and theory-driven approaches examined by example. *Environ. Sci. Pol.* 5:91–103

Niemi GJ, Hanowski JM, Lima AR, Nicholls T, Weiland N. 1997. A critical analysis on the use of indicator species in management. *J. Wildl. Manage.* 61:1240–52

Niemi GJ, Wardrop DH, Brooks RP, Anderson S, Brady VJ, et al. 2004. Rationale for a new generation of ecological indicators for coastal waters. *Environ. Health Perspect.* In press

Noss RF. 1990. Indicators for monitoring biodiversity: a hierarchical approach. *Conserv. Biol.* 4:355–64

Noss RF. 1999. Assessing and monitoring forest biodiversity: a suggested framework and indicators. *For. Ecol. Manage.* 115:135–46

O'Connell TJ, Jackson LE, Brooks RP. 2000. Bird guilds as indicators of ecological condition in the Central Appalachians. *Ecol. Appl.* 10:1706–21

O'Connor RJ, Walls TE, Hughes RM. 2000. Using multiple taxonomic groups to index the ecological condition of lakes. *Environ. Monit. Assess.* 61:207–28

Odum EP. 1963. *Ecology*. New York: Holt, Rinehart, & Winston

Olsen AR, Sedransk J, Edwards D, Gotway CA, Liggett W, et al. 1999. Statistical issues for monitoring ecological and natural resources in the United States. *Environ. Monit. Assess.* 54:1–45

O'Neill RV, Krummel JR, Gardner RH, Sugihara G, Jackson B, et al. 1988. Indices of landscape pattern. *Landsc. Ecol.* 1:153–62

Ostrom E, Burger J, Field CB, Norgarrd RB, Policansky D. 1999. Revisiting the commons: local lessons, global challenges. *Science* 284:278–82

Paerl HW, Valdes LM, Pinckney JL, Piehler MF, Dyble J, Moisander PH. 2003. Phytoplankton photopigments as indicators of estuarine and coastal eutrophication. *BioScience* 53:953–64

Paine RT. 1969a. A note on trophic complexity and community stability. *Am. Nat.* 103:91–93

Paine RT. 1969b. The *Pisaster-Tegula* interaction: prey patches, predator food preference, and intertidal community structure. *Ecology* 50:950–61

Parsons DJ. 2004. Supporting basic ecological research in US national parks: challenges and opportunities. *Ecol. Appl.* 14:5–13

Paul JF. 2003. Developing and applying an index of environmental integrity for the US Mid-Atlantic region. *J. Environ. Manage.* 67:175–85

Pérez-López M, Alonso J, Nóvia-Valiñas MC, Melgar MJ. 2003. Assessment of heavy metal contamination of seawater and marine limpet, *Patella vulgata* L., from Northwest Spain. *J. Environ. Sci. Health*, Part A: *Tox. Hazard. Subst. Environ. Eng.* 38:2845–56

Pimentel D, Westra L, Noss RF, eds. 2000. *Ecological Integrity: Integrating Environment, Conservation, and Health*. Washington, DC: Island

Rakocinski CF, Brown SS, Gaston GR, Heard RW, Walker WW, Summers JK. 1997. Macrobenthic responses to natural and contaminant-related gradients in northern Gulf of Mexico estuaries. *Ecol. Appl.* 7:1278–98

Rapport DJ. 1992. Evolution of indicators of ecosystem health. See McKenzie et al. 1992, 1:121–34

Rapport D, Costanza R, Epstein PR, Gaudet C, Levins R. 1998. *Ecosystem Health*. Malden, MA: Blackwell Sci. 72 pp.

Ratcliffe DA. 1980. *The Peregrine Falcon*. Calton, UK: Poyser

Reynoldson TB, Norris RH, Resh VH, Day KE, Rosenberg DM. 1997. The reference condition: a comparison of multimetric and multivariate approaches to assess water-quality impairment using benthic macroinvertebrates. *J. North Am. Benthol. Soc.* 16:833–52

Roberge J, Angelstam P. 2004. Usefulness of the umbrella species concept as a conservation tool. *Conserv. Biol.* 18:76–85

Rolstad J, Gjerde I, Gundersen VS, Sætersdal M. 2002. Use of indicator species to assess forest continuity: a critique. *Conserv. Biol.* 16:253–57

Root RB. 1967. The niche exploitation pattern of the blue-gray gnatcatcher. *Ecol. Monogr.* 37:317–50

Rubinoff D. 2001. Evaluating the California gnatcatcher as an umbrella species for conservation of southern California coastal sage scrub. *Conserv. Biol.* 15:1374–83

Rublee PA, Kempton JW, Schaefer EF, Allen C, Harris J, et al. 2001. Use of molecular probes to assess geographic distribution of *Pfiesteria* species. *Environ. Health Perspect.* 109:765–67

Rudi K, Skulberg OM, Skulberg R, Jakobsen KS. 2000. Application of sequence-specific labeled 16S rRNA gene oligonucleotide probes for genetic profiling of cyanobacterial abundance and diversity by array hybridization. *Appl. Environ. Microbiol.* 66:4004–11

Sauer JR, Hines JE, Fallon J. 2003. *The North American Breeding Bird Survey, Results and Analysis 1966–2002. Version 2003.1*, USGS Patuxent Wildl. Res. Cent., Laurel, MD

Savidge JA. 1984. Guam: paradise lost for wildlife. *Biol. Conserv.* 30:305–17

Schiller A, Hunsaker CT, Kane MA, Wolfe AK, Dale VH, et al. 2001. Communicating ecological indicators to decision makers and the public. *Conserv. Ecol.* 5:19

Seegert G. 2001. The development, use, and misuse of biocriteria with an emphasis on index of biotic integrity. *Environ. Sci. Pol.* 3:51–58

Semlitsch RD. 2003. *Amphibian Conservation*. Washington, DC: Smithsonian Inst.

Shannon CE, Weaver W. 1949. *The Mathematical Theory of Communication*. Urbana: Univ. Illinois Press

Sierszen ME, Frost TM. 1990. Effects of lake acidification on zooplankton feeding rates and selectivity. *Can. J. Fish. Aquat. Sci.* 47: 772–79

Simberloff D. 1998. Flagships, umbrellas, and keystones: Is single-species management passé in the landscape era? *Biol. Conserv.* 83:247–57

Simon TP, ed. 2003. *Biological Response Signatures: Indicator Patterns Using Aquatic Communities.* Boca Raton, FL: CRC

Simon TP, Emery EB. 1995. Modification and assessment of an index of biotic integrity to quantify water resource quality in Great Rivers. *Regul. Rivers: Res. Manage.* 11:283–98

Simon TP, Jankowski R, Morris C. 2000. Modification of an index of biotic integrity for assessing vernal ponds and small palustrine wetlands using fish, crayfish, and amphibian assemblages along southern Lake Michigan. *Aquat. Ecosyst. Health Manage.* 3:407–18

Simon TP, Stewart PM, Rothrock PL. 2001. Development of an index of biotic integrity for plant assemblages (P-IBI) in southern Lake Michigan. *Aquat. Ecosys. Health Manage.* 4:293–309

Smeets E, Weterings R. 1999. Environmental indicators: typology and overview. *Tech. Rep. 25*, Eur. Environ. Agency, Copenhagen, Den. http://reports.eea.eu.int:80/TEC25/en/tech_25_text.pdf

Stockwell D, Peterson AT. 2003. Comparison of resolution of methods used in mapping biodiversity patterns from point-occurrence data. *Ecol. Indic.* 3:213–21

Stolte KW, Mangis DR. 1992. Identification and use of plant species as ecological indicators of air pollution stress in national park units. See McKenzie et al. 1992, 1:373–92

Suter GW II. 1993. A critique of ecosystem health concepts and indices. *Environ. Toxicol. Chem.* 12:1533–39

Suter GW II, Norton SB, Cormier SM. 2002. A methodology for inferring the causes of observed impairments in aquatic ecosystems. *Environ. Toxicol. Chem.* 21:1101–11

ten Brink BJE, Hosper SH, Colijn F. 1991. A quantitative method for description and assessment of ecosystems: the AMOEBA approach. *Mar. Pollut. Bull.* 23:265–70

Tilman D. 1999. Diversity by default. *Science* 283:495–96

Turner MG, Hargrove WH, Gardner RH, Romme WH. 1994. Effects of fire on landscape heterogeneity in Yellowstone National Park, Wyoming. *J. Veg. Sci.* 5:731–42

Underwood RJ, Roth RR. 2002. Demographic variables are poor indicators of wood thrush productivity. *Condor* 104:92–102

Urquhart NS, Paulsen SG, Larsen DP. 1998. Monitoring for policy-relevant regional trends over time. *Ecol. Appl.* 8:246–57

US Dep. Inter. 1980. Standards for the development of habitat suitability index models. *Ecol. Serv. Manual No. 103.* Div. Ecol. Serv., US Dep. Inter. Fish Wildl. Serv., Washington, DC

US EPA. 1997. An ecological assessment of the United States Mid-Atlantic region. *EPA/620/R-97/130*, Washington, DC

US EPA. 1998. Condition of the Mid-Atlantic estuaries. *EPA 600-R-98–147*, Washington, DC

US EPA. 2000a. Evaluation guidelines for ecological indicators. *EPA/620/R-99/005*, Research Triangle Park, NC.109 pp. http://www.ecosystemindicators.org/wg/publication/Jackson_Kurtz_Fisher.pdf

US EPA. 2000b. Mid-Atlantic highlands streams assessment. *EPA/903/R-00/015*, Washington, DC. 74 pp. http://www.epa.gov/maia/pdf/MAHAStreams.pdf

US EPA. 2002a. Environmental monitoring and assessment program research strategy. *EPA 620/R-02/002*, Research Triangle Park, NC. 78 pp. http://www.epa.gov/nheerl/emap/files/emap_research_strategy.pdf

US EPA. 2002b. A SAB report: a framework for assessing and reporting on ecological condition. *EPA-SAB-EPEC-02–009*, Washington, DC. 142 pp. http://www.epa.gov/sab/pdf/epec02009.pdf

US EPA. 2002c. Biological indicator variability and stream program integration: a Maryland case study. *EPA/903/R-02/008*, Washington, DC. 92 pp. http://www.epa.gov/bioindicators/pdf/biological_indicator_md_.pdf

US EPA. 2002d. Mid-Atlantic integrated assessment (MAIA) estuaries 1997–1998. *EPA/620/R-02–003*, Washington, DC

US EPA. 2003a. EPA's draft report on the environment—technical document. *EPA/600-R-03–050*, Washington, DC. 453 pp. http://www.epa.gov/indicators

US EPA. 2003b. Developing biological indicators: lessons learned from Mid-Atlantic streams. *EPA/903/R-03/003*, Washington, DC. 52 pp. http://www.epa.gov/bioindicators/pdf/MAIA_lessons_learned_biology.pdf

US EPA. 2003c. Response of surface water chemistry to Clean Air Act amendments of 1990. *EPA/620/R-03/001*, Washington, DC. 74 pp.

US EPA. 2004. Draft national coastal condition report II. *EPA 620/R-03/002*, Washington, DC. 362 pp. http://www.epa.gov/owow/oceans/nccr2/downloads.html

Van Biema D. 1995. Prophet of poison. *Time* 145:26–33

Van Horne B. 1983. Density as a misleading indicator of habitat quality. *J. Wildl. Manage.* 47:893–901

Venier LA, Pearce JE, McKee JE, McKenney DW, Niemi GJ. 2004. Climate and satellite-derived land cover for predicting breeding bird distribution in the Great Lakes basin. *J. Biogeogr.* 31:315–31

Verner J. 1984. The guild concept applied to management of bird populations. *Environ. Manage.* 8:1–14

Whittier TR, Paulsen SG, Larsen DP, Peterson SA, Herlihy AT, Kaufmann PR. 2002. Indicators of ecological stress and their extent in the population of northeastern lakes: a regional assessment. *BioScience* 52:235–47

Wolter PT, White MA. 2002. Recent forest cover type transitions and landscape structural changes in northeast Minnesota. *Landsc. Ecol.* 17:133–55

World Economic Forum. 2002. *2002 Environmental sustainability index.* Annu. Meet., Global Leaders for Tomorrow Task Force, World Econ. Forum, Yale Univ. Cent. Environ. Law Policy, and Columbia Univ. Cent. Int. Earth Sci. Inf. Netw., New York http://www.ciesin.org/indicators/ESI/downloads.html

Yoccoz NG, Nichols JD, Boulinier T. 2001. Monitoring of biological diversity in space and time. *Trends Ecol. Evol.* 16:446–53

ECOLOGICAL IMPACTS OF DEER OVERABUNDANCE

Steeve D. Côté,[1] Thomas P. Rooney,[2] Jean-Pierre Tremblay,[1] Christian Dussault,[1] and Donald M. Waller[2]

[1]Chaire de Recherche Industrielle CRSNG-Produits forestiers Anticosti, Département de Biologie and Centre d'études nordiques, Université Laval, Québec G1K 7P4, Canada; email: steeve.cote@bio.ulaval.ca, jean-pierre.tremblay@bio.ulaval.ca, christian.dussault@fapaq.gouv.qc.ca
[2]Department of Botany, University of Wisconsin, Madison, Wisconsin 53706; email: tprooney@facstaff.wisc.edu, dmwaller@wisc.edu

Key Words browsing, Cervidae, forest regeneration, herbivory, plant-herbivore interactions

■ **Abstract** Deer have expanded their range and increased dramatically in abundance worldwide in recent decades. They inflict major economic losses in forestry, agriculture, and transportation and contribute to the transmission of several animal and human diseases. Their impact on natural ecosystems is also dramatic but less quantified. By foraging selectively, deer affect the growth and survival of many herb, shrub, and tree species, modifying patterns of relative abundance and vegetation dynamics. Cascading effects on other species extend to insects, birds, and other mammals. In forests, sustained overbrowsing reduces plant cover and diversity, alters nutrient and carbon cycling, and redirects succession to shift future overstory composition. Many of these simplified alternative states appear to be stable and difficult to reverse. Given the influence of deer on other organisms and natural processes, ecologists should actively participate in efforts to understand, monitor, and reduce the impact of deer on ecosystems.

INTRODUCTION

Deer have excited the interest of ecologists since the birth of our discipline. Interest in managing game populations fostered the development of ecology, particularly the emergence of wildlife ecology (Leopold 1933). Deer management began with understanding which habitat conditions were most favorable for deer. Later, ecologists became interested in the effects of predators and hunters on deer and in the effects of deer on plant populations and habitat conditions. Ironically, within a century, deer management has reversed course from a preoccupation with augmenting population growth through habitat protection, hunting regulations, and predator control to serious concerns about how best to limit deer densities and the consequent impacts of these animals on other ecosystem constituents and functions (Garrott et al. 1993).

Overabundance is a value judgment that has a clear meaning only when placed in a specific context (McShea et al. 1997b). Caughley (1981) proposed a series of definitions to summarize the ecological and nonecological values upon which overabundance diagnostics have been based: Animals are overabundant when they (*a*) threaten human life or livelihood, (*b*) are too numerous for their "own good," (*c*) depress the densities of economically or aesthetically important species, or (*d*) cause ecosystem dysfunction. Here, we follow this sequence and explore some of the human-deer conflicts implicit in points (*a*) and (*c*). We then emphasize point (*d*) throughout the review and show that negative effects of abundant deer occur at various densities in different habitats. The density-dependent effects on life-history traits implicit in point (*b*) are not addressed here, but see McCullough (1979, 1999) for more information.

We review some historic studies of the impact of overabundant deer and summarize how shifts in habitat conditions and levels of predation have boosted deer population growth in many temperate ecosystems. We explore how overabundant deer affect human health, forestry, and agriculture and describe the various methods used to evaluate how deer affect tree seedlings, shrubs, and herbaceous plants. We consider how deer alter interactions among competing plants; patterns of forest regeneration; succession; populations of insects, birds, and other mammals; ecosystem processes; and overall community structure. The number and significance of these effects make clear that deer can tip forest ecosystems toward alternative states by acting as "ecosystem engineers" or "keystone herbivores," greatly affecting the structure and functioning of temperate and boreal forests (McShea & Rappole 1992, Stromayer & Warren 1997, Waller & Alverson 1997). These profound impacts lead us to ponder how ecology might inform approaches to mitigating the effects of overabundant deer. We discuss how ecological research might be extended and linked more tightly to deer management. Because space and our expertise are limited, we focus our attention on interactions between deer (family Cervidae) and temperate/boreal forests, primarily in Europe and North America.

HISTORICAL INTEREST IN DEER IMPACTS ON PLANT COMMUNITIES AND ECOSYSTEM STRUCTURE

By the nineteenth century, natural historians recognized that overabundant deer could exclude certain plants from the landscape (Watson 1983). Systematic studies of deer overabundance, however, did not occur until after the emergence of wildlife ecology, developed by Aldo Leopold. Based on his experiences with the dangers of deer overabundance, Leopold was the first to discuss threats posed by growing deer herds (Leopold 1933, Leopold et al. 1947). Leopold's warnings sparked an initial period of concern in the 1940s and 1950s, mainly in the midwestern United States, which prompted the construction of exclosures to demonstrate the influence of native deer on forest regeneration (Beals et al. 1960, Pimlott 1963, Stoeckler et al. 1957, Webb et al. 1956). Interest in deer impacts expanded in the 1970s, primarily in the Midwest and the Allegheny region of Pennsylvania (Anderson & Loucks 1979, Behrend et al. 1970, Harlow & Downing 1970), but with added attention to the introduced Sitka black-tailed deer (*Odocoileus hemionus sitkensis*) in the Queen Charlotte Islands of Canada (Pojar et al. 1980). Concerns about the impact of native deer populations in Europe (Dzieciolowski 1980) and introduced deer in New Zealand (Caughley 1983, Stewart & Burrows 1989) developed at the same time.

Seminal experiments on the population dynamics of white-tailed deer (*Odocoileus virginianus*) on the George Reserve in Michigan were conducted in the 1970s (McCullough 1979). The introduction of deer into a fenced area demonstrated that, because deer have such a high potential rate of increase, they can easily overwhelm the carrying capacity of their environment and consequently have strong and persistent negative impacts on vegetation (McCullough 1979, 1997).

In North America, the study of deer impacts soon broadened to include birds (Casey & Hein 1983), interactions with weeds (Horsley & Marquis 1983), and long-term effects on forest composition (Frelich & Lorimer 1985) and sapling-bank diversity (Whitney 1984). By the late 1980s and early 1990s, the impacts resulting from high densities of deer were being tallied in review articles (Alverson et al. 1988; Gill 1992a,b; McShea & Rappole 1992; Miller et al. 1992). Broad considerations of deer impacts also emerged in the 1994 conference hosted by the Smithsonian Institution (McShea et al. 1997b) and a 1997 special topics issue of the *Wildlife Society Bulletin* (Vol. 25, No. 2). Similar recent review issues of *Forestry* (2001, Vol. 74, No. 3) and *Forest Ecology and Management* (2003, Vol. 181, No. 2–3) focused mostly on how deer affect European forests.

CAUSES OF DEER OVERABUNDANCE

Overexploitation in the second half of the nineteenth century led to major declines in deer numbers and range. Subsequent protection of deer via restricted seasons and game laws then led to rapid population increases across Europe and North America over the past 75 to 150 years (Fuller & Gill 2001, Jedrzejewska et al. 1997, Leopold et al. 1947, McShea et al. 1997b, Mysterud et al. 2000). In Virginia, white-tailed deer increased from an estimated 25,000 animals in 1931 to 900,000 animals by the early 1990s (Knox 1997). Although whether North American deer are currently more abundant than before European colonization is not known, the evidence suggests that current deer numbers are unprecedented (McCabe & McCabe 1997).

Deer populations in North America have grown rapidly since the 1960s to 1970s in response to changes in their environment and reduction of hunting pressure (McShea et al. 1997b). The number of moose (*Alces alces*) in Scandinavia has similarly increased three to five times since the 1970s (Skolving 1985, Solberg et al. 1999). Deer densities above 10/km^2 are now common throughout temperate zones (Fuller & Gill 2001, Russell et al. 2001). In North America, deer have been reintroduced in many states (McShea et al. 1997b) and introduced to islands free of predators (e.g., Anticosti, PQ, Canada) (Côté et al. 2004). These introductions contributed to the recovery and subsequent overabundance of deer populations (Knox 1997).

The most obvious factor contributing to the rapid growth of deer populations is increased forage. Widespread agricultural and silvicultural activities considerably improved deer habitat throughout the twentieth century (Alverson et al. 1988, Fuller & Gill 2001, Porter & Underwood 1999). Tree planting after logging and early successional forested landscapes provide abundant, high-quality food that increases deer habitat carrying capacity (Bobek et al. 1984, Fuller & Gill 2001, Sinclair 1997). Forest harvesting and the resulting interspersion of habitats provide good cover and abundant forage for deer (Diefenbach et al. 1997). Many openings are also intentionally managed to boost forage quality and population growth (Waller & Alverson 1997).

Reductions in hunting and natural predators across Europe and North America have also contributed to increasing deer populations. Since the 1920s, strict hunting regulations in North America have favored deer population increases, especially on some private lands and in parks where hunting was banned (Brown et al. 2000, Diefenbach et al. 1997, Porter & Underwood 1999). Even where hunting is allowed, game laws favor the killing of males, increasing female survival and, thus, population growth (Ozoga & Verme 1986, Solberg et al. 1999). In recent decades, the pressure has increased to reform game laws to allow hunting of more does and fawns in response to overabundant herds. Hunters, however, have been reluctant to embrace such reforms (Riley et al. 2003). The number of deer hunters has also stabilized or decreased with declines in the social acceptability of hunting (Brown et al. 2000, Enck et al. 2000, Riley et al. 2003). At the same time, land owners and municipalities increasingly prohibit hunting in response to safety concerns (Kilpatrick et al. 2002), which further diminishes hunting pressure (Brown et al. 2000).

By the middle of the twentieth century, wolves (*Canis lupus*) had disappeared from continental Europe and most areas south of the North American boreal forests (Boitani 1995, Paquet & Carbyn 2003). Mountain lions (*Puma concolor*) were also extirpated in eastern North America (McCullough 1997). Without predators, ungulate populations increase rapidly to (or beyond) the carrying capacity of available forage (McCullough 1997, Messier 1994, Potvin et al. 2003, Sæther et al. 1996). Their high intrinsic rate of population increase may also allow deer to escape predator control while making overshoot of habitat carrying capacity and fluctuations in population size more likely. Moderate climates as experienced recently may also contribute to deer overabundance (Forchhammer et al. 1998, Solberg et al. 1999). Mild winters increase deer body mass (Mysterud et al. 2001) and winter survival (Loison et al. 1999), which favor population growth.

SOCIAL AND ECONOMIC CONSEQUENCES OF DEER OVERABUNDANCE

Impacts on Human Activities

Deer generate both positive and negative economic values, and negative values increase as deer become overabundant (Conover 1997). Browsing of tree seedlings by deer reduces economic value, ecological stability, and species diversity of forests, in addition to reducing tree growth, which, in turn, diminishes protection from erosion and floods (Reimoser 2003). The total cost of deer damage to the forest industry is difficult to estimate. The loss of young trees, for example, results in long-term economic losses only if the composition and quality of the final stand are affected. Despite the apparent severity of deer damage to agriculture and forestry in Britain, the economic significance is considered negligible or small in many cases (Putman 1986, Putman & Moore 1998). In contrast, deer damage is considered a major problem in the United States and in Austria, where their annual impacts are estimated at more than $750 million (Conover 1997) and more than €220 million (Reimoser 2003), respectively. In northern temperate forests, saplings 30 to 60 cm tall are most vulnerable to browsing (Andren & Angelstam 1993, Gill 1992a, Kay 1993, Welch et al. 1991). Browsing by deer can kill seedlings or reduce height growth, which results in lower-density stands and requires longer stand rotations (Kullberg & Bergström 2001). Stands subjected to heavy browsing of seedlings and saplings exhibit a size structure biased toward medium and large stems (Anderson & Loucks 1979, Potvin et al. 2003, Stromayer & Warren 1997, Tilghman 1989). When the terminal bud is browsed, the tree develops multiple leaders (Putman & Moore 1998), which decreases its commercial value. Lavsund (1987) indicated that the proportion of quality stems dropped from 63% to 18% in a stand subjected to heavy browsing by moose in Sweden. Bark stripping may kill trees but often decreases quality by girdling, growth reduction, and increased risk of fungal infections (Gill 1992b, Putman & Moore 1998).

Reimoser (2003) suggested that the severity of damage to trees depends more on forest attractiveness to deer than on deer abundance. Stands become more susceptible to deer damage with (*a*) a low density of alternate food plants (Gill 1992a, Partl et al. 2002, Welch et al. 1991), (*b*) a low density of seedlings (Andren & Angelstam 1993, Lyly & Saksa 1992, Reimoser & Gossow 1996), (*c*) abundant nitrogen in the foliage or soil (Gill 1992a), (*d*) hiding cover (Gill 1992a, Kay 1993, Partl et al. 2002), and (*e*) the presence of edges (Kay 1993, Lavsund 1987, Reimoser & Gossow 1996). On larger scales, deer impacts on vegetation are greater in fragmented landscapes (Hornberg 2001, Reimoser 2003) or low-productivity habitats (Danell et al. 1991).

White-tailed deer damage many agricultural crops in the United States (Conover 2001). In 1996, 14% of nursery owners in the northeastern United States reported damages exceeding $10,000 (Lemieux et al. 2000). Deer damage to corn fields in the United States was estimated at 0.23% of the total production ($26 million) in

1993 (Wywialowski 1996). Abundant deer also damage gardens and ornamentals (McCullough et al. 1997, West & Parkhurst 2002). Deer damage to households and agriculture in the United States totaled $351 million in 1991 (Conover 1997).

A primary cost to society of deer overabundance is increased vehicle accident rates, now a serious problem in Europe, the United States, and Japan. Deer-vehicle collisions increase as deer density and traffic volume increase (Groot Bruinderink & Hazebroek 1996). Groot Bruinderink & Hazebroek (1996) estimated that 507,000 collisions between vehicles and ungulates occur annually in Europe (excluding Russia) and result in 300 deaths, 30,000 injuries, and $1 billion in material damage. In the United States, such accidents increased from 200,000 in 1980 to 500,000 in 1991 (Romin & Bissonette 1996) and cost more than $1 billion in 1991 (Conover 1997). Many airports in Canada and the United States also experience deer-aircraft problems (Bashore & Bellis 1982, Fagerstone & Clay 1997).

Transmission of Wildlife Diseases and Zoonoses

In general, high population densities of deer favor the transmission of infectious agents (Davidson & Doster 1997). Increased deer densities appear to increase the transmission of tick-borne zoonoses directly by increasing tick (*Ixodes* spp.) abundance (Ostfeld et al. 1996, Wilson & Childs 1997). In North America, two tick-borne diseases threaten human health: Lyme disease and ehrlichiosis (<5% mortality in humans) (Telford III 2002). Lyme disease has quickly become the most common vector-borne disease in the United States (13,000 cases in 1994; Conover 1997) and is also found in Europe and Asia (Steere 1994). The incidence of Lyme disease appears to track deer density in the eastern United States (Telford III 2002; Wilson et al. 1988, 1990).

Deer transmit infectious agents directly to other deer, to livestock, and to humans, especially if deer density is high. Bovine tuberculosis (*Mycobacterium bovis*) causes mortality in deer, livestock, other wildlife species, and humans (Schmitt et al. 1997). *M. bovis* affects deer populations of New Zealand and Europe to various degrees (Clifton-Hadley & Wilesmith 1991). It has been rare in North America, but incidence could increase as deer densities increase (Schmitt et al. 1997). A recent outbreak in Michigan led to concern that it would spread to domestic cattle and to a ban on deer feeding (Miller et al. 2003).

Chronic wasting disease (CWD) is a transmissible spongiform encephalopathy similar to "mad cow" disease (Williams et al. 2002). The disease was first noticed in 1967 in mule deer (*Odocoileus hemionus*) and has now spread to elk (*Cervus elaphus*), white-tailed deer, and black-tailed deer across a broad region (Figure 1) (Williams et al. 2002). The pattern of spread suggests that the disease may be transmitted from farm-raised herds (25 identified with CWD by 2002) to wild animals (Williams et al. 2002). Although it can be transmitted within and among cervid species (Gross & Miller 2001), transmission to humans or noncervid species appears unlikely (Raymond et al. 2000). Because it develops slowly, it would not appear to limit population growth greatly, but some experts express concern that it could cause population extinctions (Williams et al. 2002). Concerns over potential

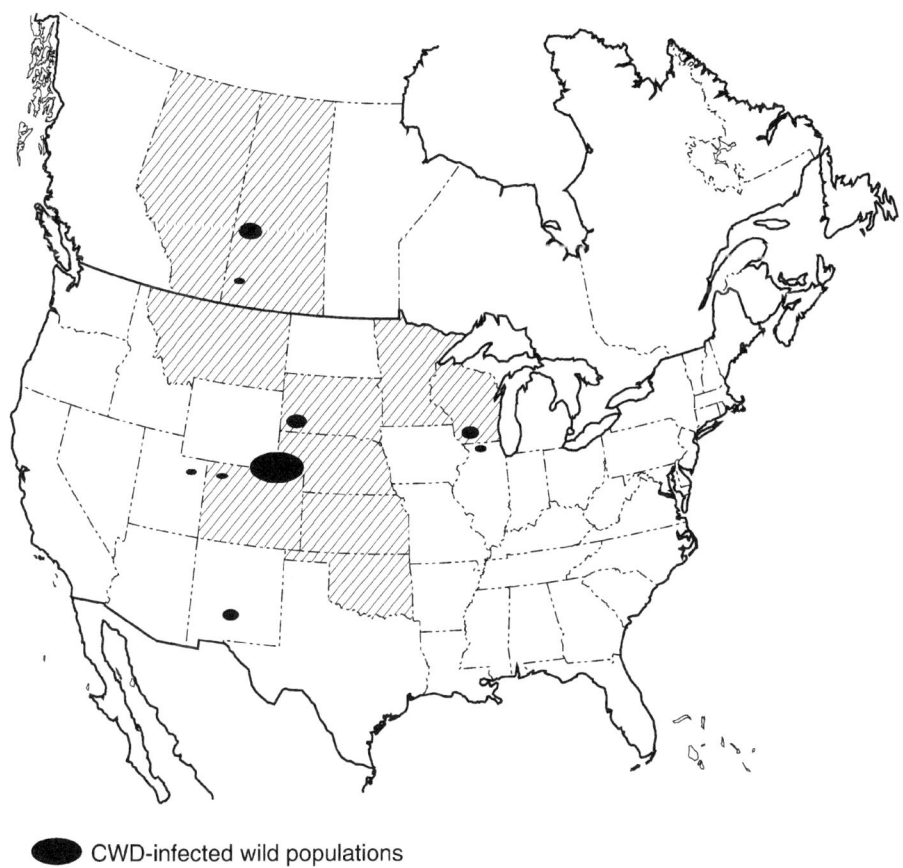

● CWD-infected wild populations
▨ States and Provinces with CWD in captive animal populations

Figure 1 Map showing states and provinces where chronic wasting disease (CWD) has been found in wild deer or elk populations or in captive herds across North America. Note the association between captive animals with CWD and escape into the wild.

human health risks from CWD could also substantially reduce hunter efforts, which already appear too low to control deer populations effectively (see the Management Issues section).

ASSESSING ECOLOGICAL EFFECTS OF DEER OVERABUNDANCE

Through most of the twentieth century, research focused on how deer affected particular species of interest (often trees) or specific areas of concern. Because site-specific management concerns drove research programs, pseudoreplication was a common feature of early research (Hurlbert 1984). There has been a gradual shift toward understanding overabundance within a stronger scientific framework. Despite this shift, the most common approaches for assessing deer impacts have not changed. Following Diamond (1983), we distinguish among natural, field, and laboratory experimental approaches.

In natural experiments, researchers select sites and collect data where spatial variation in deer abundance can be exploited. Spatial variation in deer densities allows the creation of discrete or continuous independent variables. Discrete variation arises in island-mainland systems. Deer may be absent on some islands but overabundant on others; both states offer a contrast to populations on the mainland (Balgooyen & Waller 1995, Beals et al. 1960, Côté et al. 2004, Vourc'h et al. 2001). Discrete variation may also appear in mainland systems if management varies starkly across ownership boundaries. Hunting bans on private lands and, particularly, on public lands can cause population densities to exceed those in the surrounding landscape (Nixon et al. 1991, Porter & Underwood 1999). The presence of ungulate predators can have the opposite effect; that is, reducing deer densities and impacts (Ripple & Beschta 2003, White et al. 2003). Within habitats, cliffs, boulder tops, and other physical features of the environment can create ungulate-free refuges for plants (Long et al. 1998, Rooney 1997). Such variation creates opportunities to study deer impacts by using discrete variation. Deer abundance also varies across landscapes in response to predation pressure (Lewis & Murray 1993, Martin & Baltzinger 2002) and habitat quality (Alverson et al. 1988, Reimoser & Gossow 1996), and this variation can be used to analyze ecosystem responses across gradients in deer density (Alverson & Waller 1997; Didier & Porter 2003; Rooney et al. 2000, 2002; Takada et al. 2001; Waller et al. 1996). The drawbacks of this approach are the difficulty in establishing replicates and the problem of confounding site factors (such as productivity) that themselves affect deer densities or responses to herbivory (Bergström & Edenius 2003).

The effects of overabundant deer on plants can also be studied across time. Vila et al. (2001, 2003), for example, tied browsing scars and historical variation in growth rates to fluctuating deer densities on the Queen Charlotte Islands, Canada. Before-and-after or snapshot-type studies have also been used to infer how species respond to fluctuating browsing pressure when baseline data exist (Husheer et al. 2003, Rooney & Dress 1997, Sage et al. 2003, Whitney 1984). Many such studies reflect conspicuous "signatures" of deer browsing as community composition shifts toward browse-tolerant or unpalatable species (Husheer et al. 2003). Long-term monitoring can, thus, provide powerful insights into how deer drive changes in plant communities, particularly when combined with exclosures or direct observations of which plants deer preferentially consume.

In field experiments, researchers manipulate deer densities or vegetation to study deer impacts. The use of fencing (exclosures) to exclude deer from study plots is a venerable experimental approach (Daubenmire 1940). Despite all the insights that exclosure studies bring to our understanding of deer-forest interactions, they are limited to binary treatments: They allow researchers to infer what alternate trajectory a site would take in the absence of deer. Controlled grazing experiments that utilize known deer density in enclosures appear more realistic and can be used to infer whole-community responses to manipulated deer densities (Côté et al. 2004, deCalesta 1994, Hester et al. 2000, Horsley et al. 2003, McShea & Rappole 2000, Tilghman 1989). Deer densities can also be manipulated through culling. Researchers can take advantage of culling efforts in parks and natural areas by monitoring vegetation or other response variables (Cooke & Farrell 2001). Direct manipulations of density through localized management can also be conducted under scientific objectives (Côté et al. 2004). Alternatively, vegetation can be subjected to experimental treatment. Simulated browsing treatments reveal how plants respond to defoliation in natural environments (Bergström & Danell 1995, Rooney & Waller 2001). Experimental plantings in conjunction with exclosures more accurately compare the effects of deer browsing on plant growth and mortality (Alverson & Waller 1997, Fletcher et al. 2001b, Ruhren & Handel 2003).

Laboratory experiments give researchers a high degree of control over experimental systems. Defoliation experiments can be conducted under a range of controlled environmental conditions in greenhouses or growth chambers to investigate the mechanisms of plant responses (Canham et al. 1994). Simulation models also allow researchers to forecast how deer might affect ecosystems under a broad range of deer-population and forest-management scenarios (Tremblay et al. 2004).

Each of these approaches has its strengths and weaknesses. Stronger inferences can be drawn when they are combined. Waller & Alverson (1997), for example, combined experimental plantings, exclosures, and geographic variation in deer densities to examine the effects of deer browsing on *Tsuga canadensis* growth and survival rates across a broad region. Augustine et al. (1998) combined exclosures, geographic variation in deer densities, and a simple plant-herbivore functional response model to predict time-to-extinction of forest herb populations as a function of initial abundance. Balgooyen & Waller (1995) and Martin & Balzinger (2002) compared plant responses across islands that varied in deer abundance because of hunting and introductions, both currently and historically. Meta-analysis can similarly strengthen our inferences. Gill & Beardall (2001) combined data from 13 studies to examine the effects of ungulate browsing on richness and diversity of tree species in British woodlands.

ECOLOGICAL CONSEQUENCES OF DEER OVERABUNDANCE

Plant Tolerance and Resistance to Herbivory

Deer directly affect the growth, reproduction, and survival of plants by consuming leaves, stems, flowers, and fruits. Plants defend themselves against herbivores in various ways that affect which plants are attacked, how they respond to those attacks, how herbivore individuals and populations respond to those defenses, and, ultimately, how herbivores affect ecosystem productivity and rates of nutrient cycling. Plants are often classified according to the degree to which they either resist herbivory or tolerate it. Resistant plants have traits that reduce plant selection (such as chemical defenses or low digestible content) or traits that reduce intake rates (such as leaf toughness or morphological defenses). Tolerant species can endure some defoliation with little change in growth, survival, or reproduction, whereas intolerant species are more sensitive to defoliation. In addition, woody plants often reduce their chemical and physical defenses as they grow beyond the range of mammal browsing (Bryant & Raffa 1995).

In environments with herbivores, natural selection should favor enhanced morphological and chemical defenses in plants with low tolerance. Takada et al. (2001) examined populations of the shrub *Damnacanthus indicus* (Rubiaceae) in areas with and without deer. Individual plants in areas with deer increased allocation to thorns: Both spine thickness and density were greater where deer were present. Induced and constitutive chemical defenses can make plants less palatable to deer. Red deer (*Cervus elaphus*) tend to avoid *Picea sitchensis* saplings that have higher concentrations of monoterpenes in their foliage (Duncan et al. 2001). Vourc'h et al. (2001) demonstrated that *Thuja plicata* saplings growing on islands without deer had evolved lower concentrations of foliar monoterpenes than mainland saplings growing in areas with deer. The rapid evolution of reduced defenses in cases like these strongly implies that anti-herbivore defenses are costly in terms of energy (or fitness) in situations where herbivores are scarce or absent. In environments without herbivores, undefended plants outperform defended plants (Gomez & Zamora 2002). However, selection will rarely occur quickly enough to rescue palatable populations faced with sustained overabundant deer, especially in trees where reproducing individuals are not subjected to browsing.

Tolerance to herbivory differs among species and among individuals within species. It depends on the timing and intensity of herbivory (Doak 1992, Saunders & Puettmann 1999), individual plant genotype (Hochwender et al. 2000), specific growth strategies (Canham et al. 1994, Danell et al. 1994), history of past defoliation or other stress (Cronin & Hay 1996, Gill 1992b), the density of competitors, and the degree to which the plant is under nutrient or moisture stress (Canham et al. 1994, Maschinski & Whitham 1989). Plants that lose only a small fraction of their leaves or flowers, store resources underground, hide their meristems (as in grasses), or regrow quickly via indeterminate growth tolerate deer herbivory better (Augustine & McNaughton 1998). Such species include many annuals, graminoids, deciduous trees, and shrubs and many herbs and forbs that mature in late summer. Some of the browse-tolerant species even appear to gain more biomass (or more flowers and seeds) over the course of a season than undefoliated control plants (Hobbs 1996, McNaughton 1979, Paige & Whitham 1987). Increases in final biomass yield could reflect shifts in either allocation and growth form, increased photosynthetic rates, or both. Browsing alters plant growth forms when a single terminal leader is removed, apical dominance is broken, and axillary buds give rise to a profusion of branches. Photosynthetic rates rise when changes in the water balance of residual leaves lead to an increase in stomatal conductance and foliar concentrations of carboxylating enzymes (McNaughton 1983). Although such overcompensation might be temporary, plants such as graminoids no doubt thrive under repeated grazing. Other plants can compensate at low to moderate levels of defoliation but decline once herbivore densities are high (Bergelson & Crawley 1992). Plants may also reallocate resources to grow taller or shorter when browsed (Bergström & Danell 1995, Canham et al. 1994, Edenius et al. 1993, Saunders & Puettmann 1999) Compensatory growth, however, can limit radial growth and rarely appears under repeated and heavy browsing pressure. Trees with a history of browsing also appear more susceptible to new browsing, reflecting reduced reserves, changes in tree morphology, or both (Bergqvist et al. 2003, Danell et al. 1994, Palmer & Truscott 2003, Welch et al. 1992). Deer, however, often avoid previously browsed twigs, perhaps because of induced defenses (Duncan et al. 1998).

In general, slow-growing plants will tolerate browsing less, particularly if such browsing is repeated. Shady forest understory plants, including shade-tolerant shrubs and tree seedlings, may thus be particularly vulnerable to deer browsing. Small spring ephemeral and early summer forest herbs that lose all their leaves or flowers in a single bite and cannot regrow also tolerate herbivory poorly (Augustine & McNaughton 1998, Augustine & DeCalesta 2003). Browse-intolerant species such as *Trillium* regularly suffer low or negative growth after defoliation (Rooney & Waller 2001).

Browsing directly affects reproduction in many plants, particularly if deer preferentially forage on reproductive plants or consume flowers (Augustine & Frelich 1998). Individuals of some species may not flower again for several seasons after defoliation (Whigham 1990). Where deer are abundant, browse-intolerant herbs tend to be smaller, less likely to flower, and less likely to survive relative to plants in exclosures (Anderson 1994; Augustine & Frelich 1998; Fletcher et al. 2001a; Ruhren & Handel 2000, 2003). Over time, the density of such intolerant plants tends to decline, and populations may be extirpated (Rooney & Dress 1997). Palatable herbs and shrubs such as *Taxus canadensis* remain susceptible to deer browsing throughout their lives and usually become more vulnerable to browsing as they grow larger. Deer forage selectively on the larger *Trillium grandiflorum* plants (Anderson 1994, Knight 2003). This foraging does not kill these plants because they have large, below-ground storage organs. However, defoliation often takes tall flowering stems and may cause the plants to regress in size (Knight 2003, Rooney & Waller 2001). Thus, populations subjected to abundant deer become both scarcer and dominated by small, often nonreproductive plants (Anderson 1994, Knight 2003).

Trees are obviously most vulnerable to herbivory as seeds (e.g., *Quercus* acorns), seedlings, or small saplings (Potvin et al. 2003). *Tsuga canadensis* seedlings and saplings have become scarce across much of their range in the upper Midwest in apparent response to deer browsing (Alverson & Waller 1997, Anderson & Katz 1993, Frelich & Lorimer 1985, Rooney et al. 2000, Waller et al. 1996). *Thuja occidentalis* is also disappearing from most sites in this region because deer have eliminated nearly every sapling taller than 30 cm (Rooney et al. 2002). Persistent mature trees could repopulate sites with new seedlings and saplings if browsing declined for some window of time, but this window may be as long as 70 years for slow-growing understory species such as *Tsuga* (Anderson & Katz 1993). Evergreen conifers may be particularly intolerant of browsing because they invest heavily in leaves, retain them, and do not retranslocate nutrients to stems and roots as much as deciduous species do (Ammer 1996). In addition, deer focus their browsing on evergreens in winter as other food becomes scarce.

Effects on Plant Community Structure and Interspecific Competition

Because deer forage selectively, they strongly affect competitive relationships among plant species. These shifts, in turn, may either increase or decrease overall cover and diversity. The result depends on whether or not deer primarily consume dominant species. Selective foraging on tall dominant plants in an alpine meadow favored short-statured plants, which caused species richness to increase (Schütz et al. 2003). On Isle Royale, Risenhoover & Maass (1987) attributed the higher diversity of woody vegetation in moose-browsed areas to increased light in the understory. Deer play a similar keystone role on other Lake Superior islands, where they can either enhance herbaceous plant cover and diversity (by removing *Taxus canadensis* cover) or reduce this cover and diversity as they become overabundant (Judziewicz & Koch 1993). Declines in plant cover and species richness usually occur once resistant or browse-tolerant species become dominant. Overabundant deer also commonly cause tree diversity to decline (Gill & Beardall 2001, Horsley et al. 2003, Kuiters & Slim 2002). We summarize contemporary browse-related compositional shifts in boreal and temperate forests in Table 1.

TABLE 1 Compositional shifts in dominant tree species induced by deer browsing in boreal and temperate forests

Former dominant	New dominant	Source
Balsam fir (*Abies balsamea*)	White spruce (*Picea glauca*)	Brandner et al. 1990, McInnes et al. 1992, Potvin et al. 2003
Birch (*Betula* spp.)	Norway spruce (*Picea abies*)	Engelmark et al. 1998
Eastern hemlock (*Tsuga canadensis*)	Sugar maple (*Acer saccharum*)	Alverson & Waller 1997, Anderson & Loucks 1979, Frelich & Lorimer 1985, Rooney et al. 2000
Mixed hardwoods	Black cherry (*Prunus serotina*)	Horsley et al. 2003, Tilghman 1989
Oak (*Quercus* spp.)	Savanna type system	Healy et al. 1997
Scots pine (*Pinus sylvestris*)	Hardwoods and Norway spruce	Gill 1992b

Although deer browsing can enhance ground cover and diversity, research in Pennsylvania demonstrates that indirect effects of browsing can also act against tree seedlings and herb cover. In that study, openings were often invaded by the thorny shrub *Rubus allegheniensis*, which promotes the establishment of tree seedlings (Horsley & Marquis 1983). However, deer prefer this species and, thus, reduce its abundance. This circumstance favors a competitor, the hay-scented fern *Dennstaedia punctilobula*, which deer avoid. As this species becomes more abundant, it inhibits the establishment of tree seedlings (George & Bazzaz 1999, Horsley & Marquis 1983) and excludes smaller-stature herbs (Rooney & Dress 1997). Once *Dennstaedia* is established, cessation of browsing rarely results in a recovery by *Rubus* or other species. Thus, browsing by deer shifts the forest understory to an alternate stable state that is resistant to invasion by originally dominant species (Stromayer & Warren 1997).

The extent to which deer deplete a plant population often depends on plant as well as deer abundance. Augustine et al. (1998) documented that deer have a Holling type II functional response to variable densities of the herb *Laportea canadensis*. This response results in alternative states: only moderate impacts of deer when *Laportea* is common at a site but extirpation when *Laportea* is rare. Thus, we should not assume that deer impacts are simply proportional to deer density across sites and should expect extirpations to accelerate once plant populations grow sparse. These effects likely accentuate the complex deer-plant dynamics we describe below (see Dynamics and Reversibility of Deer Impacts).

Effects on Forest Succession

Contemporary models of succession include multiple directional pathways and alternative stable states that are dependent on the local abundance and colonization potential of species, competitive interactions, and disturbance regimes (Connell & Slatyer 1977, Glenn-Lewin & van der Maarel 1992). Sustained selective browsing can sway these factors enough to affect forest succession dramatically (Engelmark et al. 1998, Frelich & Lorimer 1985, Hobbs 1996, Huntly 1991). Succession accelerates if deer break up the vegetation matrix enough to favor the establishment of later successional plants (Crawley 1997, Hobbs 1996) or if deer prefer species from early seral stages (Seagle & Liang 2001). Alternatively, succession may be stalled if browsing reduces colonization, growth, or survival in later successional species (Hobbs 1996, Ritchie et al. 1998).

Effects on Ecosystem Properties

By affecting competitive interactions among plants with varying levels of chemical defenses and by altering successional trajectories, deer alter ecosystem processes that include energy transfer, soil development, and nutrient and water cycles (Hobbs 1996, Paine 2000). When deer consume an amount of biomass that is small relative to the standing crop, as it is in grassland systems, effects on net primary productivity may be negligible or positive (Hobbs 1996). Thus, in open and productive grassland systems, grazing can increase primary production if grazing induces overcompensation in individual plants, favors more productive species, and accelerates soil processes (McNaughton 1979, 1983; Ritchie et al. 1998). Browsers accelerate nitrogen and carbon cycling if they increase the quantity and the quality of litter returned to the soil (Wardle et al. 2002). This phenomenon is more prevalent in nutrient-rich systems (Bardgett & Wardle 2003) or when deer browsing shifts the canopy composition from conifers to deciduous hardwoods (Frelich & Lorimer 1985). Browsing in early successional communities can also facilitate successional transitions toward nitrogen-fixing species such as *Alnus* sp. (Kielland & Bryant 1998). Animal excretion also increases nitrogen cycling and modifies its distribution across the landscape, which locally enhances availability (Bardgett & Wardle 2003, Singer & Schoenecker 2003). In some cases, the relative contribution of this source of nitrogen may be small compared with the adverse effects of browsing (Pastor & Naiman 1992, Pastor et al. 1993).

With an overabundant deer population, the biomass deer consume becomes large relative to standing crops, particularly in low-productivity environments such as forest understories (Brathen & Oksanen 2001). Thus, we generally expect deer to reduce productivity and decelerate nutrient cycling in forest ecosystems. Here, compensation is uncommon, growth rates are low, and deer browsing decreases the quality and quantity of litter inputs (e.g., Ritchie et al. 1998). Browsed forest plots generally show reductions in understory and woody biomass accumulation (Ammer 1996, Riggs et al. 2000). Similarly, if nitrogen limits productivity, converting plant communities from palatable, deciduous, nitrogen-rich species to species with low tissue nitrogen and more chemical defenses (e.g., conifers) will decelerate nutrient cycling as the quantity and quality of litter available to decomposers decline (Bardgett & Wardle 2003, Pastor & Naiman 1992, Pastor et al. 1993, Ritchie et al. 1998). Browsing has also been shown to reduce ectomycorrhizal infections, which amplifies reductions in nutrient intake (Rossow et al. 1997).

Cascading Effects on Animal Species

Deer exert cascading effects on animals both by competing directly for resources with other herbivores and by indirectly modifying the composition and physical structure of habitats (Fuller 2001, Stewart 2001, van Wieren 1998). For example, browsing by deer affects the population and community composition of many invertebrates, birds, and small mammals (Table 2). Maximum diversity within a stand often appears to occur at moderate browsing levels (deCalesta & Stout 1997, Fuller 2001, Rooney & Waller 2003, Suominen et al. 2003, van Wieren 1998). Heavier browsing reduces vegetative cover and complexity in the understory, which often leads to reduced habitat availability for animals. Invertebrate and bird communities are sensitive to changes in forest understory, especially foliage density (McShea & Rappole 1997, Miyashita et al. 2004). Ungulates also disrupt associations of plants and pollinators by shifting patterns of relative flower abundance (Vázquez & Simberloff 2003). Few studies have experimentally manipulated deer densities,

TABLE 2 Summary of studies addressing the effects of deer browsing on community structure of invertebrates, birds, ard small mammals, using either experimental manipulation of deer browsing pressure (including exclosure studies) or field experiments with adequate replications

Taxon/source	Forest type and site	Cervid species	Results	Comments
Invertebrates				
Bailey & Whitham 2002	*Populus tremuloides* grasslands (Arizona, US)	*Cervus elaphus*	Increase by 30% in arthropod species richness and 40% increase in abundance after intermediate-severity fire and browsing exclusion; 69% and 72% declines in richness and abundance, respectively, after high-severity fire and heavy browsing (n = 3)	
Baines et al. 1994	*Pinus sylvestris* coniferous forest (Scotland, UK)	*Cervus elaphus*	Higher abundance of most taxa in ungrazed sites (n = 8); 83% of variation in number of lepidopterous larvae explained by two indices of grazing intensity, mean annual rainfall, altitude, and tree density	
Danell & Huss-Danell 1985	*Betula pendula*, *Betula pubescens* boreal forest (Sweden)	*Alces alces*	Higher abundance of leaf-eating insects on moderately browsed birches	High moose density; effect of browsing on plant community composition
Suominen et al. 1999a	*Pinus sylvestris* coniferous forest (Sweden)	*Alces alces*, *Capreolus capreolus*	Lower abundance and higher diversity of ground-dwelling insects in grazed sites in a productive location (n = 5); no consistent differences in abundance, species richness, and diversity between grazed and ungrazed sites (n = 4) in an unproductive location	

(Continued)

TABLE 2 (Continued)

Taxon/source	Forest type and site	Cervid species	Results	Comments
Suominen et al. 1999b	*Salix* sp.—*Populus balsamifera* early successional boreal forest (Alaska, US)	*Alces alces*	Trends toward higher abundance and species richness of ground-dwelling insects in browsed sites (n = 7), except for specialized herbivores (Curculionidae)	Moderate moose density
Suominen et al. 2003	*Pinus sylvestris* or *Betula pubescens* or *Picea abies* boreal forest (Finland)	*Rangifer tarandus*	Higher abundance, species richness, and diversity of ground-dwelling beetles in grazed sites (n = 15 in four locations), except for unproductive sites where diversity was lower than in grazed sites	
Wardle et al. 2001	Southern temperate forest (New Zealand)		Lower abundance of microarthropods and macrofaunal groups in grazed sites (n = 30)	Large geographical extent
Birds				
deCalesta 1994	*Prunus serotina, Acer rubrum, A. saccharum, Fagus grandifolia* northern hardwoods (Pennsylvania, US)	*Odocoileus virginianus*	Declines of 27% and 37% in species richness and abundance of intermediate canopy nesters between lowest and highest deer densities; no effect on ground and canopy nesters; density threshold between 7.9 and 14.9 deer/km^2	Controlled grazing experiment with four simulated densities

DeGraaf et al. 1991	*Quercus* sp. dominated northern hardwoods (Massachusetts, US)	*Odocoileus virginianus*	Lower species richness and abundance of canopy feeders at higher deer density; lower migratory species richness and higher resident species richness in thinned stands with high browsing; no difference in omnivorous, insectivorous, and ground-feeding species richness and abundance (n = 12)	
McShea & Rappole 2000	*Quercus* sp. dominated mixed hardwoods (Virginia, US)	*Odocoileus virginianus*	Increased abundance of ground nesters and intermediate canopy nesters as understory vegetation resumed growth in exclosures (n = 4), but no increase in diversity because of species replacement	
Moser & Witmer 2000	*Pinus ponderosa* coniferous forest (Oregon, US)	*Cervus elaphus*	No difference in abundance, species richness and diversity between ungrazed (n = 3) and grazed (n = 3) sites	Exclosures of 20 to 40 ha
Small mammals				
McShea 2000	*Quercus* sp. dominated mixed hardwoods (Virginia, US)	*Odocoileus virginianus*	Interaction between deer browsing and previous year acorn crop: higher *Tamias striatus* and *Peromyscus leucopus* abundance in exclosures (n = 4) after low-mast years, but no difference after good-mast years	
Moser & Witmer 2000	*Pinus ponderosa* coniferous forest (Oregon, US)	*Cervus elaphus*	Higher abundance, species richness, and diversity in ungrazed (n = 3) than in grazed (n = 3) sites	Exclosures of 20 to 40 ha

which makes drawing strong inferences about the relationship between animal diversity and deer density difficult. A notable exception is the study by deCalesta (1994) of songbirds, in which a controlled grazing experiment (Horsley et al. 2003) was used to demonstrate negative and nonlinear relationships between bird diversity and deer abundance.

By modifying species abundance and diversity, deer can modify trophic interactions among species. For example, deer potentially change the interactions between mast availability, small mammals, birds, and insects (McShea 2000, McShea & Schwede 1993, Ostfeld et al. 1996). Effects on interactions within the food web may be particularly important in ecosystems where several species of large herbivores coexist, such as in western North America, Spain, or the United Kingdom.

Dynamics and Reversibility of Deer Impacts

Large herbivores have the ability to act as "biological switches" that move forest communities toward alternative successional pathways and distinct stable states (Hobbs 1996, Laycock 1991, Schmitz & Sinclair 1997). Models of forest dynamics also demonstrate how browsing by deer can alter the rate of succession (Seagle & Liang 2001), forest structure and composition (Kienast et al. 1999), successional pathways (Jorritsma et al. 1999, Tester et al. 1997), and ultimate stable states (Kramer et al. 2003). In classical succession models, the relation between deer browsing and plant abundance is gradual (Figure 2a) or sudden (Figure 2b) but in both cases, reversible. Unlike succession, however, alternative stable states are not readily reversible when the browsing pressure is reduced (Scheffer et al. 2001, Westoby et al. 1989). In Figure 2c, the system may not appear to change much as deer densities gradually increase. Then, a sudden transition may occur that sharply reduces plant population levels (or overall system diversity or productivity). Even dramatic declines in deer density at this point have little effect; recovery only occurs if deer densities remain low through some extended period of time and interventions favoring vegetation recovery are applied (May 1977, Scheffer et al. 2001, Schmitz & Sinclair 1997). By analogy with physical systems, such lags and history dependence are termed "ecological hysteresis." Such nonlinear dynamics have been described in rangeland pastures (May 1977, Laycock 1991, Lockwood & Lockwood 1993), savanna-woodland systems (Dublin 1995, Scheffer et al. 2001), and temperate and boreal forests (Augustine et al. 1998, Pastor et al. 1993).

Interactions with Predators

The role of predators in controlling ungulate populations remains uncertain, at least in some systems. Particular examples exist where the introduction of a predator did not, by itself, control ungulate populations. Wolves moving onto Isle Royale did not prevent moose overpopulation, food depletion, and a subsequent crash caused by starvation (Peterson 1999).

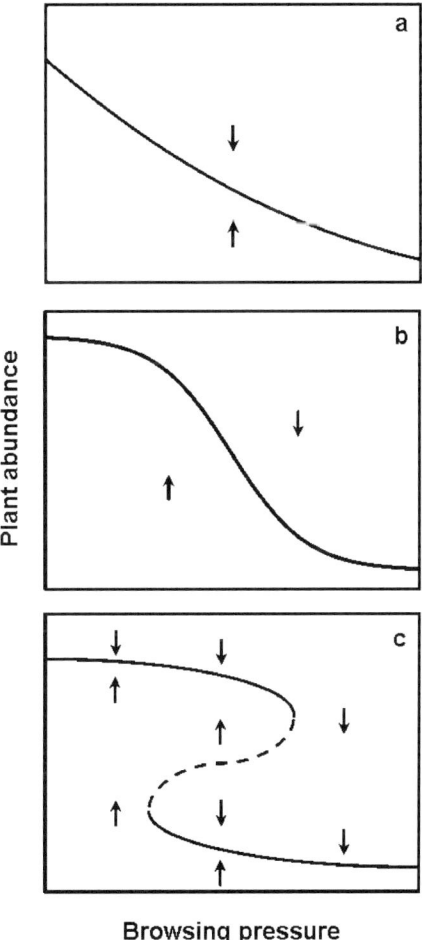

Figure 2 Three hypothetical relationships between the abundance of a forage plant and deer browsing pressure. (*a*) Deer have only modest and monotonic effects on the population. (*b*) A reversible threshold exists beyond which plant abundance drops precipitously. (*c*) Browsing beyond a certain threshold point causes a nonlinear decline that is not simply reversible. The plant population requires a large (or prolonged) reduction in browsing as well as a disturbance factor that promotes an increase of its abundance to recover. This requirement indicates an "alternate stable state." Arrows indicate dynamic changes at various points. Modified from Scheffer et al. (2001).

Recent research suggests, however, that large predators play important ecological roles. They appear to control the abundance of the "mesopredators" [e.g., raccoon (*Procyon lotor*), skunk (*Mephitis mephitis*), etc.] that prey on birds and small mammals (Crooks & Soulé 1999, Terborgh 1988). The presence of two or more predator species in the same region could work synergistically to exert significantly more population control on ungulates than either alone could exert (e.g., Gasaway et al. 1992). In the Glacier National Park area, a study by Kunkel & Pletscher (1999) concluded that combined predation from cougar and wolves is the primary factor that limits deer and elk populations. Analyzing results from 27 studies across North America, Messier (1994) used functional and numerical responses of wolves to moose to conclude that equilibrial moose densities would decline (from $2.0/km^2$ to $1.3/km^2$) in the presence of wolves. Furthermore, if habitat quality deteriorates or mortality from another predator increases, wolves are predicted to hold moose to a much lower equilibrium (0.2 to 0.4 moose/km^2). Predation effects are often nonlinear (Noy-Meir 1975) and involve lags in the manner illustrated in Figure 2*b* and 2*c* (substitute deer for plant abundance on the *y*-axis and predation for browsing pressure on the *x*-axis). Indeed, under a combined scenario, a functional guild of large predators might keep deer populations down to densities compatible with the upper curve of plant abundance in Figure 2*c*. Loss of predators could then flip the system to the alternate state represented by the bottom curve.

RESEARCH NEEDS

Whereas some species benefit from overabundant deer populations (Fuller & Gill 2001, Russell et al. 2001), overabundant deer annihilate many taxa, which disrupts community composition and ecosystem properties (Table 2) (deCalesta & Stout 1997, McShea & Rappole 1997). Between these extremes, we face much uncertainty. Ecologists should now work to identify threshold densities at which substantial impacts occur and devise effective strategies to limit deer impacts and sustain ecosystem integrity, i.e., the capacity of an ecosystem to preserve all its components and the functional relationships among those components following an external perturbation (sensu De Leo & Levin 1997; see also Hester et al. 2000, Scheffer & Carpenter 2003). Which species are affected by deer and at what densities? How fast do impacts occur? How quickly do plant populations, forest structure, and ecosystem processes recover? To what extent are deer populations and impacts constrained by food resources, predators, diseases, or hunting, and how do these limiting factors interact? This uncertainty places ecologists in an awkward position when they try to make deer management recommendations (see final section, (How) Can We Limit Deer Impacts?). Because forest communities can suffer long-term effects that are difficult to reverse, ecologists should make precautionary recommendations.

Given potential threshold effects and alternative stable states, how should we design our research? We need more controlled experiments that directly manipulate deer densities and other factors known to influence forest dynamics (e.g., logging) (Bergström & Edenius 2003, Fuller 2001, Healy et al. 1997, Hester et al. 2000, Hobbs 1996, Rooney & Waller 2003). Such experiments should span different forest types, which would allow us to predict how forest types will respond to variable deer densities (Hjalten et al. 1993, Riggs et al. 2000). We should also monitor both immediate and delayed effects and track dynamic responses to both increases and decreases in deer density. Results from such manipulations would allow us to identify what windows of low deer density are needed across space and time to allow deer-sensitive plants to persist or recover in the landscape (Sage et al. 2003, Westoby et al. 1989). Eventually, results from such experiments will allow ecologists to make specific recommendations at the right scales, such as 10 years of fewer than 7 deer/km^2 over areas of at least 60 km^2 (Hobbs 2003, Weisberg et al. 2004).

Deer management must move beyond a population-based approach to an approach that considers whole-ecosystem effects (McShea et al. 1997b). Fuller & Gill (2001) suggest that we quantify the relationships between community composition across taxa and deer at various abundances to understand the full range of deer impacts on biodiversity. Knowing how deer affect the moss layer, herbs, shrubs, saplings, trees, invertebrates, small mammals, and birds at low, intermediate, and high grazing intensities would be a major step forward. In the absence of fenced-in areas with known numbers of deer, such approaches will require that we improve our ability to estimate local deer abundances. Indicators based on vegetation measurements increase our capacity to implement localized management programs and to monitor progress toward specific management goals (Augustine & DeCalesta 2003, Augustine & Jordan 1998, Balgooyen & Waller 1995, McShea & Rappole 2000). Applied research extends to include the selection of species, varieties, and genotypes more resistant to browsing (Gill 1992b) and evaluating the risks of epidemics associated with high deer densities.

We must also learn more about how forage conditions, predator populations, and human hunting interact to affect deer population dynamics. We should seek to understand the potentially complex dynamics of tritrophic-level interactions. We need more data from a variety of systems on when predators can, alone or in combination with other factors, control deer densities. Likewise, we need to learn more about the "ecology of fear" (Brown et al. 1999), that is, how predators might influence browsing behavior even before they are numerous enough to reduce population growth appreciably (Ripple & Beschta 2003). We also have more to learn about sport hunting. We cannot yet predict, for example, how local hunting of philopatric females influences subsequent local deer densities (Côté et al. 2004, McNulty et al. 1997, but see Oyer & Porter 2004).

Finally, ecologists should work to integrate the results of individual studies into models capable of forecasting deer populations and impacts accurately enough to provide managers with sound guidance when they make decisions. Such models should integrate deer population dynamics with forest dynamics and deer hunter impacts (Tester et al. 1997). They should also incorporate the uncertainty that underpins interactions between management and science (Bergström & Edenius

2003, Bugmann & Weisberg 2003, Tremblay et al. 2004). Such models, and the research previously mentioned, have a logical place in hunter education programs and revised programs of deer management.

MANAGEMENT ISSUES

Historically, game managers strove to augment and protect deer populations, and hunters learned to limit takes and favor bucks. Today, such precepts are outmoded, but unlearning old lessons and reversing this cultural momentum has proved difficult.

The management of deer and the management of vegetation remains divorced, and this situation hampers our ability to manage them jointly (Healy et al. 1997). Their management commonly occurs in different agencies with contrasting goals and paradigms. Even the scales are different; deer density is usually estimated regionally, whereas forest managers operate on individual stands. In contrast, adaptive management seeks to merge research with management by using management prescriptions as experimental manipulations, with appropriate control areas, and by regularly incorporating research results into revised management practices (Holling 1978, Walters 1986). Ecosystem management is a further extension of conventional management that emphasizes historical patterns of abundance and disturbance and ecosystem dynamics at various scales (Christensen et al. 1996). Such approaches emphasize the importance of managing deer as part of a complex system. That promise has yet to be fully realized. Nevertheless, ecologists and wildlife managers are beginning to integrate biodiversity concerns into deer management (deCalesta & Stout 1997, Rooney 2001).

(How) Can We Limit Deer Impacts?

Foresters exploit a variety of techniques to control deer impacts locally. Keeping sapling stem density high through thinning or planting and increasing hunting pressure, for example, can allow a greater proportion of stems to escape browsing (Lyly & Saksa 1992, Martin & Baltzinger 2002, Welch et al. 1991, Reimoser 2003). Evidence indicates that within species, individual seedlings differ genetically in their susceptibility to browsing (Gill 1992b, Roche & Fritz 1997, Rousi et al. 1997, Vourc'h et al. 2002), which suggests that selection for more resistant saplings might be possible. Individual plastic tubes and wire fencing efficiently exclude deer but are costly, which limits their use to valuable seedlings or stands (Côté et al. 2004, Lavsund 1987). Electric fences are less effective but are also less expensive (Hygnstrom & Craven 1988). Repellents are also available. The most efficient repellents create fear (e.g., predator urine) (Nolte 1998, Nolte et al. 1994, Swihart et al. 1991, Wagner & Nolte 2001). The effectiveness of repellents increases with their concentration (Andelt et al. 1992, Baker et al. 1999) but decreases with (*a*) time since application (Andelt et al. 1992, Nolte 1998), (*b*) attractiveness of the food (Nolte 1998, Swihart et al. 1991, Wagner & Nolte 2001), (*c*) deer hunger

(Andelt et al. 1992), and (*d*) rainfall (Sayre & Richmond 1992). Similar methods are often employed to prevent accidents near airfields and highways (Groot Bruinderink & Hazebroek 1996, Putman 1997). Reflectors (Groot Bruinderink & Hazebroek 1996) and sound devices (Bomford & O'Brien 1990), such as gas exploders, appear ineffective in deterring deer for long periods unless the devices are activated by motion sensors (Belant et al. 1996).

Sport hunting and relocation are two methods available for controlling deer populations. Most wildlife managers consider sport hunting to be the most efficient and cost-effective method of controlling deer over large areas (Brown et al. 2000). Relocation is expensive, and relocated deer do not remain in the area of release. They also suffer high mortality (Beringer et al. 2002, McCullough et al. 1997). Sport hunting is often limited, however. For example, sport hunting cannot take place on private lands posted against hunting, in remote locations, or in urban and suburban areas. The number of hunters is also declining (Enck et al. 2000). Hunters rarely focus on young animals or hunt throughout the year as other predators do. Thus, the effectiveness of hunters is reduced. These trends, combined with growing deer populations, suggest that deer may have surpassed the point where sport hunting can reliably control their numbers (Brown et al. 2000, Giles & Findlay 2004). "Quality deer management" programs constitute an important countertrend. These programs emphasize killing doe and young animals to reduce densities, which favors the growth of large trophy bucks (Miller & Marchinton 1995).

The need for intentional culling will continue for the foreseeable future as deer populations continue to increase worldwide (McIntosh et al. 1995, McLean 1999). Hunting antlerless deer generally reduces abundance on a local scale because social groups of females usually remain in the same area from year to year (Kilpatrick et al. 2001, McNulty et al. 1997, Sage et al. 2003). This behavior prevents a rapid recolonization of the hunted area (Oyer & Porter 2004). Some affluent suburban neighborhoods employ sharpshooters working at night with low-light optics and silencers to control deer. Others have begun to experiment with birth control methods. Various fertility control and immunocontraceptive techniques can limit reproduction in deer (McShea et al. 1997a, Turner et al. 1992, Waddell et al. 2001). However, these methods are labor intensive and disrupt normal reproductive behavior (Nettles 1997); thus, their application is expensive and difficult to scale up (McCullough et al. 1997, McShea et al. 1997a, Turner et al. 1992).

Deer control efforts to date have focused on redirecting sport hunting, applying hunts specifically to reduce deer numbers, and a few high-cost techniques aimed at protecting small areas that are typically of high value. All these methods have proved inadequate thus far in preventing deer from overpopulating broad areas. Some hunters and deer managers dispute that we have any problem associated with high deer density. Still others argue that such problems are temporary or local. Even where we have agreement on the need to control deer, we see little consensus on how to achieve it. No new hunter ethos emphasizing the ecological role of hunters in limiting deer numbers and impacts has yet emerged.

Experimental hunting sites with longer seasons, liberalization of bag limits (especially for antlerless deer), and increased hunter participation could help reduce local deer density (Brown et al. 2000, Côté et al. 2004, Martin & Baltzinger 2002). Because hunters rarely fully understand deer effects on ecosystems (Diefenbach et al. 1997), scientists should provide them and society with specific goals, strategies, and actions to conserve ecosystems better.

Given divergent opinions and uncertainty, what should ecologists recommend to wildlife and land managers? The answer clearly depends on local situations and what is known about them. We urge ecologists to promote a precautionary approach. Because overabundant deer can cause severe, long-term impacts that are difficult to reverse, ecologists should persuade managers to reduce deer numbers before and not after such impacts become evident. Although research results and active involvement by ecologists may not change attitudes quickly, they play crucial long-term roles in redirecting people's attitudes and patterns of management.

ACKNOWLEDGMENTS

We thank D. Simberloff for the invitation to write this review and S. de Bellefeuille for precious help with the literature review. Research by S.D.C. and J.P.T. on deer-forest ecology is supported by the Natural Sciences and Engineering Research Council of Canada, Produits forestiers Anticosti Inc., the Fonds québécois de la recherche sur la nature et les technologies, Laval University, the Centre d'études nordiques, and the Ministère des Ressources Naturelles, de la Faune et des Parcs du Québec. Research by D.M.W. and T.P.R. is supported by the National Science Foundation Ecology Program (DEB-023633) and the United States Department of Agriculture (NRI-CRIS-2003–02472).

The *Annual Review of Ecology, Evolution, and Systematics* is online at
http://ecolsys.annualreviews.org

LITERATURE CITED

Alverson WS, Waller DM. 1997. Deer populations and the widespread failure of hemlock regeneration in northern forests. See McShea et al. 1997b, pp. 280–97

Alverson WS, Waller DM, Solheim SL. 1988. Forests too deer: edge effects in northern Wisconsin. *Conserv. Biol.* 2:348–58

Ammer C. 1996. Impact of ungulates on structure and dynamics of natural regeneration of mixed mountain forests in the Bavarian Alps. *For. Ecol. Manag.* 88:43–53

Andelt WF, Baker DL, Burnham KP. 1992. Relative preference of captive cow elk for repellent-treated diets. *J. Wildl. Manag.* 56:164–73

Anderson RC. 1994. Height of white-flowered trillium (*Trillium grandiflorum*) as an index of deer browsing intensity. *Ecol. Appl.* 4:104–9

Anderson RC, Katz AJ. 1993. Recovery of browse-sensitive tree species following release from white-tailed deer *Odocoileus virginianus* zimmerman browsing pressure. *Biol. Conserv.* 63:203–8

Anderson RC, Loucks OL. 1979. White-tail deer (*Odocoileus virginianus*) influence on

structure and composition of *Tsuga canadensis* forests. *J. Appl. Ecol.* 16:855–61
Andren H, Angelstam P. 1993. Moose browsing on Scots pine in relation to stand size and distance to forest edge. *J. Appl. Ecol.* 30:133–42
Augustine DJ, DeCalesta D. 2003. Defining deer overabundance and threats to forest communities. from individual plants to landscape structure. *Écoscience* 10:472–86
Augustine DJ, Frelich LE. 1998. Effects of white-tailed deer on populations of an understory forb in fragmented deciduous forests. *Conserv. Biol.* 12:995–1004
Augustine DJ, Frelich LE, Jordan PA. 1998. Evidence for two alternate stable states in an ungulate grazing system. *Ecol. Appl.* 8:1260–69
Augustine DJ, Jordan PA. 1998. Predictors of white-tailed deer grazing intensity in fragmented deciduous forests. *J. Wildl. Manag.* 62:1076–85
Augustine DJ, McNaughton SJ. 1998. Ungulate effects on the functional species composition of plant communities: herbivore selectivity and plant tolerance. *J. Wildl. Manag.* 62:1165–83
Bailey JK, Whitham TG. 2002. Interactions among fire, aspen, and elk affect insect diversity: reversal of a community response. *Ecology* 83:1701–12
Baines D, Sage RB, Baines MM. 1994. The implications of red deer grazing to ground vegetation and invertebrate communities of Scottish native pinewoods. *J. Appl. Ecol.* 31:776–83
Baker DL, Andelt WF, Burnham KP, Shepperd WD. 1999. Effectiveness of Hot Sauce and Deer Away repellents for deterring elk browsing of aspen sprouts. *J. Wildl. Manag.* 63:1327–36
Balgooyen CP, Waller DM. 1995. The use of *Clintonia borealis* and other indicators to gauge impacts of white-tailed deer on plant communities in northern Wisconsin, USA. *Nat. Areas J.* 15:308–18
Bardgett RD, Wardle DA. 2003. Herbivore-mediated linkages between aboveground and belowground communities. *Ecology* 84:2258–68
Bashore TL, Bellis ED. 1982. Deer on Pennsylvania airfields: problems and means of control. *Wildl. Soc. Bull.* 10:386–88
Beals EW, Cottam G, Vogel RJ. 1960. Influence of deer on vegetation of the Apostle Islands, Wisconsin. *J. Wildl. Manag.* 24:68–80
Behrend DF, Mattfeld GF, Tierson WC, Wiley JE III. 1970. Deer density control for comprehensive forest management. *J. For.* 68.695–700
Belant JL, Seamans TW, Dwyer CP. 1996. Evaluation of propane exploders as white-tailed deer deterrents. *Crop Prot.* 15:575–78
Bergelson J, Crawley MJ. 1992. The effects of grazers on the performance of individuals and populations of scarlet gilia, *Ipomopsis aggregata*. *Oecologia* 90:435–44
Bergqvist G, Bergström R, Edenius L. 2003. Effects of moose (*Alces alces*) rebrowsing on damage development in young stands of Scots pine (*Pinus sylvestris*). *For. Ecol. Manag.* 176:397–403
Bergström R, Danell K. 1995. Effects of simulated summer browsing by moose on leaf and shoot biomass of birch, *Betula pendula*. *Oikos* 72:132–38
Bergström R, Edenius L. 2003. From twigs to landscapes—methods for studying ecological effects of forest ungulates. *J. Nat. Conserv.* 10:203–11
Beringer J, Hansen LP, Demand JA, Sartwell J, Wallendorf M, Mange R. 2002. Efficacy of translocation to control urban deer in Missouri: costs, efficiency, and outcome. *Wildl. Soc. Bull.* 30:767–74
Bobek B, Boyce MS, Kosobucka M. 1984. Factors affecting red deer (*Cervus elaphus*) population density in southeastern Poland. *J. Appl. Ecol.* 21:881–90
Boitani L. 1995. Ecological and cultural diversities in the evolution of wolf-human relationships. In *Ecology and Conservation of Wolves in a Changing World*, ed. LN Carbyn, SH Fritz, DR Seip, pp. 3–11. Edmonton, Alberta: Can. Circumpolar Inst., Occas. Publ. 35. 642 pp.
Bomford M, O'Brien PH. 1990. Sonic deterrents in animal damage control: a review

of device tests and effectiveness. *Wildl. Soc. Bull.* 18:411–22

Brandner TA, Peterson RO, Risenhoover KL. 1990. Balsam fir on Isle Royale: effects of moose herbivory and population density. *Ecology* 71:155–64

Brathen KA, Oksanen J. 2001. Reindeer reduce biomass of preferred plant species. *J. Veg. Sci.* 12:473–80

Brown JS, Laundré JW, Gurung M. 1999. The ecology of fear: optimal foraging, game theory, and trophic interactions. *J. Mamm.* 80:385–99

Brown TL, Decker DJ, Riley SJ, Enck JW, Lauber TB, et al. 2000. The future of hunting as a mechanism to control white-tailed deer populations. *Wildl. Soc. Bull.* 28:797–807

Bryant JP, Raffa KF. 1995. Chemical antiherbivore defense. In *Plant Stems: Physiology and Functional Morphology*, ed. BL Gartner, 1:365–81. San Diego, CA: Academic. 440 pp.

Bugmann H, Weisberg PJ. 2003. Forest-ungulate interactions: monitoring, modeling and management. *J. Nat. Conserv.* 10:193–201

Canham CD, McAninch JB, Wood DM. 1994. Effects of the frequency, timing, and intensity of simulated browsing on growth and mortality of tree seedlings. *Can. J. For. Res.* 24:817–25

Casey D, Hein D. 1983. Effects of heavy browsing on a bird community in deciduous forest. *J. Wildl. Manag.* 47:829–36

Caughley G. 1981. Overpopulation. In *Problems in Management of Locally Abundant Wild Mammals*, ed. PA Jewell, S Holt, 1:7–20. New York: Academic. 361 pp.

Caughley G. 1983. *The Deer Wars: The Story of Deer in New Zealand.* Auckland, NZ: Heinemann. 187 pp.

Christensen NL, Bartuska AM, Carpenter S, D'Antonio C, Francis R, et al. 1996. The report of the Ecological Society of America Committee on the scientific basis for ecosystem management. *Ecol. Appl.* 6:665–91

Clifton-Hadley RS, Wilesmith JW. 1991. Tuberculosis in deer: a review. *Vet. Rec.* 129:5–12

Connell JH, Slatyer RO. 1977. Mechanisms of succession in natural communities and their role in community stability and organization. *Am. Nat.* 111:1119–44

Conover MR. 1997. Monetary and intangible valuation of deer in the United States. *Wildl. Soc. Bull.* 25:298–305

Conover MR. 2001. Effect of hunting and trapping on wildlife damage. *Wildl. Soc. Bull.* 29:521–32

Cooke AS, Farrell L. 2001. Impact of muntjac deer (*Muntiacus reevesi*) at Monks Wood National Nature Reserve, Cambridgeshire, eastern England. *Forestry* 74:241–50

Côté SD, Dussault C, Huot J, Potvin F, Tremblay JP, et al. 2004. High herbivore density and boreal forest ecology: introduced white-tailed deer on Anticosti Island. In *Lessons from the Islands: Introduced Species and What They Tell Us About How Ecosystems Work. Proc. Res. Group Introd. Species 2002 Conf.*, ed. AJ Gaston, TE Golumbia, JL Martin, ST Sharpe. Queen Charlotte City, BC: Can. Wildl. Serv., Ottawa. In press

Crawley MJ. 1997. Plant-herbivore dynamics. In *Plant Ecology*, ed. MJ Crawley, 2:401–74. Oxford: Blackwell Sci. 717 pp.

Cronin G, Hay ME. 1996. Susceptibility to herbivores depends on recent history of both the plant and animal. *Ecology* 77:1531–43

Crooks KR, Soulé ME. 1999. Mesopredator release and avifaunal extinctions in a fragmented system. *Nature* 400:563–66

Danell K, Bergström R, Edenius L. 1994. Effects of large mammalian browsers on architecture, biomass, and nutrients of woody plants. *J. Mamm.* 75:833–44

Danell K, Huss-Danell K. 1985. Feeding by insects and hares on birches earlier affected by moose browsing. *Oikos* 44:75–81

Danell K, Niemela P, Varvikko T, Vuorisalo T. 1991. Moose browsing on Scots pine along a gradient of plant productivity. *Ecology* 72:1624–33

Daubenmire RF. 1940. Exclosure technique in ecology. *Ecology* 21:514–15

Davidson WR, Doster GL. 1997. Health characteristics and white-tailed deer population density in southeastern United States. See McShea et al. 1997b, pp. 164–84

deCalesta DS. 1994. Effect of white-tailed deer on songbirds within managed forests in Pennsylvania. *J. Wildl. Manag.* 58:711–18

deCalesta DS, Stout SL. 1997. Relative deer density and sustainability: a conceptual framework for integrating deer management with ecosystem management. *Wildl. Soc. Bull.* 25:252–58

DeGraaf RM, Healy WM, Brooks RT. 1991. Effects of thinning and deer browsing on breeding birds in New England oak woodlands. *For. Ecol. Manag.* 41:179–91

De Leo GA, Levin S. 1997. The multifaceted aspects of ecosystem integrity. *Conserv. Ecol.* 1. http://www.consecol.org/vol1/iss1/art3

Diamond JM. 1983. Laboratory, field, and natural experiments. *Nature* 304:586–87

Didier KA, Porter WF. 2003. Relating spatial patterns of sugar maple reproductive success and relative deer density in northern New York State. *For. Ecol. Manag.* 181:253–66

Diefenbach DR, Palmer WL, Shope WK. 1997. Attitudes of Pennsylvania sportsmen towards managing white-tailed deer to protect the ecological integrity of forests. *Wildl. Soc. Bull.* 25:244–51

Doak DF. 1992. Lifetime impacts of herbivory for a perennial plant. *Ecology* 73:2086–99

Dublin HT. 1995. Vegetation dynamics in the Serengeti-Mara ecosystem: the role of elephants, fire and other factors. In *Serengeti II: Dynamics, Management and Conservation of an Ecosystem*, ed. ARE Sinclair, P Arcese, 1:71–90. Chicago: Univ. Chicago Press. 666 pp.

Duncan AJ, Hartley SE, Iason GR. 1998. The effect of previous browsing damage on the morphology and chemical composition of Sitka spruce (*Picea sitchensis*) saplings and on their subsequent susceptibility to browsing by red deer (*Cervus elaphus*). *For. Ecol. Manag.* 103:57–67

Duncan AJ, Hartley SE, Thurlow M, Young S, Staines BW. 2001. Clonal variation in monoterpene concentrations in Sitka spruce (*Picea sitchensis*) saplings and its effect on their susceptibility to browsing damage by red deer (*Cervus elaphus*). *For. Ecol. Manag.* 148:259–69

Dzieciolowski R. 1980. Impact of deer browsing upon forest regeneration and undergrowth. *Ekol. Pol.* 28:583–99

Edenius L, Danell K, Bergström R. 1993. Impact of herbivory and competition on compensatory growth in woody plants: winter browsing by moose on Scots pine. *Oikos* 66:286–92

Enck JW, Decker DJ, Brown TL. 2000. Status of hunter recruitment and retention in the United States. *Wildl. Soc. Bull.* 28:817–24

Engelmark O, Hofgaard A, Arnborg T. 1998. Successional trends 219 years after fire in an old *Pinus sylvestris* stand, northern Sweden. *J. Veg. Sci.* 9:583–92

Fagerstone KA, Clay WH. 1997. Overview of USDA animal damage control efforts to manage overabundant deer. *Wildl. Soc. Bull.* 25:413–17

Fletcher JD, McShea WJ, Shipley LA, Shumway D. 2001a. Use of common forest forbs to measure browsing pressure by white-tailed deer (*Odocoileus virginianus* Zimmerman) in Virginia, USA. *Nat. Areas J.* 21:172–76

Fletcher JD, Shipley LA, McShea WJ, Shumway D. 2001b. Wildlife herbivory and rare plants: the effects of white-tailed deer, rodents, and insects on growth and survival of Turk's cap lily. *Biol. Conserv.* 101:229–38

Forchhammer MC, Stenseth NC, Post E, Langvatn R. 1998. Population dynamics of Norwegian red deer: density-dependence and climatic variation. *Proc. R. Soc. London Ser. B* 265:341–50

Frelich LE, Lorimer CG. 1985. Current and predicted long-term effects of deer browsing in hemlock forests in Michigan, USA. *Biol. Conserv.* 34:99–120

Fuller RJ. 2001. Responses of woodland birds to increasing numbers of deer: a review of evidence and mechanisms. *Forestry* 74:289–98

Fuller RJ, Gill RMA. 2001. Ecological impacts of increasing numbers of deer in British woodland. *Forestry* 74:193–99

Garrott RA, White PJ, White CAV. 1993. Overabundance: an issue for conservation biologists? *Conserv. Biol.* 7:946–49

Gasaway WC, Boertje RD, Grangaard DV, Kelleyhouse DG, Stephenson RO, Larsen DG. 1992. The role of predation in limiting moose at low densities in Alaska and Yukon and implications for conservation. *Wildl. Monogr.* 120:1–59

George LO, Bazzaz FA. 1999. The fern understory as an ecological filter: emergence and establishment of canopy-tree seedlings. *Ecology* 80:833–45

Giles BG, Findlay CS. 2004. Effectiveness of a selective harvest system in regulating deer populations in Ontario. *J. Wildl. Manag.* 68:266–77

Gill RMA. 1992a. A review of damage by mammals in north temperate forests: 1. Deer. *Forestry* 65:145–69

Gill RMA. 1992b. A review of damage by mammals in north temperate forests: 3. Impact on trees and forests. *Forestry* 65:363–88

Gill RMA, Beardall V. 2001. The impact of deer on woodlands: the effects of browsing and seed dispersal on vegetation structure and composition. *Forestry* 74:209–18

Glenn-Lewin DC, van der Maarel E. 1992. Patterns and processes of vegetation dynamics. In *Plant Succession: Theory and Prediction. Popul. Community Biol. Ser. 11*, ed. DC Glenn-Lewin, RK Peet, TT Veblen, 1:11–59. New York: Chapman & Hall. 352 pp.

Gomez JM, Zamora R. 2002. Thorns as mechanical defense in a long-lived shrub (*Hormathophylla spinosa*, Cruciferae). *Ecology* 83:885–90

Groot Bruinderink GWTA, Hazebroek E. 1996. Ungulate traffic collisions in Europe. *Conserv. Biol.* 10:1059–67

Gross JE, Miller MW. 2001. Chronic wasting disease in mule deer: disease dynamics and control. *J. Wildl. Manag.* 65:205–15

Harlow RF, Downing RL. 1970. Deer browsing and hardwood regeneration in the southern Appalachians. *J. For.* 68:298–300

Healy WM, deCalesta DS, Stout SL. 1997. A research perspective on white-tailed deer overabundance in the northeastern United States. *Wildl. Soc. Bull.* 25:253–63

Hester AJ, Edenius L, Buttenschøn RM, Kuiters AT. 2000. Interactions between forests and herbivores: the role of controlled grazing experiments. *Forestry* 73:381–91

Hjalten J, Danell K, Ericson L. 1993. Effects of simulated herbivory and intraspecific competition on the compensatory ability of birches. *Ecology* 74:1136–42

Hobbs NT. 1996. Modification of ecosystems by ungulates. *J. Wildl. Manag.* 60:695–713

Hobbs NT. 2003. Challenges and opportunities in integrating ecological knowledge across scales. *For. Ecol. Manag.* 181:223–38

Hochwender CG, Marquis RJ, Stowe KA. 2000. The potential for and constraints on the evolution of compensatory ability in *Asclepias syriaca*. *Oecologia* 122:361–70

Holling CS. 1978. *Adaptive Environmental Assessment and Management.* New York: Wiley. 377 pp.

Hornberg S. 2001. Changes in population density of moose (*Alces alces*) and damage to forests in Sweden. *For. Ecol. Manag.* 149:141–51

Horsley SB, Marquis DA. 1983. Interference by weeds and deer with Allegheny hardwood reproduction. *Can. J. For. Res.* 13:61–69

Horsley SB, Stout SL, deCalesta DS. 2003. White-tailed deer impact on the vegetation dynamics of a northern hardwood forest. *Ecol. Appl.* 13:98–118

Huntly N. 1991. Herbivores and the dynamics of communities and ecosystems. *Annu. Rev. Ecol. Syst.* 22:477–503

Hurlbert SH. 1984. Pseudoreplication and the design of ecological field experiments. *Ecol. Monogr.* 54:187–211

Husheer SW, Coomes DA, Robertson AW. 2003. Long-term influences of introduced deer on the composition and structure of New Zealand *Nothofagus* forests. *For. Ecol. Manag.* 181:99–117

Hygnstrom SE, Craven SR. 1988. Electric fences and commercial repellents for reducing deer damage in cornfields. *Wildl. Soc. Bull.* 16:291–96

Jedrzejewska B, Jedrzejewski W, Bunevich AN, Milkowski L, Krasinski ZA. 1997. Factors shaping population densities and increase rates of ungulates in Bialowieza Primeval Forest (Poland and Belarus) in the 19th and 20th centuries. *Acta Theriol.* 42:399–451

Jorritsma ITM, van Hees AFM, Mohren GMJ. 1999. Forest development in relation to ungulate grazing: a modeling approach. *For. Ecol. Manag.* 120:23–34

Judziewicz EJ, Koch RG. 1993. Flora and vegetation of the Apostle Islands National Lakeshore and Madeline Island, Ashland and Bayfield Counties, Wisconsin. *Mich. Bot.* 32:43–189

Kay S. 1993. Factors affecting severity of deer browsing damage within coppiced woodlands in the south of England. *Biol. Conserv.* 63:217–22

Kielland K, Bryant JP. 1998. Moose herbivory in taiga: effects on biogeochemistry and vegetation dynamics in primary succession. *Oikos* 82:377–83

Kienast F, Fritschi J, Bissegger M, Abderhalden W. 1999. Modeling successional patterns of high-elevation forests under changing herbivore pressure—responses at the landscape level. *For. Ecol. Manag.* 120:35–46

Kilpatrick HJ, LaBonte AM, Seymour JT. 2002. A shotgun-archery deer hunt in a residential community: evaluation of hunt strategies and effectiveness. *Wildl. Soc. Bull.* 30:478–86

Kilpatrick HJ, Spohr SM, Lima KK. 2001. Effects of population reduction on home ranges of female white-tailed deer at high densities. *Can. J. Zool.* 79:949–54

Knight TM. 2003. Effects of herbivory and its timing across populations of *Trillium grandiflorum* (Liliaceae). *Am. J. Bot.* 90:1207–14

Knox WM. 1997. Historical changes in the abundance and distribution of deer in Virginia. See McShea et al. 1997b, pp. 27–36

Kramer K, Groen TA, van Wieren SE. 2003. The interacting effects of ungulates and fire on forest dynamics: an analysis using the model FORSPACE. *For. Ecol. Manag.* 181:205–22

Kuiters AT, Slim PA. 2002. Regeneration of mixed deciduous forest in a Dutch forest-heathland, following a reduction of ungulate densities. *Biol. Conserv.* 105:65–74

Kullberg Y, Bergström R. 2001. Winter browsing by large herbivores on planted deciduous seedlings in southern Sweden. *Scand. J. For. Res.* 16:371–78

Kunkel K, Pletscher DH. 1999. Species-specific population dynamics of cervids in a multipredator ecosystem. *J. Wildl. Manag.* 63:1082–93

Lavsund S. 1987. Moose relationships to forestry in Finland, Norway and Sweden. *Swed. Wildl. Res.* 1(Suppl.):229–44

Laycock WA. 1991. Stable states and thresholds of range condition on North American rangelands: a viewpoint. *J. Range Manag.* 44:427–33

Lemieux N, Maynard BK, Johnson WA. 2000. A regional survey of deer damage throughout Northeast nurseries and orchards. *J. Environ. Hort.* 18:1–4

Leopold A. 1933. *Game Management*. New York: Scribner's. 481 pp.

Leopold A, Sowls LK, Spencer DL. 1947. A survey of overpopulated deer range in the United States. *J. Wildl. Manag.* 11:162–77

Lewis MA, Murray JD. 1993. Modelling territoriality and wolf-deer interactions. *Nature* 366:738–40

Lockwood JA, Lockwood DR. 1993. Catastrophe theory: a unified paradigm for rangeland ecosystem dynamics. *J. Range Manag.* 46:282–88

Loison A, Langvatn R, Solberg EJ. 1999. Body mass and winter mortality in red deer calves: disentangling sex and climate effects. *Ecography* 22:20–30

Long ZT, Carson WP, Peterson CJ. 1998. Can disturbance create refugia from herbivores: an example with hemlock regeneration on treefall mounds. *J. Torrey Bot. Soc.* 125:165–68

Lyly O, Saksa T. 1992. The effect of stand

density on moose damage in young *Pinus sylvestris* stands. *Scand. J. For. Res.* 7:393–403
Martin JL, Baltzinger C. 2002. Interaction among deer browse, hunting, and tree regeneration. *Can. J. For. Res.* 32:1254–64
Maschinski J, Whitham TG. 1989. The continuum of plant responses to herbivory: the influence of plant association, nutrient availability, and timing. *Am. Nat.* 134:1–19
May RM. 1977. Thresholds and breakpoints in ecosystems with a multiplicity of stable states. *Nature* 269:471–77
McCabe TR, McCabe RE. 1997. Recounting whitetails past. See McShea et al. 1997b, pp. 11–26
McCullough DR. 1979. *The George Reserve Deer Herd*. Ann Arbor: Mich. Univ. Press. 271 pp.
McCullough DR. 1997. Irruptive behavior in ungulates. See McShea et al. 1997b, pp. 69–98
McCullough DR. 1999. Density dependence and life-history strategies of ungulates. *J. Mammal.* 80:1130–46
McCullough DR, Jennings KW, Gates NB, Elliott BG, Didonato JE. 1997. Overabundant deer populations in California. *Wildl. Soc. Bull.* 25:478–83
McInnes PF, Naiman RJ, Pastor J, Cohen Y. 1992. Effects of moose browsing on vegetation and litter of the boreal forest, Isle Royale, Michigan, USA. *Ecology* 73:2059–75
McIntosh R, Burlton FEW, McReddie G. 1995. Monitoring the density of a roe deer *Capreolus capreolus* population subjected to heavy hunting pressure. *For. Ecol. Manag.* 79:99–106
McLean C. 1999. The effect of deer culling on tree regeneration on Scaniport Estate, Inverness-shire: a study by the Deer Commission for Scotland. *Scot. For.* 53:225–30
McNaughton SJ. 1979. Grazing as an optimization process: grass-ungulate relationships in the Serengeti. *Am. Nat.* 113:691–703
McNaughton SJ. 1983. Compensatory plant growth as a response to herbivory. *Oikos* 40:329–36

McNulty SA, Porter WF, Mathews NE, Hill JA. 1997. Localized management for reducing white-tailed deer populations. *Wildl. Soc. Bull.* 25:265–71
McShea WJ. 2000. The influence of acorn crops on annual variation in rodent and bird populations. *Ecology* 81:228–38
McShea WJ, Monfort SL, Hakim S, Kirkpatrick J, Liu I, et al. 1997a. The effect of immunocontraception on the behavior and reproduction of white-tailed deer. *J. Wildl. Manag.* 61:560–69
McShea WJ, Rappole JH. 1992. White-tailed deer as keystone species within forested habitats of Virginia. *Va. J. Sci.* 43:177–86
McShea WJ, Rappole JH. 1997. Herbivores and the ecology of forest understory birds. See McShea et al. 1997b, pp. 298–309
McShea WJ, Rappole JH. 2000. Managing the abundance and diversity of breeding bird populations through manipulation of deer populations. *Conserv. Biol.* 14:1161–70
McShea WJ, Schwede G. 1993. Variable acorn crops: responses of white-tailed deer and other mast consumers. *J. Mammal.* 74:999–1006
McShea WJ, Underwood HB, Rappole JH, eds. 1997b. *The Science of Overabundance: Deer Ecology and Population Management*. Washington, DC: Smithson. Inst. Press. 402 pp.
Messier F. 1994. Ungulate population models with predation: a case study with the North American moose. *Ecology* 75:478–88
Miller KV, Marchinton RL. 1995. *Quality Whitetails: The Why and How of Quality Deer Management*. Mechanicsburg, PA: Stackpole Books. 320 pp.
Miller R, Kaneene JB, Fitzgerald SD, Schmitt SM. 2003. Evaluation of the influence of supplemental feeding of white-tailed deer (*Odocoileus virginianus*) on the prevalence of bovine tuberculosis in the Michigan wild deer population. *J. Wildl. Dis.* 39:84–95
Miller SG, Bratton SP, Hadidian J. 1992. Impacts of white-tailed deer on endangered plants and threatened vascular plants. *Nat. Areas J.* 12:67–74

Miyashita T, Takada M, Shimazaki A. 2004. Indirect effects of herbivory by deer reduce abundance and species richness of web spiders. *Écoscience* 11:74–79

Moser BW, Witmer GW. 2000. The effects of elk and cattle foraging on the vegetation, birds, and small mammals of the Bridge Creek Wildlife Area, Oregon. *Int. Biodeter. Biodegrad.* 45:151–57

Mysterud A. 2000. The relationship between ecological segregation and sexual body size dimorphism in large herbivores. *Oecologia* 124:40–54

Mysterud A, Stenseth NC, Yoccoz NG, Langvatn R, Steinheim G. 2001. Nonlinear effects of large-scale climatic variability on wild and domestic herbivores. *Nature* 410:1096–99

Nettles VF. 1997. Potential consequences and problems with wildlife contraceptives. *Reprod. Fertil. Dev.* 9:137–43

Nixon CM, Hansen LP, Brewer PA, Chelsvig JE. 1991. Ecology of white-tailed deer in an intensively farmed region of Illinois. *Wildl. Monogr.* 118:1–77

Nolte DL. 1998. Efficacy of selected repellents to deter deer browsing on conifer seedlings. *Int. Biodeter. Biodegrad.* 42:101–7

Nolte DL, Mason JR, Epple G, Aronov E, Campbell DL. 1994. Why are predator urines aversive to prey? *J. Chem. Ecol.* 20:1505–16

Noy-Meir I. 1975. Stability of grazing systems: an application of predator-prey graphs. *J. Ecol.* 63:459–81

Otsfeld RS, Jones CG, Wolff JO. 1996. Of mice and mast: ecological connections in eastern deciduous forests. *BioScience* 46:323–30

Oyer AM, Porter WF. 2004. Localized management of white-tailed deer in the central Adirondack Mountains, New York. *J. Wildl. Manag.* 68:257–65

Ozoga JJ, Verme LJ. 1986. Relation of maternal age to fawn-rearing success in white-tailed deer. *J. Wildl. Manag.* 50:480–86

Paige KN, Whitham TG. 1987. Overcompensation in response to mammalian herbivory: the advantage of being eaten. *Am. Nat.* 129:407–16

Paine RT. 2000. Phycology for the mammalogist: marine rocky shores and mammal-dominated communities—how different are the structuring processes? *J. Mammal.* 81:637–48

Palmer SCF, Truscott AM. 2003. Browsing by deer on naturally regenerating Scots pine (*Pinus sylvestris* L.) and its effects on sapling growth. *For. Ecol. Manag.* 182:31–47

Paquet PC, Carbyn LN. 2003. Gray wolf. In *Wild Mammals of North America: Biology, Management, and Conservation*, ed. GA Feldhamer, BC Thompson, JA Chapman, 2:482–510. Baltimore/London: Johns Hopkins Univ. Press. 1216 pp.

Partl E, Szinovatz V, Reimoser F, Schweiger-Adler J. 2002. Forest restoration and browsing impact by roe deer. *For. Ecol. Manag.* 159:87–100

Pastor J, Dewey B, Naiman RJ, McInnes PF, Cohen Y. 1993. Moose browsing and soil fertility in the boreal forests of Isle Royale National Park. *Ecology* 74:467–80

Pastor J, Naiman RJ. 1992. Selective foraging and ecosystem processes in boreal forests. *Am. Nat.* 139:690–705

Peterson RO. 1999. Wolf-moose interaction on Isle Royale: the end of natural regulation? *Ecol. Appl.* 9:10–16

Pimlott DH. 1963. Influence of deer and moose on boreal forest vegetation in two areas of eastern Canada. In *Transactions of the Sixth Congress of the International Union of Game Biologists, Bournemouth*, ed. TH Blank, pp. 105–16. London: Nat. Conserv.

Pojar J, Lewis T, Roemer H, Wilford DJ. 1980. *Relationships Between Introduced Black-Tailed Deer and the Plant Life in the Queen Charlotte Islands, B.C.* Smithers: B.C. Minist. For. 63 pp.

Porter WF, Underwood HB. 1999. Of elephants and blind men: deer management in the U.S. national parks. *Ecol. Appl.* 9:3–9

Potvin F, Beaupré P, Laprise G. 2003. The eradication of balsam fir stands by white-tailed deer on Anticosti Island, Québec: a 150 year process. *Ecoscience* 10:487–95

Putman RJ. 1986. Foraging by roe deer in

agricultural areas and impact on arable crops. *J. Appl. Ecol.* 23:91–99

Putman RJ. 1997. Deer and road traffic accidents: options for management. *J. Environ. Manag.* 51:43–57

Putman RJ, Moore NP. 1998. Impact of deer in lowland Britain on agriculture, forestry and conservation habitats. *Mamm. Rev.* 28:141–64

Raymond GJ, Bossers A, Raymond LD, O'Rourke KI, McHolland LE, et al. 2000. Evidence of a molecular barrier limiting susceptibility of humans, cattle and sheep to chronic wasting disease. *EMBO J.* 17:4425–30

Reimoser F. 2003. Steering the impacts of ungulates on temperate forests. *J. Nat. Conserv.* 10:243–52

Reimoser F, Gossow H. 1996. Impact of ungulates on forest vegetation and its dependence on the silvicultural system. *For. Ecol. Manag.* 88:107–19

Riggs RA, Tiedemann AR, Cook JG, Ballard TM, Edgerton PJ, et al. 2000. Modification of mixed-conifer forests by ruminant herbivores in the Blue Mountains Ecological Province. *Res. Pap. PNW-RP-527*. USDA For. Serv., Portland, OR. 77 pp.

Riley SJ, Decker DJ, Enck JW, Curtis PD, Lauber TB, Brown TL. 2003. Deer populations up, hunter populations down: implications of interdependence of deer and hunter population dynamics on management. *Écoscience* 10:455–61

Ripple WJ, Beschta RL. 2003. Wolf reintroduction, predation risk, and cottonwood recovery in Yellowstone National Park. *For. Ecol. Manag.* 184:299–313

Risenhoover KL, Maass SA. 1987. The influence of moose on the composition and structure of Isle Royale forests. *Can. J. For. Res.* 17:357–64

Ritchie ME, Tilman D, Knops JMH. 1998. Herbivore effects on plant and nitrogen dynamics in oak savanna. *Ecology* 79:165–77

Roche BM, Fritz RS. 1997. Genetics of resistance of *Salix sericea* to a diverse community of herbivores. *Evolution* 51:1490–98

Romin LA, Bissonette JA. 1996. Deer-vehicle collisions: status of state monitoring activities and mitigation efforts. *Wildl. Soc. Bull.* 24:276–83

Rooney TP. 1997. Escaping herbivory: refuge effects on the morphology and shoot demography of the clonal forest herb *Maianthemum canadense*. *J. Torrey Bot. Soc.* 124:280–85

Rooney TP. 2001. Deer impacts on forest ecosystems: a North American perspective. *Forestry* 74:201–8

Rooney TP, Dress WJ. 1997. Species loss over sixty-six years in the ground-layer vegetation of Heart's Content, an old-growth forest in Pennsylvania USA. *Nat. Areas J.* 17:297–305

Rooney TP, McCormick RJ, Solheim SL, Waller DM. 2000. Regional variation in recruitment of hemlock seedlings and saplings in the upper Great Lakes, USA. *Ecol. Appl.* 10:1119–32

Rooney TP, Solheim SL, Waller DM. 2002. Factors affecting the regeneration of northern white cedar in lowland forests of the upper Great Lakes region, USA. *For. Ecol. Manag.* 163:119–30

Rooney TP, Waller DM. 2001. How experimental defoliation and leaf height affect growth and reproduction in *Trillium grandiflorum*. *J. Torrey Bot. Soc.* 128:393–99

Rooney TP, Waller DM. 2003. Direct and indirect effects of deer in forest ecosystems. *For. Ecol. Manag.* 181:165–76

Rossow LJ, Bryant JP, Kielland K. 1997. Effects of above-ground browsing by mammals on mycorrhizal infection in an early successional taiga ecosystem. *Oecologia* 110:94–98

Rousi M, Tahvanainen J, Henttonen H, Herms DA, Uotila I. 1997. Clonal variation in susceptibility of white birches (*Betula* spp.) to mammalian and insect herbivores. *For. Sci.* 43:396–402

Ruhren S, Handel SL. 2000. Considering herbivory, reproduction, and gender when monitoring plants: a case study of Jack-in-the-pulpit (*Arisaema triphyllum* [L.] Schott). *Nat. Areas J.* 20:261–66

Ruhren S, Handel SL. 2003. Herbivory constrains survival, reproduction, and mutualisms when restoring nine temperate forest herbs. *J. Torrey Bot. Soc.* 130:34–42

Russell FL, Zippin DB, Fowler NL. 2001. Effects of white-tailed deer (*Odocoileus virginianus*) on plants, plant populations and communities: a review. *Am. Midl. Nat.* 146:1–26

Sæther BE, Andersen R, Hjeljord O, Heim M. 1996. Ecological correlates of regional variation in life history of the moose *Alces Alces*. *Ecology* 77:1493–500

Sage RW, Porter WF, Underwood HB. 2003. Windows of opportunity: white-tailed deer and the dynamics of northern hardwood forests of the northeastern US. *J. Nat. Conserv.* 10:213–20

Saunders MR, Puettmann KJ. 1999. Effects of overstory and understory competition and simulated herbivory on growth and survival of white pine seedlings. *Can. J. For. Res.* 29:536–46

Sayre RW, Richmond ME. 1992. Evaluation of a new deer repellent on Japanese yews at suburban homesites. *Proc. East. Wildl. Damage Control Conf.* 5:38–43

Scheffer M, Carpenter SR. 2003. Catastrophic regime shifts in ecosystems: linking theory to observation. *Trends Ecol. Evol.* 18:648–56

Scheffer M, Carpenter SR, Foley JA, Folke C, Walker B. 2001. Catastrophic shifts in ecosystems. *Nature* 413:591–96

Schmitt SM, Fitzgerald SD, Cooley TM, Bruning-Fann CS, Sullivan L, et al. 1997. Bovine tuberculosis in free-ranging white-tailed deer from Michigan. *J. Wildl. Dis.* 33:749–58

Schmitz OJ, Sinclair RE. 1997. Rethinking the role of deer in forest ecosystem dynamics. See McShea et al. 1997b, pp. 201–23

Schütz M, Risch AC, Leuzinger E, Krüsi BO, Achermann G. 2003. Impact of herbivory by red deer (*Cervus elaphus* L.) on patterns and processes in subalpine grasslands in the Swiss National Park. *For. Ecol. Manag.* 181:177–88

Seagle S, Liang S. 2001. Application of a forest gap model for prediction of browsing effects on riparian forest succession. *Ecol. Model.* 144:213–29

Sinclair ARE. 1997. Carrying capacity and the overabundance of deer: a framework for management. See McShea et al. 1997b, pp. 380–94

Singer FJ, Schoenecker KA. 2003. Do ungulates accelerate or decelerate nitrogen cycling? *For. Ecol. Manag.* 181:189–204

Skolving H. 1985. Traffic accidents with moose and roe-deer in Sweden: report of research, development and measures. *Actes du Colloque Routes et Faune Sauvage*, pp. 317–25. Strasbourg: Conseil Eur. 406 pp.

Solberg EJ, Sæther B-E, Strand O, Loison A. 1999. Dynamics of a harvested moose population in a variable environment. *J. Anim. Ecol.* 68:186–204

Steere AC. 1994. Lyme disease: a growing threat to urban populations. *Proc. Natl. Acad. Sci. USA* 91:2378–83

Stewart AJA. 2001. The impact of deer on lowland woodland invertebrates: a review of the evidence and priorities for future research. *Forestry* 74:259–70

Stewart GH, Burrows LE. 1989. The impact of white-tailed deer *Odocoileus virginianus* on regeneration in the coastal forests of Stewart Island, New Zealand. *Biol. Conserv.* 49:275–93

Stoeckler JH, Strothmann RO, Kreftint LW. 1957. Effect of deer browsing on reproduction in the northern hardwood-hemlock type in northeastern Wisconsin. *J. Wildl. Manag.* 21:75–80

Stromayer KAK, Warren RJ. 1997. Are overabundant deer herds in the eastern United States creating alternate stable states in forest plant communities? *Wildl. Soc. Bull.* 25:227–34

Suominen O, Danell K, Bergström R. 1999a. Moose, trees, and ground-living invertebrates: indirect interactions in Swedish pine forests. *Oikos* 84:215–26

Suominen O, Danell K, Bryant JP. 1999b. Indirect effects of mammalian browsers on vegetation and ground-dwelling insects in an Alaskan floodplain. *Écoscience* 6:505–10

Suominen O, Niemelä J, Martikainen P, Niemelä P, Kojola I. 2003. Impact of reindeer grazing on ground-dwelling Carabidae and Curculionidae assemblages in Lapland. *Ecography* 26:503–13

Swihart RK, Pignatello JJ, Mattina MJI. 1991. Aversive response of white-tailed deer, *Odocoileus virginianus*, to predator urines. *J. Chem. Ecol.* 17:767–77

Takada M, Asada M, Miyashita T. 2001. Regional differences in the morphology of a shrub *Damnacanthus indicus*: an induced resistance to deer herbivory? *Ecol. Res.* 16: 809–13

Telford SR III. 2002. Deer tick-transmitted zoonoses in the eastern United States. In *Conservation Medicine: Ecological Health in Practice*, ed. A Aguirre, RS Ostfeld, GM Tabor, C House, MC Pearl, pp. 310–24. New York: Oxford Univ. Press. 407 pp.

Terborgh J. 1988. The big things that run the world—a sequel to E.O. Wilson. *Conserv. Biol.* 2:402–3

Tester JR, Starfield AM, Frelich LE. 1997. Modeling for ecosystem management in Minnesota pine forests. *Biol. Conserv.* 80: 313–24

Tilghman NG. 1989. Impacts of white-tailed deer on forest regeneration in northwestern Pennsylvania. *J. Wildl. Manag.* 53:524–32

Tremblay JP, Hester A, McLeod J, Huot J. 2004. Choice and development of decision support tools for the sustainable management of deer-forest systems. *For. Ecol. Manag.* 191:1–16

Turner JW, Liu IKM, Kirkpatrick JF. 1992. Remotely delivered immunocontraception in captive white-tailed deer. *J. Wildl. Manag.* 56:154–57

Van Wieren SE. 1998. Effects of large herbivores upon the animal community. In *Grazing and Conservation Management*, ed. MF Wallis DeVries, JP Bakker, SE Van Wieren, pp. 185–214. Boston: Kluwer Acad. 390 pp.

Vázquez DP, Simberloff D. 2003. Changes in interaction biodiversity induced by an introduced ungulate. *Ecol. Lett.* 6:1077–83

Vila B, Keller T, Guibal F. 2001. Influence of browsing cessation on *Picea sitchensis* radial growth. *Ann. For. Sci.* 58:853–59

Vila B, Torre F, Guibal F, Martin JL. 2003. Growth change of young *Picea sitchensis* in response to deer browsing. *For. Ecol. Manag.* 180:413–24

Vourc'h G, Martin JL, Duncan P, Escarré J, Clausen TP. 2001. Defensive adaptations of *Thuja plicata* to ungulate browsing: a comparative study between mainland and island populations. *Oecologia* 126:84–93

Vourc'h G, Vila B, Gillon D, Escarré J, Guibal F, et al. 2002. Disentangling the causes of damage variation by deer browsing on young *Thuja plicata. Oikos* 98:271–83

Waddell RB, Osborn DA, Warren RJ, Griffin JC, Kesler DJ. 2001. Prostaglandin F2 alpha-mediated fertility control in captive white-tailed deer. *Wildl. Soc. Bull.* 29:1067–74

Wagner KK, Nolte DL. 2001. Comparison of active ingredients and delivery systems in deer repellents. *Wildl. Soc. Bull.* 29:322–30

Waller DM, Alverson WS. 1997. The white-tailed deer: a keystone herbivore. *Wildl. Soc. Bull.* 25:217–26

Waller DM, Alverson WS, Solheim S. 1996. Local and regional factors influencing the regeneration of Eastern Hemlock. In *Hemlock Ecology and Management, Conf. Proc.*, ed. G Mroz, J Martin, pp. 73–90. Houghton: Mich. Technol. Univ.

Walters CJ. 1986. *Adaptive Management of Renewable Resources*. New York: Macmillan. 374 pp.

Wardle DA, Barker GM, Yeates GM, Bonner KI, Ghani A. 2001. Introduced browsing mammals in New Zealand natural forests: aboveground and belowground consequences. *Ecol. Monogr.* 71:587–614

Wardle DA, Bonner KI, Barker GM. 2002. Linkages between plant litter decomposition, litter quality, and vegetation responses to herbivores. *Funct. Ecol.* 16:585–95

Watson A. 1983. Eighteenth century deer numbers and pine regeneration near Braemar, Scotland. *Biol. Conserv.* 25:289–305

Webb WL, King RT, Patric EF. 1956. Effect of

white-tailed deer on a mature northern hardwood forest. *J. For.* 54:391–98

Weisberg PJ, Coughenour MB, Bugmann, H. 2004. Modeling of large herbivore-vegetation interactions in a landscape context. In *Large Herbivore Ecology and Ecosystem Dynamics*, ed. K Danell, R Bergstom, P Duncan, H Olff. Cambridge, UK: Cambridge Univ. Press. In press

Welch D, Staines BW, Scott D, French DD. 1992. Leader browsing by red and roe deer on young Sitka spruce trees in western Scotland. II. Effects on growth and tree form. *Forestry* 65:309–30

Welch D, Staines BW, Scott D, French DD, Catt DC. 1991. Leader browsing by red and roe deer on young Sitka spruce trees in western Scotland: I. Damage rates and the influence of habitat factors. *Forestry* 64:61–82

West BC, Parkhurst JA. 2002. Interactions between deer damage, deer density, and stakeholder attitudes in Virginia. *Wildl. Soc. Bull.* 30:139–47

Westoby M, Walker B, Noy-Meir I. 1989. Opportunistic management for rangelands not at equilibrium. *J. Range Manag.* 42:266–74

Whigham DF. 1990. The effect of experimental defoliation on the growth and reproduction of a woodland orchid, *Tipularia discolor. Can. J. Bot.* 68:1812–16

White CA, Feller MC, Bayley S. 2003. Predation risk and the functional response of elk-aspen herbivory. *For. Ecol. Manag.* 181:77–98

Whitney GG. 1984. Fifty years of change in the arboreal vegetation of Heart's Content, an old-growth hemlock-white pine-northern hardwood stand. *Ecology* 65:403–8

Williams ES, Miller MW, Kreeger TJ, Kahn RH, Thorne ET. 2002. Chronic wasting disease of deer and elk: a review with recommendations for management. *J. Wildl. Manag.* 66:551–63

Wilson ML, Childs JE. 1997. Vertebrate abundance and the epidemiology of zoonotic diseases. See McShea et al. 1997b, pp. 224–48

Wilson ML, Ducey AM, Litwin TS, Gavin TA, Spielman A. 1990. Microgeographic distribution of immature *Ixodes dammini* ticks correlated with that of deer. *Med. Vet. Entomol.* 4:151–59

Wilson ML, Telford SR, Piesman J, Spielman A. 1988. Reduced abundance of immature *Ixodes dammini* (Acari: Isodidae) following elimination of deer. *J. Med. Entomol.* 25:224–28

Wywialowski AP. 1996. Wildlife damage to field corn in 1993. *Wildl. Soc. Bull.* 24:264–71

ECOLOGICAL EFFECTS OF TRANSGENIC CROPS AND THE ESCAPE OF TRANSGENES INTO WILD POPULATIONS

Diana Pilson and Holly R. Prendeville

School of Biological Sciences, University of Nebraska, Lincoln, Nebraska 68588-0118; email: dpilson1@unl.edu, hrp@unlserve.unl.edu

Key Words genetically engineered crop, GM crop

■ **Abstract** Ecological risks associated with the release of transgenic crops include nontarget effects of the crop and the escape of transgenes into wild populations. Nontarget effects can be of two sorts: (*a*) unintended negative effects on species that do not reduce yield and (*b*) greater persistence of the crop in feral populations. Conventional agricultural methods, such as herbicide and pesticide application, have large and well-documented nontarget effects. To the extent that transgenes have more specific target effects, transgenic crops may have fewer nontarget effects. The escape of transgenes into wild populations, via hybridization and introgression, could lead to increased weediness or to the invasion of new habitats by the wild population. In addition, native species with which the wild plant interacts (including herbivores, pathogens, and other plant species in the community) could be negatively affected by "transgenic-wild" plants. Conventional crop alleles have facilitated the evolution of increased weediness in several wild populations. Thus, some transgenes that allow plants to tolerate biotic and abiotic stress (e.g., insect resistance, drought tolerance) could have similar effects.

INTRODUCTION

Tomato with delayed ripening and canola with altered oil content were, in 1994, the first commercially released transgenic crops in the United States. In the past ten years, the number of hectares planted with crops containing genetic material derived from unrelated species has increased dramatically. Worldwide in 2003 transgenic crops were planted on more than 67.7 million hectares (James 2003). The United States accounted for 63% of this total, growing transgenic crops on 42.8 million hectares. Other large producers are Argentina, Canada, Brazil, China, and South Africa, and together with the United States, these countries plant about 99% of all transgenic crops worldwide. In the United States in 2003, 40% of the corn, 81% of the soybeans, and 73% of the cotton was transgenic, and these crops together account for the vast majority of transgenic plantings. Other transgenic crops planted in the United States in 2003 include canola, squash, and papaya (James 2003).

Although more than 40 crop phenotypes have been approved for commercial release in the United States (ISB 2004a), herbicide tolerance (in soybean, cotton, corn, and canola) and insect resistance [conferred by toxin genes derived from *Bacillus thuringiensis* (*Bt*) in cotton and corn] account for nearly all the transgenic hectareage in this country (James 2003). Transgenic virus-resistant squash and papaya are also currently planted, but together they are planted on only a few thousand hectares (although more than half of U.S. papaya crop consists of transgenic varieties) (Gianessi et al. 2002). Other crops that have been approved for commercial sale, but that have been withdrawn from the market or are rarely adopted by farmers, include *Bt* potatoes, *Bt* and herbicide-tolerant sweet corn, herbicide-tolerant sugar beets, canola with altered seed oil content, and tomatoes with various quality traits.

Three types of potential risks are associated with the commercial release of transgenic crops: food safety, agronomic, and ecological. Food safety risks include, for example, the potential for allergenicity and decreased food quality of transgenic crops relative to their nontransgenic progenitors. Food safety issues have been reviewed by Kaeppler (2000), the Royal Society (2002), GM Science Review Panel (2003, 2004), Kok & Kuiper (2003), and Thomson (2003). Agronomic risks include crop-to-crop gene flow and the evolution of insecticide resistance in insect pests and herbicide resistance in weeds. These issues have been reviewed by Tabashnik (1994), Gould (1998), Shelton et al. (2000), Tabashnik et al. (2003), Martinez-Ghersa et al. (2003), and Mellon & Rissler (2004). The focus of this review is the potential for ecological risks resulting from the commercial release of transgenic crops. Over the past several years, ecological risks have been discussed at considerable length in the literature [Tiedje et al. 1989; Rissler & Mellon 1996; Snow & Moran-Palma 1997; Wolfenbarger & Phifer 2000; Dale et al. 2002; Letourneau & Burrows 2002; National Resource Council (NRC) 2000, 2002a; GM Science Review Panel 2003, 2004; Snow et al. 2004]. Also, many websites provide a wealth of information on transgenic crops. A particularly useful site is maintained by Information Systems for Biotechnology (ISB) at Virginia Tech University (www.isb.vt.edu); links to other sites can be found at the ISB site.

Ecological risks can be divided into two types. First, nontarget effects of transgenic crops occur when the expression of a transgene in a crop has negative effects on nontarget species. For example, corn engineered to express a *Bt* toxin gene will have intended negative effects on lepidopteran pests, but it may also have unintended direct or indirect effects on native nonpest species. Nontarget effects could also occur if the crop becomes more persistent in nonagricultural habitats. Second, transgenes might escape into wild populations through the hybridization of crop plants with their wild relatives. Transgenes that increase to high frequency in wild populations might affect seed production, population size, or habitat use in the wild species. In addition, transgenes for insect resistance that establish in wild populations could have negative effects on native herbivores as well as on species with which the native herbivores interact. Horizontal gene transfer, via recombination, from transgenic plants to viruses or bacteria is also possible.

NONTARGET EFFECTS OF GENETICALLY MODIFIED CROPS

The many hypothesized nontarget effects of transgenic crops fall into two general categories. First, nontarget species could be affected, either directly or indirectly, by the transgenic product. For example, the abundance of nonpest herbivores, predators and parasitoids of target species, or pollinators might be reduced on transgenic crops. Similarly, the soil microbial community might be altered under transgenic plants. Second, a transgene that alters the habitat requirements of a crop plant could allow cultivation of crops or persistence of feral plants in previously unsuitable habitats. This could lead to a reduction in the quantity or quality of native habitat.

Species Effects of Transgenic Crops

DIRECT EFFECTS ON NONTARGET HERBIVORES Researchers estimate that insect herbivory reduces crop yields by 30–40% worldwide (Oerke et al. 1994). In response, insecticides, traditionally bred insect resistance, and now transgenic resistance are used to reduce yield losses. Each of these management tools may have unintended effects on nontarget species. For example, effects of insecticides have been well documented (Benbrook 1996). By contrast, little work has evaluated nontarget effects of traditionally bred resistance. To the extent that transgenic resistance is specific to a particular group of pests (e.g., *Bt*s), effects on nontarget herbivores of transgenic crops might be less common.

A direct effect on nontarget species occurs when the transgenic trait in the crop negatively affects herbivores that do not reduce crop yield. These nontarget species generally feed on the crop, but they may also feed on plants in and adjacent to the crop field. Several studies have examined the direct effects of transgenic plants or artificial diet with transgenic protein on nontarget herbivores (see Poppy 2000 for review), although it is not clear that all these species would naturally feed on transgenic crops (see Supplemental Table 1 and nontarget entries in Supplemental Table 3: Follow the Supplemental Material link from the Annual Reviews home page at http://www.annualreviews.org). Most of these experiments evaluated the effect of a *Bt* toxin on a nonlepidopteran herbivore, and most studies found no effect. Just four studies evaluated the effects of transgenes other than *Bt*, and in all four a negative effect was found. Without knowing more about patterns of host use in these nontarget herbivores or about the effects of conventional management practices on nontarget species, it is difficult to evaluate the ecological importance of these results.

Less commonly, a transgenic plant could directly affect herbivores feeding on adjacent plants. Perhaps the best-known example is the case of *Bt* corn pollen and the monarch butterfly. Losey et al. (1999) reported that monarch larvae fed milkweed leaves dusted with *Bt* corn pollen in the laboratory had decreased survival. The logical conclusion of this work was that if milkweed plants adjacent to *Bt* corn frequently accumulate large amounts of *Bt* pollen, and if such plants constitute a large proportion of the milkweed available to monarchs, then monarch populations would be negatively affected. Since the original Losey et al. (1999) study, and an additional study by Jesse & Obrycki (2000), a huge effort has been made to evaluate the risk to monarchs presented by *Bt* corn. In particular, a research consortium funded by the U.S. Department of Agriculture (USDA) and an industry group completed several studies on different aspects of the monarch-milkweed-*Bt*-corn interaction and found the risk to be negligible (Sears et al. 2001, Wraight et al. 2000, Hellmich et al. 2001, Oberhauser et al. 2001, Pleasants et al. 2001, Stanley-Horn et al. 2001). Although high concentrations of *Bt* corn pollen clearly have negative effects on monarch larvae, only a very small proportion of milkweed plants accumulate any *Bt* pollen, let alone pollen in sufficient concentration to have negative effects (see Obrycki et al. 2001 for review). Few other cases of potential nontarget effects on adjacent plants have been examined (Wraight et al. 2000, Zangerl et al. 2001).

EFFECTS ON POLLINATORS The effect of transgenic pollen on foraging, learning, and life history characteristics of honeybees and bumblebees has been evaluated in several laboratory studies (Supplemental Table 2). The general conclusion from this work is that proteinase inhibitors have negative effects on bees, but that *Bt* does not (Supplemental Table 2; and for review, Malone & Pham-Delegue 2001). However, interpretation of these results is difficult because the concentrations of transgenic product used in these laboratory tests, often with artificial diet, may be greater than typically encountered in the field, and the importance of nontransgenic varieties and other species in the diet is unknown.

EFFECTS ON THE THIRD TROPHIC LEVEL Natural enemies of crop pests can greatly affect pest abundance and are often used in integrated pest management programs (Bellows & Hassell 1999). Conversely, broad-spectrum insecticides generally reduce natural enemy abundance (Croft 1990). Thus, a potential advantage of transgenic crops is that relatively specific resistance to particular insect pests might leave natural enemies unaffected. However, Groot & Dicke (2002) argue that little is known about how changes in herbivore abundance or quality might affect the food web.

A variety of studies have evaluated effects of transgenic crops on the natural enemies of crop-feeding herbivores (Supplemental Table 3; see also Schuler et al. 1999 for review). In these studies, all combinations of effects have been found. The prey species can be either affected or unaffected by eating the transgenic product. In turn, predators can be either affected or unaffected by eating prey that have eaten transgenic product. It appears that effects on the predator (third-trophic-level effects) are most likely when the prey species is affected by eating the transgenic crop (or transgenic product in an artificial diet). This result contrasts with the suggestion by Groot & Dicke (2002) that unaffected prey may not modify the transgenic product, which retains its toxicity to the predator. Thus, although direct toxicity is sometimes observed (e.g., Hilbeck et al. 1998), this effect may be uncommon.

In addition to effects of the transgenic product or prey quality on predator performance, reductions in prey abundance may also lead to a reduction in predator abundance in transgenic fields. For example, a specialist predator of Colorado potato beetle was less abundant in transgenic potato fields, presumably because its prey was also less abundant (Riddick et al. 1998). By contrast, the abundance of a generalist predator (that feeds on herbivores not targeted by *Bt* potato, as well as on Colorado potato beetle) was not affected by transgenic potato (Riddick et al. 1998). Similarly, in other field studies (Hoffmann et al. 1992, Pilcher et al. 1997, Orr & Landis 1997), the densities of predators were generally not reduced in transgenic fields.

EFFECTS ON THE SOIL COMMUNITY: MICROORGANISMS AND MACROFAUNA The composition and activity of the soil microbial community are profoundly affected by the plant community (through both root exudates and litter quality and quantity), as well as by characteristics of the soil, climate, and, in agricultural communities, management practices. For example, plant species, and even crop variety, can have very large effects on rhizosphere microbial communities (Kourtev et al. 2003; Wardle 2002, Chapter 3). It is against this backdrop of natural variation that researchers must evaluate the potential effect of transgenic crop varieties. Studies of the effects of transgenic crops on the soil microbial community are summarized in Supplemental Table 4. Kowalchuk et al. (2003) have reviewed these effects and conclude that changes induced by transgenic plants are generally small relative to the effects of plant community and ecosystem properties. Similarly, transgenic varieties generally do not have negative effects on soil macrofauna (Supplemental Table 5).

Transgenic crops could affect the soil community in one of two ways. First, transgenic products could exude from roots and directly affect soil organisms; this possibility has been evaluated in a number of studies (e.g., Saxena et al. 1999, Saxena & Stotsky 2001). Second, transgenic crops could affect plant tissue quality (e.g., lignin or cellulose content) (Escher et al. 2000; Hopkins et al. 2001; for review, see Halpin & Boerjan 2003) and therefore affect tissue decomposition rates. Effects on tissue quality are most likely to have important ecological consequences (Kowalchuk et al. 2003). However, because links between microbial community structure and functional consequences (such as C and N cycling) are only poorly understood (Wardle 2002), these effects are difficult to evaluate.

Habitat Effects of Transgenic Crops

TRANSGENIC CROPS MAY BECOME FERAL If a cultivar contains a transgene that enhances the fitness of crop plants in weedy environments, the crop might become feral or might persist longer in feral populations. The potential for increased weediness in feral crops has been discussed widely (Snow & Moran-Palma 1997, Rissler & Mellon 1996, Warwick et al. 1999, Marvier 2001) but has received relatively little empirical attention. Parker & Kareiva (1996) and Bullock (1999) advocate the use of population matrix models to evaluate this risk. This approach involves gathering demographic data on nontransgenic varieties of crops in appropriate habitats (such as fields and roadsides), using these data to construct projection matrices, then performing elasticity analyses to determine which demographic transitions have the largest effects on λ, the annual rate of population increase. By comparing the transitions affected by particular transgenes with the transitions important for population growth, the potential weediness of a transgenic variety can be evaluated.

Few studies have taken such an explicitly demographic approach to evaluate the potential for transgenic crops to persist in feral populations. Parker & Kareiva (1996) found that transgenic oil-modified canola varieties had λ's that were not significantly different from those of nontransgenic isolines. Crawley et al. (1993, 2001) found no differences in persistence between transgenic herbicide-tolerant and nontransgenic varieties of oilseed rape, maize, beet, or potato. However, because the varieties used in Crawley et al.'s experiments were herbicide tolerant, and persistence was evaluated in the absence of herbicide, these results are not especially informative. It seems more likely that resistance to insects, pathogens, or other environmental stresses will enhance persistence in feral populations. For example, oilseed rape persists on road verges in France (Pessel et al. 2001), and time to extinction of feral populations might increase if plants contained transgenic insect resistance or drought tolerance.

REDUCTION IN WEED OR PEST POPULATIONS MAY NEGATIVELY AFFECT SPECIES USING AGRICULTURAL HABITATS   Crop yields can be substantially reduced by insects, other pests, and weeds. To maintain yield, farmers spray herbicides and pesticides, cultivate fields, alter planting times, and plant conventionally bred varieties that are resistant to pests or are tolerant of herbicides. This intensification of farm management practices has led to a reduction in farmland biodiversity, including reductions in weed, invertebrate, and bird populations (reviewed in Champion et al. 2003). Transgenic crops represent a further intensification of agriculture, and the effects of this technology on remaining farmland biodiversity are of some concern. For example, Watkinson et al. (2000) modeled the effect of herbicide-resistant transgenic sugar beet on weed biodiversity and bird populations using farmland in Great Britain. Depending on the particular management practices considered, their model predicts large decreases in weed biodiversity. Because insects feed on these weeds, and some farmland birds depend on these insects, decreased weed biodiversity and abundance results in decreased bird populations.

In response to this concern, the Farm Scale Evaluations (FSE) were initiated in Great Britain in 2000. These studies involve transgenic and conventional varieties of four crops, each planted in a split-field design at 60–70 sites across Great

Britain (Firbank et al. 2003, Squire et al. 2003, Champion et al. 2003). The transgenic crops included in this study are glyphosate-resistant sugar beet, glufosinate-resistant field corn, and glufosinate-resistant spring- and winter-sown oilseed rape (canola); results for winter-sown rape are not yet published. Dicot and monocot weed populations, including seedbank populations (Heard et al. 2003a,b); above- and below-ground invertebrate biodiversity, including pollinators (Brooks et al. 2003, Haughton et al. 2003); and higher-trophic-level effects (Hawes et al. 2003) were monitored over the following three years in the fields and at field margins (Roy et al. 2003). The larger objective of this work is to determine if genetically modified crops (and the management practices associated with these crops) affect farmland biodiversity differently than the management practices associated with conventional agriculture.

Data presented in papers published in 2003 (cited in the previous paragraph) suggest that transgenic varieties of sugar beet and spring-sown oilseed rape have reduced above- and below-ground (seedbank) populations of weeds, and reduced weed populations have led to generally reduced insect populations. In contrast, transgenic corn had higher weed and weed seedbank densities and generally higher insect populations than did the conventional corn fields. This is presumably because atrazine, the conventional herbicide used in corn, was more effective than glufosinate. A detailed interpretation of the FSE results and an in-depth review of other studies examining effects of transgenic herbicide-resistant crops is found in Squire et al. (2003), and Andow (2003) provides a succinct graphical summary of the results published to date. Similarly, in a comparison of transgenic and conventional soybeans, Buckelew et al. (2000) reported that weed management systems that allowed more weeds to persist supported larger insect populations.

The FSE results have been interpreted by some as evidence that transgenic herbicide-resistant beet and oilseed rape are bad for the environment, whereas transgenic herbicide-resistant maize is good for the environment (for discussion, see Dewar et al. 2004). However, another interpretation might be that transgenic varieties with improved weed control will allow higher yields and thus permit the return of marginal agricultural land to natural vegetation. In addition, application of generally less toxic herbicides (e.g., glyphosate) may be environmentally beneficial. Thus, the negative effects of transgenic crops on farmland biodiversity must be weighed against the potential benefits associated with increased yield and decreased herbicide use.

ADDITIONAL LAND MAY BE PLACED IN CULTIVATION   Transgenes that affect agronomic properties (e.g., salt tolerance, Zhang et al. 2001; water stress, Hsieh et al. 2002; low soil iron availability, Takahashi et al. 2001) could allow cultivation of currently marginal land that is nonetheless good native habitat. Conversely, through irrigation farmland often becomes too saline to sustain economically viable crops (Ghassemi et al. 1995), and transgenic salt tolerance could allow this land to remain in cultivation. These effects are currently hypothetical and have received little consideration.

MECHANISMS AND RISK OF ESCAPE OF TRANSGENES INTO WILD POPULATIONS

There are two mechanisms through which transgenes could move into wild populations: horizontal gene transfer and hybridization. Each of these mechanisms is discussed below.

Horizontal Gene Transfer

VIRUSES Transgenic virus resistance in crops released thus far has been achieved by insertion of a viral coat protein, and there is concern that recombination between infecting viruses and the transgene could result in new virus genotypes with altered host range, transmissibility, or virulence. Accumulating evidence suggests that virus/transgene recombination is likely, although most (but not all; see Aaziz & Tepfer 1999) studies have been done in laboratory conditions favoring recombination (for review, see Hammond et al. 1999, Rubio et al. 1999, Power 2002, Tepfer 2002). As discussed by Hammond et al. (1999), the ecological risks of such recombination must be evaluated against the effects of natural recombination between viruses in mixed infections, which presumably occurs at similar rates. Because such recombination can have large effects on the population biology of viruses, and in fact is responsible for much viral evolution (Roossinck 1997), the potential ecological consequences (e.g., on the effects of virus infection in wild plants) deserve further attention (Tepfer 2002, Power 2002).

Another risk associated with virus coat proteins in transgenic crops is the possibility that viral RNA will be encapsidated with transgenic coat protein. Transencapsidation is known to occur in natural mixed infections, and it can alter the transmissibility of a virus. However, because the viral genome is unaltered, this effect does not persist, and for this reason ecological risks associated with transencapsidation are thought to be low (Hammond et al. 1999). Finally, increased disease severity may occur if there are synergistic interactions between invading viruses, and by extension, potentially between invading viruses and transgenic viral genes (Tepfer 2002, Power 2002).

SOIL BACTERIA Horizontal gene transfer could also occur by natural transformation, a process through which DNA that is free in the environment can be stably integrated into the genome of competent bacteria (Nielsen et al. 1998, Bertolla & Simonet 1999). Although natural transformation is probably rare in nature (Nielsen et al. 1998, Gebhard & Smalla 1999, Nielsen et al. 2001), it is generally observed at low frequency in favorable experimental conditions (Schlüter et al. 1995; Nielsen et al. 1997, 2000a,b; Kay et al. 2002).

Hybridization with Wild Relatives

The ecological risks associated with the movement of transgenes into wild populations via hybridization must be evaluated by sequentially addressing three questions: (*a*) Is the transgenic variety of the cultivated plant sexually compatible with wild relatives? If the crop has no compatible wild relatives, then no crop genes can move into wild populations, and there is no risk of transgene escape. However, if the crop does have compatible relatives, the next question is, (*b*) Will the transgene increase in frequency in the wild population (either by demographic swamping or by natural selection)? If the transgene is not expected to increase in frequency, then the ecological risks of transgene escape are minimal. Finally, if the transgene increases the fitness of the wild relative, and therefore the transgene is expected to increase in frequency in wild populations by natural selection, the third question is, (*c*) What are the ecological consequences of the escape of the transgene into a wild population? We address the first two questions in the next two sections. In the Consequences of the Escape of Transgenes into Wild Populations section below, we discuss the third question.

IS THE TRANSGENIC VARIETY OF THE CULTIVATED PLANT SEXUALLY COMPATIBLE WITH WILD RELATIVES? All crop species were derived from one (or more) wild species, and most crops are planted sympatrically with compatible wild relatives somewhere in the world (Ellstrand 2003, Ellstrand et al. 1999). In addition, although the fertility of F1 crop-wild hybrids varies dramatically, both among crops and among crosses within crops, frequently fertility is restored in F2, BC1, and later generations. For example, in sunflower, F1 crop-wild hybrids have fertilities ranging from 1% to 100% of wild fertility, depending on the cross (Snow et al. 1998, Cummings et al. 2002), and the frequency of cultivar-specific alleles drops in later generations (Cummings et al. 2002). However, several studies have demonstrated that cultivar alleles persist in wild and weedy sunflower populations (Arias & Rieseberg 1994, Linder et al. 1998, Whitton et al. 1997). In fact, hybridization with wild relatives has been documented for 22 of the 25 most important crop species (by worldwide area planted) (Ellstrand 2003, and see Eastham & Sweet 2002), and it is likely that crop genes have introgressed into these wild populations. In addition to the overwhelming empirical evidence of gene flow between crops and their wild relatives (e.g., Warwick et al. 2003), theoretical models also suggest that introgression of transgenes is likely (e.g., Meagher et al. 2003, Thompson et al. 2003).

Because it is generally expected that crops will hybridize with their wild relatives, a variety of genetic methods of reducing the probability of introgression are under consideration and development (reviewed in Daniell 2002, NRC 2004). These include placing transgenes on chromosomes or chromosome segments that are less likely to introgress into wild populations (e.g., Rieseberg et al. 1999, 2000), closely linking transgenes with genes for domestication traits that are expected to have low fitness in wild populations, and controlling the viability or fertility of hybrid offspring using gene use–restriction technology.

Another way to reduce the risk of crop-wild hybridization is to restrict transgenic crops to areas where wild relatives do not occur. For example, wild relatives of corn and soybeans are not native to the United States, and thus, in this country, transgenes from these crops cannot move into wild relatives. Similarly, in the United States transgenic cotton is restricted to areas where wild relatives of cotton do not grow naturally (Mendelsohn et al. 2003). By contrast, transgenic virus-resistant squash (*Cucurbito pepo*) is planted within the native range of wild squash, also *Cucurbito pepo* (Wilson 1993). Although cultivated alleles have been found in wild populations (Kirkpatrick & Wilson 1988, Decker-Walters et al. 2002), little is known about either the importance of viruses in wild populations (Spencer & Snow 2001) or the frequency of hybridization between cultivated and wild squash.

Because humans move seeds around the globe, both intentionally and accidentally, restricting transgenic crops to areas outside the native range will only temporarily delay the movement of transgenes into wild relatives. For example, even though cultivation of transgenic corn has been prohibited in Mexico since 1998, transgenic material has been found in land races of maize in remote mountainous regions of Oaxaca (Quist & Chapela 2001, 2002; and see Biotech InfoNet 2002, Mann 2002). The ecological effect of a *Bt* gene that introgresses into the land races will clearly depend on the importance of lepidopteran herbivores in these populations, which is unknown. If pleiotropic effects of the *Bt* gene are weak or absent, and if linkage to other cultivar-derived traits is reduced by recombination, the *Bt* gene will only increase in frequency in the land races if it leads to the reduction of lepidopteran damage and, therefore, to an increase in seed production. Moreover, even if selection favoring the *Bt* gene in land races is strong, genetic diversity in the land races will be reduced only at loci tightly linked to the *Bt* gene itself (Maynard Smith & Haigh 1974). Thus, overall genetic diversity in the land races (and in other wild populations into which transgenes introgress) is unlikely to be affected by the escape of transgenes. In fact, at least initially, genetic diversity in wild populations hybridizing with crops may be increased (Ellstrand 2003, table 9.1).

WILL THE TRANSGENE INCREASE IN FREQUENCY IN THE WILD POPULATION? In contrast to the huge literature evaluating the potential for hybridization between transgenic crops and their wild relatives, there are few studies that have attempted to evaluate the fitness effects of transgenes once they have entered wild populations (Letourneau et al. 2003). Moreover, many authors seem to assume that if hybridization is rare (or can be made less common by the use of appropriate genetic technologies), then the consequences of hybridization do not need to be considered (e.g., Stewart et al. 2003). Similarly, many authors have suggested that if crop-wild hybrids have low fitness, the probability of introgression of a transgene into the wild population is low. This assertion is contradicted by theoretical work (Barton 1986) indicating that the rate of introgression of an allele from one population to another through hybrids can be very rapid even if the selective advantage is small, as well as by the frequent occurrence of transgressive hybrids (Rieseberg et al. 2003). In addition, Rieseberg & Wendel (1993) have documented the importance of introgression in the evolution of plant populations. Thus, evaluating the conditions under which transgenes will increase in frequency in wild populations is crucial.

A transgene could increase in frequency by either demographic swamping or by natural selection. Demographic swamping occurs with continual migration from a large source population (e.g., a crop) into a smaller recipient population (e.g., a wild relative). In a theoretical model, Haygood et al. (2003) found that with continual migration, as would occur when the crop is continuously planted, crop alleles can rapidly become fixed in wild populations, even when they are deleterious. In particular, alleles that reduce fitness can be fixed if the migration rate exceeds the selection coefficient, and when this occurs demographic swamping can lead to reduced population size and possibly local extinction. These effects (see also Huxel 1999, Wolf et al. 2001, Ferdy & Austerlitz 2002) could lead to extinction by hybridization (e.g., Levin et al. 1996, Rhymer & Simberloff 1996) and to wild populations that are endangered because of hybridization with crops (Ellstrand 2003, table 9.4).

If, instead, the effects of natural selection are expected to be more important than the effects of migration, then we must evaluate the fitness effects of the transgene in wild plants. The appropriate way to evaluate these effects is to compare the fitness of transgenic plants with the fitness of plants of the same genetic background but without the transgene. From this sort of experiment it is possible to estimate s, the selection coefficient, which quantifies the strength of selection either favoring or not favoring genotypes that include the transgene (Hartl & Clark 1997). Statistically significant selection favoring transgenic plants implies that the transgene will increase in frequency in wild populations. Moreover, if the phenotype of transgenic plants is the phenotype predicted by the transgene (e.g., lepidopteran resistance in the case of a *Bt* gene), then increased (or decreased) fitness of transgenic plants can more confidently be ascribed to the transgene, rather than to an idiosyncratic position effect. Ideally, these experiments should be done in more than one environment and year, so that the results can be more easily generalized. In addition, if the benefit of the transgene is predicted to vary across environments (e.g., because the environments vary in herbivore pressure) then the strength of selection favoring the transgene should vary as well.

If the transgene is of no benefit (or is always costly) to the wild plant, genetic drift (or purifying selection) will determine its fate in the wild population. For example, transgenes controlling traits such as ease of harvest, fruit ripening, and product shelf life, as well as newer generation traits such as pharmaceutical or industrial chemical production, are likely to be neutral or costly in wild populations. Similarly, although weeds with transgenic herbicide resistance may be more difficult to control with herbicides, these genes are likely to be neutral or costly in wild populations that are not sprayed (Snow et al. 1999, Gueritaine et al. 2002, Zhang et al. 2003).

In contrast to costly or neutral characters, transgenes for characters such as insect or pathogen resistance and drought tolerance may benefit wild populations, and therefore may increase in frequency in wild populations by natural selection. For example, damage by herbivores is generally detrimental to plant fitness in wild populations (e.g., Marquis 1992, Crawley 1997, Letourneau et al. 2002), and similarly viruses and fungal pathogens also commonly reduce fitness of wild plants (e.g., Burdon 1987, Jarosz & Davelos 1995, Maskell et al. 1999). These results suggest that transgenic herbivore and pathogen resistance would be favored by natural selection in many wild relatives of crop plants. The evolutionary dynamics of each of these traits will be determined by the balance between the benefit of the trait (in the presence of the selective agent, such as herbivores) and the cost of the trait (the reduction in fitness of transgenic individuals in the absence of the selective agent) (Simms & Rausher 1987).

Only one study has evaluated the fitness effects of a transgene in a wild genetic background in an ecologically relevant environment. In field-planted BC1 and BC3 *Helianthus annuus* with a *Bt* transgene, lepidopteran damage was reduced to near zero, and seed production was increased 15–55% (relative to plants without the transgene), depending on the year and field site (Snow et al. 2003, Pilson et al. 2004). Because lepidopteran damage is known to reduce fitness in wild *Helianthus annuus* populations (Pilson 2000, Pilson & Decker 2002), this result is not surprising. Moreover, when BC1 plants were grown in the greenhouse (in the absence of herbivores), there was neither a benefit nor a cost of the transgene (Snow et al. 2003). These results suggest that, if a *Bt* gene were to escape into wild sunflower populations, it would increase in frequency by natural selection, and the rate of increase would vary in space and time as a function of the abundance of lepidopteran herbivores.

A few additional studies have evaluated effects of transgenes in F1 or BC plants. In a cage experiment replicated in three states in the United States, Burke & Rieseberg (2003) experimentally inoculated BC3 sunflower segregating for a transgene conferring resistance to white mold. They found that both the effect of the transgene on infection frequency and the effect of infection on seed production depended on the location of the experiment. Furthermore, because the transgene had no effect on infection in locations where the pathogen had large effects on fitness, in this experiment the transgene provided no fitness benefit. Fuchs & Gonsalves (1997, 1999) found that transgenic virus-resistant F1 and BC1 squash inoculated with virus do not display symptoms and do produce fruit. BC1 and BC2 *Brassica rapa* plants (with a *Bt* gene introgressed from *Brassica napus*) produced more reproductive biomass than nontransgenic plants when inoculated with high, but not low, densities of diamondback moth larvae (Mason et al. 2003). Furthermore, in the absence of moth larvae there was no effect of the *Bt* gene on plant growth, indicating that the *Bt* gene is not costly to the plant (Mason et al. 2003). However, interpreting these studies is difficult because the incidences of naturally occurring white mold in wild sunflower populations, virus in wild squash populations, and diamondback moth in weedy *B. rapa* populations are not reported.

Because it is known that conventionally bred crop alleles have introgressed into wild populations, one way to evaluate the potential effects of transgenes in these populations is by examining the effects of these conventional alleles. Ellstrand (2003) documents 16 cases in which crop-wild hybridization has led either to the evolution of a new taxon (Ellstrand 2003, table 9.2) or to a wild population with changed ecological properties (Ellstrand 2003, table 9.3). In most of these cases the newly derived taxon or population is weedier than its wild parent. Clearly, crop alleles can sometimes enhance weediness in wild plants, but whether transgenes are more likely than conventional alleles to have this effect is an open question. Genes that affect environmental tolerances (e.g., resistance to biotic and abiotic stress) would seem the most likely candidates.

Genetic and ecological models provide another way to evaluate the potential effects of transgene movement into wild populations. Muir & Howard (1999, 2001, 2002; Howard et al. 2004) have developed and parameterized a model for transgenic fish that is now being modified for application to plant populations (ISB 2004b). Their original model predicts that a transgene that increases growth rate and adult size can (paradoxically) cause the extinction of wild populations. Extinction occurs because the transgene has opposite effects on different components of fitness, and this phenomenon has been called the Trojan gene hypothesis. Specifically, larger fish (with the transgene) get most of the matings, but because their transgenic offspring have lower viability, few survive to maturity, and population size declines. Evaluating the effects of a transgene on various components of fitness in plant populations is clearly of value (e.g., to measure λ; see Parker & Kareiva 1996, Bullock 1999). However, for plants it is not clear how frequently transgenes might have such different effects on mating success and offspring viability.

CONSEQUENCES OF THE ESCAPE OF TRANSGENES INTO WILD POPULATIONS

Although the movement of a transgene into a wild population, and its subsequent increase in frequency, are necessary, they are not sufficient to predict the environmental consequences of transgene escape. Specifically, these processes are only important to the extent that they lead to the alteration of existing ecological interactions between the wild plant and its biotic and abiotic environment. Thus, it is necessary to answer the third question raised above: What are the ecological consequences of the escape of the transgene into a wild population? A transgene that increases in frequency by natural selection in a wild population does so, by definition, because it increases survival or fecundity, and one ecological risk is the effect of increased individual fitness on population size, dynamics, and habitat use in the wild plant. In addition, transgenes that confer resistance to herbivores and pathogens will have direct effects on native species using the wild plant as a host. Clearly, these questions must be the crux of any ecological risk assessment. However, very little work has been done in these areas.

Population Dynamics and Habitat Use in the Wild Relative

A much discussed potential consequence of the escape of transgenes into wild populations is that transgenic-wild plants will become weedier (Darmency 1994, Snow & Moran-Palma 1997, Warwick et al. 1999). The size of transgenic-wild populations could increase in the wild plant's original habitat, or broader environmental tolerances could allow transgenic-wild plants to invade previously unsuitable habitat. These risks are difficult to evaluate because little is known about factors controlling population size or range in plants, or about characteristics of invading species and the receiving community that allow invasion (Hoffmann & Blows 1994, Lonsdale 1999, Kolar & Lodge 2001, Gerlach & Rice 2003).

INCREASE IN POPULATION SIZE   A transgenic-wild population could increase in size if either fecundity (e.g., Snow et al. 2003) or competitive ability (e.g., Damgaard & Jenson 2002) increase as a result of expressing the transgene, and if population size was previously limited by characters affected by the transgene. However, this possibility is difficult to evaluate because little is known about factors controlling population size in natural plant populations. However, changes in fecundity can have large effects on population growth rate (λ; Silvertown et al. 1993), suggesting that transgenes that increase fecundity could lead to increases in population size. Fecundity in sunflowers is limited by herbivory (Pilson 2000), and experimental sunflower populations in western Nebraska appear to be seed-limited, suggesting that a reduction in herbivory would lead to larger populations. In contrast, in similar populations in eastern Kansas population size is affected more by density-dependent processes (such as competition for resources) than by seed production (D. Pilson, H. Alexander, J. Moody-Weis, A. Snow, manuscript in preparation). Seed production by individual thistle plants and thistle populations can be limited by insect herbivory (Louda & Potvin 1995, Guretzky & Louda 1997). Weedy roadway populations of oilseed rape were seed-limited at the landscape scale (Crawley & Brown 1995). Small experimental populations of *Arabidopsis thaliana* were not seed-limited (Bergelson 1994).

Within a community, a likely consequence of an increase in population size of one species is the decrease in size of others. For example, in experimental plots in which goldenrod plants were sprayed with insecticide, goldenrod abundance increased while the abundance of other species decreased (Carson & Root 1999, 2000). This effect was attributed to increased competitive ability when plants were not subject to insect attack. These results suggest that if insect-resistant transgenic-wild plants (or plants with any character that increases competitive ability) are present, the abundance of other species could decline. Changes in community structure in natural habitats are thus one potential consequence of the incorporation of transgenes into wild populations.

INVASION OF PREVIOUSLY UNSUITABLE HABITAT   In addition to increased population size in their original habitat, transgenic-wild plants might also be able to invade previously unsuitable habitats. For example, one reason introduced plants are believed to become invasive is that they have escaped from their natural enemies (Mack et al. 2000, Sakai et al. 2001, NRC 2002b, Louda et al. 2003, Mitchell & Power 2003, Callaway et al. 2004). Escape from natural enemies might allow invasion directly, or reduced natural enemy attack might free resources and thus allow the evolution of increased competitive ability (Blossey & Nötzold 1995; but see Willis et al. 1999, 2000). In addition, hybridization has been shown to provide the genetic variation necessary for the evolution of invasiveness in many taxa (Ellstrand & Schierenbeck 2000), and transgressive segregation is common in crop and wild populations (Rieseberg et al. 2003). Taken together these considerations suggest that crop-wild hybrids could facilitate the evolution of broader habitat tolerances, and perhaps invasiveness, in wild populations.

Insect and Pathogen Community Structure

If transgenes for resistance to herbivores or pathogens increase to high frequency in wild populations, there will be immediate, and negative, effects on those native species that were responsible for selection to increase the frequency of the transgene in the first place. If these species are specialists, their populations could decline dramatically, perhaps resulting in local extinction. By contrast, if these species have additional host plants, the effect of the transgene will depend on how important the transgenic-wild plant is in the diet of the herbivore or pathogen. For example, although many moths feed on more than one *Helianthus* sp., *Helianthus annuus* is the most important host for many lepidopteran herbivores (Charlet et al. 1992, 1997). This fact suggests that if a *Bt* gene were to enter wild *H. annuus* populations (Snow et al. 2003, Pilson et al. 2004), the abundance of *H. annuus*–feeding lepidopterans would decrease. Furthermore, because one of the most common moths feeding on *H. annuus* has negative competitive effects on a seed midge and a seed weevil, reduction in lepidopteran abundance could result in increased seed midge and seed weevil populations (M. Paulsen & D. Pilson, manuscript in preparation). Effects at the next trophic level might include increased abundance of seed midge and seed weevil parasitoids and decreased abundance of lepidopteran parasitoids. Community-level consequences of the removal of a few species are likely to be quite complex and unpredictable (Denno et al. 1995, Polis & Strong 1996, Dunne et al. 2002, Groot & Dicke 2002).

Changes in community structure that occur because of transgenic-wild plants will not be static. Specifically, transgenic herbivore or pathogen resistance will impose strong selection on the affected species to evolve counter-resistance, similar to the selection pressure imposed by conventional insecticides. Crop pests have evolved resistance to virtually all chemical insecticides applied to crops (NRC 1986), and thus there is every reason to expect that herbivores in natural systems will evolve resistance to transgenic pesticides. The speed with which this happens will depend on the strength of selection on the transgenic host, the presence of alternative hosts, and pre-existing variation for resistance.

CURRENT TRANSGENES VERSUS TRANSGENES OF THE FUTURE

Most of our discussion has focused on transgenic traits that are likely to enhance the fitness or environmental tolerances of wild plants (e.g., insect and pathogen resistance, drought tolerance), because these traits are most likely to affect natural plant and animal populations. Furthermore, much of our discussion of insect resistance has focused on *Bt* crops because these have been released commercially and extensively evaluated (Shelton et al. 2002, Mendelsohn 2003). In the future, crops with other types of insect resistance are likely (Schuler et al. 1999, Moar 2003), as are crops with transgenic resistance to pathogens and nematodes (Atkinson et al. 2003). The sorts of ecological risks discussed here will likely be relevant to these newer transgenic products as well. By contrast, the next generation of transgenic products will probably also include plants with enhanced product quality (e.g., increased yield, altered ripening time or nutritional content, reduced lignin content in trees), as well as plants producing industrial and pharmaceutical chemicals (Dunwell 1999, Fischer & Emans 2000, Jaworski & Cahoon 2003, Ma et al. 2003, Sinclair et al. 2004). Although some of these genes could present very large agronomic or food safety risks, they seem less likely to have important ecological effects. This is because these traits are less likely to increase the fitness of wild plants and so are less likely to increase in frequency in wild populations.

POTENTIAL BENEFITS OF TRANSGENIC CROPS

Although the focus of this review is the ecological risks associated with the commercial release of transgenic crops, it is important to recognize that there may be ecological benefits as well (Wolfenbarger & Phifer 2000; Hails 2000). One potential benefit is a reduction in pesticide and herbicide use. For example, in the United States insecticide use has decreased on corn and cotton since the release of *Bt* varieties of these crops (Benbrook 2003). Similarly, in China planting *Bt* cotton has resulted in dramatic reductions in pesticide use (with both environmental and human health benefits) and increased yields (Pray et al. 2002). In addition, the use of *Bt* crops, rather than broad spectrum insecticides, could allow larger populations of beneficial insects and nonpest herbivores to persist in crop fields. However, herbicide use on herbicide-tolerant crops, especially soybean, has increased since the release of transgenic varieties, although this increase has largely been the result of increases in relatively benign herbicides such as glyphosate (Benbrook 2003). Another potential benefit of transgenic herbicide-tolerant crops is an increase in no-till or other conservation tillage practices, which lead to reduced soil erosion and run-off to streams, reduced fuel use (Fawcett & Towery 2003), and increased sequestration of atmospheric carbon. Evaluating the ecological benefits of transgenic crops is not straightforward, and no comprehensive reviews have been published.

CONCLUSIONS

We draw three general conclusions. First, transgenes that affect plant response to biotic and abiotic stress (e.g., insect and pathogen attack, drought and salt tolerance) are more likely to have negative ecological effects than are transgenes for traits affecting product quality or industrial and pharmaceutical chemical production. Second, escape of transgenes into wild populations, via hybridization and introgression, is more likely to result in negative ecological effects than are nontarget effects of the transgenic crop itself. The escape of transgenes into wild populations could lead to increased weediness or the invasion of new habitats by the wild population. In addition, native species with which the wild plant interacts (including herbivores, pathogens, and other plant species in the community) could be negatively affected by transgenic-wild plants. Conventional crop alleles have allowed the evolution of increased weediness in several wild populations. Thus, there is reason to believe that some transgenes (e.g., for insect resistance, drought tolerance) could have similar effects. Finally, there are relatively few data available with which to evaluate the potential for increased weediness in wild relatives of crop plants. A better understanding of factors controlling population size, dynamics, and range limits in weedy plants is necessary before a full ecological risk assessment can be made.

ACKNOWLEDGMENTS

Kjärstin Carlson did a tremendous job locating references and organizing the reference list. Allison Snow commented on an earlier version of the manuscript. Discussions with Allison Snow and Helen Alexander have greatly influenced our perspective on ecological risks. Our work in this area has been supported by the USDA Biotechnology Risk Assessment Program.

The *Annual Review of Ecology, Evolution, and Systematics* is online at
http://ecolsys.annualreviews.org

LITERATURE CITED

Aaziz R, Tepfer M. 1999. Recombination between genomic RNAs of two cucumoviruses under conditions of minimal selection pressure. *Virology* 263:282–89

Andow DA. 2003. UK farm-scale evaluations of transgenic herbicide-tolerant crops. *Nat. Biotechnol.* 21:1453–54

Arias DM, Rieseberg LH. 1994. Gene flow between cultivated and wild sunflowers. *Theor. Appl. Genet.* 89:655–60

Atkinson HJ, Urwin PE, McPherson MJ. 2003. Engineering plants for nematode resistance. *Annu. Rev. Phytopathol.* 41:615–39

Barton NH. 1986. The effects of linkage and density-dependent regulation on gene flow. *Heredity* 57:415–26

Bellows TS, Hassell MP. 1999. Theories and mechanisms of natural population regulation. In *Handbook of Biological Control*, ed. TS Bellows, TW Fisher, pp. 17–44. San Diego, CA: Academic

Benbrook CM. 1996. *Pest Management at the Crossroads*. Yonkers, NY: Consumers Union

Benbrook CM. 2003. *Impacts of genetically engineered crops on pesticide use in the United States: the first eight years.* http://www.biotech-info.net/Technical_Paper_6.pdf

Bergelson J. 1994. Changes in fecundity do not predict invasiveness: a model study of transgenic plants. *Ecology* 75:249–52

Bertolla F, Simonet P. 1999. Horizontal gene transfers in the environment: natural transformation as a putative process for gene transfers between transgenic plants and microorganisms. *Res. Microbiol.* 150:375–84

Biotech InfoNet. 2002. *Bt corn gene flow in Mexico.* http://www.biotechinfo.net/mexican_bt_flow.html#overview

Blossey B, Nötzold R. 1995. Evolution of increased competitive ability in invasive nonindigenous plants: a hypothesis. *J. Ecol.* 83:887–89

Brooks DR, Bohan DA, Champion GT, Haughton AJ, Hawes C, et al. 2003. Invertebrate responses to the management of genetically modified herbicide-tolerant and conventional spring crops. I. Soil-surface-active invertebrates. *Philos. Trans. R. Soc. London Ser. B* 358:1847–62

Buckelew LD, Pedigo LP, Mero HM, Owen MDK, Tylka GL. 2000. Effects of weed management systems on canopy insects in herbicide-resistant soybeans. *J. Econ. Entomol.* 93:1437–43

Bullock JM. 1999. Using population matrix models to target GMO risk assessment. *Aspects Appl. Biol.* 53:205–12

Burdon JJ. 1987. *Diseases and Plant Population Biology.* New York: Cambridge Univ. Press. 208 pp.

Burke JM, Rieseberg LH. 2003. Fitness effects of transgenic disease resistance in sunflowers. *Science* 300:1250

Callaway RM, Thelen GC, Rodriguez A, Holben WE. 2004. Soil biota and exotic plant invasion. *Nature* 427:731–33

Carson WP, Root RB. 1999. Top-down effects of insect herbivores during early succession: influence on biomass and plant dominance. *Oecologia* 121:260–72

Carson WP, Root RB. 2000. Herbivory and plant species coexistence: community regulation by an outbreaking phytophagous insect. *Ecol. Monogr.* 70:73–99

Champion GT, May MJ, Bennett S, Brooks DR, Clark SJ, et al. 2003. Crop management and agronomic context of the Farm Scale Evaluations of genetically modified herbicide-tolerant crops. *Philos. Trans. R. Soc. London Ser. B* 358:1801–18

Charlet LD, Brewer GJ, Beregovoy VH. 1992. Insect fauna of the heads and stems of native sunflowers (Asterales: Asteraceae) in eastern North Dakota. *Environ. Entomol.* 21:493–500

Charlet LD, Brewer GJ, Franzmann BA. 1997. Sunflower insects. In *Sunflower Technology and Production*, ed. AA Schneiter, pp. 183–261. Madison, WI: Soil Sci. Soc. Am.

Crawley MJ. 1997. Plant-herbivore dynamics. In *Plant Ecology*, ed. MJ Crawley, pp. 401–74. Oxford, UK: Blackwell Sci.

Crawley MJ, Brown SL. 1995. Seed limitation and the dynamics of feral oilseed rape on the M25 motorway. *Proc. R. Soc. London Ser. B* 259:49–54

Crawley MJ, Brown SL, Hails RS, Kohn DD, Rees M. 2001. Transgenic crops in natural habitats. *Nature* 409:682–83

Crawley MJ, Hails RS, Rees M, Kohn D, Buxton J. 1993. Ecology of transgenic oilseed rape in natural habitats. *Nature* 363:620–23

Croft BA. 1990. *Arthropod Biological Control Agents and Pesticides.* New York: Wiley. 723 pp.

Cummings CL, Alexander HM, Snow AA, Rieseberg LH, Kim MJ, Culley TM. 2002. Fecundity selection in a sunflower crop-wild study: Can ecological data predict crop allele changes? *Ecol. Appl.* 12:1661–71

Dale PJ, Clarke B, Fontes EMG. 2002. Potential for the environmental impact of transgenic crops. *Nat. Biotechnol.* 20:567–74

Damgaard C, Jensen BD. 2002. Disease resistance in *Arabidopsis thaliana* increases the competitive ability and the predicted probability of long-term ecological success under disease pressure. *Oikos* 98:459–66

Daniell H. 2002. Molecular strategies for gene containment in transgenic crops. *Nat. Biotechnol.* 30:581–86

Darmency H. 1994. The impact of hybrids between genetically modified crop plants and their related species: introgression and weediness. *Mol. Ecol.* 3:37–40

Decker-Walters DS, Staub JE, Chung SM, Nakata E, Quemada HD. 2002. Diversity in free-living populations of *Cucurbita pepo*

(Cucurbitaceae) as assessed by random amplified polymorphic DNA. *Syst. Bot.* 27:19–28

Denno RF, McClure MS, Ott JR. 1995. Interspecific interactions in phytophagous insects—competition reexamined and resurrected. *Annu. Rev. Entomol.* 40:297–331

Dewar AM, May MJ, Pidgeon JD. 2004. *Environmental impact of GM herbicide-tolerant crops: the UK Farm Scale Evaluations and proposal for mitigation.* http://www.isb.vt.edu/news/2004/news04.Feb.html

Dunne JA, Williams RJ, Martinez ND. 2002. Network structure and biodiversity loss in food webs: robustness increases with connectance. *Ecol. Lett.* 5:558–67

Dunwell JM. 1999. Transgenic crops: the next generation, or an example to 2020 vision. *Ann. Bot.* 84:269–77

Eastham K, Sweet J. 2002. *Genetically modified organisms (GMOs): the significance of gene flow through pollen transfer.* Luxembourg: Off. Off. Publ. Eur. Communities. 75 pp.

Ellstrand NC. 2003. *Dangerous Liaisons? When Cultivated Plants Mate with Their Wild Relatives.* Baltimore, MD: Johns Hopkins Univ. Press. 244 pp.

Ellstrand NC, Prentice HC, Hancock JF. 1999. Gene flow and introgression from domesticated plants into their wild relatives. *Annu. Rev. Ecol. Syst.* 30:539–63

Ellstrand NC, Schierenbeck KA. 2000. Hybridization as a stimulus for the evolution of invasiveness in plants? *Proc. Natl. Acad. Sci. USA* 97:7043–50

Escher N, Käch B, Nentwig W. 2000. Decomposition of transgenic *Bacillus thuringiensis* maize by microorganisms and woodlice *Porcellio scaber* (Crustacea: Isopoda). *Basic Appl. Ecol.* 1:161–69

Fawcett R, Towery D. 2003. *Conservation tillage and plant biotechnology: how new technologies can improve the environment by reducing the need to plow.* http://www.ctic.purdue.edu/CTIC/BiotechPaper.pdf

Ferdy J-B, Austerlitz F. 2002. Extinction and introgression in a community of partially cross-fertile plant species. *Am. Nat.* 160:74–86

Firbank LG, Heard MS, Woiwod IP, Hawes C, Haughton AJ, et al. 2003. An introduction to the Farm-Scale Evaluations of genetically modified herbicide-tolerant crops. *J. Appl. Ecol.* 40:2–16

Fischer R, Emans N. 2000. Molecular farming of pharmaceutical proteins. *Transgenic Res.* 9:279–99

Fuchs M, Gonsalves D. 1997. Risk assessment of gene flow associated with the release of virus-resistant transgenic crop plants. In *Virus Resistant Transgenic Plants: Potential Ecological Impact*, ed. M Tepfer, E Balazs, pp. 114–19. New York: Springer

Fuchs M, Gonsalves D. 1999. Risk assessment of gene flow from a virus-resistant transgenic squash into a wild relative. In *Methods for Risk Assessment of Transgenic Plants: III. Ecological Risks and Prospects of Transgenic Plants, Where Do We Go From Here? A Dialogue Between Biotech Industry and Science*, ed. K Ammann, Y Jacot, V Simonsen, G Kjellsson, pp. 141–43. Basel, Switz.: Birkhäuser-Verlag

Gebhard F, Smalla K. 1999. Monitoring field releases of genetically modified sugar beets for persistence of transgenic plant DNA and horizontal gene transfer. *FEMS Microbiol. Ecol.* 28:261–72

Gerlach JD, Rice KJ. 2003. Testing life history correlates of invasiveness using congeneric plant species. *Ecol. Appl.* 13:167–79

Ghassemi F, Jakeman AJ, Nix HA. 1995. *Salinisation of Land and Water Resources: Human Causes, Extent, Management, and Case Studies.* Oxfordshire, UK: CAB Int. 526 pp.

Gianessi LP, Silvers CS, Sankula S, Carpenter JE. 2002. *Plant biotechnology: current and potential impact for improving pest management in U.S. agriculture: an analysis of 40 case studies.* http://www.ncfap.org/40CaseStudies.htm

GM Sci. Rev. Panel. 2003. *GM Science Review (First Report): An open review of the science relevant to GM crops and food based on interests and concerns of the public.*

http://www.gmsciencedebate.org.uk/report/pdf/gmsci-report1-full.pdf

GM Sci. Rev. Panel. 2004. *GM Science Review (Second Report): An open review of the science relevant to GM crops and food based on interests and concerns of the public.* http://www.gmsciencedebate.org.uk/report/pdf/gmsci-report2-full.pdf

Gould F. 1998. Sustainability of transgenic insecticidal cultivars: integrating pest genetics and ecology. *Annu. Rev. Entomol.* 43:701–26

Groot AT, Dicke M. 2002. Insect-resistant transgenic plants in a multi-trophic context. *Plant J.* 31:387–406

Gueritaine G, Sester M, Eber F, Chevre AM, Darmency H. 2002. Fitness of backcross six of hybrids between transgenic oilseed rape (*Brassica napus*) and wild radish (*Raphanus raphanistrum*). *Mol. Ecol.* 11:1419–26

Guretzky JA, Louda SM. 1997. Evidence for natural biological control: insects decrease survival and growth of a native thistle. *Ecol. Appl.* 7:1330–40

Hails RS. 2000. Genetically modified plants—the debate continues. *Trends Ecol. Evol.* 15:14–18

Halpin C, Boerjan W. 2003. Stacking transgenes in forest trees. *Trends Plant Sci.* 8:363–65

Hammond J, Lecoq H, Raccah B. 1999. Epidemiological risks from mixed virus infections and transgenic plants expressing viral genes. *Adv. Virus Res.* 54:189–314

Hartl DL, Clark AG. 1997. *Principles of Population Genetics.* Sunderland, MA: Sinauer

Haughton AJ, Champion GT, Hawes C, Heard MS, Brooks DR, et al. 2003. Invertebrate responses to the management of genetically modified herbicide-tolerant and conventional spring crops. II. Within-field epigeal and aerial arthropods. *Philos. Trans. R. Soc. London Ser. B* 358:1863–77

Hawes C, Haughton AJ, Osborne JL, Roy DB, Clark SJ, et al. 2003. Responses of plants and invertebrate trophic groups to contrasting herbicide regimes in the Farm Scale Evaluations of genetically modified herbicide-tolerant crops. *Philos. Trans. R. Soc. London Ser. B* 358:1899–913

Haygood R, Ives AR, Andow DA. 2003. Consequences of recurrent gene flow from crops to wild relatives. *Proc. R. Soc. London Ser. B* 270:1879–86

Heard MS, Hawes C, Champion GT, Clark SJ, Firbank LG, et al. 2003a. Weeds in fields with contrasting conventional and genetically modified herbicide-tolerant crops. I. Effects on abundance and diversity. *Philos. Trans. R. Soc. London Ser. B* 358:1819–32

Heard MS, Hawes C, Champion GT, Clark SJ, Firbank LG, et al. 2003b. Weeds in fields with contrasting conventional and genetically modified herbicide-tolerant crops. II. Effects on individual species. *Philos. Trans. R. Soc. London Ser. B* 358:1833–46

Hellmich RL, Siegfried BD, Sears MK, Stanley-Horn DE, Daniels MJ, et al. 2001. Monarch larvae sensitivity to *Bacillus thuringiensis* purified proteins and pollen. *Proc. Natl. Acad. Sci. USA* 98:11925–30

Hilbeck A, Moar WJ, Pusztai-Carey M, Filippini A, Bigler F. 1998. Toxicity of *Bacillus thuringiensis* Cry1Ab toxin to the predator *Chrysoperla carnea* (Neutoptera: Chrysopidae). *Environ. Entomol.* 27:1255–63

Hoffman AA, Blows MW. 1994. Species borders: ecological and evolutionary perspectives. *Trends Ecol. Evol.* 9:223–27

Hoffman MP, Zalom FG, Wilson LT, Smilanick JM, Malyj LD, et al. 1992. Field evaluation of transgenic tobacco containing genes encoding *Bacillus thuringiensis* delta-endotoxin or cowpea trypsin inhibitor: efficacy against *Helicoverpa zea* (Lepidoptera: Noctuidae). *J. Econ. Entomol.* 85:2516–22

Hopkins DW, Webster EA, Chudek JA, Halpin C. 2001. Decomposition in soil of tobacco plants with genetic modifications to lignin biosynthesis. *Soil Biol. Biochem.* 33:1455–62

Howard RD, DeWoody JA, Muir WM. 2004. Transgenic male mating advantage provides opportunity for Trojan gene effect in a fish. *Proc. Natl. Acad. Sci. USA* 101:2934–38

Hsieh TH, Lee JT, Charng YY, Chan MT. 2002.

Tomato plants ectopically expressing *Arabidopsis* CBF1 show enhanced resistance to water deficit stress. *Plant Physiol.* 130:618–26

Huxel GR. 1999. Rapid displacement of native species by invasive species: effects of hybridization. *Biol. Conserv.* 89:143–52

ISB. 2004a. Crops no longer regulated by USDA: approved and pending. http://www.isb.vt.edu/cfdocs/biopetitions1.cfm

ISB. 2004b. *Workshop Report: Extending the net fitness model to considerations of crop gene flow.* http://www.isb.vt.edu/news/2004/Jan04.pdf

James C. 2003. *Preview: Global Status of Commercialized Transgenic Crops: 2003. ISAAA Briefs* No. 30. Ithaca, NY: ISAAA

Jarosz AM, Davelos AL. 1995. Effects of disease in wild plant-populations and the evolution of pathogen aggressiveness. *New Phytol.* 129:371–87

Jaworski J, Cahoon EB. 2003. Industrial oils from transgenic plants. *Curr. Opin. Plant Biol.* 6:178–84

Jesse LCH, Obrycki JJ. 2000. Field deposition of *Bt* transgenic corn pollen: lethal effects on the monarch butterfly. *Oecologia* 125:241–48

Kaeppler HF. 2000. Food safety assessment of genetically modified crops. *Agron. J.* 92:793–97

Kay E, Vogel TM, Bertolla F, Nalin R, Simonet P. 2002. In situ transfer of antibiotic resistance genes from transgenic (transplastomic) tobacco plants to bacteria. *Appl. Environ. Microbiol.* 68:3345–51

Kirkpatrick KJ, Wilson HD. 1988. Interspecific gene flow in Cucurbita: *C. texana* vs. *C. pepo. Am. J. Bot.* 75:519–27

Kok EJ, Kuiper HA. 2003. Comparative safety assessment for biotech crops. *Trends Biotechnol.* 21:439–44

Kolar CS, Lodge DM. 2001. Progress in invasion biology: predicting invaders. *Trends Ecol. Evol.* 16:199–204

Kourtev PS, Ehrenfeld JG, Haggblom M. 2003. Experimental analysis of the effect of exotic and native plant species on the structure and function of soil microbial communities. *Soil Biol. Biochem.* 35:895–905

Kowalchuk GA, Bruinsma M, van Veen JA. 2003. Assessing responses of soil microorganisms to GM plants. *Trends Ecol. Evol.* 18:403–10

Letourneau DK, Burrows BE, eds. 2002. *Genetically Engineered Organisms: Assessing Environmental and Human Health Effects.* Boca Raton, FL: CRC Press. 438 pp.

Letourneau DK, Hagen JA, Robinson GS. 2002. *Bt*-crops: Evaluating benefits under cultivation and risks from escaped transgenes in the wild. See Letourneau & Burrows 2002, pp. 33–98

Letourneau DK, Robinson GS, Hagen JA. 2003. Bt crops: predicting effects of escaped transgenes on the fitness of wild plants and their herbivores. *Environ. Biosaf. Res.* 2:219–46

Levin DA, Francisco-Ortega J, Jansen RK. 1996. Hybridization and the extinction of rare plant species. *Conserv. Biol.* 10:10–16

Linder CR, Taha I, Seiler GJ, Snow AA, Rieseberg LH. 1998. Long-term introgression of crop genes into wild sunflower populations. *Theor. Appl. Genet.* 96:339–47

Lonsdale WM. 1999. Global patterns of plant invasions and the concept of invasibility. *Ecology* 80:1522–36

Losey JE, Rayor LS, Carter ME. 1999. Transgenic pollen harms monarch larvae. *Nature* 399:214

Louda SM, Pemberton RW, Johnson MT, Follet PA. 2003. Non-target effects: the Achilles heel of biological control? Retrospective analyses to reduce risk associated with biocontrol introductions. *Annu. Rev. Entomol.* 48:365–96

Louda SM, Potvin MA. 1995. Effect of inflorescence-feeding insects on the demography and lifetime fitness of a native plant. *Ecology* 76:229–45

Ma JK-C, Pascal MWD, Christou P. 2003. The production of recombinant pharmaceutical proteins in plants. *Nat. Rev. Genet.* 4:794–805

Mack RN, Simberloff D, Lonsdale WM, Evans H, Clout M, Bazzaz FA. 2000. Biotic invasions: causes, epidemiology, global consequences, and control. *Ecol. Appl.* 10:689–710

Malone LA, Pham-Delègue M-H. 2001. Effects of transgene products on honey bees (*Apis mellifera*) and bumblebees (*Bombus* sp.). *Apidologie* 32:1–18

Mann CC. 2002. Has GM corn 'invaded' Mexico? *Science* 295:1617–18

Marquis RJ. 1992. The selective impact of herbivores. In *Plant Resistance to Herbivores and Pathogens: Ecology, Evolution, and Genetics*, ed. RS Fritz, EL Simms, pp. 301–25. Chicago: Univ. Chicago Press

Martinez-Ghersa MA, Worster CA, Radosevich SR. 2003. Concerns a weed scientist might have about herbicide-tolerant crops: a revisitation. *Weed Technol.* 17:202–10

Marvier MA. 2001. Ecology of transgenic crops. *Am. Sci.* 89:160–67

Maskell LC, Raybould AF, Cooper JI, Edwards ML, Gray AJ. 1999. Effects of turnip mosaic virus and turnip yellow mosaic virus on the survival, growth and reproduction of wild cabbage (*Brassica oleracea*). *Ann. Appl. Biol.* 135:401–7

Mason P, Braun L, Warwick SI, Zhu B, Stewart CN Jr. 2003. Transgenic *Bt*-producing *Brassica napus*: *Plutella xylostella* selection pressure and fitness of weedy relatives. *Environ. Biosaf. Res.* 2:236–76

Maynard Smith J, Haigh J. 1974. The hitchhiking effect of a favourable gene. *Genet. Res.* 23:23–35

Meagher TR, Belanger FC, Day PR. 2003. Using empirical data to model transgene dispersal. *Philos. Trans. R. Soc. London Ser. B* 358:1157–62

Mellon M, Rissler J. 2004. *Gone to Seed: Transgenic Contaminants in the Traditional Seed Supply*. Cambridge, MA: Union Concerned Sci. 70 pp.

Mendelsohn M, Kough J, Vaituzis Z, Matthews K. 2003. Are *Bt* crops safe? *Nat. Biotechnol.* 21:1003–9

Mitchell CE, Power AG. 2003. Release of invasive plants from fungal and viral pathogens. *Nature* 421:625–27

Moar WJ. 2003. Breathing new life into insect-resistant plants. *Nat. Biotechnol.* 21:1152–54

Muir WM, Howard RD. 1999. Possible ecological risks of transgenic organism release when transgenes affect mating success: sexual selection and the Trojan gene hypothesis. *Proc. Natl. Acad. Sci. USA* 96:13853–56

Muir WM, Howard RD. 2001. Fitness components and ecological risk of transgenic release: a model using Japanese Medaka (*Oryzias latipes*). *Am. Nat.* 158:1–16

Muir WM, Howard RD. 2002. Assessment of possible ecological risks and hazards of transgenic fish with implications for other sexually reproducing organisms. *Transgenic Res.* 11:101–14

Natl. Res. Counc. (NRC). 1986. *Pesticide Resistance: Strategies and Tactics for Management*. Washington, DC: Natl. Acad. Press. 471 pp.

Natl. Res. Counc. (NRC). 2000. *Genetically Modified Pest-Protected Plants: Science and Regulation*. Washington, DC: Natl. Acad. Press. 292 pp.

Natl. Res. Counc. (NRC). 2002a. *Environmental Effects of Transgenic Plants: The Scope and Adequacy of Regulation*. Washington, DC: Natl. Acad. Press. 342 pp.

Natl. Res. Counc. (NRC). 2002b. *Predicting Invasions of Nonindigenous Plants and Plant Pests*. Washington, DC: Natl. Acad. Press. 194 pp.

Natl. Res. Counc. (NRC). 2004. *Biological Confinement of Genetically Engineered Organisms*. Washington, DC: Natl. Acad. Press. 236 pp.

Nielsen KM, Bones AM, Smalla K, van Elsas JD. 1998. Horizontal gene transfer from transgenic plants to terrestrial bacteria—a rare event? *FEMS Microbiol. Rev.* 22:79–103

Nielsen KM, Gebhard F, Smalla K, Bones AM, van Elsas JD. 1997. Evaluation of possible horizontal gene transfer from transgenic plants to the soil bacterium *Acinetobacter calcoaceticus* BD413. *Theor. Appl. Genet.* 95:815–21

Nielsen KM, Smalla K, van Elsas JD. 2000a. Natural transformation of *Acinetobacter* sp.

strain BD413 with cell lysates of *Acinetobacter* sp., *Pseudomonas fluorescens*, and *Burkholderia cepacia* in soil microcosms. *Appl. Environ. Microbiol.* 66:206–12

Nielsen KM, van Elsas JD, Smalla K. 2000b. Transformation of *Acinetobacter* sp. strain BD413(pFG4 *nptII*) with transgenic plant DNA in soil microcosms and effects of kanamycin on selection of transformants. *Appl. Environ. Microbiol.* 66:1237–42

Nielsen KM, van Elsas JD, Smalla K. 2001. Dynamics, horizontal transfer and selection of novel DNA in bacterial populations in the phytosphere of transgenic plants. *Ann. Microbiol.* 51:79–94

Oberhauser KS, Prysby MD, Mattila HR, Stanley-Horn DE, Sears MK, et al. 2001. Temporal and spatial overlap between monarch larvae and corn pollen. *Proc. Natl. Acad. Sci. USA* 98:11913–18

Obrycki JJ, Losey JE, Taylor OR, Jesse LCH. 2001. Transgenic insecticidal corn: beyond insecticidal toxicity to ecological complexity. *BioScience* 51:353–61

Oerke E-C, Dehne H-W, Schönbeck F, Weber A. 1994. *Crop Production and Crop Protection: Estimated Losses in Major Food and Cash Crops*. Amsterdam: Elsevier Sci. B.V. 808 pp.

Orr DB, Landis DA. 1997. Oviposition of European corn borer (Lepidoptera: Pyralidae) and impact of natural enemy populations in transgenic versus isogenic corn. *J. Econ. Entomol.* 90:905–9

Parker IM, Kareiva P. 1996. Assessing the risks of invasion for genetically engineered plants: Acceptable evidence and reasonable doubt. *Biol. Conserv.* 78:193–203

Pessel FD, Lecomte J, Emeriau V, Krouti M, Messean A, Gouyon PH. 2001. Persistence of oilseed rape (*Brassica napus* L.) outside of cultivated fields. *Theor. Appl. Genet.* 102:841–46

Pilcher CD, Obrycki JJ, Rice ME, Lewis LC. 1997. Preimaginal development, survival, and field abundance of insect predators on transgenic *Bacillus thuringiensis* corn. *Environ. Entomol.* 26:446–54

Pilson D. 2000. Herbivory and natural selection on flowering phenology in wild sunflower, *Helianthus annuus*. *Oecologia* 122:72–82

Pilson D, Decker KL. 2002. Compensation for herbivory in wild sunflower: response to simulated damage by the head-clipping weevil. *Ecology* 83:3097–107

Pilson D, Snow AA, Rieseberg LH, Alexander HM. 2004. A protocol for evaluating the ecological risks associated with gene flow from transgenic crops into their wild relatives: the case of cultivated sunflower and wild *Helianthus annuus*. In *Introgression from Genetically Modified Plants into Wild Relatives*, ed. HCM den Nijs, D Bartsch, J Sweet, pp. 219–33. Oxfordshire, UK: CAB Int.

Pleasants JM, Hellmich RL, Dively GP, Sears MK, Stanley-Horn DE, et al. 2001. Corn pollen deposition on milkweeds in and near cornfields. *Proc. Natl. Acad. Sci. USA* 98:11919–24

Polis GA, Strong DR. 1996. Food web complexity and community dynamics. *Am. Nat.* 147:813–46

Poppy G. 2000. GM crops: environmental risks and non-target effects. *Trends Plant Sci.* 5:4–6

Power AG. 2002. Ecological risks of transgenic virus-resistant crops. See Letourneau & Burrows 2002, pp. 125–42

Pray CE, Huang JK, Hu R, Rozelle S. 2002. Five years of *Bt* cotton in China—the benefits continue. *Plant J.* 31:423–30

Quist D, Chapela IH. 2001. Transgenic DNA introgressed into traditional maize landraces in Oaxaca, Mexico. *Nature* 414:541–43

Quist D, Chapela IH. 2002. Reply. *Nature* 416:602

Rhymer JM, Simberloff D. 1996. Extinction by hybridization and introgression. *Annu. Rev. Ecol. Syst.* 27:89–109

Riddick EW, Dively G, Barbosa P. 1998. Effect of a seed-mix deployment of Cry3A-transgenic and nontransgenic potato on the abundance of *Lebia grandis* (Coleoptera: Carabidae) and *Coleomegilla maculata* (Coleoptera: Coccinellidae). *Ann. Entomol. Soc. Am.* 91:647–53

Rieseberg LH, Baird SJE, Gardner KA. 2000. Hybridization, introgression, and linkage evolution. *Plant Mol. Biol.* 42:205–24

Rieseberg LH, Kim MJ, Seiler GJ. 1999. Introgression between cultivated sunflower and a sympatric wild relative, *Helianthus petiolaris* (Asteraceae). *Int. J. Plant Sci.* 160:102–8

Rieseberg LH, Wendel JF. 1993. Introgression and its consequences in plants. In *Hybrid Zones and the Evolutionary Process*, ed. R Harrison, pp. 70–109. New York: Oxford Univ. Press

Rieseberg LH, Widmer A, Arntz MA, Burke JM. 2003. The genetic architecture necessary for transgressive segregation is common in both natural and domesticated populations. *Philos. Trans. R. Soc. London Ser. B* 358:1141–47

Rissler J, Mellon M. 1996. *The Ecological Risks of Engineered Crops.* Cambridge, MA: MIT Press. 168 pp.

Roossinck MJ. 1997. Mechanisms of plant virus evolution. *Annu. Rev. Phytopathol.* 35:191–209

Roy DB, Bohan DA, Haughton AJ, Hill MO, Osborne JL, et al. 2003. Invertebrates and vegetation of field margins adjacent to crops subject to contrasting herbicide regimes in the Farm Scale Evaluations of genetically modified herbicide-tolerant crops. *Philos. Trans. R. Soc. London Ser. B* 358:1879–98

Royal Society. 2002. *Genetically modified plants for food use and human health—an update.* http://www.royalsoc.ac.uk/files/statfiles/document-165.pdf

Rubio T, Borja M, Scholthof HB, Jackson AO. 1999. Recombination with host transgenes and effects on virus evolution: an overview and opinion. *Mol. Plant Microbe. Interact.* 12:87–92

Sakai AK, Allendorf FW, Holt JS, Lodge DM, Molofsky J, et al. 2001. The population biology of invasive species. *Annu. Rev. Ecol. Syst.* 32:305–32

Saxena D, Flores S, Stotzky G. 1999. Insecticidal toxin in root exudates from *Bt* corn. *Nature* 402:480

Saxena D, Stotzky G. 2001. *Bacillus thuringiensis* (*Bt*) toxin released from root exudates and biomass of *Bt* corn has no apparent effect on earthworms, nematodes, protozoa, bacteria, and fungi in soil. *Soil Biol. Biochem.* 33:1225–30

Schlüter K, Fütterer J, Potrykus I. 1995. "Horizontal" gene transfer from a transgenic potato line to a bacterial pathogen (*Erwinia chrysanthemi*) occurs—if at all—at an extremely low frequency. *BioTechnology* 13:1094–98

Schuler TH, Poppy GM, Kerry BR, Denholm I. 1999. Potential side effects of insect-resistant transgenic plants on arthropod natural enemies. *Trends Biotechnol.* 17:210–16

Sears MK, Hellmich RL, Stanley-Horn DE, Oberhauser KS, Pleasants JM, et al. 2001. Impact of *Bt* corn pollen on monarch butterfly populations: a risk assessment. *Proc. Natl. Acad. Sci. USA* 98:11937–42

Shelton AM, Tang JD, Roush RT, Metz TD, Earle ED. 2000. Field tests on managing resistance to *Bt*-engineered plants. *Nat. Biotechnol.* 18:339–42

Shelton AM, Zhao JZ, Roush RT. 2002. Economic, ecological, food safety, and social consequences of *Bt* transgenic plants. *Annu. Rev. Entomol.* 47:845–81

Silvertown J, Franco M, Pisanty I, Mendoza A. 1993. Comparative plant demography—relative importance of life-cycle components to the finite rate of increase in woody and herbaceous perennials. *J. Ecol.* 81:465–76

Simms EL, Rausher MD. 1987. Costs and benefits of plant resistance to herbivory. *Am. Nat.* 130:570–81

Sinclair TR, Purcell LC, Sneller CH. 2004. Crop transformation and the challenge to increase yield potential. *Trends Plant Sci.* 9:70–75

Snow AA, Andersen B, Jørgensen RB. 1999. Costs of transgenic herbicide resistance introgressed from *Brassica napus* into weedy *B. rapa*. *Mol. Ecol.* 8:605–15

Snow AA, Andow DA, Gepts P, Hallerman EM, Power A, et al. 2004. *Genetically engineered organisms and the environment: current status and recommendations.* http://www.esa.org/pao/esaPositions/Papers/geo_position.htm

Snow AA, Moran-Palma P. 1997. Commercialization of transgenic plants: potential ecological risks. *BioScience* 47:86–96

Snow AA, Moran-Palma P, Rieseberg LH, Wszelakl A, Seiler GJ. 1998. Fecundity, phenology, and seed dormancy of F-1 wild-crop hybrids in sunflower (*Helianthus annuus*, Asteraceae). *Am. J. Bot.* 85:794–801

Snow AA, Pilson D, Rieseberg LH, Paulsen MJ, Pleskac N, et al. 2003. A *Bt* transgene reduces herbivory and enhances fecundity in wild sunflowers. *Ecol. Appl.* 13:279–86

Spencer LJ, Snow AA. 2001. Fecundity of transgenic wild-crop hybrids of *Cucurbita pepo* (Cucurbitaceae): implications for crop-to-wild gene flow. *Heredity* 86:694–702

Squire GR, Brooks DR, Bohan DA, Champion GT, Daniels RE, et al. 2003. On the rationale and interpretation of the Farm Scale Evaluations of genetically modified herbicide-tolerant crops. *Philos. Trans. R. Soc. London Ser. B* 358:1779–99

Stanley-Horn DE, Dively GP, Hellmich RL, Mattila HR, Sears MK, et al. 2001. Assessing the impact of Cry1Ab-expressing corn pollen on monarch butterfly larvae in field studies. *Proc. Natl. Acad. Sci. USA* 98:11931–36

Stewart CN Jr, Halfhill MD, Warwick SI. 2003. Transgene introgression from genetically modified crops to their wild relatives. *Nat. Rev. Genet.* 4:806–17

Tabashnik BE. 1994. Evolution of resistance to *Bacillus thuringiensis*. *Annu. Rev. Entomol.* 39:47–79

Tabashnik BE, Carrière Y, Dennehy TJ, Morin S, Sisterson MS, et al. 2003. Insect resistance to transgenic *Bt* crops: lessons from the laboratory and field. *J. Econ. Entomol.* 96:1031–38

Takahashi M, Nakanishi H, Kawasaki S, Nichizawa NK, Mori S. 2001. Enhanced tolerance of rice to low iron availability in alkaline soils using barley nicotianamine aminotransferase genes. *Nat. Biotechnol.* 19:466–69

Tepfer M. 2002. Risk assessment of virus-resistant transgenic plants. *Annu. Rev. Phytopathol.* 40:467–91

Thompson CJ, Thompson BJP, Ades PK, Cousens R, Garnier-Gere P, et al. 2003. Model-based analysis of the likelihood of gene introgression from genetically modified crops into wild relatives. *Ecol. Model.* 162:199–209

Thomson J. 2003. Genetically modified food crops for improving agricultural practice and their effects on human health. *Trends Food Sci. Technol.* 14:210–12

Tiedje JM, Colwell RK, Grossman YL, Hodson RE, Lenski RE, et al. 1989. The planned introduction of genetically engineered organisms: ecological considerations and recommendations. *Ecology* 70:298–315

Wardle DA. 2002. *Communities and Ecosystems: Linking the Aboveground and Belowground Components.* Princeton, NJ: Princeton Univ. Press. 392 pp.

Warwick SI, Beckie HJ, Small E. 1999. Transgenic crops: new weed problems for Canada? *Phytoprotection* 80:71–84

Warwick SI, Simard M-J, Légère A, Beckie HJ, Braun L, et al. 2003. Hybridization between transgenic *Brassica napus* L. and its wild relatives: *Brassica rapa* L., *Raphanus raphanistrum* L., *Sinapis arvensis* L., and *Erucastrum gallicum* (Willd). O.E. Schulz. *Theor. Appl. Genet.* 107:528–39

Watkinson AR, Freckleton RP, Robinson RA, Sutherland WJ. 2000. Predictions of biodiversity response to genetically modified herbicide-tolerant crops. *Science* 289:1554–57

Whitton J, Wolf DE, Arias DM, Snow AA, Rieseberg LH. 1997. The persistence of cultivar alleles in wild populations of sunflowers five generations after hybridization. *Theor. Appl. Genet.* 95:33–40

Willis AJ, Memmott J, Forrester RI. 2000. Is there evidence for the post-invasion evolution of increased size among invasive plant species? *Ecol. Lett.* 3:275–83

Willis AJ, Thomas MB, Lawton JH. 1999. Is the increased vigour of invasive weeds explained by a trade-off between growth and herbivore resistance? *Oecologia* 120:632–40

Wilson HD. 1993. Free-living *Cucurbita pepo* in the United States: viral resistance, gene flow, and risk assessment. USDA Anim. Plant Health Insp. Serv., Hyattsville, MD

Wolf DE, Takebayashi N, Rieseberg LH. 2001. Predicting the risk of extinction through hybridization. *Conserv. Biol.* 15:1039–53

Wolfenbarger LL, Phifer PR. 2000. The ecological risks and benefits of genetically engineered plants. *Science* 290:2088–93

Wraight CL, Zangerl AR, Carroll MJ, Berenbaum MR. 2000. Absence of toxicity of *Bacillus thuringiensis* pollen to black swallowtails under field conditions. *Proc. Natl. Acad. Sci. USA* 97:7700–3

Zangerl AR, McKenna D, Wraight CL, Carroll M, Ficarello P, et al. 2001. Effects of exposure to event 176 *Bacillus thuringiensis* corn pollen on monarch and black swallowtail caterpillars under field conditions. *Proc. Natl. Acad. Sci. USA* 98:11908–12

Zhang H-X, Hodson JN, Williams JP, Blumwald E. 2001. Engineering salt-tolerant *Brassica* plants: characterization of yield and seed oil quality in transgenic plants with increased vacuolar sodium accumulation. *Proc. Natl. Acad. Sci. USA* 98:12832–36

Zhang NY, Linscombe S, Oard J. 2003. Outcrossing frequency and genetic analysis of hybrids between transgenic glufosinate herbicide-resistant rice and the weed, red rice. *Euphytica* 130:35–45

MUTUALISMS AND AQUATIC COMMUNITY STRUCTURE: The Enemy of My Enemy Is My Friend

Mark E. Hay,[1] John D. Parker,[1] Deron E. Burkepile,[1] Christopher C. Caudill,[1,2] Alan E. Wilson,[1] Zachary P. Hallinan,[1] and Alexander D. Chequer[1]

[1]*School of Biology, Georgia Institute of Technology, Atlanta, Georgia 30332-0230; email: mark.hay@biology.gatech.edu*
[2]*Fish Ecology Research Laboratory, Department of Fish and Wildlife, University of Idaho, Moscow, Idaho 83844-1141*

Key Words coevolution, foundation species, indirect effects, marine, positive interactions

■ **Abstract** Mutualisms occur when interactions between species produce reciprocal benefits. However, the outcome of these interactions frequently shifts from positive, to neutral, to negative, depending on the environmental and community context, and indirect effects commonly produce unexpected mutualisms that have community-wide consequences. The dynamic, and context dependent, nature of mutualisms can transform consumers, competitors, and parasites into mutualists, even while they consume, compete with, or parasitize their partner species. These dynamic, and often diffuse, mutualisms strongly affect community organization and ecosystem processes, but the historic focus on pairwise interactions decoupled from their more complex community context has obscured their importance. In aquatic systems, mutualisms commonly support ecosystem-defining foundation species, underlie energy and nutrient dynamics within and between ecosystems, and provide mechanisms by which species can rapidly adjust to ecological variance. Mutualism is as important as competition, predation, and physical disturbance in determining community structure, and its impact needs to be adequately incorporated into community theory.

INTRODUCTION

Ecologists have made significant advances in understanding community structure and function by focusing on negative interactions such as predation, competition, and physical disturbance (Bertness et al. 2001). However, positive interactions, such as facilitation and mutualism, also play pivotal roles in organizing communities, and incorporating positive interactions into ecological theory can fundamentally alter our understanding of the processes and mechanisms that shape communities (Stachowicz 2001, Bruno et al. 2003). In this review, we focus on mutualisms in aquatic communities and demonstrate that (*a*) mutualisms are surprisingly widespread, (*b*) mutualists in one ecological setting can be adversaries in another setting, (*c*) conversely, interactions traditionally viewed as antagonistic can be mutualistic, depending on environmental and community settings, (*d*) mutualists need not be coevolved or consistently coupled in space or time, and (*e*) mutualisms have large effects on community structure and function. Our treatment extends the traditional view of coevolved, obligate mutualisms between species pairs into a broader community context, where indirect interactions among many species are common and context dependent and can dampen or reverse direct effects as well as play significant roles in affecting community and ecosystem organization (Berlow 1999).

Mutualisms are frequently viewed as obligate, coevolved interactions uniformly benefiting both partners, such as the mutualism between yucca plants and their moth pollinators. This unconditional, pairwise focus is commonly presented in textbooks (e.g., Purves et al. 2001) and theoretical treatments of mutualism (Doebeli & Knowlton 1998, Holland et al. 2002; but see Bacher & Friedli 2002) and drives research focused on the evolution and stabilization of mutualisms (Bronstein 1994, Connor 1995, Knowlton & Rohwer 2003, Stanton 2003, Thomson 2003). However, recent experimental and theoretical investigations (e.g., Stanton 2003, Thomson 2003) have noted that well-studied mutualisms, such as those between plants and pollinators, are usually conditional and can vary tremendously, depending on which pollinators occur locally. A pollinator can switch from mutualist to parasite because of the addition of a single, more efficient pollinator to the community; this dramatic change can occur without alteration in the behavior of either initial partner (Thomson 2003). Thus, between-species associations that are mutually beneficial in one ecological setting may become neutral or harmful in another (Bronstein 1994, Connor 1995, van Baalen & Jansen 2001, Knowlton & Rohwer 2003). In this review, we argue that the consideration of mutualisms within a community context not only underscores the dynamic nature of mutualisms but also illuminates the pivotal role that mutualisms play in community-level and ecosystem-level processes (van Baalen & Jansen 2001, Gomulkiewicz et al. 2003, Stanton 2003, Thomson 2003).

Although some authors have argued that mutualists must exhibit coevolved traits, there is no compelling rationale for such a restriction, and we view mutualisms less restrictively. Like Bronstein (1994), we use mutualism to refer to interspecific interactions in which the benefits exceed the costs for both participants. We define mutualisms by the outcome of the interaction, rather than by assumptions about the coevolution of the interaction. Interactions that seem coevolved because of reciprocal traits of the participants can occur without a coevolutionary history (Steneck 1992, Vermeij 1992), thus casting doubt on our ability to distinguish interactions that are coevolved from those that are simply fortuitous. Therefore, in this review, we make no assumptions about evolutionary history, and we recognize that ecologically important mutualisms can occur fortuitously when each partner receives "by-product" benefits from the other, as when a trait of species A incidentally benefits species B, and vice versa (Connor 1995). Damselfish, for example, escape predators by sheltering in branching corals; these corals grow faster because of the nutrients excreted by the sheltering damselfish (Liberman et al. 1995). The branching morphology of the coral is intended to meet the feeding and hydrodynamic needs of the coral, yet it provides the fish with shelter; damselfish excretion is a physiological necessity for the fish, yet it benefits the coral.

Here, we explore how mutualisms structure the communities in which they occur and how community context can alter the outcomes of interactions to the extent that apparent antagonists can function as mutualists. We focus on aquatic systems because fewer mutualisms are broadly known and well investigated in these communities and because aquatic patterns provide useful contrasts with terrestrial mutualisms.

MUTUALISMS THAT SUPPORT FOUNDATION SPECIES

Foundation species such as corals, kelps, seagrasses, and marsh plants provide structure and definition to entire ecosystems (Dayton 1975, Jones et al. 1997, Bruno et al. 2003). Interactions between foundation species and their residents are not unidirectional, as residents often benefit foundation species. When foundation species support organisms that reciprocate with critical benefits, these mutualisms can be pivotal to the persistence and function of entire ecosystems.

Deep-Sea Hydrothermal Vent and Hydrocarbon-Seep Communities

Hydrothermal vent communities represent oases of high production and physical structure in a featureless landscape with few energetic resources. Unknown until 1977, these systems are examples of extreme biotic adaptation, much of it based on mutualisms between foundation species and the endosymbiotic, chemosynthetic microbes that provide their nutrition. Physical rigors at vents are extreme; fluids exiting some vents may be 400°C, have a pH of 2.8, and contain high concentrations of toxic hydrogen sulfide and other unusual cations (Tunnicliffe 1992). These hot fluids enter water that is only 2°C, but chemoautotrophic bacteria from these systems can grow at temperatures of up to 110°C, which allows them to cope with these rigors and capture the energy that powers these communities.

Foundation species such as bivalves (*Bathymodiolus* spp.) and the giant (1 m long) tubeworm *Riftia pachyptila* have neither mouth nor gut; they acquire their nutrition from sulfur-oxidizing chemoautotrophic bacteria that live within their bodies (Fiala-Medioni & Felbeck 1990). The host delivers carbon dioxide, oxygen, and hydrogen sulfide to bacteria held in special body tissues, and the bacteria use hydrogen sulfide as an electron donor to synthesize organic compounds that are utilized by the host. This mutualism enables the biomass at vent communities to become 500 to 1000 times greater than in the surrounding deep sea and forms the base of an extensive food web supported entirely by chemosynthesis (Tunnicliffe 1992, Micheli et al. 2002). Similar assemblages of hosts and endosymbiotic sulfide-oxidizing or methane-oxidizing bacteria are found at hydrocarbon seeps (Brooks et al. 1987). Stable-isotope studies indicate that some mobile predators acquire nearly 100% of their nutrition from seep production, and even vagrant predators can be substantially subsidized by this production (MacAvoy et al. 2002).

The host-microbe mutualisms of these deep-sea habitats allow a method of energy acquisition and produce a community of organisms that are among the most novel on Earth. This mutualism not only supports the foundation species of these communities but also captures biologically novel sources of nutrition in ecosystems where such high densities of macroorganisms would be impossible with only photosynthetic inputs. Of the 300 species that occur at vents, 97% were previously unknown; they represent 1 new class, 3 new orders, 22 new families, and 96 new genera (Tunnicliffe 1992).

Coral Reefs

Corals are foundation species that provide the ecological infrastructure for one of the most biologically diverse ecosystems on Earth. Perhaps the most widely studied marine mutualisms are those between corals and their photosynthetic dinoflagellate symbionts (*Symbiodinium* spp.), known as zooxanthellae (Muller-Parker & D'Elia 1997). Zooxanthellae donate carbohydrates derived from photosynthesis to the coral, while receiving nutrients in the form of nitrogenous wastes derived from the prey of the carnivorous corals. Photosynthesis by zooxanthellae provides up to 95% of the coral's carbon budget and enhances coral calcification and growth; this leads to carbonate accretion and massive amounts of reef framework in tropical seas. In the absence of zooxanthellae-assisted growth, corals could not grow fast enough to stay in well-lit surface waters during periods of rapid sea-level rise, and slow-growing reefs would drown (Neumann & MacIntyre 1985). Thus, without the coral-zooxanthellae mutualism, coral reef ecosystems would be unlikely to persist.

Corals were initially thought to coevolve pairwise mutualisms with a single species of zooxanthellae, but recent investigations suggest the existence of several distinct genetic clades and many species of zooxanthellae; these studies also show that some zooxanthellae can be rejected and new strains acquired when corals are exposed to changing environmental conditions (LaJeunesse 2002, Baker 2003). Zooxanthellae clades differ in their photosynthetic capacity, tolerance of light, and production of photoprotective compounds such as mycosporine-like amino acids (Banaszak et al. 2000, Savage et al. 2002), which makes them differentially useful to their hosts as light levels change. Baker (2001) demonstrated that several species of Caribbean corals replaced resident zooxanthellae clades with new zooxanthellae clades when moved from low-light to high-light environments but not when moved from high-light to low-light environments. Corals that did not bleach or acquire new symbionts died more frequently than corals that bleached but then acquired new symbionts. This suggests that a failure of corals to change symbionts when encountering new environmental conditions can be fatal. Such alterations in the coral-zooxanthellae mutualism may allow corals greater flexibility in adapting to global climate change (Baker 2002, Hoegh-Guldberg et al. 2002). This observation illustrates the context dependence of the coral-zooxanthellae mutualism that ultimately contributes to the success of corals as ecosystem engineers.

Although the coral-zooxanthellae mutualism drives the growth of corals that create coral reefs, a different, more diffuse and indirect, mutualism between corals and herbivores may be critical for the development and maintenance of coral reefs on both ecological and evolutionary time scales. Strong interactions between herbivorous fishes and reef corals suggest that their relationship constitutes a by-product mutualism. By feeding on seaweeds that are competitively superior to corals, herbivorous fishes both clear the substrate for settling coral larvae and prevent seaweed overgrowth of established corals (Lewis 1986, Hughes 1994). In return, the biogenic structure and topographic complexity of reef corals indirectly benefit herbivorous reef fishes by providing both habitat and food. The porous hard substrate allows rapidly growing filamentous algae (the preferred food of many reef fishes) to persist and maintain rapid rates of production, despite the constant removal of greater than 90% of algal production by reef herbivores (Carpenter 1986). When reef fishes are removed experimentally (Lewis 1986, Carpenter 1986) or by overfishing (Hughes 1994, Jackson et al. 2001), seaweeds replace corals and the biogenic structure of the reef degrades. Both reductions in coral structure and increases in seaweeds are associated with losses of herbivorous reef fishes (Jones 1991, McClanahan et al. 2000).

In fact, the diffuse mutualisms between corals and herbivorous reef fishes may have allowed the evolution of massive coral reefs. Intense herbivory probably did not exist on coral reefs until after the diversification of herbivorous fishes roughly 40 to 50 million years ago (Mya) (Vermeij 1977, Streelman et al. 2002). The reductions in seaweed biomass that followed may have been the tipping point that allowed widespread formation of coral reefs. Large scleractinian-dominated reefs first appeared 25 to 35 Mya during the Oligocene epoch, roughly 15 to 20 million years after the diversification of herbivores but almost 200 million years after the zooxanthellate corals evolved (Veron 1995, Hallock 1997). The rise of modern coral reefs may be a direct result of the evolution of intense herbivory by fishes and invertebrates, which shifted tropical benthic communities from seaweed-dominated to coral-dominated ecosystems (Wood 1998). If this hypothesis is true, then the diffuse mutualism between herbivorous fishes and corals may have been critical in allowing the development of modern reef communities.

Fishes and corals participate in other mutualisms in which fishes benefit by refuging among coral branches and corals benefit from nutrient supplements provided by the fish. For example, carnivorous grunts (*Haemulidae*) forage in seagrass beds overnight but school around stands of coral such as *Acropora palmata* and

Porites furcata during the day as a refuge from predators (Ehrlich 1975, Ogden & Ehrlich 1977). Coral stands that harbor fish schools receive nutrient supplements from fish excretion (Meyer et al. 1983, Meyer & Schultz 1985a), grow up to 23% faster, and have more nitrogen and zooxanthellae per unit area than do corals without resident fishes (Meyer & Schultz 1985b). This process of nutrient concentration and transfer also occurs on smaller spatial scales when planktivorous fishes shelter among the branches of individual corals and increase coral growth rates by as much as 40% (Shpigel & Fishelson 1986, Liberman et al. 1995). Thus, fishes that have no direct trophic link with corals collect nutrients from other communities (seagrass beds or the plankton) and concentrate these nutrients near their host coral. This nutrient subsidy facilitates coral growth and enhances the coral's value as a refuge for these fishes and for other reef organisms.

Mangroves

Mangroves dominate shallow coastlines throughout the tropics and are important habitats for a host of marine and terrestrial species. Their submerged prop roots serve as necessary hard substrate for a variety of seaweeds and invertebrates, with sponges being especially common on mangrove roots. Challenges to mangrove growth include low-nutrient soils and failure of prop roots during storms. Both challenges are lessened through mutualistic interactions with sponges such as *Tedania ignis* and *Haliclonia implexiformis*. When sponges are transplanted onto clean prop roots, fine rootlets proliferate, and nitrogen moves from the sponge to the tree while carbon moves from the tree to the sponge (Ellison et al. 1996). This exchange increases sponge growth by 40% to 100% and root growth by 100% to 300%, which is predictive of increased leaf production and net aboveground primary productivity (Farnsworth & Ellison 1996). Roots covered by sponges are also less frequently attacked by the boring isopod *Sphaeroma terebrans*, and if sponge-covered roots are attacked, they suffer less decline in growth than do attacked roots without a sponge cover (Ellison et al. 1996). Isopod attack reduces root growth by more than 50% (Perry 1988, Ellison & Farnsworth 1990) and can lead to mangroves toppling in storms and the shrinkage of mangrove islands (Rehm & Humm 1973, Svavarsson et al. 2002). Thus, protection by mutualist sponges has significant value. Furthermore, fragmentation of mangrove habitats can cause local extinctions of resident species (Grant & Grant 1997), decrease nutrient exports to other coastal ecosystems (Marshall 1994, Robertson & Alongi 1995), and reduce ecological resources such as fish and timber that are economically important to local human populations (Alongi 2002).

Seagrass and Saltmarsh Communities

Seagrasses support diverse communities of benthic invertebrates and fishes by providing biogenic structure and high productivity to otherwise featureless sand and mud flats. Seagrass blades are rapidly fouled by attached algae, sessile invertebrates, and detritus that decrease seagrass growth through shading and can cause whole plants to be ripped from the substratum through increased leaf drag (Jernakoff et al. 1996). Nutrient eutrophication generally favors epibionts over seagrasses (Jernakoff et al. 1996), and regional declines in seagrass abundance—and the fisheries they sustain—are correlated with increased nutrient supply and the deleterious impacts of fouling epibiota (Orth & Moore 1983).

Small consumers (mesograzers), such as amphipods, isopods, snails, shrimps, and crabs live on seagrass, consume fouling epibiota from its surfaces (Orth & van Montfrans 1984) and, by doing so, increase seagrass growth up to 200% (Duffy et al. 2001). The seagrass provides mesograzers with substrate for food and with a structural refuge that decreases susceptibility to predation (Leber 1985). Meadows of seagrasses support dense assemblages of mesograzers that are the primary trophic link fueling production of fishes, shrimps, crabs, and other larger animals in seagrass beds (Edgar & Shaw 1995). Therefore, the health and persistence of seagrass beds, which are critical nursery grounds for many coastal fishes, may depend on the mutualistic interactions between seagrasses and their surface-associated mesograzers.

The strong interaction between seagrasses and mesograzers is conditional and can change from mutual, to neutral, to antagonistic, depending on the environmental setting. When high water temperatures and nutrient loads favor epibiota, seagrass growth is enhanced by the addition of a mixed-species assemblage of mesograzers, but when cool water temperatures and low nutrient loads favor seagrasses over epibiota, mesograzer feeding does not enhance seagrass growth (Neckles et al. 1993). When epibiota become scarce, mesograzers may ingest living seagrass tissues and sometimes consume entire plants (Duffy et al. 2001, Duffy et al. 2003). The mutualist thus becomes a parasite under these altered environmental conditions.

Like seagrasses, saltmarsh vegetation provides biogenic structure, high production, and substrate stability on intertidal muds and sands. Saltmarsh production fuels coastal food webs, sustains fisheries, and provides nursery areas for juveniles of many coastal species. Both seagrasses and saltmarsh plants form facultative mutualisms with suspension-feeding bivalves (Peterson & Heck 2001b, Bertness 1984). Bivalves benefit rooted macrophytes by (a) harvesting plankton from the water column and enriching the substrate via feces and pseudofeces (Bertness 1984; Peterson & Heck 2001a,b), (b) stabilizing the substrate (Bertness 1984), (c) decreasing light limitation of macrophytes by filtering phytoplankton and increasing water clarity (Jackson 2001), and (d) indirectly reducing epiphytic loads on macrophytes by harboring epiphyte grazers (Peterson & Heck 2001a). In return, macrophytes provide bivalves with increased food supplies and a refuge from predators (Irlandi & Peterson 1991, Peterson & Heck 2001a).

Kelp Forests

Temperate rocky reefs with kelp forests have greater primary production, species diversity, and food-web complexity, and they export more energy to adjacent communities than do nearby reefs without kelps (Duggins et al. 1989, Estes & Duggins 1995). In Alaska, indirect interactions between sea otters and kelps form a by-product mutualism that causes dramatic shifts from rocky reefs dominated by sea urchins to high-diversity kelp forests that can support higher densities of otters (Estes 1990, Estes & Duggins 1995). By consuming large quantities of herbivorous sea urchins, sea otters diminish herbivory, allowing kelps to flourish. Productive kelps increase habitat complexity and fuel local food webs, thus increasing the abundance of nutritious prey (especially the fish, *Hexagrammos lagocephalus*) upon which Alaskan sea otters feed. High kelp production and diverse, persistent prey associated with kelp beds are positive feedbacks that increase local carrying capacity and stability of sea otter populations (Estes 1990). The excess production from kelps is exported to nearby ecosystems, where it enhances growth of other organisms, thus producing not only local but also regional effects on marine food webs (Duggins et al. 1989).

OTHER MUTUALISMS WITH COMMUNITY CONSEQUENCES

Community Impacts of Nitrogen Fixation by Symbionts

Mutualisms between nitrogen-fixing microbes and pelagic hosts can be critical to the functioning of many pelagic ecosystems. Nearly all nitrogen is energetically inaccessible to most organisms; only nitrogen-fixing microorganisms can break the triple bond of N_2 molecules and convert this nitrogen into a biologically available form. Thus, open-ocean ecosystems are commonly nitrogen limited, and nitrogen additions from nitrogen-fixing microorganisms (i.e., bacteria, cyanobacteria, and Archaea) can profoundly influence large-scale patterns of productivity by providing from less than 1% to as much as 82% of the biologically available nitrogen in pelagic ecosystems (Howarth et al. 1988, Zehr et al. 2000).

Nitrogen-fixing plankton have broad distributions and influence oceanic food-web structure and function, community composition, fisheries yield, biogeochemical cycles, and even global carbon budgets (Karl et al. 1999). When bioavailable nitrogen is limiting, planktonic diatoms and dinoflagellates adopt nitrogen fixers as symbionts to provide organic nitrogen, vitamins, growth factors, and nitrogen-rich defensive chemistry. In return, the nitrogen-fixing symbiont gains a reliable carbon source to drive nitrogen-fixation. For example, blooms of the diatom, *Hemiaulus haukii*, and its nitrogen-fixing endosymbiont, *Richelia intracelluaris*, can (*a*) be 2500 km across, (*b*) account for 25% of the total nitrogen demand of this system, (*c*) increase primary production 270% compared with areas dominated by nonmutualistic phytoplankters and cyanobacteria, and (*d*) contribute about 0.5 Tg of new nitrogen to the euphotic zone (Carpenter et al. 1999). To visualize the magnitude of this nitrogen addition, think of it as the mass of 1.5 million Harley-Davidson Road King motorcycles, 7.3 million undergraduate students, or 157 million laptop computers. When nitrogen is abundant, phytoplankton abandon their nitrogen-fixing symbionts (Gordon et al. 1994), which shows the conditional nature of these mutualisms and suggests a cost to such relationships.

Cleaner-Client Mutualisms

Cleaner fishes on tropical reefs remove parasites, mucus, and dead or infected tissue from cooperative fishes (i.e., clients) (Cote 2000). Reef-based cleaner fish are found at specific cleaning stations, usually situated on prominent portions of the reef. On the Great Barrier Reef, individual clients can visit the cleaner fish *Labroides dimidiatus* up to 144 times a day (Grutter 1995), and *L. dimidiatus* may clean up to 2300 individuals (Grutter 1996) and 132 different species (Grutter & Poulin 1998) a day. The cleaner benefits by food delivery to its territory (Grutter 1996, Grutter 1997), while the client reduces its ectoparasite load. When *L. dimidiatus* are experimentally removed from patches of reef, reef-fish diversity declines (Bshary 2003, Grutter et al. 2003), with the decline being more pronounced for highly mobile clients that visit several reef patches than for less mobile clients that stay on a single reef patch. Cleaner fishes can, thus, have a strong effect on parasite loads in their client fish and on fish usage patterns across patchy reef environments. A similar cleaner-client relationship occurs between an oligochaete worm that cleans the branchial chamber of freshwater crayfish, thereby enhancing crayfish growth and survivorship (Brown et al. 2000).

Host-Microbe Mutualisms

Mutualistic microbes can defend their host by producing chemicals that deter enemies of the host. Healthy embryos of the shrimp *Palaemon macrodactylus* are completely covered by the bacteria *Alteromonas* sp., which produce 2,3-indolinedione as a defense against the fungus *Lagenidium callinectes*, which is pathogenic to the shrimp embryos (Gil-Turnes et al. 1989). If the mutualistic bacteria are removed from the embryos with antibiotics, the embryos are killed by the fungus. If the symbiotic bacteria or the compound 2,3-indolinedione are added back to the embryos following antibiotic treatment, the embryos are protected from fungal attack. In this mutualism, the embryos provide a preferred surface for bacterial growth, and the bacteria chemically defends this resource, enhancing the survivorship of its host.

Microbial gut endosymbionts can also enhance host fitness by facilitating digestion and providing important growth factors for their hosts. In return, hosts provide their microbes with food and a predictable environment (Hungate 1975). Much like terrestrial ruminants, marine herbivorous fishes use gut microflora to help digest algal material (Choat & Clements 1998). Fishes in the genus *Kyphosus* house gut symbionts in a hindgut cecum that acts as a digestive chamber. The cecum keeps microbes under anaerobic conditions and allows fermentation of carbohydrates from the seaweed diet (Rimmer & Wiebe 1987, Seeto et al. 1996). Rates of digestive fermentation in these herbivorous fishes are comparable to those of many terrestrial vertebrate herbivores, which suggests that microbial symbionts play an important role in the digestive process (Mountfort et al. 2002) and that endosymbionts may facilitate the large impacts that herbivores have on marine communities. In addition to aiding digestion directly, microbes are known for their ability to detoxify bioactive chemicals (Iranzo et al. 2001), such as those that seaweeds produce as chemical defenses against herbivores (Hay & Fenical 1988). Therefore, herbivore gut symbionts could act as bioremediators to detoxify seaweed chemical defenses. However, this hypothesis has yet to be investigated.

Sea urchins, such as *Strongylocentrotus droebachiensis*, benefit from endosymbiotic bacteria that fix nitrogen in the digestive tract (Guerinot & Patriquin 1981). When food quality is poor, microbial nitrogen-fixation rates increase, which suggests that bacteria subsidize the nitrogen requirements of sea urchins (Guerinot & Patriquin 1981). This nutritional subsidy could provide urchins with a competitive advantage over other grazers during periods when foods are scarce, and it helps urchins avoid mortality and maintain large populations despite limited food availability (Levitan 1988). This mutualism may facilitate the dramatic impacts urchins have on benthic community structure and their ability to sustain urchin barrens (Estes & Duggins 1995).

Bodyguards

Because sessile invertebrates have a limited behavioral capacity to fend off enemies, they often recruit mutualist bodyguards to ward off attackers. Along the coast of Pacific Panama, branching pocilloporid corals are abundant and commonly harbor the crab *Trapezia ferruginea* and the shrimp *Alpheus lottini*. These crustacean mutualists shelter in corals, where they feed on energy-rich coral mucus and are only 5% as likely to be consumed by predators as are crabs that are not sheltered in a coral host; in return, the crustaceans protect corals from attack by the crown-of-thorns starfish *Acanthaster planci* (Glynn 1976). As starfish mount the coral to feed, the symbionts drive them off by nipping at their tube feet. These crustaceans significantly alter *Acanthaster* feeding preference among coral species and thus alter the impact of coral predation on reef community structure (Glynn 1976). If massive corals are surrounded by barriers of branching pocilloporid corals, crown-of-thorns starfish are prevented from attacking the palatable, massive species by the crustacean symbionts in the encircling pocilloporids (Glynn 1985). When elevated temperatures kill large areas of pocilloporids, and crustacean symbionts abandon their hosts, *Acanthaster* crosses these former barriers and almost completely consumes massive corals that had been a successful part of the coral community for 190 years (Glynn 1985). Although pocilloporids and massive corals regularly compete for space, these pocilloporids and their bodyguards protect massive corals from *Acanthaster* predation, demonstrating the dynamic and context-dependent nature of mutualisms and showing how indirect effects of mutualisms can cascade to impact other community members.

Temperate corals are especially at risk for algal overgrowth because temperate reefs lack herbivorous fishes that prevent seaweeds from overgrowing corals

(Miller 1998). Some temperate corals solve this problem by recruiting herbivorous crabs to shelter among their branches and remove competing seaweeds (Stachowicz & Hay 1999). The branching coral *Oculina arbuscula* harbors the crab *Mithrax forceps* that feeds on all encroaching seaweeds, even those that are chemically defended from other herbivores. Corals with resident crabs grow and survive in well-lit habitats, whereas those without crabs die because of overgrowth (Stachowicz & Hay 1999). Crabs in live coral have significantly enhanced growth and survivorship compared with crabs in dead coral or crabs with access to no coral. This structurally complex but competitively inferior coral provides a biogenic habitat that is used by as many as 309 other local species, and as many as 161 species occur on a single *Oculina* head (McCloskey 1970). Thus, this mutualism not only allows both the coral and the crab to persist in areas where neither could survive alone, but it also produces a biogenic habitat used by many other species.

DANGEROUS LIAISONS (NEGATIVE PAIRWISE INTERACTIONS THAT BECOME POSITIVE IN A COMMUNITY CONTEXT)

Previous authors have noted that antagonistic interactions can evolve into mutualisms (Thompson 1982, van Baalen & Jansen 2001). This finding may be expected, given the close ecological and evolutionary connections between interacting consumers and prey, or parasites and hosts. Because traits for tolerating enemies are more likely to spread through prey populations than are traits for resisting enemies (Roy & Kirchner 2000), enemy-host interactions may commonly evolve toward more benign relationships. In the following sections, we do not discuss antagonistic interactions that have evolved into mutualisms, but rather we discuss antagonistic interactions that are still antagonistic from a pairwise perspective but become mutualistic when imbedded within the nexus of community interactions.

Consumer-Prey Mutualisms

SEAWEED-HERBIVORE MUTUALISMS   Encrusting coralline algae are heavily calcified seaweeds that resemble paint on a rock. Although their hardness and morphology make them resistant to attack by many herbivores (Steneck 1986), they are often fed on by gastropods that have hardened radulae that can scrape into the alga. Littler et al. (1995) demonstrated that about 50% of the diet of the herbivorous chiton *Choneplax lata* consisted of its preferred host coralline, *Porolithon pachydermum*. This alga covers a substantial portion of reef crest habitats along the Belizian Barrier Reef, and feeding by the chiton produces excavations and burrows in the alga. When this herbivore was experimentally removed from its coralline prey, the prey became fouled by epiphytic algae that attracted powerful-jawed, deep-biting parrotfishes. These parrotfishes consumed not only the palatable epiphytes, but also the coralline host. The deep bites of the parrotfish caused much more damage to the coralline host than had been done by the chiton. With the chiton present, the coralline built up carbonate framework at a rate of 1 to 2 mm/y. When the chiton was removed, reef crest areas dominated by the coralline experienced net erosion because of heavy fish grazing. Thus, removal of the herbivorous chiton increased, rather than decreased, grazing damage to the coralline. Additionally, removal of the coralline caused a steep decline in the chiton population. In this mutualism, the grazer keeps its host free of fouling seaweeds, thus reducing attacks by damaging fishes. In return, the plant provides its consumer with a predictable food resource and a complex habitat in which to escape its own consumers. Neither the coralline nor the chiton are obligate mutualists. Each can be found without the other, but each is generally much more abundant and persistent when found together. This mutualism produces an important structural habitat that caps and protects many Caribbean reef crests and serves as a habitat for a diverse assemblage of other invertebrates (Littler et al. 1995). Steneck (1982, 1992) described a similar mutualism between the temperate coralline *Clathromorphum circumscriptum* and its major grazer, the limpet *Acamea testudinalis*. In these mutualisms, herbivore removal of some host mass in return for preventing greater loss is not fundamentally different from conventional mutualisms such as ant-plant, plant-pollinator, or seed-disperser interactions in which plants lose energy to their mutualists in the form of nectar, food bodies, or appreciable portions of their pollen or seeds in return for some offsetting benefit.

Herbivorous damselfishes form similar mutualisms with some seaweeds on tropical reefs. Through aggressive defense of the algal mats on which they feed, territorial damselfish create patches of intermediate grazing intensity where algal species richness, evenness, and diversity are increased relative to areas that are available to all grazers and relative to caged areas where all larger herbivores are excluded (Hixon & Brostoff 1983, 1996). Several algae are locally distributed only within damselfish territories. Although the rapidly growing filamentous algae in the fish's territory are its prey, they are also dependent on the territorial behavior of the fish to protect them from being grazed to local extinction by other groups of reef herbivores. If the territorial fish is removed, its algal lawn is completely consumed within hours (Hixon & Brostoff 1996, Ceccarelli et al. 2001). The presence of the territorial fish also increases algal productivity (both per unit area and per unit algal biomass), but the mechanisms that produce this effect are uncertain (Ceccarelli et al. 2001). Because of the density and ecological importance of damselfish territories to coral reefs, this consumer-prey mutualism can substantially augment reefwide production, algal biomass, and species richness (Hixon & Brostoff 1996, Ceccarelli et al. 2001).

PHYTOPLANKTON-ZOOPLANKTON MUTUALISMS Some phytoplankton, such as *Sphaerocystis schroeteri*, pass through zooplankton guts (*Daphnia*) with minimal digestion, and their growth after gut passage is dramatically elevated compared with uneaten cells (Porter 1976). Although initially described as a mutualism (Porter 1976) in which the phytoplankton received nutrients to enhance growth and the zooplankton received some gelatinous sheath material from the alga, the advantage to the grazer became less certain when it was realized that some phytoplankton species that withstood gut passage did not have gelatinous coverings and thus no covering to sacrifice (Epp & Lewis 1981). However, phytoplankton may continue appreciable photosynthesis while in zooplankton guts. The resistant algae are possibly trading photosynthate to zooplankton in return for viable gut passage (Epp & Lewis 1981), which would make this relationship a mutualism. Effects at the community level could be considerable; in grazing experiments, densities of species resistant to gut passage increase as more consumers are added (Porter 1976).

FUNGAL FARMERS   Just as leaf-cutter ants harvest tree leaves on which to culture their fungal gardens and damselfish kill corals as sites on which to culture their algal gardens, the salt marsh periwinkle, *Littoraria irrorata*, damages fresh blades of the marsh grass, *Spartina alterniflora*, as a substrate on which to culture its fungal food. The periwinkle commonly consumes ascomycete fungi in the genera *Phaeosphaeria* and *Mycosphaerella* that grow on blades of *S. alterniflora*. Although senescent blades of *S. alterniflora* are often colonized by fungi, younger blades are relatively resistant to fungal attack when undamaged. However, the periwinkle thwarts this defense by scraping surficial cells from *Spartina* and depositing fungal-rich and nutrient-rich feces onto the wounds (Silliman & Newell 2003). This process enhances fungal growth on *Spartina* by as much as 170%. Thus, the fungus is continually introduced to new resource patches that are less available without the scraping activity of the snails, and the snails harvest fungi from their expanded fungal farm (Silliman & Newell 2003). Although the snails directly damage only modest amounts of *Spartina*, their mutualism with fungi produces a large top-down effect on *Spartina* abundance. Areas of marsh where periwinkle densities are high can be nearly denuded of *Spartina* (Silliman & Zieman 2001, Silliman & Bertness 2002).

CONSUMERS AS DISPERSAL AGENTS   Consumers such as waterfowl and fishes have long been known to disperse aquatic propagules, just as terrestrial frugivores disperse seeds (Darwin 1859). In both cases, the prey lose some reproductive effort to consumers but gain the movement of their offspring over a greater area than would otherwise be possible. In ephemeral freshwater habitats, this consumer-mediated dispersal may critically affect both community dynamics (Shurin 2000) and the genetic structure of populations (De Meester et al. 2002).

Laboratory and field observations show that (*a*) 7% of seeds consumed by mallard ducks are viable after gut passage, (*b*) gut passage can increase germination success, (*c*) individual ducks may transport 5,000 to 10,000 intact seeds among wetlands during annual migration, and (*d*) many seeds are transported 13 to 75 km during short-range dispersal and up to 300 to 1400 km during "grand-passage" migrations (Figuerola & Green 2002, Holt Mueller & van der Valk 2002). Additional benefits to waterfowl could accrue because they deposit seeds of forage species in newly available or newly disturbed sites that they visit repeatedly during migrations, thus eventually reaping what they sow.

Aquatic vertebrates such as fishes also can play a strong role in flood-plain forest dynamics by eating fruits, dispersing seeds, and thereby affecting tree recruitment and nutrient cycling (Goulding 1980). Goulding (1983, in Horn 1997) argued that a greater mass of fruits and seeds were consumed by fishes than by monkeys and birds in many flood-plain forests. Horn (1997) demonstrated that germination of fig seeds (*Ficus glabrata*) was unaffected after gut passage through the characid fish *Brycon guatemalensis*; he estimated that 500 million fig seeds were consumed and dispersed by *B. guatemalensis* annually along a 6-km river segment in Costa Rica.

For some marine macroalgae, spore dispersal increases when invertebrate herbivores consume nutritious reproductive fronds. This activity is analogous to Janzen's (1984) "foliage is the fruit" hypothesis for terrestrial herbivores, where large herbivores consume and disperse seeds inadvertently while foraging on foliage. The amphipod *Hyale media* preferentially grazes reproductive versus nonreproductive fronds of the red alga *Iridaea laminarioides* and, while feeding, releases up to nine times more algal spores than are released from ungrazed individuals (Buschmann & Santelices 1987). Few of the eaten spores survive gut passage, but those that do survive grow faster than uneaten spores, possibly because of nutrient absorption during gut passage or being deposited within nutrient-rich feces. Preferential feeding on reproductive fronds, along with the presence of viable algal spores and vegetative fragments in herbivore fecal pellets (Paya & Santelices 1989, Santelices & Paya 1989), can result in greater algal recruitment and persistence at sites with high herbivore damage to reproductive blades (Gaines 1985, Buschmann & Vergara 1993). The positive correlation between levels of herbivory and algal success in some species suggests that grazers could have a net positive impact via dispersal despite the negative impacts of grazing on adult tissues. Additionally, seaweeds adapted to rapid colonization and growth are the species whose spores are most likely to survive gut passage (Santelices & Ugarte 1987), which suggests that these species may profit from being deposited in areas where herbivores have removed other competitors.

Competitors as Mutualists

Interspecific competition is traditionally regarded as a negative interaction (−,−) for both participants. This assessment is true when only the two competing species are considered, but in a community context, the effects that a nearby competitor may have in lessening physical stresses or preventing successful attacks by enemies can counteract the negative effects of competition (Hay 1986, Stachowicz 2001, Bruno et al. 2003). Among coral reef sponges, it is more common to find morphologically similar species growing intermingled in multispecific groups of up to 12 individuals than it is to find a sponge colony growing alone (Wulff 1997). In field experiments, growth rates of the sponges *Iotrochota birotulata* and *Aplysina fulva* and survivorship of *Iotrochota birotulata, Amphimedon rubens,* and *Aplysina fulva* were higher when the sponges were grown with heterospecifics than when they were grown alone or with conspecifics (Wulff 1997). Although the mechanisms conferring advantages to heterospecific groups are uncertain, these sponge species differ idiosyncratically in their susceptibility to predation, pathogens, and physical disturbance. The summed traits of the sponge consortia may enable participants to survive environmental challenges that would be insurmountable for any of them growing alone.

As a second example, congeneric species of scale-eating cichlids may benefit each other even though they share the same prey. These predatory fishes consume the scales of other living fish using a species-specific approach and attack sequence. In two congeneric species, attack success was greater when in the presence of the congeneric, but not conspecific, scale eaters (Hori 1987). Presumably, prey fish were unable to be as vigilant against multiple attack strategies. Thus, two species using a similar resource (scales on a given fish) facilitated, rather than interfered with, each other's success. Mutualism also may occur among individuals within a species, as exemplified by frequency-dependent selection in the scale eater *Perissodus microlepis*. Individual *P. microlepis* have asymmetrical mouthparts and corresponding attack strategies: "right-handed" individuals have mouthparts oriented to the right and attack the left side of their prey; "left-handed" individuals have mouthparts oriented to the left and attack the right side of their prey. Deviations from an even ratio of morphs within a population resulted in lower attack success in the dominant morph (Hori 1993). These observations suggest that these two morphs act mutually to increase attack success by decreasing prey-fish alertness for attacks from one side or the other.

Parasite-Host Mutualisms

Parasites, by definition, have direct negative effects on their hosts. Being parasitized could, however, be advantageous if the parasite also infects and has even stronger negative impacts on the host's competitors or predators (Thomas et al. 2000). The potential for parasites to influence community structure via such interactions is particularly relevant to species invasions, as introduced species can harbor pathogens that are much more virulent to native species that have never encountered the parasite. *Myxobolus cerebralis*, for example, is a protozoan parasite of salmonids that causes whirling disease in rainbow trout, cutthroat trout, brook trout, and several species of salmon, yet brown trout—thought to be the ancestral hosts of the parasite—are rarely symptomatic (Bartholomew & Reno 2002, Gilbert & Granath 2003). Where exotic brown trout co-occur with other salmonids in the western United States, the disease has been blamed for catastrophic declines of other trout populations and resulting increases in brown trout (Nehring & Walker 1996). Thus, the high virulence of *M. cerebralis* in North American salmonids may provide a net benefit to brown trout by excluding or reducing populations of competing salmonids. Furthermore, *M. cerebralis* may have benefited brown trout within its native European range by preventing the widespread establishment of rainbow trout, despite many attempted introductions (Lever 1996).

Parasites also might benefit their final host by manipulating the behavior of their intermediate host to make it more susceptible to predation by the final host (Lafferty 1992, Lafferty & Morris 1996). By definition, parasites have direct negative effects on their final host; however, if the value of the greater ease with which the final host can feed on the intermediate host outweighs the costs of parasitism for the final host, then the result can be a net mutualism. The host gets more food, and the parasite is able to complete its life cycle. For example, killifish infected with larval trematodes are 25 times more likely to be eaten by seabirds than are uninfected killifish, probably because parasitized fish frequently expose themselves to predators via conspicuous behaviors at the water's surface (Lafferty & Morris 1996). Larval trematodes are hypothesized to have negligible effect on bird fitness (Martin 1950), so the net gain associated with eating infected killifish could be substantially greater than the negative impacts of trematode infection. Although we know of no data that indicate trematode parasites and seabirds together can affect community structure, both seabirds (Micheli 1997) and killifish (Vince et al. 1976) have strong impacts on marine communities.

DIFFUSE MUTUALISMS

Although mutualisms generally are between species coupled in space and time, reciprocally beneficial interactions can also arise between individuals and species that interact only from afar or through multiple trophic links. Because diffuse mutualisms can occur across ecosystem boundaries or have significant time lags, they are rarely noticed. However, recent studies suggest that strong reciprocal interactions can occur between species that appear to interact weakly (Berlow 1999), if at all, because of their spatial separation. Here, we discuss a mutualism between marine fishes and the terrestrial trees that line the rivers and streams in which these fish spawn and die—in some cases, more than 1000 km inland.

Juvenile salmonids spend up to two years feeding in freshwater streams and rivers before migrating to marine waters, where they mature and gain nearly all of their biomass, after which they return to their natal habitats to spawn and die. Trees subsidize production in these streams with the input of nutrients, leaf litter, and woody debris that supports higher populations of aquatic invertebrates, the main food source for juvenile salmon (Everett & Ruiz 1993, Wallace et al. 1999, Helfield & Naiman 2002). At the landscape scale, forested streams typically support up to three times more salmon than unforested streams (Pess et al. 2002). Salmon, thus, benefit from living in streams surrounded by trees, but the benefit is not unidirectional.

Spawning salmon migrations inject huge amounts of marine-derived nitrogen, carbon, and phosphorous into relatively nutrient-starved systems. These massive inputs of marine-derived nutrients can be detected in the surrounding watershed, and streamside trees derive up to 26% of their nitrogen directly from spawning salmon (Helfield & Naiman 2001). As a result, annual forest growth per unit area can be up to three times higher in forests adjacent to salmon spawning sites (Helfield & Naiman 2001). Furthermore, this subsidy of nutrients may alter the competitive balance among tree species. Whereas Sitka spruce dominate nitrogen-rich stream flood plains around salmon spawning sites, western hemlock dominate similar nitrogen-poor areas upstream of spawning sites (Helfield & Naiman 2001, 2003).

Salmon often return to the same stream in which they were spawned, and salmon from particular reaches of streams are more likely to be related to one another than they are to salmon from other reaches, which suggests a heritable basis for spawning site selection (Bentzen et al. 2001). Thus, when adult salmon return to a stream and die, their carcasses effectively fertilize not only their own offspring but potentially the offspring of relatives as well. In essence, adult salmon leave a bodily inheritance, the biomass accrued during a lifetime at sea, which fertilizes the resource base that feeds their offspring.

SUMMARY

Mutualisms are more than biological oddities that enhance our understanding of evolution. They are pivotal in affecting the organization, structure, and function of communities, yet this broader-scale importance of mutualisms is rarely appreciated. Many ecologically important mutualisms are not coevolved or obligate; they are instead conditional and provide partner species with novel options for adjusting to changing physical and biotic environments. Mutualisms commonly support the foundation species that define entire ecosystems, and they can play critical roles in moving energy and nutrients across ecosystem borders—as when mutualistic nitrogen fixers convert atmospheric N_2 into bioavailable nitrogen in oceanic gyres, when bivalves or fishes move planktonic nutrients to benthic communities, and when marine fishes and riparian forests produce reciprocal benefits. Through indirect interactions within a community setting, benefits gained can offset the negative effects of competition, predation, parasitism, and physical stresses to the extent that potential enemies interact as mutualists. These interactions may allow persistence within communities or the invasion of new ones. Historically, ecology has been dominated by the study of negative interactions (e.g., competition, predation, or parasitism). However, when mutualisms are investigated within their community matrix, positive interactions become as important as negative interactions in affecting community organization and thus deserve full incorporation into basic community theory.

ACKNOWLEDGMENTS

The manuscript was improved by comments from M. Bertness, C.D. Harvell, J. Kubanek, H. Pavia, J. Stachowicz, G. Toth, and the 2004 Aquatic Ecology class students at Georgia Tech.

The *Annual Review of Ecology, Evolution, and Systematics* is online at
http://ecolsys.annualreviews.org

LITERATURE CITED

Alongi DM. 2002. Present state and future of the world's mangrove forests. *Environ. Conserv.* 29:331–49

Bacher S, Friedli J. 2002. Dynamics of a mutualism in a multi-species context. *Proc. R. Soc. London Ser. B* 269:1517–22

Baker AC. 2001. Ecosystems-reef corals bleach to survive change. *Nature* 411:765–66

Baker AC. 2002. Is bleaching really adaptive? Reply to Hoegh-Guldberg et al. *Nature* 415:602

Baker AC. 2003. Flexibility and specificity in coral-algal symbiosis: diversity, ecology, and biogeography of *Symbiodinium*. *Annu. Rev. Ecol. Evol. Syst.* 34:661–89

Banaszak AT, LaJeunesse TC, Trench RK. 2000. The synthesis of mycosporine-like amino acids (MAAs) by cultured, symbiotic dinoflagellates. *J. Exp. Mar. Biol. Ecol.* 249:219–33

Bartholomew JL, Reno PW. 2002. The history and dissemination of whirling disease. *Am. Fish. Soc. Symp.* 29:3–24

Bentzen P, Olsen JB, McLean JE, Seamons TR, Quinn TP. 2001. Kinship analysis of Pacific salmon: insights into mating, homing, and timing of reproduction. *J. Hered.* 92:127–36

Berlow EL. 1999. Strong effects of weak interactions in ecological communities. *Nature* 398:330

Bertness MD. 1984. Ribbed mussels and *Spartina alterniflora* production in a New England salt marsh. *Ecology* 65:1794–807

Bertness MD, Gaines SD, Hay ME, eds. 2001. *Marine Community Ecology*. Sunderland, MA: Sinauer. 550 pp.

Bronstein JL. 1994. Conditional outcomes in mutualistic interactions. *Trends Ecol. Evol.* 9:214–17

Brooks JM, Kennicutt MC II, Fischer CR, Macko SA, Cole K, et al. 1987. Deep-sea hydrocarbon seep communities: evidence for energy and nutritional carbon sources. *Science* 238:1138–42

Brown BL, Creed RP, Dobson WE. 2002. Brachiobdellid annelids and their crayfish hosts: Are they engaged in a cleaning symbiosis? *Oecologia* 132:250–55

Bruno JF, Stachowicz JJ, Bertness MD. 2003. Inclusion of facilitation into ecological theory. *Trends Ecol. Evol.* 18:119–25

Bshary R. 2003. The cleaner wrasse, *Labroides dimidiatus*, is a key organism for reef fish diversity at Ras Mohammed National Park, Egypt. *J. Anim. Ecol.* 72:169–76

Buschmann AH, Vergara PA. 1993. Effect of rocky intertidal amphipods on algal recruitment—a field study. *J. Phycol.* 29:154–59

Buschmann A, Santelices B. 1987. Micrograzers and spore release in *Iridaea laminarioides* Bory (Rhodophyta: Gigartinales). *J. Exp. Mar. Biol. Ecol.* 108:171–79

Carpenter EJ, Montoya JP, Burns J, Mulholland MR, Subramaniam A, et al. 1999. Extensive bloom of a N_2-fixing diatom/cyanobacterial association in the tropical Atlantic Ocean. *Mar. Ecol. Prog. Ser.* 185:273–83

Carpenter RC. 1986. Partitioning herbivory and its effects on coral-reef algal communities. *Ecol. Monogr.* 56:345–63

Ceccarelli DM, Jones GP, McCook LJ. 2001. Territorial damselfish as determinants of the structure of benthic communities on coral reefs. *Oceanogr. Mar. Biol. Annu. Rev.* 39:355–89

Choat JH, Clements KD. 1998. Vertebrate herbivores in marine and terrestrial environments: a nutritional ecology perspective. *Annu. Rev. Ecol. Syst.* 29:375–403

Connor RC. 1995. The benefits of mutualism—a conceptual framework. *Biol. Rev. Camb. Philos. Soc.* 70:427–57

Cote IM. 2000. Evolution and ecology of cleaning symbioses in the sea. *Oceanogr. Mar. Biol.* 38:311–55

Darwin C. 1859. *On the Origin of Species by Means of Natural Selection*. London: John Murray. 540 pp.

Dayton PK. 1975. Experimental evaluation of ecological dominance in a rocky intertidal algal community. *Ecol. Monogr.* 45:137–59

De Meester L, Gomez A, Okamura B, Schwenk K. 2002. The monopolization hypothesis and the dispersal-gene flow paradox in aquatic organisms. *Acta Oecol.* 23:121–35

Doebeli M, Knowlton N. 1998. The evolution of interspecific mutualisms. *Proc. Natl. Acad. Sci. USA* 95:8676–80

Duffy JE, Richardson JP, Canuel EA. 2003. Grazer diversity effects on ecosystem functioning in seagrass beds. *Ecol. Lett.* 6:637–45

Duffy JE, Macdonald KS, Rhode JM, Parker JD. 2001. Grazer diversity, functional redundancy, and productivity in seagrass beds: an experimental test. *Ecology* 82:2417–34

Duggins DO, Simenstad CA, Estes JA. 1989. Magnification of secondary production by kelp detritus in coastal marine ecosystems. *Science* 245:170–73

Edgar GJ, Shaw C. 1995. The production and trophic ecology of shallow-water fish assemblages in southern Australia. 3. General relationships between sediments, seagrasses, invertebrates and fishes. *J. Exp. Mar. Biol. Ecol.* 194:107–31

Ehrlich PR. 1975. Population biology of coral-reef fishes. *Annu. Rev. Ecol. Syst.* 6:211–46

Ellison AM, Farnsworth EJ. 1990. The ecology of Belizean mangrove-root fouling communities. I. Epibenthic fauna are barriers to isopod attack of red mangrove roots. *J. Exp. Mar. Biol. Ecol.* 142:91–104

Ellison AM, Farnsworth EJ, Twilley RR. 1996. Facultative mutualism between red mangroves and root-fouling sponges in Belizean mangal. *Ecology* 77:2431–44

Epp RW, Lewis WM Jr. 1981. Photosynthesis in copepods. *Science* 214:1349–50

Estes JA. 1990. Growth and equilibrium in sea otter populations. *J. Anim. Ecol.* 95:385–401

Estes JA, Duggins DO. 1995. Sea otters and kelp forests in Alaska: generality and variation in a community ecological paradigm. *Ecol. Monogr.* 65:75–100

Everett RA, Ruiz GM. 1993. Coarse woody debris as a refuge from predation in aquatic communities—an experimental test. *Oecologia* 93:475–86

Farnsworth EJ, Ellison AM. 1996. Scale-dependent spatial and temporal variability in biogeography of mangrove root epibiont communities. *Ecol. Monogr.* 66:45–66

Fiala-Medioni A, Felbeck H. 1990. Autotrophic processes in invertebrate nutrition: bacterial symbiosis in bivalve molluscs. *Comp. Physiol.* 5:49–69

Figuerola J, Green AJ. 2002. Dispersal of aquatic organisms by waterbirds: a review of past research and priorities for future studies. *Freshw. Biol.* 47:483–94

Gaines SD. 1985. Herbivory and between-habitat diversity: the differential effectiveness of defenses in a marine plant. *Ecology* 66:473–85

Gil-Turnes MS, Hay ME, Fenical W. 1989. Symbiotic marine-bacteria chemically defend crustacean embryos from a pathogenic fungus. *Science* 246:116–18

Glynn PW. 1976. Some physical and biological determinants of coral community structure in the eastern Pacific. *Ecol. Monogr.* 46:431–56

Glynn PW. 1985. El Niño–associated disturbance to coral reefs and post-disturbance mortality by *Acanthaster planci*. *Mar. Ecol. Prog. Ser.* 26:295–300

Gomulkiewicz R, Nuismer SL, Thompson JN. 2003. Coevolution in variable mutualisms. *Am. Nat.* 162(Suppl.):S80–93

Gordon N, Angel DL, Neori A, Kress N, Kimor B. 1994. Heterotrophic dinoflagellates with symbiotic cyanobacteria and nitrogen limitation in the Gulf of Aqaba. *Mar. Ecol. Prog. Ser.* 107:83–8

Goulding M. 1980. *The Fishes and the Forest: Explorations in Amazonian Life History*. Berkeley: Univ. Calif. Press. 280 pp.

Goulding MH. 1983. The role of fishes in seed dispersal and plant distribution in

Amazonian floodplain ecosystems. *Sonderbd. Naturwiss. Ver. Hamburg.* 7:271–83

Grant PR, Grant BR. 1997. The rarest of Darwin's finches. *Conserv. Biol.* 11:119–26

Grutter AS. 1995. Relationship between cleaning rates and ectoparasite loads in coral reef fishes. *Mar. Ecol. Prog. Ser.* 118:51–58

Grutter AS. 1996. Parasite removal rates by the cleaner wrasse *Labroides dimidiatus. Mar. Ecol. Prog. Ser.* 130:61–70

Grutter AS. 1997. Spatiotemporal variation and feeding selectivity in the diet of the cleaner fish *Labroides dimidiatus. Copeia* 2:346–55

Grutter AS, Murphy JM, Choat H. 2003. Cleaner fish drives local fish diversity on coral reefs. *Curr. Biol.* 13:64–67

Grutter AS, Poulin R. 1998. Cleaning of coral reef fishes by the wrasse *Labroides dimidiatus*: influence of client body size and phylogeny. *Copeia* 120:120–27

Guerinot ML, Patriquin DG. 1981. The association of N_2-fixing bacteria with sea urchins. *Mar. Biol.* 62:197–207

Hallock P. 1997. Reefs and reef limestones in Earth history. In *Life and Death of Coral Reefs.* ed. C Birkeland, pp. 13–42. New York: Chapman & Hall. 560 pp.

Hay ME. 1986. Associational plant defenses and the maintenance of species diversity: turning competitors into accomplices. *Am. Nat.* 128:617–41

Hay ME, Fenical W. 1988. Marine plant-herbivore interactions—the ecology of chemical defense. *Annu. Rev. Ecol. Syst.* 19:111–45

Helfield JM, Naiman RJ. 2001. Effects of salmon-derived nitrogen on riparian forest growth and implications for stream productivity. *Ecology* 82:2403–9

Helfield JM, Naiman RJ. 2002. Salmon and alder as nitrogen sources to riparian forests in a boreal Alaskan watershed. *Oecologia* 133:573–82

Helfield JM, Naiman RJ. 2003. Effects of salmon-derived nitrogen on riparian forest growth and implications for stream productivity: reply. *Ecology* 84:3399–401

Hixon MA, Brostoff WN. 1983. Damselfish as keystone species in reverse: intermediate disturbance and diversity of reed algae. *Science* 220:511–13

Hixon MA, Brostoff WN. 1996. Succession and herbivory: effects of differential fish grazing on Hawaiian coral-reef algae. *Ecol. Monogr.* 66:67–90

Hoegh-Guldberg O, Jones RJ, Ward S, Loh WK. 2002. Is bleaching really adaptive? *Nature* 415:601–2

Holland JN, DeAngelis DL, Bronstein JL. 2002. Population dynamics and mutualism: functional responses of benefits and costs. *Am. Nat.* 159:231–44

Holt Mueller M, van der Valk AG. 2002. The potential role of ducks in wetland seed dispersal. *Wetlands* 22:170–78

Hori M. 1987. Mutualisms and commensalism in a fish community in Lake Tanganyika. In *Evolution and Coadaptation in Biotic Communities*, ed. S Kawano, JH Connell, T Hidaka, pp. 219–39. Tokyo: Univ. Tokyo Press

Hori H. 1993. Frequency-dependent natural selection in the handedness of scale-eating cichlid fish. *Science* 260:216–19

Horn MH. 1997. Evidence for dispersal of fig seeds by the fruit-eating characid fish *Brycon guatemalensis* Regan in a Costa Rican tropical rain forest. *Oecologia* 109:259–64

Howarth RW, Marino R, Lane J, Cole JJ. 1988. Nitrogen-fixation in freshwater, estuarine, and marine ecosystems. I. Rates and importance. *Limnol. Oceanogr.* 33:669–87

Hughes TP. 1994. Catastrophes, phase shifts, and large-scale degradation of a Caribbean coral reef. *Science* 265:1547–51

Hungate RE. 1975. Rumen microbial ecosystem. *Annu. Rev. Ecol. Syst.* 6:39–66

Iranzo M, Sainz-Pardo I, Boluda R, Sanchez J, Mormeneo S. 2001. The use of microorganisms in environmental remediation. *Ann. Microbiol.* 51:135–43

Irlandi EA, Peterson CH. 1991. Modification of animal habitat by large plants—mechanisms by which seagrasses influence clam growth. *Oecologia* 87:307–18

Jackson JBC. 2001. What was natural in the

coastal oceans? *Proc. Nat. Acad. Sci. USA* 98:5411–18

Jackson JBC, Kirby MX, Berger WH, Bjorndal KA, Botsford LW, et al. 2001. Historical overfishing and the recent collapse of coastal ecosystems. *Science* 293:629–38

Janzen DH. 1984. Dispersal of small seeds by big herbivores—foliage is the fruit. *Am. Nat.* 123:338–53

Jernakoff P, Brearley A, Nielsen J. 1996. Factors affecting grazer-epiphyte interactions in temperate seagrass meadows. *Oceanogr. Mar. Biol.* 34:109–62

Jones GP. 1991. Postrecruitment processes in the ecology of coral reef fish populations: a multifactorial perspective. In *The Ecology of Fishes on Coral Reefs*, ed. PF Sale, pp. 294–328. San Diego, CA: Academic

Jones CG, Lawton JH, Shackak M. 1997. Positive and negative effects for organisms as ecosystem engineers. *Ecology* 78:1946–57

Karl DM, Bidigare RR, Letelier RM. 1999. Long-term changes in phytoplankton community structure and productivity in the North Pacific Subtropical Gyre: the phase shift hypothesis. *Deep Sea Res. II* 48:1449–70

Knowlton N, Rohwer F. 2003. Multispecies microbial mutualisms on coral reefs: the host as a habitat. *Am. Nat.* 162(Suppl.):S51–62

Lafferty KD. 1992. Foraging on prey that are modified by parasites. *Am. Nat.* 140:854–67

Lafferty KD, Morris AK. 1996. Altered behavior of parasitized killifish increases susceptibility to predation by bird final hosts. *Ecology* 77:1390–97

LaJeunesse TC. 2002. Diversity and community structure of symbiotic dinoflagellates from Caribbean coral reefs. *Mar. Biol.* 141:387–400

Leber KM. 1985. The influence of predatory decapods, refuge, and microhabitat selection on seagrass communities. *Ecology* 66:1951–64

Lever C. 1996. *Naturalized Fishes of the World*. San Diego, CA: Academic. 408 pp.

Levitan DR. 1988. Density-dependent size regulation and negative growth in the sea urchin *Diadema antillarum* Philippi. *Oecologia* 76:627–29

Lewis SM. 1986. The role of herbivorous fishes in the organization of a Caribbean reef community. *Ecol. Monogr.* 56:183–200

Liberman T, Genin A, Loya Y. 1995. Effects on growth and reproduction of the coral *Stylophora pistillata* by the mutualistic damselfish *Dascyllus marginatus*. *Mar. Biol.* 121:741–46

Littler MM, Littler DS, Taylor PR. 1995. Selective herbivore increases biomass of its prey: a chiton-coralline reef-building association. *Ecology* 76:1666–81

MacAvoy SE, Carney RS, Fisher CR, Macko SA. 2002. Use of chemosynthetic biomass by large, mobile, benthic predators in the Gulf of Mexico. *Mar. Ecol. Prog. Ser.* 225:65–78

Marshall N. 1994. Mangrove conservation in relation to overall environmental considerations. *Hydrobiologia* 285:303–9

Martin WE. 1950. *Euhaplorchis californiensis* n.g., n.sp., Heterophyidae, Trematoda, with notes on its life cycle. *Trans. Am. Microsc. Soc.* 69:194–209

McClanahan TR, Bergman K, Huitric M, McField M, Elfwing T, et al. 2000. Response of fishes to algae reduction on Glovers Reef, Belize. *Mar. Ecol. Prog. Ser.* 206:273–82

McCloskey LR. 1970. The dynamics of the community associated with a marine scleractinian coral. *Int. Rev. Gesamt. Hydrobiol.* 55:13–81

Meyer JL, Schultz ET. 1985a. Migrating haemulid fishes as a source of nutrients and organic matter on coral reefs. *Limnol. Oceanogr.* 30:146–56

Meyer JL, Schultz ET. 1985b. Tissue condition and growth rate of corals associated with schooling fish. *Limnol. Oceanogr.* 30:157–66

Meyer JL, Schultz ET, Helfman GS. 1983. Fish schools—an asset to corals. *Science* 220:1047–49

Micheli F. 1997. Effects of predator foraging behavior on patterns of prey mortality in marine soft bottoms. *Ecol. Monogr.* 67:203–24

Micheli F, Peterson CH, Mullineaux LS, Fisher CR, Mills SW, et al. 2002. Predation structures communities at deep-sea hydrothermal vents. *Ecol. Monogr.* 72:365–82

Miller MW. 1998. Coral/seaweed competition and the control of reef community structure within and between latitudes. *Oceanogr. Mar. Biol. Annu. Rev.* 36:65–96

Mountfort DO, Campbell J, Clements KD. 2002. Hindgut fermentation in three species of marine herbivorous fish. *Appl. Environ. Microbiol.* 68:1374–80

Muller-Parker G, D'Elia CF. 1997. Interactions between corals and their symbiotic algae. In *Life and Death of Coral Reefs.* ed. C Birkeland, pp. 96–133. New York: Chapman & Hall. 560 pp.

Neckles HA, Wetzel RL, Orth RJ. 1993. Relative effects of nutrient enrichment and grazing on epiphyte-macrophyte (*Zostera marina* L.) dynamics. *Oecologia* 93:285–95

Nehring RB, Walker PG. 1996. Whirling disease in the wild: the new reality in the intermountain West. *Fisheries* 21:28–30

Neumann AC, MacIntyre I. 1985. Reef response to sea level rise: keep-up, catch-up, or give-up. *Proc. Fifth Int. Coral Reef Cong.* 3:105–10

Ogden JC, Ehrlich PR. 1977. Behavior of heterotypic resting schools of juvenile grunts (Pomadasyidae). *Mar. Biol.* 42:273–80

Orth RJ, Moore KA. 1983. Chesapeake Bay: an unprecedented decline in submerged aquatic vegetation. *Science* 222:51–53

Orth RJ, van Montfrans J. 1984. Epiphyte-seagrass relationships with an emphasis on the role of micrograzing: a review. *Aquat. Bot.* 18:43–69

Paya I, Santelices B. 1989. Macroalgae survive digestion by fishes. *J. Phycol.* 25:186–88

Perry DM. 1988. Effects of associated fauna on growth and productivity in the red mangrove. *Ecology* 69:1064–75

Pess GR, Montgomery DR, Steel EA, Bilby RE, Feist BE, et al. 2002. Landscape characteristics, land use, and coho salmon (*Oncorhynchus kisutch*) abundance, Snohomish River, Washington, USA. *Can. J. Fish. Aquat. Sci.* 59:613–23

Peterson BJ, Heck KL. 2001a. An experimental test of the mechanism by which suspension feeding bivalves elevate seagrass productivity. *Mar. Ecol. Prog. Ser.* 218:115–25

Peterson BJ, Heck KL. 2001b. Positive interactions between suspension-feeding bivalves and seagrass—a facultative mutualism. *Mar. Ecol. Prog. Ser.* 213:143–55

Porter KG. 1976. Enhancement of algal growth and productivity by grazing zooplankton. *Science* 192:1332–34

Purves WK, Orians GH, Heller HC, Sadava D, eds. 2001. *Life, the Science of Biology.* Cranbury, NJ: Freeman. 1044 pp.

Rehm A, Humm HJ. 1973. *Sphaeroma terebrans*—threat to mangroves of southwestern Florida. *Science* 182:173–74

Rimmer DW, Wiebe WJ. 1987. Fermentative microbial digestion in herbivorous fishes. *J. Fish Biol.* 31:229–36

Robertson AI, Alongi DM. 1995. Role of riverine mangrove forests in organic carbon export to the tropical coastal ocean: a preliminary mass balance for the Fly Delta (Papua New Guinea). *Geo-Marine. Lett.* 15:134–39

Roy BA, Kirchner JW. 2000. Evolutionary dynamics of pathogen resistance and tolerance. *Evolution* 54:51–63

Santelices B, Paya I. 1989. Digestion survival of algae: some ecological comparisons between free spores and propagules in fecal pellets. *J. Phycol.* 25:693–99

Santelices B, Ugarte R. 1987. Algal life-history strategies and the resistance to digestin. *Mar. Ecol. Prog. Ser.* 35:267–75

Savage AM, Trapido-Rosenthal H, Douglas AE. 2002. On the functional significance of molecular variation in *Symbiodinium*, the symbiotic algae of Cnidaria: photosynthetic response to irradiance. *Mar. Ecol. Prog. Ser.* 244:27–37

Seeto GS, Veivers PC, Clements KD, Slaytor M. 1996. Carbohydrate utilisation by microbial symbionts in the marine herbivorous fishes *Odax cyanomelas* and *Crinodus lophodon*. *J. Comp. Phys. B* 165:571–79

Shpigel M, Fishelson L. 1986. Behavior and physiology of coexistence in 2 species of

Dascyllus (Pomacentridae, Teleostei). *Environ. Biol. Fish.* 17:253–65

Shurin JB. 2000. Dispersal limitation, invasion resistance, and the structure of pond zooplankton communities. *Ecology* 81:3074–86

Silliman BR, Bertness MD. 2002. A trophic cascade regulates salt marsh primary production. *Proc. Natl. Acad. Sci. USA* 99:10500–5

Silliman BR, Newell SY. 2003. Fungal farming in a snail. *Proc. Natl. Acad. Sci. USA* 100: 15643–48

Silliman BR, Zieman JC. 2001. Top-down control of *Spartina alterniflora* production by periwinkle grazing in a Virginia salt marsh. *Ecology* 82:2830–45

Stachowicz JJ, Hay ME. 1999. Mutualism and coral persistence: the role of herbivore resistance to algal chemical defense. *Ecology* 80:2085–101

Stachowicz JJ. 2001. Mutualism, facilitation, and the structure of ecological communities. *Bioscience* 51:235–46

Stanton ML. 2003. Interacting guilds: moving beyond the pairwise perspective on mutualisms. *Am. Nat.* 162(Suppl.):S10–23

Steneck RS. 1982. A limpet-coralline alga association: adaptations and defenses between a selective herbivore and its prey. *Ecology* 63:507–22

Steneck RS. 1986. The ecology of coralline algal crusts—convergent patterns and adaptive strategies. *Annu. Rev. Ecol. Syst.* 17:273–303

Steneck RS. 1992. Plant-herbivore coevolution: a reappraisal from the marine realm and its fossil record. In *Plant-Animal Interactions in the Marine Benthos*, ed. DM John, SJ Hawkins, JH Price, pp. 477–91. Oxford: Clarendon

Streelman JT, Alfaro M, Westneat MW, Bellwood DR, Karl SA. 2002. Evolutionary history of the parrotfishes: biogeography, ecomorphology, and comparative diversity. *Evolution* 56:961–71

Svavarsson J, Osore MKW, Olafsson E. 2002. Does the wood-borer *Sphaeroma terebrans* (Crustacea) shape the distribution of the mangrove *Rhizophora mucronata*? *Ambio* 31:574–79

Thomas F, Poulin R, Guégan JF, Michalakis Y, Renaud F. 2000. Are there pros as well as cons to being parasitized? *Parasitol. Today* 16:533–36

Thompson JN. 1982. *Interaction and Coevolution*. New York: Wiley. 179 pp.

Thomson J. 2003. When is it mutualism? *Am. Nat.* 162(Suppl.):S1–9

Tunnicliffe V. 1992. Hydrothermal-vent communities of the deep sea. *Am. Sci.* 80:115–28

van Baalen M, Jansen VAA. 2001. Dangerous liaisons: the ecology of private interest and common good. *Oikos* 95:211–24

Vermeij GJ. 1977. Patterns in crab claw size—geography of crushing. *Syst. Zool.* 26:138–51

Vermeij GJ. 1992. Time of origin and biogeographical history of specialized relationships between northern marine plants and herbivorous mollusks. *Evolution* 46:657–64

Veron JEN. 1995. *Corals in Space and Time*. Ithaca, NY: Cornell Univ. Press. 321 pp.

Vince S, Valiela I, Backus N, Teal JM. 1976. Predation by salt-marsh killifish *Fundulus heteroclitus* in relation to prey size and habitat structure—consequences for prey distribution and abundance. *J. Exp. Mar. Biol. Ecol.* 23:255–66

Wallace JB, Eggert SL, Meyer JL, Webster JR. 1999. Effects of resource limitation on a detrital-based ecosystem. *Ecol. Monogr.* 69: 409–42

Wood R. 1998. The ecological evolution of reefs. *Annu. Rev. Ecol. Syst.* 29:179–206

Wulff JL. 1997. Mutualisms among species of coral reef sponges. *Ecology* 78:146–59

Zehr JP, Carpenter EJ, Villareal TA. 2000. New perspectives on nitrogen-fixing microorganisms in tropical and subtropical oceans. *Trends Microbiol.* 8:68–73

OPERATIONAL CRITERIA FOR DELIMITING SPECIES

Jack W. Sites, Jr., and Jonathon C. Marshall
Department of Integrative Biology and M.L. Bean Life Science Museum, Brigham Young University, Provo, Utah 84602-5181; email: Jack_Sites@byu.edu, Jonathon_C_Marshall@hotmail.com

Key Words species criteria, species delimitation, systematics, biodiversity, speciation

■ **Abstract** Species are routinely used as fundamental units of analysis in biogeography, ecology, macroevolution, and conservation biology. A large literature focuses on defining species conceptually, but until recently little attention has been given to the issue of empirically delimiting species. Researchers confronted with the task of delimiting species in nature are often unsure which method(s) is (are) most appropriate for their system and data type collected. Here, we review twelve of these methods organized into two general categories of tree- and nontree-based approaches. We also summarize the relevant biological properties of species amenable to empirical evaluation, the classes of data required, and some of the strengths and limitations of each method. We conclude that all methods will sometimes fail to delimit species boundaries properly or will give conflicting results, and that virtually all methods require researchers to make qualitative judgments. These facts, coupled with the fuzzy nature of species boundaries, require an eclectic approach to delimiting species and caution against the reliance on any single data set or method when delimiting species.

> No one definition has as yet satisfied all naturalists; yet every naturalist knows vaguely what he means when he speaks of a species.
>
> Darwin (1859/1964)

INTRODUCTION

Indigenous folk taxonomies spanning widely divergent cultures suggest shared and possibly innate cognitive mechanisms for recognizing biological categories (Atran et al. 1999, Hey 2001). The challenge for biologists is to go beyond subjective judgments and develop operational methods of delimiting species. Species are usually considered fundamental to studies of ecology, evolution, systematics, and conservation biology, yet the literature on the empirical methods of delimiting species has until recently been meager relative to the extensive literature on species concepts and on theory and methods of phylogenetic analysis (Wiens 1999). This is an odd state of affairs given that two frequently stated empirical goals of systematic biology are: (*a*) to discover monophyletic groups at higher levels, and (*b*) to discover lineages (i.e., species) at lower levels (Wheeler & Meier 2000). Cracraft (2002) claims that the question, "What is a species?" remains the most crucial of the "seven great questions of systematic biology." Today, the empirical issue of species delimitation is receiving increased attention, and several novel methods have recently been proposed for delimiting species in a statistically rigorous framework (Puorto et al. 2001, Templeton 2001, Wiens & Servedio 2000, Wiens & Penkrot 2002). Nontree-based methods delimit species on the basis of gene flow assessments (Sites & Marshall 2003), whereas tree-based methods delimit species as historical lineages (Baum & Donoghue 1995, Goldstein & DeSalle 2000, Wheeler & Meier 2000).

Defining species conceptually has been contentious, as is evident by the plethora of species concepts (summaries in Zink 1997; Mayden 1997; Harrison 1998, table 2.1; Sluys & Hazevoet 1999), and some researchers have argued that there is no clear conceptual distinction between species and higher categories (Ereshefsky 2002, Mishler 2003). We share the view that species are spatio-temporally bounded entities (rather than classes defined by some common property; Baum 1998) that differ from higher categories in that they can originate by any number of mechanisms (Turelli et al. 2001) as a consequence of descent-with-modification (the unitary process of Frost & Kluge 1994). We anticipate that different data sets and different methods of delimiting species may give ambiguous or conflicting results (Hey et al. 2003, Sites & Marshall 2003) as a consequence of multiple evolutionary processes operating within and between populations across varying spatio-temporal scales (Harrison 1998, Lee 2003). Thus, the failure of any given discovery method in a particular case does not negate the reality of the species in nature (de Queiroz 1998, Wiley & Mayden 2000).

IMPORTANCE OF EMPIRICAL DELIMITATION OF SPECIES

Species delimitation is essential because species are used as basic units of analysis in several areas of biogeography, ecology, and macroevolution (Brown et al. 1996, Blackburn & Gaston 1998, Brooks & McLennan 1999, Barraclough & Nee 2001). Species are also the currency for global biodiversity assessments (Caldecott et al. 1996) and are therefore important to conservation biology (Agapow et al. 2004). At a more fine-grained level, species are often extensively subdivided (Hughes et al. 1997, Bohonak 1999), and their boundaries and population structures usually define the rates of, and limits within which, many evolutionary processes operate (Coyne et al. 1997). Empirical inference of species boundaries may be difficult when: (*a*) one or more characters acquires an independent geographic distribution within a species (which is likely given sufficient subdivision; Wade & Goodnight 1998) and therefore is/are not good whole-genome markers for the boundaries of the species (Porter 1990); or (*b*) a hybrid zone acts as a partial sieve (Martinsen et al. 2001) between two independent genealogical lineages that undergo some intercrossing at points of parapatric contact (which may have several possible outcomes; Burke & Arnold 2001). Although these phenomena occasionally mislead delimitation of species, an understanding of evolutionary processes and mechanisms requires that we attempt to define objectively the arena within or across which these processes can act (see also Cracraft 2000) with rigorous empirical methods.

SPECIES CONCEPTS AND OPERATIONAL CRITERIA

Frost & Kluge (1994) and later Mayden (1997) distinguished between primary and secondary species concepts. The primary concept defines the entities believed to be species [the evolutionary species concept (ESC; Wiley 1978) was favored by these authors as the conceptual, or ontological definition], whereas secondary concepts are the operational methods for discovery of entities in accord with the primary concept. De Queiroz (1998) extended this idea further and noted that all modern species definitions either explicitly or implicitly equated species with "segments of population level evolutionary lineages," which he termed the general lineage concept (GLC) of species. These authors all made the same distinction between the ESC ontological definition and criteria used to delimit species in nature (empirical discovery methods).

This distinction is useful. These authors' perspectives capture the meaning of lineages (see Pigliucci 2003 for a similar perspective), and although lineage splitting (speciation) may result in species that can be empirically discovered (via fixation of a character, loss of a polymorphism, attainment of reproductive isolation, etc.), the processes causing the new species eventually to become distinct are usually unpredictable. A GLC (i.e., one defining species as ancestor-descendent populations) allows investigators to test species boundaries from different philosophical perspectives. Population biologists, for example, are likely to favor biological species criteria (giving primacy to gene flow), whereas systematists usually favor phylogenetic criteria (distinctness of lineages). Regardless, the delimitation of species requires that one have clearly defined operational criteria by which individuals can be tested for species membership, and the criteria must be understood within the context of what kind of entity (interbreeding versus historical lineage) each method is designed to test. In this review, we discuss 12 methods of species delimitation and consider the kinds of suitable data for, and the strengths and limitations of, each method (Table 1).

Nontree-Based Methods

Boundaries of sexually reproducing species (sensu Mayr 1942) are conceptually defined on the basis of reproductive compatibility within, or reproductive isolation between, species, and these boundaries can occasionally be tested by direct studies of crossability (Dettman et al. 2003b). This testing is not possible for many sexual species, however, and a number of methods are available for indirectly estimating gene flow within and between hypothesized species.

TABLE 1 Empirical methods for delimiting species in the context of properties/criteria, classes of data, generality (asexual versus sexual), and some important assumptions and/or limitations of each

Method[a]	Relevant biological properties/criteria[b]	Classes of data suitable to method	Assumptions/limitations
HZB[1]	Limited or no gene flow across hybrid zone	Nuclear genes with codominance	F_{ST}-based Nm estimator; assumes drift–gene flow equilibrium, with isolation-by-distance model
GenD$_{GW}$[1]	Gene flow within but not between species	Multilocus allele frequency data	Assumes drift-gene flow equilibrium
GenD$_H$[1]	Time-dependent emergence of reproductive isolation	Multilocus allele frequency data	Assumes a molecular clock correlated with a genomic basis for reproductive isolation
FFR[1]	Recombination within nuclear loci limited by extent of gene flow	Nuclear genes with codominance	Requires identification of all alleles segregating at a locus, and no gene flow between species
PAA[1,2]	Lineage isolation sufficient for fixation of character states	Allozymes, chromosomes, morphology, binary (presence/absence) of data	Assumes conspecificity of individuals from same locality; character fixation difficult to show at conventional levels ($\alpha = 0.05$) of confidence
Corr-D[1]	Correlated divergence in morphology and gene sequence mark discontinuity	Morphology, molecular markers (both converted to pairwise distances)	DNA and morphology must be available for same specimens; test may be circular if putative species are in sympatry
M/GC[1]	Morphological discontinuities or reduction/absence or gene flow mark boundaries species limits	Morphological or genetic characters	Between species variability is greater than within species variability; introgression limited or absent
PCT[1,2]	Lineage isolation sufficient for character divergence	Allozymes, chromosomes, morphology, DNA sequences	Some versions reduce speciation to single character substitutions, others require strong a priori rejection of some modes of speciation or anagenesis
CHA[1,2]	Lineage isolation sufficient for coalescence to monophyly of haplotypes at one locus	DNA haplotypes for one locus	Equates nonrecombinant haplotype clades to species
EXCL[1]	Lineage isolation sufficient for allele coalescence to exclusivity at unlinked loci	DNA haplotypes for multiple loci	Requires unspecified number of unlinked genes with divergence profiles matched to timing of speciation events

(*Continued*)

TABLE 1 (*Continued*)

Method[a]	Relevant biological properties/criteria[b]	Classes of data suitable to method	Assumptions/limitations
WP[1]	Lineage isolation sufficient for geographical character divergence	DNA haplotypes, morphology, etc.	Assumes no gene flow between species; no interspecific recombination between haplotypes
TTC[1,2]	Lineage isolation sufficient for attainment of ecological or allopatric character divergence	Genetic, ecological, morphological, or physiological data; with DNA haplotypes	Inference key can be misled by insufficient sampling density; H_2 can never be completely falsified; choice of candidate traits may be subjective

[a]Abbreviations: cladistic haplotype aggregation (CHA), exclusivity criterion (EXCL), field for recombination (FFR), genetic distance Good & Wake (GenD$_{GW}$), genetic distance Highton (GenD$_H$), hybrid zone barrier (HZB), population aggregation analysis (PAA), Templeton's tests for cohesion (TTC), Wiens & Penkrot methods (WP), correlated distance matrixes (Corr-D), morphological–genotypic cluster (M/GC), phylogenetic/composite tree-based methods (PCT); for each of these methods, the superscripts 1 and 2 denote suitability of method to sexual and asexual taxa, respectively.

[b]These are deliberately general because multiple properties (criteria) are manifested during the speciation process, but both the order of their appearance and their relevance to testing species boundaries depend on many idiosyncratic conditions and mechanisms associated with a particular speciation event, and at what point along an evolutionary trajectory extant populations are sampled (de Queiroz 1998).

HYBRID ZONE BARRIER   Porter (1990) used gene flow statistics with allozyme loci to test species boundaries in two hybridizing North American butterflies (*Limenitis lorquini, L. weidemeyerii*). The method derives from Wright's (1931) equilibrium relationship between gene flow and genetic differentiation and uses the following expression to estimate gene flow (number of migrants):

$$Nm_{ST} \approx (1/F_{ST} - 1)/4,$$

where the subscript S represents subpopulations, T represents the total set of populations, N is the effective population size, and m is the effective proportion of migrants between populations, under an island model of population structure. Porter used the estimator θ (Weir & Cockerham 1984) in place of F_{ST} and an isolation-by-distance model in place of Wright's (1931) island model, as justified by simulation studies (Slatkin & Barton 1989).

The method proceeds in two steps. First, Nm estimates are calculated between sympatric sibling species pairs to assess the strength of biases in the Nm estimator. Because this comparison is between sets of populations not currently connected by gene flow, a result showing $Nm > 0$ must be attributable to other factors that contribute to genetic similarity between populations. As a control, Nm estimates are also calculated among subpopulations within species and are expected to reflect gene flow at a level that presumably prevents fixation of alleles in the absence of selection ($Nm \geq 1.0$). The factors most likely to bias Nm in the direction >0 (giving the appearance of gene flow when there is none) include balancing selection, sister taxa characterized by large N coupled with recent divergence, or the presence of undetected cryptic alleles (hidden heterogeneity; Coyne 1982). Porter (1990, pp. 134–35) gives details on calculating confidence intervals on Nm, as well as on what other kinds of information can be used to judge which of these biases is most likely in a given case.

The second step is to estimate Nm across a parapatric hybrid zone between nominal species under a null hypothesis (H_O) that the hybridizing populations are genetically isolated. Estimates of gene flow that are not significantly greater than those typical of sympatric sibling species would support the null hypothesis; i.e., these populations maintain separate gene pools despite some hybridization. The empirical examples presented in Porter (1990, table 1) suggest that the Nm estimator provides a reasonable estimate of gene flow except when it falls below ~0.5.

If $Nm > \sim 1.0$, then ongoing gene flow is likely important in promoting genetic similarity between hybridizing groups, and they should be considered conspecific. If $\sim 0.5 < Nm < \sim 1.0$, then gene flow is weak but probably sufficient to permit rapid exchange of selectively favored alleles; this pattern would also reject H_O in favor of conspecificity. In these cases, investigators should examine any relevant secondary evidence that could inform interpretation of the Nm result to determine what proportion of genetic similarity can be attributed to factors other than gene flow. If $0.0 < Nm < 0.5$, then the hybridizing populations are almost or fully isolated genetically; gene flow is unimportant relative to other processes, and the hybridizing populations would be considered separate species.

GENETIC DISTANCE: GOOD & WAKE   Good & Wake (1992) described a method for qualitatively assessing gene flow as a function of geographic isolation over a range of spatial scales. The method involves plotting genetic distances (in the particular case in Nei 1978, distances summed over allozyme loci) against geographic distances for all pairwise comparisons made in a particular study. A regression line is then fitted to a set of points representing samples from a priori defined subsets of the total (i.e., all samples from a single river basin or some other geographically or taxonomically defined entity). If the regression line for such a sample passes through the origin of the graph, such a pattern is most readily explained by gene flow with isolation-by-distance, and populations are interpreted to be conspecific (or at least to represent part of a single genetically cohesive unit). Conversely, the regression for a sample of populations that includes genetically isolated groups will deviate significantly from a 0 origin because genetic divergence among samples is expected to be independent of their degree of geographical separation. The method is illustrated in its original application in the salamander genus *Rhyacotriton* (Good & Wake 1992) and has been used in studies of several other genera of salamanders, including *Desmognathus* (Tilley & Mahoney 1996), *Ensatina* (Jackman & Wake 1994), and *Plethodon* (Highton & Peabody 2000, Highton 2000).

GENETIC DISTANCE: HIGHTON   Following a different line of reasoning, Highton (1989, 1990) suggested that, for some groups characterized by extremely slow rates of morphological evolution (e.g., salamanders of the family Plethodontidae), species boundaries are most easily identified by multilocus allozyme data from which genetic distances (D of Nei 1972, 1978) are calculated and used to infer the distance that correlates with the origin of intrinsic reproductive isolation (refined arguments presented in Highton 1995, 1998, 2000). Specifically, Highton suggested that groups of samples differing by a Nei $D \geq 0.15$ should be considered distinct species; he recognized that this value was arbitrary, but he cited a review of the allozyme literature (Thorpe 1982) in which most (97%) genetic identities (I) between well-defined species of vertebrates are < 0.85, whereas most ($>98\%$) I values within a species are >0.85 (a Nei I of ≈ 0.85 is the equivalent of a Nei D of ≈ 0.16). These patterns are general enough across nonavian vertebrates to suggest that, as a rule of thumb, the divergence usually needed to complete speciation is correlated with a D of ≈ 0.15–0.16.

The method is implemented by plotting a histogram of D value frequencies for pairwise comparisons between populations (Highton 1998), and under a null model of conspecificity among all populations, the distribution should be approximately unimodal (presumably owing to isolation-by-distance within a species) with most values clumping below $D \approx 0.15$ (Figure 1). If the samples comprise different species, then the distribution of D values is expected to be bimodal, with a second peak well above $D \approx 0.15$ (Figure 1). The protocol includes two criteria: (*a*) that a $D > 0.15$ approximates that point at which reproductive isolation is completed, and (*b*) that pairwise D values will be bimodally distributed if two or more species are included in the sample (Figure 1). Only the second of these criteria is independent of the taxon (salamanders of the family Plethodontidae) upon which the method was based. Highton recommended that genetic data be corroborated with morphological and/or distributional data before making strong inferences about species boundaries (see Highton & Peabody 2000).

CORRELATED DISTANCE MATRICES   A method described by Puorto et al. (2001) uses molecular and morphological data in combination to delimit species, specifically a combination of mtDNA haplotypes, a multivariate statistical summary of morphological variation, and numerical hypothesis-testing techniques, based on an example with taxonomically problematic pitvipers of the *Bothrops atrox* complex in Brazil. First, morphological variation was summarized as a matrix of Euclidean distances between specimens; this matrix is the set of dependent variables, and Mantel tests are then used to evaluate sequentially the potential alternative causal factors for morphological similarities or differences. The alternative factors tested included sex [matrix values were either 0 (between individuals of the same sex) or 1 (opposite sex)], geographic distances (straight-line distances between all paired combinations of individuals), and patristic distances of mtDNA haplotypes collected for the same individuals used to obtain Euclidian distances. Puorto et al. (2001) then tested whether morphology was associated with mtDNA distances, with the effects of the other factors regressed out; morphology was used as the only observed (dependent) variable. The expectation was that if two species

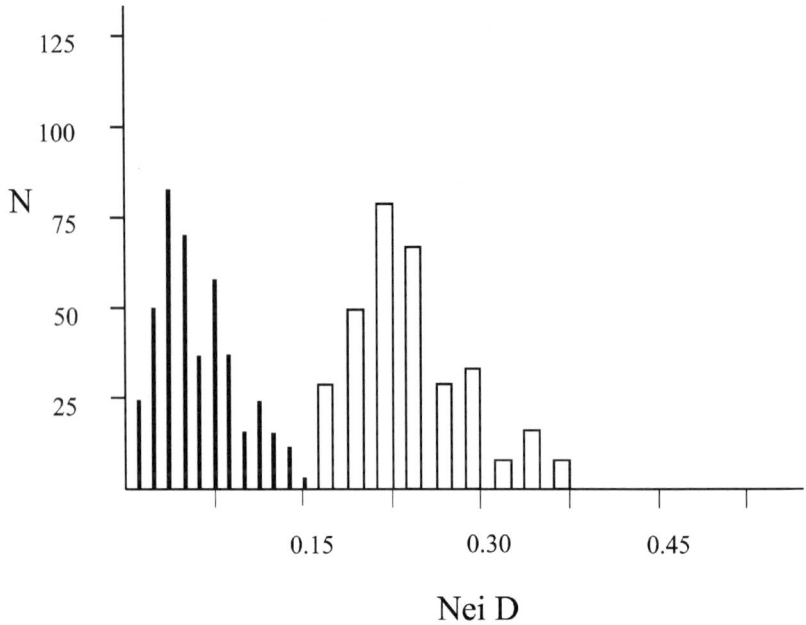

Figure 1 Histogram showing distributions of pairwise genetic distance values among multiple samples whose species status is of interest (Highton 1989, 1990, 2000). A unimodal distribution in which all D coefficients are <0.15 (Nei 1978) reflects divergence among populations within a single species (*solid bars*), and a bimodal distribution in which a substantial subset of pairwise D values are >0.15 (*both sets of bars*) represents between-species comparisons (see text for details).

are present in a sample, they should be characterized by possession of different mtDNA clusters, and specimens belonging to either species should show patterns of morphological variation congruent with species identity as revealed by mtDNA sequence variation (after regressing out the effects of sex and geographic distance). The alternative is that two (or more) mtDNA clusters may exist within a single species, in which case researchers would not expect patterns of morphological and mtDNA variation to correlate with each other.

The empirical results of this study revealed two overlapping mtDNA haplotype clusters of *Bothrops atrox* (Puorto et al. 2001, figure 3), a result compatible with a hypothesis of sympatry of two species, or conspecificity and the co-mingling of different genotypes in a single species. Mantel tests did not reveal any significant association between morphological variation and mtDNA affinities of the specimens, and in this case the authors failed to reject the single-species hypothesis.

MORPHOLOGICAL METHODS   Until the 1940s, most biologists recognized species solely on the basis of morphological differences (Coyne 1994), following a methodology approximately as described by Cronquist (1978): "Species are the smallest groups that are consistently and persistently distinct, and distinguishable by ordinary means." The implication is that the existence of "gaps in the pattern of visually observable phenetic diversity" is taken as evidence for reproductive isolation, and species are then delimited along the boundaries of morphological discontinuities. Cronquist recognized that opinions might differ on how to define the criteria for "consistent," "persistent," and "ordinary" means, but he insisted that any group of organisms that did not meet all three of these criteria to some reasonable degree should not be recognized as a species. Statistical rigor has been applied to quantify the criterion of phenetic discontinuity, most notably by numerical clustering approaches (Sokal & Sneath 1963); multivariate statistical methods identify samples divisible by phenetic gaps resulting from concordant differences in character states (i.e., minimally distinguishable in multivariate space), and these samples are often interpreted as species (see also Sokal 1973, Sneath 1976).

Mallet (1995) extended this reasoning to a genotypic cluster approach to delimit species as "clusters of monotypic or polytypic biological entities, identified using morphology or genetics, forming groups of individuals that have few or no intermediates when in contact." The morphological and genotypic clustering approaches are conceptually similar (statistical grouping algorithms, visual inspection to determine if clustering is significant), but neither is a discrete method in the sense of the others described in this review. However, both morphological and molecular characters can be coded in ways that permit implementation of some of the methods described here, particularly the population aggregation analysis (PAA).

POPULATION AGGREGATION ANALYSIS   The PAA method was formulated by Davis & Nixon (1992) on the basis of two principles: (*a*) all individuals sampled from a local population are assumed to be conspecific, and (*b*) identical character attributes shared among individuals drawn from two or more populations provide evidence for conspecificity. The PAA is a formalized protocol for traditional approaches to identifying one or more diagnostic morphological characters for species delimitation (see also Nixon & Wheeler 1990); it codifies a way to identify the morphological discontinuities considered essential by Cronquist (1978), although Davis & Nixon (1992) recognized its applicability to many other kinds of data. The PAA requires a summary of character states for all individuals in a sample to estimate a population profile for those states, and then it combines all samples with identical profiles for all character states. This process is continued iteratively until the only remaining sample aggregates are those separated from each other by fixed character state differences; these samples are the smallest diagnosable units (Cracraft 1983, Nixon & Wheeler 1990) and are taken to be species. The method was illustrated in the original paper (Davis & Nixon 1992) and was recently used by Benavides et al. (2002).

FIELD FOR RECOMBINATION   A method described by Doyle (1995) delimits species by using Mendelian loci (originally allozymes) to define a field for genetic recombination (FFR, Carson 1957), which should be coincident with the boundaries of sexually reproducing species in which constituent populations are interconnected by gene flow. Because sexual reproduction defines the FFR for Mendelian loci, Doyle (1995) argued that the distribution of alleles would more faithfully define species boundaries than the phylogenies of those same alleles; the latter may not be equivalent to species boundaries because of idiosyncrasies of lineage sorting, between-locus mutation rate heterogeneity, and intralocus recombination (Avise & Wollenberg 1997, Hare 2001). Doyle's method seeks to identify natural discontinuities between different allelic states for a given locus on the basis of overlapping sets of heterozygous individuals. The presence of two alleles in a single individual is taken as evidence that these two alleles belong to the same allele pool, and so for a single locus, sampling of multiple individuals from multiple localities will permit grouping individuals together, or separating them into distinct groups, on the basis of overlapping or nonoverlapping sets of heterozygous genotypes. Better resolution of a gene pool boundary will likely be afforded by extending this procedure to additional loci. At the organismal level, the multilocus genotype provides evidence of gene flow among populations, and individuals possessing alleles belonging to a multilocus gene pool are part of the same FFR. Individuals not sharing alleles at any loci are considered to be members of a different FFR, and species are delimited on the basis of discontinuities in multilocus FFRs. This method was illustrated in the original description (Doyle 1995) and has been implemented in at least one system (J.C. Marshall, E. Arévalo, E. Benavides, J.L. Sites & J.W. Sites, Jr., submitted manuscript).

Tree-Based Methods

Systematists typically favor phylogenetic methods to delimit species, including several versions of the phylogenetic species concept (PSC; Rosen 1979, Mishler & Brandon 1987, Baum & Donoghue 1995), as well as more recent extensions and related discovery methods. Some of these methods were reviewed by Sites & Marshall (2003) and are again covered here but in the context of several versions of the PSC not covered by that review.

PHYLOGENETIC/COMPOSITE TREE-BASED METHODS   The terminology of Brooks & McLennan (1999) is useful in recognizing tree-based methods originating from various versions of the PSC. These authors denoted an apomorphy-based method (Rosen 1979) as PSC-1, a lineage-splitting method as PSC-2, and a node-based composite species concept (CSC; see Kornet 1993). These methods differ from each other primarily with respect to how species are delimited on a phylogenetic tree and how ancestral species are handled. All rely on the reconstruction of a phylogenetic hypothesis from a data set for which discrete characters can be coded and polarized by standard cladistic methods followed by mapping character transitions onto the topology (see Brooks & McLennan 1999). These methods can be distinguished most clearly by comparing the number of species recognized and the criteria for their recognition under PSC-1, PSC-2, and CSC for the same number of terminal units (i.e., potential species).

Figure 2 compares three hypothetical trees with identical topologies characterized by the same number of terminal taxa (seven species, A–G) but differing in the number or distribution of apomorphic transitions. Figure 2A presents a tree in which all terminal and interior branches are supported by one apomorphy each; here all three methods would recognize twelve species: all seven terminal (A–G) and all five internal (= ancestral) branches. The number of species delimited would be different if a data set produced the same topology but did not recover apomorphies for some terminal branches. In Figure 2B, for example, terminals B, C, and G lack apomorphies, and a strict application of the apomorphy-based

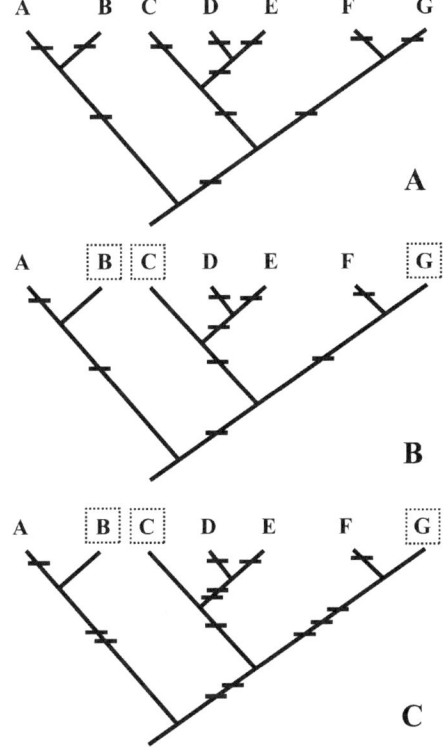

Figure 2 Hypothetical phylogenies with identical topologies for seven terminals (A–G), in which horizontal bars represent single apomorphic character transitions. (A) Tree in which each terminal and each interior branch are defined by single apomorphies; (B) tree in which each interior and all but three terminal branches are defined by single apomorphies; and (C) tree identical to B except that some internal branches are supported by two or three apomorphies. Terminals enclosed in squares in trees B and C are those not supported by apomorphies [modified from Brooks & McLennan's (1999) figure 1 with permission].

phylogenetic method (PSC-1) and the node-based CSC would recognize only nine species (terminals B, C, and G are considered species but are not distinguished from their common ancestors). The lineage-splitting PSC-2 would still recognize terminals B, C, and G as distinct species because it is based only on the branching structure of the tree and is not concerned with character evolution. Finally, in Figure 2C four internal nodes are characterized by more than one (two or three) apomorphic character transitions. All three methods recognize the seven terminals A–G as distinct species, but again the apomorphy method (PSC-1) collapses terminals B, C, and G with their common ancestors, whereas the CSC recognizes fourteen species—one for each apomorphic character on the tree. It thus permits recognition of multiple species on a single branch on the premise that the origin and fixation of each apomorphic trait required permanent lineage splitting; each apomorphy delimits the present or prior existence of a distinct species (Brooks & McLennan 1999).

CLADISTIC HAPLOTYPE AGGREGATION Brower (1999) extended the population aggregation analysis to encompass sequence data and cladistic analysis in an effort to take full advantage of the information content of the evidence. The use of DNA data requires one to decide how attributes are defined; one can either (*a*) use a well-defined region of DNA as a single attribute (Doyle 1995), in which haplotypes are scored as nonadditive alternative allelic states; or (*b*) atomize the sequence to the level of resolving single bases as attributes. Brower proposed to call these alternatives the PAA1 (sequence-as-a-single attribute) and PAA2 (string-of-attributes) methods and used a number of contrived and empirical data sets to compare the results of each with cladistic haplotype aggregation (CHA) (discussed below).

CHA is implemented by first identifying individual organisms as representatives of local populations, collecting DNA sequences (haplotypes) of exemplars, tabulating haplotypes to determine sample profiles, and aggregating sample profiles (by either PAA1 or PAA2) that do not have fixed character differences into a single population profile (as in the PAA of Davis & Nixon 1992). One then estimates a phylogeny of the original unaggregated haplotypes (all distinct population profiles) and uses this phylogeny to corroborate or refute the tentative species delimited by either PAA1 or PAA2. Brower argued that all members of a species will form a contiguous section of an unrooted tree and will be separated from all other groups of samples by a single branch along which a character transition leading to a fixed character difference can be inferred (Figure 3A). Therefore, groups identified by PAA will be corroborated by CHA if they form a contiguous section on the tree. The results of CHA "do not imply phylogenies, but rather represent parsimonious patterns of empirical grouping that corroborate or reject specific a priori hypotheses of species boundaries" (Brower 1999, p. 202).

Brower compared the performance of PAA1 and PAA2 with haplotype aggregation using contrived and real data. Figure 3B shows a hypothetical case in which two populations fixed for alternative attribute states would be successfully identified by PAA1 but not by a cladistic analysis of the gene genealogy (modified from Davis 1996). For data sets consisting of independently segregating attributes (allozymes, chromosomes, microsatellites, morphology, etc.), alternative fixation of individual character states provides plausible evidence for separation of lineages (absence of gene flow), and PAA1 is appropriate. With sequence data, however, this distinction is not so obvious. Consider the hypothetical case illustrated in Figure 3C (for visual clarity different symbols were substituted for nucleotides). Here a homologous 13-bp sequence has been collected for six exemplars representing an outgroup and two population samples whose species status is of interest. If we apply PAA1, four distinct haplotypes are identified (in the two ingroup samples) because each sequence is considered a single character. The PAA1 approach delimits the two populations as two species because of fixed character differences [mutually exclusive allelic (haplotype) polymorphisms segregating at a single locus] without considering the phylogenetic relationship between haplotypes. Brower (1999) saw this delimitation of species as arbitrary. Alternatively, when each nucleotide is treated as an independent character, PAA2 identifies six characters (attributes 1–6) and seven traits (attributes 7–13) that are by definition not useful in delimiting species by the original PAA method of Davis & Nixon (1992). PAA2 collapses haplotypes 1 and 2 into a single population profile for sample 1, and haplotypes 3 and 4 for sample 2, and also delimits two species (Figure 3C). Brower's method attempts to makes use of all 13 attributes present to test the validity of the species delimited by PAA. In this example, he recovers the seven most parsimonious cladograms, a strict consensus of which reveals no phylogenetic structure (Figure 3C) between the two populations, and thus rejects the hypothesis of distinct species. Brower argued that, by selecting from the observed data only those attributes that support the groups already assumed a priori to exist (the populations from which samples 1 and 2 were drawn), the PAA methods (in either form) are circular. Brower (1999) extended this argument by providing another hypothetical example in which the PAA failed to recover population differentiation where it actually existed, and he suggested that the reason for both kinds of inconsistencies is that the PAA does not use character congruence in a phylogenetic context to distinguish homology from homoplasy, whereas CHA is based on this premise.

GENEALOGICAL EXCLUSIVITY   Baum & Shaw (1995) presented a different tree-based method that formalized a genealogical species concept, originally suggested by Avise & Ball (1990), in which relatedness is viewed in the context of genealogical descent of the genome. Operationally, the delimitation of genealogically exclusive species requires that: (*a*) species must be basal taxa, i.e., they must not themselves contain taxa; and (*b*) genealogical species reside at the boundary between reticulate and divergent genealogy (the tokogeny-phylogeny interface, sensu Hennig 1966). Species are therefore defined as exclusive groups; those in which all members are more closely related to each other than to any organism outside of the group. As in the above phylogenetic methods, such species can only be delimited when relationships are hierarchical, but the conceptual distinction here is that the idea of genealogical exclusivity derives from coalescent theory (Hudson 1990, Baum & Shaw 1995). Coalescent theory is concerned with tracing the genealogical histories of extant samples of genetic elements (alleles or haplotypes) and is normally applied to multiple unlinked loci that characterize sexual species (Maddison 1997). Thus, one can infer that when organismal genealogy is divergent (phylogenetic) for a sufficient amount of time (Neigel & Avise 1986), then selectively neutral, unlinked genes should attain concordant genealogical histories, whereas these histories will become discordant when the organismal genealogy is reticulate (tokogenetic; see also Taylor et al. 2000).

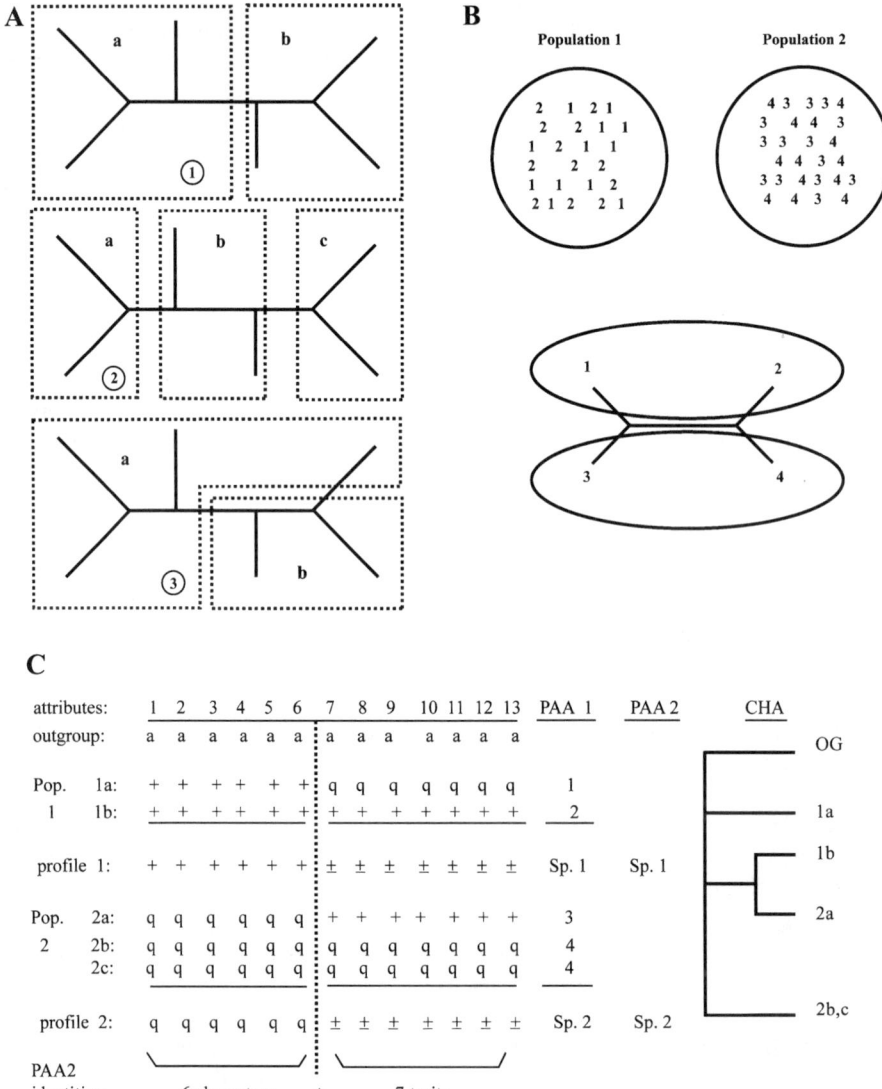

Operationally, the method requires reconstruction of genealogies for unlinked loci collected from the same exemplars and then the construction of a strict consensus of the separate gene trees to define points of concordance (resolved nodes). Species are delimited by exclusive nodes in the consensus tree. Figure 4 illustrates a contrived example in which eleven individuals are screened for four unlinked genes, which are then used to reconstruct independent gene genealogies for the eleven terminals. These four gene trees are taken to represent the coalescent histories of these loci among the organisms sampled. A strict consensus of the four genealogies defines one point of concordance among the independent gene genealogies, which shows that terminals f–k are exclusive. That is, the alleles of these terminals for all loci coalesce more recently within the group than between any member of this group and any of the other exemplars sampled. This same group is also basal—there are no other exclusive groups nested within it, and it therefore meets the criteria for recognition as a genealogical species. In contrast, exemplars a–e are not exclusive because the smallest group containing it (a–k) is not basal (it includes the exclusive group f–k; the entity a–e would be labeled a

←

Figure 3 (*A*) Hypothetical outcomes of CHA analysis (Brower 1999) in unrooted haplotype networks. Dotted lines identify distinct groups of haplotypes; two phylogenetic species are delimited in network 1 and three in network 2 because in both cases groups of haplotypes are identified by apomorphic character transitions along single branches in the network. In contrast, species-level distinctness is rejected in network 3 because the two clusters of haplotypes are connected to each other by two branches. (*B*) Hypothetical population samples characterized by fixed differences for alternate allelic states at a single locus (alleles 1 and 2, and alleles 3 and 4, in the two populations, respectively). In this example, the PAA1 of Brower (1999), as originally described by Davis & Nixon (1992), delimits the two species, whereas a gene genealogy with the topology depicted here would not delimit these populations as separate [modified from Brower's (1999) figure 2 with permission]. (*C*) Hypothetical sequence data matrix with thirteen bases collected for six exemplars (outgroup and two population samples of two and three individuals, respectively), and a comparison of outcomes of PAA1, PAA2, and CHA analyses for delimitation of species. The PAA1 (sequence region as the attribute) identifies four unique haplotypes in the five ingroup sequences given; two species are delimited, one for each sample because each segregates two mutually exclusive sets of alleles (as in panel *B*). The PAA2 (atomistic) also delimits two species (among the ingroup exemplars) on the basis of aggregated population profiles that show fixed differences at six characters (attributes 1–6), and seven polymorphic traits [attributes 7–13, following the Davis & Nixon (1992) protocol]. The CHA [a strict consensus of seven most parsimonious (MP) trees] resolves only one unambiguous clade (1b + 2a) because attributes 7–13 are polarized as synapomorphies; note that this result conflicts with a priori assumptions made about population membership [modified from Brower's (1999) figure 3 with permission].

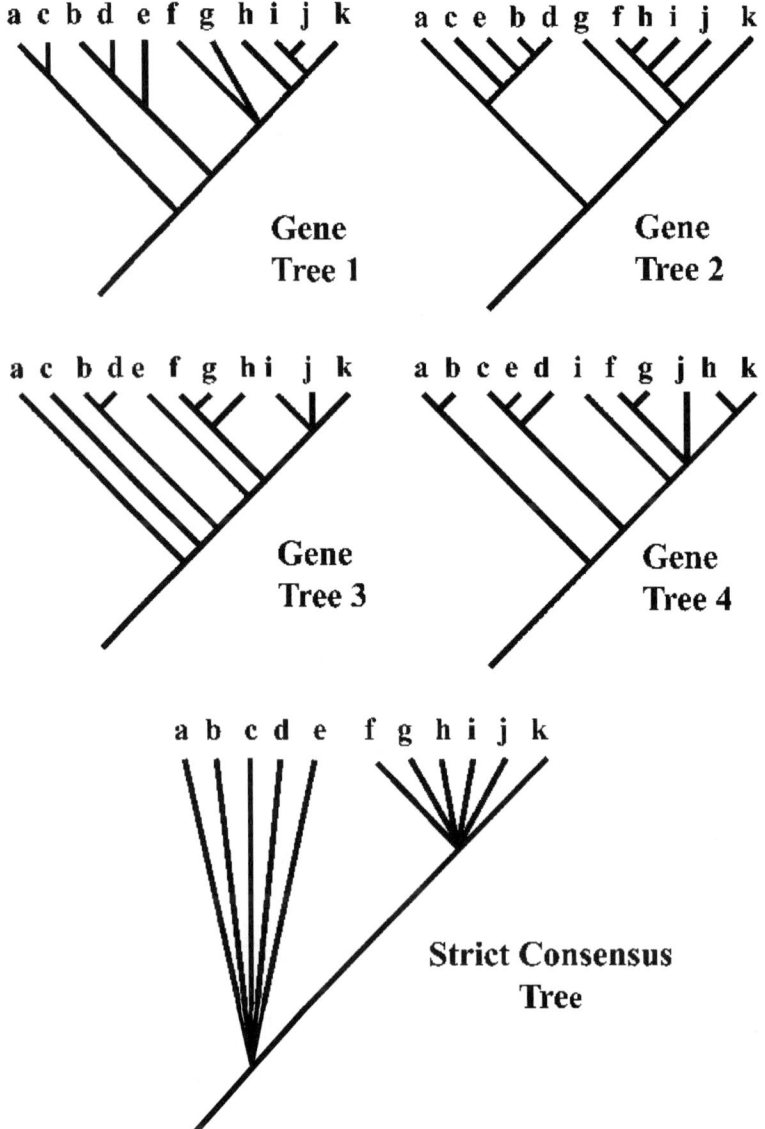

Figure 4 Hypothetical example in which gene genealogies are reconstructed from four unlinked loci for eleven individuals; the strict consensus tree recovers a single genealogical species that meets the two requirements specified by Baum & Shaw (1995): (*a*) genealogical species are basal taxa (i.e., they do not themselves contain taxa); and (*b*) these species are exclusive (i.e., all members are more closely related to each other than to any individuals outside of the group; see text for details). In this example, terminals f–k comprise a genealogical species, whereas the unresolved polytomy that includes terminals a–e does not.

SPECIES DELIMITATION METHODS C-1

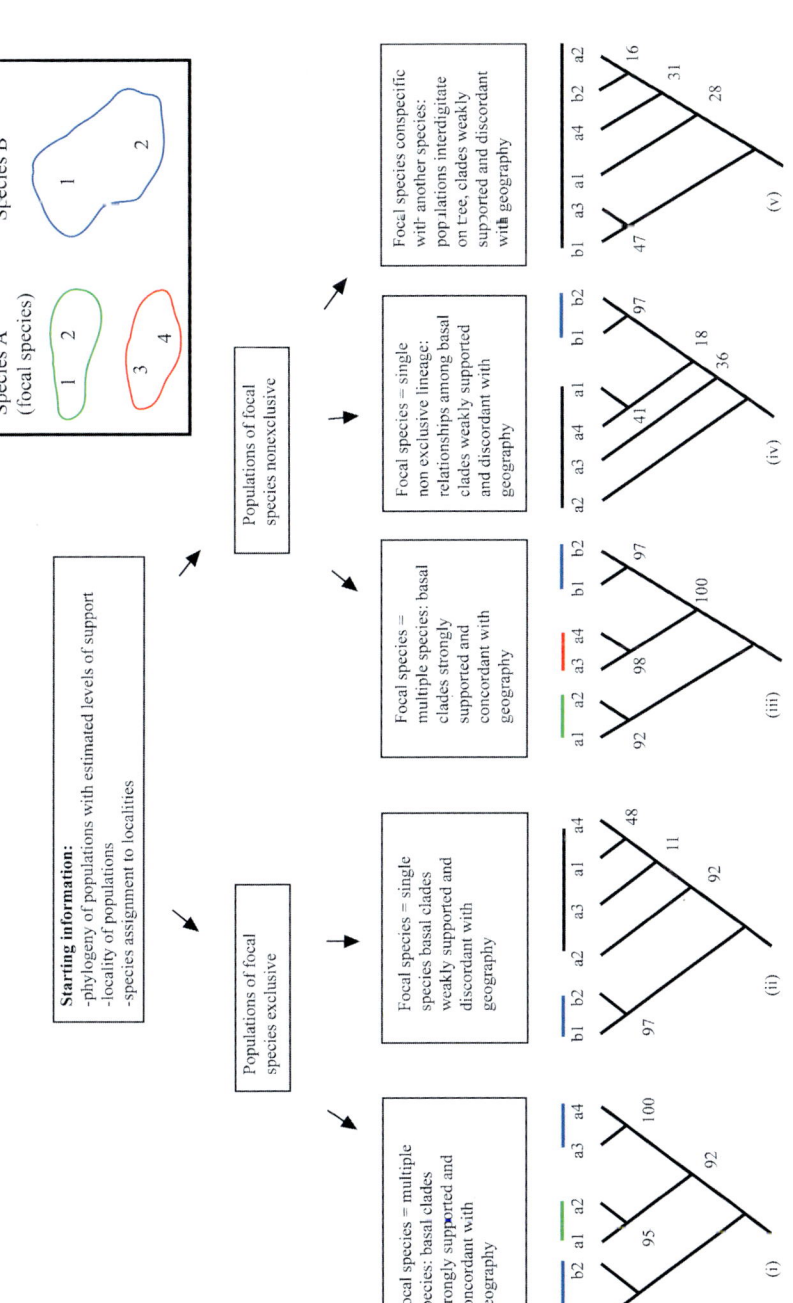

See legend on next page

Figure 5 Morphological tree-based method of species delimitation described by Wiens & Penkrot (2002). Species *a* represents the focal species in which population samples are drawn from two geographically distinct regions (indicated by *green* and *red lines*), and species *b* represents a closely related nonfocal species (delimited by *blue lines*). Phylogenetic relationships are inferred from population samples, and species are delimited on the basis of the positions of the terminal branches in the tree (the five possibilities for completely resolved trees are the topologies i–v; numbers at nodes represent bootstrap values), relative to their geographic points of origin, by following the pathways depicted here. The DNA tree-based method (not shown) is similar but uses individual haplotypes rather than population samples (see Wiens & Penkrot 2002 for more detail).

ferespecies or metaspecies by some; Graybeal 1995). This approach was recently implemented by Dettman et al. (2003a), who used four nuclear gene regions to delimit species of the fungus *Neurospora* on the basis of concordance (with strong nodal support) for three of the four single gene genealogies. They added a novel criterion of genealogical nondiscordance to recognize species when a clade was strongly supported by a single-gene genealogy and not contradicted by any other single-locus genealogy at the same level of support.

WIENS-PENKROT PHYLOGENETIC METHODS   Wiens & Penkrot (2002) recently described formal protocols for implementing molecular and morphological phylogenetic methods for delimiting species. They applied both to empirical data for *Sceloporus* lizards, along with results from a character-based morphological approach. The molecular method is intended to be used in combination with a nested clade analysis (Templeton et al. 1995) and is based on a sampling design that should ideally include: (*a*) a number of closely related reference species to test for exclusivity for the focal species (the species of interest in the study), and (*b*) sampling of at least two individuals from as many localities as possible within the focal species to increase the strength of between-population gene flow inferences. The method requires a phylogeny of nonrecombining haplotypes of known locality, and taxonomic designation (focal taxa are the testable hypotheses) and a topology that fails to recover haplotypes from a given locality as a clade is taken as evidence for potential gene flow with other populations.

The morphological method is similar but based on population, rather than individual, sampling to avoid the default treatment of all shared polymorphisms as homoplasies (rather than synapomorphies) if single individuals are used (see Wiens 2000). This method considers sets of populations that are strongly supported, exclusive, and concordant with geography as species. As with the molecular protocol, the morphological method assumes that gene flow among populations and recombination among characters will break up hierarchical patterns within species, and the absence of historical signal below the species level will permit recovery of only weakly supported trees that are discordant with geography. Both the methods are implemented using dichotomous flow charts that lead to several alternatives for making species-level decisions. Figure 5 (see color insert) diagrams the pathways for the morphological method.

TEMPLETON'S TESTS OF COHESION   Extensions of Templeton's (1989) original, general hypotheses about genotypic and phenotypic cohesion have rendered both operational (Templeton 1994, 1998, 1999; Templeton et al. 1995) and include tests of species boundaries (Templeton 2001). Hypothesized species boundaries are statistically tested through a set of nested null hypotheses structured to evaluate the correlation of genotypes and/or phenotypes with geographic location [nested clade analysis (NCA); Templeton et al. 1995]. An inference key (Templeton 1998, 2004) is then used to derive the most likely biological cause(s) for observed patterns of variation. The method tests two hypotheses:

H_1 all organisms are sampled from a single evolutionary lineage

H_2 populations of separate lineages identified by rejection of H_1 are genetically exchangeable and/or ecologically interchangeable among themselves.

Species are recognized only after rejection of H_1 and H_2 at the same levels of divergence.

An NCA inference of historical fragmentation at some nested clade level is the only inference taken as evidence for the possible existence of separate evolutionary lineages (Templeton 2001) and requires rejection of H_1. When H_1 is rejected with statistical support for historical fragmentation, the two or more lineages inferred from the first NCA may constitute different cohesion species, and the second null hypothesis (H_2) must be tested. The question here (H_2) is whether populations comprising the two (or more) lineages defined by the historical fragmentation event(s) are genetically exchangeable and/or ecologically interchangeable among themselves. H_2 can be tested by direct statistical contrasts of the lineages previously identified or through the NCA to test for a statistical concordance of candidate traits for genetic exchangeability (i.e., those associated with prevention or promotion of gene flow) or for ecological interchangeability (life-history traits, ecological requirements, physiological tolerances, etc.). H_2 is rejected only when (*a*) a significant association is detected between geography and the genetic or ecological trait associated with genetic exchangeability or ecological interchangeability, and (*b*) the phylogenetic position of this association is concordant with the previously identified historical isolation event(s) (i.e., both are identified at the same clade level). Recent attempts to test H_2 include Shaw's (1999) use of song patterns (mate recognition) in the Hawaiian cricket genus *Laupala* and Templeton's (1999) use of chromosome number in races of mole rats (*Spalax*) as candidate traits for genetic exchangeability. Templeton (1999) also used nonshivering thermogenesis, hematocrit and hemoglobin concentrations, breathing and heartbeat frequencies, and oxygen and carbon dioxide pressures in subcutaneous gas pockets as candidate traits to test for ecological interchangeability in *Spalax*. More recently, Gomez-Zurita et al. (2000) used trophic selection and altitude of habitat in leaf beetles of the *Timarcha goettingensis* complex as candidate ecological traits.

DISCUSSION

General Observations

The methods surveyed in this review vary relative to biological properties and criteria, the kinds of data suitable, and some assumptions or limitations (Table 1). Methods appear to be biased in favor of using molecular markers, but six methods also include morphological data (PAA, correlated distances, morphological/genotypic cluster, phylogenetic/composite trees, Wiens-Penkrot, and Templeton's tests for cohesion), and tree-based methods can now accommodate allele frequency data and frequency and continuous morphological characters (Wiens & Penkrot 2002).

Most methods (hybrid zone barrier, Good-Wake distance, correlated distances, PAA, FFR, morphology/genotypic cluster, exclusivity, Wiens-Penkrot, and Templeton's tests for cohesion) refer explicitly or implicitly to gene flow and are applicable to sexually reproducing species, but some nontree-based methods (PAA and CHA) are operationally general enough to apply to asexual complexes, and the various tree-based approaches (phylogenetic/composite and Templeton's) are explicit in their incorporation of asexual taxa. The hybrid zone and Good-Wake distance methods implicitly assume that populations are in approximate mutation/drift equilibrium, a condition likely rarely met in nature (Pannell & Charlesworth 1999, Whitlock & McCauley 1999).

Several methods are likely to have technological limitations for most organisms. For example, the FFR, CHA, and exclusivity methods all implicitly assume perfect sampling for any locus; that is, no heterozygotes or distantly related alleles are missed for any given population. In all these methods, homoplasy will falsely identify different alleles as identical and either underestimate topological structure (CHA and exclusivity methods) or lead to inferences of gene exchange (FFR) when gene flow is absent. Both allozyme (Murphy et al. 1996) and microsatellite loci (Li et al. 2002, Estoup et al. 2002) can be characterized by allelic homoplasy between closely related species, which will introduce an unknown error into the use of either kind of data in FFR methods. Perhaps amplified fragment length polymorphisms (Sullivan et al. 2004) or single-nucleotide polymorphisms (Brumfield et al. 2003, Morin et al. 2004) might be viable alternatives. The methods described by Porter (1990), Good & Wake (1992), and Highton (1989, 1990) may also be sensitive to resolution of alleles, but the possible biases resulting from this phenomenon have not been carefully studied. The exclusivity criterion test requires sequences of unlinked gene regions that may not be available for many taxa, although this is changing (Hare 2001, Zhang & Hewitt 2003).

Operational Assumptions and Limitations

The hybrid zone method (Porter 1990) is the most restrictive relative to conditions that must be met in nature because it requires hybrid zones and regions of sympatry between closely related species, but many such complexes are now known that could be subjected to this test (Arnold 1997, Burke & Arnold 2001). The genetic distance method of Highton (1989, 1990) assumes that reproductive isolation is due to divergence across many loci scattered over the entire genome (Coyne & Orr 1989, Orr 1996, Hollocher & Wu 1996). This assumption is the same for the presumed clock-like accumulation of genetic distances by point mutations, and although there are broad correlations between D values and reproductive isolation in some groups (Coyne & Orr 1989, Sasa et al. 1998), there are enough exceptions to inflate error terms (on D estimates) sufficiently to compromise the diagnostic value of such correlations (Ferguson 2002).

The graphical genetic distance method (Good & Wake 1992) qualitatively delimits species on the basis of a correlation of pairwise D values with geography and thus depends more on patterns of gene flow than on whether D correlates tightly with the emergence of reproductive isolation. The more quantitative correlated distance matrix method of Puorto et al. (2001) provides an explicit test of associations between independent data sets, and it discriminates between alternative hypotheses by partial correlation test to regress out factors that do not contribute to significant associations. It requires that both morphological and molecular divergence be concordant with species limits, and the tests are most powerful when both data sets are collected from the same individuals. One strength of this method is that in scenarios in which putative species are allo- or parapatric and there is an a priori basis for distinguishing these, then researchers can assign each specimen to a putative species and use such an assignment as another independent matrix in a Mantel test. A limitation is that such a test is not possible with putative species in partial sympatry because individual specimens cannot be assigned to putative species in the region of sympatry without causing the test to be circular.

The morphological methods summarized above have well-characterized limitations that need little elaboration here; they may be difficult to apply to organisms characterized by cryptic species (Mayden 1997), and when used to summarize patterns of geographic variation, phenetic clustering methods assume a causal connection between overall similarity and degree of genetic continuity (de Queiroz & Good 1997, but see Highton 2000).

Virtually all methods are sensitive to sampling design, with different implications for individual, geographic, and character sampling, and these points have been explicitly addressed in some methods. PAA (Davis & Nixon 1992) specifies the clearest relationship between fixed character differences and species boundaries; characters must be present at 100% frequency to be diagnostic of species limits. However, this criterion cannot be established with statistical confidence given finite sample sizes (Wiens & Servedio 2000). Wiens & Servedio (2000) replaced the 100% fixation criterion with one that is less restrictive (e.g., 95%) but that can be achieved with statistical confidence and may still indicate an absence of gene flow. This method enables one to evaluate the statistical confidence that a character is diagnostic (present at a frequency established by the investigator) in the context of finite sampling. The FFR method (Doyle 1995) is likely susceptible to false positives in favor of gene flow owing to retained ancestral polymorphisms (J.C. Marshall, E. Arévalo, E. Benavides, J.L. Sites & J.W. Sites, Jr., submitted manuscript).

Tree-based methods provide various criteria for species delimitation as described above, but several of these methods (the various phylogenetic/composite, CHA, and exclusivity) do not explicitly emphasize any geographic component, which will likely be relevant for inferences about gene flow in sexually reproducing species and are emphasized by other approaches (Wiens-Penkrot and Templeton methods). All methods that rely on gene genealogies can be misled by discordance if these genealogies coalesce above or below the divergence/reticulation interface because there is no basis for equating gene trees to species lineages if divergence has been recent (Hudson 1990, Avise 2000; but this may become less of a problem as time since divergence increases, see Neigel & Avise 1986). The

Wiens-Penkrot approach does emphasize the rapid sorting of the mtDNA locus in these cases, presumably because mtDNA is more likely to track recent species splits relative to nuclear loci (Moore 1995, 1997). However, stochasticity associated with population structure, recurrent mutation, selection, and other processes generate sufficiently large confidence intervals for mtDNA coalescent times that they may overlap extensively with nuclear gene coalescents (Hudson & Turelli 2003). Thus caution is required when interpreting single gene genealogies for any method.

Until recently, most DNA-based studies of species boundaries have been based on single loci, and the over-reliance on single gene regions to delimit species has been problematic because phylogenetic structure can extend below the level of species (as for asexual taxa or haploid genomes), and/or tokogenetic relationships may extend above the level of species (interspecific hybridization or reticulation; de Queiroz & Donoghue 1990, Davis 1996). If hybrids must be categorized and narrow hybrid zones are common in a particular group, then notions of diagnosability or monophyly are not useful, and other approaches must be sought. Detailed population sampling of a diversity of taxa often reveals widespread intraspecific paraphyly and polyphyly resulting from hybridization and other processes (Funk & Omland 2003) as well as mismatches between nuclear and mitochondrial genomes in cases where both have been sampled (Shaw 2002). In this case, an advantage of Templeton's tests for cohesion is that they do not require absolute properties (character fixation, exclusivity, complete absence of gene flow) if both individual and geographic sampling are adequate (Templeton 2001), but an operational limitation is the extensive sampling required for all taxa included in a study. A conceptual limitation of Templeton's methods is that negative evidence for H_2 (failure to reject) cannot be taken as evidence that different lineages belong to the same cohesion species because this outcome may simply result from failure to select the appropriate ecological or reproductive trait(s) to delimit lineages in nature that really do represent distinct species. Other concerns about the accuracy of the nested clade inference key (Knowles & Maddison 2002) can at least partly be addressed by cross-validation of inferences with other analyses based on different assumptions (Masta et al. 2003, Morando et al. 2004).

Finally, all methods that require monophyly or exclusivity (CHA, genealogical exclusivity) or that insist on the primacy of branching structure over character evidence (PSC-2 in Figure 2*B*), do not permit recognition of ancestral species that retain plesiomorphic characters relative to their descendents. This pattern is likely to occur when peripheral populations speciate from a widespread ancestor (Talbot & Shields 1996, Hedin 1997) and will result in paraphyly of the ancestral species. Adoption of these methods thus requires a priori rejection of some modes of speciation that may be important processes in nature (see also Brooks & McLennan 1999, Barraclough & Nee 2001). At another extreme, the node-based composite species method does not allow for within-species anagenesis, an assumption that also severely constrains the kinds of evolutionary processes that biologists might study in phylogenies (Brooks & McLennan 1999).

Where Do We Go from Here?

The emerging debates over empirical methods of species delimitation suggest that evolutionary biologists are giving the issue serious consideration (Brower 1999 versus Davis & Nixon 1992, Wake & Schneider 1998 versus Highton 1998). Researchers agree that speciation processes create fuzzy boundaries under which all methods will occasionally fail or be discordant with each other and that this is an unavoidable consequence of the many combinations of deterministic and stochastic processes associated with any speciation event (de Queiroz 1998, Harrison 1998, Frost 2000). Empirical examples bear out this expectation: In salamanders of the *Ensatina eschscholtzii* complex, one (Wake & Schneider 1998), two (Frost & Hillis 1990), seven (Graybeal 1995), or eleven (Highton 1998) species can be recognized on the basis of different methods. This result reflects both the multiple properties that change at different rates and in different order during speciation and the fact that different methods were often designed to detect different kinds of entities (distinct lineages versus interbreeding populations).

When viewed in this context, all methods will sometimes fail to delimit species boundaries properly, and virtually all will require researchers to make qualitative judgments. For example, there is no objective criterion for how much morphological divergence is enough to delimit a species, what threshold frequency of intermediates is needed to delimit species by genotypic clusters (Mallet 1995), what proportion of unlinked loci are needed to delimit coalescent species (Hudson & Coyne 2002), or what frequency cutoff most appropriately indicates that no significant gene flow is occurring between populations (Wiens & Servedio 2000). Proponents of the apomorphy-based criterion for historical delimitation of species have described auxiliary ranking criteria as a means of providing independent justification to assign species rank to some monophyletic groups and not others (Mishler & Brandon 1987). These criteria may include causal processes, such as interbreeding, selective constraints, and developmental canalization, as well as practical criteria (Mishler & Theriot 2000), such as number and quality of synapomorphies, bootstrap values, etc.

The impasse in the debate surrounding species concepts and species delimitation has precipitated recent proposals to dump the species category altogether because some insist that species be defined in a manner consistent with all other Linnaean ranks (Pleijel & Rouse 2000, but see Lee 2003). Others suggest that groups of organisms be delimited by the amount that they differ from other groups, independent of the concept of species (Hendry et al. 2000, but see Avise & Walker 2000). What has been missing in most empirical studies are honest assessments of uncertainties of methods (Adams 2001, Hey et al. 2003) and cross-validation of inferences from a single method or data set. The particular features of the biology of some groups make them well suited to particular methods and data sets (see Adams 1998 and Taylor et al. 2000 for reviews of species concepts and recognition in nematodes and fungi, respectively), but even when this is not the case, the variety of methods now available (Table 1) allows investigators to select the combination of data and analyses that best matches the reality (i.e., individual or character sampling limitations) of a particular study. In most cases, researchers should be able to analyze the same data set with several methods and/or to collect multiple data sets for independent corroboration; both provide opportunity for cross-validation of hypotheses based initially on one analysis of a single data set (Agapow et al. 2004, Templeton 2004).

We think evolutionary biology would now be well served by detailed comparative studies applying multiple methods and data sets to species delimitation in natural populations. Some studies (Wiens & Penkrot 2002; Dettman et al. 2003a,b; Fukami et al. 2004) show that when different methods and data give conflicting results for the same organisms, researchers can often sort out the most likely cause(s) for the conflict. We can make few generalities about which methods are best for a variety of taxa and biological properties, but additional studies should reveal how alternative methods designed to recover the same entities (i.e., interbreeding populations) compare with each other and under what conditions the boundaries of different entities (interbreeding populations versus historically distinct lineages) will coincide with each other.

ACKNOWLEDGMENTS

We thank Paul-Michael Agapow, Lynn Bohs, David Baum, Leigh Johnson, John Taylor, John Wiens, and the BYU and University of Utah systematics discussion groups for comments on earlier drafts of this manuscript, and Brad Shaffer for feedback on the submitted manuscript; we also thank Jeff Watkins for help with graphics. J.W.S. has been supported by research funds from the American Museum of Natural History, American Philosophical Society, National Geographic Society, and the National Science Foundation. J.C.M. is supported by a National Science Foundation doctoral dissertation improvement grant and by Brigham Young University graduate research and student mentoring fellowships.

The *Annual Review of Ecology, Evolution, and Systematics* is online at
http://ecolsys.annualreviews.org

LITERATURE CITED

Adams BJ. 1998. Species concepts and the evolutionary paradigm in modern nematology. *J. Nematol.* 30:1–31

Adams BJ. 2001. The species delimitation uncertainty principle. *J. Nematol.* 33:153–60

Agapow PM, Bininda-Edmonds ORP, Crandall KA, Gittleman JL, Mace GM, et al. 2004. The impact of species concept on biodiversity studies. *Q. Rev. Biol.* 79:161–79

Arnold ML. 1997. *Natural Hybridization and Evolution.* Oxford: Oxford Univ. Press

Atran S, Medin D, Ross N, Lynch E, Coley J, et al. 1999. Folk ecology and common management in the Maya Lowlands. *Proc. Natl. Acad. Sci. USA* 96:7598–603

Avise JC. 2000. *Phylogeography: The History and Formation of Species.* Cambridge, MA: Harvard Univ. Press. 447 pp.

Avise JC, Ball RM. 1990. Principles of

genealogical concordance in species concepts and biological taxonomy. *Oxford Surv. Evol. Biol.* 7:45–67
Avise JC, Walker D. 2000. Abandon all species concepts? A response. *Conserv. Genet.* 1:77–80
Avise JC, Wollenberg K. 1997. Phylogenetics and the origin of species. *Proc. Natl. Acad. Sci. USA* 94:7748–55
Barraclough TG, Nee S. 2001. Phylogenetics and speciation. *Trends Ecol. Evol.* 16:391–99
Baum DA. 1998. Individuality and the existence of species through time. *Syst. Biol.* 47:641–53
Baum DA, Donoghue MJ. 1995. Choosing among alternative "phylogenetic" species concepts. *Syst. Bot.* 20:560–73
Baum DA, Shaw KL. 1995. Genealogical perspectives on the species problem. In *Experimental and Molecular Approaches to Plant Biosystematics*, ed. PC Hoch, AG Stephenson, pp. 289–303. St. Louis: Mo. Botan. Gard.
Benavides E, Ortiz JC, Sites JW Jr. 2002. Species boundaries among the *Telmatobius* (Anura: Leptodactylidae) of the Lake Titicaca basin: allozyme and morphological evidence. *Herpetologica* 58:31–55
Blackburn TM, Gaston KJ. 1998. Some methodological issues in macroecology. *Am. Nat.* 51:6814–83
Bohonak AJ. 1999. Dispersal, gene flow, and population structure. *Q. Rev. Biol.* 74:21–45
Brooks DR, McLennan DA. 1999. Species: turning a conundrum into a research program. *J. Nematol.* 31:117–33
Brower AVZ. 1999. Delimitation of phylogenetic species with DNA sequences: a critique of Davis and Nixon's population aggregation analysis. *Syst. Biol.* 48:199–213
Brown JH, Stevens GC, Kaufman DM. 1996. The geographic range: size, shape, boundaries, and internal structure. *Annu. Rev. Ecol. Syst.* 27:597–623
Bruce RC, Jaeger RG, Houck LD, eds. 2000. *The Biology of Plethodontid Salamanders*. New York: Kluwer Academic/Plenum

Brumfield RT, Beerli P, Nickerson DA, Edwards SV. 2003. The utility of single nucleotide polymorphisms in inference of population history. *Trends Ecol. Evol.* 18:249–56
Burke JM, Arnold ML. 2001. Genetics and the fitness of hybrids. *Annu. Rev. Genet.* 35:31–52
Caldecott JO, Jenkins MD, Johnson TH, Groombridge B. 1996. Priorities for conserving global species richness and endemism. *Biodivers. Conserv.* 5:699–727
Carson HL. 1957. The species as a field for recombination. In *The Species Problem*, ed. E Mayr, pp. 23–38. Washington, DC: AAAS
Coyne JA. 1982. Gel electrophoresis and cryptic protein variation. In *Isozymes: Curr. Top. Biol. Med. Res*, Vol. 6. New York: Liss
Coyne JA. 1994. Ernst Mayr and the origin of species. *Evolution* 48:19–30
Coyne JA, Barton NH, Turelli M. 1997. A critique of Sewall Wright's shifting balance theory of evolution. *Evolution* 51:643–71
Coyne JA, Orr HA. 1989. Patterns of speciation in *Drosophila*. *Evolution* 43:362–81
Cracraft J. 1983. Species concepts and species analysis. *Curr. Ornithol.* 1:159–87
Cracraft J. 2000. Species concepts in theoretical and applied biology: a systematic debate with consequences. In *Species Concepts and Phylogenetic Theory: A Debate*, ed. QD Wheeler, R Meier, pp. 30–43. New York: Colombia Univ. Press
Cracraft J. 2002. The seven great questions of systematic biology: an essential foundation for conservation and the sustainable use of biodiversity. *Ann. Mo. Bot. Gard.* 89:127–44
Cronquist A. 1978. Once again, what is a species? In *Biosystematics in Agriculture*, ed. JA Romberger, pp. 3–20. Montclair, NJ: Allanheld Osmun
Darwin C. 1859/1964. *On the Origin of Species*. Cambridge, MA: Harvard Univ. Press
Davis JI. 1996. Phylogenetics, molecular variation, and species concepts. *Bioscience* 46:502–11
Davis JI, Nixon KC. 1992. Populations, genetic variation, and the delimitation of phylogenetic species. *Syst. Biol.* 41:421–35

de Queiroz K. 1998. The general lineage concept of species, species criteria, and the process of speciation. In *Endless Forms: Species and Speciation*, ed. DJ Howard, SH Berlocher, pp. 57–75. New York/Oxford: Oxford Univ. Press. 470 pp.

de Queiroz K, Donoghue MJ. 1990. Phylogenetic systematics and species revisited. *Cladistics* 6:83–90

de Queiroz K, Good D. 1997. Phenetic clustering in biology: a critique. *Q. Rev. Biol.* 72:3–30

Dettman JR, Jacobson DJ, Taylor JW. 2003a. A multilocus genealogical approach to phylogenetic species recognition in the model eukaryote *Neurospora*. *Evolution* 57:2703–20

Dettman JR, Jacobson DJ, Turner E, Pringle A, Taylor JW. 2003b. Reproductive isolation and phylogenetic divergence in *Neurospora*: comparing methods of species recognition in a model eukaryote. *Evolution* 57:2721–41

Doyle J. 1995. The irrelevance of allele tree topologies for species delimitation, and a non-topological alternative. *Syst. Bot.* 20: 574–88

Ereshefsky M, 2002. Linnaen ranks: vestiges of a bygone era. *Philos. Sci.* 69:S305–15

Estoup A, Jarne P, Cornuet JM. 2002. Homoplasy and mutation at microsatellite loci and their consequences for population genetics analysis. *Mol. Ecol.* 11:1591–604

Ferguson JWH. 2002. On the use of genetic divergence for identifying species. *Biol. J. Linn. Soc.* 75:509–16

Frost D. 2000. Species, descriptive efficiency, and progress in systematics. See Bruce et al. 2000, pp. 7–29

Frost DR, Hillis DM. 1990. Species in concept and practice: herpetological applications. *Herpetologica* 46:87–104

Frost DR, Kluge AG. 1994. A consideration of epistemology in systematic biology, with special reference to species. *Cladistics* 10:259–94

Fukami H, Budd AF, Levitan DR, Jara J, Kersanach R, et al. 2004. Geographic differences in species boundaries among members of the *Montastraea annularis* complex based on molecular and morphological markers. *Evolution* 58:324–37

Funk DJ, Omland KE. 2003. Species-level paraphyly and polyphyly: frequency, causes, and consequences, with insights from animal mitochondrial DNA. *Annu. Rev. Ecol. Evol. Syst.* 34:397–423

Goldstein PZ, Desalle R. 2000. Phylogenetic species, nested hierarchies, and character fixation. *Cladistics* 16:364–84

Gomez-Zurita J, Petitpierre E, Juan C. 2000. Nested cladistic analysis, phylogeography and speciation in the *Timarcha goettingensis* complex (Coleoptera, Chrysomelidae). *Mol. Ecol.* 9:557–70

Good DA, Wake DB. 1992. Geographic variation and speciation in the torrent salamanders of the genus *Rhyacotriton* (Caudata: Rhyacotritonidae). *Univ. Calif. Pub. Zool.* 126:1–91

Graybeal A. 1995. Naming species. *Syst. Biol.* 44:237–50

Hare MP. 2001. Prospects for nuclear gene phylogeography. *Trends Ecol. Evol.* 16:700–6

Harrison RG. 1998. Linking evolutionary pattern and process: the relevance of species concepts for the study of speciation. In *Endless Forms: Species and Speciation*, ed. DJ Howard, SH Berlocher, pp. 19–31. New York/Oxford: Oxford Univ. Press. 470 pp.

Hedin MC. 1997. Speciational history in a diverse clade of habitat-specialized spiders (Araneae: Nesticidae: *Nesticus*): inferences from geographic-based sampling. *Evolution* 5:1929–45

Hendry AP, Vamosi SM, Latham SJ, Heilbuth JC, Day T. 2000. Questioning species realities. *Conserv. Genet.* 1:67–76

Hennig W. 1966. *Phylogenetic Systematics.* Chicago: Univ. Chicago Press. 263 pp.

Hey J. 2001. *Genes, Categories, and Species: The Evolution and Cognitive Causes of the Species Problem.* New York: Oxford Univ. Press. 217 pp.

Hey J, Waples RS, Arnold ML, Butlin RK, Harrison RG. 2003. Understanding and confronting species uncertainty in biology and conservation. *Trends Ecol. Evol.* 18:597–603

Highton R. 1989. Biochemical evolution in the slimy salamanders of the *Plethodon glutinosus* complex in the eastern United States. Part I. Geographic protein variation. *Ill. Biol. Monogr.* 57:1–78

Highton R. 1990. Taxonomic treatment of genetically differentiated populations. *Herpetologica* 46:114–21

Highton R. 1995. Speciation in eastern North-American salamanders of the genus *Plethodon*. *Annu. Rev. Ecol. Syst.* 26:579–600

Highton R. 1998. Is *Ensatina eschscholtzii* a ring-species? *Herpetologica* 54:254–78

Highton R. 2000. Detecting cryptic species using allozyme data. See Bruce et al. 2000, pp. 215–41

Highton R, Peabody RB. 2000. Geographical protein variation and speciation in salamanders of the *Plethodon jordani* and *Plethodon glutinosus* complexes in the southern Appalachian mountains with the description of four new species. See Bruce et al. 2000, pp. 31–94

Hollocher H, Wu CI. 1996. The genetics of reproductive isolation in the *Drosophila simulans* clade: X vs. autosomal effects and male vs. female effects. *Genetics* 143:1243–55

Hudson RR. 1990. Gene genealogies and the coalescent process. *Oxford Surv. Evol. Biol.* 7:1–44

Hudson RR, Coyne JA. 2002. Mathematical consequences of the genealogical species concept. *Evolution* 56:1557–65

Hudson RR, Turelli M. 2003. Stochasticity overrules the "three times rule": genetic drift, genetic draft, and coalescent times for nuclear loci versus mitochondrial DNA. *Evolution* 57:182–90

Hughes JB, Daily GC, Ehrlich PR. 1997. Population diversity: its extent and extinction. *Science* 278:689–92

Jackman TR, Wake DB. 1994. Evolutionary and historical analysis of protein variation in the blotched forms of salamanders of the *Ensatina* complex (Amphibia: Plethodontidae). *Evolution* 48:876–97

Knowles LL, Maddison WP. 2002. Statistical parsimony. *Mol. Ecol.* 11:2623–35

Kornet D. 1993. Permanent splits as speciation events: a formal reconstruction of internodal species concept. *J. Theor. Biol.* 164:407–35

Lee MSY. 2003. Species concepts and species reality: salvaging a Linnaean rank. *J. Evol. Biol.* 16:179–88

Li YC, Korol AB, Fahima T, Beiles A, Nevo E. 2002. Microsatellites: genomic distribution, putative functions and mutational mechanism: a review. *Mol. Ecol.* 11:2453–65

Maddison WP. 1997. Gene trees in species trees. *Syst. Biol.* 46:523–36

Mallet J. 1995. A species definition for the modern synthesis. *Trends Ecol. Evol.* 10:294–99

Martinsen GD, Whitham TG, Turek RJ, Keim P. 2001. Hybrid populations selectively filter gene introgression between species. *Evolution* 55:1325–35

Masta SE, Laurent NM, Routmann EJ. 2003. Population genetic structure of the toad *Bufo woodhousii*: an empirical assessment of the effects of haplotype extinction on nested clade analysis. *Mol. Ecol.* 12:1541–54

Mayden RL. 1997. A hierarchy of species concepts: the denouement in the saga of the species problem. In *Species: The Units of Biodiversity*, ed. M Claridge, HA Darwah, MR Wilson, pp. 381–424. London: Chapman & Hall

Mayr E. 1942. *Systematics and the Origin of Species.* New York: Columbia Univ. Press

Mishler BD. 2003. The advantage of a rank-free classification for teaching and research. *Cladistics* 19:157

Mishler BD, Brandon RN. 1987. Individuality, pluralism, and the phylogenetic species concept. *Biol. Phil.* 2:397–414

Mishler BD, Theriot EC. 2000. The phylogenetic species concept (sensu Mishler and Theriot): monophyly, apomorphy, and phylogenetic species concepts. In *Species Concepts and Phylogenetic Theory: A Debate*, ed. QD Wheeler, R Meier, pp. 44–54. New York: Columbia Univ. Press

Morando M, Avila LJ, Baker J, Sites JW Jr. 2004. Phylogeny and phylogeography of the *Liolaemus darwinii* complex (Squamata:

Liolamidae): evidence for introgression and incomplete lineage sorting. *Evolution* 58:842–61

Morin PA, Luikart G, Wayne RK, SNP Workshop Group. 2004. SNPs in ecology, evolution and conservation. *Trends Ecol. Evol.* 19:208–16

Moore WS. 1995. Inferring phylogenies from mtDNA variation: mitochondrial genes versus nuclear-gene trees. *Evolution* 49:718–26

Moore WS. 1997. Mitochondrial gene trees versus nuclear-gene trees: a reply to Hoelzer. *Evolution* 51:627–29

Murphy RW, Sites JW Jr., Buth DG, Haufler CH. 1996. Proteins: isozyme electrophoresis. In *Molecular Systematics*, ed. DM Hillis, C Moritz, BK Mable, pp. 51–120. Sunderland, MA: Sinauer. 655 pp.

Nei M. 1972. Genetic distance between populations. *Am. Nat.* 106:283–92

Nei M. 1978. Estimation of average heterozygosity and genetic distance from a small number of individuals. *Genetics* 89:583–90

Neigel JE, Avise JC. 1986. Phylogenetic relationships of mitochondrial DNA under various demographic models of speciation. In *Evolutionary Processes and Theory*, ed. S Karlin, E Neto, pp. 515–35. New York: Academic. 786 pp.

Nixon KC, Wheeler QD. 1990. An amplification of the phylogenetic species concept. *Cladistics* 6:211–23

Orr HA. 1996. Dobzhansky, Bateson, and the genetics of speciation. *Genetics* 144:1331–35

Pannell JR, Charlesworth B. 1999. Neutral genetic diversity in a metapopulation with recurrent local extinction and recolonization. *Evolution* 53:664–76

Pigliucci M. 2003. Species as family resemblance concepts: the (dis-)solution of the species problem? *BioEssays* 25:596–602

Pleijel E, Rouse GW. 2000. Least-inclusive taxonomic unit: a new taxonomic concept in biology. *Proc. R. Soc. London Ser. B* 267:627–30

Porter AH. 1990. Testing nominal species boundaries using gene flow statistics: taxonomy of two hybridizing admiral butterflies (*Limenitis*: Nymphalidae). *Syst. Zool.* 39:131–47

Puorto G, Da Graça SM, Theakston RDG, Thorpe RS, Warrell DA, et al. 2001. Combining mitochondrial DNA sequences and morphological data to infer species boundaries: phylogeography of lancehead pitvipers in the Brazilian Atlantic forest, and the status of *Bothrops pradoi* (Squamata: Serpentes: Viperidae). *J. Evol. Biol.* 14:527–38

Rosen DE. 1979. Fishes from the upland and intermontane basin of Guatemala: revisionary studies and comparative biogeography. *Bull. Am. Mus. Natl. Hist.* 162:267–376

Sasa MM, Chippindale PT, Johnson NA. 1998. Patterns of postzygotic isolation in frogs. *Evolution* 52:1811–20

Shaw KL. 1999. A nested analysis of song groups and species boundaries in the Hawaiian cricket genus *Laupala*. *Mol. Phyl. Evol.* 11:332–41

Shaw KL. 2002. Conflict between nuclear and mitochondrial DNA phylogenies of a recent species radiation: what mtDNA reveals and conceals about modes of speciation in Hawaiian crickets. *Proc. Natl. Acad. Sci. USA* 99:16122–27

Sites JW Jr, Marshall JC. 2003. Delimiting species: a Renaissance issue in systematic biology. *Trends Ecol. Evol.* 18:462–70

Slatkin M, Barton NH. 1989. A comparison of three indirect methods for estimating average levels of gene flow. *Evolution* 43:1349–68

Sluys R, Hazevoet CJ. 1999. Pluralism in species concepts: dividing nature at its diverse joints. *Species Diversity* 4:243–56

Sneath PHA. 1976. Phenetic taxonomy at the species level and above. *Taxon* 25:437–50

Sokal RR. 1973. The species problem reconsidered. *Syst. Zool.* 22:360–74

Sokal RR, Sneath PHA. 1963. *Principles of Numerical Taxonomy*. San Francisco: W.H. Freeman. 359 pp.

Sullivan JP, Lavoué S, Arnegard ME, Hopkins CD. 2004. AFLPs resolve phylogeny and reveal mitochondrial introgression within a species flock of African electric fish (Mormyroidea: Teleostei). *Evolution* 58:825–41

Talbot SL, Shields GF. 1996. Phylogeography of brown bears (*Ursus arctos*) of Alaska and paraphyly with the Ursidae. *Mol. Phylogenet. Evol.* 5:477–94

Taylor JW, Jacobsen DJ, Kroken S, Kasuga T, Geiser DM, et al. 2000. Phylogenetic species recognition and species concepts in fungi. *Fungal Genet. Biol.* 31:21–32

Templeton AR. 1989. The meaning of species and speciation. In *Speciation and Its Consequences*, ed. D Otte, JA Endler, pp. 3–27. Sunderland, MA: Sinauer

Templeton AR. 1994. The role of molecular genetics in speciation studies. In *Molecular Ecology and Evolution: Approaches and Application*, ed. B Schierwater, B Streit, GP Wagner, R Desalle, pp. 455–77. Basel, Switz.: Birkhauser Verlag

Templeton AR. 1998. Species and speciation: geography, population structure, ecology, and gene trees. In *Endless Forms: Species and Speciation*, ed. DJ Howard, SH Berlocher, pp. 32–43. New York/Oxford: Oxford Univ. Press. 470 pp.

Templeton AR. 1999. Using gene trees to infer species from testable null hypothesis: cohesion species in the *Spalax ehrenbergi* complex. In *Evolutionary Theory and Process: Modern Perspectives, Papers in Honor of Eviatar Nevo*, ed. SP Wasser, pp. 171–92. Dordrecht, Neth.: Kluwer Acad. 528 pp.

Templeton AR. 2001. Using phylogeographic analyses of gene trees to test species status and boundaries. *Mol. Ecol.* 10:779–91

Templeton AR. 2004. Statistical phylogeography: methods of evaluating and minimizing inference errors. *Mol. Ecol.* 13:789–809

Templeton AR, Routman E, Phillips CA. 1995. Separating population structure from history: a cladistic analysis of the geographical distribution of mitochondrial DNA haplotype in the tiger salamander, *Ambystoma tigrinum*. *Genetics* 140:767–82

Thorpe JP. 1982. The molecular clock hypothesis: biochemical evaluation, genetic differentiation and systematics. *Annu. Rev. Ecol. Syst.* 13:139–68

Tilley SG, Mahoney MJ. 1996. Patterns of genetic differentiation in salamanders of the *Desmognathus ochrophaeus* complex (Amphibia: Plethodontidae). *Herp. Mono.* 10:1–42

Turelli M, Barton NH, Coyne JA. 2001. Theory and speciation. *Trends Ecol. Evol.* 16:330–43

Wade MJ, Goodnight C. 1998. Perspective: the theories of Fisher and Wright in the context of metapopulations: when nature does many small experiments. *Evolution* 52:1537–53

Wake DB, Schneider CJ. 1998. Taxonomy of the plethodontid salamander genus *Ensatina*. *Herpetologica* 54:279–98

Weir BS, Cockerham CC. 1984. Estimating F-statistics for the analysis of population-structure. *Evolution* 38:1358–70

Wheeler QD, Meier R, eds. 2000. *Species Concepts and Phylogenetic Theory: A Debate*. New York: Columbia Univ. Press. 256 pp.

Whitlock MC, McCauley DE. 1999. Indirect measures of gene flow and migration: $F_{ST} \neq 1/(4Nm + 1)$. *Heredity* 82:117–25

Wiens JJ. 1999. Polymorphism in systematics and comparative biology. *Annu. Rev. Ecol. Syst.* 30:327–62

Wiens JJ. 2000. Coding morphological variation for phylogenetic analysis: polymorphism and interspecific variation in higher taxa. In *Phylogenetic Analysis of Morphological Data*, ed. J Wiens, pp. 115–45. Washington, DC: Smithsonian Inst. Press

Wiens JJ, Penkrot TA. 2002. Delimiting species using DNA and morphological variation and discordant species limits in spiny lizards (*Sceloporus*). *Syst. Biol.* 51:69–91

Wiens JJ, Servedio MR. 2000. Species delimitation in systematics: inferring diagnostic differences between species. *Proc. R. Soc. London Ser. B* 267:631–36

Wiley EO. 1978. The evolutionary species concept reconsidered. *Syst. Zool.* 27:17–26

Wiley EO, Mayden RL. 2000. The evolutionary species concept. In *Species Concepts and Phylogenetic Theory: A Debate*, ed. QD Wheeler, R Meier, pp. 70–89. New York: Columbia Univ. Press

Wright S. 1931. Evolution in Mendelian populations. *Genetics* 16:97–159

Zhang DX, Hewitt GM. 2003. Nuclear DNA analyses in genetic studies of populations: practice, problems, and prospects. *Mol. Ecol.* 12:563–84

Zink RM. 1997. Species concepts, speciation, and sexual selection. *J. Avian Biol.* 27: 1–6

THE NEW VIEW OF ANIMAL PHYLOGENY

Kenneth M. Halanych

Department of Biological Sciences, Auburn University, Auburn, Alabama 36849; email: ken@auburn.edu

Key Words Metazoa, Lophotrochozoa, Ecdysozoa, Deuterostomia, Cambrian explosion

■ **Abstract** Molecular tools have profoundly rearranged our understanding of metazoan phylogeny. Initially based on the nuclear small ribosomal subunit (SSU or 18S) gene, recent hypotheses have been corroborated by several sources of data (including the nuclear large ribosomal subunit, Hox genes, mitochondrial gene order, concatenated mitochondrial genes, and the myosin II heavy chain gene). Herein, the evidence supporting our current understanding is discussed on a clade by clade basis. Bilaterian animals consist of three clades: Deuterostomia, Lophotrochozoa, and Ecdysozoa. Each clade is supported by molecular and morphological data. Deuterostomia is smaller than traditionally recognized, consisting of hemichordates, echinoderms, chordates, and *Xenoturbella* (an enigmatic worm-like animal). Lophotrochozoa groups animals with a lophophore feeding apparatus (Brachiopoda, Bryozoa, and Phoronida) and trochophore larvae (e.g., annelids and mollusk), as well as several other recognized phyla (e.g., platyhelminthes, sipunculans, nemerteans). Ecdysozoa comprises molting animals (e.g., arthropods, nematodes, tardigrades, priapulids), grouping together two major model organisms (*Drosophila* and *Caenorhabditis*) in the same lineage. Platyhelminthes do not appear to be monophyletic, with Acoelomorpha holding a basal position in Bilateria. Before the emergence of bilateral animals, sponges, ctenophorans, cnidarians, and placozoans split from the main animal lineage, but order of divergence is less than certain. Many questions persist concerning relationships within Ecdysozoa and Lophotrochozoa, poriferan monophyly, and the placement of many less-studied taxa (e.g., kinorhynchs, gastrotrichs, gnathostomulids, and entoprocts).

INTRODUCTION

In the past decade, major new hypotheses of animal evolution have shaken traditional foundations and caused researchers to abandon long-standing hypotheses. This change certainly provoked controversy, and many are critical of these new hypotheses. Skepticism has focused on uncertainty about reliability of molecular data, apparent conflict between morphology and molecular data, lack of robust phylogenetic signal, lack of well-defined morphological synapomorphies, and apparent contradictory conclusions from the same data source. In most cases, the basis for such skepticism is limited. One must keep in mind that systematic biology is a dynamic field of research, with hypotheses constantly proposed and later falsified. In contrast, many who are interested in animal phylogeny want a well-supported (i.e., static) evolutionary framework that they can use for comparative studies or teaching purposes.

Fortunately, consensus is emerging for many regions of the metazoan tree. This review provides a conceptual framework of the current understanding of animal phylogeny in light of recent advances. In particular, this review supplements invertebrate biology texts (e.g., Brusca & Brusca 2003, Ruppert et al. 2004) that do not adequately convey the recent advances. Owing to space limitations, I do not discuss many of the important traditional hypotheses that are already well reviewed in Willmer (1990) or viewpoints based solely on morphology (Nielsen 2001). I draw mainly on recent analyses to (*a*) build a basic comparative framework of metazoan phylogeny, (*b*) discuss support, or lack thereof, of major hypotheses, and (*c*) catalog significant papers for those less familiar with the field.

Before I discuss animal relationships, a few salient points deserve mention.

1. Traditional understanding of animal phylogeny (Figure 1, see color insert) was largely based on (*a*) the concept that evolution proceeds from simple to complex, (*b*) a suite of purportedly conserved embryological features (e.g., cleavage patterns, blastopore fate, mode of coelom formation), and (*c*) overall body architecture (e.g., segmentation, type of coelom). These ideas were outlined in Libbie Hyman's (1940, p. 38) figure 5, which presented her understanding of animal body plans. When compared with genetic information (or even cladistic morphological analyses), many traditional morphological and embryological characters are more evolutionarily labile than previously thought (e.g., Halanych 1996a, Valentine 1997, Halanych & Passamaneck 2001).

2. "Phyla" are man-made constructs erected because, in part, shared features were lacking between organismal groups. The phylum concept has a long history of being equated to body plans (or Baupläne), which can be misleading as to the age and diversity of a group. For example, despite very different body plans, major lineages of the phylum Arthropoda are put together because of the presence of an exoskeleton. In contrast, the formerly recognized phylum Vestimentifera (tube worms) are highly derived annelids with limited diversity and recent origins (McHugh 1997; Halanych et al. 1998, 2001).

3. Molecular data are more objective and subject to considerably more rigor than morphological data. DNA sequence contains four easily identified and mutually exclusive character states (analyses on protein coding genes usually employ translated nucleotide data). Morphological and embryological character definitions and scoring of character states are far more subjective, and most characters have been repeatedly used without critical evaluation, calling into question the utility of morphological cladistic studies that span Metazoa (Jenner 1999, 2002, 2004). For example, a quick comparison among

morphological cladistic analyses will reveal several characters scored differently by various workers. Furthermore, evolution at nucleotide and amino acid levels of housekeeping and conserved developmental genes (i.e., those used for phylogenetics) is understood to a much better degree than evolutionary forces acting on morphology. As such, more sophisticated and accurate methods of phylogeny reconstruction are available to molecular data, whereas morphological data are generally limited to parsimony methods.

NEW METAZOAN[1] TREE

In 1988, Field and coworkers published on animal phylogeny using the 18S or nuclear small ribosomal subunit (SSU) gene. Although this paper did have some internal inconsistencies, it ushered in the era of molecular systematics for higher-level animal phylogeny. The SSU was chosen at the time because enough RNA could be obtained for sequencing, it was ubiquitous in animals, and regions of the gene were conserved enough to make "universal primers." Field et al. (1988) were also one of the first to use explicit criteria and algorithms for building a phylogenetic tree of Metazoa (but see Bergstrom 1985). Thus, it provided a means to critically test and evaluate traditional hypotheses of animal relationships.

Figure 2 (see color insert) illustrates our current understanding of animal relationships. The topology presented is a conservative interpretation of available data. Because data come from many different sources, we cannot reconstruct the tree from a single, all-encompassing analysis. Throughout the text, I discuss support for the relationships presented in Figure 2. Similarly, Table 1 summarizes some major changes in our understanding of metazoan phylogeny.

BASAL METAZOAN CLADES

Metazoa constitutes a monophyletic clade closely related to Choanoflagellata. This traditional view and supporting evidence (e.g., morphological synapomorphies, including extracellular matrix, septate junctions, and spermatozoa; and gene trees, including rDNA, heat-shock proteins, and elongation factors) have been recently covered by Cavalier-Smith et al. (1996), Eernisse & Peterson (2004), and Brooke & Holland (2003) and is not elaborated here.

Sponge Paraphyly

The question of poriferan monophyly has recently generated much discussion. Molecular analyses of SSU data suggest that sponges form a basal paraphyletic

---

[1]Note that others recognize a formal distinction between Metazoa and Animalia, using Animalia to represent a more inclusive clade of Choanoflagellata plus Metazoa (e.g., Sørensen et al. 2000, Eernisse & Peterson 2004).

TABLE 1 List of important hypotheses[a] supported by our current understanding

Hypothesis	Support[b]
Sponge paraphyly	SSU, LSU, morph.
Cnidaria sister to Bilateria	SSU, SSU 2° structure, Hox, morph.
Platyhelminthes polyphyletic (Acoelomorphs basal)	SSU, LSU, myo II
Deuterostomia	SSU, LSU, Hox, morph.
Ambulacraria (Hemichordata and Echindermata)	SSU, LSU, tRNA coding, multigene protein coding, morph.
Xenoturbella is a deuterostome	SSU, COI, COII, mtDNA gene order, codon usage
Lophotrochozoa	SSU, LSU, Hox, mtDNA gene trees, mtDNA gene order, myo II, IF gene
Annelida includes sibogliniids and echiurids	EF-1α, SSU, LSU, CO1, morph.
Lophophorate polyphyly	SSU, LSU, Hox, morph.
Syndermata (Acanthocephala within Rotifera)	SSU, morph.
Gnathifera (Syndermata, Gnathostomulida, Micrognathozoa)	SSU, morph.
Platyzoa (Platyhelminthes, Gastrotricha, Cycliphora, Entoprocta, Gnathifera)	SSU, LSU, morph.
Ecdysozoa	SSU, LSU, Hox, morph., HRP activity
Scalidophora (Kinorhyncha, Loricifera, Priapula)	SSU, LSU, morph.
Pancrustacea (Hexapoda within Crustacea)	EF-1α, SSU, LSU

[a]Hypotheses that are either novel (e.g., based on molecular data) or are different from our traditional understanding.
[b]These are the primary sources of support. See text for full details for any given node.

grade in Metazoa with Hexactinellida (glass sponges), then Demospongia (spongin fibers), and then Calcarea (calcareous sponges) branching off in order from the main metazoan lineage (Borchiellini et al. 2001). Recent combined studies provide support for the claim that Calcarea shares a more recent common ancestor with other animals, but placement of the Demospongia relative to the Hexactinellida is not clear (SSU and LSU, Medina et al. 2001; morphology and SSU, Zrzavy et al. 1998, Peterson & Eernisse 2001). The presence of long cross-striated ciliary rootlets in calcareous sponge larvae and eumetazoans to the exclusion of other sponges is consistent with the recent SSU hypothesis (Amano & Hori 1992, Nielsen 2001).

In phylogenetic analyses, taxonomic representation of sponges is often limited because of problems with obtaining quality DNA or RNA free of contamination from foreign genomes. In addition to more robust taxon sampling, future studies examining the presence or absence of certain genes in different sponge lineages will also yield valuable insight. As for morphological characters, sponges are often problematic because many phylogenetic characters applied to other animals must be coded as question marks, leading to undesirable effects in tree reconstruction programs.

The paraphyletic nature of sponges has important consequences for understanding early animal evolution. Poriferans are typically considered monophyletic on the basis of their body architecture, with a water canal system, presence of choanocytes, and equatorial polar bodies during development (Nielsen 2001). However, molecular data suggest these features should be considered symplesiomorphies (shared ancestral characters) of Metazoa and not synapomorphies of Porifera. Interestingly, Hox genes, involved in anterior/posterior body patterning, have not been easy to find in sponges despite repeated attempts (Seimiya et al. 1994, Manuel & Le Parco 2000), although Degnan et al. (1995) reports finding three divergent candidates. Thus, transition from an organism with poorly defined or lacking axes to one with an anterior and posterior seems to have been accompanied by expansion of genetic machinery (Finnerty 1998, Finnerty & Martindale 1998).

Cnidarians and Ctenophores

Coelenterata is a dated term that referred to a taxon comprising Cnidaria and Ctenophora (and originally Porifera; Hyman 1940). A Cnidaria/Ctenophora clade is not supported by SSU data (Collins 1998, Kim et al. 1999, Podar et al. 2001), LSU data (Medina et al. 2001), morphology (Eernisse et al. 1992), or combined analyses (Zrzavy et al. 1998, Peterson & Eernisse 2001). However, the hypothesized position of Ctenophora varies depending on which data are analyzed. Morphological analyses place ctenophores closer to bilaterians (Schram 1991, Eernisse et al. 1992, Zrzavy et al. 1998, Peterson & Eernisse 2001) because of sperm morphology and muscle cells, among other characters. In contrast, SSU data support sponges and ctenophores as basal lineages of Metazoa (Collins 1998, Kim et al. 1999, Podar et al. 2001, Medina et al. 2001), leaving cnidarians as sister to bilaterians. Available evidence from Hox genes (reviewed in Martindale et al. 2002), as well as SSU secondary structure (Aleshin & Petrov 2002), also supports the Cnidaria/Bilateria relationship.

The traditional dogma concerning lack of bilateral symmetry and mesoderm in cnidarians and ctenophores is problematic. The issue of symmetry is really an issue of body axis evolution. How many are there? Evolution of developmental mechanisms research has largely focused on how body axes are set up and maintained during ontogeny. Thus, our understanding and definitions of symmetry are being reshaped by a growing body of evidence that demonstrates that mechanisms used in axial pattern can be found in nonbilaterian animals (Martindale et al. 2002, Brooke & Holland 2003, Wikramanayake et al. 2003, Finnerty et al. 2004). Similarly, we are beginning to appreciate that mesoderm, or mesodermal precursors, may be present in ctenopohores and cnidarians (e.g., Martindale & Henry 1999, Spring et al. 2002, Muller et al. 2003, and references therein). Information on relationships within Cnidaria may be found in Bridge et al. (1995) and Collins (2000, 2002). Podar et al. (2001) and Harbison (1985) offer views on molecular and morphological ctenophore phylogeny, respectively.

Placozoa

Placozoans are simple organisms of great interest for understanding very early evolution in animals. As with sponges, the use of placozoans in morphological cladistic analyses can be problematic because they cannot be scored for most characters. Every conceivable placement of placozoans among nonbilaterian metazoans has been proposed. Studies including SSU data variably place placozoans within or sister to the Cnidaria (Bridge et al. 1995, Siddall et al. 1995, Kim et al. 1999, Cavalier-Smith & Chao 2003; but see Zrzavy et al. 1998, Peterson & Eernisse 2001). Reports of a placozoan/cnidarian clade seem less likely based on the circular morphology of the placozoan mtDNA molecule (most cnidarians have a linear mtDNA genome) and secondary structure of the mitochondrial LSU (Ender & Schierwater 2003). Although the exact placement of placozoans is not clear, they are near the base of Metazoa (just before or after the sponge lineages) and are currently receiving considerable attention via genomic tools (e.g., Martinelli & Spring 2003, Jakob et al. 2004).

BILATERIA

Bilateria consists of three main clades that predate the Precambrian/Cambrian boundary, 540 million years ago (Mya) (Balavoine & Adoutte 1998). Unfortunately, the events that led to the last bilaterian ancestor and subsequent diversification into deuterostomes, lophotrochozoans, and ecdysozoans are not well understood. As such, there is considerable interest in determining which extant taxon is the most basal bilaterian lineage. Currently, there are two possible candidates, acoelomorphs and myxozoans (small parasitic group). Very recent data for the chaetognaths, traditionally considered deuterostomes, suggest a fairly basal position. However, arrow worms are discussed in the deuterostome section because of the relevancy of published data.

The placement of acoelomorphs is a controversial topic that highlights some of the potential problems with both morphological and SSU rDNA data. Early SSU analyses suggested Platyhelminthes was polyphyletic, with Acoela being the most basal bilaterian lineage and separate from other flatworms (Ruiz-Trillo et al. 1999). However, the acoel in question had long-branch lengths owing to high nucleotide substitution rates, a problem that randomizes signal in data and causes long-branched taxa to be artificially placed basal in a tree reconstruction (Felsenstein 1988, Wheeler 1990). Thus, researchers were concerned that the SSU acoel result was an artifact. This long-branch problem was best illustrated in Peterson & Eernisse (2001), who graphically demonstrated that acoel branches for SSU data were so long that they effectively acted as random sequences and were probably rooting the Bilateria incorrectly (see also Giribet et al. 2000). Acoel morphology was not that helpful for higher-level phylogenetic considerations because of their simplified bodies, but within Platyhelminthes morphology supported Acoelomorpha (Acoela and Nemertodermatida) as a distinct clade considerably different from other flatworms (reviewed in Giribet et al. 2000, Ruiz-Trillo et al. 2002). Subsequent work on acoel placement with an independent marker, elongation factor (EF)–1α, supported platyhelminth monophyly (Berney et al. 2000), but this work soon came under fire because of sequence alignment issues and limited taxon sampling (Littlewood et al. 2001). Recent findings with myosin II heavy chain (Ruiz-Trillo et al. 2002) and combined SSU and LSU data (Telford et al. 2003) have independently confirmed that the Acoelomorpha are a basal lineage of Bilateria. The remaining platyhelminthes appear to be within Lophotrochozoa.

Myxozoans are small enigmatic parasites with a very simple body plan. Thus, even if they are basal bilaterians, their highly derived morphology may limit their utility for understanding the last bilaterian ancestor. Myxozoans were previously considered protozoans, but molecular data demonstrated their metazoan nature (Smothers et al. 1994). Combined SSU and morphology placed them within Cnidaria (Siddall et al. 1995), but other molecular studies placed them at or near the bilaterian root (Schlegel et al. 1996, Kim et al. 1999, Ferrier & Holland 2001). Another mysterious organism, *Buddenbrockia*, has a simple body plan and has vexed scientist since its discovery in 1850. This nondescript worm-like organism is a myxozoan (Monteiro et al. 2002).

DEUTEROSTOMIA

Before the mid-1990s, Deuterostomia was generally considered to consist of three core phyla (Echinodermata, Hemichordata, and Chordata), plus Chaetognatha and lophophorate taxa (Brachiopoda, Phoronoida, and Bryozoa[2]). Lophophorates have typically been regarded as having a mix of traditional protostome and deuterostome characters, but Zimmer (1973), among others, made convincing arguments based on developmental and nervous features (e.g., body regionalization, cleavage

---

[2]Technically, the term Bryozoa refers to a clade that includes Ectoprocta and Entoprocta (or Kamptozoa). The terms Ectoprocta and Bryozoa are equated here because one would rather be a bryozologist than an ectoproctologist!

program, intraepidermal nervous system) for deuterostome affinities. Thus, many considered some or all of them as basal to Deuterostomia sensu stricto (Willmer 1990, Nielsen 2001). Building on earlier studies that had included molecular data for a brachiopod (Field et al. 1988, Lake 1990), Halanych et al. (1995) included SSU data from all three major lophophorate lineages and formally proposed that bryozoans, brachiopods, and phoronids were derived protostomes allied to annelids and mollusks. They proposed the node-based clade Lophotrochozoa. The placement of lophophorate taxa and support for Lophotrochozoa are discussed below.

Chaetognatha has also been removed from Deuterostomia. Two independent SSU papers reported that chaetognaths were not deuterostomes, but their exact placement in animal phylogeny remained elusive (Telford & Holland 1993, Wada & Satoh 1994). Further analyses with a broader range of taxa supported a nematode-chaetognath relationship, but the issue of long-branch attraction in SSU data could not be completely ruled out (Halanych 1996b, Peterson & Eernisse 2001). The uncertainty of chaetognath's position is echoed in combined studies that place them within Ecdysozoa (Zrzavy et al. 1998, Peterson & Eernisse 2001, Eernisse & Peterson 2004) or as basal bilaterians (Giribet et al. 2000). Papillon et al. (2003) surveyed Hox genes and reported that chaetognaths are basal in metazoan phylogeny because of the presence of a chimera medial/posterior Hox orthologs. Additional posterior Hox genes have been found, supporting chaetognaths as basal protostomes (D.Q. Matus, K.M. Halanych & M.Q. Martindale, unpublished results), consistent with mtDNA gene order (Helfenbein et al. 2004). Thus, many of the deuterostome-like features of chaetognaths (enterocoelous development, tripartite body, radial cleavage, etc.) are likely bilaterian symplesiomophies rather than derived features.

Ambulacraria

Hemichordates were once placed within Chordata, but they were removed because some but not all the chordate-like features were present. Perhaps surprisingly these "half-chordates" are much more closely allied to echinoderms than to chordates (e.g., for morphology and SSU data, see Turbeville et al. 1994, Giribet et al. 2000, Peterson & Eernisse 2001; for SSU, see Halanych 1995; for changes in tRNA coding, see Telford et al. 2000; for multigene, see Cameron et al. 2000, Furlong & Holland 2002). Metschnikoff (1881), focusing on similarities between the larvae of echinoderms and enteropnuests (also known as acorn worms), referred to the echinoderm-hemichordate group as Ambulacraria, which Halanych (1995) formalized as a node-based name. Morphology supporting this grouping includes characters pertaining to the tripartite larval coeloms (absent in chordates).

Ambulacraria has profound implications for understanding chordate origins by altering interpretation of the evolution of gill slits, the nervous system, and possibly the notochord. Such chordate features may have been present in the last common ancestor of the Deuterostomia. Now we must address the possibility that echinoderms lack gill slits and a notochord-like structure because these features were lost during the evolution to a pentaradial body plan. Similarly, Bather's (1913) and later Jefferies's (1986) assertions that primitive echinoderms possessed ciliated gill slits may be true. However, objective cladistic analyses of morphology and/or molecules do not support Jefferies's notion that chordates are direct descendents of echinoderms.

Noteworthy is the large amount of genomic data coming to bear on the issue of deuterostome relationships. For example, an impressive survey of hemichordate genes with orthologs involved in chordate nervous system development demonstrates an amazing amount of conservation in expression domains despite the noncentralized nature of the hemichordate nervous system (Lowe et al. 2003). Additionally, shortly after this publication, completely sequenced genome for each major deuterostome lineage should be available in GenBank.

Tunicata

The tadpole larva is used to unify Tunicata (also known as Urochordata) with Cephalochordata and Craniata. Perhaps the most convincing characters are the notochord and neural development. Except for Larvacea (Appendicularia), commonalities between the adult body form of tunicates, cephalochordates, and craniates are lacking to the point that their inclusion in the same phylum is questioned (Nielsen 2001). Interestingly, as pointed out by Swalla and colleagues (Swalla et al. 2000, Cameron et al. 2000, Winchell et al. 2002), the phylogenetic signal supporting a monophyletic Chordata (including tunicates) clade is weak. In fact, Ambulacraria is repeatedly much more robustly supported than Chordata (e.g., Winchell et al. 2002, Bourlat et al. 2003). The placement of Tunicata is variable: sister to Craniata/Cephalochordata clade (Cameron et al. 2000, SSU of Winchell et al. 2002), sister to Craniata (Giribet et al. 2000), sister to Ambulacraria (Wada & Satoh 1994, Swalla et al. 2000), or the basal deuterostome lineage (LSU of Winchell et al. 2002, Bourlat et al. 2003). Tunicates often display long branches relative to other deuterostomes, making their placement difficult. Despite the consensus that Tunicata, Cephalochordata, and Craniata form a monophyletic clade, this issue deserves more attention. The vast differences between tunicates and other chordates suggest that they possess an evolutionarily distinct body plan. Hence, the term Tunicata is preferred over Urochordata.

Paleontological work of Shu and collaborators (1999, 2001a, 2001b, 2003) is revising our understanding of the early evolution of chordates. They have discovered ascidian (e.g., *Cheungkongella*), cephalochordate-like yunnanozoan (e.g., *Haikouella*), and agnathan (e.g., *Myllokunmingia*) fossils that push the origins of these lineages at least into the Lower Cambrian (approximately 530–540 Mya) and provide evidence of their complexity early in their evolution.

Xenoturbella

Xenoturbella is a small, morphologically nondescript flatworm-like organism that feeds on bivalve eggs and larvae, whose phylogenetic affinities have long been a mystery. Although the first available molecular data on *Xenoturbella* suggested molluscan affinities (Norén & Jondelius 1997), it appears to be a deuterostome basal to the Ambulacraria. Bourlat et al. (2003) assert that in addition to their SSU, cytochrome oxidase c subunit I (CO I) and CO II data, mitochondrial gene order, and more tentatively, codon usage support this placement. Because of its basal position, questions arise as to whether *Xenoturbella*'s simplified body plan is the result of strong selective constraints on an ancestral body form or the product of secondary simplification. *Xenoturbella* clearly illustrates that many exciting and interesting findings about animal evolution await discovery.

PROTOSTOMIA

Within Bilateria, Lophotrochozoa and Ecdysozoa form a monophyletic clade, typically called Protostomia, supported by SSU and LSU data (Halanych et al. 1995, Mallat & Winchell 2002), combined analyses (Giribet et al. 2000), and Hox genes (de Rosa et al. 1999). In cases where conflicting topologies (e.g., Ecdysozoa with Deuterostomia) have been reported (Ruiz-Trillo et al. 2002), support is weak. Continued use of the term Protostomia is less than ideal but admittedly familiar. From a Hyman-like mindset, the term implies that the ultimate fate of the blastopore is phylogenetically conservative and has traditionally been applied to coelomate animals (to the exclusion of pseudocoelomates and acoelomates). In our current understanding, blastopore fate is irrelevant for many ecdysozoans because the blastopore is usually not retained. Also, both protostome clades contain several former acoelomate or pseudocoelomate taxa. Lastly, lophotrochozoans, ecdysozoans, and deuterostomes were probably all present in Precambrian times and all experienced some degree of a rapid radiation around 580–520 Mya. Thus, it seems more natural to think about bilaterians as having three main clades, given the importance of the Cambrian explosion in metazoan history (Balavoine & Adoutte 1998).

LOPHOTROCHOZOA

Lophotrochozoa is a clade originally identified by SSU data and defined as the last common ancestor of annelids, mollusks, the three lophophorate phyla (Brachiopoda, Phoronida, and Bryozoa), and all the descendants of that ancestor (Halanych et al. 1995). The name of the clade refers to the inclusion of animals that have either a trochophore or a lophophore feeding apparatus. The major implication of this hypothesis is that lophophorate taxa are not allied to the deuterostomes but are highly derived protostomes near annelids, mollusks, and their allies, calling into question many of the classical characters used to split protostomes and deuterostomes.

The Lophotrochozoa hypothesis was contentious at first because analyses of nonmolecular data still grouped some or all lophophorate taxa with deuterostomes

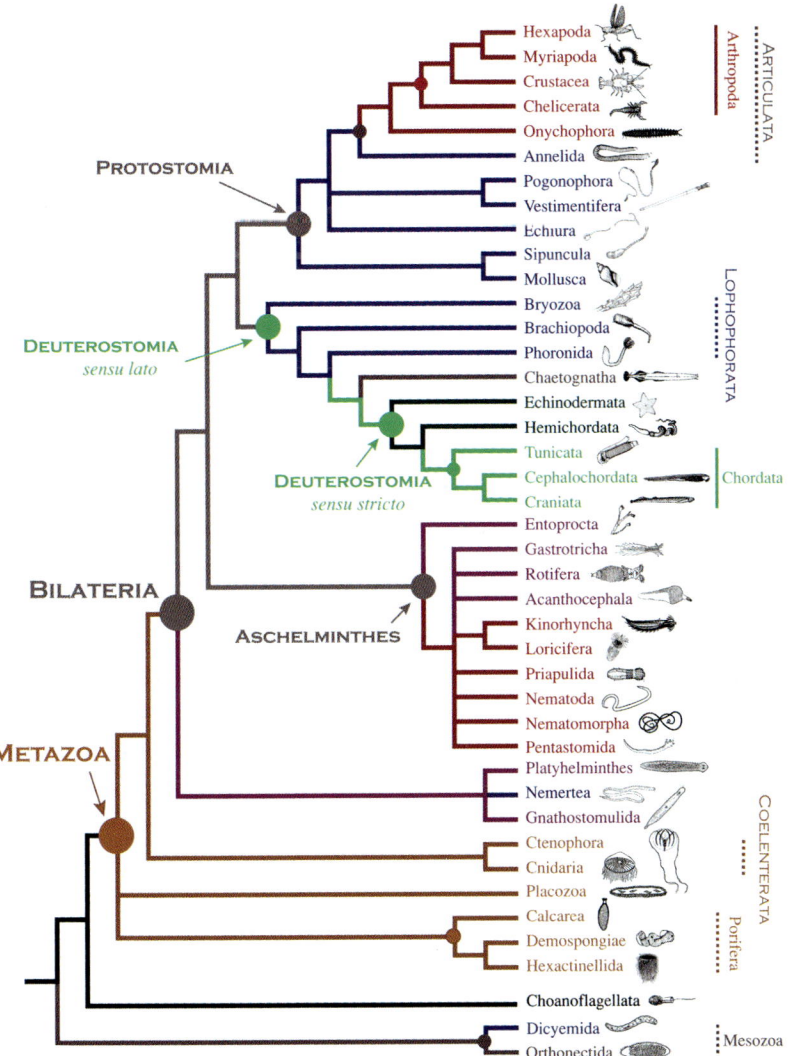

Figure 1 Traditional synthesis. The understanding of metazoan evolution prior to Field et al. (1988). This tree is drawn to illustrate major concepts. The tree is color coded to match the clades consistent with our current understanding as shown in Figure 2: Brown is Metazoa, gray is Bilateria, green is Deuterostomia, dark green is Ambulacraria, red is Ecdysozoa, blue is Lophotrochozoa, magenta is Platyzoa, and black is nonmetazoan. Researchers have proposed many variations of this general hypothesis. For example, many viewed Bryozoa, Brachiopoda, and Phoronida as forming a monophyletic Lophophorata. Also, it was widely recognized that the Aschelminthes was probably not a natural group. Note that, for illustrative purposes, not all taxa shown in Figure 2 are shown here. Filled circles correspond to labeled nodes. Dashed vertical lines indicate groups that are not monophyletic.

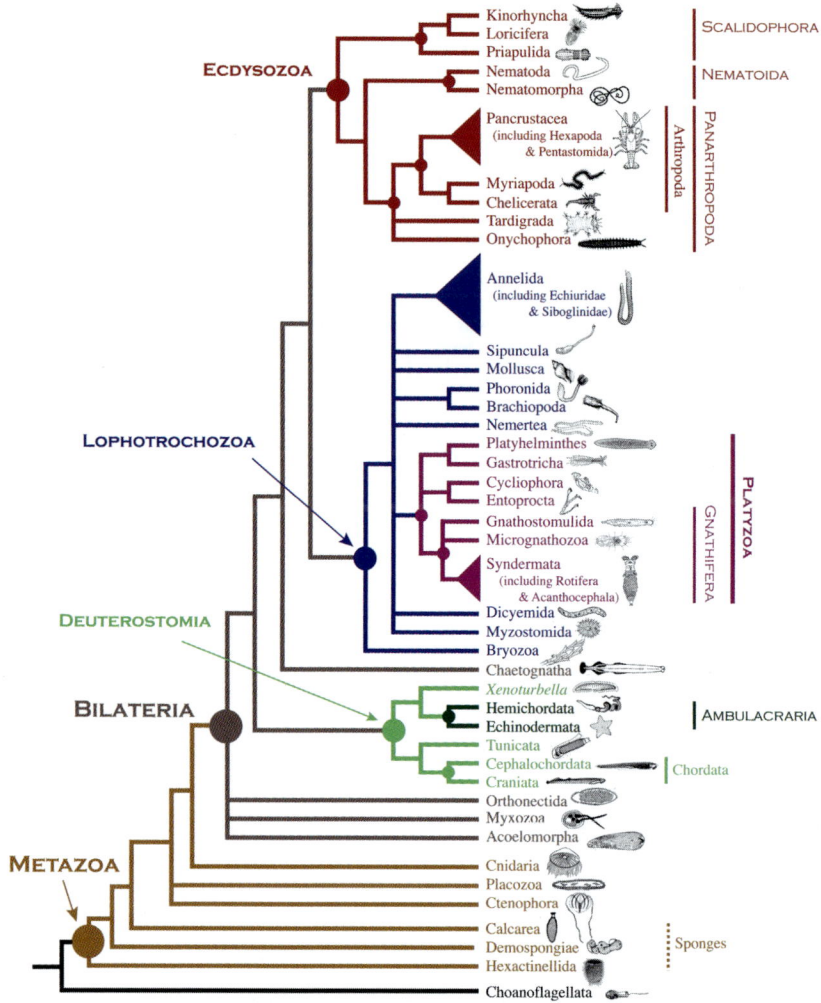

Figure 2 Modern synthesis. The new view of animal phylogeny based largely on molecular data. Details and support for various clades are discussed in the text. This figure is intended to be slightly conservative in nature, as poorly resolved issues are shown as polytomies. The tree is color coded: Brown is Metazoa, gray is Bilateria, green is Deuterostomia, dark green is Ambulacraria, red is Ecdysozoa, blue is Lophotrochozoa, magenta is Platyzoa, and black is nonmetazoan. Nested clades of one color may be inside a more inclusive clade of another color. For example, Echinodermata is dark green because it is within Ambrulacraria, but it is also a deuterostome (*green*), bilaterian (*gray*), and metazoan (*brown*). Note that Siboglinidae is the preferred name for the pogonophoran/vestimentiferan lineage, and the position of Orthonectida as a basal bilaterian needs confirmation. Filled circles correspond to labeled nodes. Dashed vertical lines indicate groups that are not monophyletic.

(Schram 1991, Eernisse et al. 1992, Zrzavy et al. 1998, Sørensen et al. 2000, Nielsen 2001). In solely molecular analyses, brachiopods, phoronids, and bryozoans have never been strongly supported with deuterostomes. Moreover, the Lophotrochozoa hypothesis has received strong support from multiple sources of molecular data, including SSU data (Halanych et al. 1995, Mackey et al. 1996, Eernisse 1997), LSU data (Mallat & Winchell 2002; Y.J. Passamaneck & K.M. Halanych, unpublished results), Hox gene data (de Rosa et al. 1999, Passamaneck & Halanych 2004), mitochondrial sequence and gene arrangement data (Stechmann & Schlegle 1999, Helfenbein et al. 2001, Helfenbein & Boore 2004), myosin II heavy chain sequence data (Ruiz-Trillo et al. 2002), and intermediate filament sequences (Erber et al. 1998). In particular, five Hox genes (*lox2*, *lox4*, *lox5*, *post1*, and *post2*) have been characterized as having lophotrochozoan-specific peptide signatures (de Rosa et al. 1999, Balavoine et al. 2002). Not surprisingly, taxonomic sampling of the above studies varies greatly, and only a few studies include all major (putative) lophotrochozoan taxa. For example, bryozoans seem to be a key taxon for defining Lophotrochozoa but have only recently been examined for data other then the SSU (see below).

Relationships within Lophotrochozoa are not well understood, and short-branch lengths between recognized phyla are a typical result. This lack of phylogenetic signal could be due to a rapid radiation of the taxa (i.e., Cambrian explosion) or indicative of the data in question. As genes other than the SSU and broader taxon sampling have been employed, lophotrochozoan interphyletic relationships have become more apparent, but support via bootstrap analyses or hypothesis testing methods (e.g., likelihood ratio tests) is often limited. For example, the terms Eutrochozoa and Trochozoa have been used to describe subsets of lophotrochozoan taxa (see Peterson & Eernisse 2001), but their weak nodal support precludes detailed discussion here. Nonetheless, within Lophotrochozoa several subclades are emerging (Figure 2).

Annelida

Annelida encompasses a greater diversity of animal body plans than traditionally recognized (McHugh 2000, Halanych et al. 2002). Both McHugh (1997) and Kojima (1998) used EF-1α data to suggest that "Polychaeta" is not monophyletic. Specifically, Vestimentifera, Pogonophora, and Echuira, which were all considered separate phyla, are within the annelid radiation. SSU (Halanych et al. 2001), CO I (Black et al. 1997), and mitochondrial genome data (Boore & Brown 2000; R.M. Jennings & K.M. Halanych, unpublished data) corroborated placement of vestimentiferans and pogonophorans within annelids. This placement has long been argued by some on the basis of morphology (van der Land & Nørrevang 1977, Southward 1988) and has been supported in a morphological cladistic analysis (Rouse & Fauchald 1997). Furthermore, Vestimentifera is clearly a clade within the recognized Pogonophora (or Frenulata) (Southward 1988; Halanych et al. 1998, 2001; Rouse 2001). McHugh (1997), and Rouse & Fauchald (1997) argue that the pogonophoran/vestimentiferan clade should revert to its original nomen Siboglinidae (Caullery 1914). Additionally, Clitellata (Oligochaeta and Hirudinea) are derived polychaetes (McHugh 2000, Rota et al. 2001, Struck et al. 2002).

McHugh's (1997) EF-1α data also placed Echiura within annelids, which was subsequently supported by morphological work on the larval nervous system (Hessling 2002) and by SSU data (Bleidorn et al. 2003). Another group that deserves brief mention is Myzostomida, commensals/parasites on echinoderms that have often been placed with annelids (Westheide & Rieger 1996, Rouse & Fauchald 1997). Available combined EF-1α and SSU data suggest an affinity to flatworms (Eeckhaut et al. 2000), but nervous system morphology (Müller & Westheide 2000) as well as larval morphology (Eeckhaut et al. 2003) are annelidlike. A recent combined analysis of SSU and morphology suggests this organism is outside the annelid radiation (Zrzavy et al. 2001). This group will receive future attention because its placement bears on issues of homology for nervous and larval characters within Lophotrochozoa.

Platyzoa

Platyzoa consists of Platyhelminthes (hereafter referred to as excluding Acoelomorpha), Rotifera, Acanthocephala, Gastrotricha, and Gnathostomulida, as recognized by Garey & Schmidt-Rhaesa (1998) and Cavalier-Smith (1998) and further supported by Giribet et al. (2000). Molecular evidence suggests Platyhelminthes are within the lophotrochozoan clade (Balavoine & Telford 1995, Balavoine 1997, de Rosa et al. 1999, Ruiz-Trillo et al. 2002), and Garey and colleagues (Garey et al. 1996b, Garey & Schmidt-Rhaesa 1998) provided strong support for acanthocephalans within rotifers, the Syndermata. The placement of gnathostomulids together with Syndermata has received support (Ahlrichs 1997, Giribet et al. 2000). Members of this group have an internal set of chitinous jaws and are therefore called the Gnathifera. Although molecular data are still wanting, the curious Micrognathozoa is likely to be in the Gnathifera (Kristensen & Funch 2000, Sørensen et al. 2000). However, recent molecular data (Giribet et al. 2004) are ambiguous, with SSU and very small fragments of LSU and Histone 3, placing Micrognathozoa near syndermatans and cycliophorans, whereas CO I placed it near entoprocts. Interestingly, the first report of Gnathostomulid SSU data (Littlewood et al. 1998) placed them next to a chaetognath-nematode clade, whereas in Giribet et al.'s (2000) analyses they clustered with platyhelminthes and gastrotrichs. Peterson & Eernisse (2001) noticed the potential long-branch issues of this group and subsequently removed the acoels, gnathostomulids, and gastrotrichs from their analyses. The LSU data support the grouping of rotifers and acanthocephalans with platyhelminthes (gnathostomulids were not included; Y.J. Passamaneck & K.M. Halanych, unpublished observations). The morphological evidence does provide support for the Platyzoa (see Giribet et al. 2000).

One key issue with Platyzoa is its placement relative to Lophotrochozoa. Although some analyses assert that it is the sister clade to Lophotrochozoa (Garey & Schmidt-Rhaesa 1998, Giribet et al. 2000), others support placing Platyzoa within Lophotrochozoa (Peterson & Eernisse 2001; Eernisse & Peterson 2004; Y.J.

Passameneck & K.M. Halanych, unpublished observations). Given that Lophotrochozoa is a node-based name, the inclusiveness of the group is determined by the position of its basal lineages, in this case Bryozoa. Thus, to determine if Platyzoa is within Lophotrochozoa, we must know its position relative to Bryozoa. The placement of Platyzoa remains an important issue because the position of platyhelminthes, with their simple body plans, shapes perceptions on the evolution of organismal complexity.

Brachipoda, Phoronida, Bryozoa

Another consistent result is that "Lophophorata" is not a monophyletic taxon, as long advocated on the basis of morphology (Nielsen 1985, 1987, 2001). Brachiopods and phoronids form a monophyletic entity, with bryozoans considerably removed. This result has been supported by SSU (Halanych et al. 1995, Mackey et al. 1996, Eernisse 1997), morphology and SSU (Zrzavy et al. 1998, Giribet et al. 2000, Peterson & Eernisse 2001), and morphology (Nielsen et al. 1996, Nielsen 2001). Lophophorate monophyly can be significantly rejected on the basis of combined SSU and LSU data (Y.J. Passamaneck & K.M. Halanych, unpublished results). In these analyses, brachiopods and phoronids tend to be placed near eutrochozoan taxa (e.g., annelids, mollusks; see de Rosa 2001, Eernisse 1997), and bryozoans are the most basal lineage of Lophotrochozoa. Whereas Giribet et al. (2000) suggested that bryozoans were basal protostomes, Hox data from bryozoans confirm their placement in Lophotrochozoa, probably as basal (Passamaneck & Halanych 2004). Cohen and coworkers (Cohen et al. 1998, Cohen 2000) suggested, on the basis of SSU data, that phoronids are derived brachiopods, but subsequent analysis suggests that they are sister taxa (Peterson & Eernisse 2001). Proposals have been put forth to recognize the new clade as either Brachiozoa (Cavalier-Smith et al. 1998), Phoronozoa (Zrzavy et al. 1998), or Lophophorata (Peterson & Eernisse 2001). The first two options represent two sides of the proverbial coin, but the last option is problematic because the term "lophophore" has a long history of being applied to brachiopods, phoronids, bryozoans, and even pterobranch hemichordates (Halanych 1996a). As such, the change in definition of Lophophorata will introduce confusion into the literature.

Other Taxa

Several other taxa deserve attention in the context of the Lophotrochozoa. As with annelids, within mollusks we do not have a good understanding of major relationships (see Haszprunar 2000, Steiner & Dreyer 2003, Medina & Collins 2003, Passamaneck et al. 2004). Although most analyses tend to place annelids and mollusks close to each other (e.g., Zrzavy et al. 1998, Giribet et al. 2000, Peterson & Eernisse 2001), we do not know how several other taxa fit in, for example scipunculans, nemerteans, and the brachiopod/phoronid clade. Examining SSU data alone in this region of the tree is problematic because they do not recover recognized phyla as monophyletic (e.g., mollusks, brachiopods, nemerteans). This situation is highlighted in Eernisse's (1997) work, which discusses the positive influence of more robust taxon sampling. A combined analysis of SSU and LSU recovers most "phyla" as monophyletic, but interphylum relationships are based on very short-branch lengths (Y.J. Passamaneck & K.M. Halanych, unpublished results). These short-branch lengths would be expected in the case of a rapid radiation such as the Cambrian explosion.

Nemerteans are also lophotrochozoans (Turbeville et al. 1992, Kmita-Cunisse et al. 1998, de Rosa et al. 1999, Balavoine et al. 2002), consistent with embryology and morphology. Sipunculans appear close to mollusks morphologically (Scheltema 1993), but recent mitochondrial gene order data place them near annelids (Boore & Staton 2002). SSU and combined analyses do not provide strong support for a sipunculid placement other than that they are within Lophotrochozoa (Zrzavy et al. 1998, Giribet et al. 2000, Peterson & Eernisse 2001). Dicymids and Orthonectids (also known as Mesozoa) were once thought of as intermediates between protists and metazoans. However, molecular evidence suggests that these two (likely independent) groups are degenerate triploblast animals (Katayama et al. 1995, Hanelt et al. 1996, Pawlowski et al. 1996). Dicymids were shown to contain the lophotrochozoan-specific peptide motif in their *lox5* homolog (Kobayashi et al. 1999). However, the position of orthonectids is less certain and may be basal in Bilateria (Hanelt et al. 1996). Entoprocts and cycliophorans also present a bit of a mystery. The placement of Entoprocta is variable but consistently within Lophotrochozoa (Mackey et al. 1996, Eernisse 1997, Giribet et al. 2000, Peterson & Eernisse 2001). Cycliophorans were first proposed to be close to entoprocts on the basis of morphological similarities (Funch & Kristensen 1995), but preliminary SSU data placed them close to rotifers (Winnepenninckx et al. 1998). This result has been supported by one combined evidence analysis (Peterson & Eernisse 2001) but refuted by two others (Zrzavy et al. 1998, Giribet et al. 2000). LSU data support the Cycliophora/Entoprocta clade (Y.J. Passamaneck & K.M. Halanych, unpublished observations), as does morphological cladistic analysis (Sørensen et al. 2000).

Noteworthy lophotrochozoan analyses using solely molecular data have not supported a clade with spiral cleavage sensu stricto (i.e., Spiralia).[3] This result is partly due to the lack of resolution currently in this region of the tree, yet we must be open to the possibility that spiral cleavage is not strictly evolutionarily conserved. In particular, phoronids and brachiopods seemed to be allied with annelids and mollusks to the exclusion of, at least, platyhelminths (as generally judged by molecules). Given that phoronids and brachiopods have radial or biradial cleavage (Zimmer 1997), we must accept that spiral cleavage, at least, has been lost in some lineages. Thus, there is no monophyletic lineage that includes all spiral cleavers to the exclusion of other cleavage patterns. Recent cell lineage work further demonstrated that phoronids show no vestiges of 4d mesoderm specification expected in spiral cleavage (Freeman & Martindale 2002). For a critical evaluation of spiral cleavage as a phylogenetic character, see Jenner (2004).

---

[3]In fact, because of the placement of platyhelminthes, most traditional treatments also do not have a monophyletic Spiralia.

ECDYSOZOA

The clade Ecdysozoa circumscribes those animals that have a cuticle shed through molting (i.e., ecdysis). Aguinaldo et al. (1997) formally proposed this clade, on the basis of SSU rDNA results, to include the last common ancestor of arthropods, tardigrades, onychorans, nematodes, nematomorphs, kinorhynchs, and priapulids, and all the descendants of that last common ancestor. Previous SSU analyses including nematodes typically found round worms clustering at the base of Bilateria (e.g., Winnepenninckx et al. 1995). Aguinaldo et al. (1997) assessed the position of nematodes using a more slowly evolving nematode sequence (in this case *Trichinella*), eliminating the potential problem of long-branch attraction. Interestingly, a previous cladistic morphological study (Eernisse et al. 1992) had also recovered the same clade of molting animals (with the exception that Priapulida was part of a basal polytomy). Initial acceptance of this radical revision was slow because it was based on a single nematode sequence. However, it has been subsequently supported by independent data sets, most notably Hox genes (*Ubx*, *abd-A*, and *Abd-B* have ecdysozoan-specific peptides; de Rosa et al. 1999, Balavoine et al. 2002), LSU rDNA data (Mallat & Winchell 2002, Mallat et al. 2004), and combined morphology and SSU (Zrzavy et al. 1998, Giribet et al. 2000, Peterson & Eernisse 2001). Additionally, Haase et al. (2001) report that Ecdysozoa show neural expression of horseradish peroxidase (HRP) immunoreactivity that is absent in other animals. They suggest that the presence of anti-HRP-reactive glycoprotein(s) is a synapomorphy for Ecdysozoa.

The presence of a multimeric form of β-thyomosin that was hypothesized as an ecdysozoan synapomorphy (Manuel et al. 2000) has been shown to be present in other metazoans (Telford 2004a). Also, genome-scale analyses have claimed to refute the Ecdysozoa hypothesis (e.g., Blair et al. 2002, Wolf et al. 2004), but unfortunately these analyses are flawed owing to limited taxon sampling (containing only three or four metazoan taxa total) and the inability to correct adequately for highly derived *Caenorhabditis elegans* sequences (i.e., long-branch issues). Wolf et al. (2004) did try to address the long-branch issue, but their effort was hampered by their limited number of metazoan taxa, three. For additional discussion concerning the problems of these papers, see Telford (2004b).

The Ecdysozoa hypothesis has had perhaps the most far-reaching effects on comparative biology because of the large amount of work on *Drosophila* and *Caenorhabditis*. Before the Ecdysozoa hypothesis, if a common genetic mechanism was found in both flies and round worms, it was presumed to be present throughout Bilateria. With the placement of these two taxa as members of the same lineage, the common machinery may be representative of only the ecdysozoan lineage, thereby limiting the inferences drawn from these model organisms.

Scalidophora and Nematoida

Researchers have limited knowledge of interrelationships within Ecdysozoa because many of the taxa are poorly studied. For example, in early 2004, only 1

GenBank entry existed for Kinorhyncha, 19 for Nematomorpha, and 26 for Priapulida. Nonetheless, there seems to be a general consensus which awaits further conformation. The Priapulida/Kinorhyncha/Loricifera clade should be referred to as Scalidophora (Lemburg 1995, Schmidt-Rhaesa 1996, Ehlers et al. 1996). Nielsen (2001) has used the name Cephalorhyncha for this clade, but previous usage of this term included the Nematomorpha (Malakhov 1980). Within this group, loriciferans are most likely sister taxon to kinorhynchs (Schram 1991, Sørensen et al. 2000, Peterson & Eernisse 2001), but no loriciferan molecular data have been collected because of difficulty in obtaining tissue. When loriciferans are not considered, priapulids and kinorhynchs form a monophyletic clade (Aguinaldo et al. 1997, Aleshin et al. 1998, Giribet et al. 2000, Peterson & Eernisse 2001).

The horsehair worms, Nematomorpha, are allied with Nematoda. This group, termed Nematoida (Schmidt-Rhaesa et al. 1998), has been supported in some analyses (Zrzavy et al. 1998; Giribet et al. 2000, combined data; Garey 2001) but not others (Giribet et al. 2000, SSU only; Peterson & Eernisse 2001). Using combined SSU and LSU data, Mallat et al. (2004) found strong support for Nematoida. The combined group of Scalidophora and Nematoida has been referred to as either the Introverta or Cycloneuralia (Nielsen 2001), but it is refuted by available data (Zrzavy et al. 1998, Peterson & Eernisse 2001, Mallat et al. 2004; but see Giribet et al. 2000). Instead, the Scalidophora is the most basal branch in the Ecdysozoa, with Nematoida and Panarthropoda (Tardigrada, Onychophora, and Arthropoda) as sister clades.

Panarthropoda

Within the Panarthropoda, the placement of tardigrades, onychophorans, and arthropods relative to each other has generated considerable debate. Early mitochondrial SSU data suggested that Onychophora were inside the Arthropoda (Ballard et al. 1992), and papers that reported a Tardigrada/Arthropoda relationship did not include an onychophoran (Garey et al. 1996a, Giribet et al. 1996). The rDNA analyses of Mallat et al. (2004) suggested an onychophoran/tardigrade clade, but the authors were tentative about this result because onychophoran rDNA appears very derived, and nodal support is weak. This result was also recovered by Giribet et al.'s (2000) combined morphology and SSU analyses.

Although of great interest, I do not discuss the relationships within the arthropods in detail. Recent work on the subject has used a variety of molecular markers (rDNA, Hox genes, mtDNA arrangement), and these are discussed elsewhere (Giribet & Ribera 2000, Giribet et al. 2001, Hwang et al. 2001, Cook et al. 2001, Mallett et al. 2004). The most notable change in arthropod phylogeny is the placement of Hexapoda within Crustacea to form Pancrustacea. This hypothesis was convincingly put forth by Regier & Shultz (1997) on the basis of EF-1α and has received considerable support (e.g., Cook et al. 2001, Mallett et al. 2004). Also the previously recognized phylum Pentastomida, parasites on vertebrates, is a derived crustacean clade (Abele et al. 1989, Lavrov et al. 2004).

Dis-Articulating

One reason many were resistant to the Lophotrochozoa and Ecdysozoa hypotheses is that they contradicted the Articulata (Arthropoda and Annelida) as a real clade. Despite the overwhelming evidence that annelids are much more closely related to mollusks than to arthropods (based on morphology as well at least five independent molecular markers), there are still attempts to maintain an Articulata-like clade (e.g., Nielsen 2003). The segmented nature of annelids and arthropods has a long history of being used to unite these taxa (Willmer 1990). Although genes involved in the segmentation program in arthropods (e.g., *Drosophila*) are well studied (for a general review, see Carroll et al. 2001), very little is known about segmentation in annelids. What we do know is that segmentation-related genes (e.g., *engrailed* and *hunchback*) in *Drosophila* appear to be doing something different in annelids (Seaver et al. 2001, Werbrock et al. 2001, but see Prud'homme et al. 2003). Seaver (2003) reviews the possibilities of independent origins of segmentation in annelids, arthropods, and chordates. As she points out, all these taxa are nested within several nonsegmented taxa. Thus, any attempt to infer that an ancestor deep in the bilaterian tree was segmented also required multiple losses of segmentation in numerous different lineages. Given that genetic machinery for segmentation does not appear to be the same, a segmented protostome or bilaterian ancestor is not likely.

Proponents of the Articulata hypothesis assert that segmentation is a very strong morphological character and thus a good phylogenetic indicator. Yet from an objective point of view there is no reason segmentation should be a better indicator of phylogenetic history than molting (in ecdysozoans) or a trochophore larva (in a subset of lophotrochozoans). The Articulata hypothesis also suggests that cleavage patterns are not immutable, as arthropods lack spiral cleavage. All these characters are intricately tied to constraints in functional morphology for which we have little understanding of the selective forces or evolutionary plasticity.

CAMBRIAN EXPLOSION

As should be clear from the previous discussion, understanding the early evolutionary events of animal history is difficult. This situation has not been made any easier by the lack of a substantial fossil record before about 570 Mya. At roughly 543 Mya, the fossil record shows a sudden diversity of animal forms that represent most of the major lineages, with some taxa displaying great diversity and derived body plans (reviewed in Grotzinger et al. 1995, Knoll & Carroll 1999, Erwin & Davidson 2002). This sudden appearance of diversity in the fossil record is called the Cambrian explosion.

Recent finds for the Lower Cambrian Chengjiang and Sirius Passet faunas (approximately 520 Mya) have provided exquisitely preserved samples of animals (Conway Morris & Peel 1995, Chen & Zhou 1997, Bengston & Zhao 1997, Shu et al. 1999). Older fossils from the Ediacaran and Doushantuo formations (up to 570 Mya) predate the Cambrian boundary and show that some animal lineages were present well before the Cambrian (Xiao et al. 1998). For the period before the

Doushantuo formation (which yielded sponges and fossil embryos; Li et al. 1998, Xiao et al. 1998), fossils for animals are generally lacking, despite the presence in the fossil record of several other eukaryotic crown groups dating back to 1200 Mya (Erwin & Davidson 2002). [The report by Seilacher et al. (1998) of ancient worm trace fossils was erroneous (Rai & Gautam 1999, Rasmussen et al. 2002).]

In contrast to the fossil data, several recent works have reported using a molecular clock to date animal diversification (Wray et al. 1996, Nikoh et al. 1997, Ayala et al. 1998, Gu 1998, Bromham et al. 1998, Lynch 1999). These works report that bilaterians diverged between 630 and 1200 Mya. Several workers (e.g., Philippe et al. 1994, Smith & Peterson, 2002; see Graur & Martin 2004 for a particularly colorful discussion) have highlighted some of the problems with using a molecular clock to date deep divergences. In particular, clock studies have been plagued by assumptions of rate homogeneity in nucleotide substitution patterns. Even when nucleotide substitution models can correct for rate variation across different positions in the same gene, available molecular tools are not sophisticated enough (yet) to deal adequately with rate variation across lineages within the same tree. Furthermore, many molecular clock analyses made the mistake of overgeneralizing their results. Specifically, they used only a limited number (3–5) of fossil calibration points, typically within craniates or other deuterostomes, whereas the Cambrian explosion was mainly lophotrochozoan and ecdysozoan in nature and involved numerous lineages (although we must recognize the diversity of echinoderms and presence of early chordates).

On the positive side, studies of molecular clocks have called our attention to a hidden history of early animal evolution. Although the fossil data suggested this hidden history was very short in nature ($\sim$30 My), molecular clock analyses in general suggested a much longer hidden history (up to 700 My; see Erwin & Davidson 2002). This last scenario seems unlikely because fossils of other crown eukaryotes are known from this period, necessitating ad hoc hypotheses to account for the dearth of animal fossils in particular.

The fact that there is any hidden history, regardless of duration, suggests that early animals were very small organisms, likely meiofaunal or small epibenthic dwellers. Such organisms would not fossilize easily (but the Chengjiang, Sirius Passet, and Doushantuo formations have proven the exception) and were probably direct developers. Another argument against a segmented last common bilaterian ancestor is that miniaturization can apparently reduce segmentation (Westheide 1997).

CONCLUSIONS

Our understanding of metazoan phylogeny is far from complete. However, in the past 15 years we have made tremendous progress toward understanding the general framework of animal evolution. Relationships among the most basal lineages of animals are not entirely clear, but poriferans are likely a paraphyletic grade that led to a Cnidarian/Bilaterian clade. Within Bilateria, we have three major clades. Lophotrochozoa is the most diverse clade in terms of body plans, and understanding their internal relationships will take considerable work. How Platyzoa fit into this group also remains to be seen. Ecdysozoa groups nematodes and arthropods and has implications for how we extrapolate information from model systems. Chaetognaths are likely to be basal to Ecdysozoa and Lophotrochozoa. Deuterostomia only contains a limited number of lineages, but the presence of the Ambulacraria (echinoderms and hemichordates) suggests that several chordate features evolved earlier than traditionally believed. At the base of Bilateria, we have the Acoelomorpha and possibly Myxozoa.

Compared with the Hyman-like concept of animal phylogeny, this new view underscores the evolutionary plasticity of embryology and functional morphology. Many of the dogmatic concepts in invertebrate biology must be questioned (e.g., cleavage patterns are immutable, evolution proceeds from simple to complex, segmentation is highly conserved).

Clearly, several relationships still need to be worked out. To promote additional research on metazoan evolution, below are ten provocative hypotheses that are likely to provide considerable insight into animal evolution when tested in a rigorous manner:

1. Placozoans branched off from the main animal lineage before sponges.
2. Mesoderm first arose in ctenophores.
3. Acoelomorphs are secondarily simplified animals.
4. Chaetognatha is the sister to the Lophotrochozoa/Ecdysozoa clade.
5. Spiral cleavers do not form a monophyletic clade exclusive of other cleavage patterns.
6. Genetic mechanisms controlling annelid segmentation are different than in arthropods.
7. The Brachiopoda/Phoronida clade is sister to Mollusca.
8. Panarthopoda evolved from a small infaunal organism.
9. Platyzoa is a derived subclade of Lophotrochozoa.
10. The hidden history of early bilaterian evolution was less than 50 My.

NOTE ADDED IN PROOF

While this work was in press, Anderson et al. (2004) published data from the sodium-potassium ATPase α-subunit gene, which also supports the Lophotrochozoa and Ecdysozoa hypotheses, but did not recover deuterostomes or Arthropods as monophyletic.

ACKNOWLEDGMENTS

I thank all those who over the years have provided fruitful discussions concerning animal evolution, including those at the Lambert residence at Friday Harbor Laboratories. Drew Harvell's support of this review is most appreciated. Comments by T. Struck and J.H. Halanych are most appreciated. C.N. Halanych and J.M.

Halanych freely shared time to support this work. H. Blasczyk assisted with figures. This work was supported by the National Science Foundation (DEB-0075618, EAR-0120646, and IBN-0333843).

The *Annual Review of Ecology, Evolution, and Systematics* is online at
http://ecolsys.annualreviews.org

LITERATURE CITED

Abele LG, Kim W, Felgenhauer BE. 1989. Molecular evidence for inclusion of the phylum Pentastomida in the Crustacea. *Mol. Biol. Evol.* 6:685–91

Aguinaldo AMA, Turbeville JM, Linford LS, Rivera MC, Garey JR, et al. 1997. Evidence for a clade of nematodes, arthropods and other moulting animals. *Nature* 387:489–93

Ahlrichs WH. 1997. Epidermal ultrastructure of *Seison nebaliae* and *Seison annulatus*, and a comparison of epidermal structures within the Gnathifera. *Zoomorphology* 117:41–48

Aleshin VV, Milyutina IA, Kedrova OS, Vladychenskaya NS, Petrov NB. 1998. Phylogeny of *Nematoda* and *Cephalorhyncha* derived from 18S rDNA. *J. Mol. Evol.* 47:597–605

Aleshin VV, Petrov NB. 2002. Molecular evidence of regression in evolution of Metazoa. *Zh. Obshch. Biol.* 63:195–208

Amano S, Hori I. 1992. Metamorphosis of calcareous sponges. I. Ultrastructure of free-swimming larvae. *Invertebr. Reprod. Dev.* 21:81–90

Anderson FE, Cordoba AJ, Thollesson M. 2004. Bilaterian phylogeny based on analyses of a region of the sodium-potassium ATPase α-subunit gene. *J. Mol. Evol.* 58:252–68

Ayala FJ, Rzhetsky A, Ayala FJ. 1998. Origin of the metazoan phyla: Molecular clocks confirm paleontological estimates. *Proc. Natl. Acad. Sci. USA* 95:606–11

Balavoine G. 1997. The early emergence of platyhelminths is contradicted by the agreement between 18S rRNA and Hox genes data. *C. R. Acad. Sci.* 320:83–94

Balavoine G, Adoutte A. 1998. One or three Cambrian radiations? *Science* 280:397–98

Balavoine G, de Rosa R, Adouette A. 2002. Hox clusters and bilaterian phylogeny. *Mol. Phylogenet. Evol.* 24:366–73

Balavoine G, Telford MJ. 1995. Identification of planarian homeobox sequences indicates the antiquity of most Hox/homeotic gene subclasses. *Proc. Natl. Acad. Sci. USA* 92:7227–31

Ballard JWO, Olsen GJ, Faith DP, Odgers WA, Rowell DM, Atkinson PW. 1992. Evidence from 12S ribosomal RNA sequences that onychophorans are modified arthropods. *Science* 258:1345–48

Bather FA. 1913. Caradocian Cystidea from Girvan. *Trans. R. Soc. Edinburgh* 49:359–529

Bengston S, Zhao Y. 1997. Fossilized metazoan embryos from the earliest Cambrian. *Science* 277:1645–48

Bergstrom J. 1985. Metazoan evolution—a new model. *Zool. Scr.* 15:189–200

Berney C, Pawlowski J, Zaninetti L. 2000. Elongation factor 1-α sequences do not support an early divergence of the Acoela. *Mol. Biol. Evol.* 17:1032–39

Black MB, Halanych KM, Maas PAY, Hoeh WR, Hashimoto J, et al. 1997. Molecular systematics of vestimentiferan tubeworms from hydrothermal vents and cold-water seeps. *Mar. Biol.* 130:141–49

Blair JE, Ikeo K, Gojobori T, Hedges SB. 2002. The evolutionary position of nematodes. *BMC Evol. Biol.* 2:7

Bleidorn C, Vogt L, Bartolomaeus T. 2003. A contribution to sedentary polychaete phylogeny using 18S rRNA sequence data. *J. Zool. Syst. Evol. Res.* 41:186–95

Boore JL, Brown WM. 2000. Mitochondrial genomes of *Galathealinum, Helobdella*, and *Platynereis*: Sequence and gene arrangement comparisons indicate that Pogonophora is not a phylum and annelida and arthropoda are not sister taxa. *Mol. Biol. Evol.* 17:87–106

Boore JL, Staton JL. 2002. The mitochondrial genome of the Sipunculid *Phascolopsis gouldii* supports its association with Annelida rather than Mollusca. *Mol. Biol. Evol.* 19:127–37

Borchiellini C, Manuel M, Alivon E, Boury-Esnault N, Vacelet J, Le Parco Y. 2001. Sponge paraphyly and the origin of Metazoa. *J. Evol. Biol.* 14:171–79

Bourlat S, Nielsen C, Lockyer A, Littlewood DT, Telford M. 2003. *Xenoturbella* is a deuterostome that eats molluscs. *Nature* 424:925–28

Bridge D, Cunningham CW, DeSalle R, Buss LW. 1995. Class-level relationships in the phylum Cnidaria: molecular and morphological evidence. *Mol. Biol. Evol.* 12:679–89

Bromham LD, Rambault A, Fortey R, Cooper A, Penny D. 1998. Testing the Cambrian explosion hypothesis by using a molecular dating technique. *Proc. Natl. Acad. Sci. USA* 95:12386–89

Brooke NM, Holland PW. 2003. The evolution of multicellularity and early animal genomes. *Curr. Opin. Genet. Dev.* 13:599–603

Brusca RC, Brusca GJ. 2003. *Invertebrates*. Sunderland, MA: Sinauer. 936 pp. 2nd ed.

Cameron CB, Garey JR, Swalla BJ. 2000. Evolution of the chordate body plan: new insights from phylogenetic analyses of deuterostome phyla. *Proc. Natl. Acad. Sci. USA* 97:4469–74

Carroll SB, Grenier JK, Weatherbee SD. 2001. *From DNA to Diversity*. London: Blackwell. 214 pp.

Caullery M. 1914. Sur les Siboglinidae, type nouveau d'invertébrés recueilli par l'expédition du Siboga. *C.R. Acad. Sci.* 158:2014–17

Cavalier-Smith T. 1998. A revised six-kingdom system of life. *Biol. Rev.* 73:203–66

Cavalier-Smith T, Allsopp M, Chao E, Boury-Esnault N, Vacelet J. 1996. Sponge phylogeny, animal monophyly, and the origin of the nervous system: 18S rRNA evidence. *Can. J. Zool.* 74:2031–45

Cavalier-Smith T, Chao EE. 2003. Phylogeny of Choanozoa, Apusozoa, and other Protozoa and early eukaryote megaevolution. *J. Mol. Evol.* 56:540–63

Chen J-Y, Zhou G-Q. 1997. Biology of the Chengjiang fauna. In *The Cambrian Explosion and the Fossil Record*, ed. J-Y Chen, Y-N Chen, H Van Iten, 10:11–116. Taiwan: Bull. Natl. Mus. Nat. Sci.

Cohen BL. 2000. Monophyly of brachiopods and phoronids: reconciliation of molecular evidence with Linnaean classification (the subphylum Phoroniformea nov.). *Proc. R. Soc. London* 267:225–31

Cohen BL, Gawthrop A, Cavaliersmith T. 1998. Molecular phylogeny of brachiopods and phoronids based on nuclear-encoded small subunit ribosomal RNA gene sequences. *Philos. Trans. R. Soc. London Ser. B* 353:2039–61

Collins AG. 1998. Evaluating multiple alternative hypotheses for the origin of Bilateria: An analysis of 18S rRNA molecular evidence. *Proc. Natl. Acad. Sci. USA* 95:15458–63

Collins AG. 2000. Towards understanding the phylogenetic history of Hydrozoa: hypothesis testing with 18S gene sequence data. *Sci. Mar.* 4:1–22

Collins AG. 2002. Phylogeny of Medusozoa and the evolution of cnidarian life cycles. *J. Evol. Biol.* 15:418–32

Conway Morris S, Peel JS. 1995. Articulated halkieriids from the lower Cambrian of north Greenland and their role in early protostome evolution. *Philos. Trans. R. Soc. London Ser. B* 347:305–58

Cook CE, Smith ML, Telford MJ, Bastianello A, Akam M. 2001. Hox genes and the phylogeny of the arthropods. *Curr. Biol.* 11:759–63

Degnan BM, Degnan SM, Giusti A, Morse DE. 1995. A hox/hom homeobox gene in sponges. *Gene* 155:175–77

de Rosa R. 2001. Molecular data indicate the protostome affinity of brachiopods. *Syst. Biol.* 50:848–59 de Rosa R, Grenier JK, Andreeva T, Cook CE, Adoutte A, et al. 1999. HOX genes in brachiopods and priapulids and protostome evolution. *Nature* 399:772–76

Eeckhaut I, Fievez L, Muller MC. 2003. Larval development of *Myzostoma cirriferum* (Myzostomida). *J. Morphol.* 258:269–83

Eeckhaut I, McHugh D, Mardulyn P, Tiedemann R, Monteyne D, et al. 2000. Myzostomida: a link between trochozoans and flatworms? *Proc. R. Soc. London* 267:1383–92

Eernisse DJ. 1997. Arthropod and annelid relationships re-examined. In *Arthropod Relationships*, ed. RA Fortey, RH Thomas, pp. 43–56. London: Chapman & Hall

Eernisse DJ, Albert JS, Anderson FE. 1992. Annelida and Arthropoda are not sister taxa: a phylogenetic analysis of spiralian metazoan phylogeny. *Syst. Biol.* 41:305–30

Eernisse DJ, Peterson KJ. 2004. The history of animals. In *Assembling the Tree of Life*, ed. J Cracraft, MJ Donoghue, pp. 197–208. New York: Oxford Univ. Press

Ehlers U, Ahlrichs W, Lemburg C, Schmidt-Rhaesa A. 1996. Phylogenetic systematization of the Nemathelminthes (Aschelminthes). *Verh. Dtsch. Zool. Ges.* 89:8

Ender A, Schierwater B. 2003. Placozoa are not derived cnidarians: evidence from molecular morphology. *Mol. Biol. Evol.* 20:130–34

Erber A, Riemer D, Bovenschulte M, Weber K. 1998. Molecular phylogeny of metazoan intermediate filament proteins. *J. Mol. Evol.* 47:751–62

Erwin DH, Davidson EH. 2002. The last common bilterian ancestor. *Development* 129:3021–32

Felsenstein J. 1988. Phylogenies from molecular sequences: inference and reliability. *Annu. Rev. Genet.* 22:521–65

Ferrier D, Holland PWH. 2001. Sipunculan ParaHox genes. *Evol. Dev.* 3:263–70

Field KG, Olsen GJ, Lane DJ, Giovannoni SJ, Ghiselin MT, et al. 1988. Molecular phylogeny of the animal kingdom. *Science* 239:748–53

Finnerty JR. 1998. Homeoboxes in sea anemones and other nonbilaterian animals: implications for the evolution of the Hox cluster and zootype. *Curr. Top. Dev. Biol.* 40:211–54

Finnerty JR, Martindale MQ. 1998. The evolution of the Hox cluster: insights from outgroups. *Curr. Opin. Genet. Dev.* 8:681–87

Freeman G, Martindale MQ. 2002. The origin of mesoderm in phoronids. *Dev. Biol.* 252:301–11

Funch P, Kristensen RM. 1995. Cycliophora is a new phylum with affinities to Entoprocta and Ectoprocta. *Nature* 378:711–14

Furlong RF, Holland PW. 2002. Bayesian phylogenetic analysis supports monophyly of Ambulacraria and of cyclostomes. *Zool. Sci.* 19:593–99

Garey JR. 2001. Ecdysozoa: the relationship between Cycloneuralia and Panarthropoda. *Zool. Anz.* 240:321–30

Garey JR, Krotec M, Nelson DR, Brooks J. 1996a. Molecular analysis supports a tardigrade-arthropod association. *Invert. Biol.* 115:79–88

Garey JR, Near TJ, Nonnemacher MR, Nadler SA. 1996b. Molecular evidence for Acanthocephala as a subtaxon of Rotifera. *J. Mol. Evol.* 43:287–92

Garey JR, Schmidt-Rhaesa A. 1998. The essential role of "minor" phyla in molecular studies of animal evolution. *Am. Zool.* 38:907–17

Giribet G, Carranza S, Baguña J, Riutort M, Ribera C. 1996. First molecular evidence for the existence of a Tardigrada + Arthropoda clade. *Mol. Biol. Evol.* 13:76–84

Giribet G, Distel DL, Polz M, Sterrer W, Wheeler WC. 2000. Triploblastic relationships with emphasis on the acoelomates and the position of Gnathostomulida, Cycliophora, Plathelminthes, and Chaetognatha: a combined approach of 18S rDNA sequences and morphology. *Syst. Biol.* 49:539–62

Giribet G, Edgecombe GD, Wheeler WC. 2001. Arthropod phylogeny based on eight molecular loci and morphology. *Nature* 413:157–61

Giribet G, Ribera C. 2000. A review of Arthropod phylogeny: new data based on ribosomal DNA sequences and direct character optimization. *Cladistics* 16:204–31

Giribet G, Sørensen MV, Funch P, Kristensen RM, Sterrer W. 2004. Investigations into the phylogenetic position of Micrognathozoa using four molecular loci. *Cladistics* 20:1–13

Graur D, Martin W. 2004. Reading the entrails of chickens: molecular timescales of evolution and the illusion of precision. *Trends Genet.* 20:80–86

Grotzinger JP, Bowring SA, Saylor BZ, Kaufman AJ. 1995. Biostratigraphic and geochronologic constraints on early animal evolution. *Science* 270:598–604

Gu X. 1998. Early metazoan divergence was about 830 million years ago. *J. Mol. Evol.* 47:369–71

Haase A, Stern M, Wachtler K, Bicker G. 2001. A tissue-specific marker of Ecdysozoa. *Dev. Genes Evol.* 211:428–33

Halanych KM. 1995. The phylogenetic position of the pterobranch hemichordates based on 18S rDNA sequence data. *Mol. Phylogeny Evol.* 4:72–76

Halanych KM. 1996a. Convergence in the feeding apparatuses of lophophorates and pterobranch hemichordates revealed by 18S rDNA: an interpretation. *Biol. Bull.* 190:1–5

Halanych KM. 1996b. Testing hypotheses of chaetognath origins: long branches revealed by 18S ribosomal DNA. *Syst. Biol.* 45:223–46

Halanych KM, Bacheller JD, Aguinaldo AMA, Liva SM, Hillis DM, Lake JA. 1995. Evidence from 18S ribosomal DNA that the lophophorates are protostome animals. *Science* 267:1641–43

Halanych KM, Dahlgren TG, McHugh D. 2002. Unsegmented annelids? Possible origins of four lophotrochozoan worm taxa. *Integr. Comp. Biol.* 42:678–84

Halanych KM, Feldman RA, Vrijenhoek RC. 2001. Molecular evidence that *Sclerolinum brattstromi* is closely related to vestimentiferans, not frenulate pogonophorans (Siboglinidae, Annelida). *Biol. Bull.* 201:65–75

Halanych KM, Lutz RA, Vrijenhoek RC. 1998. Evolutionary origins and age of vestimentiferan tube-worms. *Cah. Biol. Mar.* 39:355–58

Halanych KM, Passamaneck Y. 2001. A brief review of metazoan phylogeny and future prospects in Hox-research. *Am. Zool.* 41:629–39

Hanelt B, Van Schyndel D, Adema CM, Lewis LA, Loker ES. 1996. The phylogenetic position of *Rhopalura ophiocomae* (Orthonectida) based on 18S ribosomal DNA sequence analysis. *Mol. Biol. Evol.* 13:1187–91

Harbison GR. 1985. On the classification and evolution of the Ctenophora. In *The Origins and Relationships of Lower Invertebrates*, ed. S Conway Morris, JD George, R Gibson, HM Platt, pp. 78–100. Oxford: Oxford Univ. Press

Haszprunar G. 2000. Is the Aplacophora monophyletic? A cladistic point of view. *Am. Malacol. Bull.* 15:115–30

Helfenbein KG, Boore JL. 2004. The mitochondrial genome of *Phoronis architecta*—comparisons demonstrate that phoronids are lophotrochozoan protostomes. *Mol. Biol. Evol.* 21:153–57

Helfenbein KG, Brown WM, Boore JL. 2001. The complete mitochondrial genome of the articulate brachiopod *Terebratalia transversa*. *Mol. Biol. Evol.* 18:1734–44

Hessling R. 2002. Metameric organisation of the nervous system in developmental stages of *Urechis caupo* (Echiura) and its phylogenetic implications. *Zoomorphology* 121:221–34

Hwang UW, Friedrich M, Tautz D, Park CJ, Kim W. 2001. Mitochondrial protein phylogeny joins myriapods with chelicerates. *Nature* 413:154–57

Hyman LH. 1940. *The Invertebrates: Protozoa through Ctenophora*. New York: McGraw-Hill

Jakob W, Sagasser S, Dellaporta S, Holland P, Kuhn K, Schierwater B. 2004. The Trox-2 Hox/ParaHox gene of Trichoplax (Placozoa) marks an epithelial boundary. *Dev. Genes Evol.* 214:170–75

Jefferies RPS. 1986. *The Ancestry of the Vertebrates*. Cambridge, UK: Br. Mus. Nat. Hist.

Jenner RA. 1999. Metazoan phylogeny as a tool in evolutionary biology: current problems and discrepancies in application. *Belg. J. Zool.* 129:245–62

Jenner RA. 2002. Boolean logic and character state identity: pitfalls of character coding in metazoan cladistics. *Contrib. Zool.* 71:67–91

Jenner RA. 2004. Towards a phylogeny of the Metazoa: evaluating alternative phylogenetic positions of Platyhelminthes, Nemertea, and Gnathostomulida, with a critical reappraisal of cladistic characters. *Contrib. Zool.* 73:3–163

Katayama T, Wada H, Furuya H, Satoh N, Yamamoto M. 1995. Phylogenetic position of the dicyemid Mesozoa inferred from 18S rDNA sequences. *Biol. Bull.* 189:81–90

Kim J, Kim W, Cunningham CW. 1999. A new perspective on lower metazoan relationships form 18S rDNA sequences. *Mol. Biol. Evol.* 16:423–27

Kmita-Cunisse M, Loosli F, Bierne J, Gehring WJ. 1998. Homeobox genes in the ribbonworm *Lineus sanguineus*: Evolutionary implications. *Proc. Natl. Acad. Sci. USA* 95:3030–35

Knoll A, Carroll SB. 1999. Early animal evolution: Emerging views from comparative biology and geology. *Science* 284:2129–37

Kobayashi M, Furuya H, Holland PWH. 1999. Dicyemids are higher animals. *Nature* 401:762

Kojima S. 1998. Paraphyletic status of Polychaeta suggested by phylogenetic analysis based on the amino acid sequences of elongation factor-1-alpha. *Mol. Phylogenet. Evol.* 9:255–61

Kristensen RM, Funch P. 2000. Micrognathozoa: a new class with complicated jaws like those of Rotifera and Gnathostomulida. *J. Morphol.* 246:1–49

Lake JA. 1990. Origin of Metazoa. *Proc. Natl. Acad. Sci. USA* 87:763–66

Lavrov DV, Brown WM, Boore JL. 2004. Phylogenetic position of the Pentastomida and (pan)crustacean relationships. *Proc. R. Soc. London Ser. B Biol. Sci.* 271:537–44

Lemburg C. 1995. Ultrastructure of the introvert and associated structures of the larvae of *Halicryptus spinulosus* (Priapulida). *Zoomorphology* 115:11–29

Li C-W, Chen J-Y, Hua T-E. 1998. Precambrian sponges with cellular structures. *Science* 279:879–82

Littlewood DTJ, Olson PD, Telford MJ, Herniou EA, Riutort M. 2001. Elongation factor 1-α sequences alone do not assist in resolving the position of the Acoela within the Metazoa. *Mol. Biol. Evol.* 18:437–42

Littlewood DTJ, Telford MJ, Clough KA, Rohde K. 1998. Gnathostomulida—an enigmatic metazoan phylum from both morphological and molecular perspectives. *Mol. Phylogenet. Evol.* 9:72–79

Lowe CJ, Wu M, Salic A, Evans L, Lander E, et al. 2003. Anteroposterior patterning in hemichordates and the origins of the chordate nervous system. *Cell* 113:853–65

Lynch M. 1999. The age and relationships of the major animal phyla. *Evolution* 53:319–25

Mackey LY, Winnepennickx B, De Wachter R, Beckeljau T, Emschermann P, Garey JR. 1996. 18S rRNA suggests that Entoprocta are protostomes, unrelated to Ectoprocta. *J. Mol. Evol.* 42:552–59

Malakhov VV. 1980. Cephalorhyncha, a new type of animal kingdom uniting Priapulida, Kinorhyncha, Gordiacea, and a system of aschelminthes worms. *Zool. Zh.* 59:485–99

Mallatt J, Winchell CJ. 2002. Testing the new animal phylogeny: first use of combined large-subunit and small-subunit rRNA gene sequences to classify the protostomes. *Mol. Biol. Evol.* 19:289–301

Mallatt JM, Garey JR, Shultz JW. 2004. Ecdysozoan phylogeny and Bayesian inference: first use of nearly complete 28S and 18S rRNA gene sequences to classify the arthropods and their kin. *Mol. Phylogenet. Evol.* 31:178–91

Manuel M, Kruse M, Muller WEG, Parco YL. 2000. The comparison of β-thymosin homologues among Metazoa supports an arthropod-nematode clade. *J. Mol. Evol.* 51:378–81

Manuel M, Le Parco Y. 2000. Homeobox gene diversification in the calcareous sponge, *Sycon raphanus*. *Mol. Phylogenet. Evol.* 17:97–107

Martindale MQ, Finnerty JR, Henry JQ. 2002. The Radiata and the evolutionary origins of the bilaterian body plan. *Mol. Phylogenet. Evol.* 24:358–65

Martindale MQ, Henry JQ. 1999. Intracellular fate mapping in a basal metazoan, the ctenophore *Mnemiopsis leidyi*, reveals the origins of mesoderm and the existence of indeterminate cell lineages. *Dev. Biol.* 214:243–57

Martinelli C, Spring J. 2003. Distinct expression patterns of the two T-box homologues Brachyury and Tbx2/3 in the placozoan *Trichoplax adhaerens*. *Dev. Genes Evol.* 213:492–99

McHugh D. 1997. Molecular evidence that echiurans and pogonophorans are derived annelids. *Proc. Natl. Acad. Sci. USA* 94:8006–9

McHugh D. 2000. Molecular phylogeny of the Annelida. *Can. J. Zool.* 78:1873–84

Medina M, Collins AG. 2003. The role of molecules in understanding molluscan evolution. In *Molecular Systematics and Phylogeography of Mollusks*, ed. C Lydeard, DR Lindberg, pp. 14–44. Washington, DC: Smithsonian Inst.

Medina M, Collins AG, Silberman JD, Sogin ML. 2001. Evaluating hypotheses of basal animal phylogeny using complete sequences of large and small subunit rRNA. *Proc. Natl. Acad. Sci. USA* 98:9707–12

Metschnikoff VE. 1881. Über die systematische Stellung von *Balanoglossus*. *Zool. Anz.* 4:139–57

Monteiro AS, Okamura B, Holland PWH. 2002. Orphan worm finds a home: *Buddenbrockia* is a Myxozoan. *Mol. Biol. Evol.* 19:968–71

Müller MC, Westheide W. 2000. Structure of the nervous system of *Myzostoma cirriferum* (Annelida) as revealed by immunohistochemistry and cLSM analyses. *J. Morphol.* 245:87–98

Muller P, Seipel K, Yanze N, Reber-Muller S, Streitwolf-Engel R, et al. 2003. Evolutionary aspects of developmentally regulated helix-loop-helix transcription factors in striated muscle of jellyfish. *Dev. Biol.* 255:216–29

Nielsen C. 1985. Animal phylogeny in the light of the trochaea theory. *Biol. J. Linn. Soc.* 25:243–99

Nielsen C. 1987. Structure and function of metazoan ciliary bands and their phylogenetic significance. *Acta Zool.* 68:205–62

Nielsen C. 2001. *Animal Evolution: Interrelationships of the Living Phyla.* Oxford: Oxford Univ. Press. 561 pp. 2nd ed.

Nielsen C. 2003. Proposing a solution to the Articulata-Ecdysozoa controversy. *Zool. Scr.* 32:475–82

Nielsen C, Scharf N, Eibye-Jacobsen D. 1996. Cladistic analyses of the animal kingdom. *Biol. J. Linn. Soc.* 57:385–410

Nikoh N, Iwabe N, Kuma K, Ohno M, Sugiyama T, et al. 1997. An estimate of divergence time of Parazoa and Eumetazoa and that of Cephalochordata and Vertebrata by aldolase and triose phosphate isomerase clocks. *J. Mol. Evol.* 45:97–106

Norén M, Jondelius U. 1997. *Xenoturbella*'s molluscan relatives. *Nature* 390:31–32

Papillon D, Perez Y, Fasano L, Le Parco Y, Caubit X. 2003. Hox gene survey in the chaetognath *Spadella cephaloptera*: evolutionary implications. *Dev. Genes Evol.* 213:142–48

Passamaneck YJ, Halanych KM. 2004. Evidence from Hox genes that bryozoans are lophotrochozoans. *Evol. Dev.* 6:275–81

Passamaneck YJ, Schander C, Halanych KM. 2004. Investigation of molluscan phylogeny using large-subunit and small-subunit nuclear rRNA sequences. *Mol. Phylogenet. Evol.* 32:25–38

Pawlowski J, Montoya-Burgos JI, Fahrni JF, Wuest J, Zaninetti L. 1996. Origin of the Mesozoa inferred from 18S rRNA gene sequences. *Mol. Biol. Evol.* 13:1128–32

Peterson KJ, Eernisse DJ. 2001. Animal phylogeny and the ancestry of bilaterians: inferences from morphology and 18S rDNA gene sequences. *Evol. Dev.* 3:170–205

Philippe H, Chenuil A, Adouette A. 1994. Can the Cambrian explosion be inferred through molecular phylogeny? *Development* 1994(Suppl.):15–25

Podar M, Haddock SH, Sogin ML, Harbison GR. 2001. A molecular phylogenetic framework for the phylum Ctenophora using 18S rRNA genes. *Mol. Phylogenet. Evol.* 21:218–30

Prud'homme B, de Rosa R, Arendt D, Julien JF, Pajaziti R, et al. 2003. Arthropod-like expression patterns of *engrailed* and *wingless* in the annelid *Platynereis dumerilii* suggest a role in segment formation. *Curr. Biol.* 13:1876–81

Rai V, Gautam R. 1999. Evaluating evidence of ancient animals. *Science* 284:A1235

Rasmussen B. 2002. Discoidal impressions and trace-like fossils more than 1200 million years old. *Science* 296:1112–15

Regier JC, Shultz JW. 1997. Molecular phylogeny of the major arthropod groups indicates polyphyly of crustaceans and a new hypothesis for the origin of hexapods. *Mol. Biol. Evol.* 14:902–13

Rota E, Martin P, Erséus C. 2001. Soil-dwelling polychaetes: enigmatic as ever? Some hints on their phylogenetic relationship as suggested by a maximum parsimony analysis of 18S rRNA gene sequences. *Contrib. Zool.* 70:127–38

Rouse GW. 2001. A cladistic analysis of Siboglinidae Caullery, 1914 (Polychaeta, Annelida): formerly the phyla Pogonophora and Vestimentifera. *Zool. J. Linn. Soc.* 132:55–80

Rouse GW, Fauchald K. 1997. Cladistics and polychaetes. *Zool. Scr.* 26:139–204

Ruiz-Trillo I, Paps J, Loukota M, Ribera C, Jondelius U, et al. 2002. A phylogenetic analysis of myosin heavy chain type II sequences corroborates that Acoela and Nemertodermatida are basal bilaterians. *Proc. Natl. Acad. Sci. USA* 99:11246–51

Ruiz-Trillo I, Riutort M, Littlewood TJ, Herniou EA, Baguña J. 1999. Acoel flatworms: earliest extant bilaterian metazoans, not members of platyhelminthes. *Science* 283:1919–23

Ruppert EE, Fox RS, Barnes RD. 2004. *Invertebrate Zoology, a Functional Evolutionary Approach.* Belmont, CA: Brooks/Cole-Thomson Learn. 963 pp. 7th ed.

Scheltema AH. 1993. Aplacophora as progenetic aculiferans and the coelomate origin of mollusks as the sister taxon of Sipuncula. *Biol. Bull.* 184:57–78

Schlegel M, Lom J, Stechmann A, Bernhard D, Leipe D, et al. 1996. Phylogenetic analysis of complete small subunit ribosomal RNA coding region of *Myxidium lieberkuehni*: evidence that Myxozoa are Metazoa and related to the Bilateria. *Arch. Protistenkd.* 147:1–9

Schmidt-Rhaesa A. 1996. The nervous system of *Nectonema munidae* and *Gordius aquaticus*, with implications for the ground pattern of Nematomorpha. *Zoomorphology* 116:133–42

Schmidt-Rhaesa A, Bartolomaeus T, Lemburg C, Ehlers U, Garey JR. 1998. The position of the Arthropoda in the phylogenetic system. *J. Morphol.* 238:263–85

Schram FR. 1991. Cladistic analysis of metazoan phyla and the placement of fossil problematica. In *The Early Evolution of Metazoa and the Significance of Problematic Taxa*, ed. AM Simonetta, S Conway Morris, pp. 35–46. New York: Cambridge Univ. Press

Seaver EC. 2003. Segmentation: mono- or polyphyletic? *Int. J. Dev. Biol.* 47:583–95

Seaver EC, Paulson DA, Irvin SQ, Martindale MQ. 2001. The spatial and temporal expression of Ch-en, the *engrailed* gene in the polychaete *Chaetopterus*, does not support a role in body axis segmentation. *Dev. Biol.* 236:195–209

Seilacher A, Bose PK, Pflüger F. 1998. Triploblastic animals more than 1 billion years ago: trace fossil evidence from India. *Science* 282:80–83

Seimiya M, Ishiguro H, Miura K, Watanabe Y, Kurosawa Y. 1994. Homeobox-containing genes in the most primitive metazoa, the sponges. *Eur. J. Biochem.* 221:219–25

Shu D-G, Chen L, Han J, Zhang X-L. 2001a. An early Cambrian tunicate from China. *Nature* 411:472–73

Shu D-G, Luo H-L, Morris SC, Zhang X-L, Hu S-x, et al. 1999. Lower Cambrian vertebrates from south China. *Nature* 402:42–46

Shu D-G, Morris SC, Han J, Chen L, Zhang XL, et al. 2001b. Primitive deuterostomes from the Chengjiang Lagerstatte (Lower Cambrian, China). *Nature* 414:419–24

Shu D-G, Morris SC, Zhang ZF, Liu JN, Han J, et al. 2003. A new species of yunnanozoan with implications for deuterostome evolution. *Science* 299:1380–84

Siddall ME, Martin DS, Bridge D, Desser SS, Cone DK. 1995. The demise of a phylum of protists: phylogeny of Myxozoa and other parasitic Cnidaria. *J. Parasitol.* 81:961–67

Smith AB, Peterson KJ. 2002. Dating the time of origin of major clades: molecular clocks and the fossil record. *Annu. Rev. Earth Planet. Sci.* 30:65–88

Smothers JF, von Dohlen CD, Smith LHJ, Spall RD. 1994. Molecular evidence that the myxozoan protists are metazoans. *Science* 265:1719–21

Sørensen MV, Funch P, Willerslev E, Hansen AJ, Olesen J. 2000. On the phylogeny of the metazoa in the light of Cycliophora and Micrognathozoa. *Zool. Anz.* 239:297–318

Southward EC. 1988. Development of the gut and segmentation of newly settled stages of *Ridgeia* (Vestimentifera): implications for relationship between Vestimentifera and Pogonophora. *J. Mar. Biol. Assoc. UK* 68:465–87

Spring J, Yanze N, Josch C, Middel AM, Winninger B, Schmid V. 2002. Conservation of Brachyury, Mef2, and Snail in the myogenic lineage of jellyfish: a connection to the mesoderm of bilateria. *Dev. Biol.* 244:372–84

Stechmann A, Schlegel M. 1999. Analysis of the complete mitochondrial DNA sequence of the brachiopod *Terebratulina retusa* places Brachiopoda within the protostomes. *Proc. R. Soc. London Ser. B* 266:2043–52

Steiner G, Dreyer H. 2003. Molecular phylogeny of Scaphopoda (Mollusca) inferred from 18S rDNA sequences: support for a Scaphopoda-Cephalopoda clade. *Zool. Scr.* 32:343–56

Struck TH, Westheide W, Purschke G. 2002. Progenesis in Eunicida ("Polychaeta," Annelida)—separate evolutionary events? Evidence from molecular data. *Mol. Phylogenet. Evol.* 25:190–99

Swalla BJ, Cameron CB, Corley LS, Garey JR. 2000. Urochordates are monophyletic within the deuterostomes. *Syst. Biol.* 49:52–64

Telford MJ. 2004a. The multimeric β-thymosin found in nematodes and arthropods is not a synapomorphy of the Ecdysozoa. *Evol. Dev.* 6:90–94

Telford MJ. 2004b. Animal phylogeny: back to the coelomata? *Curr Biol* 14: R274–76

Telford MJ, Herniou EA, Russell RB, Littlewood DTJ. 2000. Changes in mitochondrial genetic codes as phylogenetic characters: two examples from the flatworms. *Proc. Natl. Acad. Sci. USA* 97:11359–64

Telford MJ, Holland PWH. 1993. The phylogenetic affinities of the chaetognaths: a molecular analysis. *Mol. Biol. Evol.* 10:660–76

Telford MJ, Lockyer AE, Cartwright-Finch C, Littlewood DT. 2003. Combined large and small subunit ribosomal RNA phylogenies support a basal position of the acoelomorph flatworms. *Proc. R. Soc. London Ser. B* 270:1077–83

Turbeville JM, Field KG, Raff RA. 1992. Phylogenetic position of phylum Nermertini, inferred from 18S rRNA sequences: molecular data as a test of morphological character homology. *Mol. Biol. Evol.* 9:235–49

Turbeville JM, Schulz JR, Raff RA. 1994. Deuterostome phylogeny and the sister group of the chordates: evidence from molecules and morphology. *Mol. Biol. Evol.* 11:648–55

Valentine JW. 1997. Cleavage patterns and the topology of the Metazoan tree of life. *Proc. Natl. Acad. Sci. USA* 94:8001–5

van der Land J, Nørrevang A. 1977. The systematic position of *Lamellibrachia* (Annelida, Vestimentifera). *Z. Zool. Syst. Evol.* 1975:85–101

Wada H, Satoh N. 1994. Details of the evolutionary history from invertebrates to vertebrates, as deduced from the sequences of 18S rDNA. *Proc. Natl. Acad. Sci. USA* 91:1801–4

Werbrock AH, Meiklejohn DA, Sainz A, Iwasa JH, Savage RM. 2001. A polychaete hunchback ortholog. *Dev. Biol.* 235:476–88

Westheide W. 1997. The direction of evolution within the Polychaeta. *J. Nat. Hist.* 31:1–15

Westheide W, Rieger R. 1996. *Spezielle Zoologie. Teil 1: Einzeller und Wirbellose Tiere.* Stuttgart: Fischer

Wheeler WC. 1990. Nucleic acid sequence phylogeny and random outgroups. *Cladistics* 6:363–68

Willmer P. 1990. *Invertebrate Relationships, Patterns in Animal Evolution.* New York: Cambridge Univ. Press. 400 pp.

Winchell CJ, Sullivan J, Cameron CB, Swalla BJ, Mallatt J. 2002. Evaluating hypotheses of deuterostome phylogeny and chordate evolution with new LSU and SSU ribosomal DNA data. *Mol. Biol. Evol.* 19:762–76

Winnepenninckx B, Backeljau T, Mackey LY, Brooks JM, De Wachter R, et al. 1995. 18S rRNA data indicate that Aschelminthes are polyphyletic in origin and consist of at least three distinct clades. *Mol. Biol. Evol.* 12:1132–37

Winnepenninckx BMH, Backeljau T, Kristensen RM. 1998. Relations of the new phylum Cycliophora. *Nature* 393:636–38

Wolf YI, Rogozin IB, Koonin EV. 2004. Coelomata and not Ecdysozoa: evidence from genome-wide phylogenetic analysis. *Genome Res.* 14:29–36

Wray GA, Levinton JS, Shapiro LH. 1996. Molecular evidence for deep Precambrian divergences among metazoan phyla. *Science* 274:568–73

Xiao S, Zhang Y, Knoll AH. 1998. Three-dimensional preservation of algae and animal embryos in a Neoproterozoic phosphorite. *Nature* 391:553–58

Zimmer RL. 1973. Morphological and developmental affinities of the lophophorates. In *Living and Fossil Bryozoa*, ed. GP Larwood, pp. 593–99. New York: Academic

Zimmer RL. 1997. Phoronids, brachiopods, and bryozoans, the lophophorates. In *Embryology, Constructing the Organism*, ed. SF Gilbert, AM Raunio, pp. 279–305. Sunderland, MA: Sinauer

Zrzavy J, Hypsa V, Tietz DF. 2001. Myzostomida are not annelids: molecular and morphological support for a clade of animals with anterior sperm flagella. *Cladistics* 17:170–98

Zrzavy J, Milhulka S, Kepka P, Bezdek A, Tietz DF. 1998. Phylogeny of the Metazoa based on morphological and 18S ribosomal DNA evidence. *Cladistics* 14:249–85

NOTE ADDED IN PROOF

Finnerty JR, Pang K, Burton P, Paulson D. Martindale MQ. 2004. Origins of bilateral symmetry; Hox and dpp expression in a sea anemone. *Science* 304:1335–37

Helfenbein KG, Fourcade HM, Vanjani RG, Boore JL, 2004. The mitochondrial genome of *Paraspadella gotoi* is highly reduced and reveals that chaetognaths are a sister group to protosotomes. *Proc. Natl. Acad. Sci. USA.* 101:10639–43

Wikramanayake AH, Hong M, Lee PN, Pang K, Byrum CA, et al. 2003. An ancient role for nuclear beta-catenin in the evolution of axial polarity and germ layer segregation. *Nature* 426:446–50

LANDSCAPES AND RIVERSCAPES: The Influence of Land Use on Stream Ecosystems

J. David Allan

School of Natural Resources and Environment, University of Michigan, Ann Arbor, Michigan 48109; email: dallan@umich.edu

Key Words catchment, disturbance, stressor response, stream health, river

■ **Abstract** Local habitat and biological diversity of streams and rivers are strongly influenced by landform and land use within the surrounding valley at multiple scales. However, empirical associations between land use and stream response only varyingly succeed in implicating pathways of influence. This is the case for a number of reasons, including (*a*) covariation of anthropogenic and natural gradients in the landscape; (*b*) the existence of multiple, scale-dependent mechanisms; (*c*) nonlinear responses; and (*d*) the difficulties of separating present-day from historical influences. Further research is needed that examines responses to land use under different management strategies and that employs response variables that have greater diagnostic value than many of the aggregated measures in current use.

> In every respect, the valley rules the stream.
> H.B.N. Hynes (1975)

INTRODUCTION

Rivers are increasingly investigated from a landscape perspective, both as landscapes in their own right (Robinson et al. 2002, Ward 1998, Wiens 1989) and as ecosystems that are strongly influenced by their surroundings at multiple scales (Allan et al. 1997, Fausch et al. 2002, Schlosser 1991, Townsend et al. 2003). River ecologists have long recognized that rivers and streams are influenced by the landscapes through which they flow (Hynes 1975, Vannote et al. 1980). However, a landscape perspective of rivers continues to evolve, owing both to the emergence of landscape ecology as a field of study (Turner et al. 2001, Wiens 1989) and to an increased focus on catchment-scale studies by freshwater ecologists.

As Wiens (2002) observed, river ecologists have been doing landscape ecology for a long time but just not calling it that. Landscape ecology places particular emphasis on habitat heterogeneity, connectivity, and scale, all of which have received considerable attention in running waters (Allan 1995). However, most earlier work was conducted at small spatial scales, often within stream reaches of a few hundred meters and their immediate surroundings; less consideration was given to the importance of larger spatial units. Our current understanding of rivers, as with other ecosystems, increasingly incorporates a conceptual framework of spatially nested controlling factors in which climate, geology, and topography at large scales influence the geomorphic processes that shape channels at intermediate scales and thereby create and maintain habitat important to the biota at smaller scales (Allen & Starr 1982, Frissell et al. 1986, Snelder & Biggs 2002). Recognizing that rivers are complex mosaics of habitat types and environmental gradients, characterized by high connectivity and spatial complexity, riverine landscapes increasingly are viewed as "riverscapes" (Fausch et al. 2002, Schlosser 1991, Ward et al. 2002), a unit that is amenable to study over a wide range of scales from a braided river and its valley (Tockner et al. 2002) to small habitat patches (Palmer et al. 2000).

Investigators increasingly recognize that human actions at the landscape scale are a principal threat to the ecological integrity of river ecosystems, impacting habitat, water quality, and the biota via numerous and complex pathways (Allan et al. 1997, Strayer et al. 2003, Townsend et al. 2003). In addition to its direct influences, land use interacts with other anthropogenic drivers that affect the health of stream ecosystems, including climate change (Meyer et al. 1999), invasive species (Scott & Helfman 2001), and dams (Nilsson & Berggren 2000). The recent increase in studies that seek to establish relationships between land use and stream condition is driven by several developments: (*a*) the widespread recognition of the extent and significance of changes in land use and land cover worldwide (Meyer & Turner 1994), (*b*) conceptual and methodological advances in landscape ecology combined with the ready availability of land use/land cover data (Turner et al. 2001), and (*c*) the increasing use of indicators of stream health to assess status and trends of rivers (Karr & Chu 2000, Norris & Thoms 1999).

Hierarchies, Habitats, and Biodiversity

An extensive literature explores the hierarchical nature of river systems, from the largest spatial scale of landscape or basin to successively smaller scales of the valley segment, channel reach, individual channel units (such as riffles and pools), and microhabitat (Figure 1) (Fausch et al. 2002, Frissell et al. 1986, Montgomery 1999). Because stream ecosystems are typically characterized by habitat and biota observed at the scale of a reach, typically $10^1 - 10^3$ m in length, and local species assemblages are strongly influenced by habitat quality and complexity, this geomorphological framework suggests how the stream environment at the local scale is influenced by the surrounding landscape. Reach-level channel morphology is influenced by valley slope and confinement, bed and bank material, and riparian vegetation, as well as by the supply from upslope of water, sediments, and wood (Montgomery & MacDonald 2002). Many features of the dynamic river channel are mutually adjusting (Church 2002), and human activities on the landscape that affect water or sediment supply or that stabilize or destabilize the existing channel shape are likely to set off a complex cascade of changes that are ultimately manifest

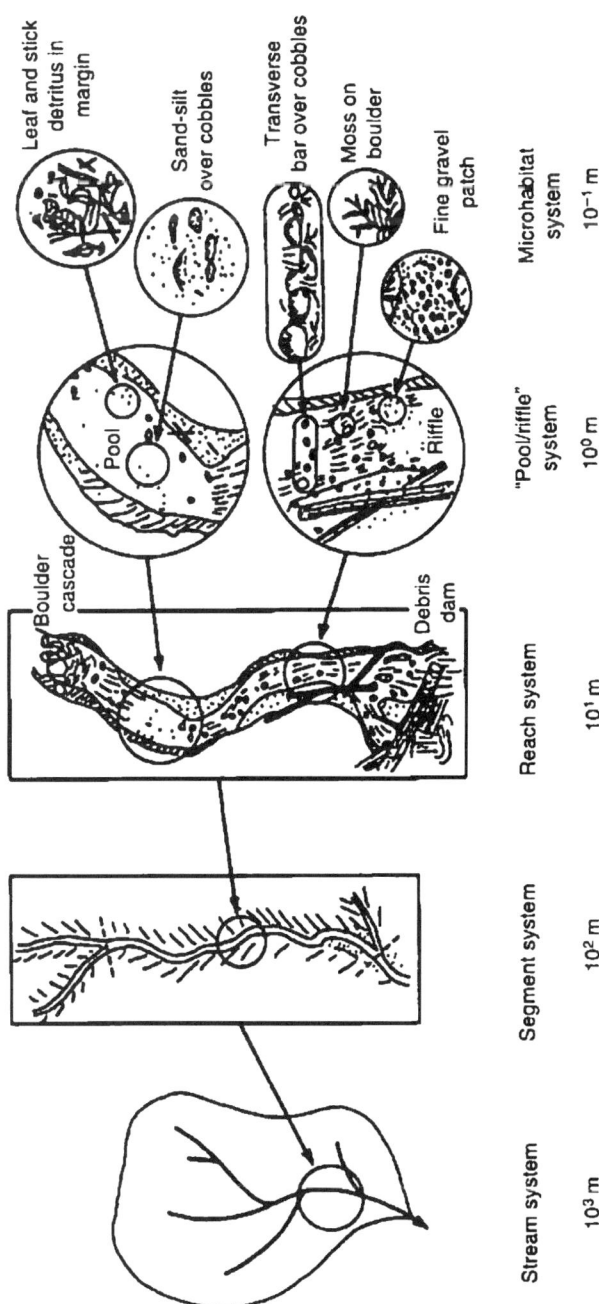

Figure 1 The hierarchical nature of river systems, from the largest spatial scale of landscape or basin to successively smaller scales of the valley segment, channel reach, individual channel units (such as riffles and pools), and microhabitat, as visualized for a small mountain stream (Frissell et al. 1986). Reprinted with permission from *Environmental Management*.

in altered and possibly degraded stream habitat. Natural hydrologic variability, and high flows in particular, move and sort sediments, and through cycles of erosion and deposition create a variety of channel features, including riffles, pools, bars, and islands; cause channel migration; maintain floodplain connectivity and other complex elements of floodplain river channels, including meander loops and side channels; and make the land-water interface both complex and dynamic (Junk et al. 1989). The resultant ever-changing mosaic of habitat patches, ecotones, and successional stages—the riverscape in all its complexity—is largely responsible for the high biodiversity of these systems (Robinson et al. 2002, Ward 1998). In this patch dynamics perspective (Townsend 1989, Hildrew & Giller 1994), the interaction between species-specific habitat needs, life histories, and dispersal ability and the ever-shifting temporal and spatial mosaic of stream habitats support greater diversity than would occur were the habitat unchanging. Thus, both the variety and the variability of habitat are important in influencing the biological diversity of streams and are linked to the larger stream system and surrounding landscape. Human actions at the landscape scale disrupt the geomorphic processes that maintain the riverscape and its associated biota and frequently result in habitat that is both degraded and less heterogeneous.

Assessment of River Health

Ecological integrity, stream condition, and river health are terms that describe the status of fluvial ecosystems and their response to human influences. Condition is defined by similarity of a test site to a set of least-impaired reference sites, whether measured by the sum of several indicators, such as the number of intolerant species and taxa richness [Index of Biotic Integrity (IBI), see Karr 1991], or by the number of observed taxa relative to expected (Rivpacs, see Wright 1995; Ausrivas, see Norris & Hawkins 2000). Additional measures include taxa richness of sensitive species; various biological and ecological traits, such as body size and shape, life history, and behavioral traits (Townsend & Hildrew 1994, Corkum 1999, Usseglio-Polatera et al. 2000b, Richards et al. 1997, Pan et al. 1999); pollution tolerance (Hilsenhoff 1988); and ecosystem processes, such as photosynthesis and respiration (Bunn et al. 1999). Habitat and water quality are also evaluated using individual variables or combined metrics (Barbour et al. 1999). Thus, a wide variety of assessment methods are available to evaluate the response of stream condition to a gradient in land use.

The shape of the relationship between a stream response variable and a measure of stress (Figure 2) likely depends mutually on the sensitivity of the response variable and mode of action of the environmental stressor. A gradual decline indicating incremental change in stream condition might be expected, for example, if a steady increase in sedimentation acted on a species assemblage that was approximately linearly ranked in their sensitivity to sediments. Nonlinear responses are expected whenever the species in question, or the majority of species, exhibit a sensitivity threshold to a particular stress, such as the frequency or magnitude of high flows. Although the response of the biota to some stressors, such as insecticides,

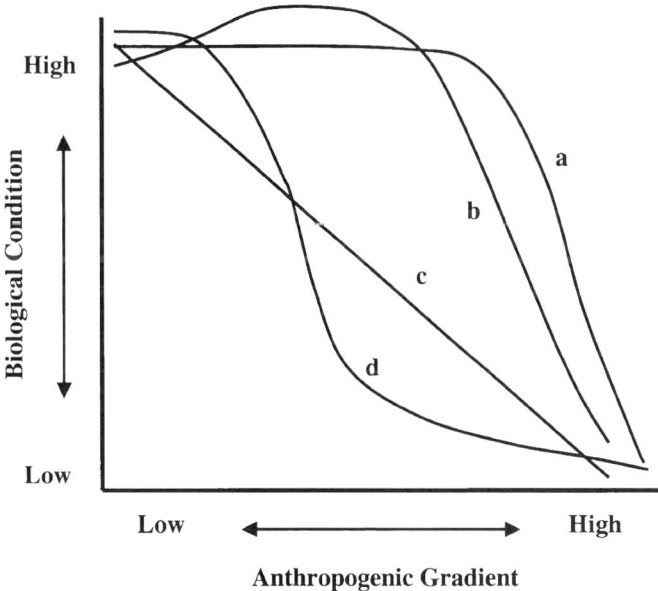

Figure 2 Hypothetical relationship depicting possible responses of stream biological condition (taxon richness, assemblage similarity, or a biological index, scaled to best attainable or reference conditions) to a gradient of increasing environmental stress, measured directly as, e.g., sedimentation, or indirectly as, e.g., agricultural land in the catchment. Possible responses include (*a*) nonlinear response occurring in the high range of the gradient, (*b*) subsidy-stress response, (*c*) linear response, and (*d*) nonlinear (threshold) response occurring in the low range of the gradient. Curves (*a*) versus (*d*) indicate low versus high sensitivity to a stressor. Modified from Norris & Thoms (1999) and Quinn (2000).

is expected to be only negative, a number of environmental stressors have positive influence at low to moderate concentrations. For example, a subsidy-stress response (Odum et al. 1979) may be a common outcome of riparian thinning and a low intensity of agriculture, in which initial increases in light, nutrients, and water temperatures increase periphyton biomass and macroinvertebrate abundance, with no apparent decline in diversity, whereas further intensification of agriculture results in loss of diversity and sensitive species (Quinn 2000).

Ultimately, the range of stream conditions from pristine to profoundly impacted reflects the system's integrated response to various human disturbances acting through the physical space of the catchment hierarchy, over short (pulse) and long (press) durations, and with cascading influences via local habitat structure and food web interactions (Quinn 2000, Townsend & Riley 1999). Different disturbances will exert their influence at different spatial scales and by different pathways (Table 1). Because streams are usually affected by multiple and interacting disturbances, matching a response to the responsible stressor can be very

TABLE 1 Principal mechanisms by which land use influences stream ecosystems

Environmental factor	Effects	References
Sedimentation	Increases turbidity, scouring and abrasion; impairs substrate suitability for periphyton and biofilm production; decreases primary production and food quality causing bottom-up effects through food webs; in-filling of interstitial habitat harms crevice-occupying invertebrates and gravel-spawning fishes; coats gills and respiratory surfaces; reduces stream depth heterogeneity, leading to decrease in pool species	Burkhead & Jelks 2001, Hancock 2002, Henley et al. 2000, Quinn 2000, Sutherland et al. 2002, Walser & Bart 1999, Wood & Armitage 1997
Nutrient enrichment	Increases autotrophic biomass and production, resulting in changes to assemblage composition, including proliferation of filamentous algae, particularly if light also increases; accelerates litter breakdown rates and may cause decrease in dissolved oxygen and shift from sensitive species to more tolerant, often non-native species	Carpenter et al. 1998, Delong & Brusven 1998, Lenat & Crawford 1994, Mainstone & Parr 2002, Niyogi et al. 2003
Contaminant pollution	Increases heavy metals, synthetics, and toxic organics in suspension associated with sediments and in tissues; increases deformities; increases mortality rates and impacts to abundance, drift, and emergence in invertebrates; depresses growth, reproduction, condition, and survival among fishes; disrupts endocrine system; physical avoidance	Clements et al. 2000, Cooper 1993, Kolpin et al. 2002, Liess & Schulz 1999, Rolland 2000, Schulz & Liess 1999, Woodward et al. 1997
Hydrologic alteration	Alters runoff-evapotranspiration balance, causing increases in flood magnitude and frequency, and often lowers base flow; contributes to altered channel dynamics, including increased erosion from channel and surroundings and less-frequent overbank flooding; runoff more efficiently transports nutrients, sediments, and contaminants, thus further degrading in-stream habitat. Strong effects from impervious surfaces and stormwater conveyance in urban catchments and from drainage systems and soil compaction in agricultural catchments	Allan et al. 1997, Paul & Meyer 2001, Poff & Allan 1995, Walsh et al. 2001, Wang et al. 2001
Riparian clearing/canopy opening	Reduces shading, causing increases in stream temperatures, light penetration, and plant growth; decreases bank stability, inputs of litter and wood, and retention of nutrients and contaminants; reduces sediment trapping and increases bank and channel erosion; alters quantity and character of dissolved organic carbon reaching streams; lowers retention of benthic organic matter owing to loss of direct input and retention structures; alters trophic structure	Bourque & Pomeroy 2001, Findlay et al. 2001, Gregory et al. 1991, Gurnell et al. 1995, Lowrance et al. 1984, Martin et al. 1999, Osborne & Kovacic 1993, Stauffer et al. 2000
Loss of large woody debris	Reduces substrate for feeding, attachment, and cover; causes loss of sediment and organic material storage; reduces energy dissipation; alters flow hydraulics and therefore distribution of habitats; reduces bank stability; influences invertebrate and fish diversity and community function	Ehrman & Lamberti 1992, Gurnell et al. 1995, Johnson et al. 2003, Maridet et al. 1995, Stauffer et al. 2000

difficult. Thus, it may be possible to determine the degree of impairment accurately without achieving the same level of certainty regarding cause (Gergel et al. 2002).

THE INFLUENCE OF LAND USE ON RIVERS

The global transition from undisturbed to human-dominated landscapes has impacted ecosystems worldwide and made the quantification of land use/land cover (hereafter, land use) a valuable indicator of the state of ecosystems (Meyer & Turner 1994). Hundreds of studies document statistical associations between land use and measures of stream condition using multisite comparisons and empirical models, and collectively these studies provide strong evidence of the importance of surrounding landscape and human activities to a stream's ecological integrity. Moreover, the extent of land use transformation is staggering. For example, before the development of pastoral agriculture in New Zealand, more than 80% of the land was forested; today, agriculture, primarily the grazing of nearly 60 million sheep and cattle, is the dominant land use in the middle and lower catchment areas of most of New Zealand's streams and rivers (Quinn 2000).

Not surprisingly, agriculture occupies the largest fraction of land area in many developed catchments, whereas urban land use is a much smaller fraction. Of some 150 major river basins of North America, agricultural land use varied from near zero in some northern river systems to 66% in the Upper Mississippi Basin (Benke & Cushing 2004). Six major river basins of the United States have more than 40% of their area in agriculture: the Lower Mississippi, Upper Mississippi, Southern Plains, Ohio, Missouri, and Colorado. Within the Upper Mississippi, the extent of agriculture in large tributary basins varies from 25% in the St. Croix and Wisconsin Rivers to 95% in the Minnesota River Basin. Comparisons of small subcatchments within a larger catchment have reported that the extent of agricultural land use varies even more widely at this smaller spatial scale, from 10% to 70% (Roy et al. 2003), 14% to 99%, (Richards et al. 1996), and 36% to 84% (Roth et al. 1996). Streams draining these landscapes can be expected to experience a wide range of human influences.

Urban land use is commonly a low percentage of total catchment area, yet it exerts a disproportionately large influence both proximately and over distance (Paul & Meyer 2001). Urban land exceeds 5% of catchment area in 29 river basins and exceeds 10% in only 10 of the 150 large basins of North America (Benke & Cushing 2004). However, a large percentage of the land area of small catchments may be urban. Among 30 small (100 km^2) subcatchments of the Etowah Basin, Georgia, combined low- and high-density urban land area averages 15%, with a maximum of 61% (Roy et al. 2003). Impervious surface area reaches as high as 51% for small streams in metropolitan subcatchments of Melbourne, Australia (Walsh et al. 2001), and urban land area is as high as 97% in small catchments of southeastern Wisconsin (Wang et al. 2001).

Other land uses affect stream condition, including forestry, mining, and recreation (Bryce et al. 1999). However, most landscape-scale studies of the influence of land use on streams have contrasted the varying extent of agricultural, urban, and natural (usually forested) land, and so these studies are the primary focus of this review.

Some important caveats apply to studies of the relationship between land use and stream condition. Because land use sums to 100%, several measures of land use may predict stream condition nearly equally well (e.g., Herlihy et al. 1998), and so the interpretation that a particular land use variable is the primary driver of stream condition must be made with caution. Comparisons of land use implicitly substitute space for time, as the often unstated assumption is that locations differing in land use are similar in essentially all other respects and can be viewed as equivalent to the progression over time of a single location experiencing the transition from natural to developed land. Forecasting changes in stream ecosystems in response to changing land use runs the risk that the relationship will change over time owing to changes in specific practices or in the environment itself. For example, revegetation of the riparian to reduce stream temperatures may be negated by future climate change; development of crops with engineered pest resistance may reduce use of pesticides, thus removing one of the pathways by which agricultural land use impacts stream biota.

Agricultural Land Use

Numerous studies have documented declines in water quality, habitat, and biological assemblages as the extent of agricultural land increases within catchments (Richards et al. 1996, Roth et al. 1996, Sponseller et al. 2001, Wang et al. 1997). Researchers commonly report that streams draining agricultural lands support fewer species of sensitive insect and fish taxa than streams draining forested catchments (Genito et al. 2002, Lenat & Crawford 1994, Wang et al. 1997). Although researchers report that row crop and other forms of intensive cultivation strongly affect stream condition, the influence of pasture agriculture may be less pronounced (Meador & Goldstein 2003, Strayer et al. 2003).

Agricultural land use degrades streams by increasing nonpoint inputs of pollutants, impacting riparian and stream channel habitat, and altering flows (Table 1). Higher inputs of sediments, nutrients, and pesticides accompany increased agricultural land use (Cooper 1993, Johnson et al. 1997, Lenat 1984, Osborne & Wiley 1988). Landscape metrics, particularly the proportion of agriculture in the catchment and forest in the riparian zone, explained 65%–84% of the variation in yields of nitrogen, dissolved phosphorus, and suspended sediments for 78 catchments across the five-state Mid-Atlantic Highlands region (Jones et al. 2001). Elevated nutrient concentrations are reported to result in greater algal production and changes in autotroph assemblage composition (Delong & Brusven 1998, Quinn 2000). However, the hypoxic conditions that high nutrient loading causes in lentic and coastal waters (Carpenter et al. 1998) are uncommon in streams and are likely to occur only in localized areas of slow-moving water. Because light levels, nutrient concentrations, and water temperature all tend to increase as riparian forest is lost, algal response may be influenced by one or more of these factors acting in concert. Changes in algal biomass and composition in the upper Roanoke Basin were primarily attributed to light and temperature because nutrients were thought to be sufficient at all sites (Sponseller et al. 2001). Another common response to lost riparian forest is increased macroinvertebrate abundance, particularly grazers, as the food web becomes increasing influenced by autochthonous rather than allochthonous energy sources (Delong & Brusven 1998, Quinn 2000).

Agricultural insecticide and herbicide runoff is likely responsible for some of the association between agricultural land use and stream biota described above (Cooper 1993, Skinner et al. 1997); however, evidence comes primarily from localized toxicity tests rather than from landscape-scale investigations. For example, field enclosures using caged amphipods and laboratory tests that exposed midge larvae to stream sediments showed pesticide toxicity in an agricultural catchment in the United Kingdom (Crane et al. 1996). Furthermore, the disappearance of 8 of the 11 most abundant invertebrate taxa from a reach of headwater stream after surface runoff from arable land was attributed to an insecticide (Schulz & Liess 1999), although most species recovered within 6–11 months, indicating a pulse disturbance. Because the concentrations of agricultural pesticides and herbicides are seldom measured in studies relating agricultural land use to stream biota, their role may be more widespread than is recognized.

Streams in highly agricultural landscapes tend to have poor habitat quality, reflected in declines in habitat indexes and bank stability (Richards et al. 1996, Roth et al. 1996, Wang et al. 1997), as well as greater deposition of sediments on and within the streambed. Sediments in runoff from cultivated land and livestock trampling (Quinn 2000, Strand & Merritt 1999) are considered to be particularly influential in stream impairment (Waters 1995). In the Piedmont region of the Chattahoochee Basin, Georgia, sediments in the channel increased with increasing agricultural land use, while heterogeneity in stream depth and the diversity of fishes associated with coarse substrate in pools declined (Walser & Bart 1999).

Changes to stream hydrology owing to increased agricultural land use are variable, depending on crop evapotranspiration rates compared with natural vegetation, changes to soil infiltration capacity, extent of drainage systems, and, if there is irrigation, whether water is extracted from the river or from groundwater. Mean annual flow of the Kankakee River, Illinois, increased during the twentieth century without any corresponding trend in precipitation, implicating land clearing and urbanization as the cause of greater runoff (Peterson & Kwak 1999). Storm flows commonly increase in magnitude and frequency, especially where runoff is enhanced owing to drainage ditches, subsurface drains, and loss of wetland area. In addition to the impact of flow extremes on erosion and habitat, high flows can eliminate taxa if such events occur during sensitive life stages or with sufficient frequency that only resistant and rapidly dispersing species can tolerate them. Macroinvertebrates that are able to withstand dislodgement or that have short and fast life cycles and good colonizing ability predominated in highly agricultural streams of Michigan (Richards et al. 1997). Alterations to flow regime affect stream fishes by downstream displacement of early life stages and disruption of spawning (Harvey 1987, Schlosser 1985). Although annual and storm flows typically increase with agricultural land use, base flows often decline owing to reduced infiltration and more episodic export of water (Poff et al. 1997). This decline results in an increased area of shallow water habitat, which usually lacks structure and is more easily warmed (Richards et al. 1996).

Wherever agriculture or other anthropogenic activity extends to the stream margin and natural riparian forest is removed, streams are usually warmer during summer and receive fewer energy inputs as leaf litter, and primary production usually increases (Quinn 2000). Bank stability may decrease, although establishment of deep-rooting grasses can stabilize banks (Davies-Colley 1997, Lyons et al. 2000), and the amount of large wood in the stream declines markedly (Johnson et al. 2003). Stable wood substrate in streams performs multiple functions, influencing channel features and local flow and habitat and providing cover for fish, perching habitat for invertebrates, and a substrate for biofilm and algal colonization (Gregory et al. 2003). Its absence can have a profound influence. For example, the presence of wood added an average of 55% and 26% to reach-level local diversity within highly agricultural catchments in Minnesota and Michigan, respectively (Johnson et al. 2003).

Urban Land Use

Substantial changes in biological assemblages are associated with increasing catchment area as urban land (Booth & Jackson 1997, Klauda et al. 1998, Lenat & Crawford 1994, May et al. 1997, Morley & Karr 2002, Tong & Chen 2002, Usseglio-Polatera & Beisel 2002, Wang et al. 2001). Urbanization is the suggested cause of the disappearance of anadromous fishes from tributaries of the Hudson River (Limburg & Schmide 1990). Change in the amount of connected impervious surface was the best single predictor of fish density, diversity, and biotic integrity across a gradient from predominantly agriculture to predominantly urban land in southeastern Wisconsin (Wang et al. 2001). Increasing urbanization among 30 sites within the Etowah Catchment, Georgia, was negatively correlated with water quality, habitat, and measures of the macroinvertebrate assemblage (Roy et al. 2003). Despite the many factors thought to potentially limit Pacific salmon populations, percentage of urban land, along with water quality and sediment flow events, explained more than 60% of the variation in Chinook salmon recruitment in the interior Columbia River Basin from 1980–1990 (Regetz 2003).

Major changes associated with increased urban land area include increases in the amounts and variety of pollutants in runoff, more erratic hydrology owing to increased impervious surface area and runoff conveyance, increased water temperatures owing to loss of riparian vegetation and warming of surface runoff on exposed surfaces, and reduction in channel and habitat structure owing to sediment inputs, bank destabilization, channelization, and restricted interactions between the river and its land margin (Table 1) (Paul & Meyer 2001).

Enhanced runoff from impervious surfaces and stormwater conveyance systems can degrade streams and displace organisms simply because of greater frequency and intensity of floods, erosion of streambeds, and displacement of sediments (Lenat & Crawford 1994). Modeled runoff within the Little Miami Basin, Ohio, was estimated to be more than 55 times greater from impervious than from pervious surfaces (Tong & Chen 2002). A comparison of streams in metropolitan areas and surrounding lands of Melbourne, Australia, found that macroinvertebrate taxa richness declined with increasing impervious surface, but streams of comparable imperviousness were markedly more degraded in the metropolitan drainage system. Flashiness of runoff was considered the primary influence throughout, but the presence of stormwater conveyance systems in the metropolitan area had the added effect of even greater flashiness and the conveyance of multiple pollutants (Walsh et al. 2001).

Whether urban land or impervious surface is a better predictor of the response of stream biota may depend on whether its primary influence is via flow alteration or also involves pollutants. Indeed, biological response measures have been better predicted by impervious area in several landscape studies of stream urbanization (Ourso & Frenzel 2003, Walsh et al. 2001, Wang et al. 2001) and by urban land area in others (Morley & Karr 2002), suggesting that hydrologic influences are primary in some studies, but the broader range of influences represented by urban area may be more important in others.

Because multiple pollutants enter urban streams, the direct influence of particular chemicals and metals is rarely demonstrated in comparisons of urban land use within catchments. Along a steep gradient of urbanization in the vicinity of Anchorage, Alaska, measured as a percentage of impervious area, macroinvertebrate taxa richness declined, and tolerant taxa replaced intolerant taxa (Ourso & Frenzel 2003). Urban land use, chemical factors, channel condition, and instream habitat all correlated with impervious area. However, stream and riparian habitat did not vary as strongly with impervious area as did water and sediment chemistry, suggesting that contaminants may have been of primary importance.

As a cause of changes in the biota, habitat degradation in response to catchment urbanization is less emphasized than other factors, particularly flow variability, although changes to bed sediments are commonly reported (Morley & Karr 2002, Roy et al. 2003). This deemphasis of habitat influence may be because some urban streams have protected corridors that maintain physical habitat but not water quality, or because habitat is relatively uniformly degraded among urbanized sites. Although catchment impervious area was the best single predictor of fish density, diversity, and biotic integrity in southeastern Wisconsin, stream habitat was not well correlated to increasing urbanization, which the authors attributed to prior habitat degradation associated with agriculture (Wang et al. 2001).

FOUR CHALLENGES

Undoubtedly, by changing the landscapes of stream catchments, human activities alter stream ecosystem in multiple ways (Table 1). However, our understanding of the relationships between anthropogenic land use and the ecological integrity of streams is complicated by covariation between anthropogenic and natural gradients, issues of scale, and uncertainties concerning the importance of legacies and thresholds. These challenges are now examined individually, although all may be of importance in a particular catchment study.

Covariation of Anthropogenic and Natural Landscape Features

Gradients of anthropogenic land use are frequently superimposed on an underlying gradient in parent geological material, soil type, topography, and other features of the natural terrain. Anthropogenic and natural factors covary because the latter influences the suitability of locations for agricultural and urban development. Sites near one another tend to be alike in both natural features and human uses, and spatial dependency can be anticipated in the distribution of organisms owing to their habitat requirements and tendency to disperse outwards from locations of high population recruitment (Corkum 1999). Whenever anthropogenic and natural gradients covary and only anthropogenic land use is assessed, the influence attributed to land use can be overestimated.

Whether natural or anthropogenic variables are found to have the stronger effect on stream condition depends substantially on the scope of the study, as well as on the adequate measurement of both types of variables. Nutrient and sediment measures often show that land use overrides natural features, particularly in agricultural lands (Johnson et al. 1997). In Lapwai Creek, an agriculturally impaired stream in northern Idaho, functional groups of macroinvertebrates were similar among sites despite expectations of differences along a river continuum, and the assemblage composition was markedly different from that found in less-impaired streams (Delong & Brusven 1998). Despite substantial variation in terrain and the extent of riparian vegetation, the relative homogeneity of the macroinvertebrate assemblages of these sites was interpreted, via increased sedimentation and the dominance of periphyton as an energy source, as evidence of the overwhelming effect of agricultural land use.

Natural factors may be of primary importance when human influence is minor, or when human influence is widespread and fairly uniform across the study region. In a study of 70 catchments within the relatively undegraded Northern Lakes and Forest Ecoregion of Wisconsin and Michigan, anthropogenic land use was not an important predictor of stream fish assemblages and attributes (Wang et al. 2003). The diversity and abundance of mollusks in the rivers of 36 catchments in Iowa were correlated with landscape factors indicative of erosional and groundwater processes, principally the average percentage of slope and percentage of land area composed of alluvial deposits (Arbuckle & Downing 2002). Possibly because the entire area is highly agricultural, anthropogenic land use was not a predictor. The distribution of mollusks throughout a catchment in southeastern Michigan was well correlated with such local habitat measures as flow stability and substrate, which in turn were more strongly related to geology than to land use (McRae et al. 2004). Although geology and anthropogenic land use both varied throughout the catchment, the former appeared to have the stronger influence over local habitat conditions.

Coequal or at least mixed influence of natural and anthropogenic variables on local stream condition is a frequent finding. Both were influential in explaining patterns in macroinvertebrate assemblages among 55 riffle sites dispersed throughout the Taieri Basin of the South Island of New Zealand (Townsend et al. 2003). Overall, the distribution and abundance of macroinvertebrates were best explained by geomorphological factors at the catchment scale, by a mixture of geomorphology and land use at the reach scale, and predominantly by land use at the bedform (local, riffle) scale. Physical habitat variables explained approximately one third to one half of variation in macroinvertebrate assemblages, depending on season, among 46 sites in the Saginaw Basin of Michigan (Richards et al. 1996). Approximately one half of the habitat variation was explained by landscape variables. Bankfull width and other channel shape measures were much more strongly influenced by geology variables and only minimally by land use. Woody debris showed the opposite partitioning, whereas bankfull depth and canopy cover were equally influenced.

Covariation among natural and anthropogenic environmental factors can make attributing relative influence difficult or impossible. For example, a study of 25 agricultural streams in eastern Wisconsin found that agriculture was a strong influence, but the interpretation was complex owing to covariance of natural and anthropogenic variables measured at multiple spatial scales (Fitzpatrick et al. 2001). Streams with more than 10% agricultural land in their buffers were almost invariably impaired, particularly as indicated by an IBI for fishes, whereas invertebrate and algal metrics were less sensitive to land use. However, because riparian vegetation, geologic conditions, and hydrologic conditions were all correlated with the response of biotic metrics to agricultural land in the catchment, and because the relationships varied with the taxonomic group assessed, researchers could not confidently separate the interrelated effects of geologic setting, catchment and buffer land cover, and base flow.

Spatial Scale

Multiscale investigations often evaluate the relationship between stream condition and land use measured at several of the following scales: (a) the local reach, described by a buffer of 100 m to several hundred meters in width on each bank, and some hundreds of meters to a kilometer in length; (b) a buffer of similar width but of greater length, often the entire upstream distance for a small stream; and (c) the entire catchment upstream of a site (Figure 3). These scales will be referred

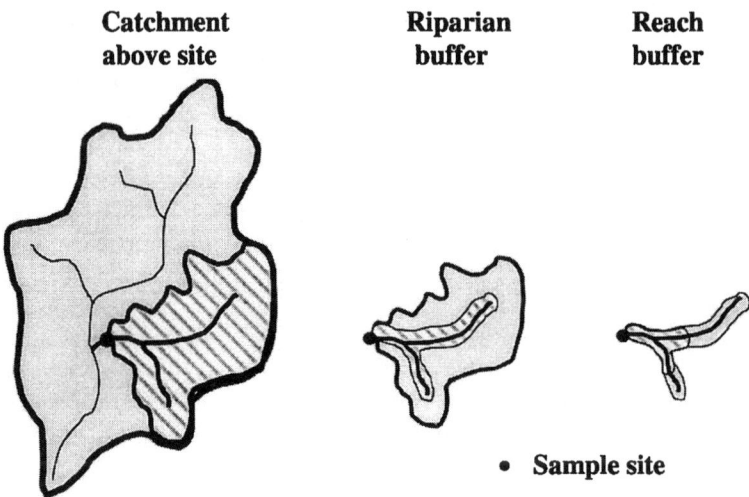

Figure 3 Three spatial scales widely used in relating landscape variables to some physical or biological measure of stream condition. The catchment typically is a subcatchment of a larger basin. Buffer widths of 100–200 m (each bank) are common. Modified from Morely & Karr (2002).

to as reach, riparian, and catchment, respectively; the former clearly is local scale, whereas the latter are two aspects of larger scale.

Different environmental variables of streams can be expected to vary in their responsiveness to large- versus local-scale environmental factors. Shade is influenced by very local patterns of riparian vegetation, water temperature responds to shading over distances of hundreds to thousands of meters, and inputs of leaf litter and wood are local but subject to downstream transport of variable distance (Allan et al. 1997, Quinn 2000). Nutrients and sediments can be transported long distances and so may be influenced by riparian conditions along a stream's entire length. Land use throughout the entire catchment governs stream hydrology through its influence over evapotranspiration, infiltration, and runoff conveyance, and land use is a strong predictor of total nutrient loading (Boyer et al. 2002). The spatial scale at which an effect is detected is influenced by how closely land use in the riparian mirrors land use throughout the catchment, by data resolution, by the interplay of anthropogenic and natural gradients, and by specifics of study design. For example, a comparison of small catchments in southeastern Michigan that spanned a large gradient in agriculture but included only minimal replication of study reaches within catchments found that variation in land use at the catchment scale is the best predictor of stream habitat and fish IBI (Roth et al. 1996). However, reach-scale variation in land use was superior to catchment-scale variation in predicting stream condition within the same river basin when the study design examined multiple reaches within just three small catchments that differed moderately in land use (Lammert & Allan 1999). Variation in land cover is often greater at the reach and riparian scales than at the catchment scale, which likely contributes to the greater influence attributed to riparian land use in many studies (e.g., Stauffer et al. 2000).

When land use at the reach and riparian scales is reported to have a strong influence over stream condition, direct local pathways are usually apparent. Pasture streams with occasional wooded reaches show marked physical and biological changes over distances of less than one kilometer. Forested reaches typically have cooler temperatures, wider channels, fewer sediments, and greater diversity of invertebrates (Abell & Allan 2002, Storey & Cowley 1997, Sweeney 1993). Fish assemblages in Ecuadorian streams changed from dominance by insectivorous and omnivorous taxa in pools with near-stream forest cover to primarily periphyton-grazers in open canopy pools, indicating a direct food web linkage (Bojsen & Barriga 2002). Near-stream connected imperviousness had a stronger influence on fish assemblages than did comparable amounts of impervious surface located farther from the stream, apparently owing to increased severity and frequency of high-flow events and lowered baseflow (Wang et al. 2001).

Even modest riparian deforestation in highly forested catchments can result in degradation of stream habitat owing to sediment inputs. A comparison of two small catchments that were less than 3% nonforested with two that were 13% and 22% nonforested found the latter to have higher concentrations of suspended sediments, higher turbidity at baseflow, five to nine times greater bedload transport, and greater embeddedness (Sutherland et al. 2002). Deforested riparian strips greater than one kilometer in length were associated with more fine sediments and less habitat diversity in a southern Appalachian stream, even though the riparian strips were vegetated and all were located within a highly forested catchment (Jones et al. 1999). Streams with reduced forest cover exhibit declines in overall fish abundance and an increase in sediment-tolerant and invasive species at the expense of those that spawn in clean gravel (Sutherland et al. 2002).

A number of studies have attributed more influence to catchment than to local land use, although pathways of influence may not be as easily detected. A composite index of habitat quality was strongly related to catchment land use and showed progressively weaker associations with riparian and reach-scale land use in southeastern Michigan, where a fish IBI was also more strongly associated with land use throughout the riparian and subcatchment than with reach-scale riparian vegetation (Roth et al. 1996). Invertebrate metrics were better predicted by catchment than by local-scale urbanization in the Puget Sound lowlands of Washington State (Morley & Karr 2002). Catchment-scale influence may be greatest when the primary mechanism is flow instability, nutrients, or some other factor related to the entire landscape.

Studies that examine a variety of measures of stream conditions in relation to land use at multiple scales report, unsurprisingly, mixed influence (Fitzpatrick et al. 2001, Richards et al. 1996, Roth et al. 1996, Stewart et al. 2001). Macroinvertebrate indexes were strongly correlated with both catchment and riparian land cover

over a range of 5%–61% total urban area and 34%–95% forest area in 100-m buffers (Roy et al. 2003). However, macroinvertebrate indexes were even more strongly predicted by environmental factors quantified at the reach-scale, including variation in substrate size and ion concentrations. Because reach-scale conditions were also associated with catchment land cover, these results are consistent with the view that large-scale landscape factors affect the biota via their influence over local-scale physical conditions.

Nonlinearities

Stream condition almost invariably responds nonlinearly to a gradient of increasing urban land or impervious area (IA). A marked decline in species diversity and IBIs with increasing urbanization has been reported from streams in Wisconsin (around 8%–12% IA, Stepenuck et al. 2002, Wang et al. 2000), Delaware (8%–15% IA, Paul & Meyer 2001), Maryland (greater than 12% IA, Klein 1979), and Georgia (15% urban land, Roy et al. 2003). Additional studies (reviewed in Paul & Meyer 2001, Stepenuck et al. 2002) provide evidence of marked changes in discharge, bank and channel erosion, and biotic condition at greater than 10% imperviousness. Although considerable evidence supports a threshold in stream health in the range of 10%–20% IA or urban land, others disagree (Bledsoe & Watson 2001, Karr & Chu 2000), and the relationship is likely too complex for a single threshold to apply. Hydrologic response is influenced by a number of catchment and stream characteristics, including slope, storage, conveyance and connectivity, and channel form (Bledsoe & Watson 2001, Walsh et al. 2001). Also, the supply of contaminants in urban storm runoff may vary independently of impervious area. In contrast to the above studies, a comparison of 45 highly urbanized sites around Seattle, Washington, reported a highly linear decline in macroinvertebrate indexes with increasing urban land and impervious area across the entire gradient ($\sim$10%–60% IA, $\sim$20%–90% urban land; Morley & Karr 2002).

Streams in agricultural catchments usually remain in good condition until the extent of agriculture is relatively high, more than 30%–50%. In previously forested catchments in New Zealand, a macroinvertebrate fauna typical of undamaged sites was retained and abundances enhanced by conversion of up to 30% of catchment area to pastoral land, but increases in agricultural land above 30% resulted in an increase in pollution-tolerant forms, illustrating a subsidy-stress relationship (Quinn 2000, Quinn & Hickey 1990). In several studies of Wisconsin streams, agricultural land use had a strong effect only when it exceeded 50% of catchment area (Wang et al. 2003, Wang et al. 1997). The response of stream condition to extent of agriculture across 172 sites from 20 major river basins throughout the United States was quite variable, and at least some sites had good fish condition even if agriculture exceeded 50% (Meador & Goldstein 2003). A study of agricultural streams in Wisconsin found indications of a decline in a fish IBI at >30% agriculture in the catchment and >10%–20% agriculture in the buffer (Fitzpatrick et al. 2001); another study reported declines in habitat quality and a fish IBI only when agriculture reached about 50% of catchment area, and some sites maintained high IBI and habitat scores at 80% agriculture (Wang et al. 1997). The wide range of responses reported from streams draining agricultural landscapes clearly indicates that extent of agriculture is not by itself sufficient to predict the strength of the response.

Legacy Effects

Legacy effects are the consequence of disturbances that continue to influence environmental conditions long after the initial appearance of the disturbance. Observing that the present-day diversity of stream macroinvertebrates and fish in forested catchments of the Appalachians, which previously had been farmed, were more similar to streams from present-day agricultural landscapes than from present-day primary forest, Harding et al. (1998) emphasized the importance of the "ghost of land use past." Interpretation of the influence of land use is further complicated when cycles of change occur, such as when agricultural land reverts to forest, or when change is sequential, as when forested land is first converted to agriculture and subsequently to urban land. The finding by Wang et al. (2001) that fish metrics but not habitat varied strongly along an urbanization gradient was interpreted as the legacy of similar habitat degradation at all sites under the common, prior influence of agriculture.

Geomorphological changes brought about by multiple human activities likely have produced lasting, complex, and often unappreciated changes in physical structure and hydrology of river systems. Landscape changes that occurred within a few decades of European settlement of New South Wales, Australia, including clearance of riparian and floodplain vegetation and draining of swamps, have fundamentally altered river structure throughout virtually the entire Bega catchment (Brierley et al. 1999). Extensive habitat transformation has resulted, including channel widening and infilling of pools in lowland sections and incision of headwater channels owing to more efficient downstream water conveyance and downstream export of sediments. Overall structural complexity has been reduced and lateral connectivity is largely lost in middle reaches but is now increased in the lowlands. Unusually, longitudinal connectivity is now greater than was likely true of the presettlement, more discontinuous system. Brierley et al. (1999) estimate that it will take thousands of years for the sediment-starved upper reaches to refill with sediments, while at the same time the oversupply of sand in the lower river is now trapped by exotic vegetation. The timescale of recovery from geomorphic channel alterations is especially long, particularly in comparison to changes in land use, and so stream habitat and channel shape may never reach equilibrium with ongoing development.

How much the channels of some large North American rivers have lost complexity has only recently been appreciated. For example, the Willamette River, Oregon, is estimated to have undergone a fourfold reduction in length of shoreline as its once expansive floodplain and backwaters have been confined to a narrower and simpler channel in response to snag removal, channel dredging, and the draining of its floodplain (Sedell & Froggatt 1984). Under the combined influence of reductions in sediment supply and construction of levees, the Cedar River, Washington, has experienced a 35% decrease in channel width and 45% decrease in channel area (Perkins 1994). Many streams of the upper Midwest have been deprived of wood by past timber harvest and the removal of existing wood in the channel by log drives (Johnson et al. 2003), and headwater regions have been transformed by the removal of beaver and consequent reduction in dams and ponds (Naiman et al. 1988). Elevated concentrations of heavy metals in the water column and in sediments of streams owing to hard-rock mining since the late 1800s were deemed responsible for reduced abundances and diversity of native fishes in northern Idaho (Maret & Maccoy 2002). Legacy effects owing to prior land clearing, channel modifications, snag removal, mining, and perhaps other human actions clearly pose a major challenge to linking present-day land use with concurrent stream condition.

MANAGEMENT APPLICATIONS

The measurement of stream health and its response to a variety of environmental stressors, including land use, requires well-tested indicators of ecological integrity. Composite measures, such as the IBI and percent assemblage similarity, are very useful in detecting overall stream degradation, but because of their aggregated nature they may be less easily interpreted than the behavior of individual response variables (Watzin & McIntosh 1999). For management and restoration actions to be effective, we must diagnose cause as well as assess harm, which requires an improved understanding of the mechanisms through which land use impacts stream ecosystems. Studies are needed that examine the response of individual species, traits, and guilds and that better connect the chain of influence from land use to stream response via studies of mechanisms.

Greater interpretability of stressor-response relationships is usually achieved when the response variables are tolerance groupings, feeding and reproductive guilds, traits, and individual taxa (e.g., Poff & Allan 1995, Townsend et al. 1997, Usseglio-Polatera et al. 2000b). In comparing the responses of several macroinvertebrate metrics with several potential stressors using a large data set from the Mid-Atlantic Highlands, Yuan & Norton (2003) found the proportional abundance of tolerant taxa to be the most sensitive indicator of nutrient enrichment and habitat degradation, whereas Ephemeroptera richness was the most sensitive indicator of high metals and ions. Both linear and nonlinear response relationships were common. Using 11 biological and 11 ecological traits for 472 invertebrate taxa from French rivers, Usseglio et al. (2000a) identified distinct ecological groupings on the basis of body size, reproductive habitat, food source, and feeding habits, an approach that appears to hold promise for bioassessment (Gayraud et al. 2003). Further examples include shifts in fish reproductive guilds in response to sediment inputs (Jones et al. 1999, Sutherland et al. 2002), changes in the relative abundance of species that feed on periphyton versus leaf litter in response to loss of riparian shade (Bojsen & Barriga 2002, Quinn 2000), and an association between taxa with multivoltine life cycles and small body sizes and the extent of shallow, slow-water habitat (Richards et al. 1997). Thus, Poff's (1997) argument that a combination of landscape and habitat filters, together with categorizing or ranking taxa by traits that determine their susceptibility to particular environmental conditions, holds much promise for a multiscale, mechanistic understanding of assemblage response to changing land use and other broad-scale disturbances.

Riparian management is particularly attractive because of the riparian zone's immediate and direct influences on stream condition via well-documented pathways (Gregory et al. 1991, Naiman & Decamps 1997) and because it promises benefits that are highly disproportionate to the land area required (Lowrance et al. 1997, Quinn et al. 2001). However, gaps in the riparian (Weller et al. 1998) as well as subsurface farm and storm drains bypass the riparian zone and diminish its effectiveness (Barton 1996, Osborne & Kovacic 1993). In addition, landscape change at the scale of entire catchments may have impacts too great for a riparian strip to moderate. Studies that evaluate the influence of landscape change across multiple spatial scales report that stream responses are complex and interacting and vary with location and landform setting. For example, nutrient concentrations often reflect catchment land use, whereas macroinvertebrate assemblages appear especially sensitive to a number of local habitat factors (Hunsaker & Levine 1995, Strayer et al. 2003).

Reversal of land use to a less-developed state at the catchment scale is rarely practical, and so improvement of stream condition more often depends on best management practices (BMPs) and improvements in landscape management and design. Some of these activities are at the catchment scale, such as conservation tillage, reduced fertilizer application, and other agricultural BMPs, as well as efforts to minimize hydrologic changes by retaining natural flow paths and infiltration capacity. Other BMPs are more proximate to the stream, such as stormwater retention ponds, managed wetlands, livestock exclusion, and maintenance of an intact riparian corridor. Evaluations of BMP benefits to stream condition commonly report improvements in physical and chemical variables, including habitat, nutrients, sediments, and turbidity (Caruso 2000, D'Arcy & Frost 2001, Lowrance et al. 1997, Strand & Merritt 1999, Wissmar & Beschta 1998). However, studies that evaluate biological responses to BMPs at the scale of the catchment are rare. One such study reported improvements in stream chemistry and streambed sediments in response to buffer practices, but the response of biological metrics was indistinct (Nerbonne & Vondracek 2001; see also Sovell et al. 2000). However, installation of riparian BMPs resulted in improvements in habitat quality and fish abundance in a Wisconsin stream (Wang et al. 2002). More studies of this kind are needed to determine whether physical improvements in stream condition are also evident in the biota.

The ecosystem functions performed by stream riparian zones vary with landform and location, as does human activity within the riparian, and so a "one size fits all" approach to riparian management is unlikely to be effective (Quinn et al. 2001, Strayer et al. 2003). Instead, knowledge of geomorphic setting and the key functions or uses of the riparian that are considered of greatest value should guide riparian management decisions. For example, Lowrance et al. (1997) estimated the amount of sediments, nitrogen, and phosphorus that forested riparian buffers would retain from runoff entering the Chesapeake Bay for each of its nine physiographic provinces. They took into consideration the differences owing to soils, slope, and hydrologic connectivity and predicted different removal efficiencies for different pollutants. Ultimately, BMPs are likely to be chosen on the basis of their demonstrated effectiveness in a particular landform and human setting and of how much society values the expected benefit to the stream ecosystem.

The demonstrated effectiveness of land use data in predicting many components of stream condition points to an expanding role for landscape analysis in catchment management (Gergel et al. 2002). At present, most current studies rely on static Geographic Information Systems (GIS) maps that may represent land cover some years displaced in time from stream condition measures. However, remotely sensed data are likely to become more widely used in the future, offering greater opportunity to synchronize the time frame of land cover and stream condition measurement and to develop new landscape indicators. One promising demonstration showed that stream chemistry, habitat, and stream fish indexes across multiple ecoregions of Nebraska, Kansas, and Missouri were correlated to various "greenness" metrics on the basis of the normalized difference vegetation index, an indicator of vegetation condition and physiological activity obtained from satellite or airborne sensors (Griffith et al. 2002). Although management to mitigate land use impacts on streams will require site-based analysis of interacting factors, detection of areas at risk and estimation of probable risk factors are important and complementary activities to site-based studies.

CONCLUSIONS

The rapidly expanding investigation of streams in the context of their catchments and landscapes clearly indicates that stream ecosystems are strongly affected by human actions across spatial scales. The impacts are numerous, both direct and indirect, and complex, owing to the various pathways by which land use influences streams and the interaction between anthropogenic gradients and the hierarchically structured influence of landform on local stream conditions. Not only does the valley rule the stream, as Hynes (1975) so aptly put it, but increasingly, human activities rule the valley. The extent of change in river health in response to future population growth and development can be anticipated from knowledge of the relationships between land use and stream condition and plausible alternative futures (Baker et al. 2004).

Our understanding of the pathways and mechanisms through which land use influences stream conditions is informed by the comparative and empirical approach that has been the focus of this review; yet, it can also be said that this knowledge at present is extremely limited, particularly for prescriptive management. Our limited understanding is due in part to the multiple effects of a particular change in land use and in part to the influence of local setting and underlying natural variation. Clearly, the influence of the surrounding landscape on a stream is manifest across multiple spatial scales and is further complicated by legacies from prior human activities. Thus, landform apparently operates mainly at the larger scale of catchment and region through its influence over geology, climate, vegetation, and topography, whereas the influence of land use operates across all scales, depending on the response variable of concern. Whether threshold responses are widespread is uncertain, owing partly to the scatter that is common in empirical relationships between land use and stream response. However, impacts of urban land use are clearly experienced at considerably lower percentages of catchment area than is true for agricultural land use, and most studies report a nonlinear response of stream condition to increasing urbanization.

Integrative measures of stream condition, including IBIs and percent similarity measures, are particularly useful for assessing overall stream health because they integrate multiple influences. However, species traits, feeding and reproductive guilds, taxa of known tolerance to particular stressors, and other less-aggregated measures are likely to prove more useful in evaluating pathways and mechanisms (Poff 1997, Usseglio-Polatera et al. 2000a). It will be particularly useful to examine the response of these more sensitive indicators to various management practices intended to offset the harmful impacts of intensive land uses. To date, the majority of catchment-scale studies has only indirectly indicated tradeoffs, as in the common finding that biological metrics are negatively associated with agricultural land in the catchment but positively associated with forested land in the riparian (Steedman 1988, Wang et al. 1997). Future studies that examine the response of more revealing measures such as trait and guild composition, within a two-dimensional matrix of varying land use and management practices, could bring new understanding to the influence of land use on stream condition.

ACKNOWLEDGMENTS

I thank Rebecca Cifaldi for her extensive contributions to all phases of the literature review. R. Abell, B. Allan, N. Bosch, P. Esselman, D. Infante, B. Kennedy, M. Khoury, L. Poff, L. Wang, J. Wiens, and L. Yuan made helpful comments on earlier drafts. JDA was supported by a David H. Smith award from The Nature Conservancy and a grant from the Office of Research and Development of the U.S. Environmental Protection Agency during preparation of this manuscript.

The *Annual Review of Ecology, Evolution, and Systematics* is online at
http://ecolsys.annualreviews.org

LITERATURE CITED

Abell RA, Allan JD. 2002. Riparian shade and stream temperatures in an agricultural catchment, Michigan, USA. *Verh. Int. Ver. Theor. Ang. Limnol.* 28:1–6

Allan JD. 1995. *Stream Ecology: Structure and Function of Running Waters*. Dordrecht, Neth.: Kluwer. 388 pp.

Allan JD, Erickson DL, Fay J. 1997. The

influence of catchment land use on stream integrity across multiple spatial scales. *Freshw. Biol.* 37:149–61

Allen TF, Starr TB. 1982. *Hierarchy: Perspectives for Ecological Complexity.* Chicago: Univ. Chicago Press

Arbuckle KE, Downing JA. 2002. Freshwater mussel abundance and species richness: GIS relationships with watershed land use and geology. *Can. J. Fish. Aquat. Sci.* 59:310–16

Baker JP, Hulse DW, Gregory SV, White D, Van Sickle J, et al. 2004. Alternative futures for the Willamette River Basin, Oregon. *Ecol. Appl.* 14:313–24

Barbour MT, Gerritsen J, Snyder BD, Stribling JB. 1999. Rapid Bioassessment Protocols for use in streams and wadeable rivers: periphyton, benthic macroinvertebrates and fish. *Second Edition. Rep. EPA/841-B-99–002*, US EPA, Off. Water, Washington, DC

Barton DR. 1996. The use of Percent Model Affinity to assess the effects of agriculture on benthic invertebrate communities in headwater streams of southern Ontario, Canada. *Freshw. Biol.* 36:397–410

Benke AC, Cushing CE, eds. 2004. *Rivers of North America.* San Diego, CA: Academic/Elsevier

Bledsoe BP, Watson CC. 2001. Effects of urbanization on channel instability. *J. Am. Water Resour. Assoc.* 37:255–70

Bojsen BH, Barriga R. 2002. Effects of deforestation on fish community structure in Ecuadorian Amazon streams. *Freshw. Biol.* 47:2246–60

Booth DB, Jackson CR. 1997. Urbanization of aquatic systems: degradation thresholds, stormwater detection, and the limits of mitigation. *J. Am. Water Resour. Assoc.* 33:311–23

Bourque CPA, Pomeroy JH. 2001. Effects of forest harvesting on summer stream temperatures in New Brunswick, Canada: an inter-catchment, multiple-year comparison. *Hydrol. Earth Syst. Sci.* 5:599–613

Boyer EW, Goodale CL, Jaworski NA, Howarth R. 2002. Anthropogenic nitrogen sources and relationships to riverine nitrogen export in the northeastern U.S.A. *Biogeochemistry* 57/58:137–69

Brierley GJ, Cohen T, Fryirs K, Brooks A. 1999. Post-European changes to the fluvial geomorphology of Bega catchment, Australia: implications for river ecology. *Freshw. Biol.* 41:839–48

Bryce SA, Larsen DP, Hughes RM, Kaufmann P. 1999. Assessing relative risks to aquatic ecosystems: a mid-Appalachian case study. *J. Am. Water Resour. Assoc.* 35:23–36

Bunn SE, Davies PM, Mosisch TD. 1999. Ecosystem measures of river health and their response to riparian and catchment degradation. *Freshw. Biol.* 41:333–46

Burkhead NM, Jelks HL. 2001. Effects of suspended sediment on the reproductive success of the tricolor shiner, a crevice-spawning minnow. *Trans. Am. Fish. Soc.* 130:959–68

Carpenter SR, Caraco NF, Howarth RW, Sharpley AN, Smith VH. 1998. Nonpoint pollution of surface waters with phosphorus and nitrogen. *Ecol. Appl.* 8:559–68

Caruso BS. 2000. Comparative analysis of New Zealand and US approaches for agricultural nonpoint source pollution management. *Environ. Manag.* 25:9–22

Church M. 2002. Geomorphic thresholds in riverine landscapes. *Freshw. Biol.* 47:541–57

Clements WH, Carlisle DM, Lazorchak JM, Johnson PC. 2000. Heavy metals structure benthic communities in Colorado mountain streams. *Ecol. Appl.* 10:626–38

Cooper CM. 1993. Biological effects of agriculturally derived surface water pollutants on aquatic systems—a review. *J. Environ. Q.* 22:402–8

Corkum LD. 1999. Conservation of running waters: beyond riparian vegetation and species richness. *Aquat. Conserv.* 9:559–64

Crane M, Delaney P, Mainstone C, Clarke S. 1996. Measurement by in situ bioassay (on site or on-the-field) of water quality in an agricultural catchment. *Water Res.* 29:2441–48

D'Arcy B, Frost A. 2001. The role of best management practices in alleviating water quality problems associated with diffuse pollution. *Sci. Total Environ.* 265:359–67

Davies-Colley RJ. 1997. Stream channels are narrower in pasture than in forest. *N. Z. J. Mar. Fresh.* 31:599–608

Delong MD, Brusven MA. 1998. Macroinvertebrate community structure along the longitudinal gradient of an agriculturally impacted stream. *Environ. Manag.* 22:445–57

Ehrman TP, Lamberti GA. 1992. Hydraulic and particulate matter retention in a 3rd-order Indiana stream. *J. N. Am. Benthol. Soc.* 11:341–49

Fausch KD, Torgersen CE, Baxter CV, Li HW. 2002. Landscapes to riverscapes: bridging the gap between research and conservation of stream fishes. *BioScience* 52:483–98

Findlay S, Quinn JM, Hickey CW, Burrell G, Downes M. 2001. Effects of land use and riparian flowpath on delivery of dissolved organic carbon to streams. *Limnol. Oceanogr.* 46:345–55

Fitzpatrick FA, Scudder BC, Lenz BN, Sullivan DJ. 2001. Effects of multi-scale environmental characteristics on agricultural stream biota in eastern Wisconsin. *J. Am. Water Resour. Assoc.* 37:1489–507

Frissell CA, Liss WJ, Warren CE, Hurley MD. 1986. A hierarchical framework for stream habitat classification: viewing streams in a watershed context. *Environ. Manag.* 12:199–214

Gayraud S, Statzner B, Bady P, Haybach A, Scholl F, et al. 2003. Invertebrate traits for the biomonitoring of large European rivers: an initial assessment of alternative metrics. *Freshw. Biol.* 48:2045–64

Genito D, Gburek WJ, Sharpley AN. 2002. Response of stream macro invertebrates to agricultural land cover in a small watershed. *J. Freshw. Ecol.* 17:109–19

Gergel SE, Turner MG, Miller JR, Melack JM, Stanley EH. 2002. Landscape indicators of human impacts to riverine systems. *Aquat. Sci.* 64:118–28

Gregory S, Boyer KL, Gurnell AM. 2003. *The ecology and management of wood in world rivers.* Am. Fisheries Soc. Symp. 37, Bethesda, MD

Gregory SV, Swanson FJ, McKee WA, Cummins KW. 1991. An ecosystem perspective of riparian zones: focus on links between land and water. *BioScience* 41:540–51

Griffith JA, Martinko EA, Whistler JL, Price KP. 2002. Interrelationships among landscapes, NDVI, and stream water quality in the US central plains. *Ecol. Appl.* 12:1702–18

Gurnell AM, Gregory KJ, Petts GE. 1995. The role of coarse woody debris in forest aquatic habitats: implications for management. *Aquat. Conserv.* 5:143–66

Hancock PJ. 2002. Human impacts on the stream-groundwater exchange zone. *Environ. Manag.* 29:763–81

Harding JS, Benfield EF, Bolstad PV, Helfman GS, Jones EBD. 1998. Stream biodiversity: the ghost of land use past. *Proc. Natl. Acad. Sci. USA* 95:14843–47

Harvey BC. 1987. Susceptibility of young-of-the-year fishes to downstream displacement by flooding. *Trans. Am. Fish. Soc.* 116:851–55

Henley WF, Patterson MA, Neves RJ, Lemly AD. 2000. Effects of sedimentation and turbidity on lotic food webs: a concise review for natural resource managers. *Rev. Fish. Sci.* 8:125–39

Herlihy A, Stoddard JL, Johnson CB. 1998. The relationship between stream chemistry and watershed land cover data in the Mid-Atlantic region, USA. *Water Air Soil Pollut.* 105:377–86

Hildrew AG, Giller PS. 1994. Patchiness, species interactions and disturbance in the stream benthos. In *Aquatic Ecology: Scale, Pattern and Process*, ed. PS Giller, AG Hildrew, DG Raffaelli, pp. 21–62. Oxford, UK: Blackwell

Hilsenhoff WL. 1988. Rapid field assessment of organic pollution with a family-level biotic index. *J. N. Am. Benthol. Soc.* 7:65–68

Hunsaker CT, Levine DA. 1995. Hierarchical approaches to the study of water-quality in rivers. *BioScience* 45:193–203

Hynes HBN. 1975. The stream and its valley. *Verh. Int. Ver. Theor. Ang. Limnol.* 19:1–15

Johnson LB, Breneman DH, Richards C. 2003. Macroinvertebrate community structure and function associated with large wood in low gradient streams. *River Res. Appl.* 19:199–218

Johnson LB, Richards C, Host GE, Arthur JW. 1997. Landscape influences on water chemistry in Midwestern stream ecosystems. *Freshw. Biol.* 37:193–208

Jones EBD, Helfman GS, Harper JO, Bolstad PV. 1999. Effects of riparian forest removal on fish assemblages in southern Appalachian streams. *J. Environ. Manag.* 13:1454–65

Jones KB, Neale AC, Nash MS, Van Remortel RD, Wickham JD, et al. 2001. Predicting nutrient and sediment loadings to streams from landscape metrics: a multiple watershed study from the United States Mid-Atlantic Region. *Landsc. Ecol.* 16:301–12

Junk WJ, Bayley PB, Sparks RE. 1989. The flood pulse concept in river-floodplain systems. In *Proc. Int. Large River Symp., Can. Spec. Publ. Fish. Aquat. Sci.*, ed. DP Dodge, 106:110–27

Karr JR. 1991. Biological integrity—a long-neglected aspect of water-resource management. *Ecol. Appl.* 1:66–84

Karr JR, Chu EW. 2000. Sustaining living rivers. *Hydrobiologia* 422/423:1–14

Klauda R, Kazyak P, Stranko S, Southerland MT, Roth NE, Chaillou J. 1998. Maryland biological stream survey: a state agency program to assess the impact on anthropogenic stress on stream habitat quality and biota. *Environ. Monit. Assess.* 51:299–316

Klein RD. 1979. Urbanization and stream water quality impairment. *Water Resour. Bull.* 15:948–63

Kolpin DW, Furlong ET, Meyer MT, Thurman EM, Zaugg SD, et al. 2002. Pharmaceuticals, hormones, and other organic wastewater contaminants in US streams, 1999–2000: a national reconnaissance. *Environ. Sci. Technol.* 36:1202–11

Lammert M, Allan JD. 1999. Assessing biotic integrity of streams: effects of scale in measuring the influence of land use/cover and habitat structure on fish and macroinvertebrates. *Environ. Manag.* 23:257–70

Lenat DR. 1984. Agriculture and stream water-quality—a biological evaluation of erosion control practices. *Environ. Manag.* 8:333–43

Lenat DR, Crawford JK. 1994. Effects of land-use on water-quality and aquatic biota of three North Carolina Piedmont streams. *Hydrobiologia* 294:185–99

Liess M, Schulz R. 1999. Linking insecticide contamination and population response in an agricultural stream. *Environ. Toxicol. Chem.* 18:1948–55

Limburg KE, Schmide RE. 1990. Patterns of fish spawning in the Hudson River tributaries: Response to an urban gradient? *Ecology* 71:1238–45

Lowrance R, Altier LS, Newbold JD, Schnabel RR, Groffman PM, et al. 1997. Water quality functions of riparian forest buffers in Chesapeake Bay watersheds. *Environ. Manag.* 21:687–712

Lowrance R, Todd R, Fail JJ, Hendrickson OJ, Leonard R, Asmussen L. 1984. Riparian forests as nutrient filters in agricultural watersheds. *BioScience* 34:374–77

Lyons J, Trimble SW, Paine LK. 2000. Grass versus trees: managing riparian areas to benefit streams of central North America. *J. Am. Water Resour. Assoc.* 36:919–30

Mainstone CP, Parr W. 2002. Phosphorus in rivers—ecology and management. *Sci. Total Environ.* 282:25–47

Maret TR, Maccoy DE. 2002. Fish assemblages and environmental variables associated with hard-rock mining in the Coeur d'Alene River Basin, Idaho. *Trans. Am. Fish. Soc.* 131:865–84

Maridet L, Wasson JG, Philippe M, Amoros C. 1995. Benthic organic-matter dynamics in three streams—riparian vegetation or bed morphology control. *Arch. Hydrobiol.* 132:415–25

Martin TL, Kaushik NK, Trevors JT, Whiteley HR. 1999. Review: denitrification in temperate climate riparian zones. *Water Air Soil Poll.* 111:171–86

May CW, Horner RR, Karr JR, Mat BW, Welch EB. 1997. Effects of urbanization on small streams in the Puget Sound Lowland Ecoregion. *Watershed Prot. Tech.* 2:485–94

McRae SE, Allan DJ, Burch JB. 2004. Reach- and catchment-scale determinants of distribution of freshwater mussels (Bivalvia: Unionidae) in south-eastern Michigan, USA. *Freshw. Biol.* 49:127–42

Meador MR, Goldstein RM. 2003. Assessing water quality at large geographic scales: relations among land use, water physicochemistry, riparian condition, and fish community structure. *Environ. Manag.* 31:504–17

Meyer JL, Sale MJ, Mulholland PJ, Poff NL. 1999. Impacts of climate change on aquatic ecosystem functioning and health. *J. Am. Water Resour. Assoc.* 35:1373–86

Meyer WB, Turner BL, eds. 1994. *Changes in Land Use and Land Cover: A Global Perspective*. New York: Cambridge Univ. Press. 537 pp.

Montgomery DR. 1999. Process domains and the river continuum. *J. Am. Water Resour. Assoc.* 35:397–410

Montgomery DR, MacDonald LH. 2002. Diagnostic approach to stream channel assessment and monitoring. *J. Am. Water Resour. Assoc.* 38:1–16

Morley SA, Karr JR. 2002. Assessing and restoring the health of urban streams in the Puget Sound Basin. *Conserv. Biol.* 16:1498–509

Naiman RJ, Decamps H. 1997. The ecology of interfaces: riparian zones. *Annu. Rev. Ecol. Syst.* 28:621–58

Naiman RJ, Johnston CA, Kelley JC. 1988. Alteration of North American streams by beaver. *BioScience* 38:753–62

Nerbonne BA, Vondracek B. 2001. Effects of local land use on physical habitat, benthic macroinvertebrates, and fish in the Whitewater River, Minnesota, USA. *Environ. Manag.* 28:87–99

Nilsson C, Berggren K. 2000. Alterations of riparian ecosystems caused by river regulation. *BioScience* 50:783–92

Niyogi DK, Simon KS, Townsend CR. 2003. Breakdown of tussock grass in streams along a gradient of agricultural development in New Zealand. *Freshw. Biol.* 48:1698–708

Norris RH, Hawkins CP. 2000. Monitoring river health. *Hydrobiologia* 435:5–17

Norris RH, Thoms MC. 1999. What is river health? *Freshw. Biol.* 41:197–209

Odum EP, Finn JT, Franz EH. 1979. Perturbation theory and the subsidy-stress gradient. *BioScience* 29:349–52

Osborne LL, Kovacic DA. 1993. Riparian vegetated buffer strips in water-quality restoration and stream management. *Freshw. Biol.* 29:243–58

Osborne LL, Wiley MJ. 1988. Empirical relationships between land use/cover and stream water quality in an agricultural watershed. *J. Environ. Manag.* 26:9–27

Ourso RT, Frenzel SA. 2003. Identification of linear and threshold responses in streams along a gradient of urbanization in Anchorage, Alaska. *Hydrobiologia* 501:117–31

Palmer MA, Swan CM, Nelson K, Silver P, Alvestad R. 2000. Streambed landscapes: evidence that stream invertebrates respond to the type and spatial arrangement of patches. *Landsc. Ecol.* 15:563–76

Pan YD, Stevenson RJ, Hill BH, Kaufmann PR, Herlihy AT. 1999. Spatial patterns and ecological determinants of benthic algal assemblages in Mid-Atlantic streams, USA. *J. Phycol.* 35:460–68

Paul MJ, Meyer JL. 2001. Streams in the urban landscape. *Annu. Rev. Ecol. Syst.* 32:333–65

Perkins SJ. 1994. The shrinking Cedar River. Channel changes following flow regulation and bank armoring. In *Effects of Human-Induced Changes on Hydrologic Systems, Proc. Am. Water Resour. Assoc. 1994 Annu. Summer Symp*, Herndon, VA, 1994, ed. RA Marston, V Haffurther, pp. 649–58

Peterson JT, Kwak TJ. 1999. Modeling the effects of land use and climate change on riverine smallmouth bass. *Ecol. Appl.* 9:1391–404

Poff NL. 1997. Landscape filters and species traits: towards mechanistic understanding and prediction in stream ecology. *J. N. Am. Benthol. Soc.* 16:391–409

Poff NL, Allan DJ. 1995. Functional organization of stream fish assemblages in relation to hydrologic variability. *Ecology* 76:606–27

Poff NL, Allan DJ, Bain MB, Karr JR, Prestegaard KL, et al. 1997. The natural flow regime: a paradigm for conservation and restoration of riverine ecosystems. *BioScience* 47:769–84

Quinn JM. 2000. Effects of pastoral development. In *New Zealand Stream Invertebrates: Ecology and Implications for Management*, ed. KJ Collier, MJ Winterbourn, pp. 208–29. Christchurch, NZ: Caxton

Quinn JM, Brown PM, Boyce W, Mackay S, Taylor A, Fenton T. 2001. Riparian zone classification for management of stream water quality and ecosystem health. *J. Am. Water Resour. Assoc.* 37:1509–15

Quinn JM, Hickey CW. 1990. The magnitude of the effects of substrate particle size, recent flooding, and catchment development on benthic invertebrates in 88 New Zealand Rivers. *N. Z. J. Mar. Fresh.* 24:411–27

Regetz J. 2003. Landscape-level constraints on recruitment of Chinook salmon (*Oncorhynchus tshawytscha*) in the Columbia River Basin, USA. *Aquat. Conserv.* 13:35–49

Richards C, Haro RJ, Johnson LB, Host GE. 1997. Catchment- and reach-scale properties as indicators of macroinvertebrate species traits. *Freshw. Biol.* 37:219–30

Richards C, Johnson LB, Host GE. 1996. Landscape-scale influences on stream habitats and biota. *Can. J. Fish. Aquat. Sci.* 53:295–311

Robinson CT, Tockner K, Ward JV. 2002. The fauna of dynamic riverine landscapes. *Freshw. Biol.* 47:661–77

Rolland RM. 2000. A review of chemically-induced alterations in thyroid and vitamin A status from field studies of wildlife and fish. *J. Wildl. Dis.* 36:615–35

Roth NE, Allan JD, Erickson DL. 1996. Landscape influences on stream biotic integrity assessed at multiple spatial scales. *Landsc. Ecol.* 11:141–56

Roy AH, Rosemond AD, Paul MJ, Leigh DS, Wallace JB. 2003. Stream macroinvertebrate response to catchment urbanisation (Georgia, USA). *Freshw. Biol.* 48:329–46

Schlosser IJ. 1985. Flow regime, juvenile abundance, and the assemblage structure of stream fishes. *Ecology* 66:1484–90

Schlosser IJ. 1991. Stream fish ecology: a landscape perspective. *BioScience* 41:704–12

Schulz R, Liess M. 1999. A field study of the effects of agriculturally derived insecticide input on stream macroinvertebrate dynamics. *Aquat. Toxicol.* 46:155–76

Scott MC, Helfman GS. 2001. Native invasions, homogenization, and the mismeasure of integrity of fish assemblages. *Fisheries* 26:6–15

Sedell JR, Froggatt JL. 1984. Importance of streamside forest to large rivers: the isolation of the Willamette River, Oregon, USA, from its floodplain by snagging and streamside forest removal. *Verh. Int. Ver. Theor. Ang. Limnol.* 22:1828–34

Skinner JA, Lewis KA, Bardon KS, Tucker P, Catt JA, Chambers BJ. 1997. An overview of the environmental impact of agriculture in the U.K. *J. Environ. Manag.* 50:111–28

Snelder TH, Biggs BJF. 2002. Multiscale river environment classification for water resources management. *J. Am. Water Resour. Assoc.* 38:1225–39

Sovell LA, Vondracek B, Frost JA, Mumford KG. 2000. Impacts of rotational grazing and riparian buffers on physicochemical and biological characteristics of southeastern Minnesota, USA, streams. *Environ. Manag.* 26:629–41

Sponseller RA, Benfield EF, Valett HM. 2001. Relationships between land use, spatial scale and stream macroinvertebrate communities. *Freshw. Biol.* 46:1409–24

Stauffer JC, Goldstein RM, Newman RM. 2000. Relationship of wooded riparian zones and runoff potential to fish community composition in agricultural streams. *Can. J. Fish. Aquat. Sci.* 57:307–16

Steedman RJ. 1988. Modification and assessment of an index of biotic integrity to quantify stream quality in southern Ontario. *Can. J. Fish. Aquat. Sci.* 45:492–501

Stepenuck KF, Crunkilton RL, Wang LZ. 2002. Impacts of urban land use on macroinvertebrate communities in southeastern Wisconsin streams. *J. Am. Water Resour. Assoc.* 38:1041–51

Stewart JS, Wang LZ, Lyons J, Horwatich JA, Bannerman R. 2001. Influences of watershed, riparian-corridor, and reach-scale characteristics on aquatic biota in agricultural watersheds. *J. Am. Water Resour. Assoc.* 37:1475–87

Storey RG, Cowley DR. 1997. Recovery of three New Zealand rural streams as they pass through native forest remnants. *Hydrobiologia* 353:63–76

Strand M, Merritt RW. 1999. Impact of livestock grazing activities on stream insect communities and the riverine environment. *Am. Entomol.* 45:13–27

Strayer DL, Beighley RE, Thompson LC, Brooks S, Nilsson C, et al. 2003. Effects of land cover on stream ecosystems: roles of empirical models and scaling issues. *Ecosystems* 6:407–23

Sutherland AB, Meyer JL, Gardiner EP. 2002. Effects of land cover on sediment regime and fish assemblage structure in four southern Appalachian streams. *Freshw. Biol.* 47:1791–805

Sweeney BW. 1993. Effects of streamside vegetation on macroinvertebrate communities of White Clay Creek in eastern North America. *Proc. Acad. Nat. Sci. Philadelphia,* Philadelphia, PA

Tockner K, Ward JV, Edwards PJ, Kollmann J. 2002. Riverine landscapes: an introduction. *Freshw. Biol.* 47:497–500

Tong STY, Chen WL. 2002. Modeling the relationship between land use and surface water quality. *J. Environ. Manag.* 66:377–93

Townsend CR. 1989. The patch dynamics concept of stream community ecology. *J. N. Am. Benthol. Soc.* 8:36–50

Townsend CR, Doledec S, Norris R, Peacock K, Arbuckle C. 2003. The influence of scale and geography on relationships between stream community composition and landscape variables: description and prediction. *Freshw. Biol.* 48:768–85

Townsend CR, Doledec S, Scarsbrook MR. 1997. Species traits in relation to temporal and spatial heterogeneity in streams: a test of habitat templet theory. *Freshw. Biol.* 37:367–87

Townsend CR, Hildrew AG. 1994. Species traits in relation to a habitat templet for river systems. *Freshw. Biol.* 31:265–76

Townsend CR, Riley RH. 1999. Assessment of river health: accounting for perturbation pathways in physical and ecological space. *Freshw. Biol.* 41:393–405

Turner MG, Gardner RH, O'Neill RV. 2001. *Landscape Ecology in Theory and Practice: Pattern and Process.* New York: Springer

Usseglio-Polatera P, Beisel JN. 2002. Longitudinal changes in macroinvertebrate assemblages in the Meuse River: anthropogenic effects versus natural change. *River Res. Appl.* 18:197–211

Usseglio-Polatera P, Bournaud M, Richoux P, Tachet H. 2000a. Biological and ecological traits of benthic freshwater macroinvertebrates: relationships and definition of groups with similar traits. *Freshw. Biol.* 43:175–205

Usseglio-Polatera P, Bournaud M, Richoux P, Tachet H. 2000b. Biomonitoring through biological traits of benthic macroinvertebrates: How to use species trait databases? *Hydrobiologia* 422:153–62

Vannote RL, Minshall WG, Cummins KW, Sedell JR, Cushing CE. 1980. The river continuum concept. *Can. J. Fish. Aquat. Sci.* 37:130–37

Walser CA, Bart HL. 1999. Influence of agriculture on in-stream habitat and fish community structure in Piedmont watersheds of the Chattahoochee River System. *Ecol. Freshw. Fish* 8:237–46

Walsh CJ, Sharpe AK, Breen PF, Sonneman JA. 2001. Effects of urbanization on streams of the Melbourne region, Victoria, Australia. I. Benthic macroinvertebrate communities. *Freshw. Biol.* 46:535–51

Wang L, Lyons J, Rasmussen P, Seelbach P, et al. 2003. Watershed, reach, and riparian influences on stream fish assemblages in the Northern Lakes and Forest Ecoregion, U.S.A. *Can. J. Fish. Aquat. Sci.* 60:491–505

Wang L, Lyons J, Kanehl P. 2001. Impacts of urbanization on stream habitat and fish across multiple spatial scales. *Environ. Manag.* 28:255–66

Wang L, Lyons J, Kanehl P. 2002. Effects of watershed best management practices on habitat and fish in Wisconsin streams. *J. Am. Water Resour. Assoc.* 38:663–80

Wang L, Lyons J, Kanehl P, Bannerman R, Emmons E. 2000. Watershed urbanization and changes in fish communities in southeastern Wisconsin streams. *J. Am. Water Resour. Assoc.* 36:1173–89

Wang L, Lyons J, Kanehl P, Gatti R. 1997. Influences of watershed land use on habitat quality and biotic integrity in Wisconsin streams. *Fisheries* 22:6–12

Ward JV. 1998. Riverine landscapes: biodiversity patterns, disturbance regimes, and aquatic conservation. *Biol. Conserv.* 83:269–78

Ward JV, Tockner K, Arscott DB, Claret C. 2002. Riverine landscape diversity. *Freshw. Biol.* 47:517–39

Waters TF. 1995. *Sediment in Streams*. Bethesda, MD: Am. Fish. Soc.

Watzin MC, McIntosh AW. 1999. Aquatic ecosystems in agricultural landscapes: a review of ecological indicators and achievable ecological outcomes. *J. Soil Water Conserv.* 54:636–44

Weller DE, Jordan TE, Correll DL. 1998. Heuristic models for material discharge from landscapes with riparian buffers. *Ecol. Appl.* 8:1156–69

Wiens JA. 1989. Spatial scaling in ecology. *Funct. Ecol.* 3:385–97

Wiens JA. 2002. Riverine landscapes: taking landscape ecology into the water. *Freshw. Biol.* 47:501–15

Wissmar RC, Beschta RL. 1998. Restoration and management of riparian ecosystems: a catchment perspective. *Freshw. Biol.* 40:571–85

Wood PJ, Armitage PD. 1997. Biological effects of fine sediment in the lotic environment. *Environ. Manag.* 21:203–17

Woodward DF, Goldstein JN, Farag AM, Brumbaugh WG. 1997. Cutthroat trout avoidance of metals and conditions characteristic of a mining waste site: Coeur d'Alene River, Idaho. *Trans. Am. Fish. Soc.* 126:699–706

Wright JF. 1995. Development and use of a system for predicting macroinvertebrates in flowing waters. *Aust. J. Ecol.* 20:181–97

Yuan LL, Norton SB. 2003. Comparing responses of macroinvertebrate metrics to increasing stress. *J. N. Am. Benthol. Soc.* 22:308–22

LONG-TERM STASIS IN ECOLOGICAL ASSEMBLAGES: Evidence from the Fossil Record*

W.A. DiMichele,[1] A.K. Behrensmeyer,[1] T.D. Olszewski,[2] C.C. Labandeira,[1] J.M. Pandolfi,[3] S.L. Wing,[1] and R. Bobe[1]

[1]*Department of Paleobiology, National Museum of Natural History, Smithsonian Institution, Washington, DC 20560; email: dimichele.bill@nmnh.si.edu, behrensmeyer.kay@nmnh.si.edu, labandeira.conrad@nmnh.si.edu, wing.scott@nmnh.si.edu, bobe.rene@nmnh.si.edu*
[2]*Department of Geology and Geophysics, Texas A&M University, College Station, Texas 77843; email: tomo@geo.tamu.edu*
[3]*Center for Marine Studies and Department of Earth Sciences, University of Queensland, St. Lucia, Queensland 4072, Australia; email: j.pandolfi@uq.edu.au*

Key Words ecosystem stability, assembly rules, paleoecology, ecological persistence, environmental tracking

■ **Abstract** Studies of plant and animal assemblages from both the terrestrial and the marine fossil records reveal persistence for extensive periods of geological time, sometimes millions of years. Persistence does not require lack of change or the absence of variation from one occurrence of the assemblage to the next in geological time. It does, however, imply that assemblage composition is bounded and that variation occurs within those bounds. The principal cause for these patterns appears to be species-, and perhaps clade-level, environmental fidelity that results in long-term tracking of physical conditions. Other factors that influence persistent recurrence of assemblages are historical, biogeographic effects, the "law of large numbers," niche differentiation, and biotic interactions. Much research needs to be done in this area, and greater uniformity is needed in the approaches to studying the problem. However, great potential also exists for enhanced interaction between paleoecology and neoecology in understanding spatiotemporal complexity of ecological dynamics.

INTRODUCTION

The objective of this paper is to review the concept of ecological persistence on the basis of evidence from the fossil record and to evaluate the extent to which such evidence supports the proposition that assemblages of organisms can remain "the

---

*The U.S. Government has the right to retain a nonexclusive, royalty-free license in and to any copyright covering this paper.

same" over thousands to perhaps millions of years in the face of environmental perturbations and species invasions. Such claims have been made on the basis of data from the marine and terrestrial fossil records, although counterarguments are equally common. In those instances that supposedly document persistence, how has it been measured, and what is indicated about the underlying cause or causes?

Contrasting views of plant and animal communities as ephemeral or persistent entities can be traced back to the works of Clements and Gleason (Golley 1993, McIntosh 1995), which pitted an "organismal" against an "individualistic" model of community organization, respectively. In more recent years, this debate has emerged in other guises. Do communities and ecosystems have emergent properties that might confer some resilience or homeostasis in the face of environmental change? Are they, instead, momentary associations that simply reflect the independent responses of local populations to changing environments and biotic interactions, and is continuity through time thus little more than a happenstance?

The role of the fossil record in our understanding of this issue has focused almost exclusively on Quaternary pollen and spores and to a lesser extent on insects, which have been cited as empirical confirmation of individualism. Here, we bring data from the fossil record, particularly that of the pre-Quaternary, to bear on the questions of long-term stasis in biotic systems. Although conceptual models founded in modern ecology can be difficult to test directly with fossil data, the fossil record may reveal patterns that do not follow from direct extrapolation of neoecological concepts. Finding ways to address empirically the differences in temporal and spatial scales between these sources of data is the challenge that could bring us to a richer and more comprehensive ecological theory.

WHAT DO PALEONTOLOGISTS WANT TO KNOW THAT IS OF RELEVANCE TO ECOLOGISTS?

Paleontologists are interested in the biological, physical, and chemical forces that have shaped evolution, as well as in the phylogenetic and biodiversity-related "what" and the genetic "how" of the history of life. Ecological questions in paleontology involve the role of environment in evolution and focus on many of the same variables that ecologists measure in modern systems (e.g., species richness and other diversity measures, climate, substrate, and biogeography), except that paleontologists and geologists can document these variables, with varying degrees of resolution, over much longer time periods and compare the resulting patterns in different places and through time. Examples of paleoecological questions include:

1. To what degree is the differential diversification of lineages related to environmental or ecological controls as opposed to intrinsic traits, and why have some lineages proliferated and diversified over geological time, whereas others have persisted with little morphological change? This question is

relevant to current ecological issues such as invasive species and draws on a rich geological and fossil record of continental connections, climatic shifts, and successes and failures of lineages over time and in different biogeographic arenas.

2. Why do some assemblages of organisms coexist for long periods of time, whereas other assemblages change rapidly and have little long-term continuity? This question is relevant to current debates about ecological assembly rules and the resilience or interdependency of organisms that co-occur in the same environments.

3. Why are some ancient ecosystems more resistant than others to externally driven biological change (e.g., invasion) or environmental stress (e.g., climate shifts)? The fossil record includes many examples of ecosystem persistence in the face of environmental change and also of thresholds where ecosystems break down and are replaced by novel floral and faunal assemblages.

4. How does ecology affect rates of evolution? Are the effects of species-environment associations on rates of cladogenesis, anagenesis, and extinction discernible? At the time scale of modern ecology, do these questions translate into the effects of environment on ranges and geographic variation of persistently co-occurring taxa, their ability to invade or colonize new habitats, and their susceptibility to habitat change? More simply, could the fossil record provide us with indicators of which species or ecomorphs are more or less resistant to short-term ecological stress or responsive to new opportunities?

ECOLOGICAL VERSUS PALEONTOLOGICAL CONCEPTS AND DATA

Differences in the nature of time, space, and resolution as perceived by ecologists and paleontologists have often led to differences in terminology. Clarification of common ground, while also recognizing irreconcilable differences, is important for future communication. Thus, we begin with a review of basic concepts and terms that also serves to highlight the different domains and perspectives of the ecologist and the paleontologist.

The most evident difference is the scale of time over which change is observed. Ecologists study the response of communities and ecosystems to change on very short time scales, whereas paleoecologists have access to a wide range of temporal scales (Figure 1). Because both migration and speciation occur over time, paleoecologists may borrow, though with caution, some of the descriptive tools developed by ecologists when appropriate for patterns in the fossil record (Miller 1993, 1996; DiMichele 1994). However, a danger of "me-too ecology" exists in which the paleontologist works too hard to serve up paleoecological evidence on a neoecological platter and not only misses important biases in the data but also

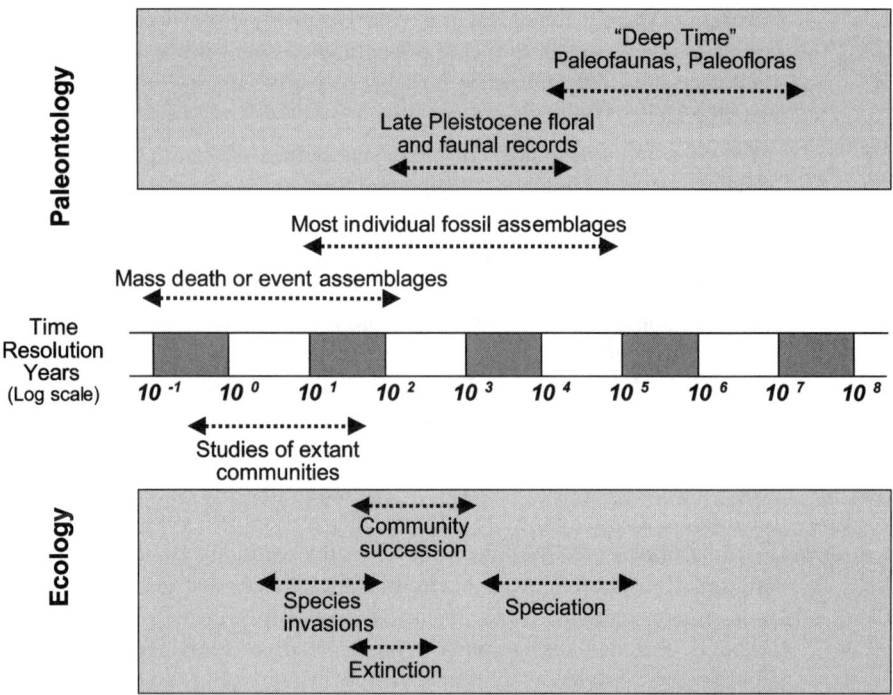

Figure 1 Comparison of the spans of temporal resolution that characterize ecological and paleontological data. Individual fossil assemblages may encapsulate varying degrees of time (i.e., time-averaging), depending on the circumstances of death and burial (taphonomy) and collection method, whereas samples of extant communities generally represent single or repeated censuses over days to decades. In paleontology, long-term floral or faunal records developed from multiple fossil assemblages are the basis for the analysis of paleocommunity stasis or change though time (*upper gray bar*). In ecology, with the exception of speciation, processes that can be observed or inferred in recent ecosystems (*lower gray bar*) occur over much shorter time intervals than can be resolved by paleontological data.

overlooks what might have been truly different about past ecosystems (Olson 1980, 1985).

The basic terms that describe long-term ecosystem behavior have been applied primarily to models in ecology and, thus, are largely conceptual (Pimm 1984). "Complexity" can be determined by the number of species in the system (richness), the number of interspecific interactions (connectance), the magnitude of the interaction among species (interaction strength), and the relative distribution of species abundances in an assemblage (evenness). Pimm (1991) described five ecological meanings for the term "stability." Each represents a pattern of change through time, independent of time scale and, therefore, applicable to a discussion of paleoecological stasis. Stability of a system is central to ecological stasis. A stable system, when at equilibrium, returns to the equilibrium condition after perturbation (Holling 1973). A stable system must be understood in terms of a range of variation around some equilibrium point. Related to stability is "persistence," an empirical measure of how long a system remains unchanged according to some defined measures of its state. "Resilience" is the rate at which a displaced variable returns to its equilibrium value. The shorter the return time, the more resilient a variable or system is. In a different formulation, resilience can be considered a measure of the amount of disturbance a system at equilibrium can absorb before moving to a new stable state (Gunderson et al. 2002). "Resistance" describes the effect on the system of a permanent change in a variable, such as the response to a disturbance. If other variables show little change, then the system is resistant to changes in the altered parameter. "Variability" is the degree to which a parameter varies over time. Although it can be measured with a statistic such as standard deviation, the duration of the examined interval must be noted so that the scale of variation can be accounted for.

The relationship between diversity/complexity and stability remains at the core of debate in ecology (May 1974, Tilman 1999, Cottingham et al. 2001). Topics of contention include the degree to which model systems capture reality, whether the focus of investigations should be on populations or on communities, and the very definition of stability itself (DeWoody et al. 2003). A corollary to these basic concepts is the process of ecosystem assembly and attendant assembly rules (Lockwood et al. 1997, Belyea & Lancaster 1999, Weiher & Keddy 2001), which can be divided broadly into rules of access to a resource and rules of coexistence on that resource.

All these concepts overtly assume that many ecosystems are capable of existing close to an equilibrium point (for a critique, see Peters 1993). They also are consistent with models of hierarchical ecosystem organization, which recognize that emergent properties (properties of the system that are not properties of any of the system's components, its species) may appear only at certain spatial or temporal scales of analysis (Allen & Hoekstra 1992, Ulanowicz 1997, Maurer 1999).

Systems also can be characterized on the basis of functional types (Simberloff & Dayan 1991, Smith et al. 1997), or ecomorphs, which are groups composed of taxa that are presumed to have similar ecological requirements and to play similar roles within an ecosystem. Systems with high levels of overlap among functional types have high levels of redundancy. High redundancy may be most common in unpredictable and variable environments (Walker 1997), although exceptions certainly abound. Coral reefs, for example, can be considered predictable environments but also can have high species diversity and high degrees of apparent redundancy.

Paleontologists and paleoecologists deal with samples of organisms and environments that are spread out in time as well as in space, and they have developed a system of new terms for unique concepts or different meanings for existing terms that reflect this domain. Fossil species, like living species, are not randomly associated with one another, and this fact allows paleoecologists to recognize biofacies, "a body of rock distinguished on the basis of its fossil content" or "a local assemblage or association of living or fossil organisms, especially one characteristic of some type of marine conditions" (Jackson & Bates 1997). Biofacies can be mapped through both time and space because different taxa are associated with different environmental conditions. The term has been used primarily in marine deposits where the composition of the biota is constrained by water temperature, depth, energy, turbidity, and chemistry, all of which may be independently recorded in physical and chemical features of the associated sediments. The term has been applied less often to terrestrial deposits, although it is equally useful there, where the flora or fauna is constrained by sedimentary characteristics.

The typical paleoecological sample is an assemblage of fossil remains collected from a single rock layer or stratum or an assemblage combined from individual samples derived from a specified thickness of strata (Behrensmeyer & Hook 1992). In certain cases, such assemblages faithfully represent original ecological communities, for example, a community that experienced a catastrophic death and burial event or one in which all the fossils are composed of rapidly degraded materials such as leaf litter (Wing & DiMichele 1995). In cases of preserved animal hard parts or accumulated resistant plant parts, the sample usually represents a period of time that is long by ecological standards, from 10^2 to 10^5 Kyr (thousands of years) (Figure 1) and, thus, may include organismal remains from more than one ancient assemblage. This incorporation of a temporal element into the sample is referred to as "time-averaging" (Johnson 1960), and is one of the major ways that paleontological faunal and some floral lists differ from ecological ones. Time-averaging also is present in some neoecological studies, such as those that assess plant damage in extant tropical forest litter, which may involve combining data from different seasons, years, or even longer intervals. For fossil assemblages composed of hard parts (e.g., shells, bones, or wood), diversity measures, rank-order of abundance, and other variables are almost always based on samples that combine the results of short-term fluctuations and provide average measures of paleocommunity characteristics over hundreds or thousands of years (Figure 1). This method can be advantageous (Kidwell 2002, Behrensmeyer et al. 2000), but significantly time-averaged paleocommunity data cannot be regarded as strictly comparable to a large proportion of census data for modern ecosystems. Thus, the short-term cycles or trends over the years or decades observed in ecological studies are largely invisible to paleontologists and paleoecologists, especially in the vertebrate or marine invertebrate fossil records. In pre-Quaternary depositional settings, where high temporal resolution can be achieved, correlation of changes in the biota at such a fine scale across a large area ordinarily is not possible. Moreover, the absence of a taxon from a particular assemblage where it might be expected can be the result of preservational sampling biases, collecting methods, or original ecological controls. A number of strategies for coping with such problems are available. These strategies include quantitative assessment of apparent versus real taxonomic ranges (i.e., origination and extinction rates) (Marshall 1997, Foote 2000, Holland 2003), holding depositional context and taphonomic situation constant through time (isotaphonomy) (Behrensmeyer & Hook 1992, Behrensmeyer & Chapman 1993), and modeling time-averaging effects on the basis of modern analogs and "taphonomic control taxa" (Kidwell & Holland 2002).

In both the marine and the nonmarine realms, "paleocommunity" has come to mean a group of taxa that form a recurring, recognizable assemblage of organisms. Criteria for defining the boundaries of biofacies or paleocommunities vary greatly. One of the most stringent definitions is that of Bambach & Bennington (1996), who restricted the term community (i.e., paleocommunity) to groups of collections that cannot be statistically distinguished on the basis of their species composition and abundance. They used the term community type (i.e., paleocommunity type) for aggregates of collections and paleocommunities that have similar but not identical taxonomic composition and occur in similar but not identical environments. This second term acknowledges that significant variation can exist among individual collections because of variation among habitats. In other words, a paleocommunity type (or a biofacies) can represent a metacommunity that varies from place to place on an ecological landscape. Watkins et al. (1973) provided an interesting distinction between Petersen communities, "a local, descriptive unit of recurring combinations of species that can be reproduced in the fossil record," (i.e., a biofacies) and Thorson parallel communities, "groups of Petersen communities related by common ancestry of contained taxa." The second concept explicitly introduces an evolutionary and historical aspect to paleocommunities, which allows them to have continuity through geologic time and global distribution, despite species differences, as long as member taxa represent related evolutionary lineages.

Much has been written about turnover and turnover events or pulses in the land mammal fossil record (e.g., Vrba 1993, McKee 2001). To paleontologists and paleoecologists, turnover means the proportion of extinctions and originations relative to the total number of taxa in a specified time interval. Turnover events are higher than normal rates of turnover relative to a clearly defined "background" rate. In ecology, turnover most typically means a change in species composition over a specified geographic or ecological boundary; that is, turnover in space rather than through time. However, numerous examples exist of turnover occurring over short time scales (years to decades), such as successional changes in old fields or after tree fall in forests, or after storms in marine systems. These uses of the term turnover touch on the much larger issue of the validity of using ecological variability on a modern time scale as an analog for temporal variability and change through the geological record.

As in studies of modern ecosystems, functional types, or ecomorphs, have also been examined in the fossil record in an attempt to reduce the effects of phylogenetic history and focus on patterns of stability or change in suites of functionally related characters. Examples include both whole communities and specific ecological features of fossil mammals (Damuth 1992, Van Valkenburgh 1995, Janis et al. 2000), plants (Wing & DiMichele 1992, DiMichele et al. 2001b), and insects (Labandeira 1998, 2002).

Application of Watkins et al.'s (1973) concept and related concepts that incorporate a historical dimension to communities has led to research in the global paleobiogeography and evolutionary history of diverse paleocommunities (Boucot 1975, Boucot & Lawson 1999). On the basis of this work, Boucot (1983) defined a series of Ecologic-Evolutionary Units (EEUs) in the Phanerozoic and emphasized the ecological context of evolution within this framework. Other workers, however, following Gleasonian traditions, maintain that paleocommunities are simply associations of taxa with similar habitat preferences and do not represent an emergent level of biotic organization (Hoffman 1979, Jablonski & Sepkoski 1996). Resolving this dispute remains an important focus in evolutionary paleobiology.

ARE PERSISTENT COMMUNITY ASSEMBLAGES POSSIBLE?

At least four hypotheses have been proposed to explain the persistence of community composition over geological time. First, species co-occurrence may reflect evolved ecological relationships that confer net mutual benefits when species occupy the same environment at the same time. Such species interactions may determine community structure and function (e.g., competition, symbiosis, or predation). Second, a significant overlap in species environmental tolerances may exist such that some species almost always occupy the same environment. Third, geographic isolation may occur because of such factors as place of evolutionary origin, vicariance patterns, and natural geographic barriers; these factors, previously set in motion, may be responsible for species co-occurrence at any given time. Fourth, the "law of large numbers" may apply, which states that the most abundant species remain most abundant because they tend to produce more offspring than the less abundant species.

Individualistic Versus Interactive Models

Central to ecological stasis is the nature of the organization of ecological assemblages. Are communities simply short-term associations of species living together because of momentarily similar responses to environmental conditions (Gleason 1926), or are such assemblages integrated in some manner that leads to emergent, system-level properties (Clements 1916)? This discussion dates to the early twentieth century (Golley 1993) but continues today in debates over assembly rules (Belyea & Lancaster 1999, Weiher & Keddy 2001), the effects of biodiversity on ecosystem stability and productivity (May 1974, Cottingham et al. 2001), and the proper construction of null models (Gotelli & Graves 1996, Hubbell 2001).

Until now, the paleontological studies that have been most widely recognized by neoecology are those that have offered support to the proposition that "communities" are transient associations of species that share similar resource requirements under a given set of climatic/environmental conditions. Most of this evidence came initially from palynological studies of the last interglacial interval in the Northern Hemisphere temperate zone (Davis 1986, Webb 1987). More recently, the records of terrestrial vertebrates (Graham 1997) and marine invertebrates (Jablonski & Sepkoski 1996, Patzkowsky & Holland 1999) have been used to document the same position, but examples to the contrary have also been presented (Pandolfi 1996, 2000). The nature of this individualism is not universally agreed upon, however, and may be dependent on scale of analysis; it may be most evident at some smaller scales within environmentally bounded species pools (biomes in the terrestrial world) but not at larger scales between species pools, where species exchange is less common.

Patterns of Stratigraphic Persistence

Olson (1952, 1958, 1980) was among the first paleontologists to document community persistence through well-defined intervals of rock strata by presenting case studies from the Permian deposits of Texas and Oklahoma. Olson established the term chronofauna for such assemblages of vertebrate taxa that recurred through long stratigraphic intervals and attempted to reconstruct food webs for the component species (Olson 1952). Numerous examples of fossil vertebrate faunas reflect some degree of continuity through time, although whether these faunas could be considered communities in the ecological sense is debatable. North American Cenozoic "Land Mammal Ages" and their equivalents on other continents are based on the persistence of suites of taxa for particular chronostratigraphic intervals, which allows correlation over wide geographic areas (Woodburne 1987, Janis et al. 1998). At a rather coarse level of resolution, the success of such schemes attests to the persistence of at least some components of mammalian faunas and the communities of which they were a part. Basically, the pattern at both regional and continental scales is similar to that observed in the marine invertebrate and plant fossil records; long intervals of relative stability are punctuated by periods of rapid change.

A similar pattern of assemblage persistence has been described by Boucot (1978, 1983) in the course of carrying out biostratigraphic and evolutionary studies of marine faunas. He recognized that invertebrate assemblages did not demonstrate patterns of continuous turnover but rather were organized into temporally coherent units between which significant differences existed (EEUs) and within which smaller units, or communities, could be recognized on the basis of recurrent patterns of composition (Sheehan 1996).

Brett & Baird (1995) introduced the term coordinated stasis to connote a pattern of bimodal evolutionary turnover within benthic marine biofacies of the Silurian and Devonian of the Appalachian Basin (Table 2). They reported a pattern in which the majority of species within a biofacies showed little, if any, evolutionary change over a period of several million years (Myr), followed by rapid extinction and replacement of those species in the same environment. To provide explicit criteria for comparison with other time periods and geographic areas, Brett et al. (1996) stated that coordinated stasis units should exceed 1 Myr in duration, during which approximately 60% of species should persist and show little morphological change. In this scheme, fewer than 40% (typically fewer than 20%) of species cross the bounding intervals, which should be no more than one tenth of the duration of the static intervals. Speciation and extinction are concentrated in the intervals of rapid turnover. The pattern of coordinated stasis need not imply strong Clementsian codependence within a community. Episodic environmental change could affect many evolutionary lineages that act independently of one another. In such a case, a pattern of coordinated stasis would reflect the timing and severity of punctuated environmental changes. Additionally, samples from a coordinated stasis unit need not all display identical species composition or abundance; room for environmental variability exists within a biofacies, as acknowledged in the definition of Bennington & Bambach's (1996) paleocommunity types (similarly by Olson 1952, 1985). Rather, the pattern is defined by coincident turnover that punctuates protracted periods of community-wide morphological stasis. Although the pattern of coordinated stasis could simply result from episodic environmental change, some authors have interpreted it as reflecting the influence of ecological interactions on the evolutionary trajectory of the lineages within a community. Morris et al. (1995) suggested a mechanism of ecological locking, "a mechanism by which ecological interactions prevent evolutionary change, resulting in long-lasting, stable systems capable of resisting some types of disturbance" (Morris et al. 1995, p. 11,272), as the necessary implication of coordinated stasis in the fossil record. Whether or not ecological locking is correct, it is not the only possible explanation of the pattern of coordinated stasis.

Coordinated stasis has been challenged because of the supposed implications of strong species interactions as a cause of the pattern (e.g., Buzas & Culver 1994, Patzkowsky & Holland 1999). In addition, strong arguments have been made regarding the need for statistical analysis of supposed patterns of persistence (Bennington & Bambach 1996, Bambach & Bennington 1996); that is, just how much variation can occur if assemblages are to be considered the same through time? Alroy (1996) carried out a statistical analysis of an extensive compilation of mammalian stratigraphic ranges and determined that the expectations of the coordinated stasis model, such as pulse-like turnover of species, did not hold.

Is the "Unified Neutral Theory of Biodiversity and Biogeography" a Link Between Rates of Taxonomic Change and Community Structure?

One of the difficulties of studying community stability in the fossil record is that most ecological models of community structure implicitly assume a static equilibrium for species richness and abundances (see review by Tokeshi 1993). An alternative approach that has received much recent attention is neutral modeling of community dynamics. Hubbell's (2001) *The Unified Neutral Theory of Biodiversity and Biogeography* assumes that local communities maintain a constant number of individuals. Consequently, new individuals can only enter the community by filling vacancies opened by the death of established individuals. Three mechanisms of replacement exist in Hubbell's model. First, a death can be offset by the birth to a parent already present in the local community. Second, an individual from the surrounding metacommunity can be recruited to fill the vacant slot and can represent a species that is not already a member of the local community. Third, a local birth could also be a speciation event, so the replacement individual represents an entirely new species. This system is neutral because all individuals have an equal probability of dying, reproducing, and speciating in a given time interval, regardless of their species identity.

The model predicts an equilibrium abundance distribution and expected richness, given the size of the community, the degree of exchange with the surrounding metacommunity, and the rate of per capita speciation (incorporated into Hubbell's fundamental biodiversity number). In contrast to statistical or niche-partitioning models (Tokeshi 1993), Hubbell's model predicts change in the taxonomic composition through time because of community drift, which reflects the constant stochastic elimination and replacement of individuals. Although common species are expected to endure longer than rare species, a consequence of large numbers effects, newborn species (or previously rare species colonizing new habitat) are not likely to rise to dominance, and previously dominant species are not likely to go extinct. The rate and likelihood of these events depends on the model parameters.

The potential usefulness of a neutral modeling approach for paleoecology lies not in whether it is right or wrong, but rather that it provides a first step toward developing a theory that ties evolutionary history to the structure of ecological communities and focuses on aspects that can be measured by use of fossil data (i.e., rates of origination and abundance distributions). Time will tell whether this potential will be fulfilled (Chave 2004), but some paleontological tests have already been formulated (Pandolfi 1996, Pandolfi & Jackson 1997, Clark & McLachlan 2003, Olszewski & Erwin 2004).

CASE STUDIES: TERRESTRIAL RECORD

The case studies mentioned in this section are summarized in Table 1.

Palynofloras: Compositional Stasis and Temporal Scale of Observation

Stasis in the composition of biotas has been observed over a wide variety of spatial and temporal scales, but many of the observations that suggest rapid and unpredictable change in community composition were derived from the past 40 Kyr of Earth history—the time during which samples can be dated by radiocarbon methods. An important question, particularly in considering the relevance of paleontological data for ecological theory, is whether changes in the temporal scale of sampling affect the detection of stasis.

To test whether temporal scale has an effect on detecting stasis in composition, we downloaded fossil pollen data from the World Data Center for Paleoclimatology (WDCP) (http://www.ngdc.noaa.gov/paleo/pollen.html). Each dataset consisted of counts of the abundances of pollen taxa in core samples. Abundance data are

TABLE 1 Summary of paleontological studies with evidence relating to persistence and punctuated change in terrestrial paleocommunities of vertebrates, invertebrates, and plants[a]

Authors	Study (age)	Duration	Resolution	Geographic scale	Taxa	Patterns through time
Barry et al. 1995, 2002	Siwalik mammals, Middle to Upper Miocene (10.7–5.7 Mya)	5 Myr	100 Kyr	Potwar Plateau, northern Pakistan	115 mammal taxa: Insectivora, Scandentia, Primates, Tubulidentata, Proboscidea, Lagomorpha, Perissodactyla, Artiodactyla, Rodentia Pholidota	Background turnover relatively high (50%–60%) but not correlated with changes in fluvial paleoenvironments; three turnover events within 100–300 Kyr intervals account for 44% of the faunal change; extinctions and appearances are not coincident in time, latter two of the three turnover events appear to be correlated with climate change
Bobe et al. 2002	Omo Shungura Formation (4.0–1.5 Mya)	2.5 Myr	3–100 Kyr	Southern Ethiopia, northern Turkana Basin	Mammals: bovids, suids, primates	Persistence of most taxa with species turnover at 2.8 Mya and after 2.0 Mya; dominant bovids stable from 3.5 to 2.0 Mya, new dominants subsequently; large mammal paleocommunity as a whole stable for 300 Kyr interval followed by 100 Kyr cyclicity in taxonomic abundances
DiMichele & Phillips 1996b, DiMichele et al. 1996, 2002	Pennsylvanian (Desmoinesian/ Westphalian D, 306–310 Mya and Missourian/ Stephanian A 303–306 Mya)	4 plus 3 Myr	Coal beds and assemblages within a coal bed: 1–10 Kyr	Southern Illinois Basin	Peat-forming plants preserved in coal balls: lycopsids, ferns, sphenopsids, pteridosperms, cordaites	Recurrent intraswamp community patterns among successive coal beds during the Westphalian; patterns terminated by a major extinction at end of Westphalian; replacement of the Desmoinesian communities by new kinds of persistent assemblages in the Missourian

Reference	Time interval	Duration	Depositional setting	Geographic area	Taxa and data	Description of stasis
DiMichele & Aronson 1992	Pennsylvanian and Early Permian	17.5 Myr	Fossiliferous clastic beds: 100 yr–1 Kyr	Europe and North America	Plants: compression floras	Two distinct biomes/species pools that persist side by side in the tropics through the Pennsylvanian and into the Early Permian with few common species (mainly weedy wetland forms)
Labandeira et al. 2002	Latest Cretaceous and earliest Paleocene (~66–64 Mya)	2.2 Myr	20 Kyr	Williston Basin, North Dakota, North America	Insect damage: 51 types documented on >13,000 dicotyledonous leaf fossils	Insect damage diverse and persistent before the Cretaceous-Paleocene boundary, then damage levels and diversity decrease particularly for host-specialist insects; this pattern persists through 800 Kyr of the early Paleocene
Wilf et al. 2001	Latest Paleocene to early middle Eocene (~56–43 Mya)	13 Myr	Individual samples probably represent <1 Kyr	Green River and Unita basins, North America	Insect damage: 40 types documented on 2435 leaf fossils from 58 host species	Three sample levels based on six quarry sites across the Eocene Continental Thermal Maximum (ECTM) interval show persistence in feeding types but changes in intensity and distribution of damage based on host-plant antiherbivore strategies
Olson 1952, 1958	Early Permian (Clear Fork Group) (~275 Mya)	3–5 Myr	Individual assemblages probably represent 100s to 1000s of years	North central Texas, North America	21 genera and 32 species: fish, amphibians, reptiles	Faunal assemblages associated with different environments (upland, stream, pond margin, and pond) can be followed through time and persist during periods of environmental change with some turnover but overall continuity in the taxonomic and ecomorphic character of the chronofauna

(Continued)

TABLE 1 (Continued)

Authors	Study (age)	Duration	Resolution	Geographic scale	Taxa	Patterns through time
Scott 1978	Late Carboniferous (Westphalian B, 311–313 Mya)	3 Myr	Fossiliferous clastic beds: 100 yr–1 Kyr	West Yorkshire coal measures, United Kingdom	Plants: compression floras	Recurrent paleoenvironmentally specific plant assemblages identified from multiple sampling horizons through time; links lithological patterns to biological patterns and shows they are recurrent (biofacies)
Pfefferkorn et al. 2000	Lower Carboniferous (Namurian A, 320–327 Mya)	6.2 Myr	Fossiliferous clastic beds: 100 yr–1 Kyr	Ostrava-Karvina Basin, Czech Republic	Plants: compression floras	Recurrent plant associations recognized for periods of about 2 Myr each, punctuated by a rapid turnover and establishment of new types of species assemblages in next interval of persistence
Pfefferkorn & Thomson 1982	Late Carboniferous (300–319 Mya)	19 Myr	Fossiliferous clastic beds: 100 yr–1 Kyr	Europe and North America	Plants: compression floras	Analysis at major group level (orders); recurrent associations of taxa and environments of deposition (biofacies) with a major change in dominance patterns occurring near the Westphalian-Stephanian boundary (paralleling coal-ball patterns)
Falcon-Lang 2003	Late Carboniferous (Westphalian A, 313–315 Mya)	1.6 Myr maximum	Fossiliferous clastic sedimentary cycles, 50–200 Kyr duration with finer resolution of individual beds	Joggins Formation, Bay of Fundy, Nova Scotia	Plants: compression floras	Eight transgressive-regressive sedimentary cycles, glacially driven; three recurrent plant assemblages: wetland, dominated by lycopsids; dryland, dominated by cordaites and sigillarian lycopsids; and coastal, dominated by progymnosperms and gymnosperms

[a]Abbreviations: Mya = millions of years ago; Myr = millions of years; Kyr = thousands of years.

desirable because over geologically short time periods, local extinction or immigration of species is rare. We also used pollen data from the Paleogene (Wing & Harrington 2001).

Change in the composition of the pollen assemblages through time was quantified by application of squared chord distance (Cd_{ij}), a dissimilarity measure that incorporates relative abundance data (Overpeck et al. 1985). Time-steps between samples (t) were calculated from age models in the WDCP pollen database (Quaternary data), or from magnetostratigraphic, biostratigraphic, and isotopic correlation to the Global Magnetic Polarity Time Scale (Paleogene data). Ages of samples between dated horizons were interpolated.

If community composition were completely static, Cd_{ij} would equal 0 between any pair of samples. In the real world, fluctuations in sample composition result from small changes in assemblage composition (e.g., tree falls or local fires) or from small changes in the way the vegetation is represented by the pollen sample. If these small, unpredictable fluctuations in composition are not directional, no correlation should exist between the Cd_{ij} and the time difference (t) between samples i and j. If change in floral composition is directional, a significant positive correlation should exist between Cd_{ij} and t.

The analysis of both Quaternary and Paleogene assemblages reveals that large changes in composition are rare, and most change in composition is fluctuating; therefore, Cd_{ij} is poorly correlated with t. Consequently, samples of similar age give high rates of compositional change (Cd_{ij}/t is large because t is small), whereas samples separated by more time yield low rates (Cd_{ij}/t is small because t is large). Rates of change in floral composition are, thus, highly inversely correlated with t, and rates measured over different intervals are difficult to compare. Short-term (e.g., less than 10 Kyr) fluctuations in composition are likely missed in deep-time datasets, which would, thus, appear to be stable. Conversely, Quaternary datasets generally are too short to capture rare extinctions or immigrations; thus, they do not show the irreversible changes in species-pool composition that are seen in many deep-time records. Differing temporal resolution thus complicates the comparison of rates of change.

Carboniferous Wetland Assemblages

The Late Carboniferous (Pennsylvanian) was a time of generally cool global climate with intervals of intense tropical precipitation that supported rainforests (Gastaldo et al. 1996) and vast peat swamps that became the coal beds of Europe and the eastern United States. Upper Carboniferous rocks appear to reflect, in part, glacial periodicity and may preserve Milankovitch cyclicity (Algeo & Wilkinson 1987). Fossil remains of the plants from peat swamp (mire) forests are preserved as "coal balls" (petrified peat), as compression-impression fossils in mudstones and sandstones, or as spores and pollen. The occurrence of plant fossils in multiple coal beds and in the intervening rocks permits the study of changes in plant composition through time under recurrent, common environmental conditions (isotaphonomy, sensu Behrensmeyer & Hook 1992). Within coal beds, fossil plant and spore-pollen samples can be collected incrementally, which permits vegetational dynamics to be resolved at time scales of less than 100 Kyr. In addition, dynamics can be examined at many sampling horizons (i.e., coal beds) and, thus, through multiple glacial-interglacial cycles and in response to both background and large-scale extinctions (Phillips et al. 1985).

Phillips and coworkers (DiMichele & Phillips 1996a,b; DiMichele et al. 1996; DiMichele et al. 2002) examined more than 50 coal beds, which represented more than 10 Myr, and reported the following basic patterns. Within any one coal bed, multiple, recurrent plant communities can be identified statistically. These communities recur in successive coals, recognized by the approximate rank-order of abundance of the dominant elements; minor taxa vary widely in abundance. A major extinction eliminated nearly two thirds of the species at the Middle–Late Pennsylvanian boundary, approximately 306 million years ago (Mya). This extinction appears to have been caused by a short pulse of global warming and drying in the tropics (Phillips & Peppers 1984, Frakes et al. 1992). After a brief interval of high variability in dominance patterns immediately after the extinction (Peppers 1996), peat-forming landscapes reorganized, and groups previously in low abundances, particularly opportunistic tree ferns, rose to dominance by replacing the prior dominants that had been eliminated by the climatic changes. This represented an internal reorganization of the ever-wet peat-substrate species pool. The pattern of vegetational persistence has been detected in Late Pennsylvanian peat-forming environments (Willard et al. 2004) by an analysis of rank-order distribution of dominant tree ferns in several successive coals in eastern Illinois. A parallel change in dominance patterns in tropical floodbasin floras occurred at approximately the same time, although the reported taxonomic resolution is at the level of families and classes (Pfefferkorn & Thomson 1982).

Patterns similar to those found in coals have been documented in floras from floodbasin sedimentary rocks (sandstones and mudstones) lying between coal beds. In deep-core samples from the beginning of the Late Carboniferous ice age taken from the Ostrava-Karvina coal basin in the Czech Republic, three to four vegetational units that conform to distinct sequential rock formations, separated by abrupt compositional turnover, were identified statistically (Pfefferkorn et al. 2000). In Westphalian B age clastic rocks from England, Scott (1978) identified recurrent patterns of species association with particular sedimentary environments (biofacies) over a 3-Myr time interval. Specific, recurrent sedimentary environments have been tied to qualitatively identified plant assemblages of distinctively different dominance and diversity patterns in the well-known Joggins section in Nova Scotia (Falcon-Lang 2003). The temporal patterns of vegetational change have been placed in a sequence-stratigraphic framework, in which rock units have been related to cycles of rising and falling sea level, presumably controlled in large part by cycles of glaciation and eustasy.

In the north temperate Angaran floristic province, Meyen (1982) documented an extensive stratigraphic sequence of Carboniferous floras at the species level. Assemblage membership persists in this stratigraphic sequence over many sampling units (beds) though time. Floristic changes occur essentially instantaneously between longer intervals of persistence.

Biome Exchange Patterns in the Late Paleozoic Tropics

During the Late Pennsylvanian and Early Permian, floras dominated by phylogenetically advanced, xeromorphic seed plants began to appear in the tropics. The fossil record shows these new plants intercalated between wetland floras in association with indicators of seasonally dry climatic conditions (Cridland & Morris 1963, Mamay & Mapes 1992, Rothwell & Mapes 1988). Ziegler (1990) referred to this biome as "summer wet." The seasonally dry/summer-wet biome and its flora ultimately replaced the ever-wet biome in the tropics as global glaciation ended in the Early Permian (see papers in Martini 1997). Broutin et al. (1990) and DiMichele & Aronson (1992) demonstrated that little floristic mixing occurred between these two species pools. Fragmentary evidence of the seasonally dry biome, in the form of transported conifer scraps, appears millions of years before the first in situ macrofossils (Lyons & Darrah 1989), which suggests growth in remote, well-drained upland areas. A biome characteristic of yet drier conditions, composed of still more phylogenetically derived plants, characteristic of the Late Permian and Mesozoic, appeared in the Early Permian in association with gypsum beds (DiMichele et al. 2001a). This unusual and unexpected flora reiterates the earlier pattern of precocious conifer occurrence and again implies strong ties of different biomes or vegetation types to different climatic-edaphic conditions.

These examples demonstrate that when similar environmental conditions can be identified over long intervals of time (in effect, removing significant environmental variation as a causative factor), species-pool composition may persist, and quantitative relationships among these species also may recur. This observation does not mean that similar environments caused the persistence. In fact, neither the persistence of species dominance-diversity patterns nor sharply defined, long-lasting boundaries between species pools are general expectations under the tenets of species individualism, given the continuously occurring effects of disturbance, migration, and low-level environmental variation.

Stability Within Versus Between Trophic-Level Interactions: Fossil Insect Patterns

Typically, most paleoecological data are collected within a trophic level, a taxonomic group, or both, such as analyses of Paleozoic corals, Mesozoic ammonites, or Cenozoic plants. Intertrophic data and comparisons usually focus on taxon-specific ecological interactions that have a relatively good and persistent fossil record, such as Paleozoic platyceratid gastropods and their crinoid and blastoid associates (Baumiller & Gahn 2002) or Cenozoic naticid gastropods and their molluscan prey (Kelley & Hansen 2001). Insect-feeding traces on fossil leaves provide evidence for the broadest effects of one trophic level on another in terms of diversity, abundance, and persistence of associations and in their implications for community-level ecological reorganization. A broad spectrum of insect damage has been documented on various plant organs, especially leaves, throughout the post-Silurian terrestrial record (Labandeira 2002).

Two case studies are relevant to establishing intertrophic inferences: the sudden end-Cretaceous extinction event, and the Early Cenozoic Thermal Maximum (ECTM) (Table 1). The Cretaceous-Paleogene and ECTM studies detect sudden to prolonged intervals of persistent insect damage patterns as well as wholesale shifts in community-wide structure that involved a cascade of floral changes that had significant impact on feeding patterns of constituent insect herbivores.

Fossil floras from the late Cretaceous and early Paleogene provide an excellent record of insect damage in lieu of a poor record of insect body fossils (Labandeira et al. 2002). In the geographically extensive Williston Basin of North Dakota, 143 floras from 106 time slices encompass a 2.2-Myr interval that straddles 183 m of the Cretaceous-Paleogene boundary (Table 1). The average geochronologic resolution is 20 Kyr, although the stratigraphic placement of samples throughout the section is not uniform. Patterns of insect herbivory were diverse and persistent before the boundary. This abundance was followed by a significant decrease in the diversity and level of insect herbivore damage at the Cretaceous-Paleogene boundary. Host-specialist insects suffered preferentially greater extinction at the boundary than generalists. Finally, no post-event rebound is evident within the 0.8-Myr interval of the earliest Paleocene, although other data (Wilf et al. 2001) indicate return several Myr later.

A profound change in the physical global environment occurred from the latest Paleocene through early-middle Eocene (Zachos et al. 2001), described as the ECTM, mentioned above. A study in which floras from the Greater Green River and Uinta Basins of the Western Interior were used (Wilf et al. 2001) documented insect herbivory from the beginning of the ECTM (56 Mya; humid warm-temperate to subtropical climate, predominantly deciduous plants) to near the maximum (53 Mya; humid subtropical climate, mixed deciduous and evergreen plants) and significantly after the peak warming (43 Mya; seasonally dry subtropical climate, mixed deciduous and thick-leaved evergreen plants). Throughout the entire 13-Myr interval, significantly more herbivory occurred on plant hosts with shorter-lived and thinner leaves than on plants with longer-lived and thicker leaves. This variance in herbivore modes was statistically separable throughout the three time slices examined and widened during the middle Eocene. These data, consistent with modern herbivore defense theory (Coley & Barone 1996), demonstrate that during these major climatic shifts, herbivory was partitioned into an accommodationist strategy that allowed for high levels on deciduous mesic hosts and an antiherbivore defensive strategy of lower levels on strongly defended, mostly evergreen plants.

These studies represent two very different modes of community-level reorganization among insects and their plant hosts. The Cretaceous-Paleogene boundary study indicates a major disruption of associational diversity that took until the Paleocene-Eocene boundary to return to predisturbance levels. As for the latest ECTM, no significant loss of insect feeding types within floras or local community-wide rearrangement of insect-plant associations occurred, but significant changes did occur in intensity and distribution of damage based on highly defended versus poorly defended plant antiherbivore strategies.

Turnover in Miocene Mammalian Faunas of Pakistan

More than 3 decades of work on the vertebrate faunas of northern Pakistan has generated a 10-Myr record of stasis and change in the mammalian community of this region and allows a comprehensive examination of evolutionary and ecological change in relation to environmental parameters (Barry et al. 2002). The intensively sampled Siwalik deposits (more than 40,000 fossil specimens) provide a temporal resolution of 100 Kyr between 10.7 and 5.7 Mya, which Barry et al. (2002) suggest may be the finest feasible level of resolution for long sequences of vertebrate-bearing strata. By use of a large suite of 115 mammal taxa, Barry et al. (2002) demonstrated a moderately high and persistent level of "background" turnover (50% to 60%) over a period of 5 Myr. Superimposed on this background are three separate short-term turnover events that also changed the character of the mammalian community. The first event involves extinction of many long-lived taxa that are recorded in southern Asia before their demise 10.3 Mya. The second and third events are separated by only about 500 Kyr and occur at 7.8 and between 7.3 and 7.0 Mya during a time of independently documented climate change toward intensified monsoons and the spread of grassland habitats (Dettman et al. 2000). Barry et al. (2002) concluded that this study supports neither the notion of coordinated stasis nor environmentally driven turnover events as the dominant mode of faunal change through time. However, turnover events documented in the Siwalik record have 3 to 13 times the expected rate relative to the background turnover (background average of 1.5 taxonomic appearances or disappearances per 100 Kyr compared with 9, 5, and 20 times the expected rate during the three turnover events). This finding provides evidence for some degree of stability of faunas in the Siwalik ecosystem over time intervals that were very long relative to modern communities. Reinforcing this notion is the fact that major changes through time in the Siwalik fluvial systems had no apparent impact on the composition of the fauna; that is, no correlation of lithologic or paleoenvironmental change with the turnover events occurred. This observation indicates that the mammalian fauna, in successive time-averaged samples of approximately 100 Kyr each, was resistant to changes in the substrate environment but sensitive to the major climate changes of the late Miocene.

Stability and Change in the East African Pliocene Mammal Record

The African late Cenozoic fossil record has been subject to intensive research because of interest in human evolution, and this research provides some of the highest resolution paleontological evidence available for mammalian community structure through time. Deposits along the lower Omo Valley in southern Ethiopia include a sequence of nearly 800 m of sediments from approximately 4 to 1 Mya (Brown 1994, de Heinzelin 1983, Feibel et al. 1989). A large and carefully documented collection of more than 40,000 fossil specimens from the Shungura Formation provides examples of faunal change at a number of different temporal scales (Bobe et al. 2003, and references therein). Analysis for ecological patterns has been done for specific groups of mammals (bovids, suids, and hominins and other primates), and time intervals can be resolved to approximately 10^3 years in parts of the sequence (de Heinzelin 1983).

The time encompassed by the Shungura Formation coincided with major climatic and environmental changes in Africa (deMenocal 1995, Vrba 1995), and some of these changes are correlated in time with changes in the mammalian fauna (Alemseged 2003, Bobe et al. 2002, Bobe & Eck 2001). Species turnover, based on first and last occurrences, is low overall between 3.5 and 2.0 Mya; the only marked turnover event occurred at 2.85 Mya and corresponds to the onset of Northern Hemisphere cooling (deMenocal & Bloemendal 1995). Examination of the relative abundances of the major mammalian taxa (4,820 specimens) provides a higher resolution record of faunal change in which a period of stability between 2.8 and 2.5 Mya (five stratigraphic sample levels) is followed by a cyclical pattern of shifting taxonomic dominance over 100-Kyr intervals up to 2.0 Mya (Bobe et al. 2002). Statistical tests indicate that the interval of stasis is unlikely to be sampling error, which provides evidence for ecological stability over several hundred thousand years, when global climates were becoming cooler and more variable (deMenocal & Bloemendal 1995). Bobe et al. (2002) speculate that the paleo-Omo River system, with its large drainage area, helped to buffer the lower riverine floodplain and gallery forest habitats from the impact of larger-scale climate change. This effect persisted for several hundred thousand years, until around 2.5 Mya, when external changes finally penetrated the local system and destabilized the ecological communities of the lower Omo Valley.

The Shungura sequence provides additional evidence for faunal stability and ecological persistence. The two most abundant species of the dominant family (Bovidae) co-occur throughout the interval from about 3.5 Mya to 2.0 Mya, with *Aepyceros shungurae* (early impala) and *Tragelaphus nakuae* (similar to bongo) alternating in first and second place (Bobe & Eck 2001). This persistent association may indicate that both species had similar tolerance limits for the wooded and moist environments of the Pliocene-Pleistocene lower Omo River. After 2.0 Mya, during a period of increased environmental change, *T. nakuae* became extinct, and *A. shungurae* became a less conspicuous element of the Omo bovid fauna.

CASE STUDIES: MARINE RECORD

The case studies mentioned in this section are summarized in Table 2.

Soft-Bottom Benthic Invertebrate Faunas

Studies of turnover through time in pre-Quaternary, marine, soft-bottom communities can be somewhat arbitrarily divided into two categories on the basis of their

TABLE 2 Examples of long-duration and high-resolution paleontological studies, with evidence relating to persistence and punctuated change in soft-bottom paleocommunities of marine invertebrates[a]

Authors	Study (age)	Duration	Resolution	Geographic scale	Taxa	Patterns through time
Long-Duration Studies						
Ludvigsen & Westrop 1983, Westrop 1996, Westrop & Cuggy 1999	Upper Cambrian (Marjuman-Sunwaptan Stages) ~495–500 Mya	10–15 Myr	~1 Myr biozones	North American continent	Trilobites: >50 supraspecific taxa	Generic-level stability punctuated by three large extinction events; generic composition persistent within intervals between extinctions; species exhibit continuous turnover; each extinction followed by rapid radiation; turnover episodes cut across biofacies; inarticulate brachiopods have longer species longevities (i.e., greater persistence) than trilobites
Patzkowsky & Holland 1997	Upper Ordovician (Mohawkian and Cincinnatian Series) ~455–480 Mya	17 Myr	Depositional sequences ~1.3 Myr each	Eastern North America	Articulate brachiopods: 96 genera, 441 species	Three periods of low turnover separated by two episodes of rapid turnover; earlier turnover episode reflects elevated extinction rates and later episode reflects invasion of new genera; turnover events correspond to extreme perturbations of regional ocean-climate system
Brett & Baird 1995	Silurian and Devonian (Llandovery through Givetian Series)	45 Myr	Evolutionary-ecological subunits 2–8 Myr each	Appalachian Basin in New York State	Corals, bryozoans, brachiopods, mollusks, trilobites, echinoderms, and other groups (hundreds of species)	Very high persistence within ecological-evolutionary subunits punctuated by episodes of very rapid turnover; extinction events closely followed by rapid radiations; nearshore assemblages show greater taxonomic persistence than open-marine assemblages

(*Continued*)

TABLE 2 (Continued)

Authors	Study (age)	Duration	Resolution	Geographic scale	Taxa	Patterns through time
Olszewski & Patzkowsky 2001	Pennsylvanian-Permian (Desmoinesian to Wolfcampian Series) ~280–310 Mya	17.5 Myr	450 Kyr cyclothems	Northern midcontinent in Kansas and Nebraska	Brachiopods: 38 genera, 102 species; bivalves: 47 genera, 94 species	Low but constant rates of background turnover punctuated by episodes of appearance or disappearance; appearance and disappearance episodes not correlated in time; turnover histories of brachiopods and bivalves not correlated
Tang & Botjer 1996	Jurassic 142–206 Mya	28 Myr	Depositional units 2 to 6 Myr	Western North America	Bivalves: 49 genera, 79 species	Episodes of elevated turnover could not be recognized; high levels of species persistence reported; origination and extinction of taxa in this region not closely correlated in time
High-Resolution Studies						
Holterhoff 1996	Upper Pennsylvanian (Stephanian)	0.5 Myr	~10^4–10^5 yr	Nebraska, Kansas, Oklahoma	Crinoids	Several biofacies recognized in both transgressive and regressive portions of a single stratigraphic cycle; one biofacies restricted to transgressive part (no analog?); collections within biofacies show great deal of variation

Study	Period	Duration	Location	Taxa	Notes
Bennington & Bambach 1996	Middle Pennsylvanian (Westphalian)	<10 Myr	0.5–2.5 Myr	Eastern Kentucky	Brachiopods and mollusks	Biofacies (paleocommunity types) are recurrent in four marine incursions; but samples (local paleocommunities) can be statistically significantly different
Olszewski & Patzkowsky 2001	Pennsylvanian-Permian (Stephanian-Wolfcampian)	~2.5 Myr	~0.5 Myr	Nebraska, Kansas	Brachiopods and bivalves	Biofacies remain distinct in each stratigraphic cycle but show a great deal of variation in the taxonomic membership of component collections
Bonuso et al. 2002	Devonian (Givetian)	15 Myr	~1–3 Myr	Central New York State	Corals, brachiopods, mollusks, trilobites, echinoderms, and other groups	Significant taxonomic variation through section; dominant taxa and ecological groups change through section; stasis not observed
Pandolfi 1996	Pleistocene	95 Kyr	~10^2–10^3 yr	35 km (PNG)	Reef corals	Persistence in taxonomic composition and species diversity through nine separate reef-building episodes

Source: Modified from Olszewski & Patzkowsky 2001.
[a]Abbreviations: Mya = millions of years ago; Myr = millions of years; Kyr = thousands of years.

temporal resolution (Table 2). High-resolution studies focus on communities at less than 1 million year resolution, whereas long-duration studies focus on changes over more than 1 million years. Although this distinction is somewhat artificial, the critical difference is whether evolutionary turnover is expected within the units being compared. High-resolution studies should show little evolutionary change within a time interval, whereas in long-duration studies, even background rates of turnover can result in observable change. In addition, high-resolution studies are usually based on individual collections, whereas long-duration studies are based on compiled lists.

High-resolution studies are noted in Table 2, grouped in the lower portion of the table. Such studies require a stratigraphic framework that allows correlation at levels finer than the time between turnover events (if these events occur in the study interval). One reason that three of the five cited studies in Table 2 (Bennington & Bambach 1996, Holterhoff 1996, Olszewski & Patzkowsky 2001) come from the Carboniferous-Permian period is the strong stratigraphic cyclicity that characterizes these rocks. Despite some of the distinct differences in these studies, none refute the observation that biofacies are recurrent (even in the face of significant environmental change). However, all five studies found that the composition of local communities within a biofacies can be quite flexible and that stability was only evident at a larger geographic scale. Also worth noting is Holterhoff's (1996) report of a nonanalog crinoid assemblage like the nonanalog Quaternary plant assemblages reported by Overpeck et al. (1992), which suggests that communities outside the Pleistocene could also respond to environmental change by reassembling rather than by extinction of member lineages.

Longer-term, soft-bottom marine studies are grouped together at the top of Table 2. These studies range in age from Cambrian to Jurassic (Table 2). These studies focus on durations of more than 10 Myr and generally consist of faunal lists from stratigraphic intervals of less than 1 Myr duration. The spatial scale of these studies usually includes multiple biofacies of an entire basin or platform. A number of generalizations can be made from examining Table 2. First, most studies do show times of elevated taxonomic turnover relative to background rates, although the rates themselves vary greatly. At least some studies (Tang & Bottjer 1996, Patzkowsky & Holland 1997, Olszewski & Patzkowsky 2001) show that episodes of elevated first appearance do not necessarily coincide with, or immediately follow, episodes of disappearance (as in Westrop 1996 and Brett & Baird 1995). In those studies that distinguished biofacies (Westrop 1996, Brett & Baird 1995, Olszewski & Patzkowsky 2001), the history of turnover appeared to differ between biofacies.

In addition, the frequency and magnitude of regional turnover episodes appear to decrease with time (Olszewski & Patzkowsky 2001). Whether this pattern will hold up to future scrutiny is not clear, but it does have implications for global Phanerozoic diversity patterns, which show decreasing rates of extinction and origination through time. This pattern may reflect decreasing susceptibility of ecological communities or their component taxa to turnover events caused by environmental perturbations of sufficient magnitude. Whether or not this interpretation is corroborated by future work, it makes the point that the fossil record can provide information on intermediate scales that connect ecological and biogeographic processes to paleontological patterns at Phanerozoic scales (Miller 1998).

Quaternary Coral Reef Dynamics

The ecological dynamics of living reef communities observed over yearly to decadal time scales is characterized by fluctuating species composition with changing environmental conditions (Connell 1978, Sale 1988, Tanner et al. 1994, Bak & Nieuwland 1995, Connell et al. 1997). Thus, community structure varies unpredictably over small temporal scales. Similarly, studies at small spatial scales (small areas on single reefs, less than 1 km) have shown convincingly that both fish (Williams 1980, Sale & Douglas 1984, Sale & Steel 1989, Doherty & Williams 1988, Sale 1988, Sale et al. 1994) and corals (Tanner et al. 1994, Connell et al. 1997) show a high degree of variability in community structure. Thus, living coral reefs appear to show a large degree of disorder in community composition at small spatial and temporal scales.

In contrast, results from several recent studies of Quaternary reefs all point to remarkable persistence in taxonomic composition and diversity during multiple episodes of global climate change over the past 500 Kyr (Jackson 1992; Hubbard et al. 1994; Stemann & Johnson 1992; Pandolfi 1996, 1999; Aronson & Precht 1997; Greenstein et al. 1998). Pandolfi (1996, 1999) assembled and analyzed data on species distribution patterns from reef coral assemblages, aged between 125 Kyr and 30 Kyr, from three sites along 35 km of the Huon Peninsula in Papua New Guinea. Persistence in species composition was found among assemblages that lived during nine successive, glacially induced, high–sea level stands (Figure 1) (Pandolfi 1996, 1999). Data on species distribution patterns from reef coral assemblages, aged between 220 Kyr and 104 Kyr, from three sites along 25 km of Barbados also showed persistence in species composition among assemblages that lived during four successive, glacially induced, high–sea level stands. These patterns were shown by use of species relative-abundance data, and the rare taxa showed similar trends to the common taxa (Pandolfi 2000).

Aronson & Precht (1997) and Aronson et al. (2002, 2004) documented a change in the dominant coral in Belize and Panama that occurred recently. In the Belize lagoon, *Acropora cervicornis*, dominant for the past 3000 years in cores, was replaced by *Agaricia tenuifolia* in the 1980s as the dominant coral species. Similarly, in the Bocas del Toro lagoon in Panamá, *Porites* spp., dominant for over 3000 years, also has now been replaced as the dominant coral by *Agaricia tenuifolia*.

These consistent species-distribution patterns demonstrate that Pleistocene reef communities comprised more predictable associations of reef coral species, over broad spatial and temporal scales, than those observed by ecologists on living reefs at smaller scales. Such broad-scale persistence in community structure is in good agreement with several studies on living reefs of fish (Ault & Johnson 1998, Robertson 1996) and corals (Geister 1977) examined at a scale of less than 1 km, as well as with a recent study that documented broad-scale predictability in living adult coral abundance for over 2,000 km of the Great Barrier Reef, Australia (Hughes et al. 1999). Patterns of persistence in space and time are, thus, well documented for coral reef communities.

Two underlying causes for persistence in coral reef community structure have been offered. Hubbell (1997) interpreted Pandolfi's reef coral data from Papua New Guinea in terms of the unified neutral theory (Hubbell 2001). He asserted that patterns of persistence in community structure could come about as easily by a "law of large numbers" as they could by niche differentiation and limited membership, as advocated by Pandolfi (1996, 1999). However, his comments were tempered by an impression that Pandolfi had used only the common corals in his analysis, when, in fact, patterns were similar whether the analyses were confined to common species or included rare species. Moreover, data analyzed from Barbados showed that patterns in persistence of common taxa were exactly mimicked in the rare taxa, further evidence that large population size is immaterial to persistence in coral reef communities (Pandolfi 2000). The data from Papua New Guinea and Barbados appear to indicate limitations on membership in reef coral associations in space and time and suggest that these associations are not random assemblies of available species.

DISCUSSION

The case studies are derived primarily from the authors' areas of research and provide only a subset of paleontological research that could be brought to bear on the question of community persistence through time. However, they serve to illustrate the following points.

Improving the Exchange Between Paleoecology and Ecology

Many aspects of communities can be described from fossil data, such as richness, species abundances, and trophic interactions, at temporal resolutions ranging from 10^0 to 10^6 years. Useful generalizations about the ecological and evolutionary processes that underlie these patterns depend on our ability to compare different studies. A good deal of confusion has been caused by workers who examined patterns at different geographic, temporal, or taxonomic scales and came to apparently contradictory conclusions. Our intention in this review is not to set down "correct" procedures or standard sampling protocols (of necessity, these must reflect the stratigraphic and taphonomic context of the study), but to clarify the information that can and should be stated explicitly in any study, theoretical or empirical, of community evolution in the fossil record. To compare the results of different studies, others need to know the spatial and environmental resolution and extent, temporal resolution and duration, and taxonomic and ecomorphic nature of the data. Many of these recommendations already are in use, which reflects broadening recognition of the issues that surround the relationship between the structure of ecological communities and rates of evolution.

First, the spatial resolution and extent of the study should be stated explicitly. How large was the area investigated? In addition, and perhaps more significant, the degree of habitat variability in the area also must be reported. The community of a large, environmentally uniform region may be more comparable to a smaller patch in a more environmentally variable region. When communities of different ages are compared, are the units of comparison individual collections, are they biofacies or paleocommunities that are based on many collections but represent a single habitat, or are they gradients or landscapes that cross multiple habitats? Have habitats been tracked geographically to assess the constancy of species composition through space and time or have they been observed only as recurrent, and temporally separated, sampling intervals within the study area? An additional issue is how biofacies or larger gradients have been identified and whether those of one study are really comparable to those of another study.

Second, the degree of environmental resolution should be considered. Is more than one habitat represented? Are environmental gradients present, and how have such gradients been sampled? What is the range in environmental variability? Is this study conducted within or between habitats?

Third, the temporal resolution and duration of the study should be noted and the degree of time-averaging in individual samples should be included (e.g., Tables 1 and 2). Even a crude estimate can be helpful in providing perspective on the appropriate scale of inferred ecological processes. Stratigraphic horizons, even in some Paleozoic successions, can be resolved at scales of less than 10^6 years, in some cases even representing in situ communities buried catastrophically. Stratigraphic resolution is critical to distinguish between intervals of slow background change or stasis and short episodes of rapid turnover. What is the duration of the study? Some Quaternary studies may represent 10^5 to 10^6 years—an interval that is below the resolution of some (not all) deep-time studies. That the results of studies at different scales may have implications for one another is not in dispute, but they cannot be compared directly as if they were equivalent.

Fourth, biological resolution of the study should be determined. Most paleontological studies use taxa—species, genera, or higher groups—as the basic unit of analysis. However, different levels of taxonomic resolution produce very different expectations for patterns of stability through time. Is the study restricted to a single clade, or does it attempt to include all available groups? Alternatively, functional diversity and ecomorphs also can be examined through time and may produce very different patterns than taxonomic analyses.

Fifth, when patterns are reported, basic descriptive data should be included, in addition to derived measures, because such data allows calculation of alternative metrics or statistics as well as assessment of the amount of data available.

Environmental Tracking

The concept of environmental tracking by floras and faunas has had a mixed reception in the ecological literature. On the basis of the data reviewed here, biotas do appear to track climates, but such tracking is certainly influenced by geographic and physicochemical barriers. Environmental tracking is important for understanding long-term ecological patterns because it removes certain high-level taphonomic megabiases in the study of the factors that may influence the patterns; thus, the range of extrinsic environmental variables that must be considered as causative factors is reduced.

Embedded within this issue is the question of species individuality. The clear recognition of Quaternary "nonanalog" floras (Overpeck et al. 1992) begs the question of scale. Miller (1993) argued that environmental tracking was a characteristic of community types rather than specific communities of marine invertebrates. In effect, a "community type" might be considered to be a "biome" in the sense of terrestrial vegetation. Biomes have been shown in both empirical studies (e.g., Ziegler 1990) and modeling studies (e.g., Kutzbach et al. 1998) to approximate climatic boundaries closely and to expand and contract in concert with climatic changes. Species individuality seems most strongly expressed within these biome-level units (DiMichele et al. 2004).

Limits to biome membership may reflect historical patterns of clade origin and distribution (Valentine 1980). Unexpectedly strong relationships between ecological niche and phylogenetic relatedness of species have been documented for extant plants (Prinzing et al. 2001) and were labeled "phylogenetic niche conservatism." Similar conservatism appears to underlie long-term patterns of within-environment, within-clade, species replacement in Permo-Carboniferous tropical ecosystems (Knoll 1984, DiMichele & Phillips 1996a), rather than such mechanisms as "ecological locking." This feature is not just a pattern found in plants. Quaternary fossil insects (Coope 1994), both individual species and communities, show strong conformance to shifting climatic conditions. Small Pleistocene mammals exhibit strong patterns of co-occurrence, with little disassociation in response to climate change (Alroy 1999). Depending on the scale of analysis, some marine invertebrate communities also show strong long-term persistence within limits of oceanographic conditions (Jackson 1994; Pandolfi 1996, 2000; Brett & Baird 1995); however, such patterns have not been widely demonstrated for many kinds of marine communities (Jablonski & Sepkoski 1996).

Strong environmental tracking does not require high levels of species interaction. However, some recent studies suggest that interspecific interactions can have strong stabilizing effects. A study of Holocene pollen from eastern North America, for example (Clark & McLachlan 2003), found that species abundances stabilized rapidly after glacial retreat and have remained stable over broad geographic scales, which counters the long-standing view of Holocene forests as randomly assembled and ever-changing species associations. Milchunas et al. (1988) have proposed that feedback in subhumid grassland ecosystems that have a long history of grazing has led to "switching mechanisms" and divergent selection that allow different species to become dominant under different stress conditions. This pattern contrasts with that in semiarid grasslands, where dominant species become yet more dominant under stress, and convergent selection reigns. The responses to grazing in grasslands with short evolutionary histories involve little feedback and are characterized by individualistic species' responses to grazing pressure. The existence of "switching mechanisms" among different species could lead to a situation in which different dominance patterns occur in an otherwise stable community. Both of these studies conflict with predictions of Hubbell's (2001) "community drift" model, which emphasizes persistent dominance patterns as simply the result of the "law of large numbers," and not a result of interspecific interactions.

Nonetheless, environmental boundaries and species conformance with them may be the single most important cause of long-term persistence of taxonomic associations in the fossil record. At least species-by-environment conformance is much more easily examined in the fossil record than are interspecific interactions. The key to the matter seems to be spatial scale, related directly to environmental scale. As noted by Zeigler et al. (2003), species distributional patterns may be constrained mainly by atmospheric and oceanic circulation patterns, which have sharp boundaries. Identifying the limits to these boundaries and the nature of their change through time is key to understanding the persistence of assemblages of organisms.

CONCLUSIONS

Above we noted that there are four underlying factors that could contribute to the persistence of species assemblages through long periods of geological time. First was the matter of evolved mutualisms. This factor is without a doubt the most difficult to identify in the fossil record (or in modern communities), but some illuminating research has been conducted on the subject. Studies of reef corals (Pandolfi 1996, 2000) suggest limited membership in certain environments, which points to selectivity, possibly the result of biotic interactions. Similarly, studies of temperate tree patterns (Clark & McLachlan 2003) and grasslands (Milchunas et al. 1988) also suggest mutualisms that may lead to community persistence. Patterns of within-clade ecomorphic replacement in late Paleozoic coal-swamp ecosystems (DiMichele and Phillips 1996a) also hint at the possibility that interspecific interactions may partly control replacement dynamics. Plant-insect associations across sudden and protracted environmental shifts indicate that levels and types of herbivory recoup from regional extinction throughout the fossil record (Wilf et al. 2001, Labandeira & Phillips 2002, Labandeira et al. 2002). Much needs to be done in this area, and most of the work will come from neoecology, with results extrapolated to the fossil record primarily through models.

Second is the issue of environmental control. Nearly all the examples given in this review point strongly to the fact that species are neither randomly nor continuously distributed across the landscape. Boundaries to species distribution clearly exist, and many species have very similar, strongly overlapping ranges and limits. This broad pattern is one that permits the recognition of evolutionary-ecological units and, at the same time, can limit the utility of local studies for global geological temporal correlation. Thus, this situation may be one in which matters of scale, spatial and temporal, can be most clearly seen as important system constraints.

Third is consideration of the effect of historical contingency, which is manifested in biogeographic patterns. Biogeographic limitations lie a level above those placed on species associations by climatic and other environmental factors. Biogeographic factors control the species pool at a fundamental evolutionary level and underlie many of the studies discussed here, without being made explicit, as indicated by the patterns of geographic restriction.

Fourth, the "law of large numbers" has been offered as a null explanation for persistence (Hubbell 2001). This model generally predicts a rapid decrease in community similarity as spatial distance and periods of time increase. The law of large numbers is invoked only when the predictions of other explanations are not borne out. Studies of coral reefs (Pandolfi 2000) and temperate Holocene forest composition (Clark & McLachlan 2003) have challenged this assertion by showing, respectively, that rare taxa can also show persistence within assemblages and that spatial decay in assemblage similarity is not the rule in some biomes. This "law" still needs to be evaluated when claims of community persistence are made.

In conclusion, the fossil record offers considerable evidence for the persistence of species assemblages at many different spatial and temporal scales. It suggests a hierarchy of ecological organization and that variance is not uniformly distributed throughout that hierarchy. Some levels show greater spatial and temporal persistence of patterns than others. The case studies and discussion presented above point to several possible causes for these patterns and provide a range of possibilities and recommendations that we hope will encourage more uniform approaches and increased exchange between paleoecologists and neoecologists in the future.

ACKNOWLEDGMENTS

We appreciate the participation and input of Alistair McGowan and members of the Paleobiology Journal Club during discussion sessions along the way to finalization of this review. This is the Evolution of Terrestrial Ecosystems (ETE) Program Publication No. 121.

The *Annual Review of Ecology, Evolution, and Systematics* is online at
http://ecolsys.annualreviews.org

LITERATURE CITED

Alemseged Z. 2003. An integrated approach to taphonomy and faunal change in Shungura Formation (Ethiopia) and its implication for hominid evolution. *J. Hum. Evol.* 44:461–78

Algeo TJ, Wilkinson BH. 1987. Periodicity of mesoscale Phanerozoic sedimentary cycles and the role of Milankovitch orbital modulation. *J. Geol.* 96:313–22

Allen TFH, Hoekstra TW. 1992. *Toward a*

Unified Ecology. New York: Columbia Univ. Press
Alroy J. 1996. Constant extinction, constrained diversification, and uncoordinated stasis in North American mammals. *Palaeogeogr. Palaeoclimatol. Palaeoecol.* 127:285–311
Alroy J. 1999. Putting North America's end-Pleistocene megafaunal extinction in context: large scale analyses of spatial patterns, extinction rates, and size distributions. In *Extinctions in Near Time: Causes, Contexts, and Consequences*, ed. RDE MacPhee, pp.105–43. New York: Plenum
Aronson RB, Macintyre IG, Precht WF, Murdoch TJT, Wapnick CM. 2002. The expanding scale of species turnover events on coral reefs in Belize. *Ecol. Monogr.* 72:233–49
Aronson RB, Macintyre IG, Wapnick CM, O'Neill MW. 2004. Phase shifts, alternative states, and the unprecedented convergence of two reef systems. *Ecol. Monogr.* In press
Aronson RB, Precht WF. 1997. Stasis, biological disturbance, and community structure of a Holocene coral reef. *Paleobiology* 23:326–46
Ault TR, Johnson CR. 1998. Spatially and temporally predictable fish communities on coral reefs. *Ecol. Monogr.* 68:25–50
Bak RPM, Nieuwland G. 1995. Long-term change in coral communities along depth gradients over leeward reefs in the Netherlands Antilles. *Bull. Mar. Sci.* 56:609–19
Bambach RK, Bennington JB. 1996. Do communities evolve? A major question in evolutionary paleoecology. In *Evolutionary Paleobiology*, ed. D Jablonski, DH Erwin, JH Lipps, pp. 123–60. Chicago: Univ. Chicago Press
Barry JC, Johnson NM, Raza SM, Jacobs LL. 1985. Neogene mammalian faunal change in southern Asia: correlations with climatic, tectonic and eustatic events. *Geology* 13:637–40
Barry JC, Morgan ME, Flynn LJ, Pilbeam D, Behrensmeyer AK, et al. 2002. Faunal and environmental change in the late Miocene Siwaliks of northern Pakistan. Memoir 3. *Paleobiology* 28:1–72

Baumiller TK, Gahn FJ. 2002. Fossil record of parasitism on marine invertebrates with special emphasis on the platyceratid-crinoid interaction. In *The Fossil Record of Predation. Paleontological Society Papers*, ed. M Kowalewski, PH Kelley, 8:195–209. Knoxville, TN: Paleontol. Soc.
Behrensmeyer AK, Chapman RE. 1993. Models and simulations of taphonomic time-averaging in terrestrial vertebrate assemblages. In *Taphonomic Approaches to Time Resolution in Fossil Assemblages. Short Courses in Paleontology No. 6*, ed. S Kidwell, AK Behrensmeyer, pp. 125–49. Knoxville, TN: Paleontol. Soc.
Behrensmeyer AK, Damuth JD, DiMichele WA, Potts R, Sues H-D, Wing SL, eds. 1992. *Terrestrial Ecosystems Through Time.* Chicago: Univ. Chicago Press
Behrensmeyer AK, Hook RW. 1992. Paleoenvironmental contexts and taphonomic modes. See Behrensmeyer et al. 1992, pp. 14–136
Behrensmeyer AK, Kidwell SM, Gastaldo RA. 2000. Taphonomy and paleobiology. In *Deep Time. Supplement to Paleobiology Volume*, ed. DH Erwin, SL Wing, 26:103–47. Knoxville, TN: Paleontol. Soc.
Belyea LR, Lancaster J. 1999. Assembly rules within a contingent ecology. *Oikos* 86:402–16
Bennington JB, Bambach RK. 1996. Statistical testing for paleocommunity recurrence: Are similar fossil assemblages ever the same? *Palaeogeogr. Palaeoclimatol. Palaeoecol.* 127:107–34
Bobe R, Behrensmeyer AK, Chapman RE. 2002. Faunal change, environmental variability and late Pliocene hominin evolution. *J. Hum. Evol.* 42:475–97
Bobe R, Behrensmeyer AK, Eck GG, Leakey LN. 2003. A comparative approach to faunal change in the Hadar and Turkana regions. *Am. J. Phys. Anthropol. Suppl.* 36:69
Bobe R, Eck GG. 2001. Responses of African bovids to Pliocene climatic change. *Paleobiol. Mem.* 27(Suppl. 2):1–47
Bonuso N, Newton CR, Brower JC, Ivany LC.

2002. Does coordinated stasis yield taxonomic and ecologic stability? Middle Devonian Hamilton Group of central New York. *Geology* 30:1055–58
Boucot AJ. 1975. *Evolution and Extinction Rate Controls.* Amsterdam: Elsevier
Boucot AJ. 1978. Community evolution and rates of cladogenesis. *Evol. Biol.* 11:545–655
Boucot AJ. 1983. Does evolution take place in an ecological vacuum? II. *J. Paleontol.* 57:1–30
Boucot AJ, Lawson JD, eds. 1999. *Paleocommunities; A Case Study from the Silurian and Lower Devonian.* Cambridge, UK: Cambridge Univ. Press
Brett CE, Baird GC. 1995. Coordinated stasis and evolutionary ecology of Silurian to Middle Devonian faunas in the Appalachian Basin. In *New Approaches to Speciation in the Fossil Record*, ed. DH Erwin, RL Anstey, pp. 285–315. New York: Columbia Univ. Press
Brett CE, Ivany LC, Schopf KM. 1996. Coordinated stasis: an overview. *Palaeogeogr. Palaeoclimatol. Palaeoecol.* 127:1–20
Broutin J, Doubinger J, Farhanel G, Freytet F, Kerp H, et al. 1990. Le renouvellement des flores au passage Carbonifère Permien: approaches stratigraphique, biologique, sédimentologique. *C. R. Acad. Sci. Paris* 311:1563–69
Brown FH. 1994. Development of Pliocene and Pleistocene chronology of the Turkana Basin, East Africa, and its relation to other sites. In *Integrative Paths to the Past*, ed. RS Corruccini, RL Ciochon, pp. 285–312. Englewood Cliffs, NJ: Prentice-Hall
Buzas MA, Culver SJ. 1994. Species pool and dynamics of marine paleocommunities. *Science* 264:1439–41
Chave J. 2004. Neutral theory and community ecology. *Ecol. Lett.* 7:241–53
Clark JS, McLachlan JS. 2003. Stability of forest biodiversity. *Nature* 423:635–38
Clements FR. 1916. *Plant Succession: An Analysis of the Development of Vegetation.* Washington, DC: Carnegie Inst.
Coley PD, Barone JA. 1996. Herbivory and plant defenses in tropical forests. *Annu. Rev. Ecol. Syst.* 27:305–35
Connell JH. 1978. Diversity in tropical rain forests and coral reefs. *Science* 199:1302–10
Connell JH, Hughes TP, Wallace CC. 1997. A 30-year study of coral abundance, recruitment, and disturbance at several scales in space and time. *Ecol. Monogr.* 67:461–88
Coope GR. 1994. The response of insect faunas to glacial-interglacial climatic fluctuations. *Philos. Trans. R. Soc. London Ser. B* 344:19–26
Cottingham KL, Brown BL, Lennon JT. 2001. Biodiversity may regulate the temporal variability of ecological systems. *Ecol. Lett.* 4:72–85
Cridland AA, Morris JE. 1963. *Taeniopteris, Walchia,* and *Dichophyllum* in the Pennsylvanian System of Kansas. *Univ. Kans. Sci. Bull.* 44:71–82
Damuth JD. 1992. Taxon-free characterization of animal communities. See Behrensmeyer et al. 1992, pp. 183–203
Davis MB. 1986. Climatic instability, time lags, and community disequilibrium. In *Community Ecology*, ed. J Diamond, TJ Case, pp. 269–84. New York: Harper & Row
de Heinzelin J. 1983. *The Omo Group.* Tervuren: Musée R. l'Afrique Central
deMenocal PB. 1995. Plio-Pleistocene African climate. *Science* 270:53–59
deMenocal PB, Bloemendal J. 1995. Plio-Pleistocene climatic variability in subtropical Africa and the paleoenvironment of hominid evolution: a combined data-model approach. In *Paleoclimate and Evolution with Emphasis on Human Origins*, ed. ES Vrba, GH Denton, TC Partridge, LH Burckle, pp. 262–88. New Haven: Yale Univ. Press
Dettman DL, Kohn MJ, Quade J, Ryerson FJ, Ojha TP, Hamidullah S. 2000. Seasonal stable isotope evidence for a strong Asian monsoon throughout the past 10.7 Ma. *Geology* 29:31–34
DeWoody YD, Swihart RK, Craig BA, Goheen JR. 2003. Diversity and stability in communities structured by asymmetric resource allocation. *Am. Nat.* 162:514–27

DiMichele WA. 1994. Ecological patterns in time and space. *Paleobiology* 20:89–92

DiMichele WA, Aronson RB. 1992. The Pennsylvanian-Permian vegetational transition: a terrestrial analogue to the onshore-offshore hypothesis. *Evolution* 46:807–24

DiMichele WA, Gastaldo RA, Pfefferkorn HW. 2004. Biodiversity partitioning in the Late Carboniferous and Early Permian and its implications for ecosystem assembly. In *Biodiversity*, ed. N Jablonski. San Francisco: Calif. Acad. Sci. In press

DiMichele WA, Mamay SH, Chaney DS, Hook RW, Nelson WJ. 2001a. An Early Permian flora with Late Permian and Mesozoic affinities from north-central Texas. *J. Paleontol.* 75:449–60

DiMichele WA, Pfefferkorn HW, Phillips TL. 1996. Persistence of Late Carboniferous tropical vegetation during glacially driven climatic and sea-level fluctuations. *Palaeogeogr. Palaeoclimatol. Palaeoecol.* 125:105–28

DiMichele WA, Phillips TL. 1996a. Clades, ecological amplitudes, and ecomorphs: phylogenetic effects and the persistence of primitive plant communities in the Pennsylvanian-age tropics. *Palaeogeogr. Palaeoclimatol. Palaeoecol.* 127:83–106

DiMichele WA, Phillips TL. 1996b. Climate change, plant extinctions, and vegetational recovery during the Middle-Late Pennsylvanian transition: the case of tropical peat-forming environments in North America, In *Biotic Recovery from Mass Extinctions. Geol. Soc. London Spec. Publ.*, ed. ML Hart, 102:201–21. London: Geol. Soc.

DiMichele WA, Phillips TL, Nelson WJ. 2002. Place vs. time and vegetational persistence: a comparison of four tropical paleomires from the Illinois Basin at the height of the Pennsylvanian ice age. *Int. J. Coal Geol.* 50:43–72

DiMichele WA, Stein WE, Bateman RM. 2001a. Ecological sorting during the Paleozoic radiation of vascular plant classes. In *Evolutionary Paleoecology*, ed. WD Allmon, DJ Bottjer, pp. 285–35. New York: Columbia Univ. Press

Doherty PJ, Williams D McB. 1988. The replenishment of coral reef fish populations. *Oceanogr. Mar. Biol. Annu. Rev.* 32:487–551

Falcon-Lang HJ. 2003. Response of Late Carboniferous tropical vegetation to transgressive-regressive rhythms at Joggins, Nova Scotia. *J. Geol. Soc. London* 160:643–48

Feibel CS, Brown FH, McDougall T. 1989. Stratigraphic context of fossil hominids from the Omo Group deposits: northern Turkana Basin, Kenya and Ethiopia. *Am. J. Phys. Anthropol.* 78:595–622

Foote M. 2000. Origination and extinction components of taxonomic diversity: general patterns. *Paleobiology* 26(Suppl.):74–102

Frakes LA, Francis JE, Syktus JI. 1992. *Climate Modes of the Phanerozoic*. Cambridge, UK: Cambridge Univ. Press

Gastaldo RA, DiMichele WA, Pfefferkorn HW. 1996. Out of the icehouse into the greenhouse: a late Paleozoic analogue for modern global vegetational change. *GSA Today* 6:1–7

Geister J. 1977. The influence of wave exposure on the ecological zonation of Caribbean coral reefs. *Proc. 3rd Int. Coral Reef Symp.* 1:23–29

Gleason HA. 1926. The individualistic concept of the plant association. *Bull. Torrey Bot. Club* 53:7–26

Golley FB. 1993. *A History of the Ecosystem Concept in Ecology*. New Haven: Yale Univ. Press

Gotelli NJ, Graves GR. 1996. *Null Models in Ecology*. Washington, DC: Smithson. Inst. Press

Graham RW. 1997. The spatial response of mammals to Quaternary climate changes. In *Past and Future Rapid Environmental Changes: The Spatial and Evolutionary Responses of Terrestrial Biota. NATO ASI Ser. 1: Global Environmental Change*, ed. B Huntley, W Cramer, AV Morgan, HC Prentice, AM Solomon, 47:153–62. New York: Springer-Verlag

Greenstein BJ, Curran HA, Pandolfi JM. 1998. Shifting ecological baselines and the demise of *Acropora cervicornis* in the western North

Atlantic and Caribbean province: a Pleistocene perspective. *Coral Reefs* 17:249–61

Gunderson LH, Pritchard L, Holling CS, Folke C, Peterson GD. 2002. A summary and synthesis of resilience in large-scale systems. In *Resilience and the Behavior of Large-scale Systems*, ed. LH Gunderson, L Prichard Jr., pp. 249–66. Washington, DC: Island

Hoffman A. 1979. Community paleoecology as an epiphenomenal science. *Paleobiology* 5:357–79

Holland SM. 2003. Confidence limits on fossil ranges that account for facies changes. *Paleobiology* 29:468–79

Holling CS. 1973. Resilience and stability of ecological systems. *Annu. Rev. Ecol. Syst.* 4:1–23

Holterhoff PF. 1996. Crinoid biofacies in Upper Carboniferous cyclothems, midcontinent North America; faunal tracking and the role of regional processes in biofacies recurrence. *Palaeogeogr. Palaeoclimatol. Palaeoecol.* 127:47–81

Hubbard DK, Gladfelter EH, Blythell JC. 1994. Comparison of biological and geological perspectives of coral-reef community structure at Buck Island, U.S. Virgin Islands. In *Proc. Colloq. Global Aspects of Coral Reefs: Health, Hazards, and History, 1993*, ed. RN Ginsburg. Miami, FL: Rosenstiel Sch. Mar. Atmos. Sci., Univ. Miami Press

Hubbell SP. 1997. A unified theory of biogeography and relative species abundance and its application to tropical rain forests and coral reefs. *Proc. 8th Int. Coral Reef Symp.* 1:33–42

Hubbell SP. 2001. *The Unified Neutral Theory of Biodiversity and Biogeography*. Princeton, NJ: Princeton Univ. Press

Hughes TP, Baird AH, Dinsdale EA, Moltschaniwskyj N, Pratchett MS, et al. 1999. Patterns of recruitment and abundance of corals along the Great Barrier Reef. *Nature* 397:59–63

Jablonski D, Sepkoski JJ Jr. 1996. Paleobiology, community ecology, and scales of ecological pattern. *Ecology* 77:1367–78

Jackson JA, Bates RL, eds. 1997. *Glossary of Geology*. Alexandria, VA: Am. Geol. Inst. 4th ed.

Jackson JBC. 1992. Pleistocene perspectives on coral reef community structure. *Am. Zool.* 32:719–31

Jackson JBC. 1994. Community unity? *Science* 264:1412–13

Janis CM, Damuth J, Theodor JM. 2000. Miocene ungulates and terrestrial primary productivity: Where have all the browsers gone? *Proc. Natl. Acad. Sci. USA* 97:7899–904

Janis CM, Scott KM, Jacobs LL, eds. 1998. *Evolution of Tertiary Mammals of North America*. Volume 1: *Terrestrial Carnivores, Ungulates, and Ungulatelike Mammals*. Cambridge, UK: Cambridge Univ. Press

Johnson RG. 1960. Models and methods for analysis of the mode of formation on fossil assemblages. *Geol. Soc. Am. Bull.* 71:1075–86

Kelley PH, Hansen TA. 2001. The role of ecological interactions in the evolution of naticid gastropods and their molluscan prey. In *Evolutionary Paleoecology*, ed. WD Allmon, DJ Bottjer, pp. 149–70. New York: Columbia Univ. Press

Kidwell SM. 2002. Time-averaged molluscan death assemblages: palimpsests of richness, snapshots of abundance. *Geology* 30:803–6

Kidwell SM, Holland SM. 2002. The quality of the fossil record: implications for evolutionary analyses. *Annu. Rev. Ecol. Syst.* 33:561–88

Knoll AH. 1984. Patterns of extinction in the fossil record of vascular plants. In *Extinctions*, ed. M Nitecki, pp. 21–65. Chicago: Univ. Chicago Press

Kutzbach J, Gallimore R, Harrison S, Behling P, Selin R, Laarif F. 1998. Climate and biome simulations for the past 21,000 years. *Q. Sci. Rev.* 17:473–506

Labandeira CC. 1998. Early history of arthropod and vascular plant associations. *Annu. Rev. Earth Planet. Sci.* 26:329–77

Labandeira CC. 2002. The history of associations between plants and animals. In *Plant-Animal Interactions*, ed. C Herrera, O

Pellmyr, pp. 26–74, 248–61. Oxford: Blackwell Sci.
Labandeira CC, Johnson KR, Wilf P. 2002. Impact of the terminal Cretaceous event on plant-insect associations. *Proc. Natl. Acad. Sci. USA* 99:2061–66
Labandeira CC, Phillips TL. 2002. Stem borings and petiole galls from Pennsylvanian tree ferns of Illinois, USA: implications for the origin of the borer and galler functional-feeding groups and holometabolous insects. *Palaeontographica A* 264:1–84
Lockwood JL, Powell RD, Nott MP, Pimm SL. 1997. Assembling ecological communities in time and space. *Oikos* 80:549–53
Ludvigsen R, Westrop SR. 1983. Trilobite biofacies of the Cambrian-Ordovician boundary interval in northern North America. *Alcheringa* 7:301–19
Lyons PC, Darrah WC. 1989. Earliest conifers in North America: Upland and/or paleoclimate indicators? *Palaios* 4:480–86
Mamay SH, Mapes G. 1992. Early Virgilian plant megafossils from the Kinney Brick Company quarry, Manzanita Mountains, New Mexico. *N. M. Bur. Mines Min. Res. Bull.* 138:61–85
Marshall CD. 1997. Confidence intervals on stratigraphic ranges with nonrandom distributions of fossil horizons. *Paleobiology* 23:165–73
Martini IP, ed. 1997. *Late Glacial and Postglacial Environmental Changes: Quaternary, Carboniferous, Permian, and Proterozoic*. New York: Oxford Univ. Press
Maurer BA. 1999. *Untangling Ecological Complexity*. Chicago: Univ. Chicago Press
May RM. 1974. *Stability and Complexity in Model Ecosystems*. Princeton, NJ: Princeton Univ. Press
McIntosh RP. 1995. H.A. Gleason's 'individualistic concept' and theory of animal communities: a continuing controversy. *Biol. Rev.* 70:317–57
McKee JK. 2001. Faunal turnover rates and mammalian biodiversity of the late Pliocene and Pleistocene of eastern Africa. *Paleobiology* 27:500–11
Meyen SV. 1982. The Carboniferous and Permian floras of Angaraland (a synthesis). *Biol. Mem.* 7:1–110
Milchunas DG, Lauenroth WK, Sala OE. 1988. A generalized model of the effects of grazing by large herbivores on grassland community structure. *Am. Nat.* 132:87–106
Miller AI. 1998. Biotic transitions in global marine diversity. *Science* 281:1157–60
Miller W III. 1993. Models of recurrent fossil assemblages. *Lethaia* 26:182–83
Miller W III. 1996. Ecology of coordinated stasis. *Palaeogeogr. Palaeoclimatol. Palaeoecol.* 127:177–90
Morris PJ, Ivany LC, Schopf KM, Brett CE. 1995. The challenge of paleoecological stasis: reassessing sources of evolutionary stability. *Proc. Natl. Acad. Sci. USA* 92:11269–73
Olson EC. 1952. The evolution of a Permian vertebrate chronofauna. *Evolution* 6:181–96
Olson EC. 1958. Faunal of the Vale and Choza: 14. Summary, review, and integration of the geology and the faunas. *Fieldiana Geol.* 10:397–448
Olson EC. 1980. Taphonomy: its history and role in community evolution. In *Fossils in the Making*, ed. AK Behrensmeyer, AP Hill, pp. 5–19. Chicago: Univ. Chicago Press
Olson EC. 1985. Vertebrate paleoecology: a current perspective. *Palaeogeogr. Palaeoclimatol. Palaeoecol.* 50:83–106
Olszewski TD, Erwin DH. 2004. Dynamic response of Permian brachiopod communities to long-term environmental change. *Nature* 428:738–41
Olszewski TD, Patzkowsky ME. 2001. Evaluating taxonomic turnover; Pennsylvanian-Permian brachiopods and bivalves of the North American Midcontinent. *Paleobiology* 27:646–68
Overpeck JT, Webb RS, Webb T III. 1992. Mapping eastern North American vegetation change of the last 18 Ka: No-analogs and the future. *Geology* 20:1071–74
Overpeck JT, Webb T III, Prentice IC. 1985. Quantitative interpretation of fossil pollen

spectra: dissimilarity coefficients and the method of modern analogs. *Q. Res.* 23:87–108

Pandolfi JM. 1996. Limited membership in Pleistocene reef coral assemblages from the Huon Peninsula, Papua New Guinea: constancy during global change. *Paleobiology* 22:152–76

Pandolfi JM. 1999. Response of Pleistocene coral reefs to environmental change over long temporal scales. *Am. Zool.* 39:113–30

Pandolfi JM. 2000. *Persistence in Caribbean coral communities over broad spatial and temporal scales.* Presented at 9th Int. Coral Reef Congr., Bali, October 23–27, 2000

Pandolfi JM, Jackson JBC. 1997. The maintenance of diversity on coral reefs: examples from the fossil record. *Proc. 8th Int. Coral Reef Symp.* 1:397–404

Patzkowsky ME, Holland SM. 1997. Patterns of turnover in Middle and Upper Ordovician brachiopods of the eastern United States: a test of coordinated stasis. *Paleobiology* 23:420–43

Patzkowsky ME, Holland SM. 1999. Biofacies replacement in a sequence stratigraphic framework: Middle and Upper Ordovician of the Nashville Dome, Tennessee, USA. *Palaios* 14:310–23

Peppers RA. 1996. Palynological correlation of major Pennsylvanian (Middle and Upper Carboniferous) chronostratigraphic boundaries in the Illinois and other coal basins. *Geol. Soc. Am. Mem.* 188:1–111

Peters RH. 1993. *A Critique for Ecology.* Cambridge, UK: Cambridge Univ. Press

Pfefferkorn HW, Thomson MC. 1982. Changes in dominance patterns in Upper Carboniferous plant-fossil assemblages. *Geology* 10: 641–44

Pfefferkorn HW, Gastaldo RA, DiMichele WA. 2000. Ecological stability during the late Paleozoic cold interval. In *Phanerozoic Terrestrial Ecosystems. Paleontological Society Papers*, ed. RA Gastaldo, WA DiMichele, 6:63–78. Knoxville, TN: Paleontol. Soc.

Phillips TL, Peppers RA. 1984. Changing patterns of Pennsylvanian coal swamp vegetation and implications of climate control on coal occurrence. *Int. J. Coal Geol.* 3:205–55

Phillips TL, Peppers RA, DiMichele WA. 1985. Stratigraphic and interregional changes in Pennsylvanian coal-swamp vegetation: environmental inferences. *Int. J. Coal Geol.* 5: 43–109

Pimm SL. 1984. The complexity and stability of ecosystems. *Nature* 307:321–26

Pimm SL. 1991. *The Balance of Nature? Ecological Issues in the Conservation of Species and Communities.* Chicago: Univ. Chicago Press

Prinzing A, Durka W, Klotz S, Brandl R. 2001. The niche of higher plants: evidence for phylogenetic conservatism. *Proc. R. Soc. London Ser. B* 268:2383–89

Robertson DR. 1996. Interspecific competition controls abundance and habitat use of territorial Caribbean damselfishes. *Ecology* 77:885–99

Rothwell GW, Mapes G. 1988. Vegetation of a Paleozoic conifer community. In *Regional Geology and Paleontology of Upper Paleozoic Hamilton Quarry Area in Southeastern Kansas. Guidebook 6*, ed. G Mapes, RH Mapes, pp. 213–23. Lawrence: Kansas Geol. Survey

Sale PF. 1988. What coral reefs can teach us about ecology. *Proc. 6th Int. Coral Reef Symp.* 1:19–31

Sale PF, Douglas WA. 1984. Temporal variability in the community structure of fish on coral patch reefs and the relation of community structure to reef structure. *Ecology* 65:409–22

Sale PF, Guy JA, Steel WJ. 1994. Ecological structure of assemblages of coral reef fishes on isolated patch reefs. *Oecologia* 98:83–99

Sale PF, Steel WJ. 1989. Temporal variability in patterns of association among fish species on coral patch reefs. *Marine Ecol. Prog. Ser.* 51:35–47

Scott AC. 1978. Sedimentological and ecological control of Westphalian B plant assemblages from West Yorkshire. *Proc. Yorkshire Geol. Soc.* 41:461–508

Sheehan PM. 1996. A new look at ecological evolutionary units (EEUs). *Palaeogeogr. Palaeoclimatol. Palaeoecol.* 127:21–31

Simberloff D, Dayan T. 1991. The guild concept and the structure of ecological communities. *Annu. Rev. Ecol. Syst.* 22:115–43

Smith TM, Shugart HH, Woodward FI, eds. 1997. *Plant Functional Types—Their Relevance to Ecosystem Properties and Global Change.* Cambridge, UK: Cambridge Univ. Press

Stemann TA, Johnson KG. 1992. Coral assemblages, biofacies, and ecological zones in the mid-Holocene reef deposits of the Enriquillo Valley, Dominican Republic. *Lethaia* 25:231–41

Tang CM, Bottjer DJ. 1996. Long-term faunal stasis without evolutionary coordination: Jurassic benthic marine paleocommunities, Western Interior, United States. *Geology* 24:815–18

Tanner JE, Hughes TP, Connell JH. 1994. Species coexistence, keystone species, and succession: a sensitivity analysis. *Ecology* 75:2204–19

Tilman D. 1999. The ecological consequences of biodiversity: a search for general principles. *Ecology* 80:1455–74

Tokeshi M. 1993. Species abundance patterns and community structure. *Adv. Ecol. Res.* 24: 111–86

Ulanowicz RE. 1997. *Ecology, The Ascendent Perspective.* New York: Columbia Univ. Press

Valentine JW. 1980. Determinants of diversity in higher taxonomic categories. *Paleobiology* 6:444–50

Van Valkenburgh B. 1995. Tracking ecology over geological time: evolution within guilds of vertebrates. *Trends Ecol. Evol.* 10:71–76

Vrba ES. 1993. Turnover-pulses, the Red Queen, and related topics. *Am. J. Sci.* 293: 418–52

Vrba ES. 1995. The fossil record of African antelopes (Mammalia: Bovidae) in relation to human evolution and paleoclimate. In *Paleoclimate and Evolution with Emphasis on Human Origins*, ed. ES Vrba, GH Denton, TC Partridge, LH Burckle, pp. 385–424. New Haven: Yale Univ. Press

Walker BH. 1997. Functional types in non-equilibrium systems. In *Plant Functional Types*, ed. TM Smith, HH Shugart, FI Woodward, pp. 91–103. Cambridge: Cambridge Univ. Press

Watkins R, Berry WBN, Boucot AJ. 1973. Why 'communities'? *Geology* 1:5–8

Webb T III. 1987. The appearance and disappearance of major vegetational assemblages: long-term vegetational dynamics in eastern North America. *Vegetation* 69:177–87

Weiher E, Keddy P. 2001. Assembly rules as general constraints on community composition. In *Ecological Assembly Rules. Perspectives, Advances, Retreats*, ed. E Weiher, P Keddy, pp. 251–71. Cambridge, UK: Cambridge Univ. Press

Westrop SR. 1996. Temporal persistence and stability of Cambrian biofacies: Sunwaptan (Upper Cambrian) trilobites of North America. *Palaeogeogr. Palaeoclimatol. Palaeoecol.* 127:33–46

Westrop SR, Cuggy MB. 1999. Comparative paleoecology of Cambrian trilobite extinctions. *J. Paleontol.* 73:337–54

Wilf P, Labandeira CC, Johnson KR, Coley PD, Cutter AD. 2001. Insect herbivory, plant defense, and early Cenozoic climate change. *Proc. Natl. Acad. Sci. USA* 98:6221–26

Willard DA, Phillips TL, Lesnikowska AD, DiMichele WA. 2004. The Calhoun Coal and the persistent organization of Late Pennsylvanian coal-swamp vegetation. *Int. J. Coal Geol.* In press

Williams D. McB. 1980. Dynamics of the pomacentrid community on small patch reefs in One Tree Lagoon (Great Barrier Reef). *Bull. Mar. Sci.* 30:159–70

Wing SL, DiMichele WA. 1992. Ecological characterization of fossil plants. See Behrensmeyer et al. 1992, pp. 139–80

Wing SL, DiMichele WA. 1995. Conflict between local and global changes in plant diversity through geological time. *Palaios* 10: 551–64

Wing SL, Harrington GJ. 2001. Floral response to rapid warming at the Paleocene/Eocene boundary and implications for concurrent faunal change. *Paleobiology* 27:539–62

Woodburne MO. 1987. *Cenozoic Mammals of North America*. Berkeley: Univ. Calif. Press

Zachos J, Pagani M, Sloan L, Thomas E, Billups K. 2001. Trends, rhythms, and aberrations in global climate 65 Ma to present. *Science* 292:686–93

Ziegler AM. 1990. Phytogeographic patterns and continental configurations during the Permian Period. *Geol. Soc. Mem.* 12:363–79

Ziegler AM, Eshel G, Rees PM, Rothfus TA, Rowley DB, Sunderlin D. 2003. Tracing the tropics across land and sea: Permian to present. *Lethaia* 36:227–54

AVIAN EXTINCTIONS FROM TROPICAL AND SUBTROPICAL FORESTS

Navjot S. Sodhi,[1] L.H. Liow,[2] and F.A. Bazzaz[3]

[1]Department of Biological Sciences, National University of Singapore, Singapore 117543, Republic of Singapore; email: dbsns@nus.edu.sg
[2]Committee on Evolutionary Biology, University of Chicago, Chicago, Illinois 60637; email: lhliow@midway.uchicago.edu
[3]Department of Organismic and Evolutionary Biology, Biological Laboratories, Harvard University, Cambridge, Massachusetts 02138; email: fbazzaz@oeb.harvard.edu

Key Words extirpation, deforestation, fragmentation, resilience, conservation

■ **Abstract** Tropical forests are being lost at an alarming rate. Studies from various tropical locations report losses of forest birds as possibly direct or indirect results of deforestation. Although it may take a century for all the sensitive species to be extirpated from a site following habitat loss, species with larger or heavier bodies and those foraging on insects, fruits, or both are particularly extinction prone. Larger- or heavier-bodied species may occur at low densities, increasing their vulnerability to habitat alterations. Insectivores are vulnerable for reasons such as the loss of preferred microhabitats, poor dispersal abilities, and/or ground nesting habits that make them susceptible to predation. The lack of year-round availability of fruits may make survival in deforested or fragmented areas difficult for frugivores. Extirpation of large predators, superior competitors, pollinators, and seed dispersers may have repercussions for tropical ecosystem functioning. Large tropical reserves that adequately protect existing forest avifauna are needed. Sound ecological knowledge of tropical forest avifauna for biodiversity-friendly forest management practices is also needed but sorely lacking.

INTRODUCTION

Despite extremely high species diversity and endemism of existing tropical forests, 16 million ha are lost annually (Achard et al. 2002). The speed of tropical deforestation is unprecedented in evolutionary history (Bierregaard et al. 1992). A large proportion of current biotas may have originated from tropical climates, so tropical extinctions have repercussions for the preservation of current as well as future biodiversity (Myers & Knoll 2001).

Globally, one in eight bird species may become extinct over the next 100 years—99% of the extinctions owing to human activities such as deforestation and hunting (BirdLife International 2000). Therefore, there is an urgent need to understand the patterns and processes of avian extinctions. We review extinctions of forest avifauna from the tropics and subtropics, lying largely between the Tropics of Cancer and Capricorn, 23.5° north and south of the equator. These forests are typically found in aseasonal or semiseasonal climates, and they harbor a disproportionate share of the planet's biodiversity (Whitmore 1980, Myers et al. 2000). Ninety-three percent of 902 threatened forest birds are found in the tropics (Birdlife International 2000). Tropical forest birds are particularly sensitive to deforestation compared with other taxa such as vascular plants (Brook et al. 2003). They may be inherently vulnerable to habitat perturbations owing to relatively low population sizes, patchy distributions, poor dispersal abilities, and high habitat specificity (Terborgh et al. 1990, Turner 1996, Laurance et al. 1997).

We do not review extinctions on the basis of fossil records, as they have been dealt with elsewhere (e.g., Milberg & Tyrberg 1993, Pimm et al. 1994). We are concerned with local extinctions (hereafter extinctions) or extirpations. Local (population) extinctions are determined by comparing results from recent surveys with those made in the past (historical approach). Alternatively, they are inferred by making comparisons among differently sized fragments (fragmentation approach). We review evidence from both empirical approaches. However, for the fragmentation approach, we include only studies that provided quantitative data on species extirpations.

We present the current state of knowledge on avian extirpations from tropical forests by exploring these questions: Are populations at the distributional edges rather than centers of their ranges more vulnerable to extinction? Are large-sized species more vulnerable to extinction than small ones? Have extinctions occurred disproportionately in certain foraging guilds? Does certain behavior(s) make a species more extinction-prone? Is there any evidence that inbreeding depression causes extinctions? Are certain families more prone to extinctions? Do extinctions affect ecosystem functioning? Can any conservation efforts reverse or halt the extinctions? We also highlight areas where future research and resources should be focused.

PATTERNS

Anthropogenically driven extinctions have occurred on tropical and subtropical islands for thousands of years (e.g., Diamond 1989, Milberg & Tyrberg 1993, Steadman 1995). Mechanisms for prehistorical extinctions may be similar to those for current extinctions: overhunting, introduced predators and diseases, and habitat destruction (Milberg & Tyrberg 1993). Further, forest maturation can render habitats unsuitable for some species and cause local extinctions (e.g., Bush & Whittaker 1991). Although "natural" or human-caused prehistoric extinctions have occurred, probably owing to accelerating habitat destruction and burgeoning human populations, extinction rates have soared recently (Balmford 1996, Kerr & Burkey 2002).

Most studies reporting bird extirpations are Neotropical (12 out of 16). Extirpations reported ranged from 1% (Costa Rica, minimum estimate; see Daily et al. 2001) to 67% (Singapore; see Castelletta et al. 2000) of the original forest avifauna. There are important patterns to note from these studies. First, the existence of refuges can probably dampen the effects of deforestation (Brash 1987). Second, habitat flexibility seems to aid survival. For example, in experimentally fragmented areas (1 and 10 ha) of Manaus (Brazil), mixed-species flocking species that used forest edges and secondary forests survived (wedge-billed woodcreeper, *Glyphorynchus spirurus*; chestnut-rumped woodcreeper, *Xiphorhynchus pardalotus*; and white-flanked antwren, *Myrmotherula axillaris*), whereas species restricted to the forest interior (e.g., cinereous antshrike, *Thamnomanes caesius*) perished (Stouffer & Bierregaard 1995a). Similarly, because of abilities to use forest edges and gaps, predominantly nectivorous understory hummingbirds suffered few negative effects from fragmentation and maintained species composition and abundance similar to those occurring prior to isolation in the same area (Stouffer & Bierregaard 1995b). Last, observed recolonizations of altered areas remain few, suggesting poor colonization abilities of most extirpated species. These poor colonization abilities may be due to variables such as low initial population density and/or reluctance to cross even small gaps (e.g., 80 m; see Bierregaard et al. 1992). Out of the eight forest-dwelling species missing from Barro Colorado Island (Panama) since the 1970s, only one (great currasow, *Crax rubra*) appeared to have recolonized (Robinson 1999). In some studies (e.g., Karr 1982a) nearby continuous forests are assumed to be a population source area. In light of possible poor dispersal abilities of most tropical bird species (see above), such an assumption needs verification.

LAG TIME IN AVIAN EXTINCTIONS

Possibly because of comparatively low reproductive effort and high environmental stability, tropical bird species should have low adult mortality. Adult mortality of only 10% to 30% per year has been estimated for birds from various tropical locations (Trinidad, Sarawak [Malaysia], and Barro Colorado Island; see Willis 1974). This fact suggests that it may take some time for a tropical bird species to disappear following habitat loss unless such habitat loss decreases adult longevity. The Puerto Rican parrot (*Amazona vittata*, longevity in captivity is 23 years), for example, had about 2000 individuals around 1900 in the Caribbean National Forest. Their numbers declined to 20 over the next 60 years, despite a halt in deforestation (Brash 1987). This result indicates that it may take decades for the extinction of some species, and once the decline starts it may be difficult to halt even if habitat degradation ceases. For such species, management actions (see below), in addition to stopping deforestation, may be required to halt or reverse the decline.

Brooks et al. (1999b) found that a plausible half-life (time taken to lose half of the species) for avifaunal extinctions in fragmented forests in Kenya was 50 years. Half-life decreased with fragment size and isolation from nearby forested areas (e.g., 23 years for a 100 ha fragment 9.0 km from the nearest patch). Because the smallest fragment was also the most isolated, it was not possible to tease apart independent effects of fragment size and isolation in this study. Smaller fragments may contain small populations, and isolation may reduce immigration and thus result in zero or minimal "rescue effect" (immigration by unrelated individuals into isolated populations; Brown & Kodric-Brown 1977). Half-lives may be further shortened in fragments by indirect factors, such as high predation, parasitism, and damage in case of catastrophic events such as fires and storms (Brooks et al. 1999b). In fact, 100 ha fragments may lose half of their species in less than 15 years (Ferraz et al. 2003). Therefore, there may be decades, even if remaining forests are large, in which to halt the extinctions of at least a large proportion of tropical forest bird species following initial habitat loss. Further, ability to tolerate habitat deterioration and dispersal capability may vary among species and affect their persistence within a disturbed landscape (Lens et al. 2002).

MECHANISMS

Ultimately, decreased survival, fecundity, or both may result in extinction of a species. Mechanisms causing extirpations may act independently or synergistically (Figure 1). Pinpointing a single cause of an extinction may be difficult (Pimm 1996, Reed 1999). Habitat loss may directly cause some extinctions, but it also facilitates colonization by invasive species and infiltration by disease vectors and parasites, and it opens up habitats for hunters. For example, in Hawaii, avian extinctions were higher than predicted by habitat loss alone (Scott et al. 1988, Pimm 1996), suggesting collective effects by a myriad of factors.

Habitat destruction is considered a major cause of species losses (e.g., Tilman et al. 1994, Pimm & Raven 2000, Gaston et al. 2003). However, some authors argue that there has been little *direct* empirical evidence that tropical deforestation causes extinctions (Heywood & Stuart 1992). Such direct evidence has come from well-designed experiments that compare data before and after deforestation, as has been done in Manaus, Brazil (Bierregaard et al. 1992). Mounting evidence also suggests that deforestation is one of the prime direct or indirect causes of reported bird extirpations. Brook et al. (2003), using the species-area equation, projected that if the current rate of deforestation continues in Southeast Asia, there will be 16% to 32% extirpations of forest birds by 2100. Similarly, Brooks et al. (1997) used deforestation data and the species-area equation to predict the number of threatened endemic species in Southeast Asia. In more detailed analyses, Brooks et al. (1999a) determined the correlation between degree of deforestation and the existence of threatened endemic species in the lowland and montane forests of Southeast Asia. Broadly, more threatened bird species were in heavily deforested areas. However, there were some regional and habitat differences. The number of threatened species in montane areas was underestimated by the species-area equation, possibly because their restricted ranges made them disproportionately more vulnerable to habitat loss (Brooks et al. 1999a). Conversely, lowland avifauna

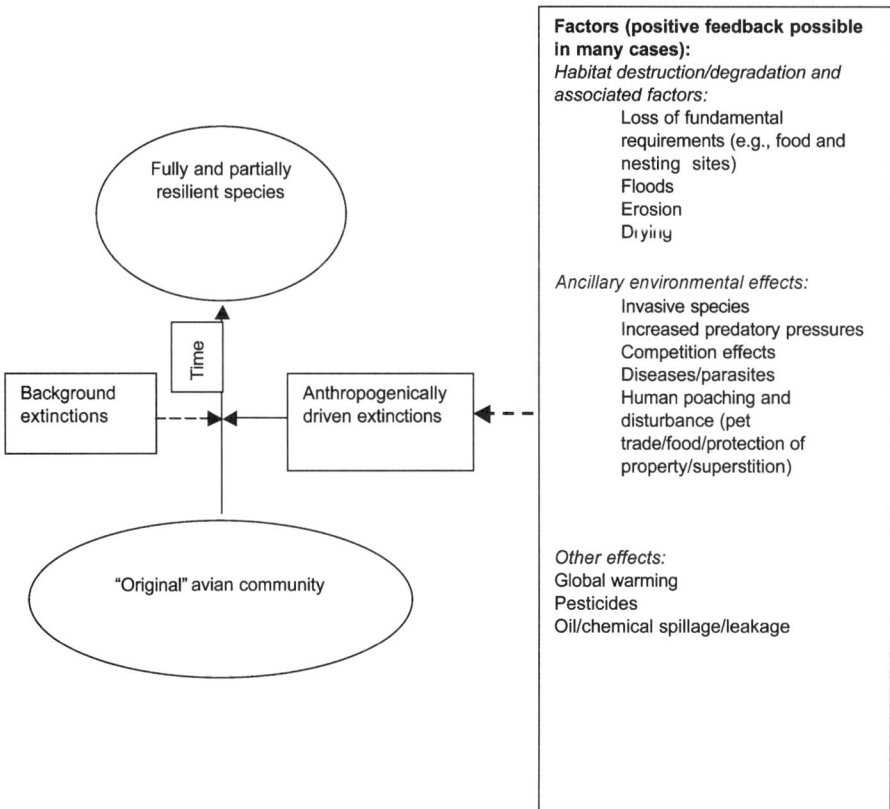

Figure 1 A generalized diagram showing the possible pathway and factors causing bird extinctions. (Background extinctions refer to extinctions from natural causes, e.g., catastrophic events such as hurricanes.)

of areas such as Lesser Sundas and Java were less threatened than predicted by the same equations. Because of centuries of habitat loss, these areas may either have lost their more sensitive endemics before any scientific surveys were conducted or they may contain inherently tolerant species. Brooks et al.'s (1999a) analyses did not demonstrate a direct link between deforestation and species endangerment but did show that deforested areas generally harbor more threatened species, because of either habitat loss or degradation or associated causes (e.g., hunting).

Deforestation can cause the loss of preferred habitats of some forest bird species. Only one or two woodpecker species occurred in fragments smaller than 2500 ha in Java, possibly because of the lack of suitable nesting trees (Van Balen 1999). Loss of forest streams in the western Andes might have caused the extinction, for instance, of the Andean cock-of-the-rock, *Rupicola peruviana*, and the crested anttanager *Habia cristata* (Kattan et al. 1994). Similarly, mesic forests are preferred by forest-dependent endemic and quasi-endemic bird species in the cerrado region of Brazil (Marini 2001), suggesting that loss of such forests may be detrimental to them.

As mentioned above, roads and trails make deforested areas more accessible to hunters (hunting for food and/or the cage; see Milner-Gulland et al. 2003). In the 1980s, 2–5 million birds were captured for the international trade (Beissinger 2001). There is little monitoring of the international bird trade, but half a million birds traded between 1991 and 1996 were listed in CITES (Convention on International Trade in Endangered Species of Wild Fauna and Flora) appendices (Beissinger 2001). Almost all these birds originated from the tropics.

In many cases, human persecution may have caused the extirpation or decline of forest species. For example, in Ecuador hunting may have partly caused the decline of species such as the great tinamou (*Tinamus major*) and crested guan (*Penelope purpurascens*; Leck 1979). Similar cases have been reported elsewhere (Brash 1987, Christiansen & Pitter 1997, Van Balen 1999, Marini 2001). Not all hunting is for meat, as some large birds are shot because it is believed they eat domestic animals (see Gillespie 2001).

Increased accessibility of forests to people may have other negative effects on birds. Diamond et al. (1987) speculated that increased human-mediated disturbance (e.g., nest predation by domestic dogs, *Canis familiaris*) might have been one factor contributing to the extinction of ground-dwelling species such as the banded pitta (*Pitta guajana*) and black-capped babbler (*Pellorneum capistratum*) in Bogor Botanical Garden (Java, Indonesia). The use of pesticides may also be detrimental to forest birds. The extirpation of two Maurisian carnivorous forest bird species, the Mauritius kestrel (*Falco punctatus*) and Mauritius cuckoo-shrike (*Coracina typica*) before the 1950s from some areas was probably due to the use of organochlorines (Safford & Jones 1997). Both species seem to have recovered since 1970, after use of organochlorines decreased. Helped by the reintroduction program and provision of nest boxes, the kestrel seems to have recolonized all the previously occupied areas. Owing to the poor dispersal abilities of the cuckoo-shrike, its reintroduction into all previously occupied areas may be needed (Safford & Jones 1997).

Deforestation may also facilitate the spread of invasives (plants and animals) into previously forested areas (e.g., Brash 1987, Diamond et al. 1987). The magnitude of the impact of invasives on extirpations remains unclear. The introduced brown tree snake (*Boiga irregularis*) was likely responsible for the loss of 12 (at least 8 forest species) of 18 native bird species from Guam (Fritts & Rodda 1998, Wiles et al. 2003). Similarly, ground-nesting birds are negatively impacted by feral cats (*Felis catus*) and domestic dogs (Leck 1979, Diamond et al. 1987, Kattan et al. 1994). However, some ground-nesting bird species may be able to counter cats because of adaptations to native ground predators (e.g., rodents and land crabs) (Mayr & Diamond 2001).

On French Polynesian islands, monarchs (*Pomarea* spp.) were rarely found on islands with the brown rat (*Rattus rattus*) (Seitre & Seitre 1992), suggesting either that they were avoiding this potential nest predator or that this rat was one of the possible reasons causing the extirpation. In a recent study, Thibault et al. (2002) reported that the rat probably caused the decline and extinction of monarchs from Polynesian islands. This decline was likely caused by nest predation and was exacerbated by a larger, superior, invasive competitor, the red-vented bulbul (*Pycnonotus cafer*).

EXTINCTION PRONENESS

Extinction proneness may depend on the level and type of threat and interactions among multiple factors (e.g., biogeographical, morphological, and behavioral).

Biogeography

Whether populations at the edges of their ranges are more vulnerable to deforestation remains unclear (Brooks 2000). Populations may be at their physiological and ecological limits at their distributional edge. In addition, because of low densities, populations at the edges of their ranges could be particularly vulnerable to genetic (e.g., inbreeding depression) and/or environmental stresses (e.g., pesticides; see Møller 1995). Nineteen of 29 species at the upper or lower limit of their elevational distributions were extirpated from the western Andes (Kattan et al. 1994). Similar instances have been reported elsewhere (Christiansen & Pitter 1997, Gillespie 2001). However, in the Colombian Andes, Renjifo (1999) found that species beyond their usual elevational limits were not disproportionately likely to become extinct, suggesting, in some cases, resilience even at distributional limits.

Rare species may be more extinction prone than common species, but rarity is not an independent factor and correlates with variables such as body size, habitat specificity, and geographic distribution (Kattan 1992, Goerck 1997). In Ecuador, some inherently rare species such as the tiny hawk (*Accipiter superciliosus*) and crane hawk (*Geranospiza caerulescens*) were possibly extirpated (Leck 1979). Small initial population size, requirement for large territories, and diet specialization may be factors in their elimination. Similarly, rare and uncommon species were more likely to be extirpated than common species in the Colombian Andes (Renjifo 1999). In tropical moist rainforests of eastern Queensland (Australia), rare species were negatively affected by fragmentation (Warburton 1997).

However, rare species are not always more extinction prone than common species. Karr (1982b) sampled the understory fauna of Limbo Hunt Club (Panama) 9 km east of Barro Colorado Island (BCI), presumed to be the source fauna for BCI. The species that were extirpated from BCI were not less abundant at Limbo Hunt Club (e.g., song wren, *Cyphorhinus phaeocephalus*). Species with higher annual population variability in the Limbo Hunt Club were the ones that disappeared from BCI. Karr hypothesized that such species track variable food resources and may require larger areas for their local movements. Similarly, apparently widespread and common species are sometimes extirpated, for example, the brown-hooded parrot

(*Pionopsitta haematotis*) and red-capped manakin (*Pipra mentalis*) in Ecuador (Leck 1979).

Extinction proneness can also be affected by the size of a species' range (Ribon et al. 2003). Small ranges may make species more vulnerable to stochastic perturbations even if local abundance is high. Proportionally, more passerines with small geographic ranges in the Americas are threatened than those with large geographic ranges (Manne et al. 1999). Deforestation may reduce habitat sizes for species requiring large home ranges. Four canopy frugivores likely requiring large home ranges (e.g., the scaly-naped parrot, *Amazona mercenaria*) disappeared from San Antonio, Ecuador (Kattan et al. 1994). Assuming that home range scales positively with body mass, Beier et al. (2002) determined that threshold patch size in Ghana for a 24 g and 920 g species was 10 ha and 8000 ha, respectively. Renjifo (1999) suggested that for some species large home ranges may not be a constraint, as such species could be adapted to track variable food resources and thus be able to travel even through the matrix (areas surrounding the forest).

Body Size

Large-bodied vertebrates are considered more extinction prone than small-bodied ones (e.g., Kattan 1992, Gaston & Blackburn 1995, Brook et al. 2003, but see Gotelli & Graves 1990 for island extinctions). A common explanation for this trend is that body size is inversely correlated with population size (Pimm et al. 1988), making large-bodied animals particularly vulnerable to environmental perturbations. The extinction proneness of large-bodied animals is further enhanced because of other correlated traits such as large area requirements, greater food intake, high habitat specificity, and low reproductive rates (Terborgh 1974, Leck 1979, Kattan 1992, Sieving & Karr 1997).

Large bird species are generally more vulnerable to human persecution, whereas small species are generally more vulnerable to habitat loss (Owens & Bennett 2000). However, we should be cautious in generalizing about the role of body size in the extinction process. Owing to a slower reproductive rate, larger parrots are more vulnerable than smaller finches, despite lesser numbers being captured for the pet trade (estimated 70% and 25% of 1.6–3.2 million birds annually, respectively) (Beissinger 2000). However, some smaller bird species (e.g., Muscicapinid flycatchers) with small population sizes can also be vulnerable owing to heavy pet trade, suggesting that in certain cases factors other than body size may enhance extinction proneness.

Many studies report the disproportionate loss of large-bodied or heavier species from avian communities (Leck 1979, Karr 1982a, Diamond et al. 1987, Christiansen & Pitter 1997, Renjifo 1999, Castelletta et al. 2000). Species that were lost from Barro Colorado Island were heavier than extant species, on the basis of sampling of the understory of nearby Limbo Hunt Club (Panama; Karr 1982b). However, there was no correlation between body mass and rarity, suggesting that larger species may not always be rare. Species extirpated between 1911 and 1997 from the Colombian Andes were heavier than extant species (Renjifo 1999). However, this relationship was not manifested for the missing species from smaller fragments, indicating resilience in some extant large-bodied species. On the same note, Castelletta et al. (2000) showed that larger species suffered more extirpations than smaller species initially, but this was not the case in recent losses (i.e., during the past 50 years). This result shows that body size may be important in early extirpations, but other factors may become more critical as the extinction process proceeds.

There may be an interaction between foraging habits and body size. Large canopy frugivores and large understory insectivores are extinction-prone guilds in the Neotropics (Willis 1979, Bierregaard & Lovejoy 1989). Extinct large canopy frugivores had significantly longer body lengths (mode 30 versus 10 cm) than extant species in San Antonio, Colombia (Kattan et al. 1994). Such species depend on fruits that typically vary spatially and temporally and may be difficult to locate in isolated forests. This was not the case for smaller canopy frugivores such as tanagers, because they supplement their diet with insects. Also, tanagers use a variety of habitats, such as the forest canopy, edges, and gaps, making them more resilient to habitat loss.

A minority of studies indicate no clear relationship between body size and extinction proneness. Newmark (1991) found that body mass was not correlated with bird occurrences. However, this result should be carefully interpreted because of relatively small variation in the body masses (7–138 g) and small sample sizes (22 out of 26 species had fewer than 30 captures). Body size also did not predict abundance in the forests of Sumba and Buru, Indonesia (Jones et al. 2001).

In addition to body size, other morphological variables may also affect extinction proneness. High investment in sexual traits may render some species less adaptable to changing environments. However, Jones et al. (2001) found that sexual dimorphism was not one of the predicators of forest bird abundance on the islands of Sumba and Buru. The effect of sexual dimorphism on extirpation of forest bird species remains unclear.

Behavioral Characteristics

Foraging habits can affect a species' ability to persist in an altered habitat. Many studies show that either frugivores or insectivores or both are more extinction prone than other guilds (Leck 1979, Willis 1979, Brash 1987, Newmark 1991, Kattan et al. 1994, Christiansen & Pitter 1997, Goerck 1997, Stratford & Stouffer 1999, Castelletta et al. 2000, Jones et al. 2001, Beier et al. 2002, Lambert & Collar 2002, Raman & Sukumar 2002, Ribon et al. 2003). The lack of year-round availability of fruit plants in smaller forests may cause the extinction of frugivores (Leck 1979).

A number of hypotheses are proposed for the disappearance of insectivores from deforested or fragmented areas. First, deforestation may impoverish the insect fauna and reduce preferred insectivore microhabitats (e.g., dead leaves). Second, insectivores may be poor dispersers and/or have near-ground nesting habits where they may be more vulnerable to nest predators. The insectivorous guild was more depauperate in small fragments (4–5 ha) than in a large fragment (227 ha) near Las Cruces (Costa Rica; Sekercioglu et al. 2002). Most large and specialized insectivorous bird species (e.g., the black-faced antthrush, *Formicarius analis*) were probably absent from small fragments. Invertebrate abundance, average length, and dry biomass were similar among the fragments. Fragment size also had little effect on the diet composition, prey biomass, and prey items per sample of most sampled bird species. This fact led Sekercioglu et al. (2002) to conclude that absence of some insectivorous bird species from small fragments may not be related to food scarcity but likely to their poor dispersal abilities.

Other guilds may be more extinction prone in some areas. In Bogor Botanic Garden, proportionally more extirpations were of carnivores than of any other foraging guild (Diamond et al. 1987). Similarly, carnivorous species were more vulnerable than other guilds in Nicaragua because of their lower population densities and susceptibility to hunting (Gillespie 2001). Are species with generalist foraging habits safe? In Singapore, mono-diet species were more vulnerable initially, but as the extinctions proceeded, multi-diet species also were extirpated, suggesting that with heavy forest loss even generalist species are affected (Castelletta et al. 2000). Two other studies showed no effect of dietary habits on extinction proneness (Karr 1982a, Renjifo 1999).

In Singapore, 30% of extinctions occurred in species that fed in the canopy (Castelletta et al. 2000). There may be an interplay of factors, as these canopy feeders were frugivores with large bodies, such as hornbills. Such species may have low population densities and may be affected by lack of adequate fruits, among other things (e.g., lack of nesting sites), in shrunken forests (Terborgh 1974).

Behaviors other than specialized foraging habits or sites also affect extinction proneness. Species forming mixed-species flocks disappeared more frequently from smaller fragments (1 and 10 ha) in Manaus (Brazil; Stouffer & Bierregaard 1995a). Some of the mixed-species flock members (e.g., the cinereous antshrike, *Thamnomanes caesius*) may have high foraging success in mixed flocks or have large territories and high territory fidelity (Munn & Terborgh 1979, Powell 1985). The latter two may be eroded in isolated fragments. Similarly, obligate armyant followers, restricted to the Neotropics, also disappeared from smaller fragments, probably because of large area requirements (Bierregaard & Lovejoy 1989, Bierregaard et al. 1992). These species (e.g., the white-rumped antbird, *Rhegmatorhina hoffmannsi*) may have home ranges between 1 and 5 km in diameter (Willis & Oniki 1978, Harper 1989, Bieregaard & Lovejoy 1989), an area not available in smaller fragments (≤ 10 ha). Harper (1989) demonstrated that obligate antfollowers (e.g., the rufous-throated antbird, *Gymnopithys rufigula*) disappeared from fragments not containing army ant colonies but were more likely to persist if such colonies were made experimentally available. In addition to mixed-species flocking and army-ant following birds, colonial nesting Icterids may also be vulnerable because they cannot form viable colonies in small fragments (Renjifo 1999).

A majority of tropical forest bird species have poor dispersal abilities (Terborgh 1974, Bierregaard et al. 1992, but see Terborgh et al. 1997). Ability to disperse may depend on morphology (e.g., wing loading) and physiology (e.g., intolerance to sunlight) (Johns 1992). Poor dispersal abilities may make certain bird species vulnerable because they cannot colonize new areas (Tilman et al. 1994, Lens et al. 2002). Indirect results show that species with superior colonization abilities and those that can exploit secondary habitats may be less extinction prone (Stouffer & Bierregaard 1995a, Gascon et al. 1999, Sekercioglu et al. 2002).

Habitat loss may also disrupt some behaviors of bird species (e.g., mating), thus making them particularly vulnerable to extinction (Reed 1999), but such data are lacking for tropical forest birds.

Life History Traits and Genetics

Habitat loss affects species regardless of their life history traits (Beissinger 2000, Purvis et al. 2000). Species with high fecundity, short generation time, low to moderate survival rate, and small body size are predicted to be vulnerable because they can succumb to large stochastic population fluctuations. Alternatively, species with low fecundity, long generation time, high survival, and large body size may also be susceptible because they would recover slowly from reductions in population size.

Attempts to understand the role of life-history traits in the extinction process of tropical birds have been rare. This is mainly because heavily concealed nests and breeding throughout the year make sound demographic data from tropical birds difficult to collect. Although artificial nest experiments do show that undisturbed areas may suffer less predation pressure than disturbed areas (Wong et al. 1998, Sodhi et al. 2003), results are difficult to generalize. For example, in the forests of the cerrado region of Brazil, the predation pressure estimated by artificial nest experiments did not vary with fragment size (Marini 2001).

Deforestation can also be one of the major factors in the loss of genetic variability within populations (Heywood & Stuart 1992). Because of poor dispersal abilities, patchy distributions, and generally low population densities, genetic diversity for some tropical bird species may be difficult to maintain. However, empirical evidence for this hypothesis is lacking. Overall, the role of inbreeding in avifaunal extinctions from isolated tropical populations remains unclear and requires scientific attention.

Taxonomy and Phylogeny

It is well known that controlling for phylogenetic effects is important in comparative behavioral ecological studies (Harvey & Pagel 1991). Lockwood (1999) correctly pointed out that phylogenetic effects are commonly measured as taxonomic effects (i.e., families or genera are taken as monophyletic). Despite the practical ease of considering taxonomy (a proxy for phylogeny) and the nonrandom nature of extinctions with respect to phylogeny (e.g., Bennett & Owens 1997, McKinney & Lockwood 1999), most studies concerning extinction vulnerability of tropical birds do not take this problem into account. A phylogenetic approach should be used more regularly for tropical birds because practical conservation measures take place in the context of local or regional avifaunas. Most of the studies concerned with phylogenetic biases have been done on a global scale (see Bennett & Owens 2002 for a review and summary), and the practical usefulness of these studies for management decisions in the tropics is not high. Another concern is that studies determining the role of phylogeny in bird extinctions rely heavily on Sibley & Ahlquist (1990), and in the absence of an independent source of information the conclusions should be drawn with caution.

Researchers have repeatedly pointed out that saving species is not enough; we must save evolutionary history (e.g., reviewed briefly in Vazquez & Gittleman 1998). But this has not been practiced in the tropics as far as we know. Poorly studied taxa are in general at higher direct risk of extinction (McKinney 1999). Because tropical areas are much less well studied than temperate areas, even for a comparatively well-studied taxon such as Aves, one can predict that tropical birds are even more at risk, owing to lower professional and public awareness (see Stutchbury & Morton 2001).

Although some authors suggest that phylogenetics should be incorporated into conservation decisions (Heard & Mooers 1999), others argue that species richness can be an adequate surrogate of phylogenetic diversity (Rodrigues et al. 2004). However, it is unclear how far theoretical considerations can go in practice, especially in the tropics. Practical tests of management of vulnerable species in the tropics need to be carefully carried out with scientific rigor to test the available (and conflicting) models (Nee & May 1997, Heard & Mooers 1999). Results thus far have been mixed. For instance, Fjeldså & Lovett (1997) pointed out that younger species of African birds are more patchily distributed and hence relatively more vulnerable to extinction. However, Lockwood et al. (2000) found that the age of a lineage does not seem to affect community homogenization (i.e., if particular lineages are lost and the community gets swamped by only a few invasive lineages).

Few specific tropical case studies deal with phylogenetic effects. Among those that exist, which primarily examine extinctions and the numbers of congeners, results are conflicting. Thiollay (1997) suggested that the presence of congeners may lead to competitive pressure, narrower niches, and lower abundance of at least some potential competitors. However, Terborgh & Winter (1980), with data from Trinidad and Venezuela, and Jones et al. (2001), with data from Sumba and Buru, suggest that congeners did not affect survivorship of their relatives.

Terborgh & Winter (1980) found that families containing disproportionately more susceptible species in Fernanado Po, Trinidad, Hainan, Sri Lanka, and Tasmania (temperate island) were: Bucerotidae, Cracidae, Falconidae, Phasianidae, Picidae, Timaliidae, Tinamidae, and Ramphastidae, although patterns for other families were mixed (containing similar numbers of both susceptible and resilient species). In the Colombian Andes, only Icteridae, as a family, was more prone to extinctions (Renjifo 1999). In Las Cruces, Sekercioglu et al. (2002) found a positive correlation between the number of species of a bird family present in nonforested habitats and their presence in small fragments. They seemed to suggest that family-level dispersal characteristics and abilities to exploit deforested habitats may assist in species persistence in small fragments.

ECOLOGICAL IMPACTS OF EXTINCTIONS

The extinction of some species (e.g., large predators and pollinators) may have higher ecological consequences than extinction of others (Terborgh 1992, Crooks & Soulé 1999). Because of the loss of larger predatory species from tropical communities, avian vulnerability to predation is often exacerbated. These larger predators (e.g., jaguars, *Panthera onca*) do not prey on birds directly but prey on smaller predators such as medium-sized and small mammals (mesopredators). In the absence of their predators, mesopredators become more abundant and prey upon adult birds, their young, and/or eggs (Terborgh 1992). Although this mesopredator release hypothesis has been applied largely to mammals, loss of birds such as raptors (e.g., the harpy eagle, *Harpia harpyja*) may cause similar effects in the ecosystem.

Does the disappearance of a competitor result in niche expansion and higher densities of subordinate species? In the undergrowth of Barro Colorado Island, the disappearance of a larger and probably superior competitor, the black-faced antthrush (*Formicarius analis*) appeared to have allowed the chestnut-backed antbird (*Myrmeciza exsul*) to become more densely and evenly distributed (Sieving & Karr 1997). Extinction of forest bird species, coupled with habitat changes (i.e., increased edge), may result in the thriving of edge species in isolated areas. Leck (1979) suggested that increased forest edge in Ecuador may have contributed to the commonness and possible increase of species such as the shiny cowbird (*Molothrus bonariensis*) and olive-crowned yellowthroat (*Geothlypis semiflava*). More than two thirds of the species that increased over time near Lagoa Santa (Brazil) live in the understory (Christiansen & Pitter 1997). These understory-dwellers are small-bodied insectivores, and their increase could be related to the increase in abundance of insects in remnants owing to an increased amount of edge.

Following deforestation, more nonforest or edge species (e.g., the rufous-collared sparrow, *Zonotrichia capensis*) can colonize new areas (Renjifo 1999). But this change is more likely due to the creation of optimal habitats than to the extinction of forest bird species. The loss of insectivorous species from experimentally isolated fragments in Manaus did not facilitate the increase of nonforest or previously uncommon species even after nine years of isolation (Stouffer & Bierregaard 1995a).

Therefore, the role of *direct* interspecific competition in the extinction process remains unclear. Similarly, how competitive release affects the extant species remains to be clearly demonstrated. The extinction of insectivorous birds from scrubs of West Indian islands correlated with the subsequent higher biomass of competing *Anolis* lizards (Wright 1981). It would have been interesting to determine whether bird reintroductions reversed the observed pattern (i.e., lowered the biomass of *Anolis* lizards).

Parasitism may also be affected by extinctions. The parasitic brush cuckoo (*Cacomantis variolosus*) disappeared from Bogor Botanic Garden, possibly because of the decline in the abundance of its likely hosts, the pied fantail (*Rhipidura javanica*) and hill blue flycatcher (*Cyornis banyumas*). In contrast, the parasitic plaintive cuckoo (*C. merulinus*) survived in this isolate because of high abundance of its primary host, the ashy tailorbird (*Orthotomus ruficeps*; Diamond et al. 1987).

Because many tropical trees produce large, lipid-rich fruits adapted for animal dispersal (Howe 1984), the demise of bird frugivores may have consequences for forest regeneration. One of the dominant fruiting trees in Puerto Rico, *Dacryodes excelsa*, failed to reestablish in areas it previously occupied, possibly because of the extinction and decline of frugivores (Brash 1987). Some late successional montane tree species (e.g., *Canarium asperum*) rely heavily on frugivorous bird species such as hornbills and fruit pigeons for seed dispersal (Hamann & Curio 1999). These bird species are under heavy hunting pressure and their likely extinction may disrupt the recruitment of the trees. A similar predicament has been found for African trees that rely heavily on birds for seed dispersal (Cordeiro & Howe 2001, 2003).

RESILIENCE

Despite our focus on tropical extinctions, not all tropical species are extinction prone. The species-area relationship leads to a rule of thumb that a 90% loss in habitat leads approximately to a 25% to 50% loss of species (Simberloff 1992a). The predictive power of this relationship may be weak because it does not account for either habitat diversity or fragmentation (Simberloff 1992b), but it is the only such existing model and is still used to make crude predictions of the extent of biotic extinctions. Thus far, tropical bird extinctions largely seem to be fewer than predicted by the species-area equation (Brooks et al. 2002). This fact indicates temporary or permanent resilience in some of the species and/or a time lag in the extinction process.

Since European colonization about 500 years ago, the Atlantic coastal rainforest of Brazil has been reduced to 12% of its approximated original size but with no recently reported avian extirpations (Brown & Brown 1992). The above results may reflect genuine resilience in some of the species. Alternatively, extinctions of species unknown to science may have already happened or there may be a time lag in extinctions following habitat loss (Balmford 1996, Brooks & Balmford 1996, Brooks et al. 1999c).

In a 15-km^2 remnant forest (0.3%) of Cebu Island (Philippines), 7 of 14 forest endemic bird species/subspecies survived (Magsalay et al. 1995). Fragmented landscape containing mature forest can retain as much as 96% of the original avifauna several decades after isolation (Renjifo 1999). Many other examples show that a proportion of forest avifauna can be found in disturbed habitats (e.g., Johns 1986, Thiollay 1992, Mitra & Sheldon 1993, Warkentin et al. 1995, Woltmann 2003). Few data are available, however, on whether persisting species in degraded areas are reproducing and surviving as well as their counterparts in pristine areas.

THE FUTURE

Establishment of appropriate protected areas is critical in halting future extinctions. About 23% of existing tropical humid forests are officially protected (Chape et al. 2003). However, it is not known if these reserves receive adequate protection (e.g., against illegal logging and poaching). In the tropics, the extent of protected areas does not correspond well with the number of threatened species (Kerr & Burkey 2002). Effective forest reserves should have viable populations of the threatened species, as mere presence cannot guarantee survival. Although population viability remains a controversial aspect, ecological data such as population size, survival, and recruitment are needed and should be collected to evaluate adequately the population viability of a given species (see Caughley & Gunn 1986, Coulsen et al. 2001, Brook et al. 2002).

Because of typically patchy distributions among tropical bird species, care should be taken in reserve selection (Diamond 1980, Van Balen 1999). Lists of threatened species must be consulted because a large reserve may fail to protect all or even most vulnerable species. For example, in the Albertine Rift Mountains in Central Africa, 9 of 15 threatened bird species do not occur in any of the protected reserves (Bibby 1994). In addition to reserve location, reserve size is also critical. Small reserves may (*a*) not contain enough resources, (*b*) facilitate the infiltration and spread of parasites and invasives from the surrounding matrix, and (*c*) contain tiny populations that may be vulnerable to environmental stochasticity, such as hurricanes. In fact, a 198 ha remnant failed to contain vulnerable species such as the black-necked aracari (*Pteroglossus aracari*) and spot-billed toucanet (*Selenidera maculirostris*) near Lagoa Santa (Brazil; Christiansen & Pitter 1997). Therefore, larger protected areas may be necessary, but there is no consensus about the minimum reserve size in tropics. Leck (1979) recommended reserves as large as 20–30 km^2 but preferred those as large as 100 km^2. Other authors also have suggested that reserves of several thousand km^2 may be needed in the tropics to halt or diminish the chances of mass extinctions (Terborgh 1974, 1992; Whitmore 1980; Myers 1986; Thiollay 1989). Such large reserves may be possible in areas such as the Congo and French Guiana where large undisturbed forests still exist (Whitmore 1997). In areas where large forested tracts are unavailable, forests around the existing reserves could be restored (Whitmore & Sayer 1992a).

The abundance of potential predators should not be unnaturally high in reserves (Terborgh et al. 1997). Ideally, mature and/or high-quality forests should be protected. Renjifo (1999) showed that fragments with mature forest can sustain an avifauna for decades after isolation. Sodhi (2002) found that a reserve with poor-quality mature forest (i.e., heath forest) contained fewer rare species than one with high-quality, similar-sized mature forest.

Habitat links (e.g., fence rows and windbreaks) between patches or reserves may facilitate dispersal and prevent local extinctions or population declines in fragmented landscapes (Lovejoy et al. 1997, Sekercioglu et al. 2002). Data on efficacy and adequacy of habitat corridors in the tropics are limited, however. Vegetation cover (both native and non-native) can enhance the abilities of corridors to attract forest species such as the short-tailed babbler, suggesting either movement through these areas or use of such edge habitats (Sodhi et al. 1999). On the negative side, corridors can be expensive to create and maintain, may not serve well in heavily fragmented landscapes, and may be counter-productive by facilitating the spread of predators and diseases (Simberloff & Cox 1987, Simberloff 1992b).

Although our review shows that a proportion of avifauna does not adapt to habitat change and will be extirpated, 26% of 144 species recorded in the countryside of Las Cruces (Costa Rica) were forest species (restricted to forest habitat including edges of the small fragments; Hughes et al. 2002). Edges with vertically complex vegetation (tree-tangle-shrub) had at least 100 species. It is unknown if the forest species in this anthropogenic habitat were maintaining sustainable populations, but this study showed adaptability in certain species (e.g., collared aracari, *Pteroglossus frantzii*). Removal of tall trees and edge habitats, even from degraded landscapes, may lower avian richness (Hughes et al. 2002, Petit & Petit 2003). These results cannot be generalized yet, as resilience will likely be species- or site-specific. For example, some endemic species, e.g., elegant sunbird (*Aethopyga duyvenbodei*), rarely use secondary forests even though congeneric species do (Whitten et al. 1987).

Habitat preservation may be a better approach as it can conserve many bird species simultaneously, but specific management may be needed to save individual species (Temple 1986). Conservation programs should not be directed solely at preventing population declines of vulnerable species but should also bring about recovery. To increase the population size of species headed toward extinction, researchers can envision habitat restoration, reintroduction, elimination of invasive predators and competitors, and captive breeding (Heywood & Stuart 1992). Eradication techniques for invasive species have been well developed over the years (Conover 2002). However, management must proceed carefully, using detailed ecological understanding to avoid failure (e.g., failed reintroduction of song wrens on Barro Colorado Island owing to high predation; Morton 1978). Above all, unsustainable forestry practices in the tropics must be curtailed (Laurance 2000). More emphasis is needed to integrate sound biology into forestry practices. For example, there is no consensus on how many years of logging rotation cycle are required to preserve the avian diversity. Research to realize biodiversity-friendly forestry practices is urgently needed.

When one considers that natural tropical forests occupy an area of 1756 million ha containing different forest types (e.g., lowland and montane; Whitmore 1997), little ground has been covered on the effects of deforestation on avifauna. Data are also insufficient to evaluate whether different types of disturbance (e.g., agriculture and urbanization) yield similar negative effects on the avifauna. Some tropical areas remain poorly surveyed and the status of their biodiversity uncertain (Whitten et al. 1987). Detailed biological understanding may be needed for proper conservation/management (Reed 1999), but such an understanding may be poor for certain tropical areas (Diamond 1987, Sodhi & Liow 2000). Systematic surveys to determine faunal vulnerability and to collect ecological data are needed throughout the tropics in order to practice educated conservation. Life-history variables (e.g., clutch size, reproductive output, and dispersal abilities) and diet of a large proportion of tropical bird species remain unknown. This lacuna makes the generalization of extinction proneness of tropical bird species ecologically shallow.

Studies determining the effects of deforestation on tropical avifauna must be better designed (Bierregaard et al. 1997, Crome 1997, Danielsen 1997). These studies must (a) obtain adequate sample sizes, (b) have adequate sampling periods with multiple visits, (c) use more than one sampling method (e.g., point counts and mist-netting) for adequate sampling, (d) estimate species abundances or densities, (e) validate local extinctions using playback surveys and present species accumulation curves (see Gotelli & Colwell 2001), and (f) have adequate controls (e.g., larger "original" forests). Data sets from the studies should be published or, better still, made available on the Internet for a wide access. We also recommend that individuals making bird lists rank species according to their abundance and clearly define their ranking system. This ranking will assist in determining whether a species is declining in abundance and whether rare species are more extinction prone than common species. Last, scientists must collect data showing whether applied conservation measures (e.g., establishment of reserves) are abating extinctions.

Our review shows that deforestation and its associated effects are permanently altering tropical forest bird communities. Extirpated species may or may not share similar characteristics across the tropics. Idiosyncrasies among species and sites and poor ecological knowledge make widely applicable extinction proneness rules for tropical forest bird species elusive. In general, biology is not properly integrated into tropical forestry practices. More ecological research is needed to evaluate the status and to increase our understanding of the biology of tropical birds.

ACKNOWLEDGMENTS

N.S.S. held the Charles Bullard Fellowship while the first draft of this manuscript was prepared. Partial research support was provided by the National University of Singapore (R-154-000-144-112). We thank Barry Brook, Tom Brooks, Lian Pin Koh, Eugene Morton, and Kevin Winker for making comments on an earlier draft.

The *Annual Review of Ecology, Evolution, and Systematics* is online at
http://ecolsys.annualreviews.org

LITERATURE CITED

Achard F, Eva HD, Stibig HJ, Mayaux P, Gallego J, et al. 2002. Determination of deforestation rates of the world's humid tropical forests. *Science* 297:999–1002

Balmford A. 1996. Extinction filters and current resilience: the significance of past selection pressures for conservation biology. *Trends Ecol. Evol.* 11:193–96

Beier P, Van Drielen M, Kankam BO. 2002. Avifaunal collapse in west African forest fragments. *Conserv. Biol.* 16:1097–111

Beissinger SR. 2000. Ecological mechanism of extinction. *Proc. Natl. Acad. Sci. USA* 97:11688–89

Beissinger SR. 2001. Trade of live birds: potentials, principles and practices of sustainable use. In *Conservation of Exploited Species*, ed. JD Reynolds, GM Mace, KH Redford, JG Robinson, pp. 182–202. Cambridge, UK: Cambridge Univ. Press

Bennett PM, Owens IPF. 1997. Variation in extinction risk among birds: Chance or evolutionary predisposition? *Proc. R. Soc. London Ser. B* 264:401–8

Bennett PM, Owens IPF. 2002. *Evolutionary Ecology of Birds*. Oxford: Oxford Univ. Press

Bibby CJ. 1994. Recent past and future extinctions in birds. *Philos. Trans. R. Soc. London Ser. B* 344:35–40

Bierregaard RO Jr., Laurance WF, Sites JW Jr., Lynam AJ, Didham RK, et al. 1997. Key priorities for the study of fragmented tropical ecosystems. See Laurance & Bierregaard 1997, pp. 515–25

Bierregaard RO Jr., Lovejoy TE. 1989. Effects of forest fragmentation on Amazonian understory bird communities. *Acta Amazonica* 19:215–41

Bierregaard RO Jr., Lovejoy TE, Kapos V, dos Santos AA, Hutchings RW. 1992. The biological dynamics of tropical rainforest fragments. *BioScience* 42:859–66

BirdLife International. 2000. *Threatened Birds of the World*. Barcelona, Spain/Cambridge, UK: Lynx Edicions/BirdLife International

Brash AR. 1987. The history of avian extinction and forest conversion on Puerto Rico. *Biol. Conserv.* 39:97–111

Brook BW, Burgman MA, Akcakaya HR, O'Grady JJ, Frankham R. 2002. Critiques of PVA ask wrong questions: throwing the heuristic baby out with the numerical bath water. *Conserv. Biol.* 16:262–63

Brook BW, Sodhi NS, Ng PKL. 2003. Catastrophic extinctions follow deforestation in Singapore. *Nature* 424:420–23

Brooks TM. 2000. Living on the edge. *Nature* 403:27–29

Brooks TM, Balmford A. 1996. Atlantic forest extinctions. *Nature* 380:115

Brooks TM, Mittermeier RA, Mittermeier CG, da Fonseca GAB, Rylands AB, et al. 2002. Habitat loss and extinction in the hotspots of biodiversity. *Conserv. Biol.* 16:909–23

Brooks TM, Pimm SL, Collar NJ. 1997. Deforestation predicts the number of threatened birds in insular Southeast Asia. *Conserv. Biol.* 11:382–94

Brooks TM, Pimm SL, Kapos V, Ravilious C. 1999a. Threat from deforestation to montane and lowland birds and mammals in insular South-east Asia. *J. Anim. Ecol.* 68:1061–78

Brooks TM, Pimm SL, Oyugi JO. 1999b. Time lag between deforestation and bird extinction in tropical forest fragments. *Conserv. Biol.* 13:1140–50

Brooks TM, Tobias J, Balmford A. 1999c. Deforestation and bird extinctions in the Atlantic forests. *Anim. Conserv.* 2:211–22

Brown JH, Kodric-Brown A. 1977. Turnover rates in insular biogeography: effect of immigration on extinction. *Ecology* 58:445–49

Brown KS Jr., Brown GG. 1992. Habitat alteration and species loss in Brazilian forests. See Whitmore & Sayer 1992b, pp. 119–42

Bush MB, Whittaker RJ. 1991. Krakatau: colonization patterns and hierarchies. *J. Biogeogr.* 18:341–56

Castelletta M, Sodhi NS, Subaraj R. 2000. Heavy extinctions of forest avifauna in Singapore: lessons for biodiversity conservation in Southeast Asia. *Conserv. Biol.* 14:1870–80

Caughley G, Gunn A. 1986. *Conservation Biology in Theory and Practice*. Cambridge, MA: Blackwell Sci.

Chape S, Blyth S, Fish L, Fox P, Spalding M. 2003. *2003 United Nations List of Protected Areas*. Gland, Switz.: IUCN & UNEP

Christiansen MB, Pitter E. 1997. Species loss in a forest bird community near Lagoa Santa in Southern Brazil. *Biol. Conserv.* 80:23–32

Conover M. 2002. *Resolving Human-Wildlife Conflicts*. Boca Raton, FL: Lewis

Cordeiro NJ, Howe HF. 2001. Low recruitment of trees dispersed by animals in African forest fragments. *Conserv. Biol.* 15:1733–41

Cordeiro NJ, Howe HF. 2003. Forest fragmentation severs mutualism between seed dispersers and an endemic African tree. *Proc. Natl. Acad. Sci. USA* 100:14052–56

Coulson T, Mace GM, Hudson E, Possingham H. 2001. The use and abuse of population viability analysis. *Trends Ecol. Evol.* 16:219–21

Crome FHJ. 1997. Researching tropical forest fragmentation: shall we keep doing what we're doing? See Laurance & Bierregaard 1997, pp. 485–501

Crooks KR, Soulé ME. 1999. Mesopredator release and avifaunal extinctions in a fragmented system. *Nature* 400:563–66

Daily GC, Ehrlich PR, Sanchez-Azofeifa GA. 2001. Countryside biogeography: use of human-dominated habitats by the avifauna of southern Costa Rica. *Ecol. Appl.* 11:1–13

Danielsen F. 1997. Stable environments and fragile communities: Does history determine the resilience of avian rain-forest communities to habitat degradation? *Biod. Conserv.* 6:423–33

Diamond JM. 1980. Patchy distributions of tropical birds. See Soulé & Wilcox 1980, pp. 57–74

Diamond JM. 1987. Extant unless proven extinct? Or, extinct unless proven extant? *Conserv. Biol.* 1:77–79

Diamond JM. 1989. The present, past and future of human-caused extinctions. *Philos. Trans. R. Soc. London Ser. B* 325:469–76

Diamond JM, Bishop KD, Van Balen S. 1987. Bird survival in an isolated Javan woodland: Island or mirror? *Conserv. Biol.* 1:132–42

Ferraz G, Russell GJ, Stouffer PC, Bierregaard RO Jr., Pimm SL, Lovejoy TE. 2003. Rates of species loss from Amazonian forest fragments. *Proc. Natl. Acad. Sci. USA* 100:14069–73

Fjeldså J, Lovett JC. 1997. Biodiversity and environmental stability. *Biod. Conserv.* 6:315–23

Fritts TH, Rodda GH. 1998. The role of introduced species in the degradation of island ecosystems: a case history of Guam. *Annu. Rev. Ecol. Syst.* 29:113–40

Gascon C, Lovejoy TE, Bierregaard RO Jr., Malcolm JR, Stouffer PC, et al. 1999. Matrix habitat and species richness in tropical forest remnants. *Biol. Conserv.* 91:223–29

Gaston KJ, Blackburn TM. 1995. Birds, body size and the threat of extinction. *Philos. Trans. R. Soc. London Ser. B* 347:205–12

Gaston KJ, Blackburn TM, Goldewijk KK. 2003. Habitat conservation and global avian biodiversity. *Proc. R. Soc. London Ser. B* 270:1293–300

Gillespie TW. 2001. Application of extinction and conservation theories for forest birds in Nicaragua. *Conserv. Biol.* 15:699–709

Goerck JM. 1997. Patterns of rarity in the birds of the Atlantic forest of Brazil. *Conserv. Biol.* 11:112–18

Gotelli NJ, Colwell RK. 2001. Quantifying biodiversity: procedures and pitfalls in the measurement and comparison of species richness. *Ecol. Lett.* 4:379–91

Gotelli NJ, Graves GR. 1990. Body size and the occurrence of avian species on land-bridge islands. *J. Biogeogr.* 17:315–25

Hamann A, Curio E. 1999. Interactions among frugivores and fleshy fruit trees in a Philippine submontane rainforest. *Conserv. Biol.* 13:766–73

Harper LH. 1989. The persistence of ant-following birds in small Amazonian forest fragments. *Acta Amazonica* 19:249–63

Harvey PH, Pagel MD. 1991. *The Comparative Method in Evolutionary Biology.* Oxford: Oxford Univ. Press

Heard SB, Mooers AO. 1999. Phylogenetically patterned speciation rates and extinction risks change the loss of evolutionary history during extinctions. *Proc. R. Soc. London Ser. B* 267:613–20

Heywood VH, Stuart SN. 1992. Species extinctions in tropical forests. See Whitmore & Sayer 1992b, pp. 91–117

Howe HF. 1984. Implications of seed dispersal by animals for tropical reserve management. *Biol. Conserv.* 30:261–81

Hughes JB, Daily GC, Ehrlich PR. 2002. Conservation of tropical forest birds in countryside habitats. *Ecol. Lett.* 5:121–29

Johns AD. 1986. Bird population persistence in Sabahan logging concessions. *Biol. Conserv.* 75:3–10

Johns AD. 1992. Species conservation in managed tropical forests. See Whitmore & Sayer 1992b, pp. 15–54

Jones MJ, Sullivan MS, Marsden SJ, Linsley MD. 2001. Correlates of extinction risk of birds from two Indonesian islands. *Biol. J. Linn. Soc.* 73:65–79

Karr JR. 1982a. Avian extinction on Barro Colorado Island, Panama: a reassessment. *Am. Nat.* 119:220–39

Karr JR. 1982b. Population variability and extinction in the avifauna of a tropical land bridge island. *Ecology* 63:1975–78

Kattan GH. 1992. Rarity and vulnerability: the birds of the Cordillera Central of Colombia. *Conserv. Biol.* 6:64–70

Kattan GH, Alvarez-Lopez H, Giraldo M. 1994. Forest fragmentation and bird extinctions: San Antonio eighty years later. *Conserv. Biol.* 8:138–46

Kerr JT, Burkey TV. 2002. Endemism, diversity, and the threat of tropical moist forest extinctions. *Biod. Conserv.* 11:695–704

Lambert FR, Collar NJ. 2002. The future for Sundaic lowland forest birds: long-term effects of commercial logging and fragmentation. *Forktail* 18:127–46

Laurance WF. 2000. Cut and run: the dramatic rise of transitional logging in the tropics. *Trends Ecol. Evol.* 15:433–34

Laurance WF, Bierregaard RO Jr., eds. 1997. *Tropical Forest Remnants: Ecology, Management, and Conservation of Fragmented Communities*. Chicago: Univ. Chicago Press

Laurance WF, Bierregaard RO Jr., Gascon C, Didham RK, Smith AP, et al. 1997. Tropical forest fragmentation: synthesis of a diverse and dynamic discipline. See Laurance & Bierregaard 1997, pp. 502–14

Leck CF. 1979. Avian extinctions in an isolated tropical wet-forest preserve, Ecuador. *Auk* 96:343–52

Lens L, Van Dongen S, Norris K, Githiru M, Matthysen E. 2002. Avian persistence in fragmented rainforest. *Science* 298:1236–38

Lockwood JL. 1999. Using taxonomy to predict success among introduced avifauna: Relative importance of transport and establishment. *Conserv. Biol.* 13:560–67

Lockwood JL, Brooks TM, McKinney ML. 2000. Taxonomic homogenization of the global avifauna. *Anim. Conserv.* 3:27–35

Lovejoy TE, Bierregaard RO Jr., Rylands AB, Malcolm JR, Quintela CE, et al. 1997. Edge and other effects of isolation on Amazon forest fragments. See Laurance & Bierregaard 1997, pp. 257–85

Magsalay P, Brooks T, Dutson G, Timmins R. 1995. Extinction and conservation on Cebu. *Nature* 373:294

Manne LL, Brooks TM, Pimm SL. 1999. Relative risk of extinction of passerine birds on continents and islands. *Nature* 399:258–61

Marini MA. 2001. Effects of forest fragmentation on birds of the Cerrado region, Brazil. *Bird Conserv. Int.* 11:13–25

Mayr E, Diamond J. 2001. *Birds of Northern Melanesia*. Oxford: Oxford Univ. Press

McKinney ML. 1999. High rates of extinction and threat in poorly studied taxa. *Conserv. Biol.* 13:1273–81

McKinney ML, Lockwood JL. 1999. Biotic homogenisation: a few winners replacing many losers in the next mass extinction. *Trends Ecol. Evol.* 14:450–53

Milberg P, Tyrberg T. 1993. Naïve birds and noble savage—a review of man-caused prehistoric extinctions of island birds. *Ecography* 16:229–50

Milner-Gulland EJ, Bennett EL, the SCB 2002 Annual Meeting Wild Meat Group. 2003. Wild meat: the bigger picture. *Trends Ecol. Evol.* 18:351–57

Mitra SS, Sheldon FH. 1993. Use of an exotic plantation by Bornean lowland forest birds. *Auk* 110:529–40

Møller AP. 1995. Patterns of fluctuating asymmetry in sexual ornaments of birds from marginal and central populations. *Am. Nat.* 145:316–27

Morton ES. 1978. Reintroducing recently extirpated birds into a tropical forest preserve. In *Endangered Birds: Management Techniques for Preserving Threatened Species*, ed. SA Temple, pp. 379–84. Madison: Univ. Wis. Press

Munn CA, Terborgh JW. 1979. Multi-species territoriality in neotropical foraging flocks. *Condor* 81:338–47

Myers N. 1986. Tropical deforestation and a mega-extinction spasm. In *Conservation Biology: The Science of Scarcity and Diversity*, ed. ME Soulé, pp. 394–409. Sunderland, MA: Sinauer

Myers N, Knoll AH. 2001. The biotic crisis and the future of evolution. *Proc. Natl. Acad. Sci. USA* 98:5389–92

Myers N, Mittermeier RA, Mittermeier CG, da Fonseca GAB, Kent J. 2000. Biodiversity hotspots for conservation priorities. *Nature* 403:853–58

Nee S, May RM. 1997. Extinction and the loss of evolutionary history. *Science* 278:692–94

Newmark WD. 1991. Tropical forest fragmentation and the local extinction of understory birds in the eastern Usambra mountains, Tanzania. *Conserv. Biol.* 5:67–78

Owens IPF, Bennett PM. 2000. Ecological basis of extinction risk in birds: habitat loss versus human persecution and introduced predators. *Proc. Natl. Acad. Sci. USA* 97:12144–48

Petit LJ, Petit DR. 2003. Evaluating the importance of human-modified lands for neotropical bird conservation. *Conserv. Biol.* 17:687–94

Pimm SL. 1996. Lessons from a kill. *Biod. Conserv.* 5:1059–67

Pimm SL, Jones H, Diamond J. 1988. On the risk of extinction. *Am. Nat.* 132:757–85

Pimm SL, Moulton MP, Justice LJ. 1994. Bird extinctions in the central Pacific. *Philos. Trans. R. Soc. London Ser. B* 344:27–33

Pimm SL, Raven P. 2000. Extinction by numbers. *Nature* 403:843–45

Powell GVN. 1985. Sociobiology and adaptive significance of interspecific foraging flocks in the neotropics. *Ornithol. Monogr.* 36:713–32

Purvis A, Jones KE, Mace GM. 2000. Extinction. *BioEssays* 22:1123–33

Raman TR, Sukumar R. 2002. Responses of tropical rainforest birds to abandoned plantations, edges and logged forest in the Western Ghats, India. *Anim. Conserv.* 5:201–16

Reed JM. 1999. The role of behaviour in recent avian extinctions and endangerments. *Conserv. Biol.* 13:232–41

Renjifo LM. 1999. Composition changes in a subandean avifauna after long-term forest fragmentation. *Conserv. Biol.* 13:1124–39

Ribon R, Simon JE, De Mattos GT. 2003. Bird extinctions in Atlantic forest fragments of the Viçosa region, Southeastern Brazil. *Conserv. Biol.* 17:1827–39

Robinson WD. 1999. Long-term changes in the avifauna of Barro Colorado Island, Panama, a tropical forest isolate. *Conserv. Biol.* 13:85–97

Rodrigues ASL, Brooks TM, Gaston KJ. 2004. Integrating phylogenetic diversity in the selection of priority areas for conservation: Does it really make a difference? In *Phylogeny and Conservation*, ed. A Purvis, JG Gittleman, TM Brooks. Cambridge, UK: Cambridge Univ. Press. In press

Safford RJ, Jones CG. 1997. Did organochlorine pesticide use cause declines in Mauritian forest birds? *Biodivers. Conserv.* 6:1445–51

Scott JM, Kepler CB, VanRipper C, Fefer SI. 1988. Conservation of Hawaii's vanishing avifauna. *BioScience* 38:238–53

Sekercioglu CH, Ehrlich PR, Daily GC, Aygen D, Goehring D, Sandi RF. 2002. Disappearance of insectivorous birds from tropical forest fragments. *Proc. Natl. Acad. Sci. USA* 99:263–67

Seitre R, Seitre J. 1992. Causes of land-bird extinction in French Polynesia. *Oryx* 26:215–22

Sibley CG, Ahlquist JE. 1990. *Phylogeny and Classification of Birds: A Study in Molecular Evolution*. New Haven, CT: Yale Univ. Press

Sieving KE, Karr JR. 1997. Avian extinction and persistence mechanisms in lowland Panama. See Laurance & Bierregaard 1997, pp. 156–70

Simberloff D. 1992a. Do species-area curves predict extinction in fragmented forest? See Whitmore & Sayer 1992b, pp. 75–89

Simberloff D. 1992b. Species-area relationships, fragmentation, and extinction in tropical forests. *Malayan Nat. J.* 45:398–413

Simberloff D, Cox J. 1987. Consequences and costs of conservation corridors. *Conserv. Biol.* 1:63–71

Sodhi NS. 2002. A comparison of bird communities of two fragmented and two continuous Southeast Asian rainforests. *Biod. Conserv.* 11:1105–19

Sodhi NS, Briffett C, Kong L, Yuen B. 1999. Bird use of linear areas of a tropical city: implications for park connector design and management. *Land. Urban Plan.* 45:123–30

Sodhi NS, Liow LH. 2000. Improving conservation biology research in Southeast Asia. *Conserv. Biol.* 14:1211–12

Sodhi NS, Peh KSH, Lee TM, Turner IM, Tan HTW, et al. 2003. Artificial nest and seed predation experiments on tropical Southeast Asian islands. *Biod. Conserv.* 12:2415–33

Soulé ME, Wilcox BA, eds. 1980. *Conservation Biology: An Evolutionary-Ecological Perspective*. New York: Sinauer

Steadman D. 1995. Prehistoric extinctions of Pacific Island birds: biodiversity meets zooarchaeology. *Science* 267:1123–31

Stouffer PC, Bierregaard RO Jr. 1995a. Use of Amazonian fragments by understory insectivorous birds. *Ecology* 76:2429–45

Stouffer PC, Bierregaard RO Jr. 1995b. Effects of forest fragmentation on understory hummingbirds in Amazonian Brazil. *Conserv. Biol.* 9:1085–94

Stratford JA, Stouffer PC. 1999. Local extinctions of terrestrial insectivorous birds in a fragmented landscape near Manaus, Brazil. *Conserv. Biol.* 13:1416–23

Stutchbury BJM, Morton ES. 2001. *Behavioral Ecology of Tropical Birds*. London: Academic

Temple SA. 1986. The problem of avian extinctions. *Curr. Ornithol.* 3:453–85

Terborgh J. 1974. Preservation of natural diversity: the problem of extinction prone species. *BioScience* 24:715–22

Terborgh J. 1992. Maintenance of diversity in tropical forests. *Biotropica* 24:283–92

Terborgh J, Lopez L, Jose Tello S. 1997. Bird communities in transition: the Lago Guri islands. *Ecology* 78:1494–501

Terborgh J, Robinson SK, Parker TA III, Munn CA, Pierpont N. 1990. Structure and organization of an Amazonian forest bird community. *Ecol. Monogr.* 60:213–38

Terborgh J, Winter B. 1980. Some causes of extinction. See Soulé & Wilcox 1980, pp. 119–33

Thibault JC, Martin JL, Penloup A, Meyer JY. 2002. Understanding the decline and extinction of monarchs (Aves) in Polynesian islands. *Biol. Conserv.* 108:161–74

Thiollay JM. 1989. Area requirements for the conservation of rain forest raptors and game birds in French Guiana. *Conserv. Biol.* 3:128–37

Thiollay JM. 1992. Influence of selective logging on bird species diversity in a Guianan rain forest. *Conserv. Biol.* 6:47–63

Thiollay JM. 1997. Distribution and abundance patterns of bird community and raptor populations in the Andaman archipelago. *Ecography* 20:67–82

Tilman D, May RM, Lehman CL, Nowak MA. 1994. Habitat destruction and the extinction debt. *Nature* 371:65–66

Turner IM. 1996. Species loss in fragments of tropical rain forest: a review of evidence. *J. Appl. Ecol.* 33:200–9

Van Balen B. 1999. Differential extinction patterns in Javan forest birds. *Trop. Res. Manag. Paper* 30:39–57

Vazquez DP, Gittleman JL. 1998. Biodiversity conservation: Does phylogeny matter? *Current Biol.* 8:R379–81

Warburton NH. 1997. Structure and conservation of forest avifauna in isolated rainforest remnants in tropical Australia. See Laurance & Bierregaard 1997, pp. 190–206

Warkentin IG, Greenberg R, Ortiz JS. 1995. Songbird use of gallery woodlands in recently cleared and older settled landscapes of the Selva Lacandona, Chiapas, Mexico. *Conserv. Biol.* 9:1095–106

Whitmore TC. 1980. The conservation of tropical rain forest. See Soulé & Wilcox 1980, pp. 303–18

Whitmore TC. 1997. Tropical forest disturbance, disappearance, and species loss. See Laurance & Bierregaard 1997, pp. 3–12

Whitmore TC, Sayer JA. 1992a. Deforestation and species extinction in tropical moist forests. See Whitmore & Sayer 1992b, pp. 1–14

Whitmore TC, Sayer JA. 1992b. *Tropical Deforestation and Species Extinction*. London: Chapman & Hall

Whitten AJ, Bishop KD, Nash SV, Clayton L. 1987. One or more extinctions from Sulawesi, Indonesia? *Conserv. Biol.* 1:42–48

Wiles GJ, Bart J, Beck RE Jr., Aguon CF. 2003. Impacts of the brown tree snake: patterns and species persistence in Guam's avifauna. *Conserv. Biol.* 17:1350–60

Willis EO. 1974. Populations and local extinctions of birds on Barro Colorado Island, Panama. *Ecol. Monogr.* 44:153–69

Willis EO. 1979. The composition of avian communities in reminiscent woodlots in Southern Brazil. *Papeis Avulsos Zool., Sao Paulo* 33:1–25

Willis EO, Oniki Y. 1978. Birds and army ants. *Annu. Rev. Ecol. Syst.* 9:243–63

Woltmann S. 2003. Bird community responses to disturbance in a forestry concession in lowland Bolivia. *Biod. Conserv.* 12:1921–36

Wong TCM, Sodhi NS, Turner IM. 1998. Artificial nest and seed predation experiments in tropical lowland rainforest remnants of Singapore. *Biol. Conserv.* 85:97–104

Wright SJ. 1981. Extinction-mediated competition: the *Anolis* lizards and insectivorous birds of the West Indies. *Am. Nat.* 117:181–92

EVOLUTIONARY BIOLOGY OF ANIMAL COGNITION

Reuven Dukas

Animal Behavior Group, Department of Psychology, McMaster University, Hamilton, Ontario, L8S 4K1, Canada; email: dukas@mcmaster.ca

Key Words behavior, ecology, evolution, genetics, learning, memory, perception, phenotypic plasticity, vision

■ **Abstract** This review focuses on five key evolutionary issues pertaining to animal cognition, defined as the neuronal processes concerned with the acquisition, retention, and use of information. Whereas the use of information, or decision making, has been relatively well examined by students of behavior, evolutionary aspects of other cognitive traits that affect behavior, including perception, learning, memory, and attention, are less well understood. First, there is ample evidence for genetically based individual variation in cognitive traits, although much of the information for some traits comes from humans. Second, several studies documented positive association between cognitive abilities and performance measures linked to fitness. Third, information on the evolution of cognitive traits is available primarily for color vision and decision making. Fourth, much of the data on plasticity of cognitive traits appears to reflect nonadaptive phenotypic plasticity, perhaps because few evolutionary analyses of cognitive plasticity have been carried out. Nonetheless, several studies suggest that cognitive traits show adaptive plasticity, and at least one study documented genetically based individual variation in plasticity. Fifth, whereas assertions that cognition has played a central role in animal evolution are not supported by currently available data, theoretical considerations indicate that cognition may either increase or decrease the rate of evolutionary change.

INTRODUCTION

Cognition can be defined as the neuronal processes concerned with the acquisition, retention, and use of information. Cognition determines behavioral traits that affect animal ecology and evolution. Examples include habitat selection, choice of food and diet breadth, predator avoidance, mate choice, social behavior, and behavioral shifts that lead to adaptive radiations (Mayr 1963, Dukas 1998a). The major cognitive traits include (*a*) perception, defined as the translation of environmental signals into neuronal representations; (*b*) learning, which is the acquisition of neuronal representations of new information, such as a new association between a stimulus and an environmental state, a new association between a stimulus and behavioral pattern, or a new motor pattern; (*c*) long-term memory, which consists of passive representations of information already learned; (*d*) working memory, consisting of a small set of neuronal representations active over some short duration; (*e*) attention, referring to the neuronal representations activated at any given time; and, finally, (*f*) decision making, involving the determination of action given the known states of relevant environmental features and experience (Dukas 1998a, Dukas 2002, Platt 2002).

Historical and social factors delayed extensive research on the evolutionary biology of animal cognition for much of the twentieth century. The major contributor to this delay was slow acceptance of the evidence that cognitive traits, like any other animal feature, are determined by a mixture of genetic and environmental factors (Richards 1987, Plomin et al. 2001). The old barriers are rapidly fading owing to increased knowledge about, and integrative research on, the genetics, neurobiology, and evolutionary biology of cognition (Dukas 1998a, Shettleworth 1998). A remaining impediment, however, is the sheer complexity of brains and other neural centers. Compared with some morphological and anatomical traits that have been subjected to intense evolutionary research (Futuyma 1998), cognitive traits are more difficult to quantify, they are typically labile, and the neurobiology underlying them is only partially understood. Thus, an evolutionary biologist tackling cognition must also be well acquainted with the relevant neurobiology and proper behavioral techniques used to quantify cognitive traits.

Modern evolutionary research has greatly benefited from thorough research programs that documented large heritable variation for various phenotypic traits, the effects of such traits on fitness, and the evolution of phenotypic traits in the wild. A mature evolutionary discipline of animal cognition can also gain from similar research that evaluates to what degree (*a*) there is genetically based individual variation in cognitive traits, (*b*) cognitive traits affect animal fitness, and (*c*) cognitive traits evolve. Below I review some of the most convincing evidence for the above requirements. Then I discuss two related topics, environmental effects on cognition and the effect of cognition on evolution. Space limitations prevent me from citing all relevant studies.

GENETIC VARIATION IN COGNITIVE TRAITS

A cognitive trait can evolve only if there is genetically based variation in this trait among individuals. Genetic variation is known for all cognitive traits, but the quantity and quality of information varies widely, with much of the nonhuman data restricted to learning, memory, and decision making.

Genetic Variation in Perception

COLOR VISION There is substantial genetically based individual variation in human color vision. Some of this variation is caused by single amino acid substitutions that cause minor shifts in color perception (Gegenfurtner & Sharpe 1999). Humans with normal color vision possess three opsin photopigments with peak sensitivities in the blue, green, and red regions. The genes for green and red opsins reside on the X chromosome. Hence, all males with recessive mutations in these genes but only females homozygous for such mutations show deficiencies in color perceptions, varying from dysfunctions to minor anomalies. For example, males lacking either a functional red opsin or a functional green opsin perceive color in two rather than the normal three dimensions; this condition is typically referred to as color blindness. Among European males, about 1% and 1.3% of individuals possess dysfunctional red and green opsins, respectively, and approximately 1% and 4.5% possess anomalous red and green opsins, respectively. There is a considerable variation in color-vision deficiency among human populations. Although approximately 7.4% of males of European descent show color-vision deficiency, only fewer than 2% of aboriginal males in Australia, Brazil, the South Pacific Islands, and North America have such deficiencies (Sharpe et al. 1999). Minor anomalies in color vision are common. For example, approximately 56% of European males have a point mutation substituting Serine for Alanine at codon 180 of the red opsin, which causes a slight red shift in color perception (Sharpe et al. 1999).

There has been extensive research on color vision in many animals (Gegenfurtner & Sharpe 1999, Briscoe & Chittka 2001), but little information exists on genetically based variation among individuals in color perception in animals other than primates. Two noted exceptions involve fish. First, Endler et al. (2001) selected for increased spectral sensitivities to either red or blue light in four lines of guppies (*Poecilia reticulata*). All four lines showed significant responses to selection and significant heritabilities for spectral sensitivity. Second, Fuller and colleagues (Fuller et al. 2004; R.C. Fuller, personal communication) documented heritable variation in retinal cone distribution in bluefin killifish (*Lucania goodie*). Fish inhabiting clear spring water had relatively higher expressions of ultra violet (UV) and violet opsins. In contrast, fish from a murky swamp had relatively higher expressions of yellow and red opsins.

TASTE The best-studied taste polymorphism involves human sensitivity to bitter substances containing the chemical group N-C = S (Fox 1932). Although much of the empirical work involved synthetic phenylthiocarbamide (PTC), N-C = S occurs naturally in various wild plants, including turnip and cabbage. Among people of European descent, approximately 65% identify PTC as bitter, whereas the rest perceive it as tasteless. In contrast, close to 100% of aboriginal Africans and North Americans identify PTC as bitter (Kalmus 1971, Prutkin et al. 2000). A somewhat similar polymorphism occurs in wild mice, which show individual variation in sensitivity to the bitter sucrose octaacetate (SOA) (Warren & Lewis 1970). Both the human and mouse polymorphisms are attributed to recessive tasteless genes inherited in a simple Mendelian fashion (Warren & Lewis 1970, Kalmus 1971).

SMELL A few studies have documented significant genetically based individual variation in response to various odors in fruit flies (Fuyama 1978, Alcorta &

Rubio 1988, Mackay et al. 1996). Becker (1970) conducted two artificial selection experiments, which yielded flies insensitive to insect repellents. These fly studies, however, did not provide direct evidence that the variation was in the perception of, rather than response to, odors. In humans, more than 89 cases of insensitivity to specific odors have been described with frequencies varying from 0.1% to 47% among specific odors and human populations (Amoore 1971, Griff & Reed 1995). At least one of these cases, insensitivity to the musk pentadecalactone, which occurred in about 7% of people tested, was attributed to a recessive allele affecting specific odor perception (Whissell-Buechy & Amoore 1973).

HEARING   Perhaps the only known naturally occurring, genetically based individual variation in hearing pertains to musical pitch recognition in humans. Drayna et al. (2001) used a test involving distorted tunes to quantify pitch recognition in monozygotic and dizygotic twins. Subjects showed wide variation in test scores, with approximately 25% having a perfect score and 40% performing poorly. These results were not correlated with peripheral hearing, suggesting that pitch recognition involved nonperipheral components of the auditory system. The estimated heritability in this study was 71%, although this figure may be an overestimate. Other human studies have also documented a genetic component underlying perfect pitch. Although various species have been used as model systems for studying hearing impairment (Steel & Kros 2001), there is apparently no data about genetically based individual variation in sound perception in nonhuman species.

CROSS-MODALITY VARIATION   In addition to individual differences within a single perceptual mode, one would expect to find variation between modes among individuals inhabiting different perceptual environments. Culver et al. (1995) documented large genetically based variation in sensory organs between adjacent populations of the amphipod *Gammarus minus*. Individuals from surface springs, which are exposed to natural light, have relatively short antennae and large eyes each consisting of about 40 ommatidia. In contrast, neighboring individuals occupying underground caves, which live in constant darkness, have relatively long antennae and tiny eyes each containing approximately five ommatidia. The external differences in morphology extend to the brain, with surface individuals having an optic lobe twice as large and olfactory lobe 25% smaller than cave individuals (Culver et al. 1995).

Genetic Variation in Learning and Memory

The best evidence for naturally occurring, genetically based individual variation in learning and memory comes from artificial selection experiments. Tolman (1924), Heron (1935), and Tryon (1940) initiated the artificial selection studies on learning. They used rats assessed for their spatial learning ability in mazes and then bred the rats to produce divergent lines of good and bad performers within a few generations. There was no evidence, however, that these early lines of "maze bright" and "maze dull" rats indeed reflected selection on spatial learning ability. Rather, apparently much of the differences could be attributed to traits such as motivation and fearfulness. In a later study, Thompson selected for rat lines of good and bad spatial learners while controlling for emotional and motivational variables (Fuller & Thompson 1978).

Further artificial selection experiments on learning used blowflies (*Formia regina*) (McGuire & Hirsch 1977) and fruit flies (*Drosophila melanogaster*) (Lofdahl et al. 1992). Both studies employed protocols consisting of individual flies learning to respond to stimuli associated with sugar solution, and they documented rapid and widely divergent changes in learning ability. For example, in the fruit fly study, the proportion of flies showing good learning increased from 19% to 77% in the "bright" line and decreased to approximately 2% in the "dull" line over 25 generations (Lofdahl et al. 1992).

Mery & Kawecki (2002) also documented rapid increase in learning ability in fruit flies (*D. melanogaster*) under strong artificial selection. Their protocol involved a training trial in which one of two types of fruit juice substrates was associated with quinine, which is strongly aversive to flies. During the successive test trial, neither substrate contained quinine, but the experimenters collected eggs for the next generation only from the substrate that had not contained quinine in the training trial. That is, only the flies that remembered to seek the substrate that had been quinine-free during training had positive fitness. Subsequent experiments indicated that, compared with the control line, flies from the "bright" line showed higher learning rates and lower rates of memory decay (Mery & Kawecki 2002).

It is unknown which genes determined the changes in learning and memory in the fly lines under artificial selection. However, research on the molecular biology of learning has established that major genes can readily alter learning and memory parameters. To date, 60 genes involved in learning and memory have been identified in fruit flies (Dubnau et al. 2003), and it is estimated that between 500 to 1000 genes are part of the fly networks of learning and memory (T. Tully, personal communication).

In humans, at least three naturally occurring genetic variants have been linked to memory ability. First, the apolipoprotein E (APOE) gene has three alleles, resulting from single nucleotide substitutions. Among Western Europeans and North Americans, the frequencies of these alleles, labeled $\varepsilon 2$, $\varepsilon 3$, and $\varepsilon 4$, are 8%, 78.5%, and 13.5%, respectively. In some populations in Scandinavia and Africa, however, the frequency of $\varepsilon 4$ is 20% or higher. There is a well-established association between the $\varepsilon 4$ allele and susceptibility to late onset Alzheimer's disease (Raber et al. 2000, Smith 2000). In addition, nondemented, middle-aged carriers of $\varepsilon 4$ perform significantly worse on learning and memory tests compared with people carrying only the $\varepsilon 2$ and $\varepsilon 3$ alleles (Flory et al. 2000, Parasuraman et al. 2002).

Another polymorphism linked to memory involves a G to A substitution in the catechol-O-methyltransferase (COMT) gene. This substitution results in a Valine (Val) to Methionine (Met) substitution at position 108/158 of the COMT enzyme, which catabolizes released dopamine. The Val and Met alleles are codominant, with Met homozygotes having $1/4$ of COMT activity of Val homozygotes. In a healthy population of 55 people, 82% had at least one Val allele. A standard test of working memory revealed highly significant effects of the COMT genotype, with the Val/Val scoring lowest and Met/Met scoring highest (Egan et al. 2001).

Finally, the Met allele of the brain-derived neurotrophic factor (BDNF) results in a single amino acid substitution from Val to Met at position 66 of the BDNF protein. In a control sample of 133 people, 32% carried at least one Met allele. People carrying the Met allele showed significantly poorer memory about details in short stories (episodic memory). Furthermore, two types of brain imaging revealed that the Met carriers had abnormal functioning of neurons in the hippocampus, the brain region involved in episodic memory (Egan et al. 2003).

Genetic Variation in Attention

Little information is available about genetic variation in attention, which has been studied only in humans. The $\varepsilon 4$ allele of the APOE, which was discussed in the section on learning and memory, has also been linked to small reductions in performance on visual attention tasks in nondemented older adults (Greenwood et al. 2000, Parasuraman et al. 2002). Attention deficit hyperactivity disorder (ADHD) is characterized by difficulty of sustaining attention on a task for more than a few minutes and by overall restlessness. In North America, ADHD has been diagnosed in about 3% of boys and 1% of girls, and twin studies indicate high heritability of ADHD (Plomin et al. 2001). A few polymorphic genes have been linked to ADHD. The most consistent finding is a positive small association between the 7-repeat allele of the dopamine D_4 receptor gene (DRD4) and ADHD (Faraone et al. 2001).

Genetic Variation in Decision Making

Traditional research on animal behavior has accumulated considerable data about heritable variation in behavior. Although much of that variation is probably attributed to decision making, many studies did not evaluate noncognitive alternatives. Hence this section focuses only on selected cases of probable variation in decision making within the three evolutionarily important categories of habitat choice, foraging, and mating behavior.

HABITAT CHOICE The two classes of studies reviewed here are taxis in fruit flies (*D. melanogaster*) and exploratory behavior in great tits (*Parus major*). Either behavior can affect the genetic makeup of populations, generate assortative mating, and lead to speciation.

Adult fruit flies (*D. melanogaster*) typically show negative geotaxis and positive phototaxis. Classic experiments in the 1960s, however, documented large individually based variation in tactic response: Experiments employing strong directional selection readily generated divergent lines of flies showing extremely negative and positive geotaxis and negative and positive phototaxis (Erlenmeyer-Kimling et al. 1962, Hadler 1964). Similar results were also obtained with *D. pseudoobscura* (Dobzhansky & Spassky 1967). The four *D. pseudoobscura* lines selected for divergent tactic responses were tested for mating preference. All six possible combinations of pairings between the negative and positive geotactic and phototactic lines were tested, and all six tests revealed significant assortative mating after 5 and 11 generations of selection (Del Solar 1966). That is, several weeks of directional selection on tactic response were associated with a change in mating preference, which could lead to incipient speciation. Toma et al. (2002) recently identified specific genes involved in geotaxis by comparing mRNA expression from the heads of fly lines selected for negative and positive geotaxis. Toma and colleagues focused on three of the identified genes and verified through mutant analyses that these genes indeed affected geotaxis.

Individuals show large genetically based variation in their tendency to explore and to seek novel situations (McClearn 1959). Two related studies evaluated the heritability of novelty seeking in great tits (*Parus major*). One study involved bi-directional artificial selection over four generations in captivity. Hand-reared juveniles were tested for their latency (*a*) to visit four out of five artificial trees in a novel environment, and (*b*) to approach two distinct objects, a small battery and a pink rubber toy. Both up and down selection lines showed large changes each generation in mean score, and the heritability (h^2) was approximately 0.5 (Drent et al. 2003). The other study consisted of successive captures, brief laboratory tests for novelty seeking, and releases of wild-caught great tits. The heritability estimates in this study, which were based on both parent-offspring and sibling analyses were between 0.2 to 0.4 (Dingemanse et al. 2002). Exploratory behavior is positively correlated with natal dispersal in great tits. Hence, variation in exploratory behavior may translate into variation in the genetic structure of great tit populations (Dingemanse et al. 2003).

FORAGING One of a few studies indicating genetic variation in feeding preference in vertebrates is Arnold's (1981a,b) work on garter snakes (*Thamnophis elegans*) in California. Much of that research focused on geographic variation in the slug-eating habit between snake populations in slug-rich and slug-poor areas. However, all the snake populations showed genetically based polymorphism for both chemoreceptive responses to slugs and the consequent behaviors of either slug feeding or slug avoidance. Populations sympatric with slugs consisted of approximately 76% slug-eating snakes. In contrast, the frequency of slug-eating individuals was only about 17% in slug-poor habitats. Genetic variation for host preference is also well known in a variety of herbivorous insects (Futuyma & Peterson 1985, Funk & Bernays 2001).

The *foraging* gene of *D. melanogaster* provides a rare example of a single locus underlying a major naturally occurring behavioral polymorphism. Fly larvae with the dominant rover allele, for^R, exhibit significantly longer foraging movements on the food medium than do larvae homozygous for the sitter allele, for^S. The two morphs, however, show no behavioral difference in the absence of food.

Density-dependent selection during the larval stage determines the frequency of the *for* alleles, with sitters having a selective advantage under low densities and rovers being more successful in crowded conditions (Sokolowski et al. 1997). Somewhat similar genetically based variation in behavior occurs in the nematode, *Caenorhabditis elegans*, in which the 215F isoform of the *npr-1* gene is associated with social feeding behavior, whereas the 215V isoform is linked to solitary behavior. Naturally occurring nematodes of the social morph move rapidly on bacterial food substrate and aggregate to feed together, whereas worms of the solitary morph move slowly and feed alone (De Bono & Bargmann 1998).

MATING BEHAVIOR   Three major components of sexual behavior in fruit flies (*D. melanogaster*), courtship song, mating latency, and mating duration, show genetically based individual variation. The courtship song of male fruit flies, which is generated by wing vibration, has strong effects on mating success. Six generations of artificial bi-directional selection on the song's interpulse interval (IPI) were sufficient to generate a short line with an average IPI 10% shorter than that of the long IPI line. That is, there was sufficient genetic variation to generate rapid divergence in the song and to increase IPI beyond the normal range of the species (Ritchie & Kyriacou 1996). IPI is affected by the *per* gene. A few amino acid substitutions in that gene determine the 50% difference in IPI between the closely related *D. melanogaster* and *D. simulans* (Wheeler et al. 1991, Tauber & Eberl 2003).

Manning (1961) applied bi-directional selection on mating latency in *D. melanogaster* by isolating and breeding separately the 10 fastest and 10 slowest pairs to mate among 50 males and 50 females placed in culture bottles. Compared with a control line, the average mating latency was half as long in the two fast lines and eight times longer in the two slow lines. Experiments with selected males and unselected females indicated that males from the fast lines initiated courtship more quickly and licked females more frequently than males from the slow lines but that the fast males exhibited less nonsexual activity than the slow males (Manning 1961). Lines of flies with short and long mating latencies have also been selected in *D. simulans* (Manning 1968) and *D. pseudoobscura* (Kessler 1968).

Copulation in *D. melanogaster* lasts an average of 20 minutes. Analyses of natural populations and bi-directional selection indicated low but significant heritabilities of mating duration, with the genetic variation attributed mostly to males. Twenty generations of divergent selection caused little increase but about 50% decrease in mating duration (MacBean & Parsons 1967).

FITNESS CONSEQUENCES OF COGNITIVE TRAITS

The other fundamental condition for the evolution of a certain cognitive trait is that the genetically based individual variation in this trait is associated with variation in fitness. Currently, a relatively small body of literature links cognitive traits to survival and reproductive success. In addition to reviewing that literature here, I also discuss key studies relating cognition to performance measures such as foraging success and growth rate, which may be positively correlated with fitness. I focus on three major categories for which data exist: perception, learning, and decision making.

Perception and Fitness

Cave and surface-spring populations of the amphipod *Gammarus minus* exhibit genetically based individual variation in eye and antenna sizes (see the subsection Cross-Modality Variation). Jones et al. (1992) computed selection gradients on the basis of individual mating success and fecundity in *Gammarus* and found significant directional selection for smaller eyes in caves and larger eyes in springs. Both populations, however, showed significant directional selection for longer antennae corrected for body size, suggesting that perhaps differential survival, which was not measured, acted against longer antennae in spring populations (Jones et al. 1992).

The genetics and evolution of color vision have been studied extensively (see sections above and below). Only recently, however, have a few studies examined the benefits of trichromatic over dichromatic vision. Most new world monkeys (platyrrhine) show polymorphism in color vision. In these species, a single polymorphic X-linked locus encodes a middle- to long-wave opsin. Together with the autosomally encoded short-wave opsin, the polymorphic locus generates dichromatic vision in all males and homozygous females and trichromatic vision in heterozygous females (Jacobs 1998). An elegant experiment in marmosets (*Callithrix geoffroyi*) took advantage of that polymorphism and documented a higher detection rate of orange food items by trichromatic than dichromatic individuals. The trichromatics, however, showed no higher detection rate of green food items compared with the dichromatics (Caine & Mundy 2000). Theoretical calculations also indicate that trichromatic vision could increase fruit detection against leaf background (Osorio & Vorobyev 1996, Regan et al. 2001). Two studies, however (Dominy & Lucas 2001, Lucas et al. 2003), indicated that trichromatic vision is crucial for detecting preferred leaves but not fruits in several primate species.

Learning and Fitness

The intuition that learning has positive effect on fitness has rarely been tested in the field. A notable exception involves the optimal timing of egg laying in great and blue tits (*Parus major* and *P. caeruleus*). Matching the timing of nestling feeding with the local peak in food abundance is positively associated with fitness (Thomas et al. 2001). Peak food abundance varies among habitats, and birds must predict the peak a few weeks in advance to time their egg laying properly. The timing of egg laying depends partially on innate population-specific responses to photoperiod (Lambrechts et al. 1997). Field evidence suggested, however, that experience in the first breeding season influenced successive timing of egg laying in great tits

(*Parus major*) (Nager & van Noordwijk 1995). A later experiment (Grieco et al. 2002) involving manipulation of food availability during nestling feeding in blue tits (*P. caeruleus*) indeed indicated that females with higher food supplies during nestling feeding laid eggs later than control females in the following year. Further experiments may critically examine the role of learning in altering the timing of egg laying.

A critical laboratory test linking learning to fitness involved grasshoppers (*Schistocera americana*) that had to choose between nutritionally balanced and nutritionally deficient food types. Grasshoppers in the learning treatment could learn to associate the balanced food with reliable taste, color, and spatial cues. In contrast, the random treatment had these cues varying randomly over time, hence preventing learning. The learning grasshoppers had significantly higher growth rates than the random grasshoppers (Dukas & Bernays 2000). Other laboratory studies indicating positive effects of learning on fitness involved (*a*) higher mating success in wild-type than learning-deficient male fruit flies (*D. melanogaster*) (Gailey et al. 1985); (*b*) increased reproductive success in parasitoid wasps (*Biosteres arizanus*) that were allowed to rely on learning for host choice (Dukas & Duan 2000); and (*c*) higher survival and reproduction of herbivorous mites (*Tetranychus urticae*) that learned to prefer one host plant species over another (Egas & Sabelis 2001).

Learning also allows individuals to anticipate and adjust a priori to events with major physiological impacts (Hollis 1997). For example, heroin-induced mortality in heroin-experienced rats was lower when the drug was injected in the environment previously associated with the drug than when it was injected in a familiar place not linked to the drug before. This difference in survival was attributed to physiological changes occurring prior to heroin injection in the predictable environment (Siegel et al. 1982). In natural settings, physiological preadjustment mediated by learning can increase (*a*) tolerance of extreme temperatures (Kissinger & Riccio 1995), (*b*) probability of winning a fight, (*c*) male reproductive success, and (*d*) the likelihood of predator avoidance (Hollis 1997).

Decision Making and Fitness

Numerous studies suggest that animal decisions influence fitness, but empirical research linking most types of decisions to fitness is scarce. This section focuses on three central categories of decisions determining fitness: foraging, antipredator behavior, and mate choice.

FORAGING A rare field study linking foraging decisions to lifetime fitness involved crab spiders (*Misumena vatia*), which are semelparous, sit-and-wait predators that hunt for insect prey on flowers. Adult females vary in their success of choosing the best hunting sites and prey types. Spiders that chose sites with higher prey densities had higher growth rates and greater lifetime egg production (Morse & Stephens 1996). A long-term study in four treatment populations of zebra finches (*Taeniopygia guttata*) manipulated the birds' rate of food intake by varying handling time while allowing for similar daily total food intake. Fecundity, survival, and the consequent population growth rate were all highly positively correlated with feeding rate (Lemon 1991).

ANTIPREDATOR BEHAVIOR   The northwest garter snake (*Thamnophis ordinoides*) shows continuous variation in color pattern (stripedness) from three complete bright stripes, to spots, to no markings at all. Survival of juvenile garter snakes in the field was higher in individuals with opposite combinations of stripedness and the behavioral tendency to perform reversal evasive movements during escape from predators. That is, survival was highest in striped snakes showing little reversal and in unstriped snakes with high reversal scores. This association between color pattern and behavior probably reflects visual predators' difficulty in judging the speed of moving striped snakes, as well as the advantage of a broken, cryptic pattern for evasive snakes (Brodie 1992).

Many animals rely on indirect cues associated with predators for antipredatory behavior (Kats & Dill 1998). Wolf spiders (*Pardosa milvina*) exposed to silk and excreta from their spider predator (*Hogna hellus*) showed a shift in space use involving vertical climbing on container walls, a behavior that was linked to higher survival when exposed to the predator. Overall, survival of wolf spiders exposed to predator silk and excreta cues and predators was higher than that of wolf spiders placed with predators but not exposed to the cues (Persons et al. 2002).

MATE CHOICE   Compared with other categories of decisions, a much larger body of literature from a wide variety of taxa indicates that mate choice decisions in both females and males have positive effects on fitness. Documented benefits of mate choice include (*a*) avoiding inviable offspring through interspecific mating; (*b*) better fertilization ability or fecundity; (*c*) better partner or offspring care; (*d*) superior territory; (*e*) lower risk of predation, disease, or injury; and (*f*) producing offspring of higher heritable quality (Andersson 1994, Moller & Alatalo 1999, Moller & Jennions 2001). A recent example relating female mate choice to fitness is work on the cricket, *Gryllus lineaticeps*, in which females reared on a low-nutrition diet had higher survivorship and more fertilized eggs when mated to males with preferred song than to other males (Wagner & Harper 2003). A recent example of a fitness effect of male mate choice is a study indicating that male house mice (*Mus domesticus*) from a feral source population fathered more offspring that had greater survivorship when they mated with their preferred female rather than with other females (Gowaty et al. 2003).

THE EVOLUTION OF COGNITIVE TRAITS

The best available information about evolution of cognitive traits is for color vision and decision making, most likely because they are easier to quantify than other traits. Other examples examined in this section mostly represent highly suggestive cases requiring further research.

Perception

COLOR VISION   Compared with other senses, color vision is more accessible to genetic and molecular analyses, which, combined with phylogenetic techniques, has allowed researchers to reconstruct key events in the evolution of color vision. Color perception is highly variable among vertebrates. The common ancestor of terrestrial vertebrates had four types of photopigments, which have been preserved in many birds and reptiles and some mammals. Most placental mammals (Eutheria), however, are dichromatic: They have only a short-wavelength sensitive photopigment (S) encoded by an autosomal gene and a long-wavelength sensitive photopigment (L) encoded by a gene on the X chromosome. Color vision is poor at low light intensities, and this fact can explain the evolutionary loss of tetra-chromatic vision in early mammals, which were presumably nocturnal. Improved color vision, however, evolved in anthropoid primates. In old world anthropoids, duplication followed by divergence of the L gene created distinct L photopigment and middle-range photopigment (M). Similar, though independent, duplication and divergence of the L gene took place in new world howler monkeys (*Alouatta* spp.). In other new world anthropoids, no duplication of the L gene has occurred. Rather, the L gene is polymorphic and encodes distinct L and M photopigments, creating trichromatic vision in heterozygous females but dichromatic vision in all other individuals (Jacobs 1998, Nathans 1999, Surridge et al. 2003). It is assumed that diurnal activity selected for trichromatic vision in anthropoid primates. Evidence indicating advantages of trichromatic over dichromatic vision in food finding is presented in Perception and Fitness, above. Excellent, though less-detailed, information pertaining to adaptive evolution of color vision in various ecological settings exists for other taxa, including fish (Yokoyama & Yokoyama 1996, Sugawara et al. 2002), birds (Hart 2001), and insects (Briscoe & Chittka 2001).

OLFACTION   The evolution of trichromatic vision in old world anthropoid primates just discussed paralleled evolutionary degradation of the vomeronasal system in these species (Zhang & Webb 2003). Many terrestrial vertebrates perceive pheromones primarily through the vomeronasal organ. Genes encoding the TRP2 ion channel and V1R pheromone receptors, which are unique to the vomeronasal pheromone transduction pathway, were impaired in a common ancestor of old world anthropoid primates (Catarrhines) approximately 23 million years ago. Further inactivation of pheromone receptor genes has been an ongoing process in these species. In contrast, TRP2 genes are functional in new world monkeys (Platyrrhines) (Zhang & Webb 2003).

Learning and Memory

The best evolutionary research on learning and memory consists of extensive work on spatial memory in birds that cache a large number of food items in scattered locations. This research program has not yet evaluated the genetic basis of spatial memory, however. Research on spatial memory has documented two major findings. First, bird species that rely more heavily on retrieval of cached food show better spatial memory than closely related noncaching species (Balda & Kamil 1989, Pravosudov & Clayton 2002). Many, though not all, the experiments documented the predicted positive association between reliance on spatial memory and performance on spatial memory tasks (reviewed in Shettleworth 2003). Overall, the positive evidence is highly compelling because it comes from several laboratories and distinct bird taxa.

The second finding from research on spatial memory was based on extensive research that indicated that the hippocampus is involved in spatial memory (Sherry & Vaccarino 1989). Analyses controlling for phylogeny indicated that in both European and North American birds, relative hippocampus volumes are larger in species that store food than in nonstoring species (Krebs et al. 1989, Sherry et al. 1989). Similar results were obtained in a within-species comparison: Individual black capped chickadees (*Poecile atricapilla*) from Alaska, which rely more heavily on stored food than intraspecific individuals from Colorado, had larger hippocampus volumes and showed better spatial memory than the Colorado birds (Pravosudov & Clayton 2002).

As already mentioned, the genetic basis of spatial memory is unknown. Work on mice has recently identified two quantitative trait loci, *Hipp1a* and *Hipp5a*, which explain 25% of the heritable variation in hippocampus size (Peirce et al. 2003). Furthermore, artificial divergent selection in mice for high and low open field activity resulted in structural hippocampal changes (Hausheer-Zarmakupi et al. 1996). Although these studies indicate genetic effects on hippocampus structure, they do not link such effects to spatial memory. In contrast, environmental effects, especially tasks requiring spatial memory, influence hippocampus size (Clayton 2001).

In sum, bird species and populations that cache food have larger hippocampus volumes, which allow them to depict better spatial memory compared with noncaching species. Some of the documented difference in hippocampus volume and spatial memory likely reflects an evolved cognitive change, but the evolutionary scenario must be augmented with relevant genetic information.

Research on song learning in oscine song birds (suborder oscines) provides another example for possible adaptive evolution of specific learning ability. In most oscines, young males learn aspects of their song from older adults. Song-repertoire size is positively associated with males' mating success (reviewed in DeVoogd 1998) and with volumes of the song control nucleus high vocal center (HVC) (DeVoogd et al. 1993, Szkeley et al. 1996). In zebra finches (*Taeniopygia guttata castanotis*), there is genetically based individual variation in HVC volume (Airey et al. 2000), which is positively associated with song-learning ability (Ward et al. 1998, Airey & DeVoogd 2000). This finding suggests that selection on larger song repertoire led to the evolution of enhanced song-learning ability and larger HVC.

Wright and colleagues (Wright et al. 1996, Marinesco et al. 2003) have taken a novel approach to examining the evolution of learning. They compared neuronal correlates of learning in the mollusc *Aplysia californica* and five related species.

Aplysia has been a prime model species for research on the neurobiology of learning (Kandel 2001). In *Aplysia*, learning has been linked to serotonin-induced increased spike duration and excitability in mechanosensory neurons of the tail-withdrawal reflex. Applications of serotonin in the five related species caused (*a*) only increased spike excitability in the two ancestral species, (*b*) both increased spike duration and excitability in two closely related species, and (*c*) neither increased duration nor increased excitability in another closely related species. These results led Wright et al. (1996) to suggest that increased spike duration in response to serotonin arose more recently in the evolution of the *Aplysia* group and that a learning-related mechanism was lost in one species. Further work with the *Aplysia* species group could evaluate the adaptive significance of the documented learning-related differences (Wright et al. 1996).

Volume of Brain Parts

In addition to research just discussed on spatial memory and song repertoire, several studies employing phylogenetic analyses have documented positive correlations between presumed enhancement of a certain cognitive ability and the volume of brain parts devoted to that trait. Examples include (*a*) presumed visual resolution as indicated by eye size and optic-lobe mass in birds (Brooke et al. 1999), (*b*) probable increased reliance on olfaction associated with nocturnal activity and enlarged olfactory bulb in birds and mammals (Healy & Guilford 1990, Barton et al. 1995), (*c*) frequency of feeding innovations (novel or unusual feeding behaviors) and the volumes of hyperstriatum ventrale and neostriatum in birds and neocortex and striatum in primates (Reader & Laland 2002, Lefebvre et al. 2004), and (*d*) larger brain size in mammals with more challenging feeding behaviors: fruit- versus insect-feeding bats (Eisenberg & Wilson 1978) and nonherbivorous versus herbivorous rodents and primates (Clutton-Brock & Harvey 1980, Mace et al. 1981). Finally, another noteworthy finding is that animal domestication was associated with large decreases in volumes of certain brain parts (Kruska 1988). All the examples above require further detailed investigations on the mechanisms generating the associations between brain and ecology and their genetic foundations.

Decision Making

The best available data on evolution of decision making exists for feeding and antipredatory behavior.

DIET IN PHYTOPHAGOUS INSECTS Many insects that feed on plant material specialize on a single host species, which typically also serves as the mating ground. Hence, host choice is a crucial decision because host shift may lead to reproductive isolation and incipient speciation (Mayr 1963, Funk et al. 2002). A well-recorded recent host shift involved adoption of introduced apples (*Malus pomila*) by apple maggot flies (*Rhagoletis pomonella*) that originally exploited native hawthorn (*Crataegus* spp.) in North America approximately 150 years ago.

Currently, the apple and hawthorn races are genetically distinct, and there is little migration of individuals between the two hosts (Feder et al. 1994). Recently, host preferences of flies reared from larvae on an artificial diet were tested using synthetic blends of apple and hawthorn volatiles. In all populations examined, flies showed strong affinities to their natal host. Moreover, flies of the hawthorn race, which were raised on apple in the laboratory for two generations, also showed a strong preference for hawthorn over apple. These results indicate that the apple race of *R. pomonella* has altered its response to fruit volatiles over the past 150 years, increasing its preference for apples and reducing its response to hawthorn (Linn et al. 2003). The evolution of host choice in phytophagous insects has been well studied in a variety of other systems as well (Funk et al. 2002).

ANTIPREDATORY BEHAVIOR   One of the best documented cases of evolution in the wild involves populations of guppies (*Poecilia reticulata*) occurring in distinct high- and low-predation pools in Trinidad. The principle guppy predators in that system are pike cichlids (*Crenicichla alta*). In addition to the natural variation among pools, there were a few experimental introductions of fish about 25 years ago. Fish from distinct predation regimes differ in various traits, including color, life history, mating, foraging, and antipredatory behaviors (reviewed in Endler 1995, Magurran et al. 1995). A laboratory study evaluated divergent evolution of antipredatory behavior in Trinidad guppies by transferring wild-caught fish to predator-free aquaria and testing a second generation of laboratory-born offspring. Offspring from both natural and introduced high-predation populations were significantly more likely to escape predation by pike cichlids than were offspring from natural and introduced low-predation populations. In the case of the guppies from pools with introduced predators, the results indicate evolutionary change in antipredatory behavior within approximately 30 generations (O'Steen et al. 2002). The exact cognitive mechanisms underlying that change have not yet been quantified.

In eastern North America, larvae of damselfly species in the genus *Enallagma* that grow in lakes with fish predators show distinct antipredatory behaviors compared with larval *Enallagma* spp. occupying fishless lakes in which dragonfly larvae are the top predators. Phylogenetic analysis suggested that fish lakes were the ancestral habitat of these damselflies, from which at least two independent lineages of dragonfly-lake species arose recently (Brown et al. 2000). The transition to dragonfly-lakes was associated with (*a*) increased activity in the absence of predators; (*b*) loss of antipredatory response to fish; and (*c*) adoption of an escape response to approaching predators, a tactic that reduces dragonfly but not fish predation (McPeek 1990, Stoks et al. 2003). Preliminary evidence suggests that much of the behavioral differences have a genetic basis, and research in progress is critically examining this issue (M. McPeek, personal communication). Dragonfly-lake damselflies (*E. boreale*), however, did learn to show antipredatory response to a fish predator (northern pike, *Esox lucius*) after simultaneous exposure to stimuli from the predator and injured conspecific damselflies (Wisenden et al. 1997).

ENVIRONMENTAL EFFECTS ON COGNITIVE TRAITS

In the sections above, I focused on genetic variation and did not explicitly discuss phenotypic plasticity. Most phenotypic traits, however, are determined by an interplay between an individual's genes and numerous internal and external factors, including nutrition, temperature, social environment, and interactions with other species. Consequently, a single genotype can have a whole set of phenotypes, termed norm of reaction, under different combinations of environmental conditions (reviewed in Futuyma 1998, Pigliucci 2001). In many anatomical and morphological characteristics, environmental effects are restricted to early developmental stages. A simple example for such irreversible plasticity is the positive association between adult body size and nutritional quality during development in many insects. Other traits can show reversible plasticity throughout ontogeny. For example, initiating a few weeks of weight lifting by a human adult would result in increased volumes of a few muscles. Quitting that exercise regime, however, would be associated with rapid atrophy of the same muscles.

The norm of reaction may be either nonadaptive, as in the case of body size and nutrition just mentioned, or adaptive, as in the change in muscle volume in response to power requirements. Most importantly, different genotypes may have distinct reaction norms. Such genetically based individual variation in the norm of reaction, or gene by environment interaction, implies that the norm of reaction can evolve. Note that evolutionary biology employs a statistical definition of "interaction": the terms "phenotypic plasticity" and "reaction norm" only imply environmental effects on the genotype. The term "gene by environment interaction" is restricted to cases where the environment has different effects on different genotypes (reviewed in Futuyma 1998, Pigliucci 2001). Most phenotypic plasticity studies have dealt with morphological characteristics. Cognition, however, can also be analyzed within the phenotypic-plasticity framework (Dukas 1998b).

Two key questions about phenotypic plasticity of cognitive traits are pertinent: (*a*) Is there evidence for adaptive phenotypic plasticity of cognitive traits in ecologically relevant settings? (*b*) Is there genetic variation for reaction norms of cognitive traits?

Adaptive Phenotypic Plasticity of Cognitive Traits in Ecologically Relevant Settings

Before discussing adaptive plasticity, I must review the extensive literature on apparently nonadaptive brain plasticity. Note, however, that little evolutionary research on cognitive plasticity has been carried out. Hence, adaptive patterns may underlie at least some cases of apparently nonadaptive cognitive plasticity. Many parts of the brain require feedback from the environment for proper development. For example, in kittens with one eye covered for two months after birth, most neurons in the striate cortex depicted normal receptive fields corresponding to the normal eye, but only a few neurons were influenced by the deprived eye. Monocular visual deprivation had a smaller negative effect in older kittens and no effect in adults (Wiesel & Hubel 1963). In natural settings, all developing individuals are typically exposed to visual feedback in both eyes. Hence, such external effects on vision can probably be regarded as nonadaptive plasticity.

Another well-studied example of apparently nonadaptive plasticity involves environmental enrichment, defined as laboratory housing conditions typically consisting of large space, exercise, various inanimate objects, and social stimulation (van Praag et al. 2000). Compared with mammals grown in the typically highly deprived laboratory setting, mammals under environmental enrichment exhibit better learning and memory and increased brain size, neuron size, dendritic branching, and synapses per neuron, as well as different expression of genes linked to neuronal structure, synaptic plasticity, and transmission (Rampon et al. 2000, van Praag et al. 2000). Similar effects of environmental enrichment on brain and behavior have been documented for fruit flies (*D. melanogaster*) (Technau 1984, Heisenberg et al. 1995, Dukas & Mooers 2003). Because the actual comparison in environmental-enrichment studies is between highly deprived and less-deprived laboratory settings, the documented plasticity probably reflects a need for proper environmental feedback for brain development and maintenance. The sections below discuss some of the available evidence for adaptive phenotypic plasticity of cognitive traits in ecologically relevant settings.

PERCEPTION   Sustained experiences that demand specific use of a narrow range of a sensory modality result in a magnified neuronal representation of that range in the cortex (Kaas 1991, Nudo et al. 1996). For example, tactile discrimination performed with a single finger in adult monkeys resulted in an enlarged neuronal map for that finger, accompanied by enhanced discrimination ability on the tactile task (Recanzone et al. 1992). Such reversible plasticity implies allocation of metabolically expensive neuronal tissue on the basis of current behavioral needs, but no research has evaluated the adaptive value of that brain plasticity.

LEARNING AND MEMORY   Learning is an example of phenotypic plasticity (Dukas 1998b), and evidence for adaptive learning was reviewed above in the section on learning and fitness. Nonetheless, a relevant question is whether there are known cases of adaptive norms of reaction for learning and memory abilities. Recent research on neurogenesis suggests such adaptive plasticity. The hippocampus and song nucleus in song birds generate new neurons throughout life. In black capped chickadees (*P. atricapillus*), hippocampal neurogenesis peaks in the fall, coinciding with high demands for spatial memory during peak caching activity. In adult male canaries (*Serinus canaria*), as well as in a few other species, peak neurogenesis in the HVC occurs during the breeding season (Barnea & Nottebohm 1994, Smulders et al. 2000, Tramontin & Brenowitz 2000, Nottebohm 2002). Similarly, neurogenesis occurs in the hippocampus of adult mammals. Several studies documented either increased neurogenesis or increased survival of new neurons in subjects under either exercise regimes or intense learning activities (Gould et al.

1999, van Praag et al. 2000). However, no direct evidence links neurogenesis to learning and memory, and some data appear to disagree with such a link. For example, neurogenesis also occurs in the HVC of adult male zebra finches and adult male song sparrows (*Melospiza melodia*) even though they learn their songs before sexual maturity (Tramontin & Brenowitz 1999, Nottebohm 2002).

In sum, there is strong evidence for both seasonal and activity-dependent plasticity in the rate of either generation or survival of new neurons. Such brain plasticity appears adaptive because it correlates with activities such as navigation, food caching, and singing. To date, however, no empirical data have either directly linked brain plasticity to learning and memory or demonstrated the adaptive significance of such plasticity.

DECISION MAKING   By definition, all decisions amount to phenotypic plasticity: individuals decide to perform activity x in one environmental state, perform activity y in response to another environmental state, or remain inactive given yet another environmental state. Evidence for adaptive decision making was discussed in the section Decision Making and Fitness, above. Here, I briefly review data on adaptive norms of reaction for decision making. Most animals exhibit apparently adaptive antipredator behavior in response to indirect and direct information about the presence of predators (Kats & Dill 1998, Lima 1998). Relyea (2003) compared age-specific responses of gray-treefrog tadpoles (*Hyla versicolor*) to odors of their dragonfly-larvae predators (*Anax longipes*). In addition to constant presence and constant absence of predatory cues, Relyea also either introduced or removed predatory cues during three stages of the 18-day experiment. Only young tadpoles responded to predation risk by hiding. By day 10, there was no difference in hiding between the predation and nonpredation treatments. Activity level under predation risk was lower than under nonpredation conditions throughout most of the tadpole ontogeny but similar at the end of the experiment. Overall, late in ontogeny, the tadpoles reduced employment of antipredatory behavior and relied mostly on morphologically plastic antipredatory responses, including greater body mass, shorter bodies, and longer tails. That is, the tadpoles exhibited an age-specific reaction norm for antipredatory behavior. The reduced use of antipredatory behaviors probably reflected their high cost in terms of reduced feeding and growth rate and perhaps the increased effectiveness of the antipredatory morphology (Relyea 2003). Other well-studied cases of apparently adaptive norms of reactions in decision include the effect of body reserves on feeding and antipredatory behavior (Cuthill & Houston 1997) and effects of experience on aggression (Yeh et al. 1996) and sexual behavior (Crews 2003).

Genetic Variation for Reaction Norms of Cognitive Traits

There is probably no data pertaining to genetic variation in cognitive plasticity except for that involving decision making. However, the lack of information probably indicates that little research on this topic has been carried out.

Many zooplankton species exhibit diurnal vertical migration, with individuals found in deeper water during the day than at night. This behavior reduces predation by visual predators such as fish (Ringelberg 1999). Clones of *Daphnia magna* show large genetic variation in their preferred water depth in daylight. Furthermore, different daphnia clones vary in their responses to the presence of fish odor in the water, with clones from habitats with high fish predation responding more strongly and moving to deeper water than clones from habitats with no fish predation (De Meester 1993, De Meester 1996). Research at a Belgian pond revealed that the large genetic variation in plasticity allowed rapid evolution of daphnia responses to temporal variation in fish predation over 30 years (Cousyn et al. 2001).

EFFECTS OF COGNITIVE TRAITS ON EVOLUTION

Learning and decision making can expose individuals to, and enhance survival in, new niches and adaptive zones. This fact suggests that cognitive traits play an important role in evolutionary change and speciation (Baldwin 1896, West-Eberhard 2003). Three recent independent analyses, however, have reached the similar conclusion that phenotypic plasticity may speed up, slow down, or have no net effect on evolutionary change (Robinson & Dukas 1999, Huey et al. 2003, Price et al. 2003). On the one hand, although an environmental change might cause extinction of a nonplastic population, plasticity could enable population persistence, allowing evolutionary change in the new environment. On the other hand, if plasticity is sufficient for maximizing fitness in the new environment, the existing genetic variation would remain hidden from natural selection. Hence, no evolutionary change would occur, and a return to the old environment would result in complete reversal to the ancestral type (Robinson & Dukas 1999, Huey et al. 2003, Price et al. 2003).

Elegant experiments by Waddington (1953, 1959) clearly illustrate that plasticity can indeed enhance evolutionary change. However, the overall effect of phenotypic plasticity in general and cognitive traits in particular on evolutionary change is still unknown. On the negative side, there is no evidence that enhanced cognitive abilities increased the rate of morphological evolution in either great apes or hominoids (Lynch & Arnold 1988). On the positive side, a few studies in birds documented positive correlations between relative brain size and (*a*) the number of species per taxon (Nicolakakis et al. 2003), (*b*) the number of subspecies per species (D. Sol, G. Stirling, N. Nicolakakis, and L. Lefebvre, unpublished manuscript), and (*c*) invasion success (Sol et al. 2002). In birds, brain size is highly positively correlated with the size of the hyperstriatum ventrale (HV), the part of the telencephalon most closely involved in multimodal sensory integration and learning. HV size is positively correlated with behavioral flexibility in birds (Lefebvre et al. 2004). Hence, at least in birds, some cognitive abilities are positively correlated with measures of evolutionary change.

CONCLUSIONS AND PROSPECTS

Relatively little research on the evolutionary biology of animal cognition has been carried out. Nevertheless, data compiled from several disciplines indicate that (*a*) there is genetically based individual variation in cognitive traits, (*b*) cognitive traits affect animal fitness, and (*c*) cognitive traits evolve. That is, cognitive traits can be subjected to evolutionary research like any other animal feature following established techniques employed in numerous studies mentioned in this review. Compared with working on morphological traits, however, studying cognitive traits involves two additional difficulties. First, cognitive traits are typically quantified through whole-animal behavioral tests. Second, many cognitive traits are more labile than morphological characteristics. Both issues can be addressed with properly controlled experimental protocols that can reliably quantify a given cognitive feature through behavior.

Cognitive phenotypes, like morphological and physiological phenotypes, are determined by genes and environment. Hence, employing tools developed for studying phenotypic plasticity mostly in the morphological domain could enhance research on the genetics and evolution of cognitive plasticity. Learning, memory, and decision making are types of phenotypic plasticity, and some information is available about their genetics, adaptive value, and evolution. We know little, however, about adaptive phenotypic plasticity of perceptual traits, including vision, hearing, and olfaction. Another important issue requiring future research involves adaptive environmental effects on cognitive traits. For example, compared with flies grown at low density, are the brains of flies developed under crowded conditions better adapted for migrating in search of new food sources (Heisenberg et al. 1995)? Finally, in spite of extensive research on brain plasticity, we know little about genetic variation in reaction norms of cognitive traits. For example, is there genetically based variation in the effect of food caching experience on hippocampal neurogenesis? And does this variation translate into future differences in memory?

The last section of this review suggests that assertions of a unique role for cognition in enhancing evolutionary change are currently not supported by sufficient critical data. Although theory indicates that cognition could either increase or decrease the rate of evolution (Robinson & Dukas 1999), further empirical work on that topic is highly desirable.

In various places throughout this review, I mention what further information is necessary to advance research on the evolutionary biology of animal cognition. In addition, a few major issues that warrant close examination are listed below. A fair amount of data exists on genetic variation in cognitive traits. How does such natural genetic variation affect behavior and fitness, and what factors help maintain this variation? There are probably no observational data on the evolution of cognitive traits in the wild, although such research is highly feasible. For example, the amphipod system studied by Culver et al. (1995) seems ideal for further investigation of the evolution of perceptual traits and their correlated brain regions. Until recently, evolutionary studies on cognition have mostly focused on benefits while ignoring costs. Various costs, however, including energy to maintain neuronal machinery and process information, and time to process complex information, can affect the fitness consequences of cognitive traits (Dukas 1999, Bernays 2001, Laughlin 2001, Mery & Kawecki 2003). Further information on this topic is needed.

In sum, this review indicates that animal cognition can readily be placed within an evolutionary framework. The evolutionary analysis highlights key issues requiring future research, which will help us understand animal cognition, its evolution, and the role of cognition in animal evolution.

ACKNOWLEDGMENTS

I thank Dan Papaj, Martin Daly, James Burns, Ana Skemp, and Emilie Snell-Rood for comments on the manuscript and the Natural Sciences and Engineering Research Council of Canada for financial support. I also apologize to the authors of all the excellent papers that I could not cite because of the strict space limit.

The *Annual Review of Ecology, Evolution, and Systematics* is online at
http://ecolsys.annualreviews.org

LITERATURE CITED

Airey DC, Castillo-Juarez H, Casella G, Pollak EJ, DeVoogd TJ. 2000. Variation in the volume of zebra finch song control nuclei is heritable: developmental and evolutionary implications. *Proc. R. Soc. London Ser. B Biol. Sci.* 267:2099–104

Airey DC, DeVoogd TJ. 2000. Greater song complexity is associated with augmented song system anatomy in zebra finches. *Neuroreport* 11:2339–44

Alcorta E, Rubio J. 1988. Genetical analysis of intrapopulational variation in olfactory response in *Drosophila melanogaster. Heredity* 60:7–14

Amoore JE. 1971. Olfactory genetics and anosmia. In *Handbook of Sensory Physiology*, Vol. IV, ed. LM Beidler, pp. 245–56. Berlin: Springer-Verlag

Andersson M. 1994. *Sexual Selection*. Princeton, NJ: Princeton Univ. Press

Arnold S. 1981a. Behavioral variation in natural populations. I. Phenotypic, genetic and environmental correlations between chemoreceptive responses to prey in the garter snake, *Thamnophis elegans. Evolution* 35:489–509

Arnold S. 1981b. Behavioral variation in natural populations. II. The inheritance of a feeding response in crosses between geographic races of the garter snake, *Thamnophis elegans. Evolution* 35:510–15

Balda RP, Kamil AC. 1989. A comparative study of cache recovery by three corvid species. *Anim. Behav.* 38:486–95

Baldwin JM. 1896. A new factor in evolution. *Am. Nat.* 30:441–51

Barnea A, Nottebohm F. 1994. Seasonal recruitment of hippocampal neurons in adult free-ranging black-capped chickadees. *Proc. Natl. Acad. Sci. USA* 91:11217–21

Barton RA, Purvis A, Harvey PH. 1995. Evolutionary radiation of visual and olfactory brain systems in primates, bats and insectivores. *Philos. Trans. R. Soc. London Ser. B* 348:381–92

Becker HJ. 1970. The genetics of chemotaxis in *Drosophila melanogaster*: selection for repellent insensitivity. *Mol. Gen. Genet.* 107:194–200

Bernays EA. 2001. Neural limitations in phytophagous insects: implications for diet breadth and evolution of host affiliation. *Annu. Rev. Entomol.* 46:703–27

Briscoe AD, Chittka L. 2001. The evolution of color vision in insects. *Annu. Rev. Entomol.* 46:471–510

Brodie ED. 1992. Correlational selection for colour pattern and antipredator behavior in the garter snake, *Thomnophis ordinoides*. *Evolution* 46:1284–98

Brooke MdL, Hanley S, Laughlin SB. 1999. The scaling of eye size with body mass in birds. *Proc. R. Soc. London Ser. B Biol. Sci.* 266:405–12

Brown JM, McPeek MA, May ML. 2000. A phylogenetic perspective on habitat shifts and diversity in the North American *Enallagma* damselflies. *Syst. Biol.* 49:697–712

Caine NG, Mundy NI. 2000. Demonstration of a foraging advantage for trichromatic marmosets (*Callithrix geoffroyi*) dependent on food colour. *Proc. R. Soc. London Ser. B Biol. Sci.* 267:439–44

Clayton NS. 2001. Hippocampal growth and maintenance depend on food-caching experience in juvenile mountain chickadees (*Poecile gambeli*). *Behav. Neurosci.* 115:614–25

Clutton-Brock TH, Harvey PH. 1980. Primates, brains and ecology. *J. Zool.* 190:309–23

Cousyn C, De Meester L, Colbourne JK, Brendonck L, Verschuren D, Volckaert F. 2001. Rapid, local adaptation of zooplankton behavior to changes in predation pressure in the absence of neutral genetic changes. *Proc. Natl. Acad. Sci. USA* 98:6256–60

Crews D. 2003. The development of phenotypic plasticity: where biology and psychology meet. *Dev. Psychobiol.* 43:1–10

Culver DC, Kane TC, Fong DW. 1995. *Adaptation and Natural Selection in Caves.* Cambridge, MA: Harvard Univ. Press

Cuthill IC, Houston AI. 1997. Managing time and energy. In *Behavioural Ecology*, ed. JR Krebs, NB Davies, pp. 97–120. Oxford: Blackwell

De Bono M, Bargmann CI. 1998. Natural variation in a neuropeptide Y receptor homolog modifies social behavior and food response in *C. elegans*. *Cell* 94:679–89

Del Solar E. 1966. Sexual isolation caused by selection for positive and negative phototaxis and geotaxis in *Drosophila pseudoobscura*. *Proc. Natl. Acad. Sci. USA* 56:484–87

De Meester L. 1993. Genotype, fish-mediated chemical, and phototactic behavior in *Daphnia magna*. *Ecology* 74:1467–74

De Meester L. 1996. Evolutionary potential and local genetic differentiation in a phenotypically plastic trait of a cyclical parthenogen, *Daphnia magna*. *Evolution* 50:1293–98

DeVoogd TJ. 1998. Causes of avian song: using neurobiology to integrate proximate and ultimate levels of analysis. In *Animal Cognition in Nature*, ed. RP Balda, IM Pepperberg, AC Kamil, pp. 337–80. London: Academic

DeVoogd TJ, Krebs JR, Healy S, Purvis A. 1993. Relations between song repertoire size and the volume of brain nuclei related to song: comparative evolutionary analyses amongst oscine birds. *Proc. R. Soc. London Ser. B Biol. Sci.* 254:75–82

Dingemanse NJ, Both C, Van Noordwijk AJ, Rutten AL, Drent PJ. 2002. Repeatability and heritability of exploratory behaviour in great tits from the wild. *Anim. Behav.* 64:929–38

Dingemanse NJ, Both C, Van Noordwijk AJ, Rutten AL, Drent PJ. 2003. Natal dispersal and personalities in great tits (*Parus major*). *Proc. R. Soc. London Ser. B Biol. Sci.* 270:741–47

Dobzhansky T, Spassky B. 1967. Effects of selection and migration on geotactic and phototactic behaviour of *Drosophila*. I. *Proc. R. Soc. London Ser. B Biol. Sci.* 168:27–47

Dominy NJ, Lucas PW. 2001. Ecological importance of trichromatic vision to primates. *Nature* 410:363–66

Drayna D, Manichaikul A, Lange Md, Snieder H, Spector T. 2001. Genetic correlates of musical pitch recognition in humans. *Science* 291:1969–72

Drent PJ, Van Oers K, Van Noordwijk AJ. 2003. Realized heritability of personalities in the great tit (*Parus major*). *Proc. R. Soc. London Ser. B Biol. Sci.* 270:45–51

Dubnau J, Chiang A-S, Tully T. 2003. Neural substrates of memory: from synapse to system. *J. Neurobiol.* 54:238–53

Dukas R. ed. 1998a. *Cognitive Ecology: The Evolutionary Ecology of Information Processing and Decision Making*. Chicago. Univ. Chicago Press

Dukas R. 1998b. Evolutionary ecology of learning. See Dukas 1998a, pp. 129–74

Dukas R. 1999. Costs of memory: ideas and predictions. *J. Theor. Biol.* 197:41–50

Dukas R. 2002. Behavioural and ecological consequences of limited attention. *Philos. Trans. R. Soc. London Ser. B* 357:1539–48

Dukas R, Bernays EA. 2000. Learning improves growth rate in grasshoppers. *Proc. Natl. Acad. Sci. USA* 97:2637–40

Dukas R, Duan JJ. 2000. Possible fitness consequences of learning in a parasitoid wasp. *Behav. Ecol.* 11:536–43

Dukas R, Mooers AØ. 2003. Environmental enrichment improves mating success in fruit flies. *Anim. Behav.* 66:741–49

Egan M, Goldberg T, Kolachana B, Callicott J, Mazzanti C, et al. 2001. Effect of COMT Val108/158 Met genotype on frontal lobe function and risk for schizophrenia. *Proc. Natl. Acad. Sci. USA* 98:6917–22

Egan MF, Kojima M, Callicott JH, Goldberg TE, Kolachana BS, et al. 2003. The BDNF val66met polymorphism affects activity-dependent secretion of BDNF and human memory and hippocampal function. *Cell* 112:257–69

Egas M, Sabelis MW. 2001. Adaptive learning of host preference in a herbivorous arthropod. *Ecol. Lett.* 4:190–95

Eisenberg JF, Wilson DE. 1978. Relative brain size and feeding strategies in the Chiroptera. *Evolution* 32:740–51

Endler JA. 1995. Multiple-trait coevolution and environmental gradients in guppies. *Trends Ecol. Evol.* 10:22–29

Endler JA, Basolo A, Glowacki S, Zerr J. 2001. Variation in response to artificial selection for light sensitivity in guppies (*Poecilia reticulata*). *Am. Nat.* 158:36–48

Erlenmeyer-Kimling L, Hirsch J, Weiss JN. 1962. Studies in experimental behavior genetics: III. Selection and hybridization analyses of individual differences in the sign of geotaxis. *J. Comp. Physiol. Psych.* 55:722–31

Faraone S, Doyle A, Mick E, Biederman J. 2001. Meta-analysis of the association between the 7-repeat allele of the dopamine D(4) receptor gene and attention deficit hyperactivity disorder. *Am. J. Psychiatry* 158:1052–57

Feder J, Opp S, Wlazlo B, Reynolds K, Go W, Spisak S. 1994. Host fidelity is an effective premating barrier between sympatric races of the apple maggot fly. *Proc. Natl. Acad. Sci. USA* 91:7990–94

Flory JD, Manuck SB, Ferrell RE, Ryan CM, Muldoon MF. 2000. Memory performance and the apolipoprotein E polymorphism in a community sample of middle-aged adults. *Am. J. Med. Genet.* 96:707–11

Fox AL. 1932. The relationship between chemical constitution and taste. *Proc. Natl. Acad. Sci. USA* 18:115–20

Fuller JL, Thompson WR. 1978. *Foundations of Behavior Genetics*. St. Louis, MO: Mosby

Fuller RC, Carleton KL, Fadool JM, Spady TC, Travis J. 2004. Population variation in opsin expression in the bluefin killifish, *Lucania goodei*: a real-time PCR study. *J. Comp. Physiol. A.* 190:147–54

Funk DJ, Bernays EA. 2001. Geographic variation in host specificity reveals host range evolution in *Uroleucon ambrosiae* aphids. *Ecology* 82:726–39

Funk DJ, Filchak KE, Feder JL. 2002. Herbivorous insects: model systems for the comparative study of speciation ecology. *Genetica* 116:251–67

Futuyma DJ. 1998. *Evolutionary Biology*. Sunderland, MA: Sinauer

Futuyma DJ, Peterson SC. 1985. Genetic variation in the use of resources by insects. *Annu. Rev. Entomol.* 30:217–38

Fuyama Y. 1978. Behavior genetics of olfactory responses in *Drosophila*. II. An odorant-specific variant in a natural population of *Drosophila melanogaster*. *Behav. Genet.* 8:399–414

Gailey DA, Hall JC, Siegel RW. 1985. Reduced reproductive success for a conditioning mutant in experimental populations of *Drosophila melanogaster*. *Genetics* 111:795–804

Gegenfurtner KR, Sharpe LT. 1999. *Color Vision: From Genes to Perception*. New York: Cambridge Univ. Press

Gould E, Beylin A, Tanapat P, Reeves A, Shors TJ. 1999. Learning enhances adult neurogenesis in the hippocampal formation. *Nat. Neurosci.* 2:260–65

Gowaty PA, Drickamer LC, Schmid-Holmes S. 2003. Male house mice produce fewer offspring with lower viability and poorer performance when mated with females they do not prefer. *Anim. Behav.* 65:95–103

Greenwood P, Sunderland T, Friz J, Parasuraman R. 2000. Genetics and visual attention: selective deficits in healthy adult carriers of the ε4 allele of the apolipoprotein E gene. *Proc. Natl. Acad. Sci. USA* 97:11661–66

Grieco F, Noordwijk van AJ, Visser M. 2002. Evidence for the effect of learning on timing of reproduction in blue tits. *Science* 296:136–38

Griff IC, Reed RR. 1995. The genetics of olfaction. *Curr. Opin. Neurobiol.* 5:456–60

Hadler NM. 1964. Genetic influence on phototaxis in *Drosophila melanogaster*. *Biol. Bull.* 126:264–73

Hart NS. 2001. The visual ecology of avian photoreceptors. *Prog. Retin. Eye Res.* 20:675–703

Hausheer-Zarmakupi Z, Wolfer DP, Leisinger-Trigona MC, Lipp HP. 1996. Selective breeding for extremes in open-field activity of mice entails a differentiation of hippocampal mossy fibres. *Behav. Genet.* 26:167–76

Healy SD, Guilford T. 1990. Olfactory bulb size and nocturnality in birds. *Evolution* 44:339–46

Heisenberg M, Heusipp M, Wanke C. 1995. Structural plasticity in the *Drosophila* brain. *J. Neurosci.* 15:1951–60

Heron WT. 1935. The inheritance of maze learning ability in rats. *J. Comp. Psychol.* 19:77–89

Hollis KL. 1997. Contemporary research on Pavlovian conditioning. A "new" functional analysis. *Am. Psychol.* 52:956–65

Huey RB, Hertz PE, Sinervo B. 2003. Behavioral drive versus behavioral inertia in evolution: a null model approach. *Am. Nat.* 161:357–66

Jacobs GH. 1998. A perspective on color vision in platyrrhine monkeys. *Vis. Res.* 38:3307–13

Jones R, Culver DC, Kane TC. 1992. Are parallel morphologies of cave organisms the result of similar selection pressures? *Evolution* 46:353–65

Kaas JH. 1991. Plasticity of sensory and motor maps in adult mammals. *Annu. Rev. Neurosci.* 14:137–67

Kalmus H. 1971. Genetics of taste. In *Taste*, ed. LM Beidler, pp. 165–79. Berlin: Springer-Verlag

Kandel ER. 2001. The molecular biology of memory storage: a dialogue between genes and synapses. *Science* 294:1030–38

Kats LB, Dill LM. 1998. The scent of death: chemosensory assessment of predation risk by prey animals. *Ecoscience* 5:361–94

Kessler S. 1968. The genetics of *Drosophila* mating behaviour. I. Organization of mating speed in *Drosophila pseudoobscura*. *Anim. Behav.* 16:485–91

Kissinger SC, Riccio DC. 1995. Stimulus conditions influencing the development of tolerance to repeated cold exposure in rats. *Anim. Learn. Behav.* 23:9–16

Krebs JR, Sherry DF, Healy SD, Perry VH, Vaccarino. 1989. Hippocampal specialization in food-storing birds. *Proc. Natl. Acad. Sci. USA* 86:1388–92

Kruska D. 1988. Mammalian domestication and its effect on brain structure and behavior. In *Intelligence and Evolutionary Biology*, ed. HJ Jerison, I Jerison, pp. 211–50. Berlin: Springer-Verlag

Lambrechts MM, Blondel J, Maistre M, Perret P. 1997. A single response mechanism is responsible for evolutionary adaptive variation in a bird's laying date. *Proc. Natl. Acad. Sci. USA* 94:5153–55

Laughlin SB. 2001. Energy as a constraint on the coding and processing of sensory information. *Curr. Opin. Neurobiol.* 11:475–80

Lefebvre L, Reader SM, Sol D. 2004. Feeding innovations and forebrain size in birds. *Brain Behav. Evol.* 63:233–46

Lemon WC. 1991. Fitness consequences of foraging behaviour in the zebra finch. *Nature (London)* 352:153–55

Lima SL. 1998. Stress and decision making under the risk of predation: recent developments from behavioral, reproductive, and ecological perspectives. *Adv. Stud. Behav.* 27:215–91

Linn C, Feder JL, Nojima S, Dambroski HR, Berlocher SH, Roelofs W. 2003. Fruit odor discrimination and sympatric host race formation in Rhagoletis. *Proc. Natl. Acad. Sci. USA* 100:11490–93

Lofdahl KL, Holliday M, Hirsch J. 1992. Selection for conditionability in *Drosophila melanogaster*. *J. Comp. Psychol.* 106:172–83

Lucas PW, Dominy NJ, Riba-Hernandez P, Stoner KE, Yamashita N, et al. 2003. Evolution and function of routine trichromatic vision in primates. *Evolution* 57:2636–43

Lynch M, Arnold SJ. 1988. The measurement of selection on size and growth. In *Size Structured Populations*, ed. B Ebenman, L Person, pp. 47–59. Berlin: Springer-Verlag

MacBean IT, Parsons PA. 1967. Directional selection for duration of copulation in *Drosophila melanogaster*. *Genetics* 56:233–39

Mace GM, Harvey PH, Clatton-Brock TH. 1981. Brain size and ecology in small mammals. *J. Zool.* 193:333–54

Mackay T, Hackett JB, Lyman RF, Wayne ML, Anholt R. 1996. Quantitative genetic variation of odor-guided behavior in a natural population of *Drosophila melanogaster*. *Genetics* 144:727–35

Magurran AE, Seghers BH, Carvalho GR, Shaw Pw. 1995. The behavioural diversity and evolution of guppy (*Poecilia reticulata*) populations in Trinidad. *Adv. Stud. Behav.* 24:155–202

Manning A. 1961. The effects of artificial selection for mating speed in *Drosophila melanogaster*. *Anim. Behav.* 9:82–92

Manning A. 1968. The effects of artificial selection for mating speed in *Drosophila simulans*. I. The behavioral changes. *Anim. Behav.* 16:108–13

Marinesco S, Duran KL, Wright WG. 2003. Evolution of learning in three aplysiid species: differences in heterosynaptic plasticity contrast with conservation in serotonergic pathways. *J. Physiol. London* 550:241–53

Mayr E. 1963. *Animal Species and Evolution*. Cambridge, MA: Harvard Univ. Press

McClearn GE. 1959. The genetics of mouse behavior in novel situations. *J. Comp. Physiol. Psych.* 52:62–67

McGuire TR, Hirsch J. 1977. Behavior genetic analysis of *Formia regina*: conditioning, reliable individual differences, and selection. *Proc. Natl. Acad. Sci. USA* 74:5193–97

McPeek MA. 1990. Behavioral differences between *Enallagma* species (Odonata) influencing differential vulnerability to predators. *Ecology* 71:1714–26

Mery F, Kawecki TJ. 2002. Experimental evolution of learning ability in fruit flies. *Proc. Natl. Acad. Sci. USA* 99:14274–79

Mery F, Kawecki TJ. 2003. A fitness cost of learning ability in *Drosophila melanogaster*. *Proc. R. Soc. London Ser. B Biol. Sci.* 270:2465–69

Moller AP, Alatalo RV. 1999. Good-genes effects in sexual selection. *Proc. R. Soc. London Ser. B Biol. Sci.* 266:85–91

Moller AP, Jennions MD. 2001. How important are direct fitness benefits of sexual selection? *Naturwissenschaften* 88:401–15

Morse DH, Stephens EG. 1996. The consequences of adult foraging success on the components of lifetime fitness in a semelparous, sit-and-wait predator. *Evol. Ecol.* 10:361–73

Nager RG, van Noordwijk AJ. 1995. Proximate and ultimate aspects of phenotypic plasticity in timing of great tit breeding in a heterogeneous environment. *Am. Nat.* 146:454–74

Nathans J. 1999. The evolution and physiology of human color vision: insights from molecular genetic studies of visual pigments. *Neuron* 24:299–312

Nicolakakis N, Sol D, Lefebvre L. 2003. Behavioural flexibility predicts species richness in birds, but not extinction risk. *Anim. Behav.* 65:445–52

Nottebohm F. 2002. Neuronal replacement in adult brain. *Brain Res. Bull.* 57:737–49

Nudo RJ, Milliken GW, Jenkins WM, Merzenich MM. 1996. Use-dependent alterations of movement representations in primary motor cortex of adult squirrel monkeys. *J. Neurosci.* 16:785–807

Osorio D, Vorobyev M. 1996. Colour vision as an adaptation to frugivory in primates. *Proc. R. Soc. London Ser. B Biol. Sci.* 263:593–99

O'Steen S, Cullum AJ, Bennett AF. 2002. Rapid evolution of escape ability in Trinidadian guppies (*Poecilia reticulata*). *Evolution* 56:776–84

Parasuraman R, Greenwood PM, Sunderland T. 2002. The apolipoprotein *E* gene, attention, and brain function. *Neuropsychology* 16:254–74

Peirce JL, Chesler EJ, Williams RW, Lu L. 2003. Genetic architecture of the mouse hippocampus: identification of gene loci with selective regional effects. *Genes Brain Behav.* 2:238–52

Persons MH, Walker SE, Rypstra AL. 2002. Fitness costs and benefits of antipredator behavior mediated by chemotactile cues in the wolf spider *Pardosa milvina* (Araneae: Lycosidae). *Behav. Ecol.* 13:386–92

Pigliucci M. 2001. *Phenotypic Plasticity: Beyond Nature and Nurture.* Baltimore, MD: Johns Hopkins Univ. Press

Platt ML. 2002. Neural correlates of decisions. *Curr. Opin. Neurobiol.* 12:141–48

Plomin R, DeFries JC, McClearn GE, McGuffin P, eds. 2001. *Behavioral Genetics.* New York: Worth

Pravosudov VV, Clayton NS. 2002. A test of the adaptive specialization hypothesis: population differences in caching, memory, and the hippocampus in black-capped chickadees (*Poecile atricapilla*). *Behav. Neurosci.* 116:515–22

Price T, Qvarnstrom A, Irwin D. 2003. The role of phenotypic plasticity in driving genetic evolution. *Proc. R. Soc. London Ser. B Biol. Sci.* 270:1433–40

Prutkin J, Duffy VB, Etter L, Fast K, Gardner E, et al. 2000. Genetic variation and inferences about perceived taste intensity in mice and men. *Physiol. Behav.* 69:161–73

Raber J, Wong D, Yu GQ, Buttini M, Mahley RW, et al. 2000. Apolipoprotein *E* and cognitive performance. *Nature* 404:352–54

Rampon C, Jiang CH, Dong H, Tang YP, Lockhart DJ, et al. 2000. Effects of environmental enrichment on gene expression in the brain. *Proc. Natl. Acad. Sci. USA* 97:12880–84

Reader SM, Laland KN. 2002. Social intelligence, innovation, and enhanced brain size in primates. *Proc. Natl. Acad. Sci. USA* 99:4436–41

Recanzone GH, Jenkins WM, Hradek GT, Merzenich MM. 1992. Progressive improvement in discriminative abilities in adult owl monkeys performing a tactile frequency discrimination task. *J. Neurophysiol.* 67:1015–30

Regan BC, Julliot C, Simmen B, Vienot F, Charles-Dominique P, Mollon JD. 2001. Fruits, foliage and the evolution of primate colour vision. *Philos. Trans. R. Soc. London Ser. B* 356:229–83

Relyea RA. 2003. Predators come and predators go: the reversibility of predator-induced traits. *Ecology* 84:1840–48

Richards RJ. 1987. *Darwin and the Emergence of Evolutionary Theories of Mind and Behavior.* Chicago: Univ. Chicago Press

Ringelberg J. 1999. The photo behaviour of *Daphnia* spp. as a model to explain diel vertical migration in zooplankton. *Biol. Rev.* 74:397–423

Ritchie MG, Kyriacou CP. 1996. Artificial selection for a courtship signal in *Drosophila melanogaster*. *Anim. Behav.* 52:603–11

Robinson B, Dukas R. 1999. The influence of phenotypic modifications on evolution: the Baldwin effect and modern perspectives. *Oikos* 85:582–89

Sharpe LT, Stockman A, Jagle H, Nathans J. 1999. Opsin genes, cone photopigments, color vision, and color blindness. See Gegenfurtner & Sharpe 1999, pp. 3–51

Sherry DF, Vaccarino AL. 1989. Hippocampus and memory for food caches in black-capped chickadees. *Behav. Neurosci.* 103:308–18

Sherry DF, Vaccarino AL, Buckenham K, Herz RS. 1989. The hippocampal complex of food-storing birds. *Brain Behav. Evol.* 34:308–17

Shettleworth SJ. 1998. *Cognition, Evolution, and Behavior*. Oxford: Oxford Univ. Press

Shettleworth SJ. 2003. Memory and hippocampal specialization in food-storing birds: challenges for research on comparative cognition. *Brain Behav. Evol.* 62:108–16

Siegel S, Hinson RE, Krank MD, McCully J. 1982. Heroin "overdose" death: contribution of drug-associated environmental cues. *Science* 216:436–37

Smith JD. 2000. Apolipoprotein *E4*: an allele associated with many diseases. *Ann. Med.* 32:118–27

Smulders TV, Shiflett MW, Sperling AJ, DeVoogd TJ. 2000. Seasonal changes in neuron numbers in the hippocampal formation of a food-hoarding bird: the black-capped chickadee. *J. Neurobiol.* 44:414–22

Sokolowski MB, Pereira HS, Hughes K. 1997. Evolution of foraging behavior in *Drosophila* by density-dependent selection. *Proc. Natl. Acad. Sci. USA* 94:7373–77

Sol D, Timmermans S, Lefebvre L. 2002. Behavioural flexibility and invasion success in birds. *Anim. Behav.* 63:495–502

Steel KP, Kros CJ. 2001. A genetic approach to understanding auditory function. *Nat. Genet.* 27:143–49

Stoks R, McPeek MA, Mitchell JL. 2003. Evolution of prey behavior in response to changes in predation regime: damselflies in fish and dragonfly lakes. *Evolution* 57:574–85

Sugawara T, Terai Y, Okada N. 2002. Natural selection of the rhodopsin gene during the adaptive radiation of East African Great Lakes Cichlid fishes. *Mol. Biol. Evol.* 19:1807–11

Surridge AK, Osorio D, Mundy NI. 2003. Evolution and selection of trichromatic vision in primates. *Trends Ecol. Evol.* 18:198–205

Szkeley T, Catchpole C, Devoogd A, Marchl Z, Devoogd T. 1996. Evolutionary changes in a song control area of the brain (HVC) are associated with evolutionary changes in song repertoire among European warblers (Sylviidae). *Proc. R. Soc. London Ser. B Biol. Sci.* 263:607–10

Tauber E, Eberl DF. 2003. Acoustic communication in *Drosophila*. *Behav. Process* 64:197–210

Technau GM. 1984. Fiber number in the mushroom bodies of adult *Drosophila melanogaster* depends on age, sex and experience. *J. Neurogen.* 1:113–26

Thomas DW, Blondel J, Perret P, Lambrechts MM, Speakman JR. 2001. Energetic and fitness costs of mismatching resource supply and demand in seasonally breeding birds. *Science* 291:2598–600

Tolman EC. 1924. The inheritance for maze-learning ability in rats. *J. Comp. Psychol.* 4:1–18

Toma DP, White KP, Hirsch J, Greenspan RJ. 2002. Identification of genes involved in *Drosophila melanogaster* geotaxis, a complex behavioral trait. *Nat. Genet.* 31:349–53

Tramontin AD, Brenowitz EA. 1999. A field study of seasonal neuronal incorporation into the song control system of a songbird that lacks adult song learning. *J. Neurobiol.* 40:316–26

Tramontin AD, Brenowitz EA. 2000. Seasonal plasticity in the adult brain. *Trends Neurosci.* 23:251–58

Tryon RC. 1940. Genetic differences in maze-learning ability in rats. *Yearb. Natl. Soc. Study Educ.* 39(Pt. 1):111–19 van Praag H, Kempermann G, Gage FH. 2000. Neural consequences of environmental enrichment. *Nat. Rev. Neurosci.* 1:191–98

Waddington CH. 1953. Genetic assimilation of an acquired character. *Evolution* 7:118–26

Waddington CH. 1959. Canalization of development and genetic assimilation of acquired characters. *Nature* 183:1654–55

Wagner WE, Harper CJ. 2003. Female life span and fertility are increased by the ejaculates of preferred males. *Evolution* 57:2054–66

Ward BC, Nordeen NJ, Nordeen KW. 1998. Individual variation in neuron number predicts differences in the propensity for avian vocal imitation. *Proc. Natl. Acad. Sci. USA* 95:1277–82

Warren RP, Lewis RC. 1970. Taste polymorphism in mice involving a bitter sugar derivative. *Nature* 227:77–78

West-Eberhard MJ. 2003. *Developmental Plasticity and Evolution.* Oxford: Oxford Univ. Press

Wheeler DA, Kyriacou CP, Greenacre ML, Yu Q, Rutila JE, et al. 1991. Molecular transfer of a species-specific behavior from *Drosophila simulans* to *Drosophila melanogaster. Science* 251:1082–85

Whissell-Buechy D, Amoore JE. 1973. Odour blindness to musk: simple recessive inheritance. *Nature* 242:271–73

Wiesel TN, Hubel DH. 1963. Single-cell responses on striate cortex of kittens deprived of vision in one eye. *J. Neurophysiol.* 26:1003–17

Wisenden BD, Chivers DP, Smith RJF. 1997. Learned recognition of predation risk by Enallagma damselfly larvae (Odonata, Zygoptera) on the basis of chemical cues. *J. Chem. Ecol.* 23:137–51

Wright WG, Kirschman D, Rozen D, Maynard B. 1996. Phylogenetic analysis of learning-related neuromodulation in molluscan mechanosensory neurons. *Evolution* 50:2248–63

Yeh SR, Fricke RA, Edwards DH. 1996. The effect of social experience on serotonergic modulation of the escape circuit of crayfish. *Science* 271:366–69

Yokoyama S, Yokoyama R. 1996. Adaptive evolution of photoreceptors and visual pigments in vertebrates. *Annu. Rev. Ecol. Sys.* 27:543–67

Zhang JZ, Webb DM. 2003. Evolutionary deterioration of the vomeronasal pheromone transduction pathway in catarrhine primates. *Proc. Natl. Acad. Sci. USA* 100:8337–41

POLLINATION SYNDROMES AND FLORAL SPECIALIZATION

Charles B. Fenster,[1] W. Scott Armbruster,[2] Paul Wilson,[3] Michele R. Dudash,[1] and James D. Thomson[4]

[1]Department of Biology, University of Maryland, College Park, Maryland 20742; email: cfenster@umd.edu; mdudash@umd.edu
[2]School of Biological Sciences, University of Portsmouth, Portsmouth, PO1 2DY, United Kingdom; Department of Biology, Norwegian University of Science and Technology, N-7491 Trondheim, Norway; Institute of Arctic Biology, University of Alaska, Fairbanks, Alaska 99775; email: ffwsa@uaf.edu
[3]Department of Biology, California State University, Northridge, California 91330-8303; email: paul.wilson@csun.edu
[4]Department of Zoology, University of Toronto, Toronto, ON M5S 3G5; email: jthomson@zoo.utoronto.ca

Key Words floral evolution, mutualism, plant-animal interaction, pollinator, pollination

■ **Abstract** Floral evolution has often been associated with differences in pollination syndromes. Recently, this conceptual structure has been criticized on the grounds that flowers attract a broader spectrum of visitors than one might expect based on their syndromes and that flowers often diverge without excluding one type of pollinator in favor of another. Despite these criticisms, we show that pollination syndromes provide great utility in understanding the mechanisms of floral diversification. Our conclusions are based on the importance of organizing pollinators into functional groups according to presumed similarities in the selection pressures they exert. Furthermore, functional groups vary widely in their effectiveness as pollinators for particular plant species. Thus, although a plant may be visited by several functional groups, the relative selective pressures they exert will likely be very different. We discuss various methods of documenting selection on floral traits. Our review of the literature indicates overwhelming evidence that functional groups exert different selection pressures on floral traits. We also discuss the gaps in our knowledge of the mechanisms that underlie the evolution of pollination syndromes. In particular, we need more information about the relative importance of specific traits in pollination shifts, about what selective factors favor shifts between functional groups, about whether selection acts on traits independently or in combination, and about the role of history in pollination-syndrome evolution.

INTRODUCTION

The paradigm that diverse floral phenotypes reflect specialization onto different groups of pollinators begins with Kölreuter's (1761) and Sprengel's (1793, 1996) descriptions of the interactions between plants and pollinators and the floral features that promote these interactions. Darwin (1862) and many others then elaborated on the view that floral-trait combinations reflect pollinator type (Müller 1883; Delpino 1868–1875; Müller & Delpino 1869; Knuth 1906, 1908; Baker 1963; Grant & Grant 1965; Fægri & van der Pijl 1966; Stebbins 1970; Johnson & Steiner 2000). When placed in an evolutionary framework, these comparative observations suggest that different pollinators promote selection for diverse floral forms that produce an array of "pollination syndromes," (e.g., Figure 1, see color insert). We define a pollination syndrome as a suite of floral traits, including rewards, associated with the attraction and utilization of a specific group of animals as pollinators. The floral traits are expected to correlate with one another across independent evolutionary events.

However, summaries of observational data show that many flowers are visited by numerous animal species (Robertson 1928; Waser et al. 1996; Ollerton 1996, 1998), which calls into question expectations from comparative biology that pollination syndromes both reflect and predict convergent selection pressures on floral traits (Fæegri & van der Pijl 1979). Furthermore, the notion that floral traits conform to a pollination syndrome and represent an adaptive response that results in specialization has been questioned because specialization is postulated to result in greater variance in reproductive success across years and, thus, ought to be selected against (Waser et al. 1996). However, comparative biology continues to highlight floral radiations onto different pollinators (Johnson et al. 1998, Goldblatt et al. 2001, Wilson et al. 2004), which likely reflects selection for specialization. How, then, can we reconcile this apparent paradox of diverse visitors at flowers with our observations of widespread convergence in floral traits?

An important first step for the study of the evolutionary relevance of pollination syndromes is to recognize that the concept implies that pollinators are clustered into functional groups (e.g., long-tongued flies or small, nectar-collecting bees) that behave in similar ways on a flower and exert similar selection pressures, which, in turn, generate correlations among floral traits (e.g., long and narrow corolla tubes, pollen presented in a certain way, or particular nectar quantities and concentrations) (Waser et al. 1996; Armbruster et al. 1999, 2000; Armbruster 2004). Such pollinator-driven floral evolution can proceed with or without the animals coevolving (Janzen 1980, Schemske 1983, Kiester et al. 1984). Here, we review the evidence that convergent selective pressures exerted by functional groups of pollinators is a prevalent underlying feature of floral diversification. We first consider the evidence for organizing diverse species of pollinators into functional groups, as well as evidence that each functional group exerts a suite of convergent selection pressures. We also evaluate whether pollinator assemblages differ in their contribution to pollination for a given plant species and, thus, may contribute differentially to the selective pressures exerted via the reproductive success of a plant. We explore whether floral traits respond differentially to selective pressures; that is, do some traits contribute more to pollination syndromes than others? Furthermore, we consider whether floral evolution represents independent or interactive selection (i.e., correlational selection) and the consequences of each type of selection. Because we emphasize a comparative approach, we evaluate the role of historical context in shaping the contemporary patterns of floral diversity. We review the patterns of character correlation that have arisen as plant lineages have shifted between pollinators (Wilson et al. 2004) and the processes that underlie this diversification (Armbruster 1992, 1993; Thomson et al. 2000; Thomson 2003). We hope to identify areas that require further study and to better quantify the contribution of pollination syndromes to our mechanistic understanding of floral diversification.

SPECIALIZATION ONTO FUNCTIONAL GROUPS

The pollination syndrome concept implies that specialization onto functional groups is a common occurrence in plant evolution. Thus, a plant has specialized pollination if it is successfully pollinated only by a subset of functionally grouped potential pollinators; such plants are also said to occupy pollination niches (Beattie 1971, Armbruster et al. 1994, Gomez & Zamora 1999). For example, some would describe *Collinsia heterophylla* as generalized because it is pollinated by some 14 species of animals, yet it is more cogently viewed as specialized onto a functional group of large-bodied, long-tongued bees in a community that contains potential pollinators of much greater functional disparity (Armbruster et al. 2002). We further illustrate this point by reexamining Robertson's (1928) dataset from the perspective of functional groups. Summarizing Robertson's (1928) observations of 15,172 visits to 441 flowering plant species found within 10 miles of Carlinville, Illinois, Waser et al. (1996) noted that the vast majority of plants received visits by many different species of potential pollinators (see their figure 1), and they concluded that 91% of the 375 native plant taxa were visited by more than one animal species and were, therefore, somewhat generalized. Waser et al. (1996) reaffirmed this conclusion from several smaller surveys of other floras. In contrast, we follow Robertson's (1928) classification of the visitors into nine functional groups (long-tongued bees, short-tongued bees, other Hymenoptera, Diptera, Coleoptera, Lepidoptera, Hemiptera, Neuroptera, and birds) and only include animal-pollinated plant species that had frequency data of the pollinators noted. By noting frequency, we could weight the relative potential importance of the different functional classes of visitors to pollination. We arbitrarily decided that if a plant species was visited three fourths or more of the time by a single functional group, then the plant manifested specialization on that functional group. We continued to add functional groups as long as the least represented functional group visited at least one half as often as the previous most frequent functional group. Robertson (1928) also noted which visitors were not pollinating, and we omitted these visitors from the list of pollinators for the particular plant species. In a number of cases, Robertson included frequency data in his original publications but did not do so in his 1928 book. We used the original references for the visitation frequency data [23 papers by Robertson between 1887 and 1924, 21 cited in Robertson (1928) and the others in Robertson (1923, 1924)]. In some cases, we could safely assess visitation in the absence of frequency data because some visitors did not pollinate or because only one type of visitor was observed.

Of the 278 animal-pollinated plant species for which we were able to perform quantitative evaluations, 150 species were pollinated by one functional group. Of the 85 species that were pollinated by two functional classes, 59 had pollination by two functional groups that probably exert very similar selection pressures and perhaps formed one more encompassing functional group. These cases include plants that are pollinated by (*a*) both long-tongued and short-tongued bees that likely form a pollen-collecting bee functional group (e.g., *Tradescantia* spp.); (*b*) long-tongued bees and bee flies that likely form a long-tongued, nectar-feeding insect functional group (e.g., *Agastache scrophlarieaefolia*); and (*c*) short-tongued bees and Diptera that form a small, pollen-collecting or nectar-feeding insect functional group (e.g., *Lepidium virginicum*). The dataset contains many biases, such as that caused by the vastly different effectiveness (discussed below) of the different pollinators. For example, long-tongued bees often work flowers at a much greater speed than do other visitors and, thus, are likely to contribute much more to pollination than is indicated by their census frequency alone. In all, we believe these biases lead to a conservative estimate of the number of functional groups that pollinate the Carlinville flora. Thus, we conclude that approximately 75% (209/278) of the flowering plant species exhibit specialization onto functional groups, a very different conclusion than that reached by Waser et al. (1996), who used the same data. Pollination of each plant species by a small subset of the available pollinators is common in other communities as well (Parrish & Bazzaz 1979; Pleasants 1980, 1990; Rathcke 1983; Armbruster 1986; Dilley et al. 2000). Furthermore, these subsets of pollinators often fall into functional groups in which the visitors likely share attributes of behavior, and these functional groups are predicted, on the basis of pollination syndrome traits (i.e., flower color, fragrance, reward, and morphology), in such divergent communities as a dipterocarp forest (Momose et al. 1998), an English meadow (Dicks et al. 2002), and *Costus* species in neotropical forests (Kay & Schemske 2003). We urge that investigators continue to organize pollinator communities by functional groups, and we emphasize here the need to understand more fully the degree to which such groups overlap in the selective pressures they exert on floral design (Ollerton & Watts 2000).

Functional groups, although sometimes difficult to delimit in practice, are clearly more relevant to specialization than are species lists. To regard *Silene vulgaris* as pollinated by one functional group of 26 nocturnal moths (Pettersson 1991) seems more informative than to consider it a generalist pollinated by 26 species of noctuids and sphingids. Understanding the functional relationships between the moth species fosters further questions. Do the moths vary in their effectiveness as pollinators? (They do.) If so, why no further specialization? Functional groups permit the diversification of flowers to be understood through adaptive evolution, not just in terms of pollinator species richness. Other possible routes of specialization in pollination ecology may involve divergence in time of day that flowers open (Armbruster 1985, Stone et al. 1998), site of pollen placement (Dressler 1968, Dodson et al. 1969, Nilsson 1987, Armbruster et al. 1994), or even homoplasy in floral traits that result in Müllerian (Schemske 1981) and Batesian mimicry (Nilsson 1983, Temeles & Kress 2003).

A number of caveats are implicit in the usage of functional groups. First, functional groupings of pollinators must be assessed by taking into account the architecture of the flower under consideration. The same pollinator (*Bombus*) may be a component of a narrow functional group that pollinates specialized zygomorphic flowers with recessed nectaries and constricted floral tubes (e.g., *Collinsia*) yet that also pollinates a highly generalized actinomorphic flower (e.g., *Rosa*) on which a broad taxonomic diversity of visitors move about in an undirected pattern. We conjecture that differences in overall structure between the two plant taxa differentially filter and focus the amount and direction of selection on floral traits. Sorting the Robertson (1928) dataset by actinomorphic versus zygomorphic flowers revealed that 52% of the 192 actinomorphic species were pollinated by one functional group, significantly less ($P < 0.01$, $\chi^2 = 11.544$) than the 61% of 86 zygomorphic species. This finding supports the notion that complex flowers reflect selection by narrower functional groups. Complexity may also lead to greater diversification rates because complexity exposes the plant to differential selective pressures exerted by different functional groups (contra Orr 2000). However, as we remark below, we have a poor comparative understanding of the types of selection pressures exerted upon plants pollinated by broad versus narrow functional groups.

Functional groups of pollinators may contain many species or only one species, and any particular species of pollinator may belong to multiple functional groups. Additionally, functional groups on which flowers specialize need not be related to pollinator taxonomy (although they often are). As an illustration of this point, Darwin (1877) described the orchid *Herminium monorchis*, which has small greenish-yellow flowers and is pollinated by taxonomically very unrelated minute insects (~ 1 mm long). These insects are compelled by the structure of the flower to behave in so similar a manner that contact with the plant's reproductive organs is associated with the same anatomical features (the outer surface of the femur of one of the front legs) of each insect. Such insects (members of the Hymenotpera, Diptera, and Coleoptera orders) are not commonly considered contributors to specialized pollination, yet the labyrinthine structure of this orchid flower imposes a uniformity of behavior and clearly fits within our notion of specialization onto a functional group, in this case comprising unrelated, minute insect taxa.

Generalization suggests that all pollinators are functionally equivalent (Gomez 2002). Thus, a generalized state may arise from the evolutionary dynamics of numerous animal species pollinating equally well and selecting for the same floral features. In other words, selection may favor adding new pollinators without losing any old ones (Aigner 2001). For example, no cost is associated with nocturnally visited flowers remaining open through the following day and being visited by diurnal pollinators, as occurs in various *Silene* spp. (Pettersson 1991; R. Reynolds, C. Fenster & M. Dudash, unpublished data) and in *Burmeistera* (Muchhala 2003). In contrast, different functional groups may be exerting quite different selective pressures, and the contemporary manifestation of generalization represents an averaging of selection over many episodes by different pollinators and functional groups (Thompson 1994, Wilson & Thomson 1996, Dilley et al. 2000). Diversifying selection also may be acting simultaneously but not toward flowers becoming more exclusive (Thompson 1999). Understanding the selective pressures responsible for the maintenance (or origin, in case of reversals) of generalization seems crucial to our understanding of the mechanisms that underlie the transition from generalization to specialization. The identity and function of the traits that contribute to this evolutionary change from generalization to specialization—perhaps color, fragrance, flowering time—are of considerable interest and deserve study. We know of only a few studies that quantify selective pressures on a generalized floral design; for example, selection for larger flowers by muscoid flies on *Ranunculus acris* (Totland 2001). Comparative studies that document selection pressures and the target traits in related species with contrasting generalized and specialized pollination systems are clearly needed. Actinomorphic and zygomorphic taxa in the Boraginaceae, Solanaceae, and Lamiales, for example, may be candidate species for study (Reeves & Olmstead 2003).

EVOLUTIONARY SPECIALIZATION VERSUS ECOLOGICAL SPECIALIZATION

Although organizing pollinators into functional groups is informative, this summary only provides an assessment of ecological specialization, the contemporary state of having pollinators mainly belonging to a single functional group (Armbruster et al. 2000). For this static variable, the reference point is either another contemporary population (such as a coflowering plant) or a theoretical state (such as the perfectly even use of resources embodied in indexes of niche breadth). In contrast, we define evolutionary specialization as evolution toward pollination by fewer functional groups, which reflects evolution toward use of fewer pollinators, less disparate pollinators, or a change in the intensity of use of a subset of preexisting pollinators (reduced evenness). For this variable, the reference point is an ancestral population or sister group in a phylogeny. Specialization may have occurred even if the resulting population appears not to be very specialized. For example, in *Asclepias*, progressively fewer pollinators and greater specialization occur in *A. solanoana* and *A. syriaca*, subgenera of the genus that appear evolutionarily derived relative to *A. incaranata* and *A. verticillata*, which have more generalized pollination (Kephart 1983, Kephart & Theiss 2004, Fishbein 1996). Evolutionary specialization implies an evolutionary response to differential selection pressures exerted by a subset of potential pollinators that often exert selection as functional groups.

One approach to studying these issues is to map measures of specialization onto plant phylogenies (Armbruster 1992, 1993; Johnson et al. 1998; Armbruster & Baldwin 1998; Wilson et al. 2004), thereby tracing the historical course of specialization and generalization on lineages. Thus, how common evolutionary specialization really is and how commonly it results in extreme ecological specialization can possibly be determined. We can also address questions about the frequency of evolutionary reversals in specialization (Armbruster & Baldwin 1998), which traits are most labile in the evolution of specialization, how the level of specialization is maintained during shifts between functional groups of pollinators, and whether evolutionary specialization is associated with floral diversification. We address these issues in a later section.

EVOLUTION IN RESPONSE TO PRINCIPAL POLLINATORS

Stebbins (1970, pp. 318–19; 1974) attempted to resolve the apparent paradox that floral diversity has arisen by divergence into pollination syndromes (evolutionary specialization) with the observation that flowers are visited by many species of animals (ecological generalization): "Since selection is a quantitative process, the characteristics of the flower will be molded by those pollinators that visit it most frequently and effectively in the region where it is evolving." Stebbins' (1970) use of the word "and" to link frequency and effectiveness implies a multiplicative relationship: The pollinator that is both relatively most effective and relatively most frequent will usually be the most important selective force. Thus, we should quantify two components of animal activity: (a) frequency of visitation during anthesis, and (b) effectiveness of pollen transfer to appropriate stigmas on each flower visit (Grant & Grant 1965; Stebbins 1970, 1974). Most studies emphasize the former because the presence of visitors is more easily observed and quantified than is the transfer of pollen (Waser et al. 1996, Dilley et al. 2000). Pollinator effectiveness has been quantified with a variety of metrics that include (a) the proportion of each species of visitor bearing pollen (Beattie 1971, Sugden 1986); (b) the rate of pollen deposition on stigmas for each species (e.g., Beattie 1971, Levin & Berube 1972, Ornduff 1975, Armbruster 1985, Herrera 1987, Fenster 1991a); (c) the number of pollen grains deposited per visit (Primack & Silander 1975, Herrera 1987, Waser & Price 1990, Fishbein & Venable 1996, Bingham & Orthner 1998, Gomez & Zamora 1999) and across sequential visits (Campbell 1985, Waser 1988); (d) the amount of pollen deposited on stigmas and pollen removed from anthers (Wolfe & Barrett 1988, Conner et al. 1995, Rush et al. 1995; Castellanos et al. 2003); (e) the frequency with which each visitor species contacts anthers and stigmas (Armbruster & Herzig 1984; Armbruster 1985, 1988,

1990); (*f*) fruit set per visit (Schemske & Horvitz 1984) and seed set per visit (Parker 1981, Motten et al. 1981, McGuire & Armbruster 1991, Olsen 1997) for each species of visitor; and (*g*) multiple components, such as pollen load, pollen removal and deposition, handling time, and potential for geitonogamy (Ivey et al. 2003). The product of per-visit probability of contacting anthers, per-visit probability of contacting stigmas, and frequency of visitation is a useful metric of pollinator importance because it incorporates both pollen-removal and pollen-deposition components of plant reproductive success and is relatively easy to measure in the field (Armbruster & Herzig 1984; Armbruster 1985, 1988, 1990). Leaving out the effectiveness component incurs the risk of misidentifying the main pollinators or misconstruing a specialized system as generalized or vice versa (Hagerup 1951; Fægri & van der Pijl 1971; Stebbins 1974; Waser & Price 1981, 1983; Schemske & Horvitz 1984; Armbruster 1985; Armbruster et al. 1989; Inouye et al. 1994).

Are we often misled by interpreting the most common visitor as the most important pollinator? The answer is frequently yes. Often, the most common visitors are poor pollinators and the least common visitors are estimated to be the best pollinators (e.g., Armbruster 1985, Armbruster et al. 1989, Schemske & Horvitz 1984, Pettersson 1991, Tandon et al. 2003). However, in other studies, the most common visitors were found to be the most important pollinators (e.g., Fishbein & Venable 1996, Olsen 1997, Fenster & Dudash 2001; and C. Fenster & M. Dudash, unpublished data).

In our view, Stebbins (1970, 1974) was correct to focus on differences in the effectiveness of pollinators, and he was correct to assert that those differences were critical to the evolution of specialization. When we examine effectiveness of different functional groups, we often observe a close match between pollinator and flower (or blossom) that is consistent with pollination syndromes. For example, hummingbirds and bees vary in their effectiveness on different *Penstemon* species (Castellanos et al. 2003), bats and hummingbirds vary in their effectiveness on alternative *Burmeistera* species, and size of bees correlates with effectiveness on *Dalehampia* taxa of different blossom sizes (Armbruster 1985, 1988, 1990). Variation in effectiveness within a functional group is frequently documented, whether we consider the group to correspond to specialization (e.g., Schemske & Horvitz 1984, 1989; Pettersson 1991; Ivey et al. 2003) or generalization (e.g., Conner et al. 1995, Rush et al. 1995).

The selective importance of a pollinator species is not a constant and likely depends on the other animals and plants in the community (Thompson 1994, 1999). In terms of pollen presentation theory (Thomson 2003), more-common but less-effective pollinators can be viewed as parasites because they remove pollen that otherwise would have been transferred to stigmas by more-effective pollinators, but this view depends critically on the specific composition of the pollinator community. Situation-specific effectiveness could, in principle, be assessed by use of simulation models to produce a "milieu analysis" (Thomson & Thomson 1992, 1998; Aigner 2001; Thomson 2003). These simulations require estimates of how a

Figure 1 (1) *Penstemon strictus*, pollinated by a variety of bees (shown here) and the wasp *Pseudomasaris vespoides* and, on occasion, hawkmoths that act as nectar thieves. It conforms to the bee-pollination syndrome in having purple flowers (bees also like yellow), nearly included anthers over a broad vestibule, a lower lip in the position of a landing platform, and the production of smaller amounts of concentrated nectar relative to hummingbird pollinated *P. barbatus* (Wilson et al. 2004). (2) *Penstemon barbatus*, visited by hummingbirds and pollen-collecting bees, conforms to the hummingbird-pollination syndrome in having red flowers, exserted anthers and stigmas, a reflexed lower lip, a position that is inclined from the horizontal, and the production of copious dilute nectar (Wilson et al. 2004). The floral tube is too long to accommodate large nectar-collecting bees. (3) *Scoliopus bigelovii*, which lives in dark, moist forests and is pollinated by fungus gnats (Mesler et al. 1980). It has lines on the sepals reminiscent of mushroom gills, and it smells like a mushroom. (4) *Ipomopsis aggregata*, visited principally by hummingbirds and conforms to the hummingbird-pollination syndrome, much like *Penstemon barbatus*. If hummingbirds are absent and nectar accumulates, bumblebees will also visit (see pictures 10 and 11) (Mayfield et al. 2001, Fenster & Dudash 2001). (5) *Ipomopsis tenuituba*, visited by hawkmoths when they are abundant and by hummingbirds (shown here), contrary to its syndrome. In keeping with the hawkmoth-pollination syndrome, it has pale pink

(*Continued*)

Figure 1 (*Continued*) flowers, very narrow tubes, short stamens and styles, and produces smaller amounts of nectar than *Ipomopsis aggregata* (Campbell et al. 1997, Melendez-Ackerman et al. 1997). (6) *Dalechampia tiliifoilia*, pollinated by a female *Eulaema cingulata* (Apidae: Euglossini), which is collecting floral resin for nest construction (Armbruster 1992). (7) *Dalechampia brownsbergensis*, pollinated by a male *Euglossa tridentata* (Apidae: Euglossini), which is mopping up liquid terpenoid fragrances from the stigmatic surface of a pistillate flower; the substances will be used later to impress (and seduce) females bees. Note that the resin gland is apparently absent; it is vestigial and hidden under a bractlet (Armbruster 1992). (8) *Silene caroliniana*, which ranges in color from pink to nearly white, presents its flowers in an upright manner and exhibits traits typical of diurnal pollination by long-tongued insects, such as narrow corolla tubes, diurnal anthesis, stigma receptivity, and reduced nectar production relative to congener hummingbird-pollinated *Silene virginica* (see picture 10) and nocturnally pollinated *Silene stellata* (see picture 13) (C. Fenster, R. Reynolds & M. Dudash, unpublished data; photograph supplied by M. Hood). Shown here pollinated by a clear-wing hawkmoth and in picture 9 by *Bombus* sp., a regular pollinator, approaching *Silene caroliniana*. The relative frequency of visitation by moths and bees to *S. caroliniana* is highly temporally variable (R. Reynolds, C. Fenster & M. Dudash, unpublished data). (10) *Silene virginica*, pollinated by its major pollinator, the hummingbird *Archilochus colubris*, has bright red, scentless flowers that are presented slightly inclined from the horizontal and that secrete copious nectar (Fenster & Dudash 2001). (11) *Silene virginica*, pollinated by *Bombus* sp. In one site of two studied and in one year of six years of observations, *Bombus* spp. were important pollinators of *S. virginica* (Fenster & Dudash 2001). (12) *Silene stellata*, white, horizontally presented flowers, becomes sexually receptive in the evening and produces relatively less nectar than hummingbird *S. virginica* or diurnally pollinated *S. caroliniana*. It is visited by nectar feeding moths during the night, of which some species also lay eggs in the flowers, and by pollen collecting *Bombus* spp., shown here the following morning (R. Reynolds, C. Fenster & M. Dudash, unpublished data).(13) *Silene stellata*, visited by a hovering Noctuid moth (R. Reynolds, C. Fenster & M. Dudash, unpublished data). (14) *Salvia mohavensis*, pollinated by huge flies in the genus *Rhaphiomidas* that have tongues approximately 18 mm long. (15–18) *Raphanus raphanistrum* exhibits traits typical of generalized pollination, including radially symmetric flowers, exposed reproductive organs and an upright flower. It is shown in picture 15 pollinated by a *Pieris* sp. collecting nectar with pollen deposited on the head and in pictures 16–18 pollinated by a halictid bee, an anthophorid bee, and a syrphid fly probing the anthers for pollen, respectively (H. Sahli & J. Conner, unpublished data).

series of animals differ in their rates of visitation, the amount of pollen they remove in each visit, and the proportion of pollen they deliver. One can also specify how these parameters might depend on floral characters, such as how exserted the anthers are or how copious the flow of nectar is. From these numbers, one can model the export of pollen from flowers and, thus, the "quality" of the various animals. The model yields the total number of pollen grains P that are delivered to stigmas by a particular mixture of pollinators. Next, one can study the effects of adding one additional visit by a particular species i of pollinator. The change in P achieved by one additional visit by pollinator species i, designated ΔP_i, is a measure of that pollinator's marginal effectiveness in the milieu under examination. By cycling through all the pollinator species and calculating ΔP_i for each, one can rank a set of covisiting pollinators by their effectiveness. This process is a Stebbinsian effectiveness in that selection acting within that particular milieu favors characteristics that increase the proportion of visits by the top-ranked pollinators and decrease visits by bottom-ranked pollinators (Castellanos et al. 2004). Selection also favors characteristics that broadly increase the ΔP_i values of animals that visit. The evolution of features that exclude less-effective pollinators may be difficult, except in the radical case of transitions between major functional classes of pollinators, which likely involves direct tradeoffs (Aigner 2004).

A constructive extrapolation of these ideas is to contrast functional groups in terms of the relationship between floral characters and plant fitness, what today we would call "selection gradients" or "fitness functions." Aigner (2001) has shown how floral characters might evolve to the net selection pressures exerted by two or more pollinators or functional groups. Hence, we can consider how selection pressures might differ both within and between functional groups. Floral traits may evolve in response to pollinators that have the most exacting and steepest selection gradients, even when those pollinators are not the most important pollinators (in terms of both frequency and effectiveness). However, these results are based on the restrictive conditions of minimal negative interactions between pollinators. When interactions are allowed, for example, the selective gradients associated with the ancestrally most important pollinators may become much steeper as additional pollinators are added to the system. Effective net stabilizing selection on the trait may then result (Aigner 2001). We are unaware of any studies that quantify the interaction among pollinators in terms of the selection they exert on floral traits. Clearly, further study is needed in this area.

Fluctuations in the pollinator milieu (e.g., Kephart 1983; Schemske & Horvitz 1984 1989; Horvitz & Schemske 1990; Bingham & Orthner 1998) and gene flow between populations with different milieus will change selection regimes and retard consistent specialization at the level of the plant species (Pettersson 1991, Wilson & Thomson 1996, Waser et al. 1996, Dilley et al. 2000). This development may be the reason functional groups include many functionally related species that, nonetheless, differ in the selective pressures they exert on floral traits. Averaged over many populations, the result of this process may well be a large functional group, even if selection for great ecological specialization onto one or a few pollinators is occurring each generation within each population (Thompson 1994).

Associating effectiveness and functional groups with floral design clearly contributes to our understanding of the relevance of pollination syndromes. Functional groups provide an intuitive biological framework on which to categorize the "tangled bank" of floral visitors into groups that may vary in their effectiveness of pollination and, hence, the selective pressures they exert on plants. Ultimately, we need to focus on the interaction between selection and traits and ask the following questions: What traits do functional groups select upon? Do different functional groups select for different traits and different trait expression? We next evaluate whether the selective pressures exerted by functional groups promote specialization into categories defined by pollination syndromes.

DO POLLINATION SYNDROMES REFLECT A RESPONSE TO SELECTION BY SPECIFIC FUNCTIONAL GROUPS?

Here, we review the evidence that specialization results from convergent selection pressures exerted by functional groups of pollinators consistent with our understanding of pollination syndromes. We consider experiments based on phenotypic-selection analysis, effects of phenotypic manipulation on pollinator discrimination, covariation in floral characters and functional groups at the ecotypic level, and phylogenetic analyses testing for correspondence between shifts in trait and shifts in functional groups of pollinators.

Phenotypic Selection

The following extended example illustrates the promise and limitations of the study of contemporary phenotypic selection on natural variation of floral traits (see also Lande & Arnold 1983; Campbell 1989, 1996; Campbell et al. 1991, 1997; Waser 1998). Studies of *Calathea ovandensis* demonstrated that two relatively short-tongued Hymenoptera species, *Rhathymus* sp. and *Bombus medius*, exerted the strongest selection of all pollinators, although they were frequently absent or in low numbers (Schemske & Horvitz 1984, 1989). These pollinators exerted selection that favored flowers with shorter corolla tubes and, thus, flowers better suited to pollination by shorter-tongued pollinators. An important conclusion from the *Calathea* study was that quantifying pollinator importance in terms of how much pollen is removed and deposited on stigmas on an absolute basis may not allow identification of the important selective agents that act on floral characteristics. Although this example demonstrates that rare pollinators can be contemporary selective agents, it does not demonstrate that short-tongued bees have been important in the origin or maintenance of the present suite of floral traits. Instead, short-tongued pollinators may act to disrupt the present suite of floral traits. Indeed, floral traits such as the long nectar tube suggest that, despite directional selection by short-tongued bees for shorter tubes, the floral morphology of *C. ovandensis* reflects selection exerted by more-frequent but less-effective long-tongued euglossine bees. The selection surface generated by euglossine bees on the study populations of *C. ovandensis* may actually be very flat, so shallow that fitness differences among individual plants are not detectable. This finding reveals the limits of phenotypic-selection studies when such studies are conducted only with natural phenotypic variation. We suggest this finding is the result of stabilizing selection averaged over the long term.

Although many other studies have documented phenotypic selection on floral traits (reviewed in Kingsolver et al. 2001), almost all studies have focused either on general floral features, such as flower number, flower size, and display height, that are likely attractive to all functional groups (e.g., Galen 1989, Johnston 1991, Maad 2000, Totland 2001) or on traits associated with breeding systems (e.g., Fenster & Ritland 1994). Studies that specifically quantify selection on floral traits that comprise specialized pollination syndromes, in pure "parental" populations, are still rare. Campbell (1989, 1996) demonstrated that hummingbirds select for stigma exsertion and wider corollas (which allow the hummingbird to enter the corolla tube more effectively). The lack of within-population phenotypic variation for floral-syndrome traits likely has been the cause of the lack of phenotypic-selection studies on these very same traits. However, our understanding of the mechanisms of pollination-syndrome evolution would be enhanced if investigators focused in the future on studying phenotypic selection on traits that contribute to pollination syndromes so that we can assess the degree to which selection is actually convergent and whether selection is acting via male or female reproductive success, or the success of both (e.g., Campbell 1989, 1996; Johnson & Steiner 1997). We also need more phenotypic-selection studies on generalized flowers to understand the selective mechanisms that underlay their origin and maintenance. For example, large syrphid flies and sweat bees exert contrasting selection on anther exsertion in *Raphanus raphanistrum*, where other traits appear to be uniformly selected by a larger suite of pollinators (H. Sahli & J. Conner, unpublished data; see Figure 1), which suggests that the functional state of generalization may simultaneously reflect balancing and uniform selection.

Species that have floral features associated with specialized pollination often do have less phenotypic variation than species that have floral features associated with more generalized pollination (Fenster 1991b, Armbruster et al. 1999, Wolfe & Kristolic 1999). In addition, floral traits most closely associated with the fit between flowers and pollinators demonstrate the least phenotypic variation (Cresswell 1998). Such comparative data are consistent with selection acting on specialized floral systems, reducing phenotypic variation in the targets of selection, and, hence, reducing our inferential powers in contemporary phenotypic selection studies.

Unusual Phenotypic Variation

One can overcome the constraint that phenotypic-selection analysis is limited to natural levels of within-population variation by surgically modifying floral characters to mimic between-species variation and then quantifying the relationship between character expression and aspects of pollinator activity thought to affect plant fitness (i.e., effect on visitation rate or pollen transfer) by introducing variation through manipulation (artificial or artificial selection). Another approach is to take advantage of situations in which phenotypic variation is greatly increased (hybrid zones) or can be artificially increased (by breeding programs). Thus, by examining situations in which phenotypic variation is inflated, one may possibly recreate the variation that was traversed through many generations of past selection or at least quantify the selective pressures responsible for the maintenance of a trait.

Phenotypic manipulations have long been a part of studies of the interaction between floral traits and pollinators (e.g., Clements & Long 1923). The most relevant studies recreate phenotypic differences similar to those known to separate closely related species that have contrasting pollination systems (Castellanos et al. 2004). In this way, the effect of the manipulated character can be isolated from all the other ways in which the species differ. Studies of hummingbird-pollinated *Aquilegia formosa* and hawkmoth-pollinated *Aquilegia pubescens* demonstrate that moths favor upright, white flowers, and although spur length had no effect on visitation by moths, it had a large effect on pollen removal (Fulton & Hodges 1999, Hodges et al. 2004). For *Silene* spp., hummingbirds favor large red flowers that are displayed high off the ground, whereas nocturnal moth pollinators discriminate on the basis of height alone relative to alternative trait expression found in sister species (C. Fenster, R. Reynolds & M. Dudash, unpublished data). Other examples of experimental manipulation include (*a*) inflorescence height of the sexually deceptive orchid *Chiloglottis trilabra* (Peakall & Handel 1993), (*b*) the structure of the lower lip in *Monarda didyma* (Temeles & Rankin 2000), (*c*) the flexibility of pedicels in *Impatiens capensis* (Hurlbert et al. 1996), (*d*) pistil height in *Brassica napus* (Creswell 2000), (*e*) the addition of nectar to the nonrewarding orchid *Anacamptis morio* (Johnson et al. 2004), and (*f*) the degree of flower stalk bending in *Pulsatilla cernua* (Huang et al. 2002). These experiments demonstrate the optimality of the natural states, whereas natural phenotypic–selection experiments might have failed to detect selection. Excision of the staminode from two bee-pollinated and two bird-pollinated *Penstemon* species demonstrated variation in function. The staminode increased pollen transfer in the bee-pollinated species but appeared to be functionless and vestigial in the bird-pollinated species (Walker-Larsen & Harder 2001). A study of the manipulation of the fit of *Impatiens* flowers around the bodies of bees found little effect on pollen transfer (Wilson 1995). Other manipulative experiments have detected directional selection on traits such as nectar-spur length of Scandinavian *Platanthera* orchids by moths (Nilsson 1988) and suggest natural selection could drive populations away from the contemporary character states. However, we suspect that the directional selection measured is just one component of stabilizing selection, and conflicting components result from selection generated by pollen thieves, herbivores, allocation tradeoffs, and other factors (e.g., Armbruster 1996a). Longitudinal phenotypic selection studies that follow cohorts through time may help quantify such conflicting selective pressures (Gustaffson & Sutherland 1988, Campbell 1997).

Trait variation may also be inflated by genetic recombination, either through controlled crosses or in natural hybrid zones. Different types of pollinators select for distinctive floral features commonly associated with pollination syndromes in an F_2-segregating hybrid population that represents a cross between principally hummingbird-pollinated *Mimulus cardinalis* and bee-pollinated *Mimulus lewisii* (Schemske & Bradshaw 1999). In the F_2 generation, bees preferred large flowers that were low in anthocyanin pigments, whereas hummingbirds favored nectar-rich flowers that were high in anthocyanins, as would be predicted by the contemporary traits that distinguish the two species. Furthermore, when flower color from one species was bred into the background of the other species, it alone seemed to result in a difference in pollinators, even while lacking covariation with nectar offerings (Bradshaw & Schemske 2003). Similarly, Campbell et al. (1997) and Meléndez-Ackerman & Campbell (1998) demonstrated that hummingbirds produce directional selection that favor traits associated with the hummingbird-pollination syndrome in a hybrid zone between red-flowered *Ipomopsis aggregata* (hummingbird syndrome) and pale-flowered *I. tenuituba* (moth syndrome). Hummingbirds selected for wide corollas (relative to the moth-pollinated species), high nectar production, and red color. Moths selected for narrower flowers but demonstrated no color preference. Preferential visitation by hummingbirds and bees to red and blue flowers, respectively, was observed in Louisiana iris hybrid zones (Wesselingh & Arnold 2000). In summary, these data demonstrate that pollinator preferences can be the source of selection for divergence of floral traits.

Comparative Data on Pollination Specialization

Associating trait shifts with shifts in functional groups is a direct test of adaptive hypotheses on which traits are selected by functional groups of pollinators. These associations can be quantified at the between-population and the among-species levels. Pollination ecotypes, populations that have genetically differentiated for traits associated with pollination (Gregory 1963–1964; Grant & Grant 1965, 1968; Breedlove 1969; Whalen 1978; Raven 1979; Miller 1981; Armbruster & Webster 1982; Armbruster 1985; Paige & Whitham 1985; Pellmyr 1986; Galen 1989; Grant & Temeles 1992; Armbruster et al. 1994; Robertson & Wyatt 1990; Johnson & Steiner 1997; Hansen et al. 2000), and floral polymorphisms within populations (Medel et al. 2003) provide strong evidence that divergent selection by different functional groups of pollinators is responsible for contemporary patterns of floral diversity within species. The lability of traits within species allows one to infer associations without necessarily taking into account ancestral and derivative relationships. If a trait such as color has diverged because of selection by one functional group (e.g., hummingbirds), then retention of a trait, even if it is an ancestral condition, is evidence that another functional group (e.g., bees) is exerting selection to maintain that trait. At the among-species level, a phylogenetic approach may allow one to identify the direction of evolution and, thus, quantify the number of shifts to different functional groups. At this level, it is conservative to consider only derived traits and functional groups because, arguably, the retention of ancestral traits and functional pollination groups also reflects selection. Here, we take the conservative approach. Examination of traits within a framework of related ecotypes or species allows testing predictions from pollination syndromes in a specific context. Rather than stating that species pollinated by hummingbirds have exserted anthers and stigmas, we state that hummingbirds tend to pollinate species that have more exserted anthers and stigmas than do closely related species pollinated by other functional groups (Thomson et al. 2000). These systematic rules remove the confounding variation in floral traits introduced by evolutionary relatedness (i.e., phylogenetic coincidence; Armbruster et al. 2002).

In Table 1, we summarize the data on trait shifts organized by reward, morphology, color, and fragrance. The data are split into ecotypic and phylogenetic cases. Several important limitations became apparent in our literature review. Traits have not been studied uniformly; color and morphology have been investigated more often than reward and fragrance. Thus, we urge that future studies quantify the full array of traits that constitute pollination syndromes (e.g., Thomson et al. 2000).

TABLE 1 Proportions of shifts in pollinators involving four kinds of floral traits[a]

Shift to:		Reward	Morphology	Color	Fragrance
Bees	Ecotype	2/6 (0/2)	9/11 (2/2)	5/7 (0/2)	1/5 (0/2)
	Phylogeny	2/6 (11/11)	6/9 (10/11)	0/6 (0/11)	1/5 (3/11)
Lepidopterans: Nocturnal	Ecotype	3/4	3/4	4/4	1/2
	Phylogeny	2/4	2/5	4/4	5/5
Lepidopterans: Diurnal	Ecotype	0/1 (0/1)	0/1 (1/1)	0/1 (0/1)	1/1 (0/1)
	Phylogeny	1/2	3/3	2/3	0/2
Flies	Ecotype	0/2	3/3 (1/1)	2/3	1/2
	Phylogeny	1/1	6/7	9/9	1/9
Hummingbird	Ecotype	1/2	4/4	4/4	0/4
	Phylogeny	3/3 (0/2)	3/3 (2/2)	3/3 (0/2)	0/3 (0/2)
		(2/2)*	(4/4)*	(0/4)*	(0/4)*
Bird	Phylogeny		2/2	2/2	
Beetle	Phylogeny	0/1	0/1	1/1	0/1
Mammals	Phylogeny	0/2	2/2	0/2	2/2
Total		28/52	59/72	38/69	14/59

[a]Denominators are the number of cases tallied from the literature; numerators are the number of cases that underwent a marked character change. In most cases, the shifts that are described are from one functional group to another (e.g., fly to moth, bee to bird), and in the remaining, noted within parentheses, the shifts are within functional groups (e.g., bee to bee, butterfly to butterfly). Asterisk (*) indicates a shift from Passerine to hummingbird pollination.

The ecotypic data likely reflects a severe bias because some workers may not have recorded the full list of pollinators but rather only those expected on the basis of the floral traits manifest in the plants under study (see Waser et al. 1996 for discussion). Consequently, many within-species pollinator shifts deserve further study in terms of quantifying visitors by functional groups, their effectiveness, and the divergent selection pressures they exert. Despite numerous plant phylogenetic studies that mention pollinator shifts, strikingly few map on the phylogeny actual data on both floral features and pollinators. Despite the formidable challenge in quantifying floral traits and pollinator-visitation data in a phylogenetic context, we are able to collate studies on 12 groups (Kurzweil et al. 1991; Armbruster 1992, 1993, 1996a,b; McDade 1992; Crisp 1994; Goldblatt & Manning 1996; Bruneau 1997; Hapeman & Inoue 1997; Johnson et al. 1998; Baum et al. 1998; Steiner 1998; Dilley et al. 2000; Beardsley et al. 2003; Patterson & Givinsh 2004), in addition to the 14 pollination ecotype studies. We believe the approaches used in these studies should be models for future research.

Table 1 allows us to make provisional comments on the role of functional groups in exerting convergent selective pressures on floral traits associated with pollination syndromes. Shifts in all four floral traits are associated with functional groups and often in ways that are predicted by traditional pollination syndromes. When tallied by trait, reward evolution appears to be strongly associated with shifts to hummingbird and nocturnal moth pollination (whether in terms of nectar composition or timing of nectar production); color appears to be associated with moth (to pale color), fly (to dark colors), and bird and hummingbird (to red or bright) pollination; and fragrance appears to be associated with nocturnal moth and mammal pollination and with shifts involving euglossine bees. Morphology responds consistently in all functional groups. Changes in overall flower size correspond with changes in the size of the members of the functional group, and changes in the size of the structure bearing the reward (e.g., tube or spur) tend to correspond with the size of the animals' probing structures. Most of the reward evolution in the dataset is associated with bees and reflects evolutionary specialization by *Dalechampia* spp. (Armbruster 1993) onto sundry bee-reward systems. The approximately 45 observed resin-reward species are pollinated almost exclusively by resin-collecting bees; 5 fragrance-reward species are pollinated exclusively by male euglossine bees; and 10 pollen-reward species are all pollinated by pollen-collecting bees, beetles, and flies (Armbruster 1988, Armbruster 1993, Armbruster & Baldwin 1998). Comparison of phylogenetic with ecotypic studies may allow us to determine if some traits are more labile than others and if syndrome evolution reflects particular trait order, especially if both studies can be done within the same group.

Mapping floral characters and pollinators onto plant phylogenies shows that the relationships between flowers and their pollinators are prone to parallelism and reversal (in addition to the previous citations, see Manning & Linder 1992, Tanaka et al. 1997, Weller et al. 1998). Some evolutionary changes narrow the spectrum of pollinators. For example, Calochortus lilies are pollinated by both beetles and pollen-collecting bees and exhibit ancestral floral traits, whereas species that are pollinated by various nectaring bees manifest derived floral traits but in turn are ancestral to species pollinated by large bees (Dilley et al. 2000, Patterson & Givnish 2003). Other evolutionary changes broaden the spectrum of pollinators. For example, in the *Dalechampia* of Madagascar, a shift from pollination by resin-collecting bees to pollination by a variety of pollen-feeding insects occurs (Armbruster & Baldwin 1998). However, most of our examples reflect shifts from one functional group of pollinators to another, such as from bee to butterfly in *Disa* orchids (Johnson et al. 1998), from bee to beetle in *Ceratandra* orchids (Steiner 1998), from fly to bee and fly to moth in *Lapeirousia* (Goldblatt & Manning 1996), and from bee to hummingbird, 14 to 25 times, in *Penstemon* (Wilson et al. 2004). If most floral-trait transitions reflect shifts from one specialized functional group of pollinators to another, then we need studies that allow us to understand the adaptive significance of such shifts. Tantalizing evidence from two cases suggests that selection for increased female reproductive success may be responsible for shifts within functional groups. For example, in *Disa* orchids, selection for longer spurs by long-tongued flies occurs in populations adapted to pollination by relatively shorter-tongued flies (Johnson & Steiner 1997), and in *Platanthera* orchids, selection for increased stigmatic area results in increasing column width and transition from proboscis to eye pollination on moths (J. Maad & L. Nilsson, unpublished data). In both cases, the "ancestral" type, shorter spurs or proboscis pollination, suffers greater pollen limitation relative to the other types in specific environments. Whether we can extrapolate these results to shifts to more disparate functional groups requires future work.

Although Table 1 suggests that all traits have responded to selection by functional groups of pollinators, the various components of syndromes (e.g., floral morphology, color, fragrance, and reward chemistry) most likely do not contribute equally to explaining variation among those animals that visit and successfully pollinate flowers (see Waser & Price 1998, Ollerton & Watts 2000). For example, flower color is an important predictor at higher taxonomic levels (e.g., between bees, flies, beetles, moths, birds, and bats), although at finer taxonomic scales, it may not perform as well (e.g., between flowers that appeal to bees collecting fragrance, nectar, or pollen) (McCall & Primack 1992; Armbruster 1996a, 2002; Waser et al. 1996; Waser & Price 1998). Similarly, reward may be the most important component of floral-trait variation associated with shifts between low-level taxa of pollinators (e.g., between different functional groups of bees), but other floral traits may be more important at higher taxonomic scales (e.g., Simpson & Neff 1983, Armbruster 1984). Patterns are likely to vary among floras around the world and among different ecological contexts. Thus, use of reward chemistry, fragrance chemistry, flower color, morphology, and other floral traits as both main and interactive effects in analyses of variance to explain portions of the variance in pollinators among plant species may be productive (Fægri & van der Pijl 1966, Armbruster et al. 2000, Wilson et al. 2004).

A reward's chemical composition, amount, and accessibility may strongly limit the functional groups of pollinators attracted, which suggests that plants have diverged in response to selection generated by the varying nutritional preferences of pollinators (e.g., Baker & Baker 1983, 1990; Pyke & Waser 1981; Simpson & Neff 1983; Bruneau 1997). For example, pollen is the reward offered by plants with poricidally dehiscent anthers [e.g., *Chamaecrista* (Caesalpinaceae), many Melastomataceae, Solanaceae, and Ericaceae], but this reward is available almost exclusively to bees that can vibrate their flight muscles to buzz the flowers. Similarly, long nectar tubes limit the kinds of animals that can access nectar.

In the tropics and subtropics, several reward systems are more restrictive in the kinds of animals they attract, and each system represents many independent evolutionary events. (*a*) Oil rewards are collected by some anthophorid and melittid bees, which pollinate hundreds of tropical and subtropical plant species, as well as a few temperate species (Cane et al. 1983, Simpson & Neff 1983, Buchmann 1987). (*b*) Fragrance is collected by pollinating male euglossine bees (Dressler 1982, Schemske & Lande 1984, Chase & Hills 1992, Armbruster 1993, Whitten et al. 1998). (*c*) Plant floral resins are used in nest building by some species of bees and wasps; floral production of resin rewards has evolved three to four times among the several hundred species of resin-reward plants that grow in most lowland tropical habitats worldwide (Armbruster 1984, 1992, 1993). (*d*) Brood-rearing site is a very specialized, but important, reward system. The relationship between figs and their seed-feeding wasp pollinators involves over 900 plant species in nearly all lowland tropical forest habitats (see Wiebes 1979, Janzen 1979, Herre & West 1997, Weiblen 2002). Additional brood-site reward relationships are known in *Yucca* (e.g., Pellmyr et al. 1996), senita cacti (Flemming & Holland 1997, Holland & Flemming 1998), and *Chaemerops* palms (Dufay & Anstett 2002), and seed predators may contribute to pollination in many *Silene* species (Pettersson 1991; S. Kephart, unpublished data; R. Reynolds, C. Fenster & M. Dudash, unpublished data).

Our review of the literature reveals that functional groups differentially exert selection pressures that can account for the convergence of floral characters into pollination syndromes. Although the broad features of the evolution of pollination syndromes are confirmed, many important details that will clarify the mechanisms that underlie floral diversification remain understudied. We need greater quantification of the contribution of specific traits to pollinator discrimination, the relative lability of traits, the order in which traits evolve, and whether the order of trait evolution determines the trajectory of subsequent pollinator shifts. We address a number of these issues in the next section.

ADAPTIVE CHARACTER COMPLEXES

In perhaps the first treatment of adaptive character complexes ("syndromes"), Simpson (1944) extended Wright's (1931) notion of adaptive topography (fitness peaks and valleys) associated with different gene combinations to an adaptive topography associated with different combinations of phenotypic traits. In discussing floral evolution, Stebbins (1950, p. 502) stressed such a role for combinations of traits: "The flower is...a harmonious unit," he wrote. "An alteration of one of its parts will immediately change the selective value of modifications in all the others." Thus, patterns of character correlation may reflect highly nonadditive interactions among traits in terms of their effect on pollination success (e.g., Fenster et al. 1997). For instance, the narrow floral tubes of hummingbird-pollinated flowers may be adaptive only when anthers and stigma are strongly exserted or anther and stigma exsertion may only be adaptive when nectar is copious. (Castellanos et al. 2004). Thus, the evolution of a whole syndrome may tend to follow the origin of certain key innovations. The evolution of nectar spurs in *Aquilegia* may have had to precede the evolution of pollination by long-tongued specialists (Hodges & Arnold 1994, 1995). The evolution of selective fruit abortion may be a key feature that predates the evolution of the mutualism between yuccas and yucca moths (Pellmyr et al. 1996, Pellmyr 1997). The evolution in *Dalechampia* of resin secretion associated with antiherbivore defense was probably a key feature that predated the evolution of pollination by resin-collecting bees (Armbruster 1997, Armbruster et al. 1997). However, certain syndromes may tend to impede subsequent pollinator shifts, retard (or in some cases promote) subsequent speciation, and, hence, result in evolutionary dead ends that are concentrated at the tips of phylogenic branches (Wilson et al. 2004). Clearly, a full understanding of the processes that underlie the evolution of pollination syndromes requires knowledge of whether the traits that constitute a syndrome can confer higher fitness (relative to the ancestral condition) independently of each other or whether the adaptive advantage depends on joint variation in floral features. Moreover, considering trait combinations may further improve predictions of functional groups by pollination syndromes.

Few studies address the interactive value of floral traits. By comparing zygomorphic species that were presumed to have more-specialized pollination and less-specialized actinomorphic species, Berg (1959, 1960) found that zygomorphic taxa had more phenotypic integration of floral traits and less correlation between floral and vegetative traits. Her results suggest that levels of covariation among floral traits respond to selection imposed by pollinators and that a selective advantage is associated with floral traits being intercorrelated with but decoupled from variation in vegetative traits (but see Herrera 1996, Armbruster et al. 1999). Attempts to describe adaptive landscapes in floral evolution by measuring patterns of interspecific and intraspecific variation have demonstrated that certain combinations of traits, such as the amount of reward, flower size, and placement of the primary sexual organs relative to the reward, conferred higher fitness than alternative combinations (Armbruster 1990, Cresswell & Galen 1991). These studies also suggested that the adaptive surface of floral traits was likely to be influenced by such factors as energetic constraints on both plants and pollinators and the physical environmental (Galen 1999, Galen & Cuba 2001). The ability of hummingbirds to feed at flowers of different length critically depends on the width of the flower (Temeles 1996, Temeles et al. 2002), and this dependency demonstrates that the joint consideration of traits can enhance our understanding of the precise relationship between plant and pollinator. Thus, we must quantify not only selection directly on floral traits but also how the evolution of floral traits interacts with the evolution of other aspects of plant morphology and life history. Furthermore, quantifying selection on trait combinations may reveal adaptive mechanisms, whereas doing so for any one trait, averaged across different trait combinations, may not (e.g., Armbruster 1990).

Phenotypic selection studies that quantify correlational selection may demonstrate selection for particular character combinations. Few multitrait studies have been conducted on natural variation in syndrome characters, and they have generally been unable to detect interpretable patterns of selection acting on trait combinations (O'Connell & Johnston 1998, Maad 2000, but see Herrera 2001). If such characters are under strong stabilizing selection, insufficient variation may exist within populations to allow detection of selection (Fenster 1991b).

Phenotypic manipulations that vary traits, both singly and together, such that they differ from the norm of the hypothesized pollination syndrome may reveal the interactive effects among traits that constitute the syndrome, thereby testing the hypothesis that pollination is maximized by certain trait combinations (Herrera 2001). Such experiments are best done by utilizing contrasting types of pollinators (moths versus bees) to quantify the role of specific visitors in the evolution of trait combinations (Castellanos et al. 2004). Phenotypic manipulation studies may allow testing of hypotheses on the order of character evolution, if they incorporate the changes in two or more characters and the interactive effects of such changes. Alternatively, one might be able to use a genetic approach: introgress traits one or more at a time across taxa pollinated by different functional groups (Bradshaw & Schemske 2003). Therefore, we may quantify how changes in one floral trait affect pollination in the context of other changes, and, thus, allow the reconstruction of the sequence of innovations.

Very few studies quantify trait interaction effects on pollinator behavior. Complementary effects of both color and floral morphology on nectar offerings have been demonstrated in *Ipomopsis* (Meléndez-Ackerman et al. 1997), although whether these effects are additive or interactive is not clear, and the studies were all done with hummingbirds, without parallel data on hawkmoths. Quantifying the interactive effects of traits may tell us more about the maintenance than the origin of syndromes (see Herrera 2001). To marshal evidence concerning the origin of syndromes, one would want to complement such experimental studies with models of the evolutionary process and with tests that utilize comparative data.

HISTORICAL STARTING POINTS

We anticipate that useful inferences about floral-trait combinations exhibited by pollination syndromes can be made by considering constraints as dictating the particular trajectory of trait evolution. Thus, the contribution of constraints to the lack of universal correspondence of floral traits to particular pollination syndromes may provide greater understanding of the observed patterns of floral variation. Floral evolution bears a strong stamp of what has been called "historicity" (Williams 1992) or "historical contingency" (Gould 1986, Futuyma 1998).

Historicity is reminiscent of Stebbins' (1974) "evolution along lines of least resistance" (Schluter 1996). History has dictated the evolutionary ability of plants to converge on pollination syndromes from a variety of starting points. The less-than-perfect correspondence of flowers into their syndromes reflects this historical effect and provides evidence of the course of floral evolution. Natural selection operates on preexisting phenotypic variation, gradually changing one form to another by making use of the structures "at hand." The details of that preexisting variation can both constrain adaptive evolution and stimulate evolutionary novelty, often in ways that are not easily predicted. For example, *Silene virginica* has presumably evolved its current floral morphology in response to selection by ruby-throat hummingbirds (Fenster & Dudash 2001), but unlike most hummingbird flowers, *S. virginica* does not technically have a tubular corolla. Instead it is polypetalous like other Caryophyllaceae, but it has a functional tube formed by the petals being enclosed by an elongated tubular calyx. Flowers usually do not contravene their lineage-specific (family-specific) traits in response to selection toward an "ideal" combination of characters, optimal or not. At the same time, historical effects can create interesting diversity, which again disrupts the conceptual unity of syndromes. In *Dalechampia* (Euphorbiaceae), most species are bee pollinated and have showy petaloid bracts. In lineages that do not deploy anthocyanins in their foliage, the floral bracts are white or pale green. In lineages that do deploy anthocyanins in their foliage, floral bracts are pink or purple, even though the same pollinator species are involved (Armbruster 1996a, 2002). Because pigmented bracts and stems appear several times simultaneously on the phylogeny, indirect selection (selection on another, genetically correlated trait, in this case, vegetative pigments) appears to have increased the diversity of bract colors (Armbruster 2002). In general, historicity makes the relationship between floral traits and pollinators more complicated than one would anticipate from a naive acceptance of pollination syndromes. Clearly, a more complete understanding of the relevance of pollination specialization requires studies that examine the interaction between history, constraints, and selective response. Assessment of systematic rules in a phylogenetic context is one effective approach to this challenge (Thomson et al. 2000).

CONCLUSIONS

We demonstrate that evolutionary specialization explains much of the striking diversity of flowers. Studies of floral specialization must continue to move from lists of pollinator species to descriptions of functional groups of pollinators and the selective pressures they exert on floral traits. Specialized floral adaptations and syndromes are often generated and maintained by selection created by functional groups of similar pollinators, whether taxonomically related or not, and only rarely by single-pollinator species. Given regional variation in the composition of pollinator communities and the role of historical contingency, it is remarkable that such dynamic complexity often converges on the traditional pollination syndromes. We advocate the continued study of both patterns of character correlation as they have arisen when plant lineages have shifted between pollinators and the processes that underlie this floral diversification in the angiosperms.

ACKNOWLEDGMENTS

We dedicate this paper to the memories of H.G. Baker, K. Fægri, and G.L. Stebbins, whose contributions to understanding floral evolution and mentoring the next generation of pollination biologists have stimulated much of the research reviewed here. We thank J. Conner, P. Goldblatt, D. Inouye, S. Kephart, A. Nilsson, J. Ollerton, M. Price, H. Sahli, D. Schemske, Ø. Totland, N. Waser, L. Wolfe, and an anonymous reviewer for sharing unpublished data and ideas and providing constructive criticisms of previous versions of this manuscript. Funding was provided by the National Science and Engineering Council of Canada, by the Norwegian Research Council, and by the National Science Foundation (USA).

The *Annual Review of Ecology, Evolution, and Systematics* is online at
http://ecolsys.annualreviews.org

LITERATURE CITED

Aigner PA. 2001. Optimality modeling and fitness trade-offs: When should plants become pollinator specialists? *Oikos* 95:177–84

Aigner PA. 2004. The evolution of specialized floral phenotypes in a fine-grained pollination environment. In *Specialization and Generalization in Plant-Pollinator Interactions*, ed. NM Waser, J Ollerton. Chicago: Univ. Chicago Press. In press

Armbruster WS. 1984. The role of resin in angiosperm pollination: ecological and chemical considerations. *Am. J. Bot.* 71:1149–60

Armbruster WS. 1985. Patterns of character divergence and the evolution of reproductive ecotypes of *Dalechampia scandens* (Euphorbiaceae). *Evolution* 39:733–52

Armbruster WS. 1986. Reproductive interactions between sympatric *Dalechampia* species: Are natural assemblages "random" or organized? *Ecology* 67:522–33

Armbruster WS. 1988. Multilevel comparative analysis of the morphology, function, and evolution of *Dalechampia* blossoms. *Ecology* 69:1746–61

Armbruster WS. 1990. Estimating and testing the shapes of adaptive surfaces: the morphology and pollination of *Dalechampia* blossoms. *Am. Nat.* 135:14–31

Armbruster WS. 1992. Phylogeny and the evolution of plant-animal interactions. *BioScience* 42:12–20

Armbruster WS. 1993. Evolution of plant pollination systems: hypotheses and tests with the neotropical vine *Dalechampia*. *Evolution* 47:1480–505

Armbruster WS. 1996a. Evolution of floral morphology and function: an integrative approach to adaptation, constraint, and compromise in *Dalechampia* (Euphorbiaceae). In *Floral Biology: Studies on Floral Evolution in Animal-Pollinated Plants*, ed. DG

Lloyd, SCH Barrett, pp. 241–72. New York: Chapman & Hall

Armbruster WS. 1996b. Exaptation, adaptation, and homoplasy: evolution of ecological traits in *Dalechampia*. In *Homoplasy: The Recurrence of Similarity in Evolution*, ed. MJ Sanderson, L Hufford, pp. 227–43. New York: Academic

Armbruster WS. 1997. Exaptations link the evolution of plant-herbivore and plant-pollinator interactions: a phylogenetic inquiry. *Ecology* 78:1661–74

Armbruster WS. 2002. Can indirect selection and genetic context contribute to trait diversification? A transition-probability study of blossom-color evolution in two genera. *J. Evol. Biol.* 15:468–86

Armbruster WS, Baldwin BG. 1998. Switch from specialized to generalized pollination. *Nature* 394:632

Armbruster WS, Di Stilio VS, Tuxill JD, Flores TC, Velasquez Runk JL. 1999. Covariance and decoupling of floral and vegetative traits in nine neotropical plants: a re-evaluation of Berg's correlation-pleiades concept. *Am. J. Bot.* 86:39–55

Armbruster WS, Edwards ME, Debevec EM. 1994. Character displacement generates assemblage structure of Western Australian triggerplants (*Stylidium*). *Ecology* 75:315–29

Armbruster WS, Fenster CB, Dudash MR. 2000. Pollination "principles" revisited: specialization, pollination syndromes, and the evolution of flowers. *Det Nor. Vidensk. Acad. I. Mat. Natur. Kl. Skr. Ny Ser.* 39:179–200

Armbruster WS, Herzig AL. 1984. Partitioning and sharing of pollinators by four sympatric species of *Dalechampia* (Euphorbiaceae) in Panama. *Ann. Mo. Bot. Gard.* 71:1–16

Armbruster WS, Howard JJ, Clausen TP, Debevec EM, Loquvam JC, et al. 1997. Do biochemical exaptations link evolution of plant defense and pollination systems? Historical hypotheses and experimental tests with *Dalechampia* vines. *Am. Nat.* 149:461–84

Armbruster WS, Keller CS, Matsuki M, Clausen TP. 1989. Pollination of *Dalechampia magnoliifolia* (Euphorbiaceae) by male euglossine bees (Apidae: Euglossini). *Am. J. Bot.* 76:1279–85

Armbruster WS, Mulder CPH, Baldwin BG, Kalisz S, Wessa B, Nute H. 2002. Comparative analysis of late floral development and mating-system evolution in tribe Collinsieae (Scrophulariaceae, s.l.). *Am. J. Bot.* 89:37–49

Armbruster WS, Webster GL. 1982. Divergent pollination systems in sympatric species of South American *Dalechampia* (Euphorbiaceae). *Am. Midl. Nat.* 108:325–37

Baker HG. 1963. Evolutionary mechanisms in pollination biology. *Science* 139:877–83

Baker HG, Baker I. 1983. Floral nectar sugar constituents in relation to pollinator type. In *Handbook of Experimental Pollination Biology*, ed. CE Jones, RJ Little, pp. 117–41. New York: Van Nostrand Reinhold

Baker HG, Baker I. 1990. The predictive value of nectar chemistry to the recognition of pollinator types. *Isr. J. Bot.* 39:157–66

Baum DA, Small RL, Wendel JF. 1998. Biogeography and floral evolution of baobabs (*Adansonia*, Bombacaceae) as inferred from multiple data sets. *Syst. Biol.* 47:181–207

Beardsley PM, Yen A, Olmstead RG. 2003. AFLP phylogeny of *Mimulus* section *Erythranthe* and the evolution of hummingbird pollination. *Evolution* 57:1397–410

Beattie AJ. 1971. Pollination mechanisms in *Viola*. *New Phytol.* 70:343–60

Berg RL. 1959. A general evolutionary principle underlying the origin of developmental homeostasis. *Am. Nat.* 93:103–5

Berg RL. 1960. The ecological significance of correlation pleiades. *Evolution* 14:171–80

Bingham RA, Orthner AR. 1998. Efficient pollination of alpine plants. *Nature* 391:238–39

Bradshaw HD, Schemske DW. 2003. Allele substitution at a flower colour locus produces a pollinator shift in monkeyflowers. *Nature* 426:176–78

Breedlove DE. 1969. *The Systematics of Fuchsia Section* Encliandra (*Onagraceae*). Berkeley: Univ. Calif. Press. 69 pp.

Bruneau A. 1997. Evolution and homology

of bird pollination syndromes in *Erythrina* (Leguminosae). *Am. J. Bot.* 84:54–71

Buchmann SL. 1987. The ecology of oil flowers and their bees. *Annu. Rev. Ecol. Syst.* 18:343–70

Campbell DR. 1985. Pollen and gene dispersal: the influence of competition for pollination. *Evolution* 39:418–31

Campbell DR. 1989. Measurements of selection in a hermaphroditic plant: variation in male and female pollination success. *Evolution* 43:318–34

Campbell DR. 1996. Mechanisms of hummingbird-mediated selection for flower width in *Ipomopsis aggregata*. *Ecology* 77:1462–72

Campbell DR. 1997. Genetic and environmental variation in life-history traits of a monocarpic perennial: a decade-long field experiment. *Evolution* 51:373–82

Campbell DR, Waser NM, Meléndez-Ackerman EJ. 1997. Analyzing pollinator-mediated selection in a plant hybrid zone: hummingbird visitation patterns on three spatial scales. *Am. Nat.* 149:295–315

Campbell DR, Waser NM, Price MV, Lynch EA, Mitchell RJ. 1991. Components of phenotypic selection: pollen export and flower corolla width in *Ipomopsis aggregata*. *Evolution* 45:1458–67

Cane JH, Eickwort GC, Wesley FR, Spielholz J. 1983. Foraging, grooming and mate-seeking behaviors of *Macropsis nuda* (Hymenoptera, Melittidae) and use of *Lysimachia ciliate* (Primulaceae) oils in larval provisions and cell linings. *Am. Midl. Nat.* 110:257–64

Castellanos MC, Wilson P, Thomson JD. 2003. Pollen transfer by hummingbirds and bumblebees, and the divergence of pollination modes in *Penstemon*. *Evolution* 57:2742–52

Castellanos MC, Wilson P, Thomson JD. 2004. 'Anti-bee' and 'pro-bird' changes during the evolution of hummingbird pollination in *Penstemon* flowers. *J. Evol. Biol.* In press

Chase MW, Hills HG. 1992. Orchid phylogeny, flower sexuality, and fragrance-seeking bees. *Bioscience* 42:43–49

Clements FE, Long FL. 1923. *Experimental Pollination: An Outline of the Ecology of Flowers and Insects*. Publ. 336. Washington, DC: Carnegie Inst. 274 pp.

Conner JK, Davis R, Rush S. 1995. The effect of wild radish morphology on pollination efficiency by 4 taxa of pollinators. *Oecologia* 104:234–45

Cresswell JE. 1998. Stabilizing selection and the structural variability of flowers within species. *Ann. Bot.* 81:463–73

Cresswell JE. 2000. Manipulation of female architecture in flowers reveals a narrow optimum for pollen deposition. *Ecology* 81:3244–49

Cresswell JE, Galen C. 1991. Frequency-dependent selection and adaptive surfaces for floral trait combinations: the pollination of *Polemonium viscosum*. *Am. Nat.* 138:1342–53

Crisp MD. 1994. Evolution of bird pollination in some Australian legumes (Fabacae). In *Phylogenetics and Ecology*, ed. P Eggelton, RI Vane-Wright, pp. 281–309. London: Academic

Darwin C. 1862. *On the Various Contrivances by Which British and Foreign Orchids Are Fertilized*. London: Murray. 365 pp.

Darwin C. 1877. *On the Various Contrivances by Which British and Foreign Orchids Are Fertilized*. New York: D. Appleton. 300 pp. 2nd ed.

Delpino F. 1868–1875. Ulteriori osservazione sulla dicogamia nel regno vegetale. *Atti della Societa Italiana di Scienze Naturali Milano*, Vols. 1 and 2

Dicks LV, Corbet SA, Pywell RF. 2002. Compartmentalization in plant-insect flower visitor webs. *J. Anim. Ecol.* 71:32–43

Dilley JD, Wilson P, Mesler MR. 2000. The radiation of Calochortus: generalist flowers moving through a mosaic of potential pollinators. *Oikos* 89:209–22

Dodson CH, Dressler RL, Hills HG, Adams RM, Williams NH. 1969. Biologically active compounds in orchid fragrances. *Science* 164:1243–49

Dressler RL. 1968. Pollination by male euglossine bees. *Evolution* 22:202–10

Dressler RL. 1982. Biology of orchid bees (Euglossini). *Annu. Rev. Ecol. Syst.* 13:373–94

Fægri K, van der Pijl L. 1966. *The Principles of Pollination Ecology*. Oxford: Pergamon. 248 pp.

Fægri K, van der Pijl L. 1971. *The Principles of Pollination Ecology*, Oxford: Pergamon. 298 pp. 2nd ed.

Fægri K, van der Pijl L. 1979. *The Principles of Pollination Ecology*, Oxford: Pergamon. 244 pp. 3rd ed.

Fenster CB. 1991a. Gene flow in *Chamaecrista fasciculata* (Leguminosae). I. Gene dispersal *Evolution* 45:398–409

Fenster CB. 1991b. Selection on floral morphology by hummingbirds. *Biotropica* 23:98–101

Fenster CB, Dudash MR. 2001. Spatiotemporal variation in the role of hummingbirds as pollinators of *Silene virginica* (Caryophyllaceae). *Ecology* 82:844–51

Fenster CB, Galloway LF, Chao L. 1997. Epistasis and its consequences for the evolution of natural populations. *Trends Ecol. Evol.* 12:282–86

Fenster CB, Ritland K. 1994. Evidence for natural selection on mating system in *Mimulus* (Scrophulariaceae). *Int. J. Plant Sci.* 155:588–96

Fishbein M, Venable DL. 1996. Diversity and temporal change in the effective pollinators of *Asclepias tuberosa*. *Ecology* 77:1061–73

Fleming TH, Holland JN. 1998. The evolution of obligate pollination mutualisms: senita cactus and senita moth. *Oecologia* 114:368–75

Fulton M, Hodges SA. 1999. Floral isolation between *Aquilegia Formosa* and *Aquilegia pubescens*. *Proc. R. Soc. London Ser. B Biol. Sci.* 266:2247–52

Futuyma DJ. 1998. *Evolutionary Biology*. Sunderland, MA: Sinauer. 810 pp. 3rd ed.

Galen C. 1989. Measuring pollinator-mediated selection on morphometric floral traits: bumble bees and the alpine skypilot, *Polemonium viscosum*. *Evolution* 43:882–90

Galen C. 1999. Why do flowers vary? The functional ecology of variation in flower size and form within natural plant populations. *Bioscience* 49:631–40

Galen C, Cuba J. 2001. Down the tube: pollinators, predators, and the evolution of flower shape in the alpine skypilot, *Polemonium viscosum*. *Evolution* 55:1963–71

Goldblatt P, Manning JC. 1996. Phylogeny and speciation in *Lapeirousia* subgenus *Lapeirousia* (Iridaceae: Ixioideae). *Ann. Mo. Bot. Gard.* 83:346–61

Goldblatt P, Manning JC, Bernhardt P. 2001. Radiation of pollination systems in *Gladiolus* (Iridaceae: Crocoideae) in southern Africa. *Ann. Mo. Bot. Gard.* 88:713–34

Gomez JM. 2002. Generalizations in the interactions between plants and pollinators. *Rev. Chil. Hist. Nat.* 75:105–16

Gomez JM, Zamora R. 1999. Generalization vs. specialization in the pollination system of *Hormathophylla spinosa* (Cruciferae). *Ecology* 80:796–805

Gould SJ. 1986. Evolution and the triumph of homology, or why history matters. *Am. Sci.* 74:60–69

Grant KA, Grant V. 1968. *Hummingbirds and Their Flowers*. New York: Columbia Univ. Press. 115 pp.

Grant V, Grant KA. 1965. *Flower Pollination in the Phlox Family*. New York: Columbia Univ. Press. 180 pp.

Grant V, Temeles EJ. 1992. Foraging ability of rufous hummingbirds on hummingbird flowers and hawkmoth flowers. *Proc. Natl. Acad. Sci. USA* 89:9400–4

Gregory DP. 1963–1964. Hawkmoth pollination in the genus *Oenothera*. *Alisio* 5:357–419

Gustaffson L, Sutherland WJ. 1988. The costs of reproduction in the collared flycatchers *Ficedula albicollis*. *Nature* 335:813–15

Hagerup O. 1951. Pollination in the faroes—in spite of rain and poverty of insects. *Den Konglige Dan. Vidensk. Selskr. Biol. Medd.* 18:1–48

Hansen T, Armbruster WS, Antonsen L. 2000. Comparative analysis of character displacement and spatial adaptations as illustrated by the evolution of *Dalechampia* blossoms. *Am. Nat.* 156(Suppl):S17–34

Hapeman JR, Inoue K. 1997. Plant-pollinator interaction and floral radiation in *Platanthera* (Orchidaceae). In *Molecular Evolution and Adaptive Radiation*, ed. TJ Givnish, KJ Sytsma, pp. 433–54. Cambridge, UK: Cambridge Univ. Press

Herre EA, West SA. 1997. Conflict of interest in a mutualism: documenting the elusive fig wasp seed trade-off. *Proc. R. Soc. London Ser. B Biol. Sci.* 264:1501–7

Herrera CM. 1987. Components of pollination "quality": comparative analysis of a diverse insect assemblage. *Oikos* 50:79–90

Herrera CM. 1996. Floral traits and plant adaptation to insect pollinators: a devil's advocate approach. In *Floral Biology: Studies on Floral Evolution in Animal-Pollinated Plants*, ed. DG Lloyd, SCH Barrett, pp. 65–87. New York: Chapman & Hall

Herrera CM. 2001. Deconstructing a floral phenotype: Do pollinators select for corolla integration in *Lavandula latifolia*? *J. Evol. Biol.* 14:574–84

Hodges SA, Arnold ML. 1994. Columbines: a geographically widespread species flock. *Proc. Natl. Acad. Sci. USA* 91:5129–32

Hodges SA, Arnold ML. 1995. Spurring plant diversification: Are floral nectar spurs a key innovation? *Proc. R. Soc. London Ser. B* 262:343–48

Hodges SA, Fulton M, Yang JY, Whittall JB. 2004. Verne Grant and evolutionary studies of *Aquilegia*. *New Phytol.* 161:113–20

Horvitz CC, Schemske DW. 1990. Spatiotemporal variation in insect mutualists of a neotropical herb. *Ecology* 71:1085–97

Huang S-Q, Takahashi Y, Dafni A. 2002. Why does the flower stalk of *Pulsatilla cernua* bend during anthesis? *Am. J. Bot.* 89:1599–603

Hurlbert AH, Hosoi SA, Temeles EJ, Ewald PW. 1996. Mobility of *Impatiens capensis* flowers: effect on pollen deposition and hummingbird foraging. *Oecologia* 105:243–46

Inouye DW, Gill DE, Dudash MR, Fenster CB. 1994. A model and lexicon for pollen fate. *Am. J. Bot.* 81:1517–30

Ivey CT, Martinez P, Wyatt R. 2003. Variation in pollinator effectiveness in swamp milkweed, *Asclepia incarnata* (Apocynaceae). *Am. J. Bot.* 90:214–25

Janzen DH. 1979. How to be a fig. *Annu. Rev. Ecol. Syst.* 10:13–51

Janzen DH. 1980. When is it coevolution? *Evolution* 34:611–12

Johnson SD, Linder HP, Steiner KE. 1998. Phylogeny and radiation of pollination systems in *Disa* (Orchidaceae). *Am. J. Bot.* 85:402–11

Johnson SD, Peter CI, Agren J. 2004. The effects of nectar addition on increased pollen removal and geitonogamy in the non-rewarding orchid *Anacamptis morio*. *Proc. Royal Soc. of London Ser. B Biol. Sci.* 271:803–9

Johnson SD, Steiner KE. 1997. Long-tongued fly pollination and evolution of floral spur length in the *Disa draconis* complex (Orchidaceae). *Evolution* 51:45–53

Johnson SD, Steiner KE. 2000. Generalization vs. specialization in plant pollination systems. *Trends Ecol. Evol.* 15:140–43

Johnston MO. 1991. Natural selection on floral traits in two species of *Lobelia* with different pollinators. *Evolution* 45:1468–79

Kay KM, Schemske DW. 2003. Pollinator assemblages and visitation rates for 11 species of Neotropical *Costus* (Costaceae). *Biotropica* 35:198–207

Kephart S. 1983. The partitioning of pollinators among three species of *Asclepias*. *Ecology* 64:120–32

Kephart S, Theiss K. 2004. Pollinator-mediated isolation in sympatric milkweeds (Asclepias): Do floral morphology and insect behavior influence species boundaries? *New Phytol.* 161:263–77

Kiester AR, Lande R, Schemske DW. 1984. Models of coevolution and speciation in plants and their pollinators. *Am. Nat.* 124:220–43

Kingsolver JG, Hoekstra HE, Hoekstra JM, Berrigan D, Vigneri SN, et al. 2001. The strength of phenotypic selection in natural populations. *Am. Nat.* 157:245–61

Knuth P. 1906. *Handbook of Flower Pollination.* Vol. I. Transl. JR Ainsworth Davis. Oxford: Clarendon. 382 pp.

Knuth P. 1908. *Handbook of Flower Pollination.* Vol. II. Transl. JR Ainsworth Davis. Oxford: Clarendon. 705 pp.

Kölreuter JG. 1761. *Vorläufige Nachrichten von einigen das Geschlecht der Pflanzen betreffenden Versuchen und Beobachtungen.* Leipzig: Gleditschischen Handlung

Kurzweil H, Linder HP, Chesselet P. 1991. The phylogeny and evolution of the *Pterygodium-Corycium* complex (*Coryciinae*, Orchidaceae). *Plant Syst. Evol.* 175:161–223

Lande R, Arnold SJ. 1983. The measurement of selection on correlated characters. *Evolution* 37:1210–26

Levin DA, Berube DE. 1972. *Phlox* and *Colias*: the efficiency of a pollination system. *Evolution* 26:242–50

Maad J. 2000. Phenotypic selection in hawkmoth pollinated *Platanthera bifolia*: targets and fitness surfaces. *Evolution* 54:112–23

Mayfield MM, Waser NM, Price MV. 2001. Exploring the 'most effective pollinator principle' with complex flowers: bumblebees and *Ipomopsis aggregata*. *Ann. Bot.* 88:591–96

McDade LA. 1992. Pollinator relationships, biogeography, and phylogenetics. *BioScience* 42:21–26

McGall C, Primack RB. 1992. Influence of flower characteristics, weather, time of day, and season on insect visitation rates in three plant communities. *Am. J. Bot.* 79:434–42

McGuire AD, Armbruster WS. 1991. An experimental test for the reproductive interactions between two sequentially blooming *Saxifraga* species. *Am. J. Bot.* 78:214–19

Manning JC, Linder HP. 1992. Pollinators and evolution in *Disperis* (Orchidaceae), or why are there so many species? *S. Afr. J. Sci.* 88:38–49

Medel R, Botto-Mahan C, Kalin-Arroyo M. 2003. Pollinator-mediated selection on the nectar guide phenotype in the Andean monkey flower, *Mimulus luteus*. *Ecology* 84:1721–32

Meléndez-Ackerman E, Campbell DR. 1998. Adaptive significance of flower color and inter-trait correlations in an *Ipomopsis* hybrid zone. *Evolution* 52:1293–303

Meléndez-Ackerman E, Campbell DR, Waser NM. 1997. Hummingbird behavior and mechanisms of selection on flower color in *Ipomopsis*. *Ecology* 78:2532–41

Mesler MR, Ackerman JD, Lu KL. 1980. The effectiveness of fungus gnats as pollinators. *Am J. Bot.* 67:564–67

Miller RB. 1981. Hawkmoths and the geographic patterns of floral variation in *Aquilegia caerulea*. *Evolution* 35:763–74

Momose K, Yumoto T, Nagamitsu T, Kato M, Nagamasu H, et al. 1998. Pollination biology in a lowland dipterocarp forest in Sarawak, Malaysia. I. Characteristics of the plant-pollinator community in a lowland dipterocarp forest. *Am. J. Bot.* 85:1477–501

Motten AF, Campbell DR, Alexander DE, Miller HL. 1981. Pollination effectiveness of specialist and generalist visitors to a North Carolina population of *Claytonia virginica*. *Ecology* 62:1278–87

Muchhala N. 2003. Exploring the boundary between pollination syndromes: bats and hummingbirds as pollinators of *Burmeistera cyclostigmata* and *B. Tenuiflora* (Campanulaceae). *Oecologia* 134:373–80

Müller H. 1883. *The Fertilization of Flowers.* Transl. D'Arcy W. Thompson. London: Macmillan. 669 pp.

Müller H, Delpino F. 1869. Application of the Darwinian theory to flowers and the insects which visit them. Transl. RL Packard. 1871, in *Am. Nat.* 5:271–97

Nilsson LA. 1983. Mimesis of bellflower (*Campanula*) by the red helleborine orchid *Cephalanthera rubra*. *Nature* 305:799–800

Nilsson LA. 1987. Angraecoid orchids and hawkmoths in central Madagascar: specialized pollination systems and generalist foragers. *Biotropica* 19:310–18

Nilsson LA. 1988. The evolution of flowers with deep corolla tubes. *Nature* 334:147–49

O'Connell LM, Johnston MO. 1998. Male and female pollination success in a deceptive orchid, a selection study. *Ecology* 79:1246–60

Ollerton J. 1996. Reconciling ecological processes with phylogenetic patterns: the apparent paradox of plant-pollinator systems. *J. Ecol.* 84:767–69

Ollerton J. 1998. Sunbird surprise for syndromes. *Nature* 394:726–27

Ollerton J, Watts S. 2000. Phenotype space and floral typology—towards an objective assessment of pollination syndromes. *Det. Nor. Vidensk. Acad. I. Mat. Natur. Kl. Skr. Ny Ser.* 39:149–59

Olsen KM. 1997. Pollination effectiveness and pollinator importance in a population of *Heterotheca subaxillaris* (Asteraceae). *Oecologia* 109:114–21

Ornduff R. 1975. Complementary roles of halictids and syrphids in the pollination of *Jepsonia heterandra* (Saxifragaceae). *Evolution* 29:371–73

Orr HA. 2000. Adaptation and the cost of complexity. *Evolution* 54:13–20

Paige KN, Whitham TG. 1985. Individual and population shifts in flower color by scarlet gilia—a mechanism for pollinator tracking. *Science* 227:315–17

Parker FD. 1981. How efficient are bees in pollinating sunflowers? *J. Kans. Entomol. Soc.* 54:61–67

Parrish JAD, Bazzaz F. 1979. Differences in pollination niche relationships in early- and late- successional plant communities. *Ecology* 60:597–610

Patterson TB, Givinish TJ. 2004. Geographic cohesion, chromosomal evolution, parallel adaptive radiations, and consequent floral adaptations in *Calochortus* (Calochortaceae): evidence from a cpDNA phylogeny. *New Phytol.* 161:253–64

Peakall R, Handel SN. 1993. Pollinators discriminate among floral heights of a sexually deceptive orchid: implications for selection. *Evolution* 47:1681–87

Pellmyr O. 1986. Three pollination morphs in *Cimicifuga simplex*: incipient speciation due to inferiority in competition. *Oecologia* 68:304–7

Pellmyr O. 1997. Pollinating seed eaters: Why is active pollination so rare? *Ecology* 78:1655–60

Pellmyr O, Thompson JN, Brown JM, Harrison RG. 1996. Evolution of pollination and mutualism in the yucca moth lineage. *Am. Nat.* 148:827–47

Pettersson MW. 1991. Pollination by a guild of fluctuating moth populations: options for unspecialization in *Silene vulgaris*. *J. Ecol.* 79:591–604

Pleasants JM. 1980. Competition for bumblebee pollinators in Rocky Mountain plant communities. *Ecology* 61:1446–59

Pleasants JM. 1990. Null-model tests for competitive displacement: the fallacy of not focusing on the whole community. *Ecology* 71:1078–84

Primack RB, Silander JA. 1975. Measuring the relative importance of different pollinators to plants. *Nature* 255:143–44

Pyke GH, Waser NM. 1981. The production of dilute nectars by hummingbird and honeyeater flowers. *Biotropica* 13:260–70

Rathcke BJ. 1983. Competition and facilitation among plants for pollination. In *Pollination Biology*, ed. LA Real, pp. 305–29. New York: Academic

Raven PH. 1979. A survey of reproductive biology in Onagraceae. *N. Z. J. Bot.* 17:575–93

Reeves PA, Olmstead RG. 2003. Evolution of the TCP gene family in Asteridae: cladistic and network approaches to understanding regulatory gene family diversification and its impact on morphological evolution. *Mol. Biol. Evol.* 20:1997–2009

Robertson C. 1923. Flowers and insects. XXII. *Bot. Gaz.* 75:60–74

Robertson C. 1924. Flowers and insects. XXIII. *Bot. Gaz.* 78:68–74

Robertson C. 1928. *Flowers and Insects. Lists of Visitors of Four Hundred and Fifty-Three Flowers*. Carlinville, IL: Charles Robertson. 221 pp.

Robertson JL, Wyatt R. 1990. Evidence for pollination ecotypes in the yellow-fringed orchid, *Platanthera ciliaris*. *Evolution* 44: 121–33

Rush S, Conner JK, Jennetten P. 1995. The effects of natural variation in pollinator visitation on rates of pollen removal in wild radish, *Raphanus raphanistrum* (Brassicaceae). *Am. J. Bot.* 82:1522–26

Schemske DW. 1981. Floral convergence and pollinator sharing in two bee-pollinated tropical herbs. *Ecology* 62:946–54

Schemske DW. 1983. Limits to specialization and coevolution in plant-animal mutualisms. In *Coevolution*, ed. MH Nitecki, pp. 67–110. Chicago: Univ. Chicago Press. 392 pp.

Schemske DW, Bradshaw HD. 1999. Pollinator preference and the evolution of floral traits in monkeyflowers (*Mimulus*). *Proc. Natl. Acad. Sci. USA* 96:11910–15

Schemske DW, Horvitz CC. 1984. Variation among floral visitors in pollination ability: a precondition for mutualism specialization. *Science* 225:519–21

Schemske DW, Horvitz CC. 1989. Temporal variation in selection on a floral character. *Evolution* 43:461–65

Schemske DW, Lande R. 1984. The evolution of self-fertilization and inbreeding depression in plants. II. Empirical observations. *Evolution* 39:41–52

Schluter D. 1996. Adaptive radiation along genetic lines of least resistance. *Evolution* 50:1766–74

Simpson BB, Neff JL. 1983. Evolution and diversity of floral rewards. In *Handbook of Experimental Pollination Ecology*, ed. CE Jones, RJ Little, pp. 277–93. New York: Van Nostrand Reinhold. 558 pp.

Simpson GG. 1944. *Tempo and Mode in Evolution*. New York: Columbia Univ. Press. 237 pp.

Sprengel CK. 1793. *Das entdeckte Geheimniss der Natur im Bau und in der Befruchtung der Blumen*. Berlin: Vieweg

Sprengel CK. 1996. Discovery of the secret of nature in the structure and fertilization of flowers. In *Floral Biology: Studies on Floral Evolution in Animal-Pollinated Plants*, ed. DG Lloyd, SCH Barrett, pp. 3–43. Transl.

P. Haase. New York: Chapman & Hall. 410 pp.

Stebbins GL. 1950. *Variation and Evolution In Plants*. New York: Columbia Univ. Press. 643 pp.

Stebbins GL. 1970. Adaptive radiation of reproductive characteristics in angiosperms. I: Pollination mechanisms. *Annu. Rev. Ecol. Syst.* 1:307–26

Stebbins GL. 1974. *Flowering Plants. Evolution Above the Species Level*. Cambridge, MA: Harvard Univ. Press. 397 pp.

Steiner KE. 1998. The evolution of beetle pollination in a South African orchid. *Am. J. Bot.* 85:1180–93

Stone G, Willmer PG, Rowe JA. 1998. Partitioning of pollinators during flowering in an African *Acacia* community. *Ecology* 79:2808–27

Sugden EA. 1986. Anthecology and pollination efficacy of *Styrax officinale* subsp. *redivivum* (Styraceaceae). *Am. J. Bot.* 73:919–30

Tanaka N, Setoguchi H, Murata J. 1997. Phylogeny of the family Hydrocharitaceae inferred from rbcL and matK gene sequence data. *J. Plant Res.* 110:329–37

Tandon R, Shivanna KR, Mohan Ram HY. 2003. Reproductive biology of *Butea monosperma* (Fabaceae). *Ann. Bot.* 92:715–28

Temeles EJ. 1996. A new dimension to hummingbird-flower relationships. *Oecologia* 105:517–23

Temeles EJ, Kress WJ. 2003. Adaptation in a plant-hummingbird association. *Science* 300:630–33

Temeles EJ, Linhart YB, Masonjones M, Masonjones HD. 2002. The role of flower width in hummingbird bill length–flower length relationships. *Biotropica* 34:68–80

Temeles EJ, Rankin AG. 2000. Effect of the lower lip of *Monarda didyma* on pollen removal by hummingbirds. *Can. J. Bot.* 78:1164–68

Thompson JN. 1994. *The Coevolutionary Process*. Chicago: Univ. Chicago Press. 376 pp.

Thompson JN. 1999. Specific hypotheses on the geographic mosaic of coevolution. *Am. Nat.* 153(Suppl.):S1–14

Thomson BA, Thomson JD. 1998. BeeVisit for Windows (Visual Basic interactive simulation package for pollination). In *BioQUEST Library Vol. V, The BioQUEST Curriculum Consortium.* New York: Academic

Thomson JD. 2003. When is it mutualism? 2001 Presidential Address, American Society of Naturalists. *Am. Nat.* 162:S1–9

Thomson JD, Thomson BA. 1992. Pollen presentation and viability schedules in animal-pollinated plants: consequences for reproductive success. In *Ecology and Evolution of Plant Reproduction: New Approaches*, ed. R. Wyatt, pp. 1–24. New York: Chapman & Hall

Thomson JD, Wilson P, Valenzuela M, Malzone M. 2000. Pollen presentation and pollination syndromes, with special reference to *Penstemon*. *Plant Species Biol.* 15:11–29

Totland O. 2001. Environment-dependent pollen limitation and selection on floral traits in an alpine species. *Ecology* 82:2233–44

Walker-Larson J, Harder LD. 2001. Vestigial organs as opportunities for functional innovation: the example of the *Penstemon* staminode. *Evolution* 55:477–87

Waser NM. 1988. Comparative pollen and dye transfer by pollinators of *Delphinium nelsonii*. *Funct. Ecol.* 2:41–48

Waser NM. 1998. Pollination, angiosperm speciation, and the nature of species boundaries. *Oikos* 81:198–201

Waser NM, Chittka L, Price MV, Williams NM, Ollerton J. 1996. Generalization in pollination systems, and why it matters. *Ecology* 77:1043–60

Waser NM, Price MV. 1981. Pollinator choice and stabilizing selection for flower color in *Delphinium nelsonii*. *Evolution* 35:376–90

Waser NM, Price MV. 1983. Pollinator behavior and natural selection for flower color in *Delphinium nelsonii*. *Nature* 302:422–24

Waser NM, Price MV. 1998. What plant ecologists can learn from zoology. *Perspect. Plant Ecol. Evol. Syst.* 1:137–50

Weiblen GD. 2002. How to be a fig wasp. *Annu. Rev. Entomol.* 47:299–330

Weller SG, Sakai AK, Rankin AE, Golonka A, Kutcher B, Ashby KE. 1998. Dioecy and the evolution of pollination systems in *Schiedea* and *Alsinidendron* (Caryophyllaceae: Alsinoideae) in the Hawaiian Islands. *Am. J. Bot.* 85:1377–88

Wesselingh RA, Arnold ML. 2000. Pollinator behaviour and the evolution of Louisiana iris hybrid zones. *J. Evol. Biol.* 13:171–80

Whalen MD. 1978. Reproductive character displacement and floral diversity in *Solanum* section *Androceras*. *Syst. Bot.* 3:77–86

Whitten WM, Hills HG, Williams NH. 1998. Occurrence of ipsdienol in floral fragrances. *Phytochemistry* 27:2759–60

Wiebes JT. 1979. Co-evolution of figs and their insect pollinators. *Annu. Rev. Ecol. Syst.* 10:1–12

Williams GC. 1992. *Natural Selection: Domains, Levels, and Challenges*. Oxford: Oxford Univ. Press. 208 pp.

Wilson P. 1995. Selection for pollination success and the mechanical fit of *Impatiens* flowers around bumblebee bodies. *Biol. J. Linn. Soc.* 55:355–83

Wilson P, Castellanos MC, Hogue JN, Thomson JD, Armbruster WS. 2004. A multivariate search for pollination syndromes among penstemons. *Oikos* 104:345–61

Wilson P, Castellanos MC, Wolfe A, Thomson JD. 2004. Shifts between bee- and bird-pollination among penstemons. In *Specialization and Generalization in Plant-Pollinator Interactions*, ed. N. Waser, J. Ollerton. Chicago: Univ. Chicago Press. In press

Wilson P, Thomson JD. 1996. How do flowers diverge? In *Floral Biology: Studies on Floral Evolution in Animal-Pollinated Plants*, ed. DG Lloyd, SCH Barrett, pp. 88–111. New York: Chapman & Hall

Wolfe LM, Barrett SCH. 1988. Temporal changes in the pollinator fauna of tristylous *Pontederia-cordata*, an aquatic plant. *Can. J. Zool.* 66:1421–24

Wolfe LM, Kristolic JL. 1999. Floral symmetry and its influence on variance in flower size. *Am. Nat.* 154:484–88

Wright S. 1931. Evolution in mendelian populations. *Genetics* 16:97–59

ON THE ECOLOGICAL ROLES OF SALAMANDERS*

Robert D. Davic[1] and Hartwell H. Welsh, Jr.[2]

[1]Ohio Environmental Protection Agency, Northeast District Office, Twinsburg, Ohio 44087; email: robert.davic@epa.state.oh.us
[2]USDA Forest Service, Pacific Southwest Research Station, Redwood Sciences Laboratory, Arcata, California 95521; email: hwelsh@fs.fed.us

Key Words amphibians, forested ecosystems, detritus-litter, succession, keystone species

■ **Abstract** Salamanders are cryptic and, though largely unrecognized as such, extremely abundant vertebrates in a variety of primarily forest and grassland environments, where they regulate food webs and contribute to ecosystem resilience-resistance (= stability) in several ways: (a) As mid-level vertebrate predators, they provide direct and indirect biotic control of species diversity and ecosystem processes along grazer and detritus pathways; (b) via their migrations, they connect energy and matter between aquatic and terrestrial landscapes; (c) through association with underground burrow systems, they contribute to soil dynamics; and (d) they supply high-quality and slowly available stores of energy and nutrients for tertiary consumers throughout ecological succession. Salamanders also can provide an important service to humans through their use as cost-effective and readily quantifiable metrics of ecosystem health and integrity. The diverse ecological roles of salamanders in natural areas underscore the importance of their conservation.

INTRODUCTION

Salamanders (Amphibia: Caudata) are ancient vertebrates that have evolved extensive ecological diversification for at least the past 150–200 million years (Gao & Shubin 2001, Schoch & Carroll 2003). They are widely distributed in North, Central, and South America, Europe, and temperate eastern Asia (Duellman 1999), with more than 400 species in 59 genera and 10 families (Zug et al. 2001). Their adaptive radiation of life history traits has resulted in exploitation of moist forest leaf litter, grasslands, underground retreats, tree canopies, talus slopes, headwater streams, riparian ecotones, swamps, caves, ponds, and seasonally inundated pools (Petranka 1998). Within these varied environments, salamanders perform many ecological roles or "key ecological functions" (Marcot & Vander Hayden 2001).

---

*The U.S. Government has the right to retain a nonexclusive, royalty-free license in and to any copyright covering this paper.

Key ecological functions refer to the primary ways that species use, influence, regulate, and alter biotic and abiotic environments—a concept recommended for multispecies planning, biodiversity conservation, and management of wildlife-habitat relationships (Johnson & O'Neil 2001). In this paper, we review literature on key ecological functions of salamanders in terrestrial and aquatic environments of North America. We offer suggestions for future research by noting basic gaps in knowledge. Nomenclature follows Collins & Taggart (2002).

This review is particularly timely because natural areas are becoming increasingly modified by destabilizing factors such as habitat alteration, toxic chemicals, loss of wetlands, and introduction of exotic species (Aber et al. 2000). Nearly three fourths of forested ecosystems in North America are considered endangered because of threats to their integrity (Noss 1999). The decline in amphibian species, many associated with forests, is now well documented (Alford & Richards 1999, Houlahan et al. 2000, Kiesecker et al. 2004). Although most attention has been given to anurans, salamander populations also are declining (Welsh 1990, Petranka et al. 1993, Wheeler et al. 2003), with unknown consequences to ecosystem processes. Of the 234 identified salamander taxa in the United States, 67 (29%) have a conservation status rank of "imperiled or critically imperiled" in at least part of their range (NatureServe 2003), yet only 13 species are protected or proposed for protection under the United States Endangered Species Act (Semlitsch 2003a). Habitat modifications are cited most often as the causes for salamander declines (Dodd & Smith 2003), with estimated losses of salamanders in some habitats in the millions (Petranka et al. 1993). In addition, zoogeographic evidence suggests that salamander faunas globally are being impacted (Duellman 1999). It is both disturbing and fortuitous that these declines are being reported at a time when salamanders are increasingly being recommended for use as bio-indicators to assess the ecological health and integrity of natural areas (Parent 1992, Welsh & Ollivier 1998, Simon et al. 2000, Welsh & Droege 2001, Micacchion 2002). Our hope is that this review will serve as a stimulus for much needed additional research on the important ecological roles of these abundant but often neglected vertebrate species.

PATTERNS OF SALAMANDER STRUCTURAL DOMINANCE

Making predictions about sustainability of ecosystems requires information on how dominant biotic and abiotic structures vary over time and space (Bailey 1996). In this section, we review literature on the structural dominance (e.g., density, biomass, calories) of salamander species in North American ecosystems at different levels of ecological organization.

Terrestrial Habitats

Numerical dominance of salamanders in the terrestrial landscape was first reported from the southern Appalachian Mountains by Hairston (1949). Consistent results

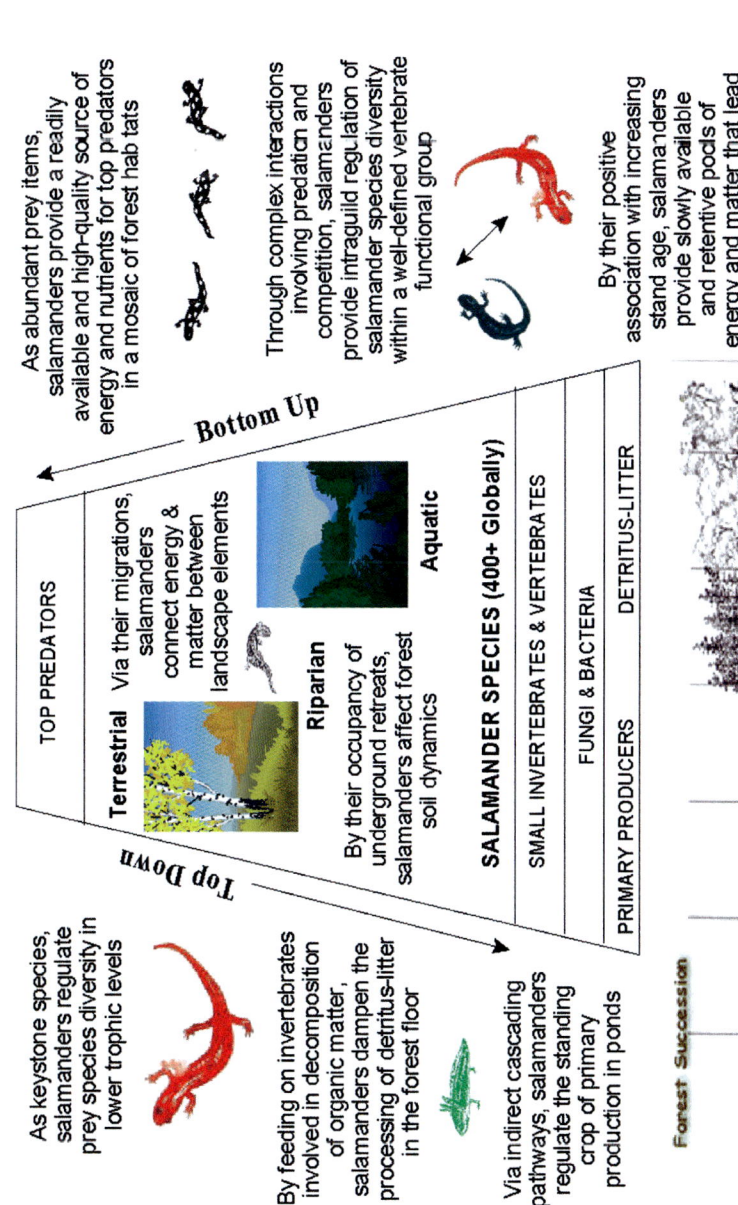

Figure 2 Key ecological functions of salamanders in ecosystems. Forest succession image used with permission of the Forest Historical Society. Salamander images courtesy of Travelin' Naturalist, Lenz Design, and US EPA (Mark Erelli, artist).

of long-term observations of high numbers along vertical transects led him to conclude that salamander species from the family Plethodontidae were the numerically dominant members of the forest vertebrate fauna, although no comparative data for other vertebrates were reported.

Burton & Likens (1975a,b) first quantified both density and biomass of a salamander guild at a watershed scale. Working in the Hubbard Brook experimental forest of New Hampshire, they estimated that five salamander species had a combined average density of 2950 salamanders/ha ($0.29/m^2$) and a biomass of 1770 g/ha wet weight. This value was 2.6 times the combined wet-weight biomass of all birds living in the watershed at the peak of bird breeding season and at least equal to that of small mammals such as shrews and mice. The nutrient pool of phosphorus in salamanders (7.79 g/ha) was greater than that in birds (4.27 g/ha) and small mammals (0.21 to 0.41 g/ha) combined. Burton & Likens (1975a,b) are often cited as evidence that salamanders are the most abundant vertebrates in mature forests; however, it does not follow that salamanders compose the greatest amount of vertebrate biomass. Not included in their biomass calculations are large herbivorous mammals, such as deer, and other vertebrates, such as fish, reptiles, or frogs. Salamanders cannot have the highest vertebrate standing crop in forests because white-tailed deer alone contain on average 1.30 $kcal/m^2$ caloric energy (Ricklefs 1979), more than the 1.165 $kcal/m^2$ estimated by Hairston (1987) for a southern Appalachian salamander guild. Hairston (1987) clarified the issue by suggesting that salamanders are the dominant "vertebrate predators" (e.g., carnivores) in forests, thus linking the ecological relevance of salamander abundance to a critical link in the trophic dynamics of food webs. To put the estimated 1.165 $kcal/m^2$ caloric contribution of these southern Appalachian salamander populations into perspective the annual average human harvest of the world's marine fishery was reported as 0.3 $kcal/m^2$ (Odum 1971).

Numerous studies expand the findings of Burton & Likens (1975a,b). Citing research from an oak woodland/redwood forest in California, Stebbins & Cohen (1995) reported that the combined density of the salamander guild was "close to the values of Burton and Likens." Comparable data were reported for a single species, *Plethodon elongatus*, in a Douglas-fir dominated stand in northwest California (Welsh & Lind 1992). In the southern Appalachian Mountains, Hairston (1987) estimated salamander guild abundance across a mosaic of habitats as 0.6 to $1.0/m^2$ (5961 to 9935/ha), more than three times the density reported by Burton & Likens (1975a) for New Hampshire. Similarly, Petranka et al. (1993) reported 10,000 salamanders/ha ($1.0/m^2$), representing 12 species in 34 mature forest stands in western North Carolina.

Researchers have often observed that a single salamander species will dominate the terrestrial habitat of a local salamander guild (Table 1). In western North America, this dominance is known to shift between Ensatina (*Ensatina eschscholtzii*), two *Plethodon* species (*P. elongatus* and *P. vehiculum*), and *Batrachoseps attenuatus* in relation to region, forest type, and seral stage (Bury et al. 1991; Welsh & Lind 1988, 1991; Cooperrider et al. 2000). The environmental factors

TABLE 1 Salamander guild species richness and evenness in five forested landscapes with a minimum of 1000 captures (see also Figure 3)[a]

Taxon Rank	Ash 1997	Welsh & Lind 1988	Ross et al. 2000	Mitchell et al. 1997	Ford et al. 2002
1	1234	1097	1967	756	2556
2	69	213	1025	182	626
3	40	72	430	70	392
4	10	38	318	28	184
5	2	21	144	25	59
6	—	20	83	24	45
7	—	5	31	10	30
8	—	4	17	5	21
9	—	1	7	2	17
10	—	1	2	1	7
11	—	—	1	—	—
Total no.	1355	1472	4026	1104	3937
Capture Method	Area search	Pitfall traps	Area search	Pitfall traps	Pitfall traps
Dom.[b] sp.	*Plethodon jordani*	*Ensatina eschscholtzii*	*Plethodon cinereus*	*Plethodon cinereus*	*Plethodon glutinosus*
Reg.[c] slope	−1.48 lower diversity	−0.73	−0.72	−0.63	−0.60 higher diversity

[a]Data represent captures across forest stands in varying stages of disturbance.
[b]Numerically dominant species within guild.
[c]Regression slopes from Figure 3, see text for explanation.

responsible for these shifts are mostly unknown. In eastern forests, pioneering surveys by Shelford (1913) identified the Northern redback salamander (*Plethodon cinereus*) as a dominant vertebrate within the leaf litter of late-successional beech-maple stands, and a variety of forest types that converge toward the beech-maple climax state. Contemporary studies confirm that *P. cinereus* numerically dominates salamander guilds in many forest types in the eastern United States (Burton & Likens 1975a,b; Carfioli et al. 2000; see also Table 1). *Plethodon* species other than *P. cinereus* are also known to dominate terrestrial salamander assemblages. For instance, *Plethodon glutinosus* populations dominated four stands of yellow poplar–northern red oak–white oak (15 to >85 years of age) in the Chattahoochee National Forest in Georgia (Ford et al. 2002; see also Houze & Chandler 2002). Jordan's redcheek salamander (*Plethodon jordani*), which is endemic to a small geographic area of the Blue Ridge Mountains in the southern Appalachians (Petranka 1998), has been shown to be the dominant salamander species in relatively dry terrestrial habitats within its range (Harper & Guynn 1999, Bartman et al. 2001). These observations suggest that not all narrowly distributed endemic salamander species are necessarily rare as some, such as *P. jordani*, can be numerically dominant and potentially provide important biotic control over ecosystem dynamics within isolated geographic areas.

Grassland Habitats

Although the vast majority of salamander species in North America are forest specialists and require relatively intact forest stands to complete at least part of their life history, a number of species are known to occupy mostly grassland habitats as adults. The California tiger salamander (*Ambystoma californiense*) is endemic to grasslands and can be the dominant vertebrate predator in ephemeral ponds; densities as high as 325 males and 216 females from a single 3660 m^2 breeding pond have been reported (Trenham et al. 2001). Mammal burrows are critical limited resources for both juveniles and adults of *A. californiense*, and loss of grassland burrow habitat and associated ephemeral breeding ponds have been associated with decline in local populations (Fisher & Shaffer 1996). Other species known to migrate into grassland habitats as adults include the various demic populations of *Ambystoma tigrinum* (see Shaffer & McKnight 1996), the most widely distributed salamander in North America (Petranka 1998), and three species of slender salamanders (*Batrachoseps nigriventris*, *B. attenuatus*, and *B. pacificus*). However, the extent of salamander use of grasslands and adult population density in relation to other habitat types are largely unknown and represent important venues for future research.

Riparian Habitats

The riparian ecotone between aquatic and terrestrial environments provides unique habitats for salamanders, and some researchers have suggested that this landscape structure has its own ecological identity for amphibians (Bury 1988, Krzysik 1998, Sheridan & Olson 2003). Thirty-five percent of the salamander genera of North America use riparian habitats to complete their life history (Krzysik 1998). Within the Humid-Temperate-Domain ecoregion of eastern North America (Bailey 1996), 47 salamander species use stream or pond riparian corridors for reproduction, foraging, and shelter (Pauley et al. 2000).

Densities of salamanders in riparian areas can exceed those found in upland terrestrial environments. Talus riparian habitats on Vancouver Island support as many as 11,600 *Plethodon vehiculum* salamanders/ha (Ovaska & Gregory 1989), more than three times the salamander density reported by Burton & Likens (1975a) for the entire Hubbard Brook watershed. In the southern Appalachians, salamander density was estimated as 18,486 individuals/ha (1.8/m^2) from riparian habitats alone (Petranka & Murray 2001), a value 7 times higher than reported by Burton & Likens; biomass was 14 times higher (16.53 kg/ha). Riparian areas along headwater streams in second-growth Douglas-fir forest (southwestern Washington) contained large numbers of salamanders of the genus *Plethodon*, which occurred adjacent to 93% of streams surveyed (Wilkins & Peterson 2000). In contrast, Waters et al. (2001) studied abundances of amphibians and small mammals along small, intermittent headwater streams in northern California and found the riparian zone to be dominated by small mammals [Allen's chipmunk (*Tamias senex*) and deer mouse (*Peromyscus maniculatus*)], not salamanders. The low numbers of salamanders in these riparian environments may be associated with unpredictable hydroperiods because nearby upland forest habitats supported high numbers of *Ensatina eschscholtzii* (J.R. Waters & H.H. Welsh, unpublished data). Removal of riparian vegetation can have detrimental effects on salamander densities, and is of particular concern for endemic species with patchy distribution (Williams et al. 2002).

Aquatic Habitats

Numerous studies document that salamanders, not fish, dominate the vertebrate community in the headwater habitats of watersheds (Murphy & Hall 1981, Petranka 1983, Resetarits 1997, Wilkins & Peterson 2000, Lowe & Bolger 2002). For example, giant salamanders (*Dicamptodon*) replace fish as the dominant vertebrate predator in headwater streams from the Pacific Northwest, contributing 99% of the total predator biomass in certain areas (Murphy & Hall 1981). Diller & Wallace (1996) reported populations of the cold water adapted *Rhyacotriton variegatus* in 80.3 % of randomly surveyed headwater streams in Northern California. Conceptually, these low-order stream habitats (sensu Strahler 1964) represent a salamander-dominated region in the upper reaches of the river continuum (Vannote et al. 1980). Salamanders can move higher into headwater streams than fish because physical attributes such as intermittent hydrology, size and depth of pools, and cascades and waterfalls limit the ability of fish to access these areas. Headwater streams likely offered an attractive ecological niche free from fish predation during the lower Paleozoic to upper Mesozoic (360–200 millions of years before present) when fish-tetrapod-salamander evolution occurred (Schoch & Carroll 2003). This hypothesis is evidenced by the widespread adaptive radiation of extant salamander taxa above the species level in headwater regions of watersheds across biomes.

Seven salamander genera in North America are specifically adapted to conditions found in headwater streams, including *Desmognathus, Dicamptodon, Eurycea, Gyrinophilus, Pseudotriton, Rhyacotriton,* and *Stereochilus*; some *Ambystoma* and *Taricha* species also breed in low-order stream environments (Petranka 1998, Corn et al. 2003). When larval age classes are included in the tally, total salamander density and biomass in headwater streams can be high compared with average densities of approximately $1.0/m^2$ reported in terrestrial habitats. For example, Nussbaum & Tait (1977) estimated densities of *Rhyacotriton* populations from $12.9/m^2$ to $41.2/m^2$ in Oregon. Davic (1983) reported high seasonal density and biomass for a complete aquatic salamander guild, including larvae (*Desmognathus, Eurycea, Gyrinophilus*), from a fishless stream in North

TABLE 2 Seasonal changes in the density and biomass of an aquatic salamander guild from a fishless and spring-fed headwater stream in North Carolina (1980)

	June #/m² (g/m²)	August #/m² (g/m²)	October #/m² (g/m²)
Desmognathus quadramaculatus			
Larvae	2.8 (5.6)	2.3 (4.0)	1.7 (1.2)
Juveniles	1.0 (1.6)	1.6 (3.7)	1.4 (2.9)
Adults	0.2 (0.1)	0.55 (0.7)	0.45 (1.0)
Eurycea wilderae			
Larvae	4.7 (0.25)	10.1 (1.3)	8.6 (0.9)
Adults	0.1 (0.15)	0.0 (0.0)	0.08 (0.01)
Gyrinophilus porphyriticus danielsi			
Larvae	0.4 (0.1)	0.2 (0.05)	0.08 (0.01)
Salamander Guild Totals	9.2 (7.8)	14.7 (9.75)	12.3 (6.1)

Source: Unpublished data from Davic (1983) with corrected biomass values.

Carolina (Table 2). Little variation in salamander guild structure was noted over seasonal time in this study, likely because of the stable environmental conditions provided by a spring-fed environment. Huang & Sih (1991a) reported exceptional densities of *Ambystoma barbouri* larvae in fishless headwater stream pool habitats, on average 20–30/m² with values as high as 50/m². Biomass of coastal giant salamander larvae in Caspar Creek on the northern California coast reached 10.4 g/m² (Nakamoto 1998). Welsh & Lind found a wide range of salamander larval densities in streams throughout northwestern California, ranging from 0.1 to 5.0/m² for *Rhyacotriton variegatus* (Welsh & Lind 1996), to 0.03 to 1.61/m² for *Dicamptodon tenebrosus* (Welsh & Lind 2002). Ultimate reasons for natural variation in salamander numbers in headwater stream environments are largely unknown; however, strong association between biotic and/or abiotic factors and salamander density has been reported for a variety of species (Petranka 1983; Davic & Orr 1987; Welsh & Lind 1996, 2002; Diller & Wallace 1996; Welsh & Ollivier 1998; Lowe & Bolger 2002; Barr & Babbitt 2002).

Large river systems support large-bodied salamanders from the genera *Necturus* and *Cryptobranchus*. Multi-year (1989–1991) mark-recapture surveys of the common mudpuppy (*Necturus maculosus*) in Ohio resulted in 382 salamanders collected along a 700 × 50 m stream reach (Matson 1998). Petranka (1998) reports *Cryptobranchus alleganiensis* densities as high as six individuals per 100 m². Given the large body size of *Necturus* and *Cryptobranchus* adults, they may rival the biomass of predatory fish species in localized stream reaches, but we are unaware of any studies that quantify salamander density or biomass in relation to other vertebrates in large rivers.

It has long been recognized that salamanders from the genera *Ambystoma*, *Notophthalmus*, and *Siren* are dominant vertebrate predators in seasonal pools and ponds (see reviews in Morin 1983, 1995; Wilbur 1997; Walls & Williams 2001). Salamander densities in these lentic habitats can be extremely high. Adult red-spotted newt densities as high as $10/m^2$ were recorded in ponds from south-central Indiana (Cortwright 1998). A study of prairie grassland lakes in North Dakota revealed a maximum density in *Ambystoma tigrinum* of 5000 larvae/ha and a maximum biomass of 180 kg/ha (Deutschmann & Peterka 1988). Even where fish were present, the standing crop of the lesser siren (*Siren intermedia*) in a Texas pond was greater than the combined value of seven fish species (Gehlbach & Kennedy 1978). Likewise, in ponds with fish on the coastal plain of southeastern United States, Means (2000) observed that the dwarf salamander, *Eurycea quadridigitata*, especially the larvae, was the numerically dominant vertebrate predator in 25 of 38 (66%) sampled habitats.

Altering the hydrologic regime can significantly alter salamander dominance in aquatic habitats (Semlitsch 2003b). Herpetofaunal communities inhabiting streams impounded by beaver ponds were compared with unimpounded streams by Metts et al. (2001). Salamanders were dominant amphibians in unimpounded streams (1680 of 2664 captures or 63.1%), but only 8.2% of captures were from habitats impounded by beavers. Snodgrass et al. (2000) reported that wetlands with different hydroperiods contain distinct amphibian assemblages and concluded that short-hydroperiod wetlands are important in maintaining amphibian biodiversity across a landscape because they may support species not found in longer hydroperiod wetlands.

Succession

A large body of literature indicates that the density of some salamander species is closely associated with forest successional stage, with higher numbers of salamanders in older, more structurally complex systems. The California slender salamander (*Batrachoseps attenuatus*) is 10 times more abundant in old-growth redwood forest than younger regenerating stands (Bury 1983, Cooperrider et al. 2000). Welsh & Lind (1988, 1991) surveyed 54 terrestrial sites in the mixed Douglas-fir/hardwood forests of northern California ranging in age from 30 to 560 years. Three species of terrestrial salamanders were more abundant on old-growth than on younger sites. As reported by Welsh & Droege (2001), salamander abundances tracked closely with several structural attributes that model the forest chronosequence (Figure 1). One of these species, *Plethodon elongatus*, is closely associated with ecological conditions found primarily in the late seral stage of the interior mixed conifer/hardwood forests in California (Welsh & Lind 1995). Old-growth forests are known to support more salamanders than second-growth managed stands on Vancouver Island, Canada (Dupuis et al. 1995). Mitchell et al. (1997) reported significantly more salamanders, especially *Plethodon cinereus* and *Ambystoma jeffersonianum*, from eastern forests 80 to >100 years old compared with forests 2–50 years old. In Georgia forests ranging in age from 15 to >85 years old, both salamander species richness and diversity increased with age (Ford et al.

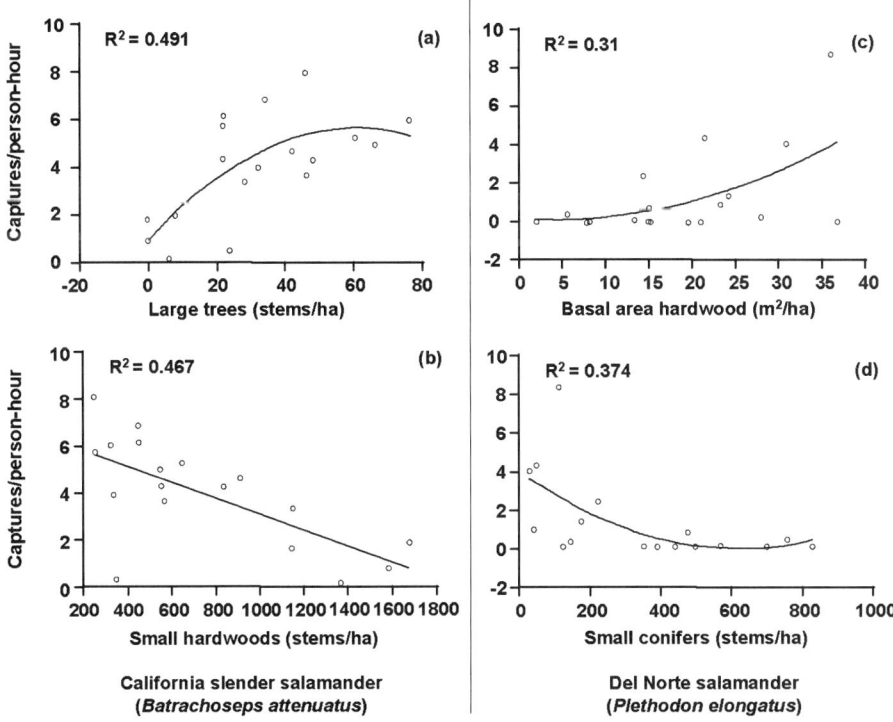

Figure 1 Bivariate scatterplots and regressions for two Pacific Northwest salamander species showing relationship to structural aspects of forest chronosequence (all coefficients significant at $p < 0.05$). Figure reprinted with slight modification from Welsh & Droege (2001) with permission.

2002). Although salamander diversity did not differ among land use categories in southern Appalachian forests, Hicks & Pearson (2003) also report that overall salamander numbers were greater in older, least altered stands. The observation that woodland salamander density increases during forest succession was verified for *P. cinereus* with a timber-harvest, GIS-based simulator model to predict both abundance and mass per unit area in forests of different ages (Gustafson et al. 2001).

Not all studies report a significant association of salamander abundances with seral stage (see review by deMaynadier & Hunter 1995). The reasons for this lack of association include differences in sampling technique, species tolerances, and/or availability of suitable microhabitat cover. For example, the threatened Cheat Mountain salamander (*Plethodon nettingi*) is most abundant in young red spruce forest stands but is rarely collected in mature forests (Brooks 1948). Metapopulation sizes of red-spotted newts are predicted to decrease as forest succession cools breeding ponds, making the pools less suitable for adults and production of efts (Cortwright 1998). Welsh (1990) and deMaynadier & Hunter (1995) both proposed that forest seral stage is an indirect measure of the age-related environmental attributes (e.g., coarse woody debris, foliage height diversity, canopy cover, litter type and depth, and cool, moist, equable microclimatic conditions) that determine whether a site is suitable for a given species. Biotic interactions of predation and competition between species may also be an important causal factor affecting salamander distribution at different stages of seral succession (see review by Hairston 1996).

PATTERNS OF SALAMANDER FUNCTIONAL DOMINANCE

Ecosystem function refers to direct and indirect interactions of biotic and abiotic components, as well as to their contribution to the performance of the ecosystem as a whole (Müller & Windhorst 2000). The extremely high densities and biomass of salamanders within a variety of discrete forest environments lend credence to a hypothesis that they can regulate ecosystem functions at many different spatial scales and seral stages. Given their low ectothermal energy demands, salamanders would be predicted to affect ecosystem processes more as density-dependent regulators (e.g., process equilibrators or modifiers) or as retentive stores of nutrients than as movers of energy (Chew 1974, Pough 1983). In this section, we review literature on the roles of salamanders as biotic regulators of ecosystem processes and their contribution to resilience-resistance pathways that contribute to ecosystem stability.

Predatory Effects on Lower Trophic Levels

All the more than 400 salamander species worldwide are obligate carnivores, with most having a polyphagus feeding strategy (Petranka 1998, Zug et al. 2001). Salamanders consume a wide variety of invertebrates and vertebrates within aquatic and terrestrial environments, although aerial prey can also be an important food resource (Davic 1991).

Hairston (1987) calculated that a salamander guild from the southern Appalachians could consume 5.80 kcal/m^2 annually, which is greater than the estimated 5.04 kcal/m^2 of all soil invertebrates. These data led Hairston to conclude that "the impact of salamanders on the soil fauna should be taken seriously." Petranka (1998) suggested that abundant woodland species of the genera *Plethodon* and *Ensatina* would be predicted to regulate the population density of forest floor invertebrates. Other genera (*Ambystoma, Amphiuma, Cryptobranchus, Desmognathus, Dicamptodon, Notophthalmus, Siren,* and *Taricha*) were identified as potential regulators of invertebrate densities in aquatic environments. A number of studies, which we review below (see also Table 3), have now demonstrated through the manipulation of salamander densities that an important linkage exists between

TABLE 3 Field experiments reporting regulatory effects by salamanders on invertebrate populations in both terrestrial and aquatic environments

Species	Affect/Habitat	Reference
Plethodon cinereus[a]	Invertebrate leaf fragmentors/Forest floor	Wyman 1998
Plethodon cinereus	Collembola via indirect predation/Forest floor	Rooney et al. 2000
Desmognathus quadramaculatus	Benthic macroinvertebrates/Headwater stream	Davic 1983
Dicamptodon tenebrosus	Benthic macroinvertebrates/Headwater stream	Parker 1992
Ambystoma barbouri	The benthic isopod, *Lirceus fontinalis*/Headwater stream	Huang & Sih 1991a,b
Notophthalmus viridescens and *Ambystoma* spp.	Zooplankton/Artificial tanks	Morin et al. 1983; Morin 1987, 1995; Leibold & Wilbur 1992
Ambystoma maculatum	Isopods, amphipods/Artificial tanks	Harris 1995
Ambystoma tigrinum	Benthic invertebrates and zooplankton/Artificial enclosures	Holomuzki et al. 1994
Ambystoma tigrinum	Caddisfly larvae/Wetlands	Wissinger et al. 1998
Ambystoma tigrinum and *Ambystoma laterale*	Zooplankton and mosquito larvae/Mesocosms and wetlands	Brodman et al. 2003

[a]Species in bold are identified by authors as potential "keystone species" (sensu Paine 1969).

salamander abundance, prey species diversity, trophic cascades, nutrient cycling, and the detritus-litter food webs of forest, grassland, and associated aquatic environments.

Wyman (1998) experimentally manipulated densities of *Plethodon cinereus* salamanders using leaf litter enclosures and concluded that these abundant salamanders are strong regulators of forest floor invertebrate populations. Rooney et al. (2000) also manipulated *P. cinereus* abundances in forest enclosures and found that salamanders indirectly enhanced the abundance of *Collembola*, a noningested prey item, by regulating the density of invertebrate predators of *Collembola*. Although we are unaware of similar experiments in western forests, the numerical dominance of *Ensatina*, *Plethodon*, and *Batrachoseps* in Douglas-fir forests suggests they also may have significant regulatory effects on invertebrate densities.

The effects of salamander predation on benthic macroinvertebrates in a lotic environment were investigated by Davic (1983) via repeated removal of biomass dominant *Desmognathus quadramaculatus* (adults, juveniles, and larvae—see Table 2) from small stream plots over a 12-month period. The results of this study showed a significant number of direct (predatory) and indirect (competitive release) impacts on prey species density and biomass where salamanders were removed, but it reported little impact on nonprey species. The collective predation pressure of the aquatic salamander guild was estimated to remove 23 benthic macroinvertebrates/day/m^2 from June to October. Parker (1992) removed *Dicamptodon tenebrosus* larvae for 96 days from a stream pool, which also showed that salamanders can have both direct and indirect effects on benthic macroinvertebrate prey populations. Larvae of *D. tenebrosus* ingested 2.2 g/m^2 of benthic prey compared with a mean standing crop of 3.1 g/m^2 of available prey, which led Parker (1994) to conclude that the regulatory effect by salamanders on invertebrate populations is intense. Huang & Sih (1991a) experimentally added *Ambystoma barbouri* larvae to isolated stream pools and found that the larvae significantly reduced the density and altered the use of microhabitats by the benthic isopod, *Lirceus fontinalis*. Individual *Ambystoma* larvae ingested on average 28.6 isopods/day, indicating that they collectively could remove roughly 5000 isopods/day from pooled areas of the stream. Laboratory experiments showed complex interactions among salamander larvae, isopods, and a top predator, the green sunfish (Huang & Sih 1991b). Subsequently, Sparkes (1996) documented that female isopods mature at larger sizes in stream pools containing *Ambystoma* larvae, thus releasing themselves from intense predation pressure by salamanders.

In contrast to the above studies, two attempts to exclude the Northern two-lined salamander (*Eurycea bislineata*) and fish predators in streams using mesh cages (Reice 1983, Reice & Edwards 1986) reported no effect on benthic macroinvertebrate prey. However, we view the results of these experiments as inconclusive. The 6.35 mm mesh used for experimental cages by Reice (1983) would not have excluded *Eurycea* larvae and small juveniles. These life stages can be abundant in streams (see Table 2) and are known to ingest a variety of benthic macroinvertebrates (Petranka 1998). In the experiment of Reice & Edwards (1986), adult *E. bislineata* were excluded, but these salamanders are known to feed extensively on terrestrial prey and migrate seasonally well away from flowing water (Petranka 1998); they are unlikely to be strong predators on aquatic benthic macroinvertebrates. However, given their wide-ranging terrestrial dispersal, abundant populations of *E. bislineata* adults may regulate invertebrate species diversity within the riparian stream environment, a hypothesis open to experimentation.

Predatory effects of salamanders are well known to reduce the population density of frog tadpoles and alter coexistence patterns of salamander species in lentic habitats (see Calef 1972, Morin 1995, Wilbur 1997, Kurzava & Morin 1998, Walls & Williams 2001, Brodman et al. 2003). Experiments in artificial ponds led Morin (1995) to conclude that adult *Notophthalmus viridescens* and larval *Ambystoma opacum* show "functional redundancy" in their nearly identical regulatory effects on anuran prey populations. Where salamander predators were abundant, frog density was reduced, leading to increased biomass of primary producers such as phytoplankton. We are unaware of experiments that have tested for indirect effects of salamander predation on macrophyte biomass or diversity.

Predatory effects by salamanders on invertebrates in pond habitats are also well documented (Table 3). For example, Morin (1987) experimentally manipulated the densities of *Notophthalmus viridescens* and *Ambystoma tigrinum* in artificial pools and found that salamander predation significantly altered patterns of seasonal succession of zooplankton. A subsequent experiment revealed that the effect of salamander predation in ponds can cascade through multiple trophic levels to increase algae production (Morin 1995). Similar experiments with *Ambystoma tigrinum* larvae showed both direct and indirect effects on zooplankton and benthic macroinvertebrate abundances in fishless pond enclosures (Holomuzki et al. 1994). Larvae of *Ambystoma tigrinum* also regulate population densities of caddisfly prey in subalpine wetlands of Colorado (Wissinger et al. 1998). Given their tendency to specialize on mollusk prey, *Siren* spp. may play an important role in structuring snail populations in ponds (Petranka 1998). In a more applied vein, Brodman et al. (2003) demonstrated the effectiveness of *Ambystoma* larvae in controlling mosquito larvae populations—mosquito larvae density was 98% lower in wetlands with salamanders compared with salamander-free wetlands.

A variety of field experiments in both terrestrial and aquatic environments have demonstrated that salamander species can function as "keystone predators" (i.e., Morin 1981, Davic 1983, Fauth & Resetarits 1991, Fauth 1999, Wissinger et al. 1998, Wyman 1998). According to the classic concept of Paine (1969), keystone species prevent dominant prey from monopolizing limited resources, thus allowing the coexistence of additional species and/or an increase in the evenness of prey species abundances within a community (see review by Menge & Freidenburg 2001). As discussed by Schulze & Mooney (1993), keystone species as a group have no redundant representation; they exert disproportionate biotic regulation within an ecosystem because their elimination causes changes in community function not performed by other species. Therefore, loss of keystone species such as salamanders could have serious negative effects on ecosystem stability by altering resilience-resistance pathways (Chapin et al. 1997). Davic (2003) proposed a modification of Paine's (1969) keystone species concept that links the a priori identity of potential keystone species to biomass dominance in ecological functional groups. This view of the keystone species concept is congruent with widespread observations that a single salamander species often dominates multispecies salamander guilds in a variety of habitat types (see Table 1), and its application offers a novel management tool for the a priori identification of potential keystone salamander species in natural areas.

Regulation of Detritus-Litter Food Webs

More than 90% of the net energy production of a temperate forest is consumed by decomposer organisms and less than 10% by herbivores (Ricklefs 1979). Those invertebrate organisms responsible for most of the decomposition and fragmentation of detritus-litter are well known for both aquatic (Wallace & Webster 1996) and terrestrial environments (Swift et al. 1979). However, the ecological roles of vertebrate predators as potential regulators of decomposer populations in ecosystems, and ecosystem processes associated with detritus-litter food webs, are poorly known (Bormann & Likens 1979, Konishi et al. 2001).

Some researchers have suggested that salamanders may provide an important indirect regulatory role in the processing of detritus-litter by ingestion of detritivore prey (Burton & Likens 1975b, Hairston 1987, Stebbins & Cohen 1995). Field experiments in lotic (Davic 1983) and terrestrial (Wyman 1998) environments support this hypothesis. Both investigations demonstrated in situ that the presence of salamanders slow the rate of detritus-litter decomposition. These findings are contrary to the suggestion of Hairston (1987) that salamanders feeding on invertebrates, which themselves feed on the bacteria and fungi in the forest floor leaf microflora, would promote a more rapid rate of leaf litter decomposition. Indirect effects on detrital processing have rarely been documented for vertebrates; however, fish predators also have been reported to slow the rate of leaf decomposition in streams (Konishi et al. 2001).

Dominant salamanders in terrestrial environments may serve to maintain resilience-resistance pathways in forests by indirectly dampening the seasonal release of essential micronutrients from leaf litter to the root systems of the flora. Leaf litter decomposes at a rate directly related to the number of invertebrate animals in the litter and the underlying soil (Perry 1994). Wyman's (1998) documentation that *Plethodon* salamanders reduce soil invertebrate numbers and indirectly dampen leaf litter processing has significant implications for the mineralization and immobilization of elements such as carbon, nitrogen, and phosphorus. Results from ongoing experiments by Wyman (2003) suggest an important linkage of *P. cinereus* salamanders to long-term retention of nitrogen compounds from leaf litter and potential regulation of the carbon-nitrogen cycles in forests. By reducing densities of invertebrates that prefer ingesting leaf sections with high nutrient value, salamander predation may allow for longer retention of nutrients in soil over time (Wyman 2003). In a headwater stream, predatory effects of *Desmognathus quadramaculatus* salamanders slow detrital processing (Davic 1983), thus potentially dampening the release of fine particulate organic matter to downstream communities. The downstream movement of organic matter is a central theme of the river continuum concept (Vannote et al. 1980) and resource spiraling (Elwood et al. 1983).

To our knowledge, no experimental studies have investigated effects of salamanders on detrital processing in lentic habitats. However, some evidence suggests that salamanders may play such a role. Wissinger et al. (1998) investigated the predatory effects of *Ambystoma tigrinum* on two species of leaf shredding limnephilid caddisfly larvae. They observed competition between caddisfly species involved in detrital processing, which resulted in strong keystone species effects on prey diversity by salamander predation. Although Wissinger et al. (1998) did not measure decay rate of detritus, their observations suggest that salamanders have the potential to regulate detrital processing in this lentic environment by predatory control of competitively dominant, leaf-shredding invertebrate species. Efford (1969) estimated the energy budget of Lake Marion, Canada, and found that detritus in the lake bottom represented 86% of the total annual carbon energy storage of 280 g of carbon/m^2/yr. The detritivore *Hyallela azteca* reduced detritus content in the bottom of the lake by 40% to 45% in one season and was a highly selected prey of the salamanders *Taricha granulosa* and *Ambystoma gracile*.

The role of salamanders in damping litter decomposition, with possible global significance, has been discussed by Wyman (1998, 2003). Forests are estimated to contain approximately three fourths of all carbon contained in living terrestrial vegetation, and a little less than one half of that is stored in soils (Perry 1994). Wyman calculated that an 11% to 17% reduction in the rate of forest floor leaf decomposition (because of the estimated regulatory effect of salamander predation) would result in 261 kg to 476 kg of carbon/ha not being released into the atmosphere annually. This cybernetic feedback between salamanders and leaf litter processing suggests that reported declines in salamander densities may be causing an increase in rate of leaf litter decomposition and concomitant increase in CO_2 release to the atmosphere. Wyman (1998, 2003) speculated that this process may contribute to global warming, but he cautions that this idea has too many assumptions to be taken as anything other than a testable hypothesis that warrants further investigation.

Coupling Aquatic and Terrestrial Habitats

By migrating between environments, consumers can affect food webs of communities at several spatial scales (Polis et al. 1996). Biological coupling of aquatic and terrestrial landscapes has only recently been investigated as this coupling relates to ecosystem integrity. Fisher et al. (1998) propose a conceptual "telescoping ecosystem model," which suggests that biological cross-links between the aquatic and terrestrial landscapes may enhance the resilience of the ecotone, although flooding is assumed to be the dominant mechanism. For instance, salmon carcasses were found to link the energy and nutrient budgets of both aquatic and terrestrial ecosystems (Cederholm et al. 1999).

Many salamander species are migratory and exhibit both short-term and short-distance movements along landscape corridors or between habitat patches, including migrations of adults to breeding sites. Hairston (1987) and Pauley et al. (2000) cite numerous examples of salamander species that migrate through forests between aquatic and terrestrial landscapes, often at night or during wet periods. Reported distances moved range from 3–1600 m for *Ambystoma* species, 3–60 m for *Desmognathus* species, 100 m for *Eurycea bislineata*, and 800 m for *Notophthalmus viridescens* (Pauley et al. 2000). Disruption of a riparian habitat can significantly alter salamander migrations (Williams et al. 2002), although land-use patterns in the upper watershed may be as important for dispersal of salamander populations as riparian habitat (Willson & Dorcas 2003).

Some salamander species function as dispersal vectors during their migrations. *Ambystoma* salamanders are known to transport mollusks (e.g., pea clam, *Pisidium adamsi*) and achenes of the bur-marigold, *Bidens cernua*, between pooled habitats during spring migrations (Lowcock & Murphy 1990). Mudpuppies (*Necturus*) serve as a migratory host for the salamander mussel (*Simpsonaias ambigua*), the only North American mussel species known to parasitize a vertebrate host other than a fish (Watters 1995). Metamorphic *Ambystoma* salamanders are responsible for dispersal of fairy shrimp (*Branchinecta coloradensis*) eggs between forest pools by feeding on female fairy shrimp in one pool and defecating in another (Bohonak & Whiteman 1999). These observations suggest a hypothesis that salamander dispersal may provide biotic control of ecosystem processes within both stream and pond ecotones as indicated by the telescoping ecosystem model (Fisher et al. 1998). Experimental verification of this hypothesis would require large-scale removal of salamanders from the riparian, coupled with long-term monitoring of changes in energy flow, nutrient cycling, and population density of salamander prey and predators.

Salamanders also may play an important role in the riparian ecotone via processes of chemical transformation. Amphibians are reported to oxidize ingested aromatic hydrocarbons followed by conjugations to glucuronides and organic sulfates (National Research Council 1981). The high efficiency at which salamanders store lipids and proteins in their tails (Burton & Likens 1975b) suggests that salamanders living within the riparian ecotone could ingest aquatic prey with high body burdens of toxic organic compounds, such as pesticides and chlorohydrocarbons, which could then be oxidized and translocated into the terrestrial environment in less toxic form during salamander migrations. Conversely, Johnson et al. (1999) found that dermal exposure of trinitrotoluene (TNT) and polychlorinated biphenyls (PCBs) resulted in bioaccumulation in tissue of *Ambystoma tigrinum* at concentrations that could affect food-web modeling. Given their relatively long life spans and high numbers in ecosystems, salamanders may be a critical food-web link in the bioaccumulation of persistent chemicals such as mercury and PCBs. Research here would provide useful information on the toxicological role of salamanders as elemental sinks, chemical transformers, and cross-links of organic molecules and heavy metal ions between aquatic and terrestrial environments (see Sparling et al. 2000).

Regulation of Salamander Diversity

Hairston (1996), Petranka (1998), and Walls & Williams (2001) provide comprehensive reviews of the literature dealing with the ecological role of salamanders as regulators of other salamander species via processes of predation, competition, or both. Studies in which salamander species were either removed from or added to experimental plots indicate that salamanders regulate the distribution and abundance of other salamander species through complex interactions of competition and predation. The studies reviewed by Hairston (1987, 1996) in terrestrial habitats, and Wilbur (1997) in artificial ponds, demonstrate the predictive power of field experiments to address complex ecological questions concerning biotic interactions. Although the relative role of competition versus predation appears to vary among different salamander guilds and in different habitats, the experimental evidence is now conclusive that both processes are important in the regulation of salamander communities.

Salamanders as Prey

Many animals are known to consume salamanders, including birds, mammals, snakes, fishes, turtles, frogs, crayfish, predatory insects, and other salamanders (Petranka 1998). The nocturnal habits of salamanders, mimicry, and the toxic skin secretions present in many species indicate that predation pressure is an important selective agent that regulates the distribution and abundance of salamander populations. Burton & Likens (1975a) concluded that salamanders in mature forests "represent a higher quality source of energy and nutrients for tertiary consumers than birds, mice, and shrews." Given their relatively small size compared with birds and mammals, salamanders can exploit small prey items not selected by larger vertebrates and convert these food sources into biomass that is then made available to larger vertebrate predators (Feder 1983, Pough 1983).

Long-term storage of salamander energy and biomass should have strong stabilizing effects on ecosystem processes. Hairston (1987) suggests that the impact of this storage by salamanders is to dampen stochastic fluctuations in the rate of energy flow. One can extend this line of thinking to the cycling of nutrients. Perry (1994) reported that healthy forests retain nutrients at a similar efficiency regardless of successional stage because trophic pathways exist that allow diversions to "slowly available nutrient pools." The low energy demand, long life span, slow growth rates, and great abundance of salamanders suggest they may well be the most important slowly available nutrient pools in forests. Margalef (1968) concluded that self-regulating ecosystems tend to conserve information by replacing ecologically equivalent system elements during succession. Salamander life histories and population dynamics fit well this holistic concept of ecosystem function. Different salamander species with similar ecological roles are found across a wide range of environments and seral stages. Given their well-documented numerical dominance in discrete macroenvironments in forests, and the tendency for the density of many species to increase during ecological succession (as herein reviewed), we suggest that salamanders can help maintain the long-term resilience-resistance of trophic pathways by providing abundant biomass and slowly available nutrient pools for top predators, at each seral stage of forest succession.

Underground Retreats

The environmental impact of underground retreats is well known and extensively reviewed (Meadows & Meadows 1991, Butler 1995). Burrows and underground passageways have ecosystem level functions beyond the increased fitness they incur to the species that make and use them. A large number of salamander species are known to occupy underground retreats (Petranka 1998). This mode of life is widespread across numerous families and genera, with obvious adaptive value to organisms susceptible to desiccation and predation (Semlitsch 1983). Although the density of salamanders in subsurface soil habitats is mostly unknown, a census of *Plethodon cinereus* in Michigan (Test & Bingham 1948) found that successive removals of salamanders from plot-strips yielded captures of 118, 146, 131, and 101 individuals over time. Failure to reduce salamander densities after repeated removals led Test & Bingham (1948) to suggest that a large percentage of the salamander population was underground in burrows and not directly beneath cover objects in the forest floor. Taub (1961) experimentally documented that *P. cinereus* spends significant time in burrows at least 12 inches deep. The availability of small mammal runways and burrows are thought to limit the population density of some *Ambystoma* species (Faccio 2003). Recently, a three-year study of *Plethodon* salamanders from the Great Smoky Mountains verified that significant proportions of terrestrial populations are subterranean (Bailey et al. 2004a).

Although use of underground retreats by salamanders is well documented, a long-standing controversy exists as to whether salamanders create their own burrows or merely take residence in burrows constructed by other animals. According to Dunn (1926), two *Desmognathus fuscus* left in a terrarium for over a year were found with many well-formed soil burrows. Dunn (1928) subsequently argued that *Desmognathus* and *Plethodon* salamanders have the ability to make their own burrows and do not merely follow crannies made by other animals as implied by Nobel (1927). A number of salamander genera have now been reported to either create or modify soil burrows including *Plethodon* (Brooks 1946, Heatwole 1960), *Ambystoma* spp. (Gruberg & Stirling 1972, Semlitsch 1983, Jennings 1996), *Siren* (Etheridge 1986), and *Phaeognathus hubrichti* (Hale & Guyer 2000). Stebbins (1951) noted that captive *Dicamptodon ensatus* are "good burrowers" and dig in gravel in an attempt to bury themselves. Marcot & Vander Hayden (2001) list 9 of 21 salamander species from the Pacific Coast that either create or modify soil burrows. Within the lungless salamander family, genera from the subfamily (Desmognathinae) retain basal morphological characters that are associated with burrowing and wedging between rocks (Titus & Larson 1996), including heavily ossified skull, flat wedge-like head, atlanto-mandibular ligaments, enlarged dorsal spinal muscles, and hind limbs relatively larger than forelimbs.

Organisms that modulate the availability of resources to other species by causing either physical or chemical changes to habitats have been referred to as "ecosystem engineers" (Jones et al. 1994). The above citations, although somewhat circumstantial, allow for a hypothesis that salamander species may serve an important ecological role in forests as ecosystem engineers of soil dynamics by creation, modification, and long-term occupancy of underground burrow systems. Regardless of how it is accomplished, either by creating burrows or using existing passages, the long-term residence of salamanders below ground suggests a number of significant ecosystem level effects: (*a*) translocation of nutrients, fungi, and other microorganisms from the forest floor to subsurface plant root systems; (*b*) deposition of excretory nutrients and organic matter for use by bacteria and fungi; and (*c*) increased dispersal of gases (e.g., dissolved oxygen, nitrogen, carbon dioxide) through the soil matrix. We are unaware of any experimental studies of these ecosystem processes, which represent an important area for future investigation in forest soil dynamics. Migrations by abundant salamanders into underground retreats during catastrophic events such as forest fires (Pilliod et al. 2003) and volcanos (Zalisko & Sites 1989) may help reset the chronosequence of forest ecosystem recovery, with surviving salamanders acting as biological legacies, both as a source of high energy prey and as predators that regulate invertebrate prey populations.

SUMMARY AND DISCUSSION

This review considers the key ecological functions (Marcot & Vander Hayden 2001) of salamanders in terrestrial, riparian, aquatic, and subterranean environments within North America ecosystems (Figure 2, see color insert). The compiled evidence supports a hypothesis that salamanders help provide fundamental biotic control of numerous ecosystem processes: (*a*) They furnish an abundant source of energy and nutrients for both terrestrial and aquatic consumers such as birds, fish, reptiles, mammals, and decomposers; (*b*) as predators of invertebrate species associated with the decomposition of organic matter, they modulate energy pathways and the release of essential minerals; (*c*) as keystone predators (sensu Paine 1969), they decrease the abundance of competitively dominant prey, thereby increasing taxa diversity in lower trophic levels; (*d*) through their complex life cycles they serve as connecting pathways for energy and matter between aquatic and terrestrial landscape elements; (*e*) by occupying and modifying underground refugia, they serve as facilitators of soil dynamics; and (*f*) by converting and storing large amounts of secondary production in the form of salamander biomass, they enhance forest resilience-resistance ($=$ stability) throughout ecological succession.

Salamanders can also provide an important service to humans as sentinels of ecosystem integrity through their use as cost-effective and readily quantifiable metrics of ecosystem resilience-resistance. Welsh & Droege (2001) report that the coefficient of variation (CV) associated with statistical sampling trends for forest dwelling plethodontid salamanders (CV $=$ 27%) is significantly lower than other vertebrates, such as passerine birds (57%), small mammals (69%), and other amphibians (37%–46%). These numbers imply that population trends for terrestrial salamanders can be detected more quickly and with fewer years of monitoring effort than other vertebrate species. Up to 20 years of stability in local population density has been reported for some plethodontid species (Hairston 1996). Population densities for migratory pond-breeding ambystomatids and salamandrids are known to vary by an order of magnitude depending on the rain-year, making these salamander species less attractive for use as long-term quantitative sentinels of ecosystem resilience and resistance (Pechmann et al. 1991, Trenham et al. 2000, Semlitsch 2003b), although Cortwright (1998) statistically demonstrated no broad shifts in density of red-spotted newts over a 10- to 11-year period from 36 breeding ponds.

Rolstad et al. (2002) concluded that the use of indicator species to assess forest health should concentrate on species that show long-term population stability and repeatedly occur in distinct habitats within old-forest stands. Many species of salamanders fit well the indicator species criteria of Rolstad et al. (2002); however, monitoring programs will need to consider variation in detection probability to track long-term trends in population density (see Hyde & Simons 2001; Bailey et al. 2004a,b). Observations on the presence-absence of sentinel salamander species also can provide useful information about existing conditions in natural areas. For example, breeding populations of spotted salamanders (*Ambystoma maculatum*) have been associated with wetlands that have high floral diversity and integrity (Micacchion 2002). Numerous Plethodontid species have been suggested to help classify aquatic life use potential under the Clean Water Act for primary headwater streams (Ohio EPA 2002). As aptly coined by Vitt et al. (1990), salamanders can provide an important ecological role as "harbingers of environmental decay" and sentinels of ecosystem condition.

Our review documents widespread observations in which at least one salamander species (rarely two) dominates the local salamander guild in different macroenvironments (see text on structural dominance and Table 1). These findings indicate that current emphasis on protection of rare species of salamanders, although clearly worthy, should be expanded to include those species that are dominant (in numbers and/or biomass) across the landscape and/or that function as keystone predators (sensu Paine 1969). Disturbances that reduce these ecologically dominant salamander species could result in profound alteration of critical ecosystem functions (see Conner 1988, Chapin et al. 1997).

This recurrent pattern of numerical dominance by a single salamander species at the landscape scale may be a fundamental aspect of the "rules-of-assembly" (Wilson 1999) of undisturbed forests. Ecologists have long recognized that closely related taxa in a community are not equally abundant but that their numbers tend to conform to mathematical patterns such as the geometric-series, log-series, log-normal, or MacArthur broken-stick distribution (reviewed by Tokeshi 1993, Brown 1995). A review of salamander abundance data from five North American forests with at least 1000 captures (Welsh & Lind 1988, Mitchell et al. 1997, Ash 1997, Ross et al. 2000, Ford et al. 2002) shows that a geometric-series model, combined with an exponential power-law regression of the semi-log data, provides a good statistical fit (Table 1, Figure 3). The simplicity of the model is appealing because variations in regression slopes (steepness and elevation), coefficients of determination (R^2), and intercepts of x-axis (estimate of species richness) and y-axis (abundance of the dominant taxon) may be useful diagnostic tools to help predict structural changes of salamander communities between disturbed and undisturbed forests. For instance, the highly negative slope of the regression curve for the salamander community studied by Ash (1997) in Figure 3 suggests that the overall species diversity of the salamander community was disturbed, with disproportional numbers of the dominant ranked species, *Plethodon jordani*. Tokeshi (1993) used geometric-series models to show greater loss of plant community diversity

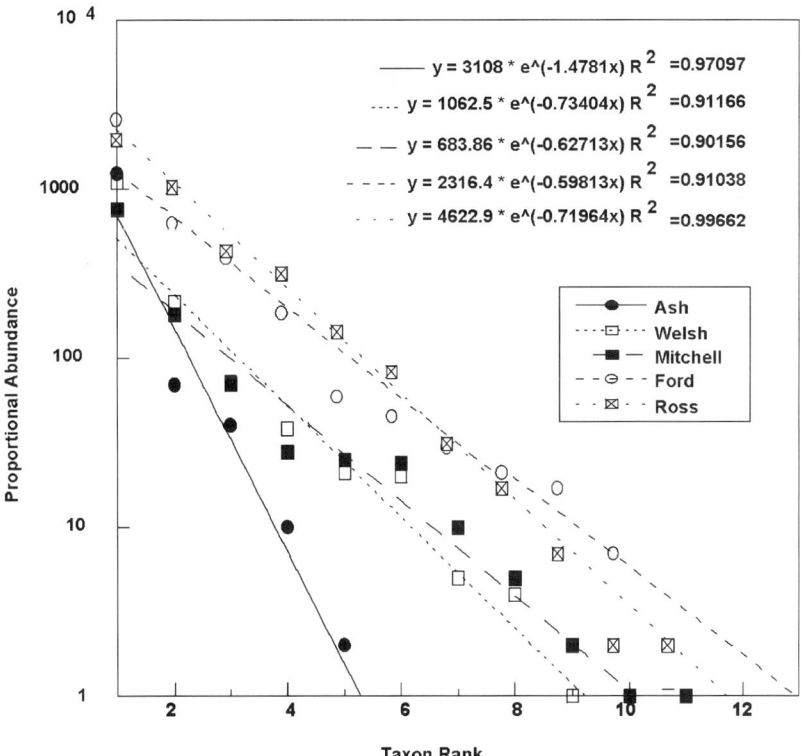

Figure 3 Geometric-series plots for salamander communities from five different forested landscapes. Regressions are exponential functions (all coefficients significant at $p < 0.01$). See Table 1 for raw data.

after long-term application of nitrogen fertilizer, and Sponseller et al. (2001) applied geometric-series regressions to study variation in benthic macroinvertebrate communities under stress from different watershed land uses. Future studies are needed to determine precisely how the mosaic pattern of forest habitats might cause deviations from a generalized geometric-series distribution for salamander communities, in a variety of undisturbed and disturbed forest areas, and at various stages of seral succession.

Concern over declining amphibian populations has produced a number of calls for greater conservation of salamander populations and their habitats (Bury et al. 1991; deMaynadier & Hunter 1995; Welsh & Lind 1995, 1996; Petranka 1998; Welsh & Droege 2001; Semlitsch & Rothermel 2003; Wyman 2003). If we view the landscape from the perspective of the salamander, it is the mosaic pattern of microenvironmental conditions present within different types of habitats (e.g., stream, pond, wetland, forest floor) that is critical to the long-term survival of salamander populations. Given the diversity of life history adaptations in salamanders and the tendency for salamanders to migrate between different habitats, protection of salamander species diversity will require attention to both micro and macrohabitat requirements (Welsh 1990; deMaynadier & Hunter 1995; Welsh & Lind 1995, 1996; Trenham 2001; Semlitsch & Rothermel 2003) in both aquatic and terrestrial environments.

ACKNOWLEDGMENTS

We thank Nancy Karraker, Lowell Orr, Thomas Pauley, Bradley Shaffer, and Clara Wheeler for helpful comments and input on various drafts of the manuscript.

The *Annual Review of Ecology, Evolution, and Systematics* is online at
http://ecolsys.annualreviews.org

LITERATURE CITED

Aber J, Christensen N, Fernandez I, Franklin J, Hidinger L, et al. 2000. Applying ecological principles to management of U.S. national forests. In *Issues in Ecology, No. 6.* Washington, DC: Ecol. Soc. Am.

Alford RA, Richards SJ. 1999. Global amphibian declines: a problem in applied ecology. *Annu. Rev. Ecol. Syst.* 30:133–65

Ash AN. 1997. Disappearance and return of plethodontid salamanders to clearcut plots in the southern Blue Ridge Mountains. *Conserv. Biol.* 11:983–89

Bailey LL, Simons TR, Pollock KH. 2004a. Estimating detection probability parameters for *Plethodon* salamanders using robust capture-recapture design. *J. Wildl. Manag.* 68:1–13

Bailey LL, Simons TR, Pollock KH. 2004b. Spatial and temporal variation in detection probability of *Plethodon* salamanders using robust capture-recapture design. *J. Wildl. Manag.* 68:14–24

Bailey RG. 1996. *Ecosystem Geography*. New York: Springer. 204 pp.

Barr GE, Babbitt KJ. 2002. Effects of biotic and abiotic factors on the distribution and abundance of larval two-lined salamanders (*Eurycea bislineata*) across spatial scales. *Pop. Ecol.* 133:176–85

Bartman CE, Parker KC, Laerm J, McCay TS. 2001. Short-term response of Jordon's salamander to a shelterwood timber harvest in western North Carolina. *Phys. Geogr.* 22:154–66

Bohonak AJ, Whiteman HH. 1999. Dispersal of the fairy shrimp *Branchinecta coloradensis* (Anostraca): effects of hydroperiod and salamanders. *Limnol. Oceanogr.* 44:487–93

Bormann FH, Likens GE. 1979. *Pattern and Process in a Forested Ecosystem*. New York: Springer. 253 pp.

Brodman R, Ogger J, Kolaczk M, Pulver RA, Long AJ, Bogard T. 2003. Mosquito control by pond-breeding salamander larvae. *Herpetol. Rev.* 34:116–19

Brooks M. 1946. Burrowing of *Plethodon jordani*. *Copeia* 1946:102

Brooks M. 1948. Notes on the Cheat Mountain salamander. *Copeia* 1948:239–44

Brown JH. 1995. *Macroecology*. Chicago: Univ. Chicago Press. 269 pp.

Burton TM, Likens GE. 1975a. Salamander populations and biomass in the Hubbard Brook experimental forest, New Hampshire. *Copeia* 1975:541–46

Burton TM, Likens GE. 1975b. Energy flow and nutrient cycling in salamander populations in the Hubbard Brook experimental forest, New Hampshire. *Ecology* 56:1068–80

Bury RB. 1983. Differences in amphibian

populations in logged and old growth redwood forest. *Northwest Sci.* 57:167–78

Bury RB. 1988. Habitat relationships and ecological importance of amphibians and reptiles. In *Streamside Management: Riparian Wildlife and Forestry Interactions*, ed. KJ Raedeke, pp. 61–76. Seattle: Univ. Wash., Inst. For. Res., Contrib. No. 59

Bury RB, Corn PS, Aubry KB. 1991. Regional patterns of terrestrial amphibian communities in Oregon and Washington. *USDA For. Serv. Gen. Tech. Rep. GTR PNW-GTR-285.* pp. 341–50

Bury RB, Corn PS, Aubry KB, Gilbert FF, Jones LLC. 1991. Aquatic amphibian communities in Oregon and Washington. *USDA For. Serv. Gen. Tech. Rep. GTR PNW-GTR-285.* pp. 326–38

Butler DR. 1995. *Zoogeomorphology: Animals as Geomorphic Agents.* New York: Cambridge Univ. Press. 231 pp.

Calef GW. 1972. Natural mortality of tadpoles in a population of *Rana aurora. Ecology* 54:741–58

Carfioli MA, Tiebout HM III, Pagano SA, Heister KM, Lutcher FC. 2000. Monitoring *Plethodon cinereus* populations. In *The Biology of Plethodontid Salamanders*, ed. RC Bruce, RG Jaeger, LD Houck, pp. 463–75. New York: Kluwer Academic/Plenum. 485 pp.

Cederholm CJ, Kunze MD, Murota T, Sibatani A. 1999. Pacific salmon carcasses: essential contributions of nutrients and energy for aquatic and terrestrial ecosystems. *Fisheries* 24:6–15

Chapin FS III, Walker BH, Hobbs RJ, Hooper DU, Lawton JH, et al. 1997. Biotic control over functioning of ecosystems. *Science* 277:500–4

Chew RM. 1974. Consumers as regulators of ecosystems: an alternative to energetics. *Ohio J. Sci.* 74:359–70

Collins JT, Taggart TW. 2002. *Standard Common and Current Scientific Names for North American Amphibians, Turtles, Reptiles, and Crocodilians.* Lawrence, KS: Cent. N. Am. Herpetol. 44 pp. 5th ed.

Conner RN. 1988. Wildlife populations: Minimally viable or ecologically functional? *Wildl. Soc. Bull.* 16:80–84

Cooperrider AR, Noss F, Welsh HH Jr., Carroll C, Zielinski W, et al. 2000. Terrestrial fauna of redwood forests. In *The Redwood Forest: History, Ecology, and Conservation*, ed. RF Noss, pp. 119–164. Covelo, CA: Island. 339 pp.

Corn PS, Bury RB, Hyde EJ. 2003. Conservation of North American stream amphibians. See Semlitsch 2003a, pp. 24–36

Cortwright SA. 1998. Ten- to eleven-year population trends of two pond-breeding amphibian species, red-spotted newts and green frogs. In *Status and Conservation of Midwestern Amphibians*, ed. MJ Lannoo, pp. 61–71. Iowa City: Univ. Iowa Press. 507 pp.

Davic RD. 1983. *An investigation of salamander guild predation in a North Carolina stream: an experimental approach.* PhD thesis. Kent State Univ. 237 pp.

Davic RD. 1991. Ontogenetic shift in diet of *Desmognathus quadamaculatus. J. Herpetol.* 25:108–11

Davic RD. 2003. Linking keystone species and functional groups: a new operational definition of the keystone species concept. *Conserv. Ecol.* 7(1):r11 http://www.consecol.org/vol7/iss1resp11

Davic RD, Orr LP. 1987. The relationship between rock density and salamander density in a mountain stream. *Herpetologica* 43:357–61

deMaynadier PG, Hunter ML Jr. 1995. The relationship between forest management and amphibian ecology: a review of the North American literature. *Environ. Rev.* 3:230–61

Deutschmann MR, Peterka JJ. 1988. Secondary production of tiger salamanders in three North Dakota prairie lakes. *Can. J. Fish Aquat. Sci.* 45:691–97

Diller LV, Wallace RL. 1996. Distribution and habitat of *Rhyacotriton variegates* in managed, young growth forests in north coastal California. *J. Herpetol.* 30:184–91

Dodd CK Jr., Smith LL. 2003. Habitat destruction and alteration: historical trends and future prospects for amphibians. See Semlitsch 2003a, pp. 94–112

Duellman WE. 1999. Global distribution of amphibians: patterns, conservation, and future challenges. In *Patterns of Distribution of Amphibians, a Global Perspective*, ed. WE Duellman, pp. 1–30. Baltimore, MD: Johns Hopkins Univ. Press. 633 pp.

Dunn ER. 1926. *Salamanders of the Family Plethodontidae*. Northhampton, MA: Smith College Anniv. Publ. 446 pp.

Dunn ER. 1928. The habitats of Plethodontidae. *Am. Nat.* 62:236–48

Dupuis LA, Smith JNM, Bunnell F. 1995. Relation of terrestrial-breeding amphibian abundance to tree-stand age. *Conserv. Biol.* 9:645–53

Efford IE. 1969. Energy transfer in Marion Lake, British Columbia; with particular reference to fish feeding. *Verh. Int. Ver. Limnol.* 17:104–8

Elwood JW, Newbold JD, O'Neill RV, Winkle WV. 1983. Resource spiraling: an operational paradigm for analyzing lotic ecosystems. In *Dynamics of Lotic Ecosystems*, ed. TD Fontaine III, SM Bartell, pp. 3–27. Ann Arbor, MI: Ann Arbor Sci. 494 pp.

Etheridge K. 1986. *Estivation in the sirenid salamanders, Siren lacertina, (Linnaeus) and Pseudobranchus striatus (Le Conte)*. PhD thesis. Univ. Florida. 74 pp.

Faccio SD. 2003. Postbreeding emigration and habitat use by Jefferson and spotted salamanders in Vermont. *J. Herpetol.* 37:479–89

Fauth JE. 1999. Identifying potential keystone species from field data—an example from temporary ponds. *Ecol. Lett.* 2:36–43

Fauth JE, Resetarits WJ. 1991. Interactions between the salamander *Siren intermedia* and the keystone predator *Notophthalmus viridescens*. *Ecology* 72:827–38

Feder ME. 1983. Integrating the ecology and physiology of plethodontid salamanders. *Herpetologica* 39:291–310

Fisher RN, Shaffer HB. 1996. The decline of amphibians in California's Great Central Valley. *Conserv. Biol.* 10:1387–97

Fisher SG, Grimm NB, Marti E, Holmes RM, Jones JB Jr. 1998. Material spiraling in stream corridors: a telescoping ecosystem model. *Ecosystems* 1:19–34

Ford WM, Chapman BR, Menzel MA, Odum RH. 2002. Stand age and habitat influences on salamanders in Appalachian cove hardwood forests. *For. Ecol. Manag.* 155:131–41

Gao K-Q, Shubin NH. 2001. Late Jurassic salamanders from northern China. *Nature* 410:574–77

Gehlbach FR, Kennedy SE. 1978. Population ecology of a highly productive aquatic salamander (*Siren intermedia*). *Southwest Nat.* 23:423–30

Gruberg ER, Stirling RV. 1972. Observations on the burrowing habits of the tiger salamander (*Ambystoma tigrinum*). *Herpetol. Rev.* 4:85–87

Gustafson EJ, Murphy NL, Crow TR. 2001. Using a GIS model to assess terrestrial salamander response to alternative forest management plans. *J. Environ. Manag.* 63:281–92

Hairston NG Sr. 1949. The local distribution and ecology of the plethodontid salamanders of the southern Appalachians. *Ecol. Monogr.* 19:49–73

Hairston NG Sr. 1987. *Community Ecology and Salamander Guilds*. New York: Cambridge Univ. Press. 230 pp.

Hairston NG Sr. 1996. Predation and competition in salamander communities. In *Long-Term Studies of Vertebrate Communities*, ed. ML Cody, JA Smallwood, pp. 161–86. San Diego, CA: Academic. 597 pp.

Hale SM, Guyer C. 2000. Effects of air temperature and depth on burrows of the red hills salamander (*Phaeognathus hubrichti*). *Herpetol. Nat. His.* 7:87–90

Harper CA, Guynn DC Jr. 1999. Factors affecting salamander density and distribution within four forest types in the southern Appalachian mountains. *For. Ecol. Manag.* 114:245–52

Harris PM. 1995. Are autecologically similar species also functionally similar? A test of pond communities. *Ecology* 76:544–52

Heatwole H. 1960. Environmental factors influencing local distribution and activity of the salamander *Plethodon cinereus*. *Ecology* 43:460–72

Hicks NG, Pearson SM. 2003. Salamander diversity and abundance in forests with alternative land use histories in the southern Blue Ridge Mountains. *For. Ecol. Manag.* 177:117–30

Holomuzki JR, Collins JP, Brunkow PE. 1994. Trophic control of fishless ponds by tiger salamander larvae. *Oikos* 71:55–64

Houlahan JE, Findlay CS, Schmidt BR, Meyer AH, Kuzmin SL. 2000. Quantitative evidence for global amphibian population declines. *Nature* 404:752–55

Houze CM Jr., Chandler CR. 2002. Evaluation of coverboards for sampling terrestrial salamanders in South Georgia. *J. Herpetol.* 36:75–81

Huang C, Sih A. 1991a. An experimental study on the effects of salamander larvae on isopods in stream pools. *Freshw. Biol.* 25:451–59

Huang C, Sih A. 1991b. Experimental studies on direct and indirect interactions in a three trophic-level stream system. *Oecologia* 85:530–36

Hyde EJ, Simons TR. 2001. Sampling plethodontid salamanders: sources of variability. *J. Wildl. Manag.* 65:624–32

Jennings MR. 1996. *Ambystoma californiense* (California tiger salamander), burrowing ability. *Herpetol. Rev.* 27:194

Johnson MS, Franke LS, Lee RB, Holladay SD. 1999. Bioaccumulation of 2,4,6-trinitrotoluene and polychlorinated biphenyls through two routes of exposure in a terrestrial amphibian: Is the dermal route significant? *Environ. Toxicol. Chem.* 18:873–78

Johnson DH, O'Neil TA. 2001. *Wildlife-Habitat Relationships in Oregon and Washington*. Corvallis: Oregon State Univ. Press. 736 pp.

Jones CG, Lawton JH, Shachak M. 1994. Organisms as ecosystem engineers. *Oikos* 69:373–86

Kiesecker JM, Belden LK, Shea K, Rubbo MJ. 2004. Amphibian decline and emerging diseases. *Am. Sci.* 92:138–47

Konishi M, Nakano S, Iwata T. 2001. Trophic cascading effects of predatory fish on leaf litter processing in a Japanese stream. *Ecol. Res.* 16:415–22

Krzysik AJ. 1998. Amphibians, ecosystems, and landscapes. In *Status and Conservation of Midwestern Amphibians*, ed. MJ Lannoo, pp. 31–44. Iowa City: Univ. Iowa Press. 507 pp.

Kurzava LM, Morin PJ. 1998. Tests of functional equivalence: complementary roles of salamanders and fish in community organization. *Ecology* 79:477–89

Leibold MA, Wilbur HM. 1992. Interactions between food-web structure and nutrients on pond organisms. *Nature* 360:341–43

Lowe WH, Bolger DT. 2002. Local and landscape-scale predictors of salamander abundance in New Hampshire headwater streams. *Conserv. Biol.* 16:183–93

Lowcock LA, Murphy RW. 1990. Seed dispersal via amphibian vectors: passive transport of Bur-marigold, *Bidens cerbua*, achenes by migrating salamanders, genus *Ambystoma*. *Can. Field Nat.* 104:298–300

Marcot BG, Vander Hayden M. 2001. Key ecological functions of wildlife species. See Johnson & O'Neil 2001, pp. 168–86

Margalef R. 1968. *Perspectives in Ecological Theory*. Chicago: Univ. Chicago Press. 111 pp.

Matson TO. 1998. Evidence for home ranges in mudpuppies and implications for impacts due to episodic applications of the lampricide TFM. In *Status and Conservation of Midwestern Amphibians*, ed. MJ Lannoo, pp. 278–87. Iowa City: Univ. Iowa Press. 507 pp.

Meadows PS, Meadows A. 1991. The environmental impact of burrowing animals and animal burrows. *Proc. Symp. Zool. Soc. London, No. 63*. Oxford: Clarendon. 349 pp.

Means DB. 2000. Southeastern U.S. coastal

plain habitats of the plethodontidae: the importance of relief, ravines, and seepage. In *The Biology of Plethodontidae Salamanders*, ed. RC Bruce, RG Jaeger, LD Houck, pp. 287–302. New York: Kluwer Academic/Plenum. 485 pp.

Menge BA, Freidenburg TL. 2001. Keystone species. In *Encyclopedia of Biodiversity*, ed. SA Levin, 3:613–31. San Diego, CA: Academic

Metts BS, Lanham JD, Russell KR. 2001. Evaluation of herpetofaunal communities on upland streams and beaver-impounded streams in the upper Piedmont of South Carolina. *Am. Midl. Nat.* 145:54–65

Micacchion M. 2002. Amphibian index of biotic integrity (AmphIBI) for wetlands. *Final Report to US EPA, Grant No. CD985875-01*, Ohio Environ. Prot. Agency, Div. Surf. Water, Columbus, OH. http://www.epa.state.oh. us/dsw/wetlands/2002_amphibian_report_final_rev.pdf

Mitchell JC, Rinehart SC, Pagels JF, Buhlmann KA, Pague CA. 1997. Factors influencing amphibian and small mammal assemblages in central Appalachian forests. *For. Ecol. Manag.* 96:65–76

Morin PJ. 1981. Predatory salamanders reverse the outcome of competition among three species of anuran tadpoles. *Science* 212:1284–86

Morin PJ. 1983. Competitive and predatory interactions in natural and experimental populations of *Notophthalmus viridescens dorsalis* and *Ambystoma tigrinum. Copeia* 1983:628–39

Morin PJ. 1987. Salamander predation, prey facilitation, and seasonal succession in microcrustacean communities. In *Predation: Direct and Indirect Impacts on Aquatic Communities*, ed. WC Kerfoot, A Sih, pp. 174–87. Hanover, NH: Univ. Press New England. 386 pp.

Morin PJ. 1995. Functional redundancy, nonadditive interactions, and supply-side dynamics in experimental pond communities. *Ecology* 76:133–49

Morin PJ, Wilbur HM, Harris RN. 1983. Salamander predation and the structure of experimental communities: responses of *Notophthalmus* and microcrustacea. *Ecology* 64:1430–36

Müller F, Windhorst W. 2000. Ecosystems as functional entities. In *Handbook of Ecosystem Theories and Management*, ed. SE Jørgensen, F Müller, pp. 33–49. Boca Raton, FL: CRC Press. 584 pp.

Murphy ML, Hall JD. 1981. Varied effects of clear-cut logging on predators and their habitat in small streams of the Cascade Mountains, Oregon. *Can. J. Fish Aquat. Sci.* 38:137–45

Nakamoto RJ. 1998. Effects of timber harvest on aquatic vertebrates and habitat in the North Fork Casper Creek. *USDA For. Serv. Gen. Tech. Rep. PSW-GTR-168*, pp. 87–95. Pac. Southwest Res. Stn., Berkeley, CA

National Research Council. 1981. *Testing for Effects of Chemicals on Ecosystems*. Washington, DC: Natl. Acad. Press. 103 pp.

NatureServe. 2003. Natural heritage central databases, Version 1.8. Arlington, VA. http://www.natureserve.org/explorer

Noble GK. 1927. The Plethodontid salamanders: some aspects of their evolution. *Am. Mus. Novitates.* 249:1–26

Noss RF. 1999. Assessing and monitoring forest biodiversity: a suggested framework and indicators. *For. Ecol. Manag.* 115:135–46

Nussbaum RA, Tait CK. 1977. Aspects of the life history of the Olympic salamander, *Rhyacotriton olympicus* (Gaige). *Am. Midl. Nat.* 98:176–99

Odum EP. 1971. *Fundamentals of Ecology*. Philadelphia, PA: W. B. Saunders. 574 pp. 3rd ed.

Ohio EPA. 2002. *Field Evaluation Manual for Ohio's Primary Headwater Habitat Streams*. Version 1.0, ed. RD Davic. Ohio Environ. Prot. Agency, Div. Surface Water, Columbus, OH. http://www.epa.state.oh.us/dsw/wqs/headwaters/PHWHManual_2002_102402.pdf

Ovaska K, Gregory PT. 1989. Population structure, growth, and reproduction in a

Vancouver Island population of the salamander *Plethodon vehiculum*. *Herpetologica* 45:133–43

Paine RT. 1969. A note on trophic complexity and community stability. *Am. Nat.* 103:91–93

Parent GH. 1992. L'utilisation des batracians et des reptiles comme bio-indicateurs. *Les Nat. Belges* 73:33–63

Parker MS. 1992. *Feeding ecology of larvae of the Pacific giant salamander* (Dicamptodon tenebrosus) *and their role as top predator in a headwater stream benthic community*. PhD thesis. Univ. Calif., Davis. 141 pp.

Parker MS. 1994. Feeding ecology of stream-dwelling Pacific giant salamander larvae (*Dicamptodon tenebrosus*). *Copeia.* 1994:705–18

Pauley TK, Mitchell JC, Buech RR, Moriarty J. 2000. Ecology and management of riparian habitats for amphibians and reptiles. In *Riparian Management in Forests of the Continental Eastern United States*, ed. ES Verry, JW Hornbeck, CA Dolloff, pp. 169–91. Boca Raton, FL: Lewis. 402 pp.

Pechmann JHK, Scott DE, Semlitsch RD, Caldwell JP, Vitt LJ, Gibbons JW. 1991. Declining amphibian populations: the problem of separating human impacts from natural fluctuations. *Science* 253:892–95

Perry DA. 1994. *Forest Ecosystems*. Baltimore, MD: Johns Hopkins Univ. Press. 649 pp.

Petranka JW. 1983. Fish predation: a factor affecting the spatial distribution of a stream-dwelling salamander. *Copeia* 1983:624–28

Petranka JW. 1998. *Salamanders of the United States and Canada*. Washington, DC: Smithson. Inst. Press. 587 pp.

Petranka JW, Eldridge ME, Haley KE. 1993. Effects of timber harvesting on southern Appalachian salamanders. *Conserv. Biol.* 7:363–70

Petranka JW, Murray SS. 2001. Effectiveness of removal sampling for determining salamander density and biomass: a case study in an Appalachian streamside community. *J. Herpetol.* 35:36–44

Pilliod DS, Bury RB, Hyde EJ, Pearl CA, Corn PS. 2003. Fire and amphibians in North America. *For. Ecol. Manag.* 178:163–81

Polis GA, Holt RD, Menge BA, Winemiller KO. 1996. Time, space, and life history: influence on food webs. In *Food Webs: Integration of Pattern and Dynamics*, ed. GA Polis, KO Winemiller, pp. 435–60. New York: Chapman & Hall. 472 pp.

Pough HF. 1983. Amphibians and reptiles as low-energy systems. In *Behavioral Energetics: The Cost of Survival in Vertebrates*, ed. WP Aspey, SI Lustick, pp. 141–88. Columbus: Ohio State Univ. Press. 300 pp.

Reice SR. 1983. Predation and substratum: factors in lotic community structure. In *Dynamics of Lotic Ecosystems*, ed. S Bartell, T Fontaine, pp. 323–345. Ann Arbor, MI: Ann Arbor Sci. 494 pp.

Reice SR, Edwards RL. 1986. The effect of vertebrate predation on lotic macroinvertebrate communities in Quebec, Canada. *Can. J. Zool.* 64:1930–36

Resetarits WJ Jr. 1997. Differences in an ensemble of streamside salamanders (Plethodontidae) above and below a barrier to brook trout. *Amphibia-Reptilia* 18:15–25

Ricklefs RE. 1979. *Ecology*. New York: Chiron. 966 pp. 2nd ed.

Rolstad J, Gjerde I, Gundersen VS, Saetersdal M. 2002. Use of indicator species to assess forest continuity: a critique. *Conserv. Biol.* 16:253–57

Rooney TP, Antolik C, Moran MD. 2000. The impact of salamander predation on collembola abundance. *Proc. Entomol. Soc. Wash.* 102:308–12

Ross B, Fredericksen T, Ross E, Hoffman W, Morrison M, et al. 2000. Relative abundance and species richness of herpetofauna in forest stands in Pennsylvania. *For. Sci.* 46:139–46

Schoch RR, Carroll RL. 2003. Ontogenetic evidence for Paleozoic ancestry of salamanders. *Evol. Develop.* 5:314–24

Schulze E-D, Mooney HA. 1993. Ecosystem function of biodiversity: a summary. In

Biodiversity and Ecosystem Function. ed. E-D Schulze, HA Mooney, pp. 497–510. Berlin: Springer-Verlag. 525 pp.

Semlitsch RD. 1983. Burrowing ability and behavior of salamanders of the genus *Ambystoma. Can. J. Zool.* 61:616–20

Semlitsch RD, ed. 2003a. *Amphibian Conservation.* Washington, DC: Smithson. Inst. 324 pp.

Semlitsch RD. 2003b. Introduction: general threats to amphibians. See Semlitsch 2003a, pp. 1–7

Semlitsch RD. 2003c. Conservation of pond-breeding amphibians. See Semlitsch 2003a, pp. 8–23

Semlitsch RD, Rothermel BB. 2003. A foundation for conservation and management of amphibians. See Semlitsch 2003a, pp. 242–60

Shaffer HB, McKnight ML. 1996. The polytypic species revisited: genetic differentiation and molecular phylogenetics of the tiger salamander *Ambystoma tigrinum* (Amphibia: Caudata) complex. *Evolution* 50:417–33

Shelford VE. 1913. Animal communities in temperate America. *Geogr. Soc. Chicago, Bull. No. 5.* Chicago: Univ. Chicago Press. 362 pp.

Sheridan CC, Olson DH. 2003. Amphibian assemblages in zero-order basins in the Oregon Coast Range. *Can. J. For. Res.* 33:1452–77

Simon TP, Jankowski R, Morris C. 2000. Modification of an index of biotic integrity for assessing vernal ponds and small palustrine wetlands using fish, crayfish, and amphibian assemblages along southern Lake Michigan. *Aquat. Ecosys. Health Manag.* 3:407–18

Snodgrass JW, Komoroski MJ, Bryan AL Jr., Burger J. 2000. Relationships among isolated wetland size, hydroperiod, and amphibian species richness: implications for wetland regulations. *Conserv. Biol.* 14:414–19

Sparkes TC. 1996. The effects of size-dependent predation risk on the interaction between behavioral and life history traits in a stream-dwelling isopod. *Behav. Ecol. Sociobiol.* 39:411–17

Sparling DW, Linder G, Bishop CA. 2000. *Ecotoxicology of Amphibians and Reptiles.* Pensacola, FL: Soc. Environ. Toxicol. Chem. 904 pp.

Sponseller RA, Benfield EF, Valett HM. 2001. Relationships between land use, spatial scale, and stream macroinvertebrate communities. *Freshw. Biol.* 46:1409–24

Stebbins RC. 1951. *Amphibians of Western North America.* Berkeley: Univ. Calif. Press. 539 pp.

Stebbins RC, Cohen N. 1995. *A Natural History of Amphibians.* Princeton, NJ: Princeton Univ. Press. 316 pp.

Strahler AN. 1964. Quantitative geomorphology of drainage basins and channel networks. In *Handbook of Applied Hydrology,* ed. VT Chow, pp. 40–74. New York: McGraw-Hill

Swift MJ, Heal OW, Anderson J. 1979. *Decomposition in Terrestrial Ecosystems.* Berkeley: Univ. Calif. Press. 372 pp.

Taub FB. 1961. The distribution of the red-backed salamander, *Plethodon c. cinereus,* within the soil. *Ecology* 42:681–98

Test FH, Bingham BA. 1948. Census of a population of the red-backed salamander (*Plethodon cinereus*). *Am. Midl. Nat.* 39:362–72

Titus TA, Larson A. 1996. Molecular phylogenetics of desmognathine salamanders (Caudata: Plethodontidae): a reevaluation of evolution in ecology, life history, and morphology. *Syst. Biol.* 45:451–72

Tokeshi R. 1993. Species abundance patterns and community structure. *Adv. Ecol. Res.* 24:111–95

Trenham PC, Shaffer HB, Koenig WD, Stromberg MR. 2000. Life history and demographic variation in the California tiger salamander (*Ambystoma californiense*). *Copeia* 2000:365–77

Trenham PC, Koenig WD, Shaffer HB. 2001. Spatially autocorrelated demography and interpond dispersal in the salamander *Ambystoma californiense. Ecology* 82:3519–30

Vannote RL, Minshall GW, Cummins KW, Sedell JR, Cushing CE. 1980. The river continuum concept. *Can. J. Fish. Aquat. Sci.* 37:130–37

Vitt LJ, Caldwell JP, Wilbur HM, Smith DC. 1990. Amphibians as harbingers of decay. *Bioscience* 40:418

Walls SC, Williams MG. 2001. The effect of community composition on persistence of prey with their predators in an assemblage of pond-breeding amphibians. *Oecologia* 128:134–41

Waters JR, Zabel CJ, McKelvey KS, Welsh HH Jr. 2001. Vegetation patterns and abundances of amphibians and small mammals along streams in a northwestern California watershed. *Northwest Sci.* 75:37–52

Watters GT. 1995. *A Guide to the Freshwater Mussels of Ohio*. Columbus, OH: Div. Wildl., Ohio Dep. Nat. Res. 3rd ed.

Wallace JB, Webster JR. 1996. The role of macroinvertebrates in stream ecosystem function. *Annu. Rev. Entomol.* 41:115–39

Welsh HH Jr. 1990. Relictual amphibians and old-growth forests. *Conserv. Biol.* 4:309–19

Welsh HH Jr., Droege S. 2001. A case for using plethodontid salamanders for monitoring biodiversity and ecosystem integrity of North American forests. *Conserv. Biol.* 15:558–69

Welsh HH Jr., Lind AJ. 1988. Old growth forests and the distribution of the terrestrial herpetofauna. *USDA For. Serv. Gen. Tech. Rep. GTR RM-166*. pp. 439–54. Rocky Mountain For. Range Exp. Stn., Ft. Collins, CO

Welsh HH Jr., Lind AJ. 1991. The structure of the herpetofaunal assemblage in the Douglas-fir/hardwood forests of northwestern California and southwestern Oregon. *USDA For. Serv. Gen. Tech. Rep. GTR PNW-GTR-285*. pp. 394–13. Pac. Northwest Res. Stn., Portland, OR

Welsh HH Jr., Lind AJ. 1992. Population ecology of two relictual salamanders from the Klamath Mountains of northwestern California. In *Wildlife 2001: Populations*, ed. DR McCulloch, RH Barrett, pp. 419–437. London: Elsevier Sci. 1163 pp.

Welsh HH Jr., Lind AJ. 1995. Habitat correlates of the Del Norte salamander, *Plethodon elongatus* (Caudata: Plethodontidea), in northwestern California. *J. Herpetol.* 29:198–210

Welsh HH Jr., Lind AJ. 1996. Habitat correlates of the southern torrent salamander, *Rhyacotriton variegatus* (Caudata: Rhyacotritonidae), in northwestern California. *J. Herpetol.* 30:385–98

Welsh HH Jr., Lind AJ. 2002. Multiscale habitat relationships of stream amphibians in the Klamath-Siskiyou Region of California and Oregon. *J. Wildl. Manag.* 66:581–602

Welsh HH Jr., Ollivier LM. 1998. Stream amphibians as indicators of ecosystem stress: a case study from California's redwoods. *Ecol. Appl.* 8:1118–31

Wheeler BA, Prosen E, Mathis A, Wilkinson RF. 2003. Population declines of a long-lived salamander: a 20+ year study of hellbenders, *Cryptobranchus alleganiensis*. *Biol. Conserv.* 109:151–56

Wilbur HM. 1997. Experimental ecology of food webs: complex systems in temporary ponds. *Ecology* 78:2279–302

Williams LR, Crosswhite DL, Williams MG. 2002. Short-term effect of riparian distribution on *Desmognathus brimleyorum* (Plethodontidae) at a natural spring in Oklahoma. *Southwest Nat.* 47:611–13

Wilkins R, Peterson NP. 2000. Factors related to amphibian occurrence and abundance in headwater streams draining second-growth Douglas-fir forests in southwestern Washington. *For. Ecol. Manag.* 139:79–91

Willson JD, Dorcas ME. 2003. Effects of habitat disturbance on stream salamanders: implications for buffer zones and watershed management. *Conserv. Biol.* 17:763–71

Wilson EO. 1999. *The Diversity of Life*. New York: Norton. 424 pp.

Wissinger SA, Whiteman HH, Sparks GB, Rouse GL, Brown WS. 1998. Foraging trade-offs along a predator-permanence gradient in subalpine wetlands. *Ecology* 80:2102–16

Wyman RL. 1998. Experimental assessment

of salamanders as predators of detrital food webs: effects on invertebrates, decomposition and the carbon cycle. *Biodivers. Conserv.* 7:641–50

Wyman RL. 2003. Conservation of terrestrial salamanders with direct development. See Semlitsch 2003a, pp. 37–52

Zalisko EJ, Sites RW. 1989. Salamander occurrence within Mt. St. Helens blast zone. *Herpetol. Rev.* 20:84

Zug GR, Vitt LJ, Caldwell JP. 2001. *Herpetology: An Introductory Biology of Amphibians and Reptiles.* San Diego, CA: Academic. 630 pp. 2nd ed.

ECOLOGICAL AND EVOLUTIONARY CONSEQUENCES OF MULTISPECIES PLANT-ANIMAL INTERACTIONS

Sharon Y. Strauss[1] and Rebecca E. Irwin[2]

[1]Section in Evolution and Ecology and Center for Population Biology, University of California, Davis, California 95616; email: systrauss@ucdavis.edu
[2]Institute of Ecology, University of Georgia, Athens, Georgia 30602; email: rirwin@uga.edu

Key Words coevolution, demography, herbivory, indirect effects, mutualism

■ **Abstract** Ecologists and evolutionary biologists are broadly interested in how the interactions among organisms influence their abundance, distribution, phenotypes, and genotypic composition. Recently, we have seen a growing appreciation of how multispecies interactions can act synergistically or antagonistically to alter the ecological and evolutionary outcomes of interactions in ways that differ fundamentally from outcomes predicted by pairwise interactions. Here, we review the evidence for criteria identified to detect community-based, diffuse coevolution. These criteria include (*a*) the presence of genetic correlations between traits involved in multiple interactions, (*b*) interactions with one species that alter the likelihood or intensity of interactions with other species, and (*c*) nonadditive combined effects of multiple interactors. In addition, we review the evidence that multispecies interactions have demographic consequences for populations, as well as evolutionary consequences. Finally, we explore the experimental and analytical techniques, and their limitations, used in the study of multispecies interactions. Throughout, we discuss areas in particular need of future research.

INTRODUCTION

A major goal in ecology and evolutionary biology is to understand how the interactions among organisms influence their abundance, phenotypes, and genotypes. Although the complexity of interactions in natural systems has been acknowledged (e.g., Billick & Case 1994, Juenger & Bergelson 1998, Paine 1992, Polis & Strong 1996, Wootton 1993), studies in terrestrial plant-animal interactions have classically focused on direct, pairwise interactions. More recently, however, an appreciation has developed of how multispecies interactions significantly alter both the ecological and evolutionary outcomes of interactions in ways that could not be predicted from an understanding of pairwise interactions alone (Miller & Travis 1996, Pilson 1996, Strauss 1991, Thompson 1999). The community context of interactions, primarily in terrestrial plant-animal systems, is the focus of this paper, and our goal is to review and highlight common themes among disparate studies of multispecies interactions. Because this area is so large, we focus this review almost solely on the responses of plants to community membership; however, the same kinds of interactions and selective effects also occur for the animals participating in these interactions.

Plants rarely interact with a single mutualistic or antagonistic species. Rather, sessile plants must integrate interactions across a suite of different mutualists and antagonists, usually simultaneously. These visitors are taxonomically diverse, use many different parts of a plant, and usually vary in their impacts on plant fitness along a continuum from positive to negative, direct to indirect. For example, when plants are attacked by enemies, a cascade of responses typically ensues. Common plant reactions to damage include altered allocation to root:shoot biomass, short-term increases in photosynthetic rate, and the induction of costly structural and chemical defenses in plant tissues. Changes in allocation to reproductive structures such as floral size and rewards after damage have been shown to alter relationships with pollinators (see section below). Moreover, induced responses are often systemic and may influence relationships with pathogens, enemies of herbivores, and other community members. Understanding how plants and animals interact and evolve in this community context is key to understanding trait evolution. Below, we consider the evidence for criteria that create conditions under which diffuse, community-dependent evolution or selection could occur. In addition, we review the evidence that these multispecies interactions have ecological demographic consequences for plant populations, as well as evolutionary consequences. Throughout, we emphasize the areas that we think offer the most promising prospects for future research.

PAIRWISE VERSUS DIFFUSE EVOLUTION

The idea that communities exert selective pressures that differ fundamentally from those imposed by multiple, pairwise combinations of species has been discussed for many years, and the term "diffuse coevolution" was introduced by Janzen (1980) in his seminal note *When Is It Coevolution?* In recent prominent papers, several authors have outlined criteria to evaluate whether evolution is diffuse (determined by interactions with many species), as opposed to pairwise (reflecting the independent interactions between pairs of species, even multiple pairs) (e.g., Hougen-Eitzman & Rausher 1994, Stinchcombe & Rausher 2002, Iwao & Rausher 1997).

In pairwise evolution, traits involved in one set of interactions evolve independently of traits involved in other interactions. The evolutionary dynamics of the two interacting species are independent of the presence or actions of other community members. For evolution to be driven by some emergent property of multispecies communities, effects of interactors should be nonadditive; that is, one could not predict the selective pressures a focal species would experience simply from knowing the selective effects of each interacting species alone. However, both additive and nonadditive effects of multiple species may lead to diffuse selection. The following criteria have been proposed to determine whether selection is diffuse. These criteria were originally proposed for herbivores only and were presented as criteria for pairwise evolution that, if violated, would provide evidence for diffuse evolution. They are paraphrased from the original papers to reflect criteria for broad communities and to reflect diffuse selection (Hougen-Eitzman & Rausher 1994, Iwao & Rausher 1997, Stinchcombe & Rausher 2002).

1. Traits important to interactions with multiple species are genetically correlated with one another; that is, selection on one trait will influence the value of traits important in other interactions. We add to this criterion the variant that there may be conflicting selection on the same trait exerted by multiple interactors.
2. The presence or absence of one community member mediates interactions with others. For example, attack by one species alters the likelihood or intensity of interactions with, or selection by, other community members.
3. The effects of multiple interactors on plant fitness are not additive. Thus, the effect of species in combination on the fitness of a focal species are not just the sum of the effects of each species separately. In considering such effects, it may be conceptually useful to divide effects of the interactors on fitness into (a) the cumulative costs (or benefits) of response to the interactors and (b) the effects of the interactors themselves (Miller & Travis 1996). As stated, this criterion addresses ecological impacts of multiple species more than evolutionary ones, because to show diffuse selection, one must focus on the nonadditive effects of species on the relationship between trait(s) and fitness, not fitness alone (see Inouye & Stinchcombe 2001, Strauss et al. 2004 for more discussion.)

What evidence do we have that bears on these criteria? Although a number of studies address criterion 2, surprisingly few studies have addressed whether genetic correlations exist among traits important in multispecies plant-animal interactions (criterion 1) and whether the impacts of multiple interactors on plants are nonadditive (criterion 3, especially for multispecies mutualisms). Here, we review evidence for each criterion from a diverse suite of multispecies plant-animal interactions.

Criterion 1: Genetic Correlations Among Traits

When traits important to interactions with multiple species are genetically correlated with one another, multiple interactors can affect the evolution of single traits. The nature of the correlation, as well as the nature of interactions with multiple species, will promote or de-emphasize the community context of such evolution. In addition, traits that have effects on multiple interactions will be shaped by the combined effects of community members.

Consider a simple scenario of a pair of genetically correlated plant traits, X1 and X2, in which increased values of X1 attract one insect species and increased values of X2 attract a different insect species. If both insects are mutualists or if both are antagonists, a positive genetic correlation between X1 and X2 could speed the rate of fixation of alleles that influence these traits. In this case, both agents drive correlated traits in the same direction, even in years when one agent is less abundant or less important in determining plant fitness. The community context will affect the speed, but not the outcome, of evolutionary change.

On the other hand, if the two insect species have opposing effects on plant fitness—for example, one is a pollinator and the other is a seed predator—then a positive genetic correlation between X1 and X2 is likely to result in selection that fluctuates, and may even change sign, with changes in insect abundance and interaction strength. In this case, fluctuating selection may prevent the fixation of alleles that optimize values of traits X1 and X2.

A negative genetic correlation between these traits would have similar effects. When both insects are antagonists (or both are mutualists), we expect fluctuating selection. However, if the two insect species have opposing effects on plant fitness, we expect fixation of alleles in these traits because both agents are driving traits in the same direction because of their correlated responses. In these scenarios, community context and the frequency and intensity of fitness impacts of multiple community members become critical to understanding trait evolution. Interactions in the field are notoriously variable in strength because species composition and abundance varies from year to year, and the strength of the interactions may be modified by the presence or absence of other community members, for example, another pollinator or seed predator (Thompson 1994) or abiotic conditions (Galen 1999).

HERBIVORE-HERBIVORE INTERACTIONS In a recent review, Rausher (1996) stated that most genetic correlations between resistance to different herbivores were either zero or positive. A correlation of zero means that defense traits are evolving in response to independent pairwise interactions (barring the existence of phenomena in criteria 2 and 3). However, Berenbaum et al. (1986) showed negative genetic correlations between the amounts of different secondary compounds in *Pastinaca sativa* (Apiaceae) and, thus, documented constraints on the evolution of resistance to different herbivores because compounds were differentially effective at deterring different herbivore species. Since the review by Rausher (1996), a few more cases of negative genetic correlations between attack by different herbivores have been documented (Juenger & Bergelson 1998, Mitchell-Olds et al. 1996, Stinchcombe & Rausher 2001), along with other studies that show positive or no genetic correlations (Tiffin & Rausher 1999).

TRAITS INVOLVED IN VARIOUS MULTISPECIES INTERACTIONS Remarkably few studies have undertaken examinations of genetic correlations between traits important in diverse, simultaneous interactions across a wide range of taxa and interactions. Siemens & Mitchell-Olds (1998) found a negative genetic correlation between resistance to the fungal pathogen *Peronospora parasitica* and plant-growth rate in *Brassica rapa* (Brassicaceae), a result that suggests the possibility of a disease resistance/plant competition tradeoff. Similarly, *Diplacus aurantiacus* (Scrophulariaceae) and *Pastinaca sativa* exhibit negative genetic correlations between the production of secondary plant metabolites that possess documented antiherbivore functions and plant growth rate (Berenbaum et al. 1986, Han & Lincoln 1995). For *Diplacus aurantiacus*, the production of 1 mg of resin content comes at a cost of 25 mg of dry-shoot biomass growth (Han & Lincoln 1995). Again, this result supports the possibility that better-defended plants may be poorer competitors for light and resources, but such growth/defense tradeoffs (Coley et al. 1985) remain to be documented in these systems. A phenomenon that might work against such a tradeoff is the possibility that increased levels of secondary compounds may also serve as allelochemicals that suppress competitors (Siemens et al. 2002). This suppression may offset the costs of production of these compounds in terms of growth rate and, thus, may not result in a defense/competition tradeoff.

To our knowledge, no studies to date document a negative genetic correlation between separate attraction and defense traits in plants, although Strauss et al. (2004) come close. After crossing plants into similar genetic background and using only greenhouse-grown plants to control for parental environment, they found a significant correlation between petal pigment, which is heritable, and foliar glucosinolate content, which serves as a defense against herbivores and also has a heritable basis (Strauss et al. 2004). Anthocyanin-dominant purple and bronze morphs of *Raphanus sativus* (wild radish; Brassicaceae) had greater induction of foliar glucosinolates than did anthocyanin-recessive yellow and white morphs. In addition, field evidence shows that pollinators prefer yellow-flowered plants (see also Stanton 1987), the least-defended genotype. Herbivores, on the other hand, generally had better performance on yellow and white genotypes and, thus, should act to select against anthocyanin-recessive morphs in the field (Irwin et al. 2003). As a result of the linked expression of these traits, herbivores may affect petal-color traits important to pollinators, and pollinators may drive variation in defense traits by favoring the least-defended, yellow genotype. Given the large amount of work that has focused on plant-animal interactions, surprisingly few studies have investigated genetic correlations between traits involved in interactions with diverse community members.

CONFLICTING SELECTION PRESSURES ON THE SAME TRAIT   Opposing selection by antagonistic and mutualistic community members may be particularly common for attractive characters, such as flowers, because mutualists and antagonists alike can use these conspicuous traits to locate plants. Such examples might be thought of as ecological pleiotropy, wherein the same trait affects multiple interactions. In some cases, community members can act upon the same trait in opposing directions, which makes fitness effects of the trait in different ecological contexts negatively genetically correlated. Community-based selection is then likely to influence the evolution of ecologically pleiotropic traits in fluctuating environments. How the community context influences selection will depend, in part, on whether community composition covaries with factors that increase or decrease the average performance of a focal species and on whether selection is hard or soft (Futuyma 1986).

Some well-known instances of conflicting selection pressures imposed by different community members on the same traits include pollinators versus seed predators (e.g., Brody 1992) and pollinators versus florivores/nectar thieves (e.g., Galen & Cuba 2001, Gómez 2003, Herrera et al. 2002). Floral traits often reflect an adaptive compromise between relationships with mutualistic pollinators and relationships with antagonists (Brody 1992, Galen & Cuba 2001). Different plant parts have typically been ascribed to serve in each function—sepals and the ovary wall protect the ovary and developing seed (Grant 1950), whereas the corolla or modified sepals serve in pollinator attraction and efficacy. However, defense and attraction functions may not be as easily partitioned as initially thought. Petals are defended against herbivores (Euler & Baldwin 1996, Strauss et al. 2004), and the degree to which nectar and pollen reward chemistry are independent of petal and leaf chemistry is unclear (Adler 2000). In *Hypericum calycinum* (Hypericaceae) flowers, the same ultraviolet pigments play a defensive role in the stamens and ovaries and an attractive role in the petals (Gronquist et al. 2001). In the *Dalechampia* clade (Euphorbiaceae), resins involved in chemical defense against herbivores and microbes have secondarily assumed a role in pollinator attraction and reward, and bracts serving to attract pollinators have been co-opted as a defense against flower-feeding herbivores (Armbruster 1997).

Galen has done some of the most complete work on how the effects of extremely diverse selective agents affect trait evolution in plants (reviewed in Galen 1999). In this case, different agents are acting in opposing directions on the same corolla flare/corolla tube-length traits, which are positively genetically correlated. Flowers of the alpine sky pilot, *Polemonium viscosum* (Polemoniaceae), that have smaller, narrow corollas experience lower risk of predation on floral parts from nectar-thieving ants than do larger, more flared flowers. However, as documented by experimental manipulations and observations in the field, narrow-tube forms, although better defended from ants, receive 47% fewer pollen grains and set 62% fewer seeds than do flared corollas. This difference in reproductive success reflects the foraging behavior of bumblebee pollinators that also seek nectar rewards. In this case, both mutualist pollinators and antagonist nectar thieves forage for the same nectar reward and use flower shape and size as a cue (Galen 1999). Moreover, in high-elevation populations, water stress favors smaller corollas that reduce water loss from the plant. Thus, the benefits of bee pollination counter costs of predation by ants and water loss in flared corolla morphs.

Several other cases are known in which herbivores and pollinators act on the same floral traits (Brody 1992, Ehrlén et al. 2002). Scape length in *Primula farinosa* (Primulaceae) appears to be under opposing selection from seed predators and pollinators (Ehrlén et al. 2002). Similarly, tall *Erysimum mediohispanicum* (Brassicaceae) plants exposed to browsers are selected against because of grazing, whereas tall plants inside ungulate exclosures are favored by pollinators (Gómez 2003). In a different correlative study across 20 species of composites (with three additional within-species comparisons), increasing capitulum size was associated with increasing incidence of infestation from bud predators and was not correlated with other variables, a result that suggests that capitulum size may be determined primarily from opposing selection from pollinators and bud predators (Fenner et al. 2002).

Opposing selection on the same trait can also be seen in tradeoffs between resistance to generalist and specialist herbivores: specialists often use volatiles to locate hosts and often have the ability to detoxify, sequester, or excrete secondary compounds. In contrast, generalist herbivores are often deterred by the presence of high concentrations of these same compounds. Thus, plants that are well defended against generalists are often consumed by specialists (Gatehouse 2002). Taken together, these examples suggest many cases in which the same trait comes under opposing selection from multiple community members.

Criterion 2: Interactions with one Species Affect the Likelihood or Intensity of Interactions with Other Species

Community context can affect the likelihood or intensity of interactions among community members and can have important ecological effects on the population size of interactors as well as important evolutionary effects if they change patterns of selection on traits. Patterns of plant use, and potential patterns of selection on plants, may be influenced by interactions through at least two common pathways: density-mediated effects and trait-mediated effects (for a review, see Wootton 2002). Density-mediated effects arise when plant-animal interactions are altered through changes in population density of community members. For example, herbivores may alter plant–seed disperser interactions through changes in seed density. Alternatively, trait-mediated effects arise when interactions with one species cause a change in trait value that subsequently affects interactions with a third species; for example, herbivores may change plant-pollinator interactions through changes in the quality of flowers produced. Shifts in ecological interactions may translate into altered selective landscapes. The presence of spider predators can cause grasshoppers to shift from eating primarily grasses to primarily forbs (Schmitz 1998) and may therefore have impacts not only on forb and grass abundance but also on the defensive traits of these co-occurring species.

One theme running through all of the studies described below is that trait-mediated effects are particularly common. Trait-mediated effects are pervasive properties of interacting species. Their effects may be nonadditive and difficult to predict in isolation (Wootton 1993). In general, most organisms have immune or defensive systems that respond dynamically and plastically to attack from pathogenic or trophic organisms. Such responses are likely omnipresent (see Table 1 for

TABLE 1 Scenarios under which plant interactions with one species could affect the likelihood or intensity of plant interactions with others[a]

Scenario	Type of interaction	Example of potential mechanism involved	Exemplar reference
Antagonist → antagonist	Herbivore → herbivore	Induction of defensive trait	Karban & Baldwin 1997
	Pathogen → herbivore	SAR pathway	Rojo et al. 2003
	Herbivore → seed predator	Change in flowering phenology	Juenger & Bergelson 1998
Mutualist → mutualist[b]	Pollinator → pollinator	Pollinator scent marks	Stout & Goulson 2001
	Ant → ant	Competition for rewards	Palmer et al. 2003
	Granivore → granivore	Change in vegetation structure	Valone et al. 1994
	Pollinator → granivore	Change in fruit/seed abundance	—[c]
	Granivore → pollinator	Change in plant demography	—
	Nectar robber → pollinator	Change in nectar availability	Irwin & Brody 2000
	Mycorrhizae → pollinator/fruit disperser	Change in nectar or fruit quality	—
	Endophytic fungi → enemies of herbivores	Change in plant quality	Omacini et al. 2001
	Endophyte → granivore	Change in seed quality	Madej & Clay 1991
Antagonist → mutualist[b]	Herbivore → pollinator	Change in floral traits	Mothershead & Marquis 2000
	Herbivore → granivore	Change in vegetation structure	Smit et al. 2001
	Herbivore → mycorrhizae	Change in plant-resource status	Eom et al. 2001
Mutualist[b] → antagonist	Pollinator → seed predator	Change in fruit/seed rewards	Herrera 2000
	Granivore → herbivore	Change in seedling demography	—
	Endophyte → herbivore	Induction of defensive trait	Clay et al. 1993
	Mycorrhizae → herbivore	Change in plant quality	Gange et al. 2002

[a] This list should be considered as a starting point for the types of multispecies interactions that could occur and is by no means exhaustive. The strength and outcome of many of these multispecies interactions are conditional on the organisms involved and on the environmental conditions in which they occur (see text).

[b] Mutualists can vary in the strength of their positive effects on plants depending on community context (reviewed in Stanton 2003). We are lumping granivores into "mutualists" here because sometimes granivores can increase seedling germination by dispersing or caching seeds; however, we fully realize that costs are associated with granivory as well, and sometimes the costs can outweigh the benefits. The same logic applies to nectar robbers and endophytes.

[c] Dashes indicate that we were unable to find a study that conclusively demonstrated the proposed interaction. These types of interactions are in need of further empirical attention.

different categories of interactions and an exemplar mechanism and citation). Below, we review the literature for interactions that have not received much prior attention and suggest review articles for other topics that have been recently reviewed.

HERBIVORE-HERBIVORE/PATHOGEN INTERACTIONS   Defenses induced by pathogen or herbivore attack in plants fit well within the trait-mediated paradigm. Karban & Baldwin (1997) document more than 24 cases in which attack by one herbivore results in altered levels of leaf damage by, attractiveness to, or mortality of, another subsequently interacting herbivore species that fed on the same plant. These examples were drawn solely from the agricultural literature, but numerous other examples suggest that these effects are ubiquitous in native systems as well (e.g., 13 examples from natural systems in Hougen-Eitzman & Rausher 1994).

In addition, a growing body of literature discusses cross-talk between induction from pathogens that cause upregulation of the salicylate (SA)-based systemic acquired resistance (SAR) and ethylene (E) pathways in plants and the jasmonate-based (JA-based) induced-response (IR) pathway stimulated by insect attack. In some cases, these pathways appear to compete with one another, such that upregulation of the SA pathway caused by disease makes plants more susceptible to insect attack (and less susceptible to disease), whereas induction of JA by insects has the reverse effect (Rojo et al. 2003, Thaler et al. 2002).

MUTUALIST-MUTUALIST INTERACTIONS   Just as herbivore feeding can influence plant resistance to subsequent enemy attack, patterns of host-plant use by one mutualist may alter subsequent plant-mutualist interactions. For example, many pollinators leave attractant and repellant scent marks that affect subsequent floral visitation (Goulson et al. 2000, Guirfa 1993). Stout & Goulson (2001) found that scent marking by *Bombus lapidarius* on the flowers of *Melilotus officinalis* (Fabaceae) deterred conspecific and heterospecific (*Apis mellifera*) floral visitors over a 40-minute period. Similarly, pollinator visitation may induce flower-color change in many plant species and have subsequent effects on floral visitation (Gori 1983). Hummingbird pollinators of *Malvaviscus arboreus* var. *mexicanus* (Malvaceae) avoid flowers that have been nectar robbed by orchard orioles in 97% of their visits because of a color change associated with damaged, older flowers (Gass & Montgomerie 1981). Conversely, floral visitation to some plant species stimulates nectar production (e.g., Gill 1988), which may increase total visitation to plants by the pollinator assemblage.

When mutualists share the same resource, competition for access to host rewards may be widespread and important in the patterns of host use by multiple species (recently reviewed in Palmer et al. 2003). Shifts in the community composition of mutualists may, in turn, exert selection on plant traits. For example, a strong dominance hierarchy among four ant species for a limiting host tree, *Acacia drepanolobium* (Mimosaceae), in Kenya affects spatial patterns of host-tree use by subordinate ant species (Palmer 2003). Africanized honey bees in French Guiana indirectly determine the abundance of native pollinators (and potentially native pollinator visitation to flowers) through exploitative competition for floral rewards (Roubik 1978, 1980). Inouye (1978) experimentally removed the bumblebee *Bombus appositus* from the flowers of its preferred host species, *Delphinium barbeyi* (Ranunculaceae), and found increased visitation by another bumblebee species, *B. flavifrons*. Conversely, removal of *B. flavifrons* from *Aconitum columbianum* (Ranunculaceae) resulted in increased visitation by *B. appositus*. These results suggest that resource utilization by one species influenced the visitation patterns of the other species. In the case of plant-pollinator/nectar-robber interactions, visits by the nectar-robbing bumblebee, *Bombus occidentalis*, decrease subsequent visitation by hummingbird pollinators (Irwin & Brody 1998) because of reduced nectar availability. In all of these studies, access to rewards or modification of rewards results in alterations in host use within multispecies mutualisms.

Interactions among other mutualists, such as seed dispersers and harvesters, may also be driven by a combination of direct and indirect effects via changes in vegetation, resource abundance, or resource distribution. For example, in an impressive three-way factorial design, Longland et al. (2001) excluded scatter-hoarder and larder-hoarder rodents and ant seed dispersers and consumers of Indian ricegrass (*Oryzopsis hymenoides* [Poaceae]). They found significantly more ant larders in plots with rodents excluded than in plots with rodents present, which suggests that rodent activity affected ant activity.

HERBIVORE-POLLINATOR INTERACTIONS   Recently, the role of trait-mediated effects of herbivore damage on the numbers and behaviors of mutualist pollinators has received experimental attention. Distinguishing among the different kinds of herbivores (i.e., foliar versus floral herbivore) is useful when these interactions are considered. Foliar herbivory has been shown to diminish the amounts or quality of floral rewards (e.g., Frazee & Marquis 1994, Lehtilä & Strauss 1999, Mutikainen & Delph 1996) and flower size (e.g., Cresswell et al. 2001, Strauss et al. 1996) and cause induction of chemicals in floral or reward chemistry (Euler & Baldwin 1996, Strauss et al. 2004). Although these studies document changes in postdamage floral traits, only a few studies show the impacts of foliar damage on subsequent pollinator visitation. Generally, pollinators tend to visit damaged plants less frequently or for shorter durations (Hamback 2001, Lehtila & Strauss 1997, Mothershead & Marquis 2000, Strauss et al. 1996), but sometimes damage has no effects on visitation patterns (Hamback 2001, Strauss et al. 2001); in one case, herbivory by wireworms caused an increase in pollination in *Sinapis arvense* (Brassicaceae) (Poveda et al. 2003). Because of self-incompatibility, stigmatic clogging, and geitonogamy, to name just a few factors, translating pollinator-foraging behavior into impacts on plant fitness can be difficult. Even fewer studies examine the effects of foliar herbivory via pollinators all the way to impacts on plant fitness (but see Mothershead & Marquis 2000, Strauss et al. 2001).

In contrast, a bit more attention has been paid to the effects of florivory on both pollination and plant reproduction. Again, florivory generally reduces subsequent pollinator visitation to damaged plants (e.g., Galen 1999, Krupnick et al. 1999, and see references in Table 2) through a number of different mechanisms, such as decreasing flower number, increasing floral asymmetry, and decreasing petal size and floral conspicuousness (e.g., Alados et al. 2002).

HERBIVORE–SEED PREDATOR/DISPERSER INTERACTIONS Herbivory not only affects plant-pollinator interactions but also plant–seed predator interactions. In some cases, these effects may be caused by changes in the traits of plants after herbivory. For example, in the monocarpic herb *Ipomopsis aggregata* (Polemoniaceae), herbivory of the flowering stalk by mule deer reduces oviposition by the predispersal seed-predator fly *Hylemya* spp. (Freeman et al. 2003, Juenger & Bergelson 1998). One likely mechanism that explains reduced seed-predator oviposition on browsed plants is that herbivory induces a delay in flowering phenology, which results in a mismatch between peak-flower and seed-predator abundance (Brody 1997, Freeman et al. 2003), although other mechanisms, such as changes in floral display size or plant or fruit chemistry, cannot be ruled out (Juenger & Bergelson 1998).

Herbivory can also affect the activity of granivores. For example, exclusion of mammalian herbivores from pine and oak woodland communities increased granivory by wood mice (*Apodemus sylvaticus*) and field voles (*Microtus agrestis*), likely because of herbivore-induced changes in vegetation structure (Smit et al. 2001). Reversing the interaction, postdispersal seed predators could affect seedling herbivory in populations where seedling herbivory is density dependent. Indirect evidence suggests that the activity of seedling predators (meadow voles, *Microtus pennsylvanicus*) and seed predators (white-footed mice, *Peromyscus leucopus*) in North American old fields is negatively correlated. Experimental manipulation of seedling predators affected the activity of seed predators (Ostfeld et al. 1997), and the opposite pattern may also hold in cases where the activity of seedling predators is dependent on seedling density, although such a scenario was not experimentally tested to our knowledge.

POLLINATOR–SEED PREDATOR INTERACTIONS Whereas herbivores can affect pollinator use of plants, pollinators, in turn, can affect the attack levels of seed and fruit feeders. Predispersal seed predators rely on successful fruit set to provision their developing larvae, and postdispersal seed predators rely on flowers to set fruit for adequate food resources. Thus, pollinator visitation to flowers may affect plant susceptibility to both predispersal and postdispersal seed predators. Leaving aside well-known instances of pollinators that also act as seed predators (i.e., yucca moths and fig wasps) (e.g., Aker & Udovic 1981, Pellmyr 1997), studies have found a positive link between pollination and predispersal seed predation (e.g., Cariveau et al. 2004, Herrera 2000), although this association is not universal (e.g., Ehrlén et al. 2002). The positive link between pollination and seed predation may be driven by two very different mechanisms. Seed predators may cue in on increased fruit and seed density associated with increased pollinator visitation, or pollinators and seed predators may use the same traits, such as flower number or

TABLE 2 Multispecies effects on plant fitness or on selection on plant traits involving multiple herbivores and herbivores and pollinators

Interactors	Plant species (family)	Fitness component	Effects on fitness additive?	Nature of nonadditivity[a]	Plant trait(s) measured	Nonadditive effects on selection?	Reference
Multiple herbivores							
Above-ground and below-ground insect herbivores	Lupinus arboreus (**Fabaceae**)	Growth and seed production	Yes		No trait examined		Maron 1998
Deer and mollusk herbivores	Lupinus chamissonis (Fabaceae)	Survival and growth rate	Yes		No trait examined		Warner & Cushman 2002
Leaf miners and insect leaf herbivores	Ipomoea purpurea (Convolvulaceae)	Seed production	Yes		Resistance to two herbivores, but never related traits to fitness		Hougen-Eitzman & Rausher 1994
Deer browsing, insect herbivore guild, fungal pathogen guild	Ipomoea hederacea (Convolvulaceae)	Seed production	Not considered		Tolerance to herbivores, pathogens, deer	Yes	Stinchcombe & Rausher 2002
Simulated root and leaf herbivory	Salix planifolia spp. planifolia (Salicaceae)	Biomass, leaf demography, and transpiration	Yes		No trait examined		Houle & Simard 1996
Eggplant flea beetle and horse nettle beetle herbivores	Solanum carolinense (Solanaceae)	Fruits and seeds, fruit size, seed mass, germination, and root mass	Yes		No trait examined		Wise & Sacchi 1996[b]
Above-ground and below-ground insect herbivores	Tripleurospermum perforatum (Asteraceae)	Plant density and flower heads	Yes		No trait examined		Mueller-Schaerer & Brown 1995
Simulated deer browsing, seed flies, and caterpillars	Ipomopsis aggregata (Polemoniaceae)	Seed, flower, fruit production	Yes = Browse/insect; No = Seed fly/caterpillar	Lesser	Flowering phenology	Yes (all three herbivores) No (just insects)	Juenger & Bergelson 1998
Spittlebug and leaf beetle herbivores	Solidago altissima (Asteraceae)	Buds, leaves, roots, shoots and biomass	No	Greater (intensity of interaction depends on the response variable)	No trait examined		Hufbauer & Root 2002

Interaction	Species	Response measured		Effect	Trait	Reference
Deer and leaf beetle herbivores and stem borers	*Rhus glabra* (Anacardiaceae)	Growth, seed production, and survivorship	No	Greater	No trait examined	Strauss 1991
Diamondback moth and flea beetle herbivores	*Brassica rapa* (Brassicaceae)	Seed production	No	Direction depends on densities of herbivores	Leaf area and resistance to the two herbivore species	Yes	Pilson 1996
Wire worm root herbivores and butterfly larval folivores	*Sinapis arvense* (Brassicaceae)	Flowers, fruits, seeds and pollinator visitation	No	Lesser	No measured traits were related to fitness		Poveda et al. 2003
Spittlebugs and plume moth herbivores	*Erigeron glaucus* (Asteraceae)	Flower number	No	Lesser	No trait examined		Karban & Strauss 1993
Lodgepole pine cone borer moth, crossbills and squirrels	*Pinus contorta* var. *latifolia* (Pinaceae)	Seeds	Not addressed		Cone traits, resistance to predation	Selection on cone traits differed in habitats with differing seed predator communities, but nonadditivity could not be addressed	Siepielski & Benkman 2004
Herbivore-pollinator interactions						
Insect florivores and supplemental hand-pollination[c]	*Castilleja indivisa* (Scrophulariaceae)	Seed number	Yes		No measured traits were related to fitness		Adler 2003
Floral herbivores and insect pollinators	*Helleborus foetidus* (Ranunculaceae)	Recruits in the next generation in the field	No	Detrimental effects of herbivores only in presence of pollinators; positive effects of pollinators only in the absence of herbivores	No trait examined		Herrera et al. 2002
Insect and ungulate fruit-feeders and insect pollinators	*Paeonia broteroi* (Paeoniaceae)	Seed set in the field	No	Detrimental effects of herbivores only in presence of pollinators; positive effects of pollinators only in the absence of herbivores	No trait examined		Herrera 2000

(*Continued*)

TABLE 2 (Continued)

Interactors	Plant species (family)	Fitness component	Effects on fitness additive?	Nature of nonadditivity[a]	Plant trait(s) measured	Nonadditive effects on selection?	Reference
Insect and ungulate fruit-feeders and insect pollinators	*Erysimum mediohispanicaum* (Brassicaceae)	Seed set in the field	Not addressed		Plant height, floral shape, flower number	Selective effects of pollinators only present when browsers absent; selective effects of browsing reverse direction of selection on some traits; non-additivity not addressed	Gómez 2003
Simulated deer browsing and supplemental hand-pollination[d]	*Ipomopsis aggregata* (Polemoniaceae)	Seed set, flower and fruit production	No	Effects of herbivory on fitness mediated by pollination intensity	Flowering phenology, plant size, plant height		Juenger & Bergelson 1997
Simulated browsing and emasculation to prevent self-pollination[d]	*Ipomopsis aggregata* (Polemoniaceae)	Seed set, flower and fruit production	No	Benefits of emasculation only in the absence of damage	No traits examined		Juenger & Bergelson 2000
Simulated leaf herbivory and supplemental bee hive pollinator additions	*Curcurbita melo* (Cucurbitaceae)	Fruit number and mass	No	Effects of herbivory on fitness mediated by pollination intensity	No trait examined		Strauss & Murch 2004

[a]Expected based on additive effects of species. Lesser = impacts of interactors on plant fitness in combination less than would be predicted from pairwise interactions. Greater = impacts of interactors on plant fitness in combination greater than would be predicted from pairwise studies.
[b]In Wise & Sacchi (1996), for both root mass and seed germination, a marginal but not a statistically significant interaction occurred between the two herbivores.
[c]*Castilleja indivisa* is primarily bumblebee pollinated.
[d]*Ipomopsis aggregata* is primarily hummingbird pollinated.

flower size, to select plants (especially in cases where seed predators oviposit on flowers before or during the pollination stage) (e.g., Brody 1992). By experimentally manipulating pollination (either by excluding pollinators from flowers or by hand-pollinating flowers), one may possibly disentangle the two mechanisms.

In an experimental study with the perennial herb *Paeonia broteroi* (Paeoniaceae), Herrera (2000) found a significant interaction between pollinator and "herbivore" exclusion treatments for seed production. Mammalian fruit predators (referred to as "herbivores") fed exclusively on fruits from flowers that had been exposed to pollinators and ignored fruits produced from self-pollinated flowers. Similarly, in *Castilleja linariaefolia* (Scrophulariaceae), pollen supplementation to flowers, to mimic increased hummingbird-pollinator visitation, resulted in a marginal, although not statistically significant, increase in predispersal seed predation by plume moth larvae and fly larvae (Cariveau et al. 2004). In both of these studies, seed predators may have cued in on increased fruit and seed abundance associated with increased pollinator visitation or plant investment in fruit quality. Thus, when the effects of pollinators and seed predators are combined, seed predators may mask any plant-fitness benefits related to increased pollination (see also Gómez 2003).

In the case of granivory and frugivory, the link between increased pollination and increased seed predation has not been empirically measured, to our knowledge. Evidence suggests that seed-addition experiments increase seedling recruitment (Turnbull et al. 2000) and that granivores respond to increased seed abundance (Edwards & Crawley 1999) and strongly affect plant recruitment (Edwards & Crawley 1999, Maron & Simms 2001). Similarly, pollination also affects the quantity and quality of fleshy fruits (Gonzalez et al. 1998), and such variation can have sizable effects on the foraging behavior of frugivores (Moegenburg & Levey 2003). However, studies are rare that have manipulated pollination and measured subsequent fruit and seed risk to, or response of, granivores or frugivores.

ENDOPHYTE-HERBIVORE INTERACTIONS   Defense mutualisms, such as those between plants and endophytes, can have a variety of effects on the host plant, which include increased resistance to herbivores and pathogens (Clay 1988). The endophytes of many cool-season agronomic grasses, such as *Festuca arundinacea* (Poaceae) and *Lolium perenne* (Poaceae), produce alkaloidal mycotoxins that increase plant resistance to invertebrate and mammalian herbivores in laboratory and field trials (Cheplick & Clay 1988, Clay 1988). Endophyte-mediated herbivore resistance is also found in other nongrass plant species (e.g., Raps & Vidal 1998, Saikkonen et al. 1996). For example, the cabbage plant *Brassica oleracea* var. *gemmifera* (Brassicaceae), when inoculated with an unspecialized endophyte (*Acremonium alternatum*), supported lower growth rates and survival of the diamondback moth larvae *Plutella xylostella*, potentially because of endophyte-mediated changes in plant phytosterol metabolism (Raps & Vidal 1998). However, endophyte-mediated resistance to herbivory is not universal (e.g., Faeth & Hammon 1997), and some studies have shown that plant-endophyte interactions may even benefit herbivores (Gange 1996, Saikkonen et al. 1999). For example, the mean relative growth rate of grasshoppers was higher on the native grass *Festuca arizonica* (Poaceae) infected with *Neotyphodium* endophytes compared with uninfected *F. arizonica* (Saikkonen et al. 1999). Some of the ambiguity in endophyte effects on plant-herbivore interactions may be driven by the host plant (e.g., native versus agronomic versus exotic, grass versus tree), endophyte, and herbivore involved, as well as by environmental conditions and other species in the community (e.g., predators of herbivores) (Faeth 2002). Multiple types of fungal infections can also influence the effects of endophytes on herbivore and plant performance. For example, the beneficial effects of the foliar endophyte *Neotyphodium lolii* on perennial ryegrass (*Lolium perenne*) was reduced by mycorrhizal fungal infection (Vicari et al. 2002). Despite the contrasting results of these studies, the growing body of literature on endophytes suggests that these fungi provide an additional component of variation in the nature and strength of plant-herbivore interactions. To our knowledge, no study has examined the relationship between endophyte infection and impacts on pollination.

HERBIVORE-MYCORRHIZAL INTERACTIONS   Mycorrhizae are fungi that have hyphal associations with the roots of many plant species. The fungi receive carbon from the plant and plants receive inorganic nutrients, especially phosphorus, from the fungi (or nitrogen if the fungi are ectomycorrhizal). In infertile soils, nutrients taken up by the mycorrhizal fungi can lead to improved plant growth and reproduction, and mycorrhizal-infected plants are often better able to tolerate environmental stresses and competition than are non-mycorrhizal-infected plants (for comprehensive information, see van der Heijden & Sanders 2002). However, the nature of mycorrhizal relationships with host plants ranges from mutualistic to parasitic, and the biotic and abiotic environment usually alter the plant/fungal relationship. The complexity of interactive effects is more extreme when the plant-fungal-herbivore relationship is considered (Gange et al. 2002). For example, in low-phosphorus environments, we expect mycorrhizal fungi to increase phosphorus uptake in the plant and to increase plant performance. Herbivore performance, in turn, may also increase with infection under these conditions (Gange et al. 1999). Thus, mycorrhizae may indirectly benefit herbivores by increasing the quality of the host plant. However, under stressful conditions, both herbivores and mycorrhizae may compete for plant resources (Gange et al. 2002, Gehring et al. 1997). Such conditionality may be one reason why mycorrhizal infection does not result in any predictable effects on herbivore performance (reviewed in Gehring & Whitham 2002). Interestingly, infection with arbuscular mycorrhizal (AM) fungi is much more likely to affect herbivore performance than ectomycorrhizal (EM) fungi, although, again, the directionality of this effect is inconsistent (Gehring & Whitham 2002).

Another reason for the inconsistency in herbivore response to mycorrhizal infection may lie in the fact that infection causes changes in both nutritive and defensive leaf chemistry (Gange & West 1994, Goverde et al. 2000). These changes, in turn, may differentially affect specialist and generalist herbivores. The performance of generalist herbivores is often more sensitive to changes in defensive chemistry than is the performance of specialist herbivores, and, although sample sizes are small, Gehring & Whitham (2002) provide some evidence from a survey of studies that generalist insects may be more affected by fungal symbionts than are specialist insects. Responses of generalist and specialist herbivores to infected plants are complex, however—both the level of infection and the species composition of AM fungi influenced the performance of a specialist lepidopteran herbivore on *Lotus corniculatus* (Fabaceae) (Goverde et al. 2000). In summary, mycorrhizal infection can clearly alter the likelihood and intensity of plant interactions with herbivores, but the highly conditional nature of the plant-fungal interaction, coupled with the diverse responses of herbivores to changes in plant chemistry, mean that such effects may be idiosyncratic.

When herbivores, as opposed to fungi, are the first interactor with the plant, herbivory affects the likelihood and extent of subsequent mycorrhizal colonization of the shared host. Again, Gehring & Whitham (2002) reviewed the literature on herbivore-mycorrhizal interactions and found that in 28 of 42 cases, herbivores inhibited the colonization of mycorrhizae that shared the same host; in 11 cases, no effects occurred, and in 10 cases, herbivory facilitated the colonization of mycorrhizae. Herbivores may also affect the species composition of mycorrhizal communities (Eom et al. 2001, Gehring & Whitham 2002). To the best of our knowledge, no studies have examined the interactions between mycorrhizal fungi and pollinators. However, one could imagine that fungi could consume plant carbon that might otherwise sweeten nectar for pollinators. The converse might also hold: mycorrhizal-infected plants with greater nutrient uptake and greater growth rates might produce higher-quality nectar, larger flowers, or a larger floral display. Similar effects of mycorrhizal infection may influence fruit quality and the use of fruit by seed dispersers.

In summary, because plants are dynamic, living resources, interactions with other organisms elicit responses that influence simultaneously and subsequently interacting species of all types. Ample evidence suggests that the intensity or likelihood of interactions changes with the suite of interacting species. Plant responses to interactors may carry fitness costs (Koricheva 2002). Thus, for example, the benefits of deterrence of future herbivores may be offset by the costs of induced resistance. Only a few studies have examined the relative costs (energetic and ecological) and benefits (ecological) of induction simultaneously (Agrawal 2000, Baldwin 1998, Hare et al. 2003, Sagers & Coley 1995, Valverde et al. 2003). These costs and benefits ultimately are assessed with respect to the same bottom line: a change in plant fitness as a result of multispecies interactions.

Criterion 3: Multiple Interactors Have Nonadditive Effects on Plant Fitness or on Selection

Despite the copious evidence that patterns of host-plant use are altered by other plant-animal (or animal-animal) interactions, we have much less evidence that these altered interactions have effects on plant fitness. However, effects of multiple species on mean fitness alone in treatments cannot tell us anything about the nature of selection, because selection, by definition, is the relationship between a trait and fitness (for further discussion, see Juenger & Bergelson 1998, Stinchcombe & Rausher 2002, Strauss et al. 2004). The original wording of this criterion in Hougen-Eitzman & Rausher's (1994) seminal paper states that nonadditive effects of species on plant fitness provide evidence for diffuse selection. In this case, the authors measure two traits (resistances to two herbivores) but do not relate leaf area removed by each herbivore to plant fitness (selection on resistance); they analyze mean individual plant fitness in different communities. In a subsequent paper, Iwao & Rausher (1997) develop an extremely thorough theoretical and empirical treatment for analyzing selection gradients that relate fitness to resistance as a function of additive and nonadditive effects of community members. Thus, whereas the experimental and analytical approaches both test the effects of communities on selection and evolution, the verbal description of the criteria remained unchanged. In fact, to our knowledge, only six papers actually measure how selection changes in response to community composition: Pilson (1996), Iwao & Rausher (1997), Juenger & Bergelson (1998), Gómez (2003), Tiffin (2002), and Stinchcombe & Rausher (2002). Five of these papers document nonadditive effects of community members on selection on a focal trait or species. Because the initial criteria for evidence of diffuse evolution did not explicitly define the importance of linking traits to plant fitness, or of assessing the relative fitness of different genotypes in the different communities, many investigators have misunderstood the basic approach required to document diffuse selection or diffuse evolution (e.g., see several studies included in Table 2).

It is important to reiterate that nonadditive effects of community members on mean fitness in treatments may still have important implications for the ecological consequences of multispecies interactions through changes in population size and may affect community-level properties, such as trophic cascades (e.g., Peckarsky & McIntosh 1998) or biodiversity (Mueller-Schaerer & Brown 1995).

MEASURING THE CONSEQUENCES OF MULTIPLE-SPECIES INTERACTIONS ON PLANT FITNESS   When plant fitness is to be measured in response to multispecies interactions, one question is essential: What is the most appropriate plant response variable to measure? In part, this decision depends on the aim of the study and whether the study includes both ecological and evolutionary perspectives.

Typically, studies have focused on measuring some correlate of plant fitness, such as plant growth, survival, flower production, pollen receipt, or seed production [hereafter referred to as fitness components (Campbell 1991)]. In many cases, however, plant-animal interactions directly and indirectly affect multiple, sequential fitness components (Adler et al. 2001), and total plant fitness may be differentially sensitive to particular fitness components (Ehrlén 2003). In addition, tradeoffs may occur between the components. For example, a large increase in seed production by a perennial plant in one year may be countered by lower growth, survival, or reproduction in subsequent years (Ehrlén 2002, Primack & Hall 1990). In addition, male fitness components (seeds sired) and female fitness components (seeds produced) are not always affected similarly by the same interaction. For example, male reproductive success may be more strongly affected by herbivory (Strauss et al. 2001) or pollination (Stanton et al. 1991) than female reproductive success, although there are exceptions (e.g., Irwin & Brody 2000, Krupnick & Weis 1999).

Fitness components, such as seed production (a female component), may have ecological effects as long as a link exists between seed production and subsequent population size. In this case, multispecies interactions may drive plant population persistence, spread, and dynamics. In contrast, fitness measured through male function does not necessarily have effects on population size (a single male could sire all the seeds, but the numbers of seeds could remain constant if resource limitation, as opposed to pollination, sets bounds on seed numbers). Fitness through male function may have important evolutionary effects, however, because soft selection can occur in a population of constant size when allele frequencies within the population change. Thus, the choice of fitness components to be measured might be influenced by the motivation of the study, that is, whether the goal is to examine multispecies effects on plant population size or on selection on plant traits.

EXPERIMENTAL APPROACHES TO MEASURING THE COMBINED EFFECTS OF MULTIPLE INTERACTORS

Fully-crossed factorial designs An experimental approach is the most straightforward means of testing for nonadditive effects of multiple visitors on plant fitness or reproduction. To measure selection, trait measurements are also necessary. By experimentally manipulating each of the interacting players in a fully-crossed design, the individual and combined effects of visitors on plant fitness and on selection can be assessed. If one wants to explore selection by communities on a particular trait, one can regress values of the trait in question against the relative fitness of genotypes or plant families (or against fitness of individuals randomly assigned to treatments for phenotypic selection gradients). Such an approach allows exploration of selection gradients in treatments in which the presence or absence of interactors has been manipulated. Iwao & Rausher (1997) discuss in detail the theoretical considerations, experimental design, and statistical analyses to partition selection on a trait by a suite of interactors into diffuse versus pairwise components. These kinds of experiments give us insights into how evolution of traits may be influenced by a community context.

The demographic responses to multispecies plant-animal interactions Fully-crossed factorial designs such as those described above and below, and in which mean fitness of individuals in treatments is the sole response variable measured, in some cases may tell us about the effects of communities on the population dynamics of component species. A handful of studies demonstrate that changes in fitness or reproduction as a result of multispecies interactions translate into changes in population size. For example, in an ambitious long-term study, Herrera et al. (2002) found nonadditive effects of pollinators and herbivores on plant fitness that translated into different numbers of seedling and adult recruits in populations of *Helleborus foetidus* (Ranunculaceae) that belonged to different experimental treatments. In addition, experimental flower removal and early-season defoliation negatively affected the population growth rate of *Primula veris* (Primulaceae) (García & Ehrlén 2002). Maron & Simms (2001) found that although exclusion of rodent granivores increased seedling emergence of *Lupinus arboreus* (Fabaceae), such effects were only marginally evident after three years (because of cutworm herbivory on seedlings). These studies suggest that multispecies plant-animal interactions can scale up to have population-level consequences in some (Ackerman et al. 1996, Louda & Potvin 1995), but not all, cases (Crawley & Nachapong 1985, Eriksson & Ehrlén 1992).

The above works are commendable in their long-term study of the effects of community composition on the populations of focal species across multiple generations. How single-year, short-term effects on seed production or growth influence population dynamics is difficult to determine because a correlation between increased seed production and future population size cannot be assumed. Several processes uncouple seed numbers from future plant population size; two processes are safe-site limitation (Crawley & Nachapong 1985, Eriksson & Ehrlén 1992, Maron et al. 2002) and self-thinning (e.g., Akifumi 1996). Thus, seed inputs do not always affect levels of adult plant recruitment (reviewed in Turnbull et al. 2000). These demographic links are only now beginning to be explored in detail (e.g., García & Ehrlén 2002, Louda & Potvin 1995, Maron & Gardner 2000, Turnbull et al. 2000), and this area of plant ecology is sorely lacking in empirical data (but see Maron et al. 2002, McEvoy et al. 1993, and references above).

Several caveats to the interpretation of demographic studies should be considered. Because rare long-distance dispersal events may be important to the founding of new populations and range expansions, and because these rare events are extremely difficult to assess in field studies, how levels of seed inputs affect them is difficult to determine. Even for wind-dispersed seeds, the vast majority of seeds are deposited very near the parent plant (Augspurger & Kitajima 1992). Local dynamics are what we can see and measure. Thus, we do not know the degree to which differences in seed production affect the incidence of important rare events, such as the establishment of new populations. Yet, these events clearly have large ecological impacts and fitness benefits when they occur. Similarly, we tend to measure seedling emergence but have a much weaker grasp on the meaning of seed inputs to seed-bank populations, which may also play important roles in population persistence (e.g., Kalisz & McPeek 1992). The effects of multispecies interactions on the population dynamics of plants and on plant-range limits and distributions need much more empirical attention.

Additive versus nonadditive effects of multiple interactors on plants Factorial experiments can tell us whether the effects of species together differ from those we would expect on the basis of their separate effects. As indicated by the studies summarized in Table 2, multispecies effects on plant fitness, growth, or reproduction often cannot be predicted from knowledge of the effects of each species in isolation. For example, *Trirhabda* beetles and *Philaenius* spittlebugs feeding together on the tall goldenrod *Solidago altissima* (Asteraceae) reduce the mass of the apical bud and the foliage more than would be expected from either insect feeding alone (Hufbauer & Root 2002). Similarly, Longland et al. (2001) show that the beneficial effects of larder hoarding by rodents on seed germination of Indian ricegrass are only evident in plots that also experience scatter hoarding by Merriam's kangaroo rats (*Dipodomys merriami*). When pollinators are limiting, effects of fruit-feeding herbivores are diminished (Herrera 2000, Herrera et al. 2002). The few studies that have addressed selection in response to community composition also show possible nonadditive, selective effects of community members. In *Ipomopsis aggregata*, a self-incompatible monocarp, browsing causes increased branching and flower production. However, pollinator limitation curtails fitness responses to damage from browsers in some years, and investigators found that selection on resistance to browsing may be diminished by pollinator responses (Juenger & Bergelson 1997). Alternatively, the relationships among plants and pollinators weaken when resource limitation overrides pollinator limitation in damaged plants (Juenger & Bergelson 1997). Browsers can also negate the selective effects of pollinators on plant height and corolla shape by removing flowers and fruits (Gómez 2003).

In summary, community composition significantly affects plant fitness, growth, or reproduction in the majority of experiments that have explored this question through the use factorial experimental designs (Table 2). The results from studies on the effects of multiple herbivores are mixed; about half show independent effects of herbivores on plant fitness (e.g., additive effects of herbivores on plant reproduction). Also, when nonadditive effects on plant fitness do occur, they are not consistent in direction. For example, in some cases, joint attack results in less fitness loss than would be predicted from the separate effects of each herbivore, whereas in other cases, the reverse is true. Although sample sizes are small, five of the six studies in Table 2 that have examined cumulative effects of herbivores and pollinators on plant fitness have found nonadditive effects.

We had a difficult time finding factorial experiments that examined the separate and combined effects of multiple-pollinator or multiple-seed-disperser species on plant fitness; this outcome likely reflects the difficulty of experimental manipulation in some of these systems (see below for other approaches).

Alternative approaches to measuring multispecies effects on plants Although experimental manipulations have the advantage of disentangling the individual and combined effects of multispecies interactions by application of standard experimental tools (i.e., factorial designs), this approach has limitations. Some interactions are very difficult to manipulate, most notably interactions among plants and multiple species that have similar body size and activity patterns. For example, when pollinators have the same body size, phenology, and foraging behavior (i.e., comparing the individual versus combined effects of two bumblebee species), it is often difficult to exclude only one pollinator. In complex systems, so many species pairs may exist that fully-crossed factorial designs and replication of treatments would be logistically impossible. In such cases of intractability, there are other ways of attacking these problems.

Several approaches employed to examine interaction strengths of species in complex communities have been reviewed (Wootton 2002). Typically, the response variables for measuring interaction strengths are per capita growth rates or changes in population size. Some of these same techniques can be applied to evolutionary questions for which the response variable is the fitness of a focal species. Path analysis (reviewed in Shipley 2000) combined with structural-equation modeling (SEM, reviewed in Mitchell 1993) provides an additional approach for quantifying multispecies effects on plant fitness. Path analysis allows the dissection of complicated webs of direct and indirect effects among multiple interactors by use of a set of a priori hypotheses (Kingsolver & Schemske 1991, Mitchell 1993) and is particularly useful for generating hypotheses about the causal mechanisms of selection in systems where experimental manipulation is impractical (Grace & Pugesek 1998; but see Smith et al. 1997). Path analysis is a sequence of multiple regressions and correlations structured by a priori hypotheses regarding the causal relationships among variables (Mitchell 1993). The degree to which a path model provides an appropriate fit to the observed data can be tested by SEM. SEM tests the observed correlation structure in the data against the expected correlation structure in the path model through the use of a goodness of fit test (reviewed in Mitchell 1993).

Path analysis in combination with SEM can be used in a number of different ways in ecological and evolutionary studies. First, path analysis combined with SEM can be used to compare multiple, hypothesized causal structures in communities (Cariveau et al. 2004, Gómez & Zamora 2000). Second, given a particular causal structure, the relative strength of different direct and indirect effects of multiple agents can be separated and compared (Adler et al. 2001, Schemske & Horvitz 1988). Within the context of natural selection, path analysis combined with multiplicative fitness components are a powerful, multivariate approach by which to dissect and understand complex patterns of selection (Conner 1996), and path analytical techniques can help reduce environmental bias when estimating natural selection (reviewed in Stinchcombe et al. 2002) by including environmental "condition" variables in the path models (Scheiner et al. 2002). Alternatively, environmental bias can be reduced by estimating selection that uses genetic or family means from genetic replicates (e.g., Iwao & Rausher 1997).

Path analysis is increasingly used in studies of multispecies plant-animal interactions (e.g., Adler et al. 2001, Cariveau et al. 2004, Gómez & Zamora 2000, Juenger & Bergelson 1997, Mothershead & Marquis 2000, Schemske & Horvitz 1988) to evaluate the direct and indirect effects of multiple interactors on plant fitness. Mothershead & Marquis (2000) used experimental manipulations combined with path analysis to compare the direct effects of leaf damage to *Oenothera macrocarpa* (Onagraceae) through decreased resource availability with the indirect effects of leaf damage through changes in floral characters and plant-pollinator interactions. In this system, the magnitude of the indirect effects of herbivory on changes in plant-pollinator interactions and plant fitness outweighed the direct consumptive effects of herbivory on plant fitness. Conversely, the direct effects of bud herbivory to *Castilleja indivisa* (Scrophulariaceae) on seed production outweighed the indirect effects of herbivory through changes in floral characters and plant-pollinator interactions (Adler et al. 2001). One common theme running through both studies, however, is that the direction or magnitude of effect of one interactor (i.e., pollinators) is dependent on the direction or magnitude of the other (i.e., herbivores). Studies in multispecies plant-animal interactions also include path analysis to estimate causal mechanisms of selection. In *Castilleja linariaefolia*, the strength of selection on calyx length, flower production, and plant height was greater for pathways through seed predation than for pathways through pollination because seed predators had strong negative effects on relative seed set compared with the weak benefits of pollinators (Cariveau et al. 2004). In southeastern Spain, *Hormathophylla spinosa* (Brassicaceae) experiences positive pollinator-mediated selection for flower number per plant and flower density in populations with low ungulate herbivory. However, in populations with high ungulate herbivory, the direct negative effects of herbivores on relative plant fitness masked any beneficial effects of flower number in attracting pollinators, and in one population with high herbivore pressure, plants experienced conflicting selection pressures between maximizing pollination and minimizing plant risk to herbivory (Gómez & Zamora 2000).

Despite their advantages, path analysis and SEM have limitations (for a more complete listing of limitations, see Mitchell 1993). First, path analysis should not be used to infer causation among variables (Mitchell 1993, Wootton 1994). Rather, path analysis identifies important correlations among variables and possible targets of selection that can be further tested experimentally (Kingsolver & Schemske 1991, Petraitis et al. 1996). Second, path analysis and the magnitude of path coefficients are strongly conditional upon which variables are included in the path model. Because the coefficient estimates from path analysis depend on the causal path structure, they do not produce selection gradients that can be used to predict evolutionary response to selection (Scheiner et al. 2000). Third, as it has been currently used, path analysis does not quantify nonadditivity of interactions on plant fitness. However, cross-product terms might possibly be incorporated into a path analysis to examine nonadditive effects of interactors on plant fitness. For example, one might include the cross-product term between honeybee visitation and butterfly visitation in a multiple regression that relates visitation rate to seed set. Similarly, nonlinear effects of interactors on plant fitness could be included by use of quadratic terms in the regressions. Approaches that integrate a variety of techniques, such as experimental tests of hypotheses developed from path analysis, may provide the most promising avenues for understanding and predicting the selective effects of multispecies interactions in natural communities.

Along with path analysis, other approaches are also being explored. A recent paper has used optimality models to show that the degree to which a plant exhibits specialized traits for one interactor is bounded by the cost of this adaptation in its relationship with other interactors (Aigner 2001). Under this scenario, adaptations to uncommon or relatively ineffective interactors may occur as long as the cost is minimal. This approach speaks to the importance of experiments that elucidate costs of traits in the absence of species interactions, in addition to benefits in the presence of species interactions in factorial design. To examine evolution as a function of community membership in this context, manipulative experiments are required to exclude various species from focal plants. Fitness tradeoffs must be measured not only with respect to the mean plant phenotype but also across the whole range of phenotypic variation. The results of Aigner's (2001) model suggest that the criteria required for multiple species to affect trait evolution may be even broader than appreciated, although addressing these conditions experimentally will be at least as difficult as any of the previously discussed experimental approaches.

FUTURE DIRECTIONS

Countless studies document the multifarious and diverse responses of community members to the actions of plant associates that precede them. These responses are typically mediated by plant reactions to the effects of previous interactors. Far fewer studies, however, document whether genetic correlations exist between traits important to interactions with multiple species, and even fewer studies hunt for genetic correlations between traits important in very diverse interactions (e.g., herbivore-pollinator, herbivore-competitor). We surmise that the latter is true because a priori mechanistic links between traits involved in diverse interactions are not always obvious. However, the presence of what we call "ecological pleiotropy," when the same trait influences very diverse interactions with multiple species, suggests that traits involved with diverse interactors may be more linked than we currently appreciate. As we understand more about the impacts of a trait on an interaction, its role in, or linkages to, other important agents of selection on the focal species may be worth exploring. Many traits that appear to be very important to plant fitness still exhibit heritable variation. We must, therefore, be open to considering links between these traits and the actions of diverse community members that may prevent fixation of favorable alleles specific to a single interaction (Rudgers 2004, Thompson 1994). Our review is unabashedly phytocentric. All the responses to selection have been measured in terms of the plant. For practical reasons, assessing plant fitness, and response to interactors, is much easier than assessing the fitness of plant associates, which may also be affected by the same multispecies interactions. For example, one could imagine that selection on the ability of an herbivore species (herbivore A) to detoxify plant chemical defenses may depend on the suite of herbivores that precede it in feeding on the host. In some years or places, these predecessors may change and, by virtue of the different plant responses they elicit, they may either facilitate or impede the performance of herbivore A upon the plant. As far as we know, very little attention has been paid to the evolutionary responses of plant associates to the prior interactions a plant has had with other associates (but see Siepielski & Benkman 2004).

Another area that has received far less attention than it should is the effect of multiple interactors on selection on plant traits. Carrying interactions through to a thorough examination of fitness components requires time and energy (and is even more difficult for perennial plants that exhibit costs of reproduction in the following year). Moreover, these changes in fitness must also be related to consistent changes in trait values for us to compare selection in different community contexts. Studies that examine only fitness in response to multispecies interactions can inform us on the (potential) ecological effects of communities on population dynamics but not on differences in selection or evolution (for further discussion, see Strauss et al. 2004). Few studies have combined the measurement of traits with measures of plant fitness in this multispecies context (Table 2). Additional experimental approaches and analytical tools that will help us explore complex systems outside of factorial designs are likely to play a key role in the efforts to understand effects of communities on both ecological and evolutionary processes.

In summary, it is clear that ecological communities shape the traits of component species, not only through pairwise interactions between species but also through the joint actions of species that result in synergisms and alternate evolutionary trajectories. Multispecies effects may constrain the response of traits to selection and may be an important source of fluctuating selection that maintains genetic variation in ecologically important traits. In addition, multispecies interactions affect not only trait evolution but also, in the few studies that have examined it, the population dynamics of species within communities. We have a growing understanding that taking a community perspective will inform us of how multiple species together affect community structure and species diversity, as well as species evolution.

The *Annual Review of Ecology, Evolution, and Systematics* is online at
http://ecolsys.annualreviews.org

LITERATURE CITED

Ackerman JD, Sabat A, Zimmerman JK. 1996. Seedling establishment in an epiphytic orchid: an experimental study of seed limitation. *Oecologia* 106:162–98

Adler LS. 2000. The ecological significance of toxic nectar. *Oikos* 91:409–20

Adler LS. 2003. Host species affects herbivory, pollination, and reproduction in experiments with parasitic *Castilleja*. *Ecology* 84:2083–91

Adler LS, Karban R, Strauss SY. 2001. Direct and indirect effects of alkaloids on plant fitness via herbivory and pollination. *Ecology* 82:2032–44

Agrawal AA. 2000. Benefits and costs of induced plant defense for *Lepidium virginicum* (Brassicaceae). *Ecology* 81:1804–13

Aigner PA. 2001. Optimality modeling and fitness trade-offs: when should plants become pollinator specialists? *Oikos* 95:177–84

Aker CL, Udovic D. 1981. Oviposition and pollination behavior of the yucca moth *Tegeticula maculata* (Lepidoptera: Prodoxidae), and its relation to the reproductive biology of *Yucca whipplei* (Agavaceae). *Oecologia* 49:96–101

Akifumi M. 1996. Density regulation during the regeneration of two monocarpic bamboos: self-thinning or intraclonal regulation? *J. Veg. Sci.* 7:281–88

Alados CL, Giner ML, Dehesa L, Escos J, Barroso FG, et al. 2002. Developmental instability and fitness in *Periploca laevigata* experiencing grazing disturbance. *Int. J. Plant. Sci.* 163:969–78

Armbruster WS. 1997. Exaptations link evolution of plant-herbivore and plant-pollinator interactions: a phylogenetic inquiry. *Ecology* 78:1661–72

Augspurger CK, Kitajima K. 1992. Experimental studies of seedling recruitment from contrasting seed distributions. *Ecology* 73:1270–84

Baldwin IT. 1998. Jasmonate-induced responses are costly but benefit plants under attack in native populations. *Proc. Natl. Acad. Sci. USA* 95:8113–18

Berenbaum MR, Zangerl AR, Nitao JK. 1986. Constraints on chemical coevolution: wild parsnips *Pastinaca sativa* and the parsnip webworm *Depressaria pastinacella*. *Evolution* 40:1215–28

Billick I, Case TJ. 1994. Higher order interactions in ecological communities: What are they and how can they be detected? *Ecology* 75:1529–43

Brody AK. 1992. Oviposition choices by a predispersal seed predator (*Hylemya* sp.). 1. Correspondence with hummingbird pollinators, and the role of plant size, density and floral morphology. *Oecologia* 91:56–62

Brody AK. 1997. Effects of pollinators, herbivores, and seed predators on flowering phenology. *Ecology* 78:1624–31

Campbell DR. 1991. Effects of floral traits on sequential components of fitness in *Ipomopsis aggregata*. *Am. Nat.* 137:713–37

Cariveau D, Irwin RE, Brody AK, Garcia-Mayeya L, von der Ohe A. 2004. Direct and indirect effects of pollinators and seed predators to selection on plant and floral traits. *Oikos* 104:15–26

Cheplick GP, Clay K. 1988. Acquired chemical defenses in grasses: the role of fungal endophytes. *Oikos* 52:309–18

Clay K. 1988. Fungal endophytes of grasses: a defensive mutualism between plants and fungi. *Ecology* 69:10–16

Clay K, Marks S, Cheplick GP. 1993. Effects of insect herbivory and fungal endophyte infection on competitive interactions among grasses. *Ecology* 74:1767–77

Coley PD, Bryant JP, Chapin FS III. 1985. Resource availability and plant antiherbivore defense. *Science* 230:895–99

Conner JK. 1996. Understanding natural selection: an approach integrating selection gradients, multiplicative fitness components, and path analysis. *Ethol. Ecol. Evol.* 8:387–97

Crawley MJ, Nachapong M. 1985. The establishment of seedlings from primary and regrowth seeds of ragwort (*Senecio jacobaeae*). *J. Ecol.* 73:255–61

Cresswell JE, Hagen C, Woolnough JM. 2001. Attributes of individual flowers of *Brassica napus* L. are affected by defoliation but not by intraspecific competition. *Ann. Bot.* 88:111–17

Edwards GR, Crawley MJ. 1999. Rodent seed predation and seedling recruitment in mesic grassland. *Oecologia* 118:288–96

Ehrlén J. 2002. Assessing the lifetime consequences of plant-animal interactions for the perennial herb *Lathyrus vernus* (Fabaceae). *Perspect. Plant. Ecol. Evol. Syst.* 5:145–63

Ehrlén J. 2003. Fitness components versus total demographic effects: evaluating herbivore impacts on a perennial herb. *Am. Nat.* 162:796–810

Ehrlén J, Käck A, Ågren J. 2002. Pollen limitation, seed predation and scape length in *Primula farinosa*. *Oikos* 97:45–51

Eom A-H, Wilson GWT, Hartnett DC. 2001. Effects of ungulate grazers on arbuscular mycorrhizal symbiosis and fungal community structure in tallgrass prairie. *Mycologia* 93:233–42

Eriksson O, Ehrlén J. 1992. Seed and microsite limitation of recruitment in plant populations. *Oecologia* 91:360–64

Euler M, Baldwin IT. 1996. The chemistry of defense and apparency in the corollas of *Nicotiana attenuata*. *Oecologia* 107:102–12

Faeth SH. 2002. Are endophytic fungi defensive plant mutualists? *Oikos* 98:25–36

Faeth SH, Hammon KE. 1997. Fungal endophytes in oak trees: long-term patterns of abundance and associations with leafminers. *Ecology* 78:810–19

Fenner M, Cresswell JE, Hurley RA, Baldwin T. 2002. Relationship between capitulum size and pre-dispersal seed predation by insect larvae in common Asteraceae. *Oecologia* 130:72–77

Frazee JE, Marquis RJ. 1994. Environmental contribution to floral trait variation in *Chamaecrista fasciculata* (Fabaceae: Caesalpinioideae). *Am. J. Bot.* 81:206–15

Freeman RS, Brody AK, Neefus CD. 2003. Flowering phenology and compensation for herbivory in *Ipomopsis aggregata*. *Oecologia* 136:394–401

Futuyma DJ. 1986. *Evolutionary Biology*. Sunderland, MA: Sinauer Associates. 600 pp.

Galen C. 1999. Why do flowers vary? The functional ecology of variation in flower size and form within natural plant populations. *Bioscience* 49:631–40

Galen C, Cuba J. 2001. Down the tube: pollinators, predators, and the evolution of flower shape in the Alpine skypilot, *Polemonium viscosum*. *Evolution* 55:1963–71

Gange AC. 1996. Positive effects of endophytic infection on sycamore aphids. *Oikos* 75:500–10

Gange AC, Bower E, Brown VK. 1999. Positive effects of an arbuscular mycorrhizal fungus on aphid life history traits. *Oecologia* 120:123–31

Gange AC, Bower E, Brown VK. 2002. Differential effects of insect herbivory on arbuscular mycorrhizal colonization. *Oecologia* 131:103–12

Gange AC, West HM. 1994. Interactions between arbuscular mycorrhizal fungi and foliar-feeding insects in *Plantago lanceolata* L. *New Phytol.* 128:79–87

García MB, Ehrlén J. 2002. Reproductive effort and herbivory timing in a perennial herb: fitness components at the individual and population levels. *Am. J. Bot.* 89:1295–302

Gass CL, Montgomerie RD. 1981. Hummingbird foraging behavior: decision-making and energy regulation. In *Foraging Behavior: Ecological, Ethological, and Psychological Approaches*, ed. AC Kamil, TD Sargent, pp. 159–94. New York: Garland Gatehouse JA. 2002. Plant resistance towards insect herbivores: a dynamic interaction. *New Phytol.* 156:145–69

Gehring CA, Cobb NS, Whitham TG. 1997. Three-way interactions among ectomycorrhizal mutualists, scale insects, and resistant and susceptible pinyon pines. *Am. Nat.* 149:824–41

Gehring CA, Whitham TG. 2002. Mycorrhizae-herbivore interactions: population and community consequences. In *Ecological Studies. Mycorrhizal Ecology*, ed. MGA van der Heijden, IR Sanders, pp. 295–320. New York: Springer-Verlag

Gill FB. 1988. Effects of nectar removal on nectar accumulation in flowers of *Heliconia imbricata* (Heliconiaceae). *Biotropica* 20:168–71

Gómez JM. 2003. Herbivory reduces the strength of pollinator-mediated selection in the Mediterranean herb *Erysimum mediohispanicum*: consequences for plant specialization. *Am. Nat.* 162:242–56

Gómez JM, Zamora R. 2000. Spatial variation in the selective scenarios of *Hormathophylla spinosa* (Cruciferae). *Am. Nat.* 155:657–68

Gonzalez MV, Coque M, Herrero M. 1998. Influence of pollination systems on fruit set and fruit quality in kiwifruit (*Actinidia deliciosa*). *Ann. Appl. Biol.* 132:349–55

Gori DF. 1983. Post-pollination phenomena and adaptive floral changes. In *Handbook* of *Experimental Pollination Biology*, ed. CE Jones, RJ Little, pp. 31–49. New York: Van Nostrand Reinhold

Goulson D, Stout JC, Langley J, Hughes WOH. 2000. The identity and function of scent marks deposited by foraging bumblebees. *J. Chem. Ecol.* 26:2897–911

Goverde M, van der Heijden MGA, Wiemken A, Sanders IR, Erhardt A. 2000. Arbuscular mycorrhizal fungi influence life history traits of a lepidopteran herbivore. *Oecologia* 125:362–69

Grace JB, Pugesek BH. 1998. On the use of path analysis and related procedures for the investigation of ecological problems. *Am. Nat.* 152:151–59

Grant V. 1950. The protection of ovules in flowering plants. *Evolution* 4:179–201

Gronquist M, Bezzerides A, Attygale A, Meinwald J, Eisner M, Eisner T. 2001. Attractive and defensive functions of the ultraviolet pigments of a flower (*Hypericum calycinum*). *Proc. Natl. Acad. Sci. USA* 98:13745–50

Guirfa M. 1993. The repellent scent-mark of the honeybee *Apis mellifera ligustica* and its role as communication cue during foraging. *Insectes Soc.* 40:59–67

Hamback PA. 2001. Direct and indirect effects of herbivory: feeding by spittlebugs affects pollinator visitation rates and seedset of *Rudbeckia hirta*. *Ecoscience* 8:45–50

Han K, Lincoln DE. 1995. The evolution of carbon allocation to plant secondary metabolites: a genetic analysis of cost in *Diplacus aurantiacus*. *Evolution* 48:1550–63

Hare JD, Elle E, van Dam NM. 2003. Costs of glandular trichomes in *Datura wrightii*: a three-year study. *Evolution* 57:793–805

Herrera CM. 2000. Measuring the effects of pollinators and herbivores: evidence for non-additivity in a perennial herb. *Ecology* 81:2170–76

Herrera CM, Medrano M, Rey Pedro J, Sanchez-Lafuente AM, Garcia MB, et al. 2002. Interaction of pollinators and herbivores on plant fitness suggests a pathway for correlated evolution of mutualism- and antagonism-related traits. *Proc. Natl. Acad. Sci. USA* 99:16823–28

Hougen-Eitzman D, Rausher MD. 1994. Interactions between herbivorous insects and plant-insect coevolution. *Am. Nat.* 143:677–97

Houle G, Simard G. 1996. Additive effects of genotype, nutrient availability and type of tissue damage on the compensatory response of *Salix planifolia* spp. *planifolia* to stimulated herbivory. *Oecologia* 107:373–78

Hufbauer RA, Root RB. 2002. Interactive effects of different types of herbivore damage: *Trirhabda* beetle larvae and *Philaenus* spittlebugs on goldenrod (*Solidago altissima*). *Am. Midl. Nat.* 147:204–13

Inouye BD, Stinchcombe JR. 2001. Relationships between ecological interaction modifications and diffuse coevolution: similarities, differences, and causal links. *Oikos* 95:353–60

Inouye DW. 1978. Resource partitioning in bumblebees: experimental studies of foraging behavior. *Ecology* 59:672–78

Irwin RE, Brody AK. 1998. Nectar robbing in *Ipomopsis aggregata*: effects on pollinator behavior and plant fitness. *Oecologia* 116:519–27

Irwin RE, Brody AK. 2000. Consequences of nectar robbing for realized male function in a hummingbird-pollinated plant. *Ecology* 81:2637–43

Irwin RE, Strauss SY, Storz S, Emerson A, Guibert G. 2003. The role of herbivores in the maintenance of a flower color polymorphism in wild radish. *Ecology* 84:1733–43

Iwao K, Rausher MD. 1997. Evolution of plant resistance to multiple herbivores: quantifying diffuse coevolution. *Am. Nat.* 149:316–35

Janzen DH. 1980. When is it coevolution? *Evolution* 34:611–12

Juenger T, Bergelson J. 1997. Pollen and resource limitation of compensation to herbivory in scarlet gilia, *Ipomopsis aggregata*. *Ecology* 78:1684–95

Juenger T, Bergelson J. 1998. Pairwise versus diffuse natural selection and the multiple herbivores of scarlet gilia, *Ipomopsis aggregata. Evolution* 52:1583–92

Juenger T, Bergelson J. 2000. Does early season browsing influence the effect of self-pollination in scarlet gilia? *Ecology* 81:41–48

Kalisz S, McPeek MA. 1992. Demography of an age-structured annual: resampled projection matrices, elasticity analyses, and seed bank effects. *Ecology* 73:1082–93

Karban R, Baldwin IT. 1997. *Induced Responses to Herbivory*. Chicago: Univ. Chicago Press. 319 pp.

Karban R, Strauss SY. 1993. Effects of herbivores on growth and reproduction of their perennial host, *Erigeron glaucus. Ecology* 74:39–46

Kingsolver JG, Schemske DW. 1991. Path analyses of selection. *Trends Ecol. Evol.* 6:276–80

Koricheva J. 2002. Meta-analysis of sources of variation in fitness costs of plant antiherbivore defenses. *Ecology* 83:176–90

Krupnick GA, Weis AE. 1999. The effect of floral herbivory on male and female reproductive success in *Isomeris arborea. Ecology* 80:135–49

Krupnick GA, Weis AE, Campbell DR. 1999. The consequences of floral herbivory for pollinator service to *Isomeris arborea. Ecology* 80:125–34

Lehtila K, Strauss SY. 1997. Leaf damage by herbivores affects attractiveness to pollinators in wild radish, *Raphanus raphanistrum. Oecologia* 111:396–403

Lehtilä K, Strauss SY. 1999. Effects of foliar herbivory on male and female reproductive traits of wild radish, *Raphanus raphanistrum. Ecology* 80:116–24

Longland WS, Jenkins SH, Vander Wall SB, Veech JA, Pyare S. 2001. Seedling recruitment in *Oryzopsis hymenoides*: Are desert granivores mutualists or predators? *Ecology* 82:3131–48

Louda SM, Potvin MA. 1995. Effect of inflorescence-feeding insects on the demography and lifetime fitness of a native plant. *Ecology* 76:229–45

Madej CW, Clay K. 1991. Avian seed preference and weight-loss experiments—the effect of fungal endophyte-infected tall fescue seeds. *Oecologia* 88:296–302

Maron JL. 1998. Insect herbivory above- and belowground: individual and joint effects on plant fitness. *Ecology* 79:1281–93

Maron JL, Gardner SN. 2000. Consumer pressure, seed versus safe-site limitation, and plant population dynamics. *Oecologia* 124:260–69

Maron JL, Combs JK, Louda SM. 2002. Convergent demographic effects of insect attack on related thistles in coastal vs. continental dunes. *Ecology* 83:3382–92

Maron JL, Simms EL. 2001. Rodent-limited establishment of bush lupine: field experiments on the cumulative effect of granivory. *J. Ecol.* 89:578–88

McEvoy PB, Rudd NT, Cox CS, Huso M. 1993. Disturbance, competition, and herbivory effects on ragwort *Senecio jacobaea* populations. *Ecol. Monogr.* 63:55–75

Miller TE, Travis J. 1996. The evolutionary role of indirect effects in communities. *Ecology* 77:1329–35

Mitchell RJ. 1993. Path analysis: pollination. In *Design and Analysis of Ecological Experiments*, ed. SM Scheiner, J Gurevitch, pp. 211–31. New York: Chapman and Hall

Mitchell-Olds T, Siemens D, Pedersen D. 1996. Physiology and costs of resistance to herbivory and disease in *Brassica. Ent. Exp. Appl.* 80:231–37

Moegenburg SM, Levey DJ. 2003. Do frugivores respond to fruit harvest? An experimental study of short-term responses. *Ecology* 84:2600–12

Mothershead K, Marquis RJ. 2000. Fitness impacts of herbivory through indirect effects on plant-pollinator interactions in *Oenothera macrocarpa. Ecology* 81:30–40

Mueller-Schaerer H, Brown VK. 1995. Direct and indirect effects of above- and belowground insect herbivory on plant density and performance of *Tripleurospermum perforatum* during early plant succession. *Oikos* 72:36–41

Mutikainen P, Delph LF. 1996. Effects of herbivory on male reproductive success in plants. *Oikos* 75:353–58

Omacini M, Chaneton EJ, Ghersa CM, Muller CB. 2001. Symbiotic fungal endophytes control insect host-parasite interaction webs. *Nature* 409:78–81

Ostfeld RS, Manson RH, Canham CD. 1997. Effects of rodents on survival of tree seeds and seedlings invading old fields. *Ecology* 78:1531–42

Paine RT. 1992. Food-web analysis through field measurement of per capita interaction strength. *Nature* 355:73–75

Palmer TM. 2003. Spatial habitat heterogeneity influences competition and coexistence in an African acacia ant guild. *Ecology* 84:2843–55

Palmer TM, Stanton ML, Young TP. 2003. Competition and coexistence: exploring mechanisms that restrict and maintain diversity within mutualist guilds. *Am. Nat.* 162:S63–S79

Peckarsky BL, McIntosh AR. 1998. Fitness and community consequences of avoiding multiple predators. *Oecologia* 113:565–76

Pellmyr O. 1997. Pollinating seed eaters: Why is active pollination so rare? *Ecology* 78:1655–60

Petraitis PS, Dunham AE, Niewiarowski PH. 1996. Inferring multiple causality: the limitations of path analysis. *Funct. Ecol.* 10:421–31

Pilson D. 1996. Two herbivores and constraints on selection for resistance in *Brassica rapa*. *Evolution* 50:1492–500

Polis GA, Strong DR. 1996. Food web complexity and community dynamics. *Am. Nat.* 147:813–46

Poveda K, Steffan-Dewenter I, Scheu S, Tscharntke T. 2003. Effects of below- and above-ground herbivores on plant growth, flower visitation and seed set. *Oecologia* 135:601–5

Primack RB, Hall P. 1990. Costs of reproduction in the pink lady's slipper orchid: a four-year experimental study. *Am. Nat.* 136:638–56

Raps A, Vidal S. 1998. Indirect effects of an unspecialized endophytic fungus on specialized plant-herbivorous insect interactions. *Oecologia* 114:541–47

Rausher MD. 1996. Genetic analysis of coevolution between plants and their natural enemies. *Trends Genet.* 12:212–17

Rojo E, Solano R, Sanchez-Serrano Jose J. 2003. Interactions between signaling compounds involved in plant defense. *J. Plant. Growth Regul.* 22:82–98

Roubik DW. 1978. Competitive interactions between neotropical pollinators and africanized honeybees. *Science* 201:1030–32

Roubik DW. 1980. Foraging behavior of competing africanized honeybees and stingless bees. *Ecology* 61:836–45

Rudgers JA. 2004. Enemies of herbivores can shape plant traits: selection in a facultative ant-plant mutualism. *Ecology* 85:192–205

Sagers CL, Coley PD. 1995. Benefits and costs of defense in a neotropical shrub. *Ecology* 76:1835–43

Saikkonen K, Helander M, Faeth SH, Schulthess F, Wilson D. 1999. Endophyte-grass-herbivore interactions: the case of *Neotyphodium* endophytes in Arizona fescue populations. *Oecologia* 121:411–20

Saikkonen K, Helander M, Ranta H, Neuvonen S, Virtanen T, et al. 1996. Endophyte-mediated interactions between woody plants and insect herbivores? *Ent. Exp. Appl.* 80:269–71

Scheiner SM, Donohue K, Dorn LA, Mazer SJ, Wolfe LM. 2002. Reducing environmental bias when measuring natural selection. *Evolution* 56:2156–67

Scheiner SM, Mitchell RJ, Callahan HS. 2000. Using path analysis to measure natural selection. *J. Evol. Biol.* 13:423–33

Schemske DW, Horvitz CC. 1988. Plant-animal interactions and fruit production in a neotropical herb: a path analysis. *Ecology* 69:1128–37

Schmitz OJ. 1998. Direct and indirect effects of predation and predation risk in old-field interaction webs. *Am. Nat.* 151:327–42

Shipley B. 2000. *Cause and Correlation in Biology*. Cambridge, UK: Cambridge Univ. Press

Siemens DH, Garner SH, Mitchell-Olds T, Callaway RM. 2002. Cost of defense in the context of plant competition: *Brassica rapa* may grow and defend. *Ecology* 83:505–17

Siemens DH, Mitchell-Olds T. 1998. Evolution of pest-induced defenses in *Brassica* plants: tests of theory. *Ecology* 79:632–46

Siepielski AM, Benkman CW. 2004. Interactions among moths, crossbills, squirrels, and lodgepole pine in a geographic selection mosaic. *Evolution* 58:95–101

Smit R, Bokdam J, den Ouden J, Olff H, Schot-Opschoor H, Schrijvers M. 2001. Effects of introduction and exclusion of large herbivores on small rodent communities. *Plant Ecol.* 155:119–27

Smith FA, Brown JH, Valone TJ. 1997. Path analysis: a critical evaluation using long-term experimental data. *Am. Nat.* 149:29–42

Stanton ML. 1987. Reproductive biology of petal color variants in wild populations of *Raphanus sativus* I. Pollinator response to color morphs. *Am. J. Bot.* 74:178–87

Stanton ML. 2003. Interacting guilds: moving beyond the pairwise perspective on mutualisms. *Am. Nat.* 162:S10–S23

Stanton ML, Young HJ, Ellstrand NC, Clegg JM. 1991. Consequences of floral variation for male and female reproduction in experimental populations of wild radish, *Raphanus sativus* L. *Evolution* 45:268–80

Stinchcombe JR, Rausher MD. 2001. Diffuse selection on resistance to deer herbivory in the ivyleaf morning glory, *Ipomoea hederacea*. *Am. Nat.* 158:376–88

Stinchcombe JR, Rausher MD. 2002. The evolution of tolerance to deer herbivory: modifications caused by the abundance of insect herbivores. *Proc. R. Soc. London Ser. B Biol. Sci.* 269:1241–46

Stinchcombe JR, Rutter MT, Burdick DS, Tiffin P, Rausher MD, Mauricio R. 2002. Testing environmentally induced bias in phenotypic estimates of natural selection: theory and practice. *Am. Nat.* 160:511–23

Stout JC, Goulson D. 2001. The use of conspecific and interspecific scent marks by foraging bumblebees and honeybees. *Anim. Behav.* 62:183–89

Strauss SY. 1991. Direct, indirect, and cumulative effects of three native herbivores on a shared host plant. *Ecology* 72:543–58

Strauss SY, Conner JK, Lehtila KP. 2001. Effects of foliar herbivory by insects on the fitness of *Raphanus raphanistrum*: damage can increase male fitness. *Am. Nat.* 158:496–504

Strauss SY, Conner JK, Rush SL. 1996. Foliar herbivory affects floral characters and plant attractiveness to pollinators: implications for male and female plant fitness. *Am. Nat.* 147:1098–107

Strauss SY, Irwin RE, Lambrix VM. 2004. Optimal defence theory and flower petal colour predict variation in the secondary chemistry of wild radish. *J. Ecol.* 92:132–41

Strauss SY, Murch P. 2004. Towards an understanding of the mechanisms of tolerance: compensating for herbivore damage by enhancing a mutualism. *Ecol. Entomol.* 29:234–39

Strauss SY, Sahli H, Conner JK. 2004. Towards a more trait-centered approach to diffuse (co)evolution. *New Phytol.* In press

Thaler JS, Karban R, Ullman DE, Boege K, Bostock RM. 2002. Cross-talk between jasmonate and salicylate plant defense pathways: effects on several plant parasites. *Oecologia* 131:227–35

Thompson JN. 1994. *The Coevolutionary Process*. Chicago: Univ. Chicago Press. 376 pp.

Thompson JN. 1999. The evolution of species interactions. *Science* 284:2116–18

Tiffin P. 2002. Competition and time of damage affect the pattern of selection acting on plant defense against herbivores. *Ecology* 83:1981–90

Tiffin P, Rausher MD. 1999. Genetic constraints and selection acting on tolerance to herbivory in the common morning glory *Ipomoea purpurea*. *Am. Nat.* 154:700–16

Turnbull LA, Crawley MJ, Rees M. 2000. Are plant populations seed-limited? A review of seed sowing experiments. *Oikos* 88:225–38

Valone TJ, Brown JH, Heske EJ. 1994. Interactions between rodents and ants in the Chihuahuan Desert: an update. *Ecology* 75:252–55

Valverde PL, Fornoni J, Nunez-Farfan J. 2003. Evolutionary ecology of *Datura stramonium*: equal plant fitness benefits of growth and resistance against herbivory. *J. Evol. Biol.* 16:127–37

van der Heijden MGA, Sanders IR. 2002. *Ecological Studies. Mycorrhizal Ecology.* New York: Springer-Verlag. 469 pp.

Vicari M, Hatcher PE, Ayres PG. 2002. Combined effect of foliar and mycorrhizal endophytes on an insect herbivore. *Ecology* 83:2452–64

Warner PJ, Cushman JH. 2002. Influence of herbivores on a perennial plant: variation with life history stage and herbivore species. *Oecologia* 132:77–85

Wise MJ, Sacchi CF. 1996. Impact of two specialist insect herbivores on reproduction of horse nettle, *Solanum carolinense*. *Oecologia* 108:328–37

Wootton JT. 1993. Indirect effects and habitat use in an intertidal community: interaction chains and interaction modifications. *Am. Nat.* 141:71–89

Wootton JT. 1994. Predicting direct and indirect effects: an integrated approach using experiments and path analysis. *Ecology* 75:151–65

Wootton JT. 2002. Indirect effects in complex ecosystems: recent progress and future challenges. *J. Sea Res.* 48:157–72

SPATIAL SYNCHRONY IN POPULATION DYNAMICS*

Andrew Liebhold,[1] Walter D. Koenig,[2] and Ottar N. Bjørnstad[3]

[1]Northeastern Research Station, USDA Forest Service, Morgantown, West Virginia 26505; email: aliebhold@fs.fed.us
[2]Hastings Reservation, University of California, Berkeley, Carmel Valley, California 93924; email: wicker@berkeley.edu
[3]Departments of Entomology and Biology, Pennsylvania State University, University Park, Pennsylvania 16802; email: onb1@psu.edu

Key Words dispersal, Moran effect, stochastic dynamics, spatial dynamics, autocorrelation

■ **Abstract** Spatial synchrony refers to coincident changes in the abundance or other time-varying characteristics of geographically disjunct populations. This phenomenon has been documented in the dynamics of species representing a variety of taxa and ecological roles. Synchrony may arise from three primary mechanisms: (a) dispersal among populations, reducing the size of relatively large populations and increasing relatively small ones; (b) congruent dependence of population dynamics on a synchronous exogenous random factor such as temperature or rainfall, a phenomenon known as the "Moran effect"; and (c) trophic interactions with populations of other species that are themselves spatially synchronous or mobile. Identification of the causes of synchrony is often difficult. In addition to intraspecific synchrony, there are many examples of synchrony among populations of different species, the causes of which are similarly complex and difficult to identify. Furthermore, some populations may exhibit complex spatial dynamics such as spiral waves and chaos. Statistical tests based on phase coherence and/or time-lagged spatial correlation are required to characterize these more complex patterns of spatial dynamics fully.

INTRODUCTION

The quest to understand mechanisms behind the spatiotemporal dynamics of natural populations has been a force motivating studies of animal and plant populations for more than a century. Intensive studies on individual species have yielded information useful for describing the important roles that trophic interactions and exogenous forces play in specific systems. However, for most species we still lack a predictive understanding of the causes of population fluctuations.

---

*The U.S. Government has the right to retain a nonexclusive, royalty-free license in and to any copyright covering this paper.

Because of the difficulty in determining the importance of specific ecological processes, many recent studies have attempted to infer mechanisms indirectly via the analysis of population time series (Kendall et al. 1999, Royama 1981, Turchin 1990). These studies characterize temporal patterns (measured by autocorrelation functions, periodograms and similar types of analytical techniques) and use this information to infer the relative role of biotic and abiotic forces in temporal dynamics. The recent flurry of studies on population synchrony may be seen as advancing this line of inquiry to include spatiotemporal dynamics. Population synchrony, as measured by correlation in abundance (or changes in abundance), was extensively studied starting in the 1990s (reviewed in Bjørnstad et al. 1999a, Koenig 1999). These types of studies have continued to advance in scope, and recently there have been important innovations to the analysis of spatial dynamics via the application of phase analyses. These methods have expanded our characterization of spatial dynamics beyond detection of simple synchrony to include more complex patterns of spatial dynamics such as waves.

A perpetual limitation in the analysis of dynamics of multiple populations has been the overwhelming complexity of analyzing large quantities of spatially referenced time series data. However, with the advent of geographical information systems, faster and cheaper computers, more powerful spatial statistics, and greater access to data, dealing with such large and complex datasets is no longer the insurmountable problem it was a few decades ago. Results from many such analyses indicate that different populations of the same species often fluctuate more or less synchronously, even when populations are quite distant from each other. Specifically, increases and decreases in abundance of one population tend to occur simultaneously with increases and decreases of other populations. As the distance separating populations increases, synchrony typically declines. This phenomenon may reflect two different patterns: for populations that fluctuate in a nonperiodic fashion, the pattern of fluctuation may become increasingly dissimilar with distance. For populations that fluctuate in a cyclic or quasiperiodic fashion, the populations' cycle phase may diverge with separation distance.

The world around us is a spatially autocorrelated place (Legendre 1993). In any biotic or abiotic property, such as elevation, temperature, population density, or reproductive rate, values at nearby locations tend to be similar (Rossi et al. 1992). However, spatial synchrony is a special type of spatial autocorrelation in that synchrony refers to spatial autocorrelation of variation through time. This special type of space-time variation sets it apart from other spatially autocorrelated properties.

Simultaneous with the advent of analytical studies of space-time patterns in real populations has been a blossoming of theoretical studies of spatial dynamics, both with respect to spatially extended populations (e.g., Bascompte & Sole 1998a) and metapopulations (e.g., Hanski 1998). Population models have been useful instruments for testing hypotheses concerning the mechanisms behind spatial synchrony. In any model in which two or more populations are allowed to fluctuate according to some mixture of density-dependent and density-independent processes, three mechanisms may cause synchrony: (*a*) dispersal among populations; (*b*) synchronous stochastic effects, often referred to as the "Moran effect"; and (*c*) trophic interactions with other species that are themselves either synchronized or mobile. As in many other ecological problems, linking theory to reality is often challenging. Although it is possible to identify these synchronizing processes in models, identification of the dominant synchronizing processes in field populations is often more difficult because all three mechanisms may produce nearly identical signatures of synchrony among populations. Ecologists are often frustrated in their efforts to identify the cause of synchrony, but this quest poses an exciting challenge to population ecologists, and many methods have been described for drawing mechanistic inferences.

THE UBIQUITOUS PRESENCE OF SPATIAL SYNCHRONY IN POPULATION DYNAMICS

Spatial synchrony of abundance has been found in populations representing virtually all major taxa and ecological roles (Table 1). Given that any two subunits of the same population are, in theory, perfectly synchronized if there is no distance separating them (in which case in a practical sense they should be considered the same population), spatial synchrony is, not surprisingly, common among nearby populations. However, researchers have observed considerable variation in the geographical range of synchrony in studies of various organisms, only some of which can be attributed to differences in scale intrinsic to the organism itself. The observed range of spatial synchrony is a few centimeters in microcosm experiments with protozoa to hundreds or thousands of kilometers in numerous "free-ranging" systems (Table 1). Synchrony usually decreases as the distance between populations increases (Ranta et al. 1995a, Sutcliffe et al. 1996, Bjørnstad et al. 1999a); however, traveling population waves can lead to U-shaped (Ranta & Kaitala 1997) or cyclic (Bjørnstad & Bascompte 2001) relationships between synchrony and distance.

In addition to synchrony of abundance, specific demographic properties, such as reproduction or mortality, and even population characteristics, such as mean size and age distributions, may be spatially correlated. In particular, many plant species exhibit a phenomenon, known as "masting" or "mast seeding," in which temporal variation in reproduction is highly synchronized among individual plants (Koenig et al. 1999, Kelly & Sork 2002, Liebhold et al. 2004). Comparable patterns of spatially synchronous reproductive dynamics have been observed in animals as well (Ims 1990, Myers et al. 1995). More recently, other kinds of demographic factors, such as the sex ratio of goshawk (*Accipiter gentiles*) nestlings (Byholm et al. 2002) and age- and size-distributions of California tiger salamanders (*Ambystoma californiense*) in ponds (Trenham et al. 2001) have been found to exhibit significant spatial synchrony. Mortality may also be synchronized among spatially disjunct populations as a consequence of spatially synchronous disease dynamics (Viboud et al. 2004).

TABLE 1 Summary of records of intraspecific spatially synchronous population dynamics among all taxa

Taxa	Geographical extent of synchrony	References
Protista: ciliophora	10–500 cm (microcosm)	Holyoak & Lawler 1996
Fungal plant pathogen	0.5–3 km	Thrall et al. 2001
Viral human pathogen	1–1000 km	Bolker & Grenfell 1996, Rohani et al. 1999, Viboud et al. 2004
Insect detritivores	5–20 m	Tobin & Bjørnstad 2003
Insect herbivores	1–1000 km	Hanski & Woiwod 1993; Hawkins & Holyoak 1998; Liebhold et al. 1996; Liebhold & Kamata 2000; Myers 1990, 1998; Peltonen & Hanski 1991; Peltonen et al. 2002; Pollard 1991; Raimondo et al. 2004; Rossi & Fowler 2003; Shepherd et al. 1988; Sutcliffe et al. 1996; Tenow 1972; Williams & Liebhold 1995, 2000b; Zhang & Alfaro 2003
Insect predators and parasitoids	10 m–400 km	Baars & Van Dijk 1984, Rossi & Fowler 2003, Satake et al. 2004, Tobin & Bjørnstad 2003
Fish	10–500 km	Fromentin et al. 2000; Myers et al. 1995, 1997; Ranta et al. 1995a
Amphibians	0.2–100 km	Trenham et al. 2001, 2003
Birds	5–2000 km	Bellamy et al. 2003; Bock & Lepthien 1976; Cattadori et al. 1999; Jones et al. 2003; Small et al. 1993; Ranta et al. 1995a,b; Koenig 1998, 2001, 2002; Moss et al. 2000; Paradis et al. 1999, 2000; Watson et al. 2000
Mammals	10–1000 km	Bjørnstad et al. 1999b; Christiansen 1983; Elton & Nicholson 1942; Grenfell et al. 1998; Ims & Andreassen 2000; Mackin-Rogalska & Nabaglo 1990; Moran 1953b; Post & Forchhammer 2002; Ranta et al. 1995a,b, 1997a,b, 1998; Small et al. 1993; Smith 1983; Swanson & Johnson 1999
Mollusks	2–30 km	Burrows et al. 2002

These examples of population synchrony all involve geographically disjunct populations of the same species. However, there have been numerous observations of synchronous fluctuations among different species as well. Many examples of interspecific synchrony can be found between pairs of species that have direct trophic interactions, such as predators and their prey. Numerical responses seen in trophic interactions often lag each other in time, which can result in time-lagged synchrony. Note, for instance, that the classic Lotka-Volterra models produce cycles in which the predator peak lags that of the prey by one-quarter-cycle length. Alternative prey species that share a common predator, a situation referred to as "apparent competition," may as a result be forced into synchrony (Bulmer 1975, Small et al. 1993, de Roos et al. 1998, Jones et al. 2003). Similar synchronization may occur among species that share a fluctuating food resource (Bock & Lepthien 1976, Koenig 2001, Jones et al. 2003). Numerical tracking of prey populations by predator populations can lead to synchrony, as illustrated by interspecific synchrony between mast-seeding plants and their seed predators (Curran & Webb 2000, Schauber et al. 2002, Satake et al. 2004). Synchrony has also been observed among populations of sympatric species that are not directly linked through trophic interactions, such as herbivorous forest insects (Miller & Epstein 1986, Hawkins & Holyoak 1998, Myers 1998, Raimondo et al. 2004) and allopatric large herbivores (Post & Forchhammer 2002).

MEASURING SYNCHRONY

Early studies relied on visual inspection of data to assess synchrony (e.g., Elton 1942). The current tradition, however, is to rely on formal statistical analyses. This, in turn, requires strict numerical definition of synchrony. Several alternative and/or complementary definitions exist. Crudely speaking, these methods may be divided into those measuring synchrony and those measuring phase synchrony (sometimes called phase coherence). An obvious measure of synchrony is the correlation between two time series of abundance. Correlation is here usually measured (albeit not necessarily) by the Pearson product-moment correlation coefficient or lag-0 cross-correlation coefficient. Hanski & Woiwod (1993), however, argued that synchrony should refer to covariation in rates of change, resulting in definitions involving correlation of time series of ratios of abundance in successive years ($R_t = N_t/N_{t-1}$) or differences in log-abundance ($r_t = \log N_t - \log N_{t-1}$) (Steen et al. 1996, Bjørnstad et al. 1999a). The latter two address the coincidence in the direction of population change (i.e., increasing or decreasing) between two populations (Buonaccorsi et al. 2001). Practically speaking, the three measures of synchrony exhibit only slight differences in numerical behavior, but they all pertain to the same dynamical property and are fraught with similar statistical complications (reviewed below). As an intermediate between the measures of synchrony and phase synchrony, coincidence of peaks has occasionally been used as a definition of synchrony; different methods have been applied to test the significance of such coincidence (Liebhold et al. 1996, Steen et al. 1996, Myers 1998). As a modification of this technique, phase synchrony is a powerful concept that has recently received strict definition in ecological dynamics (Haydon & Greenwood 2000, Grenfell et al. 2001, Cazelles & Stone 2003). This concept is particularly relevant for cyclic (or quasiperiodic) populations for which the relative phase in the cycle can be defined in the frequency domain by Fourier analysis (Bulmer 1975), by wavelet phase analysis (Grenfell et al. 2001) or in the time domain (Haydon & Greenwood 2000, Cazelles & Stone 2003). We first review the measures of synchrony and later turn to phase synchrony.

Given two time series of abundance, two important statistical issues must be considered. The first issue is how to handle long-term trends. The second issue is inference in the face of serial dependence, or temporal autocorrelation. In the presence of long-term trends in abundance, correlation that is due to common trends may obscure synchrony in short-term fluctuations (Buonaccorsi et al. 2001). To the extent that one is less interested in shared common trends, it may be wise to detrend the series by focusing on the residuals from a linear, polynomial, or local regression of abundance against time. For example, by detrending population densities of a bird species increasing throughout its range because of global change, the overall synchrony that is due to this trend is eliminated, and remaining synchrony, if any, is due to short-term changes in populations. Alternatively, one may use the time series of differenced log-abundances (r_t), which explicitly focus on rates of change and for which no additional detrending is required.

A related problem is how to conduct statistical inference on synchrony given that serial dependence (temporal autocorrelation) is ubiquitous in population time series (Royama 1992). When the significance of a correlation between two time series is tested using ordinary methods, the autocorrelation violates the assumption of independence among samples (Lennon 2000, Buonaccorsi et al. 2001). "Prewhitening," a statistical technique to remove serial dependence, has often been applied to eliminate this problem (Hanski & Woiwod 1993, Williams & Liebhold 1995, Sutcliffe et al. 1997, Paradis et al. 1999). This technique involves fitting autoregressive models to each time series, then correlating residuals from these models and conducting ordinary tests of significance. Unfortunately, by removing the autoregressive portion of these time series one may be ignoring biologically meaningful variability and synchrony (Buonaccorsi et al. 2001). Fortunately, if one has many time series, one may perform a single test for synchrony using the matrix of all pairwise correlations. For such matrices, an additional problem is that the correlation between, for example, series 1 and 2 is not independent of the correlation between series 1 and 3. However, bootstrap procedures involving resampling of locations with replacement can provide a valid test for the significance of both mean synchrony and how synchrony depends on distance (Bjørnstad et al. 1999a,b; Bjørnstad & Falck 2001). The advantage of this approach is that it does not rely on assumptions concerning independence among observations in each time series.

Technical tedium aside, a key empirical observation is how spatial synchrony often decreases as the distance between populations increases (Baars & Van Dijk

1984; Hanski & Woiwod 1993; Ranta et al. 1995a; Sutcliffe et al. 1996; Koenig & Knops 1998a; Bjørnstad et al. 1999a,b; Bjørnstad 2000, Williams & Liebhold 2000b; Peltonen et al. 2002). Various suggestions have been made for how to model this relationship between synchrony and lag distance. Of the parametric methods (i.e., methods that assume a specific functional form between synchrony and distance), the relationship has been modeled as a linear decay (Baars & Van Dijk 1984, Hanski & Woiwod 1993, Sutcliffe et al. 1996, Williams & Liebhold 2000b) or by models from geostatistical theory, such as an exponential function (Myers et al. 1995). However, because little theory exists to predict a functional form to chose (e.g., Bjørnstad & Bolker 2000), nonparametric methods that are able to capture more complex patterns of spatial correlation are often used. The two most common of these approaches is to model the relationship using smoothing splines [i.e., the nonparametric covariance function (Bjørnstad et al. 1999a,b; Bjørnstad & Bascompte 2001; Bjørnstad & Falck 2001; Peltonen et al. 2002)] or piece-wise constant 'regressograms' [cf., the modified correlogram (Koenig & Knops 1998b) or Mantel correlogram (Oden & Sokal 1986, Bjørnstad & Falck 2001)]. Figure 1 shows one parametric model and the two nonparametric methods

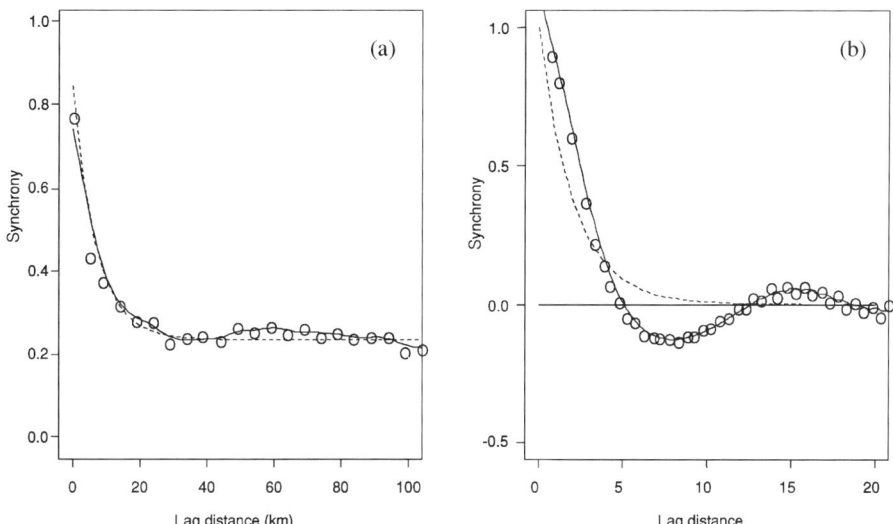

Figure 1 (*a*) Analyses of synchrony in the Japanese wood mouse (Bjørnstad et al. 1999b) using (1) a parametric geostatistical model (*dotted line*) assuming an exponential decay to a sill (the model is $y = 0.6*\exp[-x/6.9] + 0.25$); (2) The nonparametric covariance function (*solid line*), and (3) the modified correlogram (*circles*). (*b*) Synchrony in traveling waves in a spatially extended host-parasitoid model [both host and parasitoid dispersal rates are set to 0.7; see Bjørnstad & Bascompte (2001) for details]. Symbols are as in (*a*). The parametric model is $y = \exp(-x/2.1)$. Note how the nonparametric models are flexible enough to capture the cyclic nature of the correlation.

applied to the wood mice data from Hokkaido, Japan (Bjørnstad et al. 1999b), and to the traveling waves that result from spatially extended host-parasitoid models (Bjørnstad & Bascompte 2001). Bootstrap procedures are, again, useful for testing the significance of nuances in these spatial correlation functions, such as the maximum distance over which spatial synchrony extends (Koenig & Knops 1998b, Bjørnstad et al. 1999b, Bjørnstad & Falck 2001).

Approaches that focus on the correlation among series as a function of distance are in many ways analogous to the suite of geostatistical methods that quantify dependence in strictly spatial data (e.g., Rossi et al. 1992). They differ, however, in that spatial synchrony involves time and can yield absolute estimates of synchrony, whereas simple spatial autocorrelation refers to a static (nontemporal) dependence and is restricted to measuring relative synchrony. One issue that is important to both the quantification of spatial pattern in static data and to the detection of spatial synchrony in spatiotemporal data is that the scale of sampling and analysis influences the emergent patterns (Levin 1992, Dungan et al. 2002). We use the term "scale" here to refer to the magnitude of the sample unit size, the geographical extent of sampling, and the geographical resolution of the analysis. Results from one scale may not necessarily be used to draw inferences about processes operating at other scales, and therefore investigators should take care to limit generalizations from analyses to the specific scale of a study.

Correlational measures of synchrony may not always be the best representation of the synchrony-asynchrony continuum. In particular, for cyclic (or quasiperiodic) populations, subtle differences in the timing of cycles (e.g., Figure 2a) may result in low statistical correlation even when populations are closely linked. For example, the two series shown in Figure 2a appear to be superficially very correlated, yet their correlation coefficient is only $r = 0.16$. Visual inspection indicates a coincidence of peaks and troughs, although the series with open circles tends to lag behind the series with closed circles. For such dynamics it is useful to consider differences (or correlation) in cycle-phase rather than correlation in abundance (Haydon & Greenwood 2000, Grenfell et al. 2001, Cazelles & Stone 2003).

Such phase analysis is accomplished by transforming the time series N_t into a series of phase angles. This type of analysis is most easily illustrated by the method proposed by Cazelles & Stone (2003). They first identified the quasi-periodic maxima (e.g., arrows in Figure 2a) and then assigned phase angles, θ, that increase linearly between 0 and 2π as a function of time (Figure 2b). When populations are perfectly synchronized, phase differences will be near zero. However, if one series consistently lags behind the other series, then the phase differences will be consistently different from zero. Cazelles & Stone (2003) suggested a statistical test for the significance of this phase difference by comparing the frequency distribution of observed phase differences (triangles in Figure 2b) with a distribution of differences simulated for the two series under the condition of no association. However, a potential problem with the phase analysis proposed by Cazelles & Stone (2003) is that the assignment of phase angle values for each observation in a series depends entirely on identifying quasi-periodic maxima. This detail means

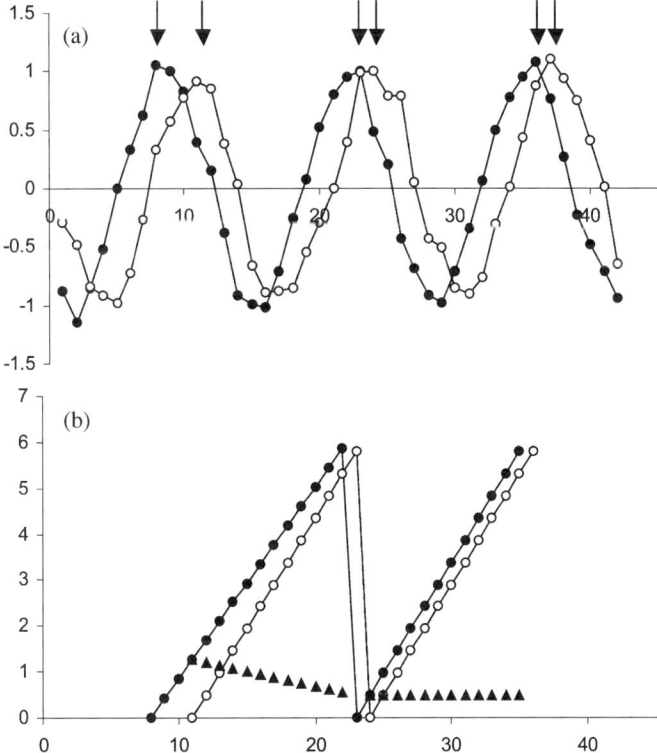

Figure 2 Calculation of phase angle differences. (*a*) Two hypothetical time series. Arrows point to quasi-periodic maxima. (*b*) Phase angles of the series shown in (*a*). Triangles are the differences between the two series.

that most of the fluctuations between the maxima are not used in measuring phase synchrony. Furthermore, identification of maxima may be difficult if the series are not periodic or if there is a great deal of high-frequency "noise" in the system. Both of these problems may be overcome by the use of wavelet phase angles (Grenfell et al. 2001), which result from a local (in time) frequency decomposition of the time series. A further advantage of wavelet phase analyses is that one may consider multiple cycle-periods (e.g., seasonal *and* multiannual periods) and average the relative phase differences across periods (Grenfell et al. 2001).

Measures of phase synchrony provide different information about spatial dynamics than do the conventional synchrony measures (above). In theory, phase angles are unaffected by the relative amplitude of time series and they are specifically designed for detecting lags in dependence between cyclic time series. As discussed below, detection of such lags may be critical for identifying waves and other forms of complex spatial dynamics. Grenfell et al. (2001) proposed a hybrid between spatial correlograms and phase analysis. Their so-called phase coherence function quantifies, in a nonparametric manner, how phase difference varies as a function of lag distance. Under this approach, all pairwise phase correlations are calculated—note here that because phases are circular, so-called circular correlations are appropriate (Jammalamadaka & SenGupta 2001)—and these phase correlations are regressed on distance. An interesting property of this function is that the increase in phase difference (i.e., the decay in phase correlation) is directly related to the speed of traveling waves. Grenfell et al. (2001) applied wavelet phase analysis to the cyclic time series of measles epidemics and found that nearby time series tended to have smaller phase differences than series located more distantly from each other. Furthermore, by analyzing how the phase difference depended on distance, they were able to map the location of apparent "foci" from which outbreaks first appeared (big cities) and spread outward at a speed of about 5 km per week.

We illustrate the wavelet phase analysis with data from the spatially extended host-parasitoid model introduced in Figure 1b. This model is well known to produce complex spatial dynamics such as spiral waves and spatial chaos (Hassell et al. 1991, Bascompte & Sole 1998b, Bjørnstad & Bascompte 2001). We performed simulations using this model in a 30 × 30 coupled map lattice with parameters set to generate spiral waves. A movie of simulated host densities can be viewed by following the Supplemental Material link from the Annual Reviews home page at http://www.annualreviews.org. Figure 3a shows the spatial distribution of relative wavelet phase angles (Grenfell et al. 2001) calculated from the time series on a 30 × 30 lattice. The figure demonstrates the existence of several primary foci of the spiral waves. Figure 3b shows the phase coherence function derived from these data. The steep decay in phase coherence testifies to rapidly moving spiral waves.

CAUSES OF SYNCHRONY

Given the frequent demonstration of population synchrony, a single cause of this behavior would be attractive. However, several possible mechanisms may cause both intraspecific or interspecific synchrony, and identifying which mechanism is more important is often difficult. The causes of population synchrony thus remain an elusive ecological question. The quest for causes is an excellent example of how simple process-oriented mathematical models can be used to test ecological hypotheses (Bascompte & Sole 1998a).

Population synchrony is closely related to a phenomenon called "phase locking" in dynamical systems theory. The existence of periodic or quasi-periodic oscillations in phenomenon ranging from animal abundance to high-speed energy waves and the prices of commodities has been recognized for hundreds of years (Huygens 1673). Phase locking between two oscillators is said to exist when $n\phi_1 - m\phi_2 = \varphi$, where n and m are integers, ϕ_1 and ϕ_2 are the phases of oscillators 1 and 2, and φ is a constant phase difference. Chaotic oscillators, in addition to periodic oscillators, may exhibit phase locking as well (Rosenblum et al. 1996).

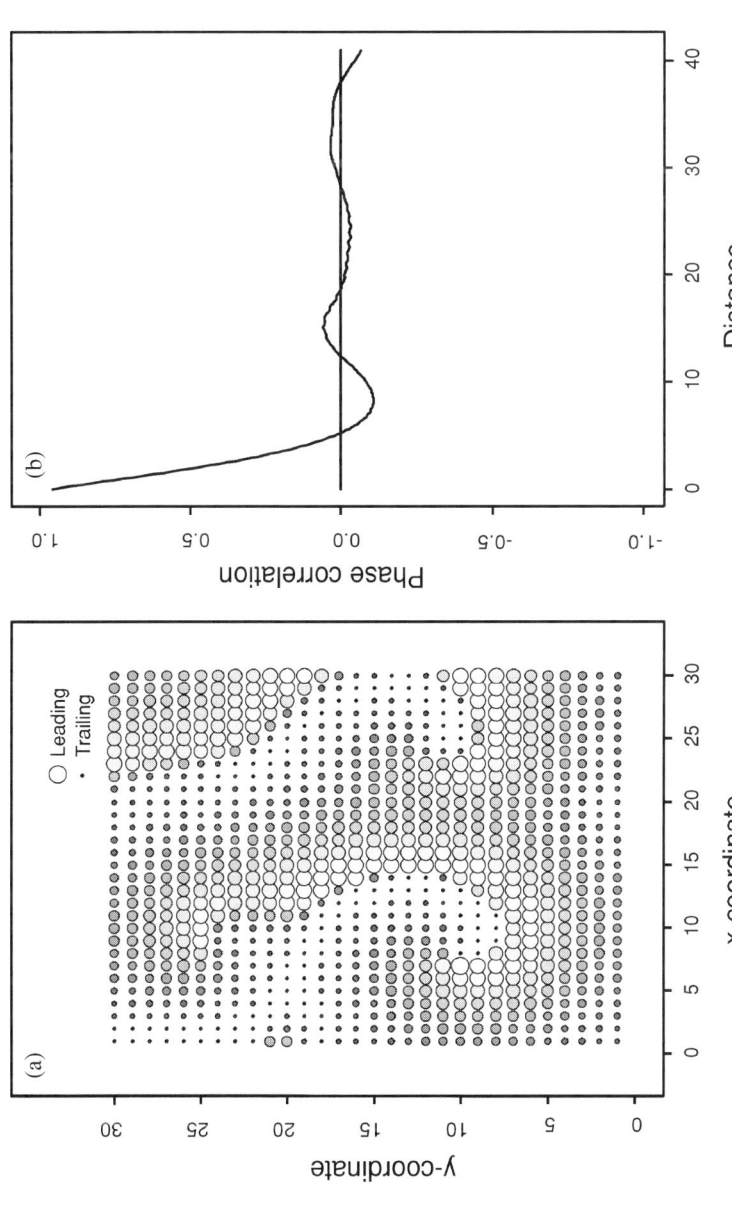

Figure 3 Wavelet phase analysis. (*a*) The relative phase differences for the 30 × 30 local populations in the spatially extended host-parasitoid model from Bjørnstad & Bascompte (2001) introduced in Figure 1. A movie of simulated values can also be viewed in the Supplemental Material. Large, light circles represent leading populations. (*b*) The phase correlation function for the data; the steep decay in phase coherence testifies to rapidly moving spiral waves.

In either case, phase locking does not require that the amplitudes of the two series are at all the same.

When one strictly cyclic oscillator even slightly influences values of the other oscillators, phase locking may result (Rosenblum et al. 1996). Alternatively, two oscillators can be synchronized by an external source of noise or random perturbation that simultaneously affects both oscillators; even a relatively small external source may produce phase locking (Stone 1991). Similarly, theoretical models in ecology indicate that two population time series may be synchronized via direct influences such as dispersal or indirect influences such as weather or shared predation. One key conclusion is that populations may operate largely independently of each other via density-dependent processes, but a small interaction between them may be sufficient to result in synchronization. Moran (1953b) recognized this fact in his seminal paper a half century ago (below).

The role of dispersal in synchronizing populations has been studied using both autoregressive (linear) models and a variety of nonlinear population models (Barbour 1990, Holmes et al. 1994, Molofsky 1994, Ranta et al. 1995a, Haydon & Steen 1997, Kendall et al. 2000, Ripa 2000, Bjørnstad & Bascompte 2001, Kaitala et al. 2001). These studies indicate that fluctuations of any two populations that are governed by the same density-dependent process can be brought into synchrony via exchange of a small number of individuals each generation, or when populations reach their maxima. Dispersal may still synchronize populations that differ slightly in driving processes, but if the density-dependent parameters vary so much that their dynamics are vastly different (for example, if they oscillate with very different periodicity), synchronization may be difficult (Barbour 1990, Ranta et al. 1998, Swanson & Johnson 1999).

Population models can also be used to demonstrate the role of exogenous effects on synchronization. Recognition of this phenomenon is attributed to Patrick Moran, an Australian statistician whose many works include several important contributions to the theory of population ecology. Among these was a simple but profound finding that emerged from his analyses of Canadian lynx trapping records, work inspired by Charles Elton, his colleague at Oxford University (Hyde 1991). Moran (1953a,b) fit a second-order autoregressive model to the lynx time series and showed that even a small synchronous, random additive effect synchronized any two series that had identical dynamics. Furthermore, he demonstrated that the correlation between two series obeying linear (or more exactly "log-linear") local dynamics should equal the correlation in the random effect, a mechanism that is now generally referred to as the "Moran effect" or "Moran's theorem." The synchronizing effect of regional stochasticity has subsequently been observed in a variety of nonlinear population models as well (Haydon & Steen 1997, Ranta et al. 1997b, Grenfell et al. 1998, Kendall et al. 2000). In most systems, the Moran effect is thought to be the result of random but synchronous weather influences acting on spatially disjunct populations (Ranta et al. 1997b; Hudson & Cattadori 1999; Koenig 1999, 2002).

Weather is not the only exogenous factor that can synchronize populations, however (Ostfeld & Keesing 2000). In some systems, mobile predators may synchronize oscillations in spatially disjunct populations (Small et al. 1993, Ims & Andreassen 2000). Some species may be synchronized by synchronous fluctuations in populations at either lower or higher trophic levels. For example Selås and colleagues (Selås 1997, Selås et al. 2001) hypothesized that population oscillations of the herbivorous autumnal moth, *Epirrita autumnata*, are primarily determined by spatially synchronous mast seeding in their host birch trees [though serious problems with this hypothesis were recently raised (Klemola et al. 2003)]; according to this hypothesis, host foliage chemistry is more suitable for population growth in years following mast events, and synchrony of masting may thus synchronize moth oscillations. In a more complex example, synchronous masting events in oaks lead to synchronous oscillations in small mammal populations; because small mammals are the dominant predators of low-density gypsy moth, *Lymantria dispar*, populations, their synchronous oscillations may result in synchronous moth oscillations (Ostfeld & Jones 1996, Liebhold et al. 2000). Satake et al. (2004) further demonstrated that the wide-scaled synchrony in fruit production among populations of *Sorbus aucuparia* throughout southern Norway causes synchronous dynamics in both the apple moth, *Argyresthia conjugella*, which feeds on Rowan berries, and its dominant predator, the parasitoid wasp (*Microgaster politus*). Synchronous dynamics of herbivorous insects have also been suggested to cause partial spatial synchrony in insectivorous birds (Jones et al. 2003). Blasius et al. (1999) simulated the spatially extended dynamics of populations embedded in a complex food web and found that under certain conditions spatial synchrony among populations of a single species may be propagated to populations of other species in the food web [as earlier conjectured by Bulmer (1975)]. This result was extended by Cazelles & Boudjema (2001), who demonstrated a theoretical propagation of synchrony among trophically linked populations, even when their density-dependent dynamics differ. Population synchrony in any species is likely affected both by synchrony in trophically related species and by regional stochasticity, as suggested by Koenig's (2001) finding that intraspecific synchrony in North American bird species varied only slightly among seed-eating and insectivorous species.

Climatic forcing may also cause interspecific synchrony. Sympatric populations of different species sometimes exhibit synchrony even though they do not have any direct trophic interactions (Miller & Epstein 1986, Post & Forchhammer 2002, Raimondo et al. 2004). Interspecific synchrony cannot be caused by dispersal, but shared stochastic effects clearly could synchronize populations (Lindström et al. 1996, Watson et al. 2000, Post & Forchhammer 2002). We note, though, that in many systems interspecific synchrony appears to be at least partially due to the simultaneous forcing effects from shared predator population fluctuations (Ydenberg 1987, Marcström et al. 1988, Ims & Steen 1990) or the synchronizing effect of shared (Swanson & Johnson 1999) predator functional responses (Small et al. 1993). With respect to synchronized reproduction, the Moran effect also has been hypothesized to be a cause of spatial synchrony in mast seeding (Koenig & Knops 1998a; Koenig et al. 1999, 2000); however, synchronization via pollen limitation may also be capable of producing synchrony over relatively large geographic distances (Isagi et al. 1997; Satake & Iwasa 2002a,b).

For most systems, theoretical evidence suggests that the nature of the density-dependent population processes affect the extent to which either dispersal or regional stochasticity results in synchrony. Although Moran's representation of density-dependent growth by a log-linear autoregressive model (Moran 1953b) provides useful insight into how stochasticity affects synchrony, most populations are thought to be dominated by nonlinear processes (Turchin 2003). Many nonlinear density-dependent processes may lead to a decrease in the level of synchrony predicted under linear dynamics (Allen et al. 1993; Grenfell et al. 1998, 2000; Jansen 1999; Ranta et al. 1998, 1999). However, nonlinear processes that enhance temporal periodicity may actually enhance synchrony (Ranta et al. 1998, Bjørnstad 2000). These findings are in general agreement with more theoretical studies of phase locking in nonlinear systems that indicate a close relationship between the propensity for phase locking and the Lyapunov exponent (that distinguishes asymptotically stable dynamics from cycles and chaos) (Pecora & Caroll 1990, Rosenblum et al. 1996, Earn et al. 2000).

Moran (1953b) assumed that spatially disjunct populations were governed by identical density-dependent processes. Similar assumptions are typically made when considering the synchronizing effect of dispersal (Barbour 1990, Ranta et al. 1995a). However, this assumption probably does not hold up in many systems (Swanson & Johnson 1999) because geographical variation in the population dynamics of a given species is well known (Bjørnstad et al. 1995, 1998; Henttonen et al. 1985; Williams & Liebhold 2000a). For example, Peltonen et al. (2002) demonstrated that geographical variation in gypsy moth dynamics can lead to a decrease in synchrony caused by regional stochasticity. Barbour (1990), using linear models, showed that the synchronizing effect of dispersal was largely unaffected by slight differences in dynamics. In general, however, there is limited information on how variation in dynamics affects synchronization via regional stochasticity or dispersal (Cazelles & Boudjema 2001).

Spatial variation in the habitat may also affect dispersal rates among populations in ways that affect the synchronizing effect of dispersal. For example, Bellamy et al. (2003) found that populations of certain woodland bird species were more synchronous when they were separated by forests than when they were separated by open land. Perhaps the greatest challenge to ecologists studying spatial synchrony is that the synchronizing effect of regional stochasticity is difficult to differentiate from the effect of dispersal or other synchronizing factors. In some systems, one can rule out dispersal (Williams & Liebhold 1995, Grenfell et al. 1998) or regional stochasticity (Holyoak & Lawler 1996) and thus identify the source of synchrony. But in most systems, these two phenomena co-occur, and the interaction between dispersal and regional stochasticity as causes of synchrony may not be simple; Kendall et al. (2000) found that overall synchrony was lower than would be expected by simply adding the synchrony expected from dispersal with that expected from regional stochasticity.

One approach to evaluating the overall importance of dispersal as a synchronizing agent is to compare species varying in their mobility. Analyses of breeding bird population time series from Great Britain (Paradis et al. 1999) indicated that species with greater dispersal capabilities are more highly synchronized, implicating dispersal as a major cause of the observed synchronous dynamics. Peltonen et al. (2002), in contrast, compared synchrony among six forest insect species that varied from immobile to highly mobile and found that intraspecific population synchrony showed little relation to mobility. They suggested that in these systems, regional stochasticity is a more important source of synchrony than dispersal. However, the lack of a relationship between synchrony and mobility may be due in part to the way in which complex spatiotemporal dynamics produce patterns of synchrony that greatly differ from the predictions from simple dispersal or Moran effects (Bjørnstad et al. 2002). The geographic scale of synchrony or the shape of the distance/synchrony relationship may also provide some clues as to the relative contribution of dispersal versus regional stochasticity. Sutcliffe et al. (1996) compared the geographical range of synchrony among various butterfly species that differed in their dispersal capabilities. They found that populations of more mobile species were generally more synchronous at a local (<5 km) scale but that at larger spatial scales all species were similar, exhibiting declining synchrony with distances up to ca. 200 km.

Population synchrony typically declines with the distance between populations, and most climatic variables (e.g., June maximum temperature) exhibit comparable patterns of synchrony as a function of distance (Koenig & Knops 2000, Williams & Liebhold 2000b, Peltonen et al. 2002). Determining which weather variable is a driving factor in producing population synchrony is nonetheless complicated by a typical lack of ecological understanding sufficient to finger which specific weather variable is most influential on population growth. Intriguingly, Koenig (2002) recently compared patterns of spatial synchrony in both temperature and precipitation using historical weather data collected at stations throughout the world and found that patterns varied little among different continents. This uniformity in patterns of weather synchrony may be an important explanation for the ubiquitous nature of population synchrony.

Evidence from a handful of systems suggests that synchrony may vary through time (Ranta et al. 1998, Koenig 2001) and space (Ranta et al. 1997a). Simulations using both linear and nonlinear models incorporating both dispersal and regional stochasticity indicate that this type of spatial and temporal variation in spatial synchrony can be expected as part of the self-organized properties of certain systems (Kaitala et al. 2001). However, we expect that this variation may also be due to spatial and temporal variation in the habitat, as it influences local dynamics (e.g., Liebhold et al. 2004). There is also evidence that changes in large-scale global weather patterns on a multi-decadal time scale, such as occurs in the North Atlantic Oscillation (Hurrell 1995), may affect the levels of observed spatial synchrony in animal populations (Post & Forchhammer 2002).

SYNCHRONY AND METAPOPULATION PERSISTENCE

The classic concept of metapopulations refers to a collection of local populations, among which the density of individuals fluctuates independently and within which there is a reasonable chance of extinction (Harrison & Quinn 1989, Hanski & Gilpin 1991). The existence of synchrony is particularly significant to such systems because synchrony is directly related to the likelihood of global extinction (Heino et al. 1997). The more spatially synchronous a metapopulation is, the shorter its expected persistence time. The reason for this is straightforward: If all subpopulations fluctuate in unison, then when one goes extinct, all others are likely to suffer the same fate; if spatial synchrony is low, some subpopulations are likely to be abundant and serve to re-establish extinct subpopulations.

The primary mechanism responsible for this phenomenon is known as the rescue effect, and since it was first defined 25 years ago (Brown & Kodric-Brown 1977), it has been invoked as an important occurrence in a variety of systems (e.g., Sinsch 1997, Martin et al. 1997). Traditionally, the metapopulation approach has focused on species in which suitable habitats are restricted to discrete patches, but evidence suggests that a variety of species whose distributions are not subdivided in any obvious way still exhibit dynamics characteristic of metapopulations, including frequent local extinctions and recolonization from other areas. For example, in an analysis of wintering North American birds, Koenig (2001) found that local extinctions were common and that 65% of species failed to exhibit significant spatial synchrony between sites <100 km apart, a distance well within the dispersal range of most of the species. Thus, although spatial synchrony is widespread, it may nonetheless be too low to affect local extinction in many species.

The spatial distribution of habitats can profoundly affect how spatial synchrony impacts metapopulation persistence (Doak et al. 1992, Adler & Nuemberger 1994, King & With 2002). Most previous studies of the effects of spatial synchrony have found that increased synchrony leads to decreased metapopulation persistence, but increased clustering of habitat patches can potentially increase persistence by increasing the immigration between nearby patches even when dispersal is costly (Adler & Nuemberger 1994). These two factors appear to act independently of each other; that is, increased synchrony always decreases metapopulation persistence, regardless of the distribution of habitat patches, whereas increased habitat clumping appears to generally increase persistence whether the landscape is dynamic or static (Johst & Drechsler 2003).

SPATIAL DYNAMICS BEYOND SYNCHRONY

Spatial synchrony is one manifestation among a broader array of space-time patterns. Theoretical ecologists have long demonstrated that relatively simple, nonlinear, spatially explicit models are capable of generating a variety of complex spatio-temporal patterns such as spiral waves, spatial chaos, and crystal lattices. Thus far, however, there are only a few examples of such complex spatial dynamics in natural systems. These examples include a likely "Turing patch" (comparable to a crystal lattice formation) in Western tussock moth populations (Hastings et al. 1997, Maron & Harrison 1997, Wilson et al. 1999); traveling waves in Canadian lynx populations (Ranta & Kaitala 1997, Ranta et al. 1997a, Haydon & Greenwood 2000); and recurrent waves in several host-enemy systems, including rabies (Bacon 1985, Murray 1993), measles (Grenfell et al. 2001), and larch budmoth outbreaks in the European Alps (Bjørnstad et al. 2002).

Complex spatial dynamics are generally considered to be the result of local reproduction of hosts and natural enemies coupled with dispersal ("reaction-diffusion"). Hassell et al. (1991) found that in a spatially extended Nicholson-Bailey model, certain dispersal rates yielded spiral wave dynamics. Consistent with the behavior of this model, predator-prey or host-parasitoid dynamics have been implicated in most examples of complex dynamics thus far described from field populations (Bascompte et al. 1997, Bjørnstad et al. 2002, Hastings et al. 1997, Haydon & Greenwood 2000, Wilson et al. 1999). The identification of complex spatial dynamics in nature is a considerable challenge, to no small extent because methods for identification of complex time-space patterns are in their infancy. One method involves using a directional additive model proposed by Lambin et al. (1998). A common indirect approach involves using spatial correlograms (Figure 1b). For example, Ranta and colleagues (Ranta & Kaitala 1997, Ranta et al. 1997a) used "U-shaped" correlograms as evidence for the existence of traveling waves in Canadian Lynx populations. Bjørnstad & Bascompte (2001) generated data known to exhibit various forms of complex spatial dynamics using a spatially explicit host-parasitoid model and analyzed them using spatial correlograms. Their results indicate that although the spatial correlation functions are unique, the difference between certain patterns (e.g., waves and spatial chaos) are subtle. Time-lagged correlation functions (Bjørnstad et al. 2002) or phase-coherence functions (Grenfell et al. 2001) appear to be diagnostic approaches for distinguishing among complex space-time patterns, but further method refinements and analyses of simulated data with known patterns are needed.

CONCLUSIONS

Although the existence of spatial synchrony is often quite clear, the mechanisms behind synchrony are often murky. In the quest to identify the causes of spatial synchrony, ecological theory and models have contributed, and are likely to continue to contribute, to clarifying how endogenous and exogenous forces interact in natural populations. Clarifying the mechanisms behind spatial synchrony represents a fascinating intellectual challenge for ecologists; it also could ultimately provide critical information for understanding and managing species conservation, pest outbreaks, and disease epidemiology.

Complex spatial dynamics may well be as common as spatial synchrony in natural systems, but detecting them and understanding their causes remains a technical challenge. With the advent of new statistical methods for identifying complex spatial dynamics, we are poised at the threshold of a new era for understanding the complex interactions that drive the spatially extended dynamics of natural systems.

ACKNOWLEDGMENTS

This work was conceived as part of the "Evolutionary causes and ecological consequences of mast seeding in plants" working group supported by the National Center for Ecological Analysis and Synthesis, a Center funded by NSF (Grant #DEB-0072909), the University of California, and the Santa Barbara campus. This work was also funded by USDA NRI grant 2002–35302-12656 to O.N.B. and A.M.L. and NSF grant IBN-0090807 to W.D.K.

The *Annual Review of Ecology, Evolution, and Systematics* is online at
http://ecolsys.annualreviews.org

LITERATURE CITED

Adler FR, Nuemberger B. 1994. Persistence in patchy, irregular landscapes. *Theor. Popul. Biol.* 45:41–75

Allen JC, Schaffer W, Rosko D. 1993. Chaos reduces species extinction by amplifying local population noise. *Nature* 364:229–32

Baars MA, Van Dijk TS. 1984. Population dynamics of two carabid beetles at a Dutch heathland. I. Subpopulation fluctuations in relation to weather and dispersal. *J. Anim. Ecol.* 53:375–88

Bacon PJ. 1985. *Population Dynamics of Rabies in Wildlife*. London: Academic

Barbour DA. 1990. Synchronous fluctuations in spatially separated populations of cyclic forest insects. In *Population Dynamics of Forest Insects*, ed. AD Watt, SR Leather, MD Hunter, NAC Kidd, pp. 339–46. Andover, UK: Intercept

Bascompte J, Solé RV. 1998a. Spatiotemporal patterns in nature. *Trends Ecol. Evol.* 13:173–74

Bascompte J, Solé RV, eds. 1998b. *Modeling Spatiotemporal Dynamics in Ecology*. New York: Springer-Verlag. 216 pp.

Bascompte J, Solé RV, Martínez N. 1997. Population cycles and spatial patterns in snowshoe hares: an individual-oriented simulation. *J. Theor. Biol.* 187:213–22

Bellamy PE, Rothery P, Hinsley SA. 2003. Synchrony of woodland bird populations: the effect of landscape structure. *Ecography* 26:338–48

Bjørnstad ON. 2000. Cycles and synchrony: two historical "experiments" and one experience. *J. Anim. Ecol.* 69:869–73

Bjørnstad ON, Bascompte J. 2001. Synchrony and second order spatial correlation in host-parasitoid systems. *J. Anim. Ecol.* 70:924–33

Bjørnstad ON, Bolker B. 2000. Canonical functions for dispersal-induced synchrony. *Proc. R. Soc. London Ser. B Biol. Sci.* 267:1787–94

Bjørnstad ON, Falck W. 2001. Nonparametric spatial covariance functions: estimation and testing. *Environ. Ecol. Stat.* 8:53–70

Bjørnstad ON, Falck W, Stenseth NC. 1995. A geographic gradient in small rodent density fluctuations: a statistical modelling approach. *Proc. R. Soc. London Ser. B* 262:127–33

Bjørnstad ON, Ims RA, Lambin X. 1999a. Spatial population dynamics: analysing patterns and processes of population synchrony. *Trends Ecol. Evol.* 14:427–31

Bjørnstad ON, Peltonen M, Liebhold AM, Baltensweiler W. 2002. Waves of larch budmoth outbreaks in the European Alps. *Science* 298:1020–23

Bjørnstad ON, Stenseth NC, Saitoh T. 1999b. Synchrony and scaling in dynamics of voles and mice in northern Japan. *Ecology* 80:622–37

Bjørnstad ON, Stenseth NC, Saitoh T, Lingjære OC. 1998. Mapping the regional transitions to cyclicity in *Clethrionomys rufocanus*: spectral densities and functional data analysis. *Res. Popul. Ecol.* 40:77–84

Blasius B, Huppert A, Stone L. 1999. Complex dynamics and phase synchronization in spatially extended ecological systems. *Nature* 399:354–59

Bock CE, Lepthien LW. 1976. Synchronous eruptions of boreal seed-eating birds. *Am. Nat.* 110:559–71

Bolker BM, Grenfell BT. 1996. Impact of vaccination on the spatial correlation and persistence of measles dynamics. *Proc. Natl. Acad. Sci. USA* 93:12648–53

Brown JH, Kodric-Brown A. 1977. Turnover rates in insular biogeography: effects of immigration on extinction. *Ecology* 58:445–49

Bulmer MG. 1975. Phase relations in the ten-year cycle. *J. Anim. Ecol.* 44:609–21

Buonaccorsi JP, Elkinton JS, Evans SR, Liebhold AM. 2001. Measuring and testing for spatial synchrony. *Ecology* 82:1668–79

Burrows MT, Moore JJ, James B. 2002. Spatial synchrony of population changes in rocky shore communities in Shetland. *Mar. Ecol. Prog. Ser.* 240:39–48

Byholm P, Ranta E, Kaitala V, Lindén H, Saurola P, Wikman M. 2002. Resource availability and goshawk offspring sex ratio variation: a large-scale ecological phenomenon. *J. Anim. Ecol.* 71:994–1001

Cattadori IM, Hudson PJ, Merler S, Rizzoli A. 1999. Synchrony, scale and temporal dynamics of rock partridge (*Alectoris graeca saxatilis*) populations in the Dolomites. *J. Anim. Ecol.* 68:540–49

Cazelles B, Boudjema G. 2001. The Moran effect and phase synchronization in complex spatial community dynamics. *Am. Nat.* 157:670–75

Cazelles B, Stone LS. 2003. Detection of imperfect population synchrony in an uncertain world. *J. Anim. Ecol.* 72:953–68

Christiansen E. 1983. Fluctuations in some small rodent populations in Norway 1971–1979. *Holarc. Ecol.* 6:24–31

Curran LM, Webb CO. 2000. Experimental tests of the spatiotemporal scale of seed predation in mast-fruiting Dipterocarpaceae. *Ecol. Monogr.* 70:129–48

de Roos AM, McCauley E, Wilson WG. 1998. Pattern formation and the spatial scale of interaction between predators and their prey. *Theor. Popul. Biol.* 53:108–30

Doak DF, Marino P, Kareiva PM. 1992. Spatial scale mediates the influence of habitat fragmentation on dispersal success: implications for conservation. *Theor. Popul. Biol.* 41:315–36

Dungan JL, Perry JN, Dale MRT, Legendre P, Citron-Pousty S, et al. 2002. A balanced view of scale in spatial statistical analysis. *Ecography* 25:626–40

Earn DJD, Levin SA, Rohani P. 2000. Coherence and conservation. *Science* 290:1360–64

Elton C, Nicholson M. 1942. The ten-year cycle in numbers of the lynx in Canada. *J. Anim. Ecol.* 11:215–44

Fromentin J-M, Gjøsæter J, Bjørnstad ON, Stenseth NC. 2000. Biological processes and environmental factors regulating the temporal dynamics of the Norwegian Skagerrak cod since 1919. *ICES J. Mari. Sci.* 57:330–38

Grenfell BT, Bjørnstad ON, Kappey J. 2001. Traveling waves and spatial hierarchies in measles epidemics. *Nature* 414:716–23

Grenfell BT, Finkenstädt BF, Wilson K, Coulson TN, Crawley MJ. 2000. Ecology—nonlinearity and the Moran effect. *Nature* 406:847

Grenfell BT, Wilson K, Finkenstädt BF, Coulson TN, Murray S, et al. 1998. Noise and determinism in synchronised sheep dynamics. *Nature* 394:674–77

Hanski I. 1998. Metapopulation dynamics. *Nature* 396:41–49

Hanski I, Gilpin M. 1991. Metapopulation dynamics: brief history and conceptual domain. *Biol. J. Linn. Soc.* 42:3–16

Hanski I, Woiwod IP. 1993. Spatial synchrony in the dynamics of moth and aphid populations. *J. Anim. Ecol.* 62:656–68

Harrison S, Quinn JF. 1989. Correlated environments and the persistence of metapopulations. *Oikos* 56:293–98

Hassell MP, Comins HN, May RM. 1991. Spatial structure and chaos in insect population dynamics. *Nature* 353:255–58

Hastings A, Harrison S, McCann K. 1997. Unexpected spatial patterns in an insect outbreak match a predator diffusion model. *Proc. R. Soc. London Ser. B* 264:1837–40

Hawkins BA, Holyoak M. 1998. Transcontinental crashes of insect populations? *Am. Nat.* 152:480–84

Haydon D, Steen H. 1997. The effects of large- and small-scale random events on the synchrony of metapopulation dynamics: a theoretical analysis. *Proc. R. Soc. London Ser. B Biol. Sci.* 264:1375–81

Haydon DT, Greenwood PE. 2000. Spatial coupling in cyclic population dynamics: models and data. *Theor. Popul. Biol.* 58:239–54

Heino M, Kaitala V, Ranta E, Lindström J. 1997. Synchronous dynamics and rates of extinction in spatially structured populations. *Proc. R. Soc. London Ser. B* 264:481–86

Henttonen H, McGuire D, Hansson L. 1985. Comparisons of amplitude and frequencies (spectral analyses) of density variations in long-term data sets of *Clethrionomys* species. *Ann. Zool. Fennici* 22:221–27

Holmes EE, Lewis MA, Banks JE, Veit RR. 1994. Partial differential equations in ecology: spatial interactions and population dynamics. *Ecology* 75:17–29

Holyoak M, Lawler SP. 1996. Persistence of an extinction-prone predator-prey interaction through metapopulation dynamics. *Ecology* 77:1867–79

Hudson PJ, Cattadori IM. 1999. The Moran effect: a cause of population synchrony. *Trends Ecol. Evol.* 14:1–2

Hurrell JW. 1995. Decadal trends in the North Atlantic Oscillation: regional temperatures and precipitation. *Science* 269:676–79

Huygens C. 1673. *Horologium Oscillatorium*. Paris: F. Muguet

Hyde CC. 1991. Patrick Alfred Pierce Moran 1917–1988. *Biographical Memoirs of Fellows of the Royal Society of London*, 37:365–79. London: R. Soc.

Ims RA. 1990. The ecology and evolution of reproductive synchrony. *Trends Ecol. Evol.* 5:135–40

Ims RA, Andreassen HP. 2000. Spatial synchronization of vole population dynamics by predatory birds. *Nature* 408:194–96

Ims RA, Steen H. 1990. Geographical synchrony in microtine population cycles: a theoretical evaluation of the role of nomadic avian predators. *Oikos* 57:381–87

Isagi Y, Sugimura K, Sumida A, Ito H. 1997. How does masting happen and synchronize? *J. Theor. Biol.* 187:231–39

Jammalamadaka SR, SenGupta A. 2001. *Topics in Circular Statistics, Section 8*. Singapore: World Sci. Press

Jansen VAA. 1999. Phase locking: another cause of synchronicity in predator-prey systems. *Trends Ecol. Evol.* 14:278–79

Johst K, Drechsler M. 2003. Are spatially correlated or uncorrelated disturbance regimes better for the survival of species? *Oikos* 103:449–56

Jones J, Doran PJ, Holmes RT. 2003. Climate and food synchronize regional forest bird abundances. *Ecology* 84:3024–32

Kaitala V, Ranta E, Lundberg P. 2001. Self-organized dynamics in spatially structured populations. *Proc. R. Soc. London Ser. B Biol. Sci.* 268:1655–60

Kelly D, Sork V. 2002. Mast seeding: patterns, causes, and consequences. *Annu. Rev. Ecol. Syst.* 33:427–47

Kendall BE, Bjørnstad ON, Bascompte J, Keitt TH, Fagan WF. 2000. Dispersal, Environmental correlation, and spatial synchrony in population dynamics. *Am. Nat.* 155:628–36

Kendall BE, Briggs CJ, Murdoch WW, Turchin P, Ellner SP, et al. 1999. Why do populations cycle? A synthesis of statistical and mechanistic modeling approaches. *Ecology* 80:1789–805

King AW, With KA. 2002. Dispersal success on spatially structured landscapes: When do spatial pattern and dispersal behavior really matter? *Ecol. Model.* 147:23–39

Klemola T, Hanhimaki S, Ruohomaki K, Senn J, Tanhuanpaa M, et al. 2003. Performance of the cyclic autumnal moth, *Epirrita autumnata*, in relation to birch mast seeding. *Oecologia* 135:354–61

Koenig WD. 1998. Spatial autocorrelation in California land birds. *Conserv. Biol.* 12:612–20

Koenig WD. 1999. Spatial autocorrelation of ecological phenomena. *Trends Ecol. Evol.* 14:22–26

Koenig WD. 2001. Synchrony and periodicity of eruptions by boreal birds. *Condor* 103:725–35

Koenig WD. 2002. Global patterns of environmental synchrony and the Moran effect. *Ecography* 25:283–88

Koenig WD, Knops JMH. 1998a. Scale of mast-seeding and tree-ring growth. *Nature* 396:225–26

Koenig WD, Knops JMH. 1998b. Testing for spatial autocorrelation in ecological studies. *Ecography* 21:423–29

Koenig WD, Knops JMH. 2000. Patterns of annual seed production by northern hemisphere trees: a global perspective. *Am. Nat.* 155:59–69

Koenig WD, Knops JMH, Carmen WJ, Stanback MT. 1999. Spatial dynamics in the absence of dispersal: acorn production by oaks in central coastal California. *Ecography* 22:499–506

Lambin X, Elston DA, Petty SJ, MacKinnon JL. 1998. Spatial asynchrony and periodic travelling wave in cyclic field vole populations. *Proc. R. Soc. London Ser. B* 265:1491–96

Legendre P. 1993. Spatial autocorrelation: Trouble or new paradigm? *Ecology* 74:1659–73

Lennon JJ. 2000. Red-shifts and red herrings in geographical ecology. *Ecography* 23:101–13

Levin SA. 1992. The problem of pattern and scale in ecology. *Ecology* 73:1943–67

Liebhold A, Elkinton J, Muzika RM. 2000. What causes outbreaks of the gypsy moth in North America? *Popul. Ecol.* 42:257–66

Liebhold A, Kamata N. 2000. Are population cycles and spatial synchrony universal characteristics of forest insect populations? *Popul. Ecol.* 42:205–9

Liebhold A, Kamata N, Jacob T. 1996. Cyclicity and synchrony of historical outbreaks of the beech caterpillar, *Quadricalcarifera punctatella* (Motschulsky) in Japan. *Res. Popul. Ecol.* 38:87–94

Liebhold A, Sork V, Peltonen M, Koenig W, Bjørnstad O, et al. 2004. Within-population spatial synchrony in mast seeding of North American oaks. *Oikos* 104:156–64

Lindström J, Ranta E, Linden H. 1996. Large-scale synchrony in the dynamics of capercaillie, black grouse and hazel grouse populations in Finland. *Oikos* 76:221–27

Mackin-Rogalska R, Nabaglo L. 1990. Geographical variation in cyclic periodicity and synchrony in the common vole, *Microtus arvalis*. *Oikos* 59:343–48

Marcström V, Kenward RE, Engren E. 1988. The impact of predation on boreal tetranoids during vole cycles: an experimental study. *J. Anim. Ecol.* 57:859–72

Maron JL, Harrison S. 1997. Spatial pattern formation in an insect host-parasitoid system. *Science* 278:1619–21

Martin K, Stacey PB, Braun CE. 1997. Demographic rescue and maintenance of population stability in grouse—beyond metapopulations. *Wildl. Biol.* 3:295–96

Miller WE, Epstein ME. 1986. Synchronous population fluctuations among moth species (Lepidoptera). *Environ. Entomol.* 15:443–47

Molofsky J. 1994. Population dynamics and pattern formation in theoretical populations. *Ecology* 75:30–39

Moran PAP. 1953a. The statistical analysis of the Canadian lynx cycle. *Aust. J. Zool.* 1:163–73

Moran PAP. 1953b. The statistical analysis of the Canadian lynx cycle. II. Synchronization and meteorology. *Aust. J. Zool.* 1:291–98

Moss R, Elston DA, Watson A. 2000. Spatial asynchrony and demographic traveling waves during red grouse population cycles. *Ecology* 81:981–89

Murray JD. 1993. *Mathematical Biology*. New York: Springer-Verlag. 2nd ed.

Myers JH. 1990. Population cycles of western tent caterpillars: experimental introductions and synchrony of fluctuations. *Ecology* 71:986–95

Myers JH. 1998. Synchrony in outbreaks of forest Lepidoptera: a possible example of the Moran effect. *Ecology* 79:1111–17

Myers RA, Mertz G, Barrowman NJ. 1995. Spatial scales of variability in cod recruitment in the North Atlantic. *Can. J. Fish. Aquat. Sci.* 52:1849–62

Myers RA, Mertz G, Bridson J. 1997. Spatial scales of interannual recruitment variations of marine, anadromous, and freshwater fish. *Can. J. Fish. Aquat. Sci.* 54:1400–7

Oden NL, Sokal RR. 1986. Directional autocorrelation: an extension of spatial correlograms to two dimensions. *Syst. Zool.* 35:608–17

Ostfeld RS, Jones CG. 1996. Of mice and mast. *BioScience* 46:323–30

Ostfeld RS, Keesing F. 2000. Pulsed resources and community dynamics of consumers in terrestrial ecosystems. *Trends Ecol. Evol.* 15:232–37

Paradis E, Baillie SR, Sutherland WJ, Gregory RD. 1999. Dispersal and spatial scale affect synchrony in spatial population dynamics. *Ecol. Lett.* 2:114–20

Paradis E, Baillie SR, Sutherland WJ, Gregory RD. 2000. Spatial synchrony in populations of birds: effects of habitat, population trend, and spatial scale. *Ecology* 81:2112–25

Pecora LM, Caroll TL. 1990. Synchronization in chaotic systems. *Phys. Rev. Lett.* 64:821–24

Peltonen A, Hanski I. 1991. Patterns of island occupancy explained by colonization and extinction rates in shrews. *Ecology* 72:1698–708

Peltonen M, Liebhold A, Bjørnstad ON, Williams DW. 2002. Variation in spatial synchrony among forest insect species: roles of regional stochasticity and dispersal. *Ecology* 83:3120–29

Pollard E. 1991. Synchrony of population fluctuations: the dominant influence of widespread factors on local butterfly populations. *Oikos* 60:7–10

Post E, Forchhammer MC. 2002. Synchronization of animal population dynamics by large-scale climate. *Nature* 420:168–71

Raimondo S, Liebhold AM, Strazanac JS, Butler L. 2004. Population synchrony within and among Lepidoptera species in relation to weather, phylogeny, and larval phenology. *Environ. Entomol.* 29:96–105

Ranta E, Kaitala V. 1997. Traveling waves in vole population dynamics. *Nature* 390:456

Ranta E, Kaitala V, Lindström J. 1997a. Dynamics of Canadian lynx populations in space and time. *Ecography* 20:454–60

Ranta E, Kaitala V, Lindström J. 1999. Spatially autocorrelated disturbances and patterns in population synchrony. *Proc. R. Soc. London Ser. B Biol. Sci.* 266:1851–56

Ranta E, Kaitala V, Lindström K, Helle E. 1997b. The Moran effect and synchrony in population dynamics. *Oikos* 78:136–42

Ranta E, Kaitala V, Lindström J, Lindén H. 1995a. Synchrony in population dynamics. *Proc. R. Soc. London Ser. B* 262:113–18

Ranta E, Kaitala V, Lundberg P. 1998. Population variability in space and time: the dynamics of synchronous populations. *Oikos* 83:376–82

Ranta E, Lindström J, Lindén H. 1995b. Synchrony in tetranoid population dynamics. *J. Anim. Ecol.* 64:767–76

Ripa J. 2000. Analysing the Moran effect and dispersal: their significance and interaction in synchronous population dynamics. *Oikos* 89:175–87

Rohani P, Earn DJ, Grenfell BT. 1999. Opposite patterns of synchrony in sympatric disease metapopulations. *Science* 286:968–71

Rosenblum MG, Pikovsky AS, Kurths J. 1996.

Phase synchronization of chaotic oscillators. *Phys. Rev. Lett.* 76:1804–7

Rossi MN, Fowler HG. 2003. The sugarcane borer, *Diatraea saccharalis* (Fabr.) (Lep., Crambidae) and its parasitoids: a synchrony approach to spatial and temporal dynamics. *J. Appl. Entomol.* 127:200–8

Rossi RE, Mulla DJ, Journel AG, Franz EH 1992. Geostatistical tools for modeling and interpreting ecological spatial dependence. *Ecol. Monogr.* 62:277–314

Royama T. 1981. Fundamental concepts and methodology for the analysis of animal population dynamics, with special reference to univoltine species. *Ecol. Monogr.* 51:473–93

Royama T. 1992. *Analytical Population Dynamics.* London: Chapman & Hall

Satake A, Bjørnstad O, Kobro S. 2004. Masting and trophic cascades: interplay between Rowan trees, apple fruit moth, and their parasitoid in Southern Norway. *Oikos* 104:540–50

Satake A, Iwasa Y. 2002a. Spatially limited pollen exchange and a long range synchronization of trees. *Ecology* 90:830–33

Satake A, Iwasa Y. 2002b. The synchronized and intermittent reproduction of forest trees is mediated by the Moran effect, only in association with pollen coupling. *J. Ecol.* 90:830–38

Schauber EM, Kelly D, Turchin P, Simon C, Lee WG, et al. 2002. Masting by 18 New Zealand plant species: the role of temperature as a synchronizing cue. *Ecology* 83:1214–25

Selås V. 1997. Cyclic population fluctuations of herbivores as an effect of cyclic seed cropping of plants: the mast depression hypothesis. *Oikos* 80:257–68

Selås V, Hogstad O, Andersson G, von Proschwitz T. 2001. Population cycles of autumnal moth, *Epirrita autumnata*, in relation to birch mast seeding. *Oecologia* 129:213–19

Shepherd R, Bennet DD, Dale JW, Tunnock S, Dolph RE, Thier RW. 1988. Evidence of synchronized cycles in outbreak patterns of Douglas-fir tussock moth, *Orgyia pseudotsugata* (McDunnough) (Lepidoptera: Lymantriidae). *Mem. Entomol. Soc. Canada, Ottawa* 146:107–21

Sinsch U. 1997. Postmetamorphic dispersal and recruitment of first breeders in a *Bufo calamita* metapopulation. *Oecologia* 112:42–47

Small RJ, Marcström V, Willebrand T. 1993. Synchronous and nonsynchronous population fluctuations of some predators and their prey in central Sweden. *Ecography* 16:360–64

Smith CH. 1983. Spatial trends in the Canadian snowshoe hare, *Lepus americanus*, population cycles. *Can. Field Nat.* 97:151–60

Steen H, Ims RA, Sonerud GA. 1996. Spatial and temporal patterns of small rodent population dynamics at a regional scale. *Ecology* 77:2365–72

Stone E. 1991. Frequency entrainment of phase coherent attractors. *Phys. Lett. A* 163:367–74

Sutcliffe OL, Thomas CD, Moss D. 1996. Spatial synchrony and asynchrony in butterfly population dynamics. *J. Anim. Ecol.* 65:85–95

Sutcliffe OL, Thomas CD, Yates TJ, Greatorex-Davies JN. 1997. Correlated extinctions, colonizations and population fluctuations in a highly connected ringlet butterfly metapopulation. *Oecologia* 109:235–41

Swanson BJ, Johnson DR. 1999. Distinguishing causes of intraspecific synchrony in population dynamics. *Oikos* 86:265–74

Tenow O. 1972. The outbreaks of *Oporinia autumnata* Bkh. and *Operophthera* spp. (Lep. Gometridae) in the Scandinavian mountain chain and northern Finland 1862–1968. *Zool. Bidr. från Upps.* (Suppl. 2):1–107

Thrall PH, Burdon JJ, Bock CH. 2001. Short-term epidemic dynamics in the *Cakile maritima—Alternaria brassicicola* host-pathogen association. *J. Ecol.* 89:723–35

Tobin PC, Bjørnstad ON. 2003. Spatial structuring and cross-correlation in a transient predator-prey system. *J. Anim. Ecol.* 72:460–67

Trenham PC, Koenig WD, Mossman MJ, Stark SL, Jagger LA. 2003. Regional dynamics of wetland-breeding frogs and toads: turnover and synchrony. *Ecol. Appl.* 13:1522–32

Trenham PC, Koenig WD, Shaffer HB. 2001. Spatially autocorrelated demography and interpond dispersal in the salamander *Ambystoma californiense*. *Ecology* 82:3519–30

Turchin P. 1990. Rarity of density dependence or population regulation with lags? *Nature* 344:660–63

Turchin P. 2003. *Complex Population Dynamics: A Theoretical/Empirical Synthesis*. Princeton, NJ: Princeton Univ. Press

Viboud C, Boëlle P-Y, Pakdaman K, Carrat F, Valleron A-J, Flahault A. 2004. Influenza epidemics in the United States, France, and Australia, 1972–1997. *Emerg. Infect. Dis.* 10:32–39

Watson A, Moss R, Rothery P. 2000. Weather and synchrony in 10-year population cycles of rock ptarmigan and red grouse in Scotland. *Ecology* 81:2126–36

Williams DW, Liebhold AM. 1995. Influence of weather on the synchrony of gypsy moth (Lepidoptera: Lymantriidae) outbreaks in New England. *Environ. Entomol.* 24:987–95

Williams DW, Liebhold AM. 2000a. Spatial scale and the detection of density dependence in spruce budworm outbreaks in eastern North America. *Oecologia* 124:544–52

Williams DW, Liebhold AM. 2000b. Spatial synchrony of spruce budworm outbreaks in eastern North America. *Ecology* 81:2753–66

Wilson WG, Harrison SP, Hastings A, McCann K. 1999. Exploring stable pattern formation in models of tussock moth populations. *J. Anim. Ecol.* 68:94–107

Ydenberg R. 1987. Nomadic predators and geographic synchrony in microtine population cycles. *Oikos* 50:270–72

Zhang Q-B, Alfaro RI. 2003. Spatial synchrony of the two-year cycle budworm outbreaks in central British Columbia, Canada. *Oikos* 102:146–54

ECOLOGICAL RESPONSES TO HABITAT EDGES:
Mechanisms, Models, and Variability Explained

Leslie Ries,[1,2] Robert J. Fletcher, Jr.,[3] James Battin,[1,2] and Thomas D. Sisk[2]

[1]*Department of Biological Sciences,* [2]*Center for Environmental Sciences and Education, Northern Arizona University, Flagstaff, Arizona 86011; email: lries@umd.edu, Thomas.Sisk@nau.edu, James.Battin@noaa.gov*
[3]*University of Montana, Avian Science Center, Division of Biological Sciences, Missoula, Montana 59812; email: Robert.Fletcher@mso.umt.edu*

Key Words ecological boundary, ecotone, edge effect, effective area model, core area model, habitat fragmentation

■ **Abstract** Edge effects have been studied for decades because they are a key component to understanding how landscape structure influences habitat quality. However, making sense of the diverse patterns and extensive variability reported in the literature has been difficult because there has been no unifying conceptual framework to guide research. In this review, we identify four fundamental mechanisms that cause edge responses: ecological flows, access to spatially separated resources, resource mapping, and species interactions. We present a conceptual framework that identifies the pathways through which these four mechanisms can influence distributions, ultimately leading to new ecological communities near habitat edges. Next, we examine a predictive model of edge responses and show how it can explain much of the variation reported in the literature. Using this model, we show that, when observed, edge responses are largely predictable and consistent. When edge responses are variable for the same species at the same edge type, observed responses are rarely in opposite directions. We then show how remaining variability may be understood within our conceptual frameworks. Finally, we suggest that, despite all the research in this area, the development of tools to extrapolate edge responses to landscapes has been slow, restricting our ability to use this information for conservation and management.

INTRODUCTION

The edges between habitat patches are often ecologically distinct from patch interiors, and understanding how ecological patterns change near edges is key to understanding landscape-level dynamics such as the impacts of fragmentation. Landscapes are often viewed as patches of habitat and nonhabitat (Figure 1a, see color insert), and thus "edge effects" have often been conceptualized as an ecological change that is due to moving away from the "core" area of a patch and not directly linked to landscape context. In fact, landscapes exist as mosaics of several different patch types (Figure 1b), so understanding the ecology of habitat edges requires understanding the complex influences that each different adjacent patch has on a focal patch. As patches become smaller and more irregularly shaped, they become increasingly dominated by edge habitat. Therefore, understanding the ecology of habitat edges is critical both for landscape ecology and for large-scale conservation and management decisions.

Because of their importance and ubiquity, habitat edges are one of the most extensively researched areas in ecology. Most edge studies measure ecological patterns, such as changes in abundance or fitness with respect to the edge. However, the literature covers a wide range of topics, including the factors that maintain or change the position of edges in a landscape (e.g., Arris & Eagleson 1989, Bowman & Fensham 1991, Allen & Breshears 1998, Callaghan et al. 2002, Klasner & Fagre 2002); the identification and measurement of edges within remotely sensed data (e.g., Fortin 1994, Choesin & Boerner 2002, Fagan et al. 2003); the impact of edges on migration through their role in mediating dispersal (e.g., Stamps et al. 1987, Haddad 1999, Ries & Debinski 2001, Schultz & Crone 2001, Schtickzelle & Baguette 2003); and the evolutionary impacts of edges on populations, including speciation (e.g., Smith et al. 1997, Schilthuizen 2000) and the maintenance of hybrid zones (Young 1996). For this review, we focus on ecological responses, particularly species abundance patterns, to the presence of habitat edges. Our goal is to understand the mechanisms that underlie those responses and to place the research into a conceptual framework that helps explain reported patterns and variability. Further, we explore how these patterns may help us understand large-scale dynamics.

Edges are generally defined as boundaries between distinct patch types, so the identification of edges depends on how patches are defined within a landscape. Patch definition can occur at a variety of scales, from patches of different species of plants within a meadow to major biomes within continents (Cadenasso et al. 2003). To limit our scope, we focus on patches of different vegetation or land-use classes (i.e., forests, meadows, scrub, agriculture, urban areas, etc.) within terrestrial landscapes. We refer to any variable that increases near edges as having a positive edge response, a variable that exhibits no pattern as having a neutral response, and a variable that decreases near edges as having a negative edge response (see Figure 2).

The earliest reference to edge-related ecology is attributed to Clements (1907), who introduced the term "ecotone." The influence of these "zones of transition" on wildlife has been noted for decades, with Leopold (1933) using the term "edge effect" to describe the increase in game species in patchy landscapes. Other early accounts of increased diversity near edges (Lay 1938, Johnston 1947) added to the general conception that edges were good for wildlife, and their creation was often recommended in management (Harris 1988, Yahner 1988). The view of edges as

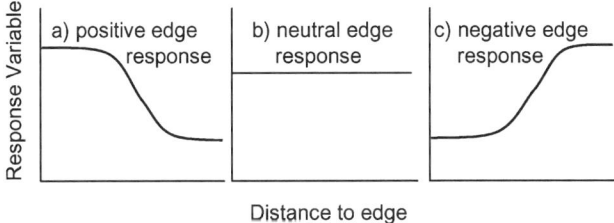

Figure 2 Three classes of ecological edge responses with respect to distance from the closest habitat edge. Responses are generally categorized as (*a*) positive edge responses, where the variable of interest increases near the edge; (*b*) neutral responses, where there is no pattern with respect to the edge; and (*c*) negative responses, where the variable decreases near the edge.

largely beneficial began to change by the late 1970s with the discovery that some birds were experiencing higher rates of nest predation and parasitism near forest edges (Gates & Gysel 1978, Chasko & Gates 1982). This pattern, coupled with increased bird numbers near edges, led many to view forest edges as "ecological traps" (Gates & Gysel 1978) and to suggest that this phenomenon may contribute to the decline of songbirds (Brittingham & Temple 1983, Wilcove 1985). In addition, edges became associated with decreased quality for habitat specialists (Mills 1995, Burke & Nol 1998) and the invasion of exotic species into habitat reserves (Morgan 1998, Honnay et al. 2002). These issues helped form the modern perception of edges, especially anthropogenic edges, as undesirable landscape features (Harris 1988, Saunders et al. 1991).

The past two decades have seen a flood of edge research on a wide range of organisms (Table 1) at an increasingly diverse number of edge types. In addition, a growing focus on mechanisms that underlie edge effects (Murcia 1995, McCollin 1998, Fagan et al. 1999, Cadenasso et al. 2003) has led to many empirical investigations of these mechanisms (e.g., Kingston & Morris 2000, Fletcher & Koford 2003, Kristan et al. 2003). However, highly variable response patterns (Murcia 1995, Villard 1998, Sisk & Battin 2002) create the impression that edges are associated with idiosyncratic responses to disparate ecological phenomena (Ehrlich 1997). Yet much of the difficulty in grappling with the edge literature stems from the lack of a framework for understanding the patterns reported (Murcia 1995, Cadenasso et al. 2003) or comparing responses at the different edges present in a landscape (Ries & Sisk 2004). Determining how edge response patterns manifest at broader scales is also difficult, limiting most studies' utility for management and conservation (Sisk & Haddad 2002, Sisk et al. 2002, Battin & Sisk 2003). To clarify these issues, we present four major sections for this review:

1. A mechanistic model of edge responses. We synthesize ideas from the literature into a conceptual model of the fundamental mechanisms underlying

TABLE 1 A summary from a subset of studies measuring responses at forest edges showing the number of times positive, negative, neutral, or mixed results were observed[a]

Taxon	Number of studies[b]	Response variable	Pos	Neg	Neutral	Mixed[c]
Birds	10	Richness/diversity	2	0	3	5
	32	Abundance by species[d]	86	49	344	28
Plants	13	Richness/diversity	7	1	2	3
	16	Abundance by species[d]	85	30	167	9
Mammals	1	Richness/diversity	0	1	0	0
	20	Abundance by species[d]	12	6	25	9
Invertebrates	8	Richness/diversity	1	0	4	3
	10	Abundance by species[d]	6	7	70	11
Herps	1	Richness/diversity	0	1	0	0
	5	Abundance by species[d]	0	6	13	2
Abiotic	20	Several[e]	22	8	21	18

[a]Only studies that measured responses within forest edges adjacent to open habitat and met certain design criteria (see text) were included. See Appendix 1a for specific studies used.
[b]Only studies that met certain design criteria were included, see text for details.
[c]Studies reported more than one response (e.g., positive and neutral) for the same response variable when analyses were stratified by some factor.
[d]Summed tallies of all species reported in each study.
[e]For example, temperature, light levels, humidity, wind, soil properties, etc.

edge response patterns and describe the empirical evidence supporting each component of that model.

2. A predictive model of edge responses. We review a model that predicts how abundances change near edges and assess the ability of that model to account for variability described within the literature.
3. Interactions affecting edge responses. We review studies that identify ecological factors that interact with edge responses and determine how that variability can be understood within the framework of the above models.
4. Extrapolating edge responses to larger scales. We explore our ability to use edge response patterns to understand the distribution of organisms in heterogeneous landscapes.

To determine the strength of evidence for suggested patterns and mechanisms, we reviewed over 900 empirical papers on terrestrial edge responses located through database searches and from citations within papers. From those papers, we selected a subset for review on the basis of the following criteria: (*a*) distance to edge was an explanatory variable for analysis, (*b*) there was a minimum of three sampling replicates, (*c*) basic habitat descriptions were given for both sides of the edge, and (*d*) the adjacent patches among replicate study sites were similar in structure (i.e., all forest or all open). We imposed these criteria to ensure that studies were both comparable and relatively rigorous. Because studies on avian nest success have been thoroughly reviewed (see Paton 1994, Hartley & Hunter 1998, Lahti 2001, Chalfoun et al. 2002b), we did not include those empirical papers but instead present a synthesis of those reviews. These criteria resulted in 263 core papers used for our review of the empirical literature.

A MECHANISTIC MODEL OF EDGE RESPONSES

A Conceptual Framework

Researchers have invoked a broad range of mechanisms to explain changes in organism abundance near edges (Wiens et al. 1985, McCollin 1998, Fagan et al. 1999, Lidicker 1999, Cadenasso et al. 2003), yet a comprehensive framework has remained elusive (Murcia 1995, Ries & Sisk 2004). Here, we present a unified model of the mechanisms underlying edge effects on the abundance of organisms. From our review of the literature, we identified four fundamental mechanisms that change organismal abundance patterns across habitat edges: (*a*) ecological flows, (*b*) access to spatially separated resources, (*c*) resource mapping, and (*d*) species interactions. Ecological flows involve the movement of material, organisms, or energy between patches (Wiens et al. 1985, Cadenasso et al. 2003). Access to spatially separated resources may be enhanced near edges for organisms whose required resources are found in multiple habitat types (Leopold 1933, Dunning et al. 1992, McCollin 1998, Fagan et al. 1999). These first two mechanisms, flows and access, fundamentally alter habitat quality of edges relative to patch interiors. Habitat edges have maximum exposure to flows from adjacent patches and are the ideal location to gain access to spatially separated resources.

The final two mechanisms, resource mapping and species interactions, represent general ecological dynamics that, although not restricted to edges, are important components in an overall framework of how edges influence distributions. Resource mapping occurs when an organism's distribution reflects that of its resources. Species interactions describe any interspecific relationship that influences one or both species. In some cases (e.g., predator-prey interactions), an interaction constitutes resource use from one organism's point of view, causing overlap between the categories of resource mapping and species interactions. To avoid confusion, we define predator dynamics as species interactions from the perspective of the prey but resource mapping from that of the predator. Mutualisms and competition are considered under the category of species interactions. Although these four mechanisms capture most dynamics that influence edge responses, there may be some taxon-specific factors that do not fit easily into this, or any, general framework. For instance, the geometry of packing bird territories into patches may cause decreases in bird abundances near edges (passive displacement) that are not based on any change in habitat quality (King et al. 1997, Ortega & Capen 1999, Fletcher & Koford 2003).

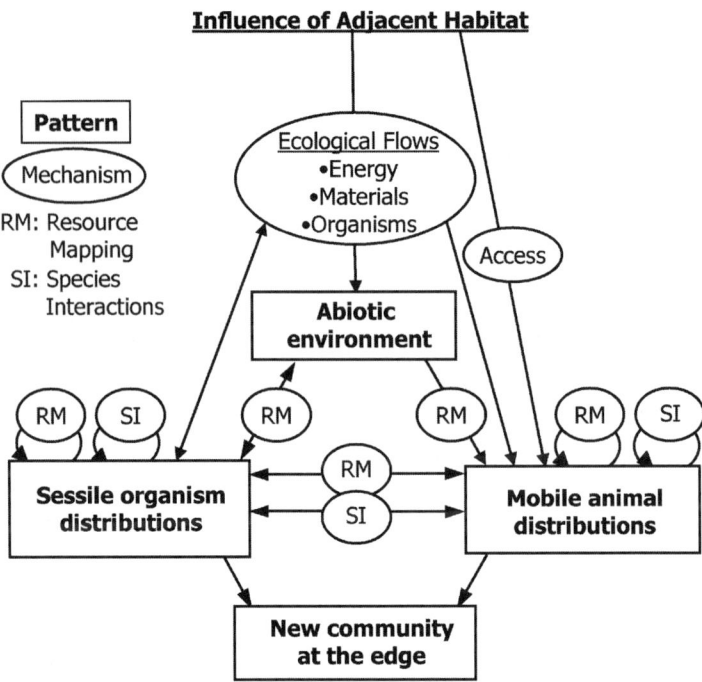

Figure 3 A mechanistic model describing the different pathways by which the distributions of organisms and, ultimately, community structure are altered near habitat edges. Patterns in the abiotic environment, mobile and sessile organism distributions, and community structure (*boxes*) are influenced by four fundamental mechanisms (*ovals*). Ecological flows of energy, material, and organisms across the edge influence the abiotic environment as well as organismal distributions. Mobile organisms whose resources are spatially separated can gain better access by being near edges. Organisms map onto changes in the distribution of their resources. Changes in species' distributions near edges can lead to novel species interactions that can further influence abundance and distributions. All these changes in species distribution lead to altered community structure near edges.

Figure 3 illustrates how these four mechanisms influence patterns in organismal distributions near habitat edges. We separate organisms into mobile and sessile groups because different pathways occur for each (specified below). Ecological flows influence mobile and sessile organisms directly by providing a source of materials and immigrants across edges. Indirect influences can also occur via organisms mapping onto changes in the abiotic environment near edges. However, only sessile organisms can feed back and directly influence ecological flows and the abiotic environment (note double-sided arrows in Figure 3 for those pathways). This feedback can occur either through changes in edge architecture that influence the permeability of the edge (Didham & Lawton 1999, Cadenasso et al. 2003) or by creating a gradient that affects the abiotic environment. Another factor that differentiates mobile from sessile organisms is that only mobile organisms are known to actively gain access to spatially separated resources in adjacent patches. Some sessile organisms may also demonstrate such responses if advantages at the edge can be realized via, for example, root or branch growth, but we are not aware of any examples demonstrating this effect. Instead, sessile organisms can gain access to resources associated with adjacent patches by being located near edges that receive flows from neighboring patches. Within the current edge literature, the categories of mobile and sessile organisms separate animals from plants and lichens, but that need not always be the case (for instance, many sessile aquatic organisms are animals). However, owing to this current dichotomy in the terrestrial literature, we formulate the bulk of our remaining discussion with respect to research on animals (mobile organisms) and plants or lichens (sessile organisms).

Finally, any change in a species' distribution may have cascading effects throughout the community (Figure 3). For example, changes in abiotic conditions near a habitat edge may lead to the establishment of new plant species, and previously absent animals may map onto these new resources, leading to novel interactions affecting multiple taxa and, ultimately, changing overall community structure (Figure 3). Below we review the evidence for each of the four mechanisms and, where possible, show how effects cascade throughout the community.

Empirical Evidence for Underlying Mechanisms

ECOLOGICAL FLOWS Ecological flows from adjacent habitat patches are a key mechanism underlying the distinction between edge and interior zones. The rate of ecological flows between patches is a function of edge permeability, the degree to which a given flow can penetrate the boundary between two patches. Edges can amplify, attenuate, or reflect ecological flows (Strayer et al. 2003). Edge permeability is strongly influenced by the architecture of the edge, largely on the basis of plant structure (Cadenasso et al. 2003). Changes in overstory cover can affect vertical penetration of light into a patch (Turton & Sexton 1996, Didham & Lawton 1999, Dignan & Bren 2003a), which may, in turn, affect temperature and humidity levels near the edge. The relative concentrations of energy, materials, and organisms on either side of the edge also affect flow rates. Materials that move passively may naturally diffuse to areas of lower concentration, but mobile organisms can move against natural gradients, making net flows more difficult to predict.

A number of studies have documented ecological flows from one patch into another. Microclimatic changes have been widely documented (e.g., Matlack 1993, Cadenasso et al. 1997, Burke & Nol 1998, Meyer et al. 2001) and are influenced by differential movement of light, heat, moisture, and wind from one patch to another (Weathers et al. 2001). The result of these differential flows is that environmental conditions near edges are often intermediate between conditions in both adjacent patch interiors. For instance, forest edges near open habitat are hotter, drier, and lighter than the forest interior (Chen et al. 1999). Conversely, open habitat near forest edges experiences increased shading, leading to lower temperatures and higher humidity (Cadenasso et al. 1997). Materials can be moved into adjacent habitat through diffusion, deposition, or the flow of animals (Cadenasso et al. 2003). Movement of animals across edges is widely documented in studies that quantify how edges mediate dispersal or influence movement direction (Sakai & Noon 1997, Haddad 1999, Ries & Debinski 2001, Schultz & Crone 2001, Matthysen 2002, Schtickzelle & Baguette 2003). The flow of plant propagules (seeds and pollen) across habitat edges has been demonstrated, but has received less study (Cadenasso & Pickett 2001, Cubiña & Aide 2001). In some cases, edges can act as relatively impermeable barriers that cause the accumulation of organisms or materials at the edge (e.g., Desrochers & Fortin 2000).

ACCESS Access is the other key mechanism that separates the quality of edge habitat from interior zones. When resources are spatially separated between two adjacent patches, edges provide maximum access to both resources (Dunning et al. 1992, McCollin 1998, Fagan et al. 1999). In his explanation of why edges should harbor higher densities of animals, Leopold (1933) used the example of bobwhite quail (*Colinus virginianus*), which use four different habitat types: forest, brushland, grassland, and agricultural fields. He suggested that quail should occur at higher densities in landscapes in which these four habitats are highly interspersed. In another early example, aspen experienced the heaviest outbreaks of a leaf-mining insect (*Lithocolletis salicifoliella*) near stands of coniferous trees, which are used as the insects' overwintering sites (Martin 1956). In perhaps the best-known example, brown-headed cowbirds (*Molothrus ater*), which parasitize the nests of forest-dwelling songbirds but forage in open pastures, often increase in abundance near forest edges (Lowther 1993). Access to different habitat types may be especially important to animals whose juvenile and adult forms have different habitat requirements (Martin 1956, Ponsero & Joly 1998). However, one must distinguish between cases in which resources in adjacent patches are different (complementary) from cases in which adjacent patches contain the same (supplementary) resources (Dunning et al. 1992), because in the latter case no edge response is expected.

RESOURCE MAPPING Resource mapping influences species distributions through more pathways than any other mechanism in our model (Figure 3). Any edge-related change in the distribution of an organism's resources may result in a concordant change in that organism's distribution. Although resource mapping is the most frequently studied mechanism, most studies are correlational and fail to establish that the variables being measured influence the abundance and distribution of the focal organism.

Both plants and animals map onto abiotic gradients, with microclimate (e.g., solar radiation, temperature, humidity, soil moisture) the most extensively documented. Several studies suggest that increased light levels near the forest edge cause changes in the plant community (Wales 1972, Honnay et al. 2002,

Watkins et al. 2003), whereas studies on plant responses to nutrient gradients are rarer. Researchers have documented relationships between animal distributions and microclimate gradients (Burke & Nol 1998, Haskell 2000, Brotons et al. 2001, Fernandez-Juricic 2001, Schlaepfer & Gavin 2001). Lichens mapping onto tree distributions and liverworts mapping onto downed log distributions (Moen & Jonsson 2003) provide rare examples of sessile organisms mapping onto the distribution of other sessile organisms.

The most frequently studied mechanism for edge effects is resource mapping by animals onto plants, most commonly the relationship between bird distributions and vegetation structure, although most studies have not found evidence to support this mechanism (e.g., Ortega & Capen 1999, Fernandez-Juricic 2001, Beier et al. 2002, Fletcher & Koford 2003). Most of these studies, however, do not independently establish which vegetation characteristics are most directly related to bird distributions before they test for a correlation with distance to the edge. This approach limits the usefulness of these studies because animals are not expected to map onto gradients that do not influence the quality of their habitat. Kristan et al. (2003) provide a rare example of a study that independently tested the importance of different vegetation characteristics on bird and mammal distributions. They examined whether changes in those vegetation features near habitat edges could account for observed edge effects. They found that the abundance of two bird species and one mammal tracked changes in vegetation with respect to the edge, while another three species showed edge responses that were not related to vegetation (Kristan et al. 2003).

Animals mapping onto the distributions of other animals is also well studied. Spotted owls (*Strix occidentalis*) illustrate a clear example of a species whose edge response patterns are directly related to those of their prey. Spotted owls show increased abundances near edges when wood rats (*Neotoma* sp.) serve as their main prey (Zabel et al. 1995, Ward et al. 1998). Wood rats are more abundant at edges because they flow into forests from neighboring clearcuts, their primary habitat (Sakai & Noon 1997). In contrast, owls show no edge effect when their main prey base is flying squirrels (*Glaucomys sabrinus*), a species that shows no edge effect (Zabel et al. 1995). These and other examples demonstrate that organisms map onto gradients in their resources, but that researchers must identify the variables that most influence habitat quality before attempting to relate those variables to edge effects.

SPECIES INTERACTIONS   Interactions in which one organism benefits at the expense of another (predation, parasitism, herbivory) are the most extensively studied class of species interactions within the edge literature, with the greatest attention paid to predation on bird nests. Edges are associated with increased nest predation and parasitism rates for birds, although most studies have found no edge effect on nest success, and a few have even found a decrease (see Paton 1994, Lahti 2001, Chalfoun et al. 2002b for reviews). These patterns may be influenced by the fact that most studies were conducted in only a few biogeographic regions (Tewksbury et al. 1998, Sisk & Battin 2002). Parasitism is also higher at edges for some amphibians (Schlaepfer & Gavin 2001) and mammals (Wolf & Batzli 2001). There is less evidence for consistent effects of predation and parasitism on invertebrates. Bird predation rates on the mantid species *Stagmomantis limbata* increased near edges, whereas parasitism rates were unaffected (Ries & Fagan 2003). Conversely, viral infections decreased near edges for the tent caterpillar *Malacosoma disstria* (Roland & Kaupp 1995), but Peltonen & Heliövaara (1999) found no edge effect in predation for bark beetles. Actively avoiding edges to escape predation pressure has been demonstrated most convincingly for mammals (Bowers & Dooley 1993, Jacob & Brown 2000, Wahungu et al. 2001, but see Morris 1997), with some evidence for amphibians (Schlaepfer & Gavin 2001). Birds show little evidence of avoiding predation pressure at edges, and some species may be unable to assess risk accurately (Gates & Gysel 1978, Schlaepfer et al. 2002, Battin 2004).

Predation on plants (herbivory and seed predation) is also well documented, with many studies finding higher predation levels at the edge (e.g., Restrepo & Vargas 1999, Roach et al. 2001, Donoso et al. 2003, Tallmon et al. 2003). For example, higher abundances of white-footed mice (*Peromyscus leucopus*) at forest-clearcut edges lead to higher seed predation rates for *Trillium ovatum*, an understory herb, decreasing recruitment and thus reducing population size near edges (Jules & Rathcke 1999, Tallmon et al. 2003). Other studies have documented the opposite pattern, showing that some herbivores avoid edges, leading to reduced seed predation rates (Ostfeld et al. 1997, Manson et al. 2001, Nickel et al. 2003).

Competitive interactions and mutualisms can also influence the distribution of organisms near edges, but empirical studies are relatively rare. Morgan (1998) found no evidence for a competitive interaction between native and exotic plants species. In contrast, Argentine ants (*Linepithema humile*) appear to outcompete the local, native ant species at some edges (Suarez et al. 1998). The noisy miner bird (*Manorina melanoecephala*) inhabits open areas and adjacent forest edges and drives many forest bird species away from the edge (Piper & Catterall 2003). In theory, edges may even reverse competitive interactions (an inferior competitor becomes the superior competitor) on the basis of the edge's permeability and of the difference in quality between adjacent patches (Cantrell et al. 1998), but there is no empirical evidence for this. Most examples of edge-related mutualisms involve animal-mediated pollination and seed dispersal. For instance, trillium recruitment is decreased near forest-clearcut edges because of lower pollination rates, which may be the result of lower pollinator abundance at edges (Jules & Rathcke 1999). The ecology of animal seed-dispersers also influences plant colonization rates and dispersal distances on the basis of the animal's movement rates (Brunet & von Oheimb 1998, Ingle 2003).

Summary

Our model presents four mechanisms (ecological flows, access, resource mapping, and species interactions) that capture most dynamics driving edge responses

(Figure 3). Whereas most studies investigate only one type of mechanistic pathway, our model clearly illustrates how edge effects may arise through multiple pathways. These different mechanisms may interact in complex ways to influence an individual species' distribution and, ultimately, community structure at an edge (Figure 3). A single edge effect may create a cascade of edge effects in a wide range of organisms (Wiens et al. 1985). For instance, alterations in light levels near an edge may increase the abundance of a plant that an herbivorous insect subsequently maps onto. This may cause a decrease in that insect's competitors and/or an increase in local predator densities. This example illustrates the potential pitfalls of using correlational studies to ascertain the actual mechanism driving observed abundance patterns. When two factors change concordantly with respect to the edge, one is often interpreted as being the mechanism for the other, when this may not be the case.

Without manipulating potential pathways, it is difficult to conclusively identify the mechanism most directly responsible for an observed edge response. However, the distribution of resources, as well as species interactions, are identified as the major drivers underlying edge responses (Figure 3). Therefore, one way to avoid spurious conclusions that may result from correlational studies is to know which resources and interactions are most important in driving the abundance patterns of each study organism. Studies of edge effects in which the variables that drive organism abundance are assessed independently from, as well as in association with, edges are much more powerful than strictly correlational studies (e.g., Ross et al. 1997, Kristan et al. 2003), but they are rare. Another way to avoid spurious conclusions is to have a priori predictions for the expected response of each organism at each edge type. We next review a predictive model based largely on the concepts and mechanisms illustrated in Figure 3 and show how it can make sense of patterns and variability reported in the edge literature.

A PREDICTIVE MODEL OF EDGE RESPONSES

Although the mechanistic model presented earlier (Figure 3) offers some clues to the conditions that lead to positive, negative, and neutral responses for a particular species at a particular edge type, it is not specifically predictive. Until recently (Brand 2004, Ries & Sisk 2004), no models had been presented that allow a researcher to predict changes in abundance near edges. Brand (2004) presents models that are specific to birds at forest edges, which we review later. Below, we focus on a more general model that predicts changes in abundance near edges for any species in any landscape (Ries & Sisk 2004). This predictive model is a simplification of the mechanisms illustrated in Figure 3 and is based primarily on resource distribution (Figure 4).

Predictions from this model (Figure 4) are based on relative habitat quality of adjacent patches and the distribution of resources within and between each patch (Ries & Sisk 2004). Habitat quality is defined by the relative availability of

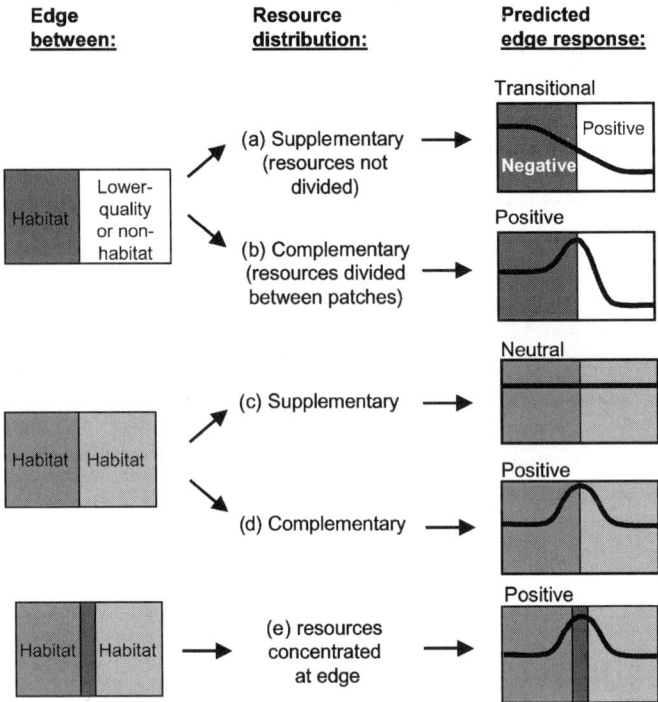

Figure 4 A predictive model of edge effects. This model is largely based on the mechanisms described in Figure 3. Edges are defined as the boundaries between patches, and patches are classified as habitat when they contain at least one resource, as nonhabitat when they contain no resources, and as lower-quality habitat when they contain fewer resources than the adjacent patch. When habitat borders nonhabitat or lower-quality habitat, where any available resources are (*a*) supplementary (the same) to those in the higher-quality patch, a transitional response is predicted. If there are resources in the lower-quality patch that are (*b*) complementary (different), then a positive edge response is predicted. When both patches contain resources, edge response predictions are based on whether the resources are (*c*) supplementary, which leads to a predicted neutral response, or (*d*) complementary in each patch, which leads to a positive prediction. When resources are (*e*) concentrated along the edge, a positive response is predicted. Reprinted with permission from Ries & Sisk (2004).

resources between the two patches. Patches are considered of lower quality when they contain fewer resources compared with the adjacent patch and nonhabitat when they contain no resources. Where organisms occur near nonhabitat or lower-quality patches, with supplementary resources (those that offer nothing different), organisms are predicted to decrease in abundance within higher-quality habitat near edges, although they may still spill over into the adjacent patch (Figure 4*a*).

(a) A simplified view of edges between habitat and nonhabitat in a "fragmented" landscape

(b) Edges are more complex when the same landscape is viewed in the context of a mosaic of different patches

Figure 1 Dark green patches show the distribution of old-growth long-leaf pine (*Pinus pilustrus*) on a section of Ft. Benning in Georgia, USA. These pine stands are habitat for the red-cockaded woodpecker (*Picoides borealis*), an endangered species restricted to the Southeast. In the top panel (*a*), habitat is illustrated in a classic "fragmentation" context, where all patches are portrayed as either habitat (*dark green*) or nonhabitat (*white*). In fact, the surrounding landscape forms a complex mosaic (*b*) that has multiple influences on each patch. Lighter greens represent younger pine age classes, yellow patches are open or brushy habitat, brown represents hardwood stands, orange is mixed hardwood-pine stands, and purple is areas developed for military use. The influence of these different types of patches, experienced largely as "edge effects," can be as varied as the types of patches found within the landscape. Forest stand maps are used with permission of Ft. Benning.

This spillover is due to flows of materials between patches (Figure 3) that tend to make edges more similar to adjacent patches and cause a gradual transition in habitat quality onto which organisms may map (Lidicker 1999). In contrast, if the lower-quality habitat contains resources that are complementary (different), then a positive response is predicted on both sides of the edge (Figure 4*b*) because access to all resources is increased near edges (Figure 3). When two adjacent patches contain relatively equal amounts of resources, no response is predicted when resources are supplementary (Figure 4*c*) because being near the edge confers no additional access to resources. Again, complementary resource distribution leads to a predicted increase near edges (Figure 4*d*) owing to increased access (Figure 3). Finally, if resources are concentrated along the edge, organisms may map onto the increase of their resources near edges (Figure 3) and are therefore predicted to increase (Figure 4*e*). One important implication of this model is that all species are predicted to show positive, neutral, and negative edge responses, depending on the edge type encountered. This may explain much of the inter- and intraspecific variability reported in the edge literature.

A Test of the Model on Several Taxa

To evaluate the general applicability of this edge response model, we compared predictions with observed responses for four taxa: birds, butterflies, mammals, and plants. Outcomes are divided into cases in which the model is correct and in which it is incorrect; however, we further divided incorrect outcomes into two distinct cases. We classified cases in which the model failed to predict a significant positive or negative edge response as a "wrong" prediction and cases in which the model predicted a positive or negative edge response, but no response was found as a "neutral" outcome (Figure 5*a*). This distinction allows separate evaluation of the factors that may have led to unpredicted neutral responses versus unpredicted positive and negative responses, outcomes that may have different underlying causes. In addition, neutral outcomes, although still incorrect, may have causes (e.g., lack of statistical power) that do not necessarily conflict with the underlying framework of the model.

Two tests of this edge response model have already been carried out. On the basis of a review of the bird literature (Ries & Sisk 2004), the model correctly predicted the direction of 83% of edge responses observed for more than 50 avian species at edges between forests and open habitat (Figure 5*b*). An intensive field study testing the predictions of the model was carried out for 15 butterfly species at 12 edge types (Ries 2003). The model performed well in all but one habitat type (which exhibited a high degree of internal heterogeneity). In the other four habitats investigated, when positive or negative predictions were made, observed edge responses were in the predicted direction 75% of the time (Figure 5*c*).

To determine how well the model performs for other taxa, we conducted a test for plants and mammals using our database of 263 papers (reptiles and amphibians

(a) Analysis of model performance

Prediction	Observation		
	-	0	+
-	Correct	Neutral	Wrong
0	Wrong	Correct	Wrong
+	Wrong	Neutral	Correct

(b) Birds

Pred	Observation		
	-	0	+
-	11	18	4
0/+	0	7	5
+	0	3	9

(c) Butterflies

Pred	Observation		
	-	0	+
-	19	55	5
0	2	46	7
+	8	61	21

(d) Mammals

Pred	Observation		
	-	0	+
-	7	0	4
0/+	1	1	9
+	0	0	3

(e) Plants

Pred	Observation		
	-	0	+
-	6	1	1
0	0	0	0
+	0	0	4

Figure 5 Analysis of the performance of the edge effect model (Figure 4). (*a*) Outcomes are separated into cases in which the model was correct and incorrect. Incorrect outcomes are further separated into two separate cases in which an unpredicted neutral response was observed and cases in which wrong positive or negative responses were observed. Results are presented for tests of the model on four taxa. (*b*) Bird results are from Ries & Sisk (2004). (*c*) Butterfly results are from Ries (2003). (*d*) Mammal and (*e*) plant results are from a review of the literature, and details for these tallies are given in Appendix 1b. (The Appendix is included as Supplemental Material. Follow the Supplemental Material link from the Annual Reviews home page at http://www.annualreviews.org.) Note that for birds and mammals, we lacked information to differentiate between positive and neutral predictions in some cases, so outcomes were identified as correct if either observation was recorded. Outcomes for which model predictions were correct are shaded in gray.

were not included because too few studies were available). This test allowed a determination of how generally applicable this model is across a wide variety of taxa and edge types. For both plants and mammals, we predicted a negative edge response when abundances were measured in the superior habitat and a positive response in lower-quality or nonhabitat (Figure 4*a*). Information on resource distribution (complementary versus supplementary) was rarely included, so for mammals, when a species was associated with both habitat types, either a neutral or positive (but not negative) response was predicted (see Figure 4c,d). This distinction was not necessary for plants because sessile organisms cannot generally gain access to resources in two separate patches. Observed edge responses were in the direction predicted by the model in 83% of cases for mammals (Figure 5d) and 91% for plants (Figure 5e). Papers used to test the model for mammals and plants are detailed in Appendix 1b (see Supplemental Material).

It is clear from these four tests (Figure 5b–e) that when edge responses are observed, they are largely predictable, even for different species at different edge types. Negative edge responses are due to individuals avoiding edges of low-quality habitat, and positive responses result when organisms gain access to resources either in adjacent patches or near edges. Some variability clearly remains unexplained; yet by using this model to isolate unpredicted responses, researchers can more easily focus on the remaining variability and attempt to account for it. Except for birds at forest edges (see below), there are no alternative predictive models with which to compare the performance of this model.

Accounting for Remaining Variation

For mammals, most unexplained variation comes from observing positive responses when negative ones were predicted (Figure 5d). This anomaly is potentially due to a lack of information on resource distribution, because a positive response would be predicted if complementary resources were known to occur in the less-preferred patch (Figure 4b) or if resources were concentrated along the edge (Figure 4e). Although resource distribution (either complementary, supplementary, or concentrated along the edge) is critical to the generation of more precise model predictions, this information is never as widely available as basic information on habitat associations. We therefore suggest that the relative distribution of critical resources on both sides of the edge should become standard information reported in edge studies.

For birds (Figure 5b) and butterflies (Figure 5c), most unexplained variation came from observing unpredicted neutral responses rather than unpredicted positive or negative "wrong" responses. One potential cause of these unpredicted neutral responses is that some species may be particularly insensitive to edges (Wiens et al. 1985, Lidicker 1999). Many species are often labeled as edge species, but this is likely due to a historical focus on a single edge type (forest edges). This labeling is especially true for birds, which are routinely classified as edge or interior species, as if such behaviors were consistent across all habitats and edge types. Although these classifications are under increasing challenge (Baker et al. 2002, Imbeau et al. 2003), they are still commonly used. Because the edge response model predicts that all species will show neutral responses at some edge types (Figure 4c), to gauge edge sensitivity accurately researchers must separate predicted neutral responses from unpredicted neutral responses. Only species that consistently fail to show edge responses where they are predicted should be considered edge insensitive. Currently, rigorous classification of any species as edge sensitive or insensitive is difficult. However, it would be useful to determine if some species are intrinsically less sensitive to habitat edges and, if so, if any life-history or ecological characteristics are associated with that insensitivity. Several characteristics, such as body size, mobility, and vulnerability to predation, have been suggested to influence overall sensitivity (Wiens et al. 1985, Lidicker 1999), although little evidence exists to substantiate these ideas.

Brand (2004) studied the ecological and life-history characteristics associated with edge responses in forest birds. In addition to habitat utilization characteristics (that closely mirror the predictions illustrated in Figure 4a,c), species are more likely to exhibit negative responses to forest-open edges if they require mesic conditions, have smaller body sizes, are less ecologically plastic, and have longer incubation and nestling periods (Brand 2004). Likewise, species are more likely to exhibit positive responses to the same forest edges if they nest in shrubs, have open cup nests, and are more ecologically plastic (Brand 2004). In the only other study of traits associated with edge sensitivity, Ries (2003) found that butterflies are more sensitive to edges if they are more vulnerable to predation and lighter in color (making them potentially more sensitive to microclimatic variation).

Another cause of unpredicted neutral responses is that most edge studies have low replication and therefore little power to detect any but the strongest responses (Murcia 1995). This problem is exacerbated by any ecological factor that interacts to weaken edge responses, a topic we explore in the following section. Finally, the predictive edge response model illustrated in Figure 4 is a necessary simplification of the complex pathways and different mechanisms presented in Figure 3. When unpredicted responses occur, it is useful to return to that underlying mechanistic framework and determine if critical resources were not considered, other species interactions were occurring, or complex pathways were not captured by the simplifying assumptions of this predictive model.

INTERACTIONS AFFECTING EDGE RESPONSES

Several studies have explored how a suite of ecological factors influence observed edge responses, and these studies allow insight into some of the unexplained variation described above. We review the four factors that have received the most attention: edge orientation, temporal effects, habitat fragmentation, and edge contrast. To explore these factors rigorously, we reviewed studies that measured edge responses for the same species at the same (or comparable) edge types while varying one of the above four factors. We then determined the percentage of times each factor influenced edge responses to cause the following four outcomes: (*a*) no interaction with edge responses; (*b*) expression, defined as a unidirectional edge response observed in some situations but not others; (*c*) strength, defined as changes in the magnitude of the effect or the distance it penetrates into the patch; or (*d*) direction, defined as an edge response switching direction from positive to negative or vice versa (see Figure 6).

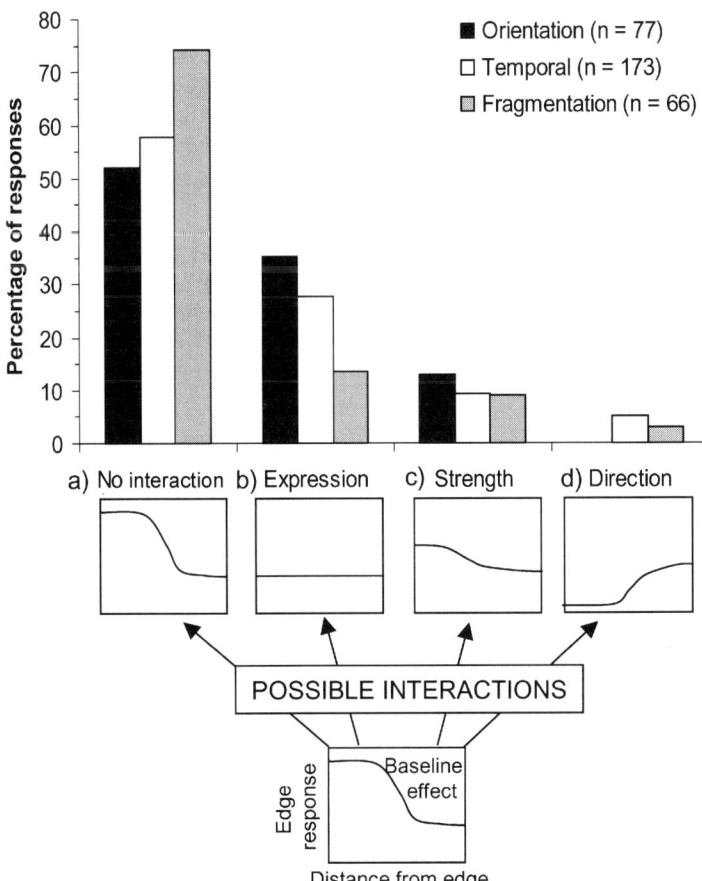

Figure 6 The influence of ecological factors that may interact to influence observed edge responses. Edge responses were compared within the same study for interactions on the basis of different orientations within the landscape, different temporal effects (daily, monthly, seasonally, or annually), and different levels of fragmentation. Outcomes of these comparisons fell into four classes. The ecological influences of orientation, time, or fragmentation either had (*a*) no interaction with observed edge responses, (*b*) caused an observed edge response not to be expressed, (*c*) changed the strength of an observed edge response, or (*d*) caused the observed edge response to change direction. The figure shows the percentage of times each outcome was observed. Details for each study are given in Appendix 1c.

Edge Orientation

The mechanistic model in Figure 3 indicates that the flow of energy is one of the fundamental drivers of edge responses. Because solar radiation is a major factor influencing the movement of energy, edges in different positions relative to the sun are likely to experience different rates, but not directions, of energetic flows (Matlack 1993, Dignan & Bren 2003b). Several studies have demonstrated that edge orientation can impact the strength or expression, but never the direction, of edge responses (Figure 6). Because edge orientation is rarely controlled for (Murcia 1995), it may be one cause of the unpredicted neutral responses reported earlier for the test of the edge response model, but not of the unpredicted positive and negative "wrong" responses (Figure 5).

If differential permeability to abiotic flows is the primary mechanism causing edge responses to weaken based on edge orientation, then we might expect two patterns. The first expected pattern is that orientation effects will be different in different latitudes. Specifically, in the northern temperate zones, south-oriented forest edges bordering open habitat should exhibit stronger edge effects (because of increased exposure to sunlight) than north-oriented edges. The converse should be true in the southern hemisphere (Kapos 1989, Young & Mitchell 1994), and little difference should be found in the tropics. These predictions may vary for other edge types, but only data on forest edges have been reported. There is strong support for the hypothesis that south-facing forest edges exhibit stronger edge responses in the northern temperate zone. When there were significant interactions in the northern hemisphere, edge responses were stronger or were expressed only at south-facing edges 42% of the time, compared with 3% at north-facing edges (n = 67). In the remaining cases, there was no interaction (Appendix 1c). Studies in the tropics and southern temperate zones were too rare to allow us to determine if patterns in those regions matched expectations.

The second expected pattern is that orientation will have a stronger influence on microclimatic patterns than on plant patterns and that the importance of orientation might continue to decrease through successively higher trophic levels. Orientation had a measurable effect on abiotic responses 63% of the time (n = 19), compared with only 42% of the time for plants (n = 50) (Appendix 1c), suggesting that damping may be occurring. Currently, there are too few studies on orientation effects on animals to determine if this damping effect occurs at higher trophic levels.

Temporal Effects

Temporal effects on edge responses have been investigated at a variety of scales, including time of day (Meyer et al. 2001), season (Young & Mitchell 1994), and year (Chalfoun et al. 2002a). Although these effects, particularly year effects, are often considered nuisance parameters, understanding their cause could help explain much of the observed variability in edge responses (Figure 6). Time most likely affects edge responses because of temporal patterns in resource distribution or use (Manson & Stiles 1998) that could vary either seasonally (Noss 1991, Young & Mitchell 1994) or throughout the day (Meyer et al. 2001). Predictable seasonal or daily change in resource use or distribution can be incorporated into the predictions of the edge response model (Figure 4) by stratifying predictions by season or time of day. Depending on how resource use or distribution changes temporally, it is even possible that the direction of an edge response could reverse, although these types of changes are rare (Figure 6d). Information on daily or seasonal changes in resource use or distribution are usually not reported, making this prediction difficult to test explicitly from the current literature. Year effects may also be driven by changes in resource distribution but are more difficult to predict than daily or seasonal effects because they likely originate from stochastic events, potentially driven by broad climatic patterns. Other temporal effects may originate from lag times in species responses, especially in sessile organisms with long generation times (e.g., Rose & Fairweather 1997) or from more gradual changes as the quality of the edge or adjacent patch changes through time.

Habitat Fragmentation

Fragmentation effects, which include patch size, isolation, and landscape composition, had the least effect on measured edge responses (Figure 6), although this pattern is strongly influenced by the results of one study (Moen & Jonsson 2003). When researchers did detect an interaction, the responses measured in one type of landscape were most commonly not expressed in another (Figure 6b). For example, brown-headed cowbirds are more likely to express edge responses in highly fragmented landscapes (Donovan et al. 1997), whereas black-tailed deer (*Odocoileus hemionus columbianus*) showed the opposite effect, with edge responses more likely when landscapes were less fragmented (Kremsater & Bunnell 1992). Many potential mechanisms are responsible for fragmentation influences on edge effects, and those mechanisms may be situation specific, making predictions difficult. Therefore, placing fragmentation interactions into a predictive framework may prove to be more difficult than for other factors discussed in this review.

Edge Contrast

Edge contrast may influence edge effects, and researchers generally assume that responses will be weaker near "soft" (low-contrast) edges than near "hard" (high-contrast) ones (Stamps et al. 1987, Duelli et al. 1990). Two main factors underlie differences in edge contrast: (*a*) differences in mean vegetation height between adjacent patches forming the edge, and (*b*) different vegetation densities within the same edge type. For both factors, differences in edge responses are likely due to differences in the permeability of the edge, resulting in different rates of ecological flows (Figure 3). Therefore, when differences in edge responses are due to edge contrast, the strength of the response is more likely to be affected than its direction. Few studies allow rigorous comparisons of edge contrast (so this factor is not included in Figure 6); however, results are similar to those presented earlier in that when a significant interaction is found, strength or expression is affected, but rarely direction.

The influence of edge contrast based on relative vegetation height is rarely rigorously explored because most studies do not control for differences in the relative qualities of the adjacent patches. This shortcoming is critical, because differences in adjacent habitat qualities are predicted to influence the strength of edge responses (Ries & Sisk 2004), and quality is often correlated with the degree of edge contrast, making these two factors particularly difficult to disentangle. Studies that have effectively controlled for habitat quality while varying edge contrast show mixed results. Fletcher & Koford (2003) found that grassland-dependent bobolinks (*Dolichonyx oryzivorus*) exhibited stronger edge avoidance near woodland edges (high contrast) than near rowcrop edges (low contrast), even though both constitute nonhabitat. Ries & Debinski (2001) found that the likelihood of a prairie endemic butterfly turning to avoid edges was stronger at high-contrast nonprairie edges than at low-contrast edges. In contrast, Ries (2003) found that edge contrast had no effect on the edge response of butterflies at several different kinds of riparian habitat edges.

The second type of edge contrast study, in which edge type is the same but vegetation structure in the focal habitat varies, is also rare but has received more rigorous treatment because the edges have been experimentally manipulated, a rarity in the edge literature. In the first example, the understory structure of a deciduous forest near an open edge was experimentally thinned. The change in vegetation density influenced whether edge responses were expressed for nutrient concentrations (Weathers et al. 2001) and rates of seedling herbivory by voles and deer (Cadenasso & Pickett 2000). The strength of edge responses varied for seed dispersal (Cadenasso & Pickett 2001). Likewise, Didham & Lawton (1999) compared relatively "open" and "closed" edges in experimentally created forest fragments in Brazil and found that edge responses were not expressed in closed-structure edges for canopy density, litter depth, or biomass, whereas edge responses changed direction for litter moisture.

Summary

The ecological factors examined rarely showed a significant interaction with edge responses (Figure 6a), but when they did, responses observed in one situation were either not expressed in another (Figure 6b) or the strength of the edge effect was modified (Figure 6c). In some cases, edge responses that were not observed may represent cases in which response strength was really affected but a lack of statistical power limited the ability to detect weaker patterns. The design of most edge studies, which often have low replication, few distance categories, and little penetration into patches, limits the detection of more subtle changes in edge response strength. In contrast, most edge studies are designed to capture differences in response direction, yet these cases are by far the least common outcome (Figure 6d). This fact suggests that changes in edge response direction for the same variable at the same edge type are rare occurrences. This pattern is encouraging and contrasts with the common conception that edge responses are intractably variable. It also suggests that, when edge responses are documented for specific species at specific edge types, managers can use this information as a generally reliable characteristic, with unforeseen shifts in the direction of edge responses unlikely to limit conservation strategies. Finally, interactions are often based on factors that are consistent with the models presented (Figures 3 and 4) and can therefore be either incorporated directly into predictions generated from the edge response model (Figure 4) or understood within the framework of our mechanistic model (Figure 3). Together, these models provide a tractable approach for understanding variability in edge responses.

EXTRAPOLATING EDGE RESPONSES TO LARGER SCALES

The main rationale cited for carrying out most edge studies is to understand better how landscape structure, usually fragmentation, influences the distribution and abundance of organisms. Although much progress has been made in describing how edges influence distributions at local scales, progress has been more limited in extrapolating measured responses to larger scales. This limited progress is unfortunate because knowledge of edge responses can contribute meaningfully to management and conservation strategies, particularly in landscapes undergoing rapid change (Sisk & Haddad 2002, Sisk et al. 2002, Battin & Sisk 2003). Few studies have tested how information on edge responses can be used to improve the understanding of distributions at larger scales (but see Temple 1986, Sisk et al. 1997, Haddad & Baum 1999).

Studies that are designed specifically to determine the role of edges as mechanisms underlying fragmentation-related patterns often demonstrate a direct link (Rosenberg & Raphael 1986, Roland 1993, Burke & Nol 1998, Didham et al. 1998, Bolger et al. 2000, Davies et al. 2001, Fletcher & Koford 2002). Furthermore, a formal meta-analysis of patch-size effects shows that species that avoid edges show increased densities in larger patches, and edge-attracted species show the opposite effect, whereas species that do not respond to edges show weak or no patch-size effects (Bender et al. 1998). This result suggests that edge responses are one of the main factors driving area sensitivity. Despite the obvious connection, three factors have limited researchers' ability to extrapolate edge responses to larger landscapes: (*a*) a lack of models and the software tools to implement them, (*b*) an incomplete knowledge of how deeply into patches edge effects extend, and (*c*) a poor understanding of how responses are influenced by the presence of multiple edges. We end this section with a brief discussion of how "scaling up" edge responses may help inform our understanding of population and community dynamics.

Models and Software Tools

Two primary models are used for "scaling up" edge responses to real landscapes: the core area model (Temple 1986, Laurance & Yensen 1991) and the effective area model (Sisk & Margules 1993, Sisk et al. 1997). Core area models use an estimated distance of edge influence to determine the amount of habitat in a patch that is not affected by edges (the core area). This model constitutes a significant advance in that it allows researchers to consider the portion of a landscape that is thought to be free of edge influences, and this model is often used when considering reserve designs (Temple & Cary 1988, Laurance 1991, Zipperer 1993, Ohman & Erikson 1998). Several software packages have the ability to measure the amount of core area in a landscape, facilitating the use of this approach. However, core area models have limited utility for species exhibiting variable responses to edges or in severely fragmented landscapes where most core habitat has already disappeared. The effective area model extends the core area approach by describing density (or other parameters) as a function of distance from edge, allowing researchers to generate quantitative predictions of distributions throughout an entire landscape. Furthermore, it allows researchers to specify different edge responses for each species at each unique edge type (Sisk et al. 1997, Sisk & Haddad 2002), thereby reflecting the complexity that exists in real landscapes (see Figure 1b). However, this model's use is currently limited by a lack of data for parameterization. Furthermore, software, although available, is at an early stage of development (Sisk et al. 2002).

Neither of the above models can incorporate variable edge responses based on landscape-level interactions, including edge orientation, patch size, or fragmentation effects (but see Zheng & Chen 2000). In the previous section, we demonstrated that these factors can interact to modify edge responses but that only strength or expression are generally affected, rarely the direction of the response (Figure 6). Edge responses are thus likely to manifest at larger scales in a predictable direction (Kolbe & Janzen 2002), but resulting magnitudes may be dampened depending on the extent to which orientation and landscape structure interact either to weaken edge responses or limit their expression.

Depth of Edge Influence

For both core area and effective area models, researchers must know the distance that edge effects extend into habitat patches. This value, often referred to as the depth of edge influence (DEI), is critical for determining the scale at which edge responses operate in larger landscapes (Laurance 2000). Although DEI values are reported for many edge responses, only recently have statistical techniques been suggested to allow for a rigorous determination of DEI (Fraver 1994, Cadenasso et al. 1997, Laurance et al. 1998, Mancke & Gavin 2000, Brand & George 2001, Harper & MacDonald 2001, Toms & Lesperance 2003). In many studies, DEI is determined by "visual inspection," and these numbers are likely influenced by study design (i.e., the length of survey transects and the number of distance categories). Without controlling for the statistical rigor of each study, abiotic and plant responses are generally reported to extend up to 50 m into patches, invertebrate responses up to 100 m, and bird responses 50–200 m (Appendix 1d). Examples of deeper DEIs exist for all classes, and Laurance (2000) suggested that some edge effects might occur over the scale of kilometers, yet most studies are not designed to capture effects at this scale. As empirical studies employ more rigorous designs

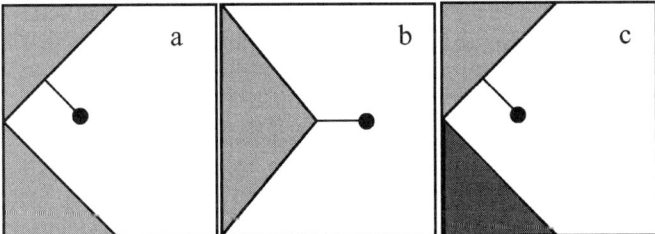

Figure 7 Complex and multiple edge effects. In most published studies, the influence of the edge is described for a point on the basis of the distance from that point to the nearest edge (shown by the dot and line), which is the same for all three panels in this figure (*a*, *b*, *c*). However, the type and strength of edge influence is likely to be different for each situation. Even at the same edge type (*a*, *b*), the geometry of the patch shape is likely to exhibit complex effects that cause edge influences to be either (*a*) greater or (*b*) lesser on points equidistant from the closest edge. A further complication that has received no empirical treatment is the convergence of multiple edge types at one point (*c*).

and statistical techniques, estimates of DEI can be more easily compared and used in landscape-level extrapolations.

Multiple Edge Effects

Another critical issue when extrapolating edge responses to landscapes is that most empirical research on edge effects has ignored the cumulative effect of being near more than one edge. Most studies (and models) use the linear distance to the closest edge as the main explanatory variable (Figure 7), and researchers generally avoid placing plots near corners or other converging edge types to limit their potential influence. However, complex patch geometry and the convergence of multiple edge types are dominant features of all real landscapes (see Figure 1*b*). If edge responses are different near multiple edges, then that factor should be incorporated into any model that extrapolates edge responses to complex landscapes. Multiple edge effects arising from complex patch geometry (Figure 7*a,b*) are poorly understood, yet they almost certainly influence the strength of edge effects by influencing both the magnitude and the depth of edge influence (Chen et al. 1995). For example, bobolinks exhibit a greater reduction in their probability of occurrence in plots located near corners compared with straight edges (Fletcher 2003). A further complication of multiple edge effects can arise when different edge types converge (Figure 7*c*), potentially causing complex edge responses on the basis of the relative influence of each adjacent habitat type (Fernandez et al. 2002). We know of no empirical studies that measure how edge responses are affected by the convergence of multiple habitat types. Mathematical approaches to deal with these issues have been suggested (Malcolm 1994, Mancke & Gavin

2000, Zheng & Chen 2000, Fernandez et al. 2002, Fletcher 2003), but the lack of empirical treatment severely limits our ability to understand how edge responses manifest in real landscapes.

Population and Community Consequences

Edge responses can also "scale up" to influence long-term population dynamics and community structure. The parameters that ultimately determine distributions (birth, death, immigration and emigration rates) are rarely modeled with respect to edges, which makes determining a cumulative influence of edges on population dynamics difficult. Edge effects may not have strong implications on population-level parameters, such as population growth rates, if the effect is weak or the response variable is not a limiting factor in population regulation. For example, Flaspohler et al. (2001) examined whether edge effects could affect population growth rates of ovenbirds. Although nest predation increased near edges, clutch size was higher, thereby offsetting predation and causing estimates of population growth rates to be similar for edge and interior habitats (see also Harris & Reed 2002). Also, the degree to which edges are important depends on the strength of the response and the relative amount of edge in a landscape (With & King 2001, Fletcher 2003). For these reasons, knowledge of edge responses may, in some cases, have limited utility in understanding population dynamics.

Although we have focused on individual species' responses to edges, our mechanistic model shows that changes in individual species' distributions ultimately culminate in an altered community at the edge (Figure 3). However, a framework for predicting community-level responses to habitat edges currently does not exist. Many studies have measured community changes near edges, with increases in species diversity the most common result (Table 1), but the causes and implications of changes in diversity near edges remain unclear. Increased diversity near edges could arise from an additive effect of species using each habitat type, increases of species using novel edge habitat, or some other emergent property. If increased diversity near edges is indeed a common and widespread pattern, the implications have received little exploration.

Summary

Despite decades of research describing how a suite of organisms respond to habitat edges, the tools to extrapolate edge responses to the landscape scale effectively and the specific data required to do so remain limited. Most studies are not designed to allow even the most basic models to be rigorously parameterized. Depth of edge influence is rarely quantified in a rigorous way, and the influence of multiple, complex edge effects has, for the most part, been ignored. This is unfortunate because understanding and predicting edge responses are clearly keys to understanding the impacts of fragmentation and other landscape-level changes. A recent review of the fragmentation literature (Fahrig 2003) admonished researchers to approach the study of fragmentation in a way that allows the impacts of habitat loss to be clearly separated from those of habitat structure. Understanding how edges impact patch quality is clearly one approach to meeting that important goal. A greater understanding of how edge responses influence population and community dynamics is also needed. Much work remains to be done to develop the knowledge and tools to accomplish this. Until these questions are more rigorously addressed, our ability to use edge response data to inform conservation and management issues remains limited.

CONCLUSIONS

Our review of the edge literature shows that the mechanisms underlying edge responses have been well described over several decades of research. The two models we present provide tools to make sense of a highly descriptive and variable literature. Our mechanistic model (Figure 3) illustrates four fundamental mechanisms underlying edge responses: ecological flows, access to spatially separated resources, resource mapping, and species interactions. These mechanisms form the basis of a general, predictive model of edge responses (Figure 4) that can be used for any species in any landscape. Our review of an extensive literature shows that edge responses, when observed, are generally predictable and consistent when species and edge type are held constant. This pattern contrasts with the generally held view that edge responses are largely idiosyncratic and based on disparate ecological phenomena. These models capture most variation, and additional variation can be explored by allowing researchers to focus on unpredicted outcomes. We identify several factors, including intrinsic sensitivity to edges, edge orientation, edge contrast, fragmentation effects, temporal shifts in resource distribution or use, and study design, that may influence observed edge responses. Finally, although many of the patterns and much of the variability in the current edge literature can be accounted for with these models, the tools necessary to extrapolate those responses to larger landscapes or to use them to understand population dynamics and community patterns have received less development, limiting our ability to use this information to inform conservation and management decisions in real, dynamic landscapes. We hope that this synthesis stimulates future research by allowing studies to be carried out under a theoretical, predictive framework. Edge responses are a key component in furthering our understanding of landscape-level dynamics, and their study continues to hold great promise in contributing even more to our growing understanding of how the spatial patterning of landscapes influences the abundance and distribution of organisms.

ACKNOWLEDGMENTS

M. Howe helped us with the creation and maintenance of our database of edge papers. The manuscript was substantially improved by comments from T.J. Fontaine, B. Kristan, K. Moloney, and D. Simberloff. Funding for much of this work came from the Strategic Environmental Research and Development Program (Project CS-1100).

The *Annual Review of Ecology, Evolution, and Systematics* is online at http://ecolsys.annualreviews.org

LITERATURE CITED

Allen CD, Breshears DD. 1998. Drought-induced shift of a forest-woodland ecotone: rapid landscape response to climate variation. *Proc. Natl. Acad. Sci. USA* 95:14839–42

Arris LL, Eagleson PS. 1989. Evidence of a physiological basis for the boreal-deciduous forest ecotone in North America. *Vegetatio* 82:55–58

Baker J, French K, Whelan RJ. 2002. The edge effect and ecotonal species: bird communities across a natural edge in southeastern Australia. *Ecology* 83:3048–59

Battin J. 2004. When good animals love bad habitats: ecological traps and the conservation of animal populations. *Conserv. Biol.* In press

Battin J, Sisk TD. 2003. Assessing landscape-level influences of forest restoration on animal populations. In *Ecological Restoration of Southwestern Ponderosa Pine Forests*, ed. P Frederici, pp. 175–90. Washington, DC: Island

Beier P, Van Drielen M, Kankam BO. 2002. Avifaunal collapse in West African forest fragments. *Conserv. Biol.* 16:1097–111

Bender DJ, Contreras TA, Fahrig L. 1998. Habitat loss and population decline: a meta-analysis of the patch size effect. *Ecology* 79:517–33

Bolger DT, Suarez AV, Crooks KR, Morrison SA, Case TJ. 2000. Arthropods in urban habitat fragments in southern California: area, age, and edge effects. *Ecol. Appl.* 10:1230–48

Bowers MA, Dooley JL. 1993. Predation hazard and seed removal by small mammals—microhabitat versus patch scale effects. *Oecologia* 94:247–54

Bowman D, Fensham RJ. 1991. Response of a monsoon forest-savanna boundary to fire protection, Weipa, northern Australia. *Aust. J. Ecol.* 16:111–18

Brand LA. 2004. *Prediction and assessment of edge response and abundance for desert riparian birds in southeastern Arizona*. PhD thesis. Colo. State Univ., Fort Collins

Brand LA, George TL. 2001. Response of passerine birds to forest edge in coast redwood forest fragments. *Auk* 118:678–86

Brittingham MC, Temple SA. 1983. Have cowbirds caused forest songbirds to decline? *BioScience* 33:31–35

Brotons L, Desrochers A, Turcotte Y. 2001. Food hoarding behaviour of black-capped chickadees (*Poecile atricapillus*) in relation to forest edges. *Oikos* 95:511–19

Brunet J, von Oheimb G. 1998. Migration of vascular plants to secondary woodlands in southern Sweden. *J. Ecol.* 86:429–38

Burke DM, Nol E. 1998. Influence of food abundance, nest-site habitat, and forest fragmentation on breeding ovenbirds. *Auk* 115:96–104

Cadenasso ML, Pickett STA. 2000. Linking forest edge structure to edge function: mediation of herbivore damage. *J. Ecol.* 88:31–44

Cadenasso ML, Pickett STA. 2001. Effect of edge structure on the flux of species into forest interiors. *Conserv. Biol.* 15:91–97

Cadenasso ML, Pickett STA, Weathers KC, Jones CG. 2003. A framework for a theory of ecological boundaries. *BioScience* 53:750–58

Cadenasso ML, Traynor MM, Pickett STA. 1997. Functional location of forest edges: gradients of multiple physical factors. *Can. J. For. Res.* 27:774–82

Callaghan TV, Crawford RMM, Eronen M,

Hofgaard A, Payette S, et al. 2002. The dynamics of the tundra-taiga boundary: an overview and suggested coordinated and integrated approach to research. *Ambio* 12:3–5

Cantrell RS, Cosner C, Fagan WF. 1998. Competitive reversals inside ecological reserves: the role of external habitat degradation. *J. Math. Biol.* 37:491–533

Chalfoun AD, Ratnaswamy MJ, Thompson FR. 2002a. Songbird nest predators in forest-pasture edge and forest interior in a fragmented landscape. *Ecol. Appl.* 12:858–67

Chalfoun AD, Thompson FR, Ratnaswamy MJ. 2002b. Nest predators and fragmentation: a review and meta-analysis. *Conserv. Biol.* 16:306–18

Chasko GG, Gates JE. 1982. Avian habitat suitability along a transmission-line corridor in an oak-hickory forest region. *Wildl. Monogr.* 82:1–41

Chen J, Franklin JF, Spies TA. 1995. Growing-season microclimatic gradients from clearcut edges into old-growth Douglas-fir forests. *Ecol. Appl.* 5:74–86

Chen J, Saunders SC, Crow TR, Naiman RJ, Brosofske KD, et al. 1999. Microclimate in forest ecosystem and landscape ecology. *BioScience* 49:288–97

Choesin D, Boerner REJ. 2002. Vegetation boundary detection: a comparison of two approaches applied to field data. *Plant Ecol.* 158:85–96

Clements FE. 1907. *Plant Physiology and Ecology*. New York: Holt

Cubiña A, Aide TM. 2001. The effect of distance from forest edge on seed rain and soil seed bank in a tropical pasture. *Biotropica* 33:260–67

Davies KF, Melbourne BA, Margules CR. 2001. Effects of within- and between-patch processes on community dynamics in a fragmentation experiment. *Ecology* 82:1830–46

Desrochers A, Fortin MJ. 2000. Understanding avian responses to forest boundaries: a case study with chickadee winter flocks. *Oikos* 91:376–84

Didham RK, Hammond PM, Lawton JH, Eggleton P, Stork NE. 1998. Beetle species responses to tropical forest fragmentation. *Ecol. Monogr.* 68:295–323

Didham RK, Lawton JH. 1999. Edge structure determines the magnitude of changes in microclimate and vegetation structure in tropical forest fragments. *Biotropica* 31:17–30

Dignan P, Bren L. 2003a. A study of the effect of logging on the understory light environment in riparian buffer strips in a southeast Australian forest. *For. Ecol. Manag.* 172:161–72

Dignan P, Bren L. 2003b. Modelling light penetration edge effects for stream buffer design in mountain ash forest in southeastern Australia. *For. Ecol. Manag.* 179:95–106

Donoso DS, Grez AA, Simonetti JA. 2003. Effects of forest fragmentation on the granivory of differently sized seeds. *Biol. Conserv.* 115:63–70

Donovan TM, Jones PW, Annand EM, Thompson FR. 1997. Variation in local-scale edge effects: mechanisms and landscape context. *Ecology* 78:2064–75

Duelli P, Studer M, Marchand I, Jakob S. 1990. Population movements of arthropods between natural and cultivated areas. *Biol. Conserv.* 54:193–207

Dunning JB, Danielson BJ, Pulliam HR. 1992. Ecological processes that affect populations in complex landscapes. *Oikos* 65:169–75

Ehrlich PR. 1997. *A World of Wounds: Ecologists and the Human Dilemma*. Oldendorf/Luhe, Ger.: Ecology Inst.

Fagan WF, Cantrell RS, Cosner C. 1999. How habitat edges change species interactions. *Am. Nat.* 153:165–82

Fagan WF, Fortin MJ, Soykan C. 2003. Integrating edge detection and dynamic modeling in quantitative analyses of ecological boundaries. *BioScience* 53:730–38

Fahrig L. 2003. Effects of habitat fragmentation on biodiversity. *Annu. Rev. Ecol. Syst.* 34:487–515

Fernández C, Acosta FJ, Abellá G, López F, Díaz M. 2002. Complex edge effect fields as additive processes in patches of ecological systems. *Ecol. Model.* 149:273–83

Fernández-Juricic E. 2001. Avian spatial segregation at edges and interiors of urban parks in

Madrid, Spain. *Biodivers. Conserv.* 10:1303–16

Flaspohler DJ, Temple SA, Rosenfield RN. 2001. Effects of forest edges on ovenbird demography in a managed forest landscape. *Conserv. Biol.* 15:173–83

Fletcher RJ. 2003. *Spatial and temporal scales of distribution and demography in breeding birds: implications of habitat fragmentation and restoration.* PhD thesis. Iowa State Univ., Ames

Fletcher RJ, Koford RR. 2002. Habitat and landscape associations of breeding birds in restored and native grasslands. *J. Wildl. Manag.* 66:1011–22

Fletcher RJ, Koford RR. 2003. Spatial responses of bobolinks (*Dolichonyx oryzivorus*) near different types of edges in northern Iowa. *Auk* 120:799–810

Fortin MJ. 1994. Edge detection algorithms for two-dimensional ecological data. *Ecology* 75:956–65

Fraver S. 1994. Vegetation responses along edge-to-interior gradients in the mixed hardwood forests of the Roanoke River basin, North Carolina. *Conserv. Biol.* 8:822–32

Gates JE, Gysel LW. 1978. Avian nest dispersion and fledging success in field-forest ecotones. *Ecology* 59:871–83

Haddad NM. 1999. Corridor use predicted from behaviors at habitat boundaries. *Am. Nat.* 153:215–27

Haddad NM, Baum KA. 1999. An experimental test of corridor effects on butterfly densities. *Ecol. Appl.* 9:623–33

Harper KA, MacDonald SE. 2001. Structure and composition of riparian boreal forest: new methods for analyzing edge influence. *Ecology* 82:649–59

Harris LD. 1988. Edge effects and conservation of biotic diversity. *Conserv. Biol.* 2:330–32

Harris RJ, Reed JM. 2002. Effects of forest-clearcut edges on a forest-breeding songbird. *Can. J. Zool.* 80:1026–37

Hartley MJ, Hunter ML. 1998. A meta-analysis of forest cover, edge effects, and artificial nest predation rates. *Conserv. Biol.* 12:465–69

Haskell DG. 2000. Effects of forest roads on macroinvertebrate soil fauna of the southern Appalachian mountains. *Conserv. Biol.* 14:57–63

Honnay O, Verheyen K, Hermy M. 2002. Permeability of ancient forest edges for weedy plant species invasion. *For. Ecol. Manag.* 161:109–22

Imbeau L, Drapeau P, Mönkkönen M. 2003. Are forest birds categorised as "edge species" strictly associated with edges? *Ecography* 26:514–20

Ingle NR. 2003. Seed dispersal by wind, birds, and bats between Philippine montane rainforest and successional vegetation. *Oecologia* 134:251–61

Jacob J, Brown JS. 2000. Microhabitat use, giving-up densities and temporal activity as short- and long-term anti-predator behaviors in common voles. *Oikos* 91:131–38

Johnston VR. 1947. Breeding birds of the forest edge in Illinois. *Condor* 49:45–53

Jules ES, Rathcke BJ. 1999. Mechanisms of reduced trillium recruitment along edges of old-growth forest fragments. *Conserv. Biol.* 13:784–93

Kapos V. 1989. Effects of isolation on water status of forest patches in the Brazilian Amazon. *J. Trop. Ecol.* 5:173–85

King DI, Griffin CR, DeGraaf RM. 1997. Effect of clearcut borders on distribution and abundance of forest birds in northern New Hampshire. *Wilson Bull.* 109:239–45

Kingston SR, Morris DW. 2000. Voles looking for an edge: habitat selection across forest ecotones. *Can. J. Zool.* 78:2174–83

Klasner FL, Fagre DB. 2002. A half century of change in alpine treeline patterns at Glacier National Park, Montana, USA. *Arct. Antarct. Alp. Res.* 34:49–56

Kolbe JJ, Janzen FJ. 2002. Spatial and temporal dynamics of turtle nest predation: edge effects. *Oikos* 99:538–44

Kremsater LL, Bunnell FL. 1992. Testing responses to forest edges: the example of black-tailed deer. *Can. J. Zool.* 70:2426–35

Kristan WB, Lynam AJ, Price MV, Rotenberry JT. 2003. Alternative causes of edge-abundance relationships in birds and small mammals of California coastal sage scrub. *Ecography* 26:29–44

Lahti DC. 2001. The "edge effect on nest predation" hypothesis after twenty years. *Biol. Conserv.* 99:365–74

Laurance WF. 1991. Edge effects in tropical forest fragments: application of a model for the design of nature reserves. *Biol. Conserv.* 57:205–19

Laurance WF. 2000. Do edge effects occur over large spatial scales? *Trends Ecol. Evol.* 15:134–35

Laurance WF, Ferreira LV, Rankin-De Merona JM, Laurance SG. 1998. Rain forest fragmentation and the dynamics of Amazonian tree communities. *Ecology* 79:2032–40

Laurance WF, Yensen E. 1991. Predicting the impacts of edge effects in fragmented habitats. *Biol. Conserv.* 55:77–92

Lay DW. 1938. How valuable are woodland clearings to birdlife? *Wilson Bull.* 50:254–56

Leopold A. 1933. *Game Management*. New York: Charles Scribner's Sons

Lidicker WZ. 1999. Responses of mammals to habitat edges: an overview. *Landsc. Ecol.* 14:333–43

Lowther PE. 1993. Brown-headed cowbird. In *The Birds of North America, No. 47*, ed. A Poole, F Gill. Philadelphia, PA: Birds N. Am.

Malcolm JR. 1994. Edge effects in central Amazonian forest fragments. *Ecology* 75:2438–45

Mancke RG, Gavin TA. 2000. Breeding bird density in woodlots: effects of depth and buildings at the edges. *Ecol. Appl.* 10:598–611

Manson RH, Ostfeld RS, Canham CD. 2001. Long-term effects of rodent herbivores on tree invasion dynamics along forest-field edges. *Ecology* 82:3320–29

Manson RH, Stiles EW. 1998. Links between microhabitat preferences and seed predation by small mammals in old fields. *Oikos* 82:37–50

Martin JL. 1956. The bionomics of the aspen blotch miner, *Lithocolletis salicifolia* Cham. (Lepidoptera: Gracillariidae). *Can. Ent.* 88:155–70

Matlack GR. 1993. Microenvironment variation within and among forest edge sites in the eastern United States. *Biol. Conserv.* 66:185–94

Matthysen E. 2002. Boundary effects on dispersal between habitat patches by forest birds (*Parus major, P. caeruleus*). *Landsc. Ecol.* 17:509–15

McCollin D. 1998. Forest edges and habitat selection in birds: a functional approach. *Ecography* 21:247–60

Meyer CL, Sisk TD, Covington WW. 2001. Microclimatic changes induced by ecological restoration of ponderosa pine forests in northern Arizona. *Rest. Ecol.* 9:443–52

Mills LS. 1995. Edge effects and isolation: redbacked voles on forest remnants. *Conserv. Biol.* 9:395–402

Moen J, Jonsson BG. 2003. Edge effects on liverworts and lichens in forest patches in a mosaic of boreal forest and wetland. *Conserv. Biol.* 17:380–88

Morgan JW. 1998. Patterns of invasion of an urban remnant of a species-rich grassland in southeastern Australia by non-native plant species. *J. Veg. Sci.* 9:181–90

Morris DW. 1997. Optimally foraging deer mice in prairie mosaics: a test of habitat theory and absence of landscape effects. *Oikos* 80:31–42

Murcia C. 1995. Edge effects in fragmented forests: implications for conservation. *Trends Ecol. Evol.* 10:58–62

Nickel AM, Danielson BJ, Moloney KA. 2003. Wooded habitat edges as refugia from microtine herbivory in tallgrass prairies. *Oikos* 100:525–33

Noss RF. 1991. Effects of edge and internal patchiness on avian habitat use in an old-growth Florida hammock. *Nat. Areas J.* 11:34–47

Ohman K, Eriksson LO. 1998. The core area concept in forming contiguous areas for long-term forest planning. *Can. J. For. Res.* 28:1032–39

Ortega YK, Capen DE. 1999. Effects of forest roads on habitat quality for ovenbirds in a forested landscape. *Auk* 116:937–46

Ostfeld RS, Manson RH, Canham CD. 1997. Effects of rodents on survival of tree seeds and seedlings invading old fields. *Ecology* 78:1531–42

Paton PWC. 1994. The effect of edge on avian nest success: How strong is the evidence? *Conserv. Biol.* 8:17–26

Peltonen M, Heliövaara K. 1999. Attack density and breeding success of bark beetles (Coleoptera, Scolytidae) at different distances from forest-clearcut edge. *Agric. For. Ent.* 1:237–42

Piper SD, Catterall CP. 2003. A particular case and a general pattern: hyperaggressive behavior by one species may mediate avifaunal decreases in fragmented Australian forests. *Oikos* 101:602–14

Ponsero A, Joly P. 1998. Clutch size, egg survival and migration distance in the agile frog (*Rana dalmatina*) in a floodplain. *Arch. Hydrobiol.* 142:343–52

Restrepo C, Vargas A. 1999. Seeds and seedlings of two neotropical montane understory shrubs respond differently to anthropogenic edges and treefall gaps. *Oecologia* 119:419–26

Ries L. 2003. *Placing edge responses into a predictive framework*. PhD thesis. North. Ariz. Univ., Flagstaff. 152 pp.

Ries L, Debinski DM. 2001. Butterfly responses to habitat edges in the highly fragmented prairies of central Iowa. *J. Anim. Ecol.* 70:840–52

Ries L, Fagan WF. 2003. Habitat edges as a potential ecological trap for an insect predator. *Ecol. Ent.* 28:567–72

Ries L, Sisk TD. 2004. A predictive model of edge effects. *Ecology*. In press

Roach WJ, Huntly N, Inouye R. 2001. Talus fragmentation mitigates the effects of pikas, *Ochotona princeps*, on high alpine meadows. *Oikos* 92:315–24

Roland J. 1993. Large-scale forest fragmentation increases the duration of tent caterpillar outbreak. *Oecologia* 93:25–30

Roland J, Kaupp WJ. 1995. Reduced transmission of forest tent caterpillar (Lepidoptera, Lasiocampidae) nuclear polyhedrosis virus at the forest edge. *Environ. Ent.* 24:1175–78

Rose S, Fairweather PG. 1997. Changes in floristic composition of urban bushland invaded by *Pittosporum undulatum* in northern Sydney, Australia. *Aust. J. Bot.* 45:123–49

Rosenberg KV, Raphael MG. 1986. Effects of forest fragmentation on vertebrates in douglas-fir forests. In *Wildlife 2000: Modeling Habitat Relationships of Terrestrial Vertebrates*, ed. J Verner, ML Morrison, CJ Ralph, pp. 263–72. Madison: Univ. Wis. Press

Ross WG, Kulhavy DL, Conner RN. 1997. Stand conditions and tree characteristics affect quality of longleaf pine for red-cockaded woodpecker cavity trees. *For. Ecol. Manag.* 91:145–54

Sakai HF, Noon BR. 1997. Between-habitat movement of dusky-footed woodrats and vulnerability to predation. *J. Wildl. Manag.* 61:343–50

Saunders DA, Hobbs RJ, Margules CR. 1991. Biological consequences of ecosystem fragmentation: a review. *Conserv. Biol.* 5:18–32

Schilthuizen M. 2000. Ecotone: speciation-prone. *Trends Ecol. Evol.* 15:130–31

Schlaepfer MA, Gavin TA. 2001. Edge effects on lizards and frogs in tropical forest fragments. *Conserv. Biol.* 15:1079–90

Schlaepfer MA, Runge MC, Sherman PW. 2002. Ecological and evolutionary traps. *Trends Ecol. Evol.* 17:474–80

Schtickzelle N, Baguette M. 2003. Behavioural responses to habitat patch boundaries restrict dispersal and generate emigration-patch area relationships in fragmented landscapes. *J. Anim. Ecol.* 72:533–45

Schultz CB, Crone EE. 2001. Edge-mediated dispersal behavior in a prairie butterfly. *Ecology* 82:1879–92

Sisk TD, Battin J. 2002. Habitat edges and avian ecology: geographic patterns and insights for western landscapes. *Stud. Avian Biol.* 25:30–48

Sisk TD, Haddad NM. 2002. Incorporating the effects of habitat edges into landscape models: effective area models for cross-boundary management. In *Integrating Landscape Ecology into Natural Resource Management*, ed. J Liu, W Taylor, pp. 208–40. Cambridge, UK: Cambridge Univ. Press

Sisk TD, Haddad NM, Ehrlich PR. 1997. Bird assemblages in patchy woodlands: modeling the effects of edge and matrix habitats. *Ecol. Appl.* 7:1170–80

Sisk TD, Margules CR. 1993. Habitat edges and restoration: methods for quantifying edge effects and predicting the results of restoration efforts. In *Nature Conservation 3: Reconstruction of Fragmented Ecosystems*, ed. DA Saunders, RJ Hobbs, PR Ehrlich, pp. 57–69. Sydney, Aust.: Surrey, Beatty & Sons

Sisk TD, Noon BR, Hampton HM. 2002. Estimating the effective area of habitat patches in heterogeneous landscapes. In *Predicting Species Occurrences: Issues of Accuracy and Scale*, ed. JM Scott, PJ Heglund, ML Morrison, pp. 713–25. Washington, DC: Island

Smith TB, Wayne RK, Girman DJ, Bruford MW. 1997. A role for ecotones in generating rainforest biodiversity. *Science* 276:1855–57

Stamps JA, Buechner M, Krishnan VV. 1987. The effects of edge permeability and habitat geometry on emigration from patches of habitat. *Am. Nat.* 129:533–52

Strayer DL, Power ME, Fagan WF. 2003. A classification of ecological boundaries. *BioScience* 53:723–29

Suarez AV, Bolger DT, Case TJ. 1998. Effects of fragmentation and invasion on native ant communities in coastal southern California. *Ecology* 79:2041–56

Tallmon DA, Jules ES, Radke NJ, Mills LS. 2003. Of mice and men and trillium: cascading effects of forest fragmentation. *Ecol. Appl.* 13:1193–203

Temple SA. 1986. Predicting impacts of habitat fragmentation on forest birds: a comparison of two models. In *Wildlife 2000: Modeling Habitat Relationships of Terrestrial Vertebrates*, ed. J Verner, ML Morrison, CJ Ralph, pp. 301–4. Madison: Univ. Wis. Press

Temple SA, Cary JR. 1988. Modeling dynamics of habitat-interior bird populations in fragmented landscapes. *Conserv. Biol.* 2:340–47

Tewksbury JJ, Hejl SJ, Martin TE. 1998. Breeding productivity does not decline with increasing fragmentation in a western landscape. *Ecology* 79:2890–903

Toms JD, Lesperance ML. 2003. Piecewise regression: a tool for identifying ecological thresholds. *Ecology* 84:2034–41

Turton SM, Sexton GJ. 1996. Environmental gradients across four rainforest-open forest boundaries in northeastern Queensland. *Aust. J. Ecol.* 21:245–54

Villard MA. 1998. On forest-interior species, edge avoidance, area sensitivity, and dogmas in avian conservation. *Auk* 115:801–5

Wahungu GM, Catterall CP, Olsen MF. 2001. Predator avoidance, feeding and habitat use in the red-necked pademelon, *Thylogale thetis*, at rainforest edges. *Aust. J. Zool.* 49:45–58

Wales BA. 1972. Vegetation analysis of north and south edges in a mature oak-hickory forest. *Ecol. Monogr.* 42:451–71

Ward JP, Gutiérrez RJ, Noon BR. 1998. Habitat selection by northern spotted owls: the consequences of prey selection and distribution. *Condor* 100:79–92

Watkins RZ, Chen J, Pickens J, Brosofske KD. 2003. Effects of forest roads on understory plants in a managed hardwood landscape. *Conserv. Biol.* 17:411–19

Weathers KC, Cadenasso ML, Pickett STA. 2001. Forest edges as nutrient and pollutant concentrators: potential synergisms between fragmentation, forest canopies, and the atmosphere. *Conserv. Biol.* 15:1506–14

Wiens JA, Crawford CS, Gosz JR. 1985. Boundary dynamics: a conceptual framework for studying landscape ecosystems. *Oikos* 45:421–27

Wilcove DS. 1985. Nest predation in forest tracts and the decline of migratory songbirds. *Ecology* 66:1211–14

With KA, King AW. 2001. Analysis of landscape sources and sinks: the effect of spatial pattern on avian demography. *Biol. Conserv.* 100:75–88

Wolf M, Batzli GO. 2001. Increased prevalence of bot flies (*Cuterebra fontinella*) on white-footed mice (*Peromyscus leucopus*) near forest edges. *Can. J. Zool.* 79:106–9

Yahner RH. 1988. Changes in wildlife communities near edges. *Conserv. Biol.* 2:333–39

Young A, Mitchell N. 1994. Microclimate and vegetation edge effects in a fragmented podocarp-broadleaf forest in New Zealand. *Biol. Conserv.* 67:63–72

Young ND. 1996. An analysis of the causes of genetic isolation in two Pacific Coast iris hybrid zones. *Can. J. Bot.* 74:2006–13

Zabel CJ, McKelvey K, Ward JP. 1995. Influence of primary prey on home-range size and habitat-use patterns of northern spotted owls (*Strix occidentalis caurina*). *Can. J. Zool.* 73:433–39

Zheng DL, Chen JQ. 2000. Edge effects in fragmented landscapes: a generic model for delineating area of edge influences (D-AEI). *Ecol. Model.* 132:175–90

Zipperer WC. 1993. Deforestation patterns and their effects on forest patches. *Landsc. Ecol.* 8:177–84

EVOLUTIONARY TRAJECTORIES AND BIOGEOCHEMICAL IMPACTS OF MARINE EUKARYOTIC PHYTOPLANKTON

Miriam E. Katz,[1] Zoe V. Finkel,[2] Daniel Grzebyk,[2] Andrew H. Knoll,[3] and Paul G. Falkowski[1,2]

[1]*Department of Geological Sciences, Rutgers University, Piscataway, New Jersey 08854; email: mimikatz@rci.rutgers.edu*
[2]*Institute of Marine and Coastal Sciences, Rutgers University, New Brunswick, New Jersey 08901; email: finkel@imcs.rutgers.edu, grzebyk@imcs.rutgers.edu, falko@imcs.rutgers.edu*
[3]*Department of Organismic and Evolutionary Biology, Harvard University, Cambridge, Massachusetts 02138; email: aknoll@oeb.harvard.edu*

Key Words coccolithophores, diatoms, dinoflagellates, phylogenetic trees, carbon cycle

■ **Abstract** The evolutionary succession of marine photoautotrophs began with the origin of photosynthesis in the Archean Eon, perhaps as early as 3.8 billion years ago. Since that time, Earth's atmosphere, continents, and oceans have undergone substantial cyclic and secular physical, chemical, and biological changes that selected for different phytoplankton taxa. Early in the history of eukaryotic algae, between 1.6 and 1.2 billion years ago, an evolutionary schism gave rise to "green" (chlorophyll *b*–containing) and "red" (chlorophyll *c*–containing) plastid groups. Members of the "green" plastid line were important constituents of Neoproterozoic and Paleozoic oceans, and, ultimately, one green clade colonized land. By the mid-Mesozoic, the green line had become ecologically less important in the oceans. In its place, three groups of chlorophyll *c*–containing eukaryotes, the dinoflagellates, coccolithophorids, and diatoms, began evolutionary trajectories that have culminated in ecological dominance in the contemporary oceans. Breakup of the supercontinent Pangea, continental shelf flooding, and changes in ocean redox chemistry may all have contributed to this evolutionary transition. At the same time, the evolution of these modern eukaryotic taxa has influenced both the structure of marine food webs and global biogeochemical cycles.

INTRODUCTION

Phytoplankton comprise a diverse, polyphyletic group of single-celled and colonial aquatic photosynthetic organisms that drift with the currents (Falkowski & Raven 1997). Fewer than 25,000 morphologically defined forms are distributed among at least eight major divisions or phyla. In contrast, nearly all photosynthetic organisms on land belong to a single clade (Embryophyta) that contains approximately 275,000 species (Table 1). Marine phytoplankton constitute less than 1% of Earth's photosynthetic biomass, yet they are responsible for more than 45% of our planet's annual net primary production (Field et al. 1998). Their evolutionary trajectories have shaped trophic dynamics and strongly influenced global biogeochemical cycles. In this paper, we examine the macroevolutionary histories of the major eukaryotic phytoplankton taxa that dominate the contemporary oceans, and consider their relationships to biogeochemical cycles.

ORIGINS OF PHOTOSYNTHESIS AND EUKARYOTIC PHYTOPLANKTON: MOLECULAR EVIDENCE

The origins of photoautotrophy are uncertain. Sedimentary microstructures as old as 3,500 Mega-annum (Ma) have been interpreted as cyanobacteria (Schopf 1993, Schopf 2002), but these reports have been questioned (Brasier et al. 2002). Carbon isotope measurements of reduced carbon in successions as old as 3,800 Ma seem to indicate widespread autotrophy in the oceans (Rosing 1999); however, the processes that led to the isotopic fractionation are not well understood. Perhaps more convincingly, fossil lipid biomarkers indicate that oxygenic photoautotrophs were present in the oceans by ca. 2,800 Ma (Brocks et al. 2003) and by ca. 2,300 Ma they had oxidized Earth's atmosphere (Bekker et al. 2004). Oxygenic photosynthesis subsequently spread to eukaryotes via cyanobacterial endosymbiotic associations that evolved to keep the process localized in membrane-bound organelles called plastids (Yoon et al. 2004). A series of primary, secondary, and possibly even tertiary endosymbiotic associations spread photoautotrophy to five of the eight major extant eukaryotic clades, giving rise to multiple new eukaryotic phyla (Figure 1, see color insert) (Baldauf 2003, Delwiche 1999, Palmer 2003, Yoon et al. 2004).

The earliest photosynthetic eukaryotes arose from one and possibly two successive endosymbiotic events in which the derived organelles were originally photosynthetic bacteria. One clade of early eukaryotes engulfed an ancestral cyanobacterium that was transformed into a plastid. More controversial is the proposal that an earlier, ancestral eukaryote incorporated the antecedent of an extant purple nonsulfur bacterium, which conferred an anoxygenic photosynthetic pathway on the host cell (Taylor 1987). The engulfed bacterium ultimately became the mitochondrion and lost its ability to photosynthesize, perhaps because of environmental oxidation. To accommodate the production of molecular oxygen within this new host-cell complex, the nascent mitochondrion would have operated its electron transport pathway in reverse (Osyczka et al. 2004), thereby giving rise to an oxygen-dependent respiratory electron transport chain with extremely high energy-conversion efficiencies. In this view, the evolution of the only two organelles that originated via endosymbiotic events were "motivated" by photosynthetic carbon-acquisition pathways.

This original symbiotic engulfment of a cyanobacterium into a host cell was a "primary" endosymbiotic process. The engulfed cyanobacterium contained a full complement of genes that allowed the organism to replicate itself. The symbiotic association was accompanied by the transfer of coding genes from the cyanobacterial genome into the eukaryotic nucleus. Assuming the genomic composition of extant cyanobacteria is representative of the ancestral symbiont genome, then 90% to 99% of the genes in the nascent plastid were subsequently lost or transferred to the host cell nucleus (Grzebyk et al. 2003, McFadden 1999, Palmer 2003). As a result, although they retain a core set of genes and limited autonomous capabilities for some housekeeping molecular processes, plastids cannot live longer than hours and cannot replicate outside of the host cell.

Three extant phytoplankton phyla (Chlorophyta, Rhodophyta, and Glaucophyta) result from primary endosymbiosis. Various datasets of the eukaryotic host cell [18S ribosomal RNA gene (Figure 1), protein genes (Baldauf et al. 2000), or mitochondria (Gray et al. 1998)] indicate that these clades were derived from a common heterotrophic ancestor. Plastid gene phylogenies, such as those inferred from 16S ribosomal genes (Figure 2, see color insert), also show the initial division of the plastid cluster according to the three types of primary plastids, typically surrounded by two membranes: cyanelles, green plastids (chloroplasts), and red plastids (rhodoplasts). Thus, despite their physiological differences, the three phyla may have originated from a single endosymbiotic event. Primary green plastids appear to have differentiated from the other two early in their evolutionary history. The accessory pigment, chlorophyll b (chl b), is the hallmark of the "green" plastid lineage, which became the forerunner of all chlorophyte algae and (subsequently) all higher plants that colonized terrestrial ecosystems in the Paleozoic. Cyanelles and rhodoplasts have many structural, biochemical, and genetic features in common; they contain chlorophyll a (chl a) only and have phycobilin pigments located in phycobilisomes, a simple carotenoid composition with zeaxanthin and α-carotene or β-carotene similar to cyanobacteria, and plastid genomes that retain numerous genes lost in chloroplasts.

Secondary endosymbiosis is a process that involves two eukaryotic cells and appears to be unique to plastid evolution. A heterotrophic cell first engulfed an alga with primary plastids. This complex mix of organelles was subsequently "enslaved," and the core photosynthetic machinery was salvaged and reduced to a new plastid. Secondary endosymbiotic plastids are present in a number of algal phyla. Two of them are characterized by secondary green plastids (Euglenophyta and Chlorarachniophyta, Figure 1). Plastids in four other algal lineages—cryptophytes, haptophytes (including coccolithophores), dinoflagellates (division Dinophyta), and heterokonts (Figures 1 and 2)—originated from a symbiotic association that appears to have differentiated from rhodophytes. Heterokonts include multiple algal classes, such as diatoms (Bacillariophyceae), brown algae (Phaeophyceae), and classes with species that form harmful algal blooms (Raphidophyceae and Pelagophyceae).

TABLE 1 The higher systematic groups of oxygenic photoautotrophs, with estimates of the approximate number of total known species, their distributions between marine and freshwater habitats,[a] and the type of plastid and major accessory pigments they contain

Taxonomic group		Known species			Plastids	
Category	Name	Total	Marine	Freshwater	Type	Accessory pigments[b]
Empire	Bacteria (= Prokaryota)					
Kingdom	Eubacteria					
Subdivision	Cyanobacteria (stricto sensu) (= Cyanophytes, blue-green algae)	1,500	150	1,350	N/A	B, Z, β, A
Subdivision	Chloroxybacteria (= Prochlorophyta)	3	2	1	N/A	dvb, Z, α, β
Empire	Eukaryota					
Kingdom	Protozoa					
Division	Euglenophyta	1,050	30	1,020		
Class	Euglenophyceae				S (green)	b, Dd, Dt, β, N, A
Division	Dinophyta (Dinoflagellates)	2,000	1,800	200		
Class	Dinophyceae				S (red)	c_2, Pe, Dd, Di, Dt, β
Kingdom	Plantae					
Subkingdom	Biliphyta					
Division	Glaucocystophyta	13	—	13		
Class	Glaucocystophyceae				P (cyan.)	B, Z, α, β
Division	Rhodophyta	6,000	5,880	120		
Class	Bangiophyceae				P (red)	B, Z, β, α
Class	Florideophyceae					
Subkingdom	Viridiplantae				P (green)	b, L, N, V, Z, β, α, A
Division	Chlorophyta					
Class	Chlorophyceae	2,500	100	2,400		
	Prasinophyceae	120	100	20		+P
	Ulvophyceae	1,100	1,000	100		
	Charophyceae	12,500	100	12,400		

Rank	Taxon			Species	Plastid	Pigments
Division	Bryophyta (mosses, liverworts)			22,000	—	1,000
Division	Lycopsida			1,228	—	70
Division	Filicopsida (ferns)			8,400	—	94
Division	Magnoliophyta (flowering plants)			240,000		
Subdivision		Monocotyledoneae		52,000	55	455
Subdivision		Dicotyledoneae		188,000	—	391
Kingdom	Chromista					
Subkingdom		Chlorechnia				
Division		Chlorarachniophyta				S (green)	b, L, N, V, β
Class			Chlorarachniophyceae	3–4	3–4	—
Subkingdom		Euchromista				
Division		Cryptophyta				S (red)	c_2, B, Al, α
Class			Cryptophyceae	200	100	100
Division		Haptophyta				S (red)	$c_1/c_2/c_3$, Fu/BF/HF, Dd, Dt, β
Class			Prymnesiophyceae	500	100	400
Division		Heterokontophyta				S (red)	
Class			Bacillariophyceae (diatoms)	10,000	5,000	5,000	c_1/c_2, Fu, Dd, Dt, β
			Chrysophyceae	1,250	800	450	c_1/c_2, Fu/BF, Dd, Dt, β
			Eustigmatophyceae	12	6	6	Va, V, β
			Phaeophyceae (brown algae)	1,500	1,497	3	$c_1/c_2/c_3$, Fu, Dc, Dt, V, β
			Raphidophyceae	27	10	17	c_1/c_2, Fu, Dd, Dt, β
			Tribophyceae (Xanthophyceae)	600	50	500	Va, Dd, Dt, β, c_1/c_2
Kingdom	Fungi					
Division		Ascomycotina (lichens)		13,000	15	20

[a] The difference between the number of marine and freshwater species, and that of known species, is accounted for by terrestrial organisms. Dashes indicate that no species are known (by us) for their particular group in this environment. Abbreviations are as follows. Plastid type: P = primary plastid (cyanelle, red, or green); S = secondary plastid (from the green or red plastid lineage). Accessory pigments: B = bilipigments; b = chlorophyll b; $c_1/c_2/c_3$ = chlorophyll c type 1, 2 or 3; dvb = divinyl chlorophyll b. Carotenoids: α = α-carotene; β = β-carotene; A = antheraxanthin; Al = Aljalloxanthin; BF = 19′-butanoylexyfucoxanthin; Dd = diadinoxanthin; Di = dinoxanthin; Dt = diatoxanthin; Fu = fucoxanthin; HF = 19′-hexanoyloxyfucoxanthin; L = lutein; N = neoxanthin; P = prasinoxanthin; Pe = peridinin; V = violaxanthin; Va = vaucheriaxanthin; Z = zeaxanthin. One-digit abbreviations indicate pigments already present in cyanobacteria or primary plastids; two-digit abbreviations indicate pigments appearing in secondary plastids.

[b] Sources: Van den Hoek et al. (1995), Jeffrey et al. (1997), Paerl et al. (2003).

There is ongoing debate over the number of endosymbiotic events in the evolutionary history of eukaryotic phytoplankton. Early molecular phylogenetic analyses suggested that the two secondary green plastid clades and four secondary red phyla were derived from independent endosymbiosis events. The nuclear and mitochondrial genomes of these two secondary green phyla are not closely related (Figure 1) (Baldauf 2003, Gray et al. 1998). Similarly, nuclear and plastid gene phylogenetic analyses indicate that the cryptophytes, haptophytes, dinoflagellates, and heterokonts arose from independent host cells. Plastid gene phylogenies indicate multiple origins of secondary plastids within rhodophytes (see Figure 2) (Bhattacharya & Medlin 1998, Müller et al. 2001, Oliveira & Bhattacharya 2000, Yoon et al. 2002a), and eukaryotic gene sequence analysis indicates that secondary host cells were unrelated or, at best, that heterokonts and alveolates share an ancient common ancestor (Figure 1) (Baldauf 2003, Baldauf et al. 2000, Bhattacharya & Medlin 1998). However, it has been proposed that secondary endosymbiosis occurred only twice: once in the green algal lineage and once in the red algal lineage (Cavalier-Smith 2003). A single endosymbiotic origin of plastids in the "green" euglenids and chlorarachniophytes seems unlikely, but some phylogenetic studies support the single-event hypothesis for phyla that contain secondary red plastids (Fast et al. 2001, Harper & Keeling 2003, Yoon et al. 2002b). This would indicate that all extant basal heterokonts and alveolates must have then lost their plastids to return to the heterotrophic nutritional mode.

Endosymbiotic events transformed both the eukaryotic host and symbiotic cells. The most noticeable impact concerned the composition of accessory photosynthetic pigments (Table 1). Most extant marine cyanobacteria contain only chl a, open-chain tetrapyrroles (phycobilins), and two carotenoids (zeaxanthin and β-carotene). However, the cyanobacterial division of Prochlorophyta also contains chl b or a divinyl derivative as a major accessory pigment. Within the Eucarya, chl b is present only in green plastids (hereafter, the "green" line), whereas phycobilins are present only in red plastids. For that reason, the ancestor of plastids was suggested to have been a cyanobacterium that contained both chl b and phycobilins (Tomitani et al. 1999). However, that hypothesis appears to be unlikely because extant prochlorophytes are genetically very distant from plastids (Hess et al. 2001). After primary endosymbiosis, rhodophytes retained the cyanobacterial pigments (but with α-carotene instead of β-carotene), and glaucophytes acquired an additional minor carotenoid. In contrast, primary green plastids greatly diversified in carotenoid pigments; they added four major pigments that absorb light in the wavelength range 400 to 520 nm (versus the less energetic wavelength range of 480 to 580 nm absorbed by rhodophytes). Whereas one may ask if the "primitive" pigment composition of rhodophytes limited the evolution of this phylum, interestingly, accessory pigments diversified in the red plastid lineage after secondary endosymbiosis. The new set of pigments, although specific to secondary red plastids, conferred similar properties to new plastids that mimic those provided by accessory pigments of primary green plastids. Chlorophyllide c and its derivatives, which are not present in any extant photosynthetic prokaryotes or in rhodophytes, absorb light in the 450 to 480 nm range, which nearly overlaps the absorption range by chl *b*. Secondary red plastid carotenoids, some of which are unique to a phytoplankton phylum, absorb light from 400 nm up to 580 nm. We refer to the secondary red plastid phyla that contain chl *c* as the "red plastid lineage."

EARLY PHYTOPLANKTON EVOLUTION: FOSSIL EVIDENCE

Fossils of probable eukaryotic origin first appear in abundance in rocks dated as approximately 1,700 to 1,900 Ma (Han & Runnegar 1992, Javaux et al. 2001, Knoll 1994). This age should be regarded as a minimum date for eukaryotic diversification because the paleontological record of older strata is sparse; molecular biomarkers suggest that both cyanobacteria and at least stem group eukaryotes existed by 2,700 Ma (Brocks et al. 1999, Summons et al. 1999), and the former, at least, may have evolved earlier (Knoll 2003). The systematic affinities of the earliest eukaryotic fossils are poorly constrained, but by 1,200 Ma, multicellular red algae lived along tidal flats, attached to hardgrounds (Butterfield 2000).

The presence of rhodophytes in mid-Proterozoic rocks places a minimum constraint on the timing of the red-green plastid schism. Reds radiated early as multicellular, benthic constituents of nearshore communities (Xiao et al. 2004). In contrast, early branching greens are planktonic, the paraphyletic "prasinophytes" that still contribute to photosynthesis in open ocean environments (Melkonian & Surek 1995). Xanthophyte fossils in uppermost Mesoproterozoic and Neoproterozoic rocks (Butterfield 2002, German 1990) indicate that secondary endosymbiosis followed relatively closely on the heels of primary events.

The functional as well as phylogenetic schism of reds and greens suggests that early members of the two lineages became adapted for different environmental regimes; the greens proliferated in oligotrophic open oceans underlain by anoxic waters, whereas the reds differentiated in coastal waters where both nutrients and fully oxygenated waters were more abundant (Anbar & Knoll 2002). Cyanobacteria appear to have dominated primary production in early to middle Proterozoic oceans, but by the end of the era (543 Ma), eukaryotic algae had become important constituents of marine ecosystems (Knoll 1989, Knoll 1992, Lipps 1993, Tappan 1980).

Many fossils of Proterozoic (and younger) protists cannot be assigned to extant taxa with any degree of certainty. Closed, organic-walled structures that resemble (at least broadly) dinocysts or prasinophyte phycomata are called "acritarchs," a grouping that makes them easier to deal with, but not to understand. Acritarchs first appeared as a minor component of the fossil record approximately 1700 to 1900 Ma (Summons et al. 1992, Zhang 1986); a moderate increase in diversity occurred approximately 800 to 900 Ma (Knoll 1994). This diversification coincides with the early stages of Rodinia rifting; however, no firm causal link has been established between the two events (Figure 3, see color insert).

In the Cambrian and Ordovician, eukaryotic phytoplankton underwent a second diversification in concert with the early Paleozoic radiations of marine invertebrates. The affinities of Early Paleozoic phytoplankton remain a subject for debate (Molyneux et al. 1996). Molecular biomarkers indicative of dinoflagellates occur in association with some Cambrian microfossils (Moldowan & Talyzina 1998), but morphological and ultrastructural analyses also show that prasinophyte green algae played a larger role in the continental shelf phytoplankton than they have in more recent times (Colbath & Grenfell 1995, Talyzina & Moczydlowska 2000). Preserved phytoplankton diversity peaked in the mid-Paleozoic and declined rapidly in the Late Devonian to Early Mississippian.

THE RISE OF THE RED LINEAGE

Mesozoic Expansion of the Red Lineage

In the early Mesozoic, the red eukaryotic phytoplankton began to assume a new ecological importance in the marine realm (e.g., Falkowski et al. 2004a,b). Marine prasinophytes declined during the Jurassic and Cretaceous, although this group of green algae is still represented in the oceans today (albeit as a minor constituent of open-ocean communities). A period of transition to red-line-dominated primary production occurred in the Triassic to Early Jurassic, ultimately resulting in the ecological dominance of coccolithophores, dinoflagellates, and diatoms in the contemporary ocean.

The first unequivocal dinoflagellates appeared in the fossil record as organic-walled cysts preserved in Middle Triassic continental margin sediments (Stover et al. 1996). Studies of molecular biomarkers indicate that dinoflagellates may have existed as far back as the Neoproterozoic (Moldowan & Talyzina 1998, Summons & Walter 1990); however, these biomarkers did not become prominent constituents of marine bitumens until the Triassic (Moldowan et al. 1996, Moldowan & Jacobson 2000), when microfossils more clearly document their expansion and radiation (Fensome et al. 1996, Stover et al. 1996) (Figure 3).

The calcareous nannoplankton (dominated by coccolithophorids) were the second group in the red lineage to radiate in the fossil record. They originated in the Late Triassic (Bown et al. 2004) (Figure 3) at about the same time that molecular biomarkers of coccolithophorids became common (Moldowan & Jacobson 2000). The earliest nannoplankton have been identified in Carnian sediments from the southern Alps (Bown 1998, Janofske 1992) and Nevada (F. Tremolada, personal communication).

The silica-encased diatoms were the last of the three major groups to emerge in the Mesozoic. The siliceous diatom frustules are highly soluble compared with other siliceous fossils (e.g., radiolaria and sponge spicules), which possibly imparts a preservational bias to the fossil record of diatom origin. Reports of diatom frustules in Jurassic sediments (Rothpletz 1896) have proved difficult to replicate by later workers, although molecular biological clock estimates (Medlin et al. 2000) and molecular biomarkers (Moldowan & Jacobson 2000) indicate that diatoms may have evolved earlier, but remained minor components in the marine realm until the Cretaceous. The first unequivocal fossil records of diatoms document radiations in the Early Cretaceous oceans (Harwood & Nikolaev 1995). Diatom morphologies in the Early Cretaceous were dominated by cylindrical and long-cyclindrical forms that show very little variation (Gersonde & Harwood 1990, Harwood & Gersonde 1990); the similarity among these early diatom morphologies argue against a pre-Mesozoic origin (Harwood & Nikolaev 1995). Late Cretaceous diatoms were dominated by discoidal and biddulphioid frustule morphologies (Harwood & Nikolaev 1995). Diatoms were present in nonmarine environments by 70 Ma (Chacon-Baca et al. 2002).

Regardless of the exact timing of the evolutionary origins of coccolithophores, dinoflagellates, and diatoms, fossil and biomarker data document the major expansion of all three groups in the Mesozoic (Bown et al. 2004, Grantham & Wakefield 1988, Harwood & Nikolaev 1995, Moldowan & Jacobson 2000, Stover et al. 1996) (Figure 3). They began their evolutionary trajectories to ecological prominence as the supercontinent Pangea began to break apart in the Late Triassic to Early Jurassic (~200 Ma), which marked the opening phase of the current Wilson cycle of continental breakup, dispersal, and reassembly (see The Role of the Wilson Cycle, below). Sea level rose as Pangea fragmented and the Atlantic Ocean basin widened, flooding continental shelves and low-lying inland areas (Figure 3). In addition, the fragmentation of the continents and creation of a new ocean basin produced an increase in the total length of coastline where many plankton lived. Nutrients (such as phosphate) that were previously locked up in the large continental interior of Pangea were transported to newly formed shallow seas. At the same time, diversities increased in the three groups of eukaryotic phytoplankton, and in marine invertebrates, paralleling a long-term increase in sea level that began in the Early Jurassic (Haq et al. 1987, Vail et al. 1977) (Figure 3). Greater nutrient availability coupled with expanded habitat area may well have contributed to the radiation of phytoplankton that lived along continental margins. Accordingly, the diversities of eukaryotic phytoplankton of the red lineage parallel sea-level rise through the Mesozoic (Figure 3). In contrast, short-term [1 million year (myr) scale] sea-level changes may have resulted in species turnover events, but had very limited impact on the large-scale evolutionary history of phytoplankton clades.

Life-cycle strategies may also have been an important component that favored the radiation of the red lineage as continental shelves flooded and epicontinental seas became widespread in the Jurassic. Dinoflagellates, coccolithophorids, and diatoms produce resting stages. After a bloom, a small fraction of the phytoplankton population becomes arrested in a specific stage of the cell's life cycle in which the production of cell armor increases, and the cell sinks to the seafloor until conditions become favorable to bloom (usually the following year). The timing of the bloom depends on environmental conditions such as ocean stratification and day length. The resting stage is associated with gamete formation and gene exchange in the planktonic portion of the life cycle. This life-cycle strategy requires shallow marine areas and also promotes genetic isolation by reducing the gene flow. Gene transfer, which can occur through sexual recombination, lateral vectors (such as viral infection), or both, is highly attenuated in benthic stages. Over relatively short periods of time, genetic isolation may have increased the tempo of evolution and phenotypic selection, processes that have been observed in contemporary phytoplankton assemblages (Medlin ct al. 1997).

Cenozoic Expansion of Diatoms

Bolide impact at the Cretaceous/Tertiary boundary (65 Ma) caused major extinctions (Alvarez et al. 1980) that are recorded in the fossil records of the coccolithophores and, to a lesser extent, the diatoms and dinoflagellates (Figure 3). The environmental factors that favored the expansion of the red lineage through the Mesozoic allowed these phytoplankton to repopulate the marine realm after the Cretaceous/Tertiary boundary (K/T) mass extinction event. Dinoflagellates and calcareous nannoplankton recovered to preextinction diversity levels by the earliest Eocene (~55 Ma), only to decline through the rest of the Cenozoic as long-term sea level began to fall in the mid-Paleogene and the extent of flooded continental areas decreased. Among fossilizable taxa, modern dinoflagellate species diversity has declined to levels comparable to the earliest Middle Jurassic, whereas modern coccolithophorid species diversity has declined to Late Jurassic levels.

In contrast to the other phytoplankton, diatom diversity has increased through the Cenozoic, despite falling sea level. Two pulses of diversification occurred in diatoms: the Eocene/Oligocene boundary interval and the middle to late Miocene. The latter was accompanied by a substantial radiation among pennate species (Strelnikova 1991). Conditions that may have favored the expansion of diatoms in the Cenozoic are discussed in the following section.

SELECTION PRESSURES AND ADAPTATIONS

Resource Acquisition

In contrast to most species of dinoflagellates and coccolithophores, diatoms frequently form extensive blooms along continental margins and in upwelling regions of the contemporary ocean. These organisms are responsible for approximately 40% of the net primary production and more than 50% of the organic carbon that is exported to the ocean interior (Falkowski et al. 2003). Planktonic diatoms have evolved a nutrient storage vacuole that retains high concentrations of nitrate and phosphate (Raven 1997). The storage vacuole allows diatoms to acquire pulses of inorganic nutrients, which can deprive competing taxa of these essential resources, overcome light-dependent nutrient uptake in mixing systems, or both. The storage capacity of the vacuole is sufficient to allow two to three cell divisions without the need for external nutrient resources. Consequently, diatoms thrive best in regions where nutrients are supplied with high pulse frequencies.

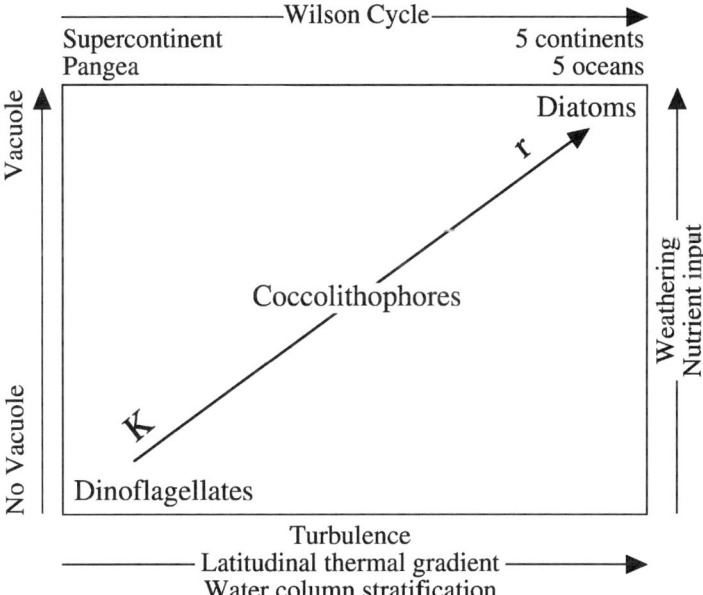

Figure 4 Models based on resource acquisition strategies (e.g., Grover 1988; Litchman E, Klausmeier CA, Miller JR, Schofield OM, Falkowski PG, in review; Tilman 1977; Tozzi et al. 2004) suggest that diatoms dominate when brief periods of water-column stability are punctuated by high turbulence, whereas coccolithophores and dinoflagellates dominate when the water column is more stably stratified. After Tozzi and others (2004).

Competition among diatoms, coccolithophores, and dinoflagellates has been modeled by application of resource acquisition strategies (e.g., Grover 1991; Litchman E, Klausmeier CA, Miller JR, Schofield OM, Falkowski PG, in review; Tilman 1977; Tozzi et al. 2004). The results of these models suggest that diatoms dominate when brief periods of water-column stability are punctuated by high turbulence, such as storm events (Figure 4). In contrast, coccolithophores and dinoflagellates dominate when the water column is more stably stratified. In theory, competitive exclusion could occur if equilibrium conditions were reached. In reality, the coexistence of two or more taxa competing for a single resource is a consequence of the dynamically unstable nature of aquatic ecosystems (Li 2002, Siegel 1998). Margalef (1994) recognized these fundamental differences in physiology and proposed that competition among the three major red lineage taxa could be related to upper-ocean turbulence and the supply of nutrients. The so-called Margalef mandala can be extended over geological time (Prauss 2000) to infer selection processes that led to the early rise of dinoflagellates and coccolithophorids in the relatively stable conditions of the Mesozoic, followed by the rise of diatoms in the Cenozoic.

Increasing latitudinal thermal gradients and decreasing deep-ocean temperatures have contributed to greater vertical thermal stratification through the latter half of the Cenozoic, which increased the importance of wind-driven upwelling and mesoscale eddy turbulence in providing nutrients to the upper ocean. The ecological dominance of diatoms under certain sporadic mixing conditions suggests that their long-term success in the Cenozoic probably can be attributed in part to an increase in event-scale turbulent energy dissipation in the upper ocean (Falkowski et al. 2004a). Sporadic nutrient influx to the euphotic zone may favor diatoms over coccolithophores and dinoflagellates, and a change in concentration and pulsing may favor small diatoms over large diatoms. As a result, a significant decrease in the average size of diatoms occurred in the Cenozoic. Periods of change were concentrated in the middle to late Eocene and early to middle Miocene (ZV Finkel, ME Katz, JD Wright, OM Schofield & PG Falkowski, in review).

Silica Bioavailability

Diatoms precipitate orthosilicic acid in a protein matrix to form extremely strong, structurally intricate biogenic opal shells called frustules. The modern oceans are undersaturated with respect to silica, which is largely the result of the evolution and ecological success of diatoms in removing this element from the dissolved phase, especially in the latter half of the Cenozoic (Conley 2002). Eighty percent of the silica in the oceans is derived from chemical and biological weathering of continental rocks (De La Rocha et al. 2000); the annual net riverine flux is 5.0 Tmol of silica per year (Tréguer et al. 1995). Sustaining the silica flux on geological timescales requires that fresh rock surfaces continuously become exposed to the soil interface, a process that is perpetuated by erosion. The rate of nutrient flux to the oceans on multimillion-year timescales is determined in part by continental elevation. Both orogeny and regression have characterized nearly all of the continents during the Cenozoic, which has increased nutrient fluxes (including silicic acid) (Maldonado et al. 1999) and facilitated diatom expansion (Falkowski et al. 2004a,b). An additional loop in the silica cycle that developed in the Cenozoic also may have accelerated the success of diatoms. The loop involves the coevolution of mammals, grasses, and diatoms. Grasses extract silicic acid from groundwater in soils and store it in opal phytoliths (Conley 2002). Silica can constitute as much as 15% of the dry weight of grasses (Alexandre et al. 1997, Bartoli 1983, Rapp & Mulholland 1992), and phytolith dissolution can release soluble silica two times more efficiently than silicate weathering (Alexandre et al. 1997).

Grasses originated in the Cretaceous, but remained sparse until the Eocene/Oligocene boundary (33.7 Ma) (Kellogg 2000, Retallack 2001), when global climates became more arid as a result of major glaciation in the Antarctic. As grasslands expanded, grazing ungulates evolved and displaced browsers (Janis & Damuth 2000). Hypsodont (high-crown) dentition in ungulates was selected over the brachydont (leaf-eating) dentition in browsing mammals, which coincided with the widespread distribution of silica-rich phytoliths and grit in grassland

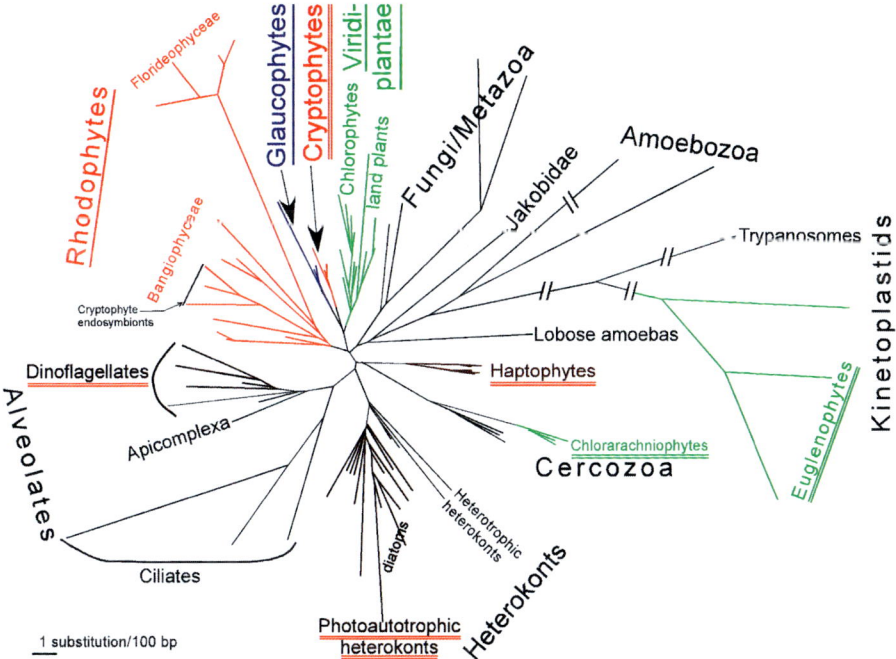

Figure 1 Relationships among photosynthetic eukaryotes inferred from 18S rRNA gene sequences (species names and sequence accession numbers are not shown for the clarity of the figure). Names underlined with one or two strokes indicate taxa that contain primary or secondary endosymbiotic plastids (*green* or *red*), respectively. Colored branches indicate taxa that contain photosynthetic plastids. Phylogenetic tree was constructed with the PHYLIP software package, in which genetic distance calculation (F84 substitution model) and the neighbor-joining method were used. Tree displayed is consistent with consensus trees obtained from 1,000 bootstrapped data sets.

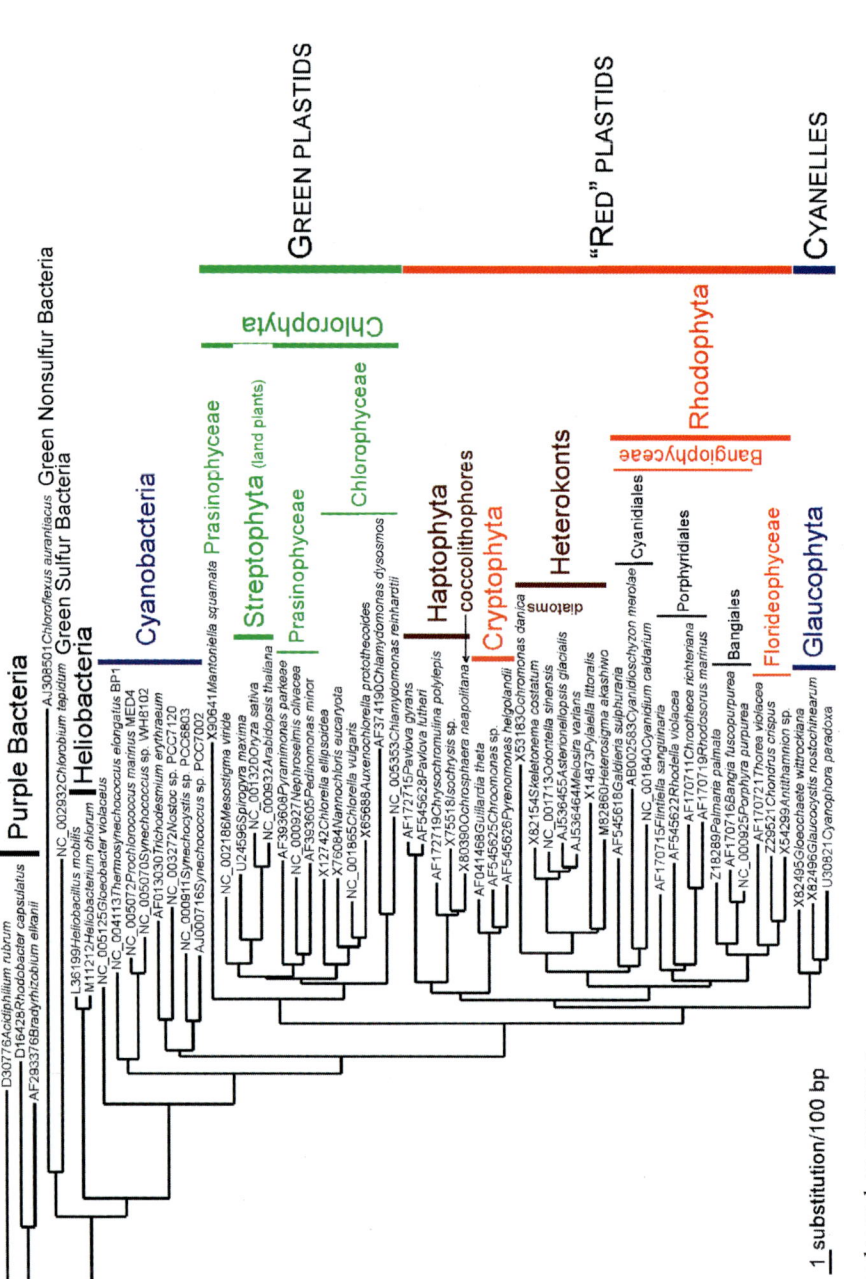

See legend on next page

EVOLUTION OF EUKARYOTIC PHYTOPLANKTON   C-3

Figure 2   Phylogenetic tree of plastids inferred from 16S rRNA gene sequences rooted with anoxygenic photosynthetic bacteria. Tree shows all plastids originating once from cyanobacteria, likely from an ancestral lineage that subsequently became extinct. Data from euglenophytes and chlororachniophytes (*secondary green plastids*) and from dinoflagellates (*two types of secondary red plastids*) have been excluded because of high genetic divergences responsible for inconsistent phylogenetic branchings. Phylogenetic tree was constructed with the PHYLIP software package, in which genetic distance calculation (F84 substitution model) and the neighbor-joining method were used. Tree displayed is consistent with consensus tree obtained from 1,000 bootstrapped data sets. Each species name is preceded by the accession number of the DNA sequence used in the phylogenetic analysis. Colors indicate groups of photosynthetic plastids.

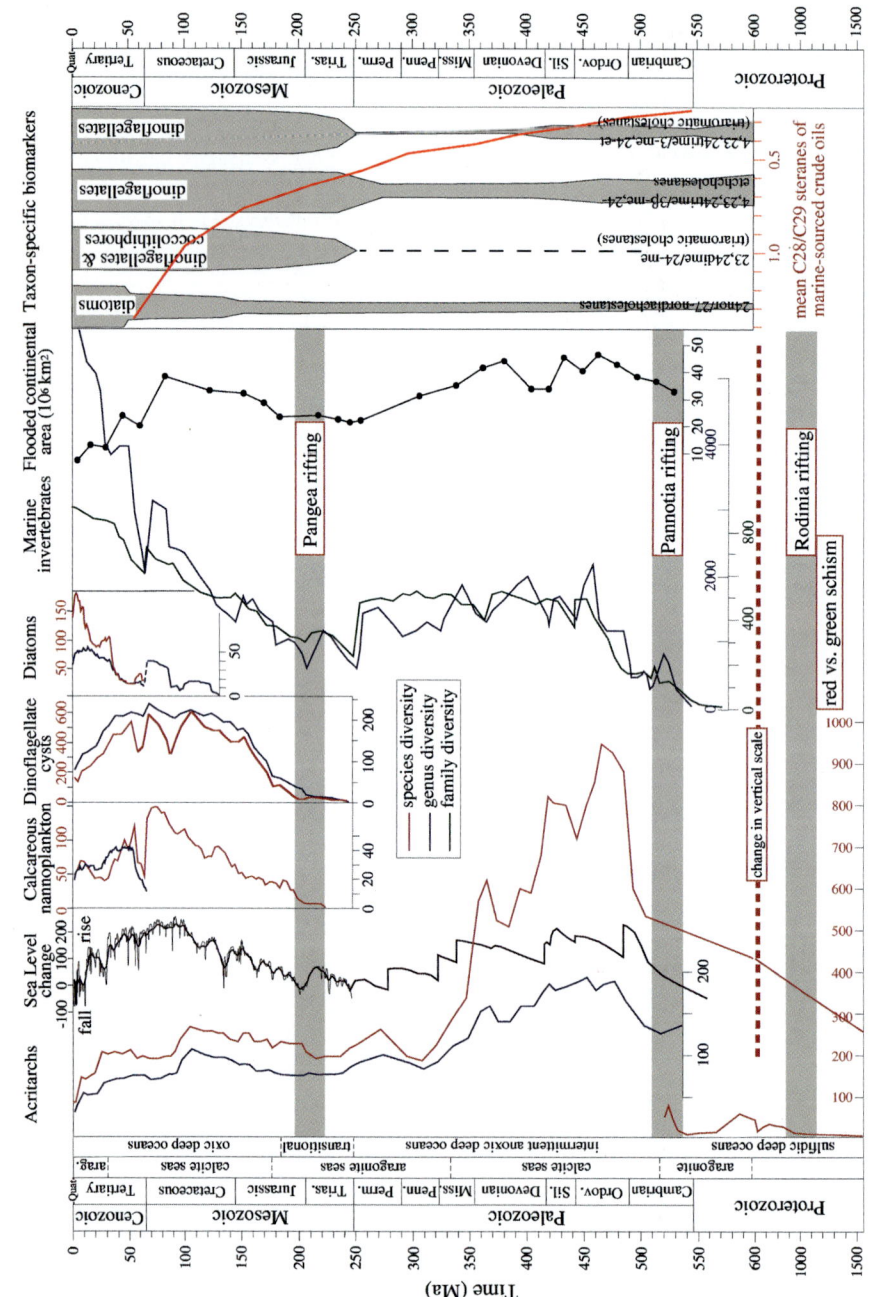

See legend on next page

Figure 3 Comparison of eukaryotic phytoplankton diversity curves (after Katz et al. 2004) with sea-level change (Mesozoic-Cenozoic [Haq et al. 1987] and Paleozoic [Vail et al. 1977]), flooded continental areas (Ronov 1994), and marine genus (*blue*) and family (*green*) invertebrate diversities (Sepkowski 1997). Phytoplankton species (*red*) diversities are from published studies (calcareous nannofossils [Bown et al. 2004], dinoflagellates [Stover et al. 1996], diatoms [Spencer-Cervato 1999], acritarchs [Proterozoic (Knoll 1994)], [Phanerozoic (R.A. MacRae, unpublished data)]). Phytoplankton genus (*blue*) diversities were compiled for this study from publicly available databases (calcareous nannofossils and diatoms [Spencer-Cervato 1999], dinoflagellates, and acritarchs [R.A. MacRae, unpublished data]. All records are adjusted to the Berggren et al. (1995) (Cenozoic), Gradstein et al. (1995) (Mesozoic), and GSA (http://rock.geosociety.org/science/timescale/timescl.htm) (Paleozoic) timescales. Taxon-specific biomarkers (Moldowan & Jacobson 2000) and C28/C29 sterane ratios (Grantham & Wakefield 1988) provide a record of increased biomass preservation of eukaryotic phytoplankton in the Mesozoic-Cenozoic. Episodes of supercontinent rifting are shaded. We base our interpretations of acritarch evolution on two compilations: a high-quality Proterozoic and Early Cambrian record of acritarch species (Knoll 1994) and a global, genera-level, Phanerozoic, quality-controlled literature compilation that may include taxonomic synonyms (R.A. MacRae, personal communication). Biological uncertainty exists in the interpretation of different genera and species within the acritarch record. Even though Knoll's (1994) record underestimates global diversity, and MacRae's record overestimates it, the long-term patterns of diversification and extinction appear consistent and robust. Data are available at http://mychronos.chronos.org/~miriamkatz/20040721

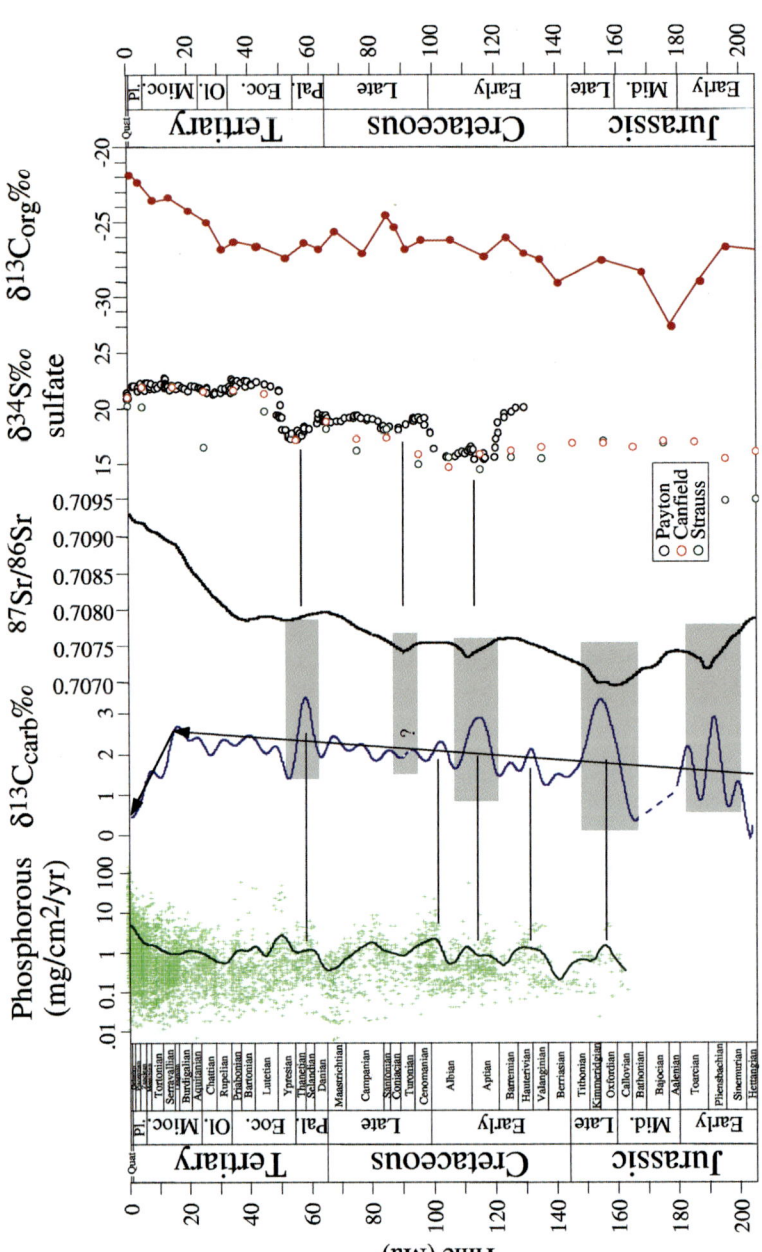

Figure 7 Geochemical proxy records showing comparisons of phosphorous flux [(Follmi 1995); curve fit (B.S. Cramer, personal communication) uses the SSA-MTM Toolkit from http://www.atmos.ucla.edu/tcd/ssa/ (Ghil et al. 2002)], bulk sediment $\delta^{13}C_{carb}$ (Katz et al., 2004), strontium isotopes (Howarth & McArthur 1997), sulfur isotopes (Canfield 1998, Payton et al. 1998, Payton et al. 2004, Strauss 1999), and bulk sediment $\delta^{13}C_{org}$ (Hayes et al. 1999). Timescales are as in Figure 3. Data are available at http://mychronos.chronos.org/~miriamkatz/20040721

forage (Retallack 2001). The rise of grazing ungulates and the radiation of grasses may have acted as a biologically catalyzed silicate weathering process (Falkowski et al. 2004a,b). Phytolith diversity and abundance has increased since the late Eocene (Jacobs et al. 1999, Retallack 2001); almost certainly, grass-mediated silica mobilization from soils increased as the grasses radiated. The subsequent transfer of silica to the oceans (primarily via riverine transport) increased the bioavailability of silica for diatom growth (Falkowski et al. 2004a,b). Accordingly, diatom species diversity, and presumably abundance, increased dramatically at the Eocene/Oligocene boundary (Figure 3). A major expansion of grasslands in the Neogene (Retallack 1997, Retallack 2001) was accompanied by a second pulse of diatom diversification at the species level (Falkowski et al. 2004a,b).

Armor

We can infer the evolutionary trajectories of all three major red lineage clades that appear to have dominated the eukaryotic phytoplankton community since the Mesozoic because they have fossilizable cell walls that also provide armor. This armor likely protected the phytoplankton from grazers long enough for the phytoplankton to form blooms (Banse 1992). Although the composition of the cell walls differs markedly among dinoflagellates, coccolithophorids, and diatoms, grazing by zooplankton may have provided a common evolutionary selection pressure. Unfortunately, virtually no fossil record exists for major modern pelagic zooplankton groups; chitinous crustaceans (e.g., copepods and euphausids) decompose rapidly in the sediments, and virtually all soft-bodied organisms such as salps decompose in the water column before they reach the sea floor. The potential role of cell walls as armor against grazing has long been debated (Smetacek 1999, 2001). The strength and flexibility of diatom frustules makes fracturing them a challenge for invertebrates; in fact, diatoms can pass through a copepod gut intact (Hamm et al. 2003). The chitinous teeth of copepods have a siliceous coating that provides significant compressive strength, and mandible muscles in these crustaceans are highly developed. This feature is a clear example of adaptive evolution between predator and prey in pelagic ecosystems (Adams 2001). However, if diatoms are consumed, zooplankton greatly facilitate the dissolution of the frustule (Bidle & Azam 1999).

Coccolithophores are grazed, but the nutritional benefit derived from ingesting a cell consisting of 30% calcium carbonate is lower than that derived from ingesting a naked cell of the same size. Indeed, zooplankton avoid coccolithophorids if presented with optional, unarmored food sources (Falkowski and Wyman, unpublished data). In addition to armor, thecate dinoflagellates (and some diatoms) have evolved a strategy of vertical migration; they obtain nutrients below the pycnocline at night and rise to the upper portion of the euphotic zone during the day to optimize photosynthesis (Kamykowski 1981, Villareal et al. 1993). This migration strategy is out of phase with that of many zooplankton grazers (Banse 1964).

Ocean Chemistry

Secular changes in seawater chemistry also appear to have influenced phytoplankton evolutionary trajectories. For example, trace elements (including iron, copper, zinc, and manganese) play essential roles in mediating critical biochemical reactions in all phytoplankton. The bioavailability of these elements in seawater is strongly dependent on redox state (Whitfield 2001). Algae with green plastids have substantially higher quotas for iron, zinc, and copper (i.e., Fe:P, Zn:P, and Cu:P ratios) than do red eukaryotes, whereas the latter have higher quotas for cadmium, cobalt, and manganese (Quigg et al. 2003). Furthermore, the redox state of the oceans may discriminate between red and green plastid lineages with respect to fixed nitrogen preferences (Anbar & Knoll 2002; Falkowski & Raven 1997; Litchman E, Klausmeier CA, Miller JR, Schofield OM, Falkowski PG, in review), as well as the availability of phosphate. Members of the green plastid lineage tend to have higher N/P ratios than members of the red plastid lineage (Falkowski et al. 2004b). Hence, the form and availability of macronutrients and trace metals have the potential to be strong selective agents that favor red- or green-plastid-containing taxa.

The Archean biosphere was anoxic, and its initial oxidation in the Paleoproterozoic Era introduced oxygen primarily into the surface mixed layer of the oceans. Complete oxygen depletion appears to have remained common beneath the pycnocline, and sulfidic deep waters were widespread (Canfield 1998) (Figure 3). Under these conditions, early algae likely competed best in the better-oxygenated coastal regions, where rivers delivered essential metals (Anbar & Knoll 2002). As noted above, the evolutionary divergence of red and green plastids took place in the context of this redox heterogeneity and may have contributed to the different evolutionary trajectories of rhodophytes and early chlorophytes. Increasing oxidation of ocean waters may have facilitated the expansion of algae across shelves in latest Neoproterozoic times; undoubtedly, interactions with evolving animals also contributed to Cambrian and Ordovician phytoplankton diversification (Butterfield 1997).

Black shales are common in Proterozoic marine successions, which indicates oxygen depletion beneath surface water masses in the oxygen-minimum zone (Shen et al. 2002, Shen et al. 2003); black shales also occur episodically through much of the Paleozoic (Arthur & Sageman 1994). Deep-water anoxia may have been particularly pronounced near the end of the Permian (Isozaki 1997). The expanded oxygen-minimum zone persisted through the Early Triassic (Twitchett 1999, Wignall & Twitchett 2002), and may have altered the distributions of many trace elements within the oceans (Whitfield 2001a). Ocean anoxia increases the availability of Fe, Mn, P, and ammonium and decreases the availability of Cd, Cu, Mo, Zn, and nitrate; hence, the green lineage may have been favored over the red lineage at times when subsurface reducing conditions prevailed.

A secular shift in ocean redox conditions in the early Mesozoic changed trace metal availability in the oceans and exerted a selective pressure that favored the red lineage by better meeting their metal requirements (Falkowski et al. 2004a,b).

The last major occurrence of large concentrations of the green algal prasinophytes was in the Early Jurassic, when black shale deposition and ocean anoxia appears to have been widespread (Falkowski et al. 2004a,b). As deep-ocean oxygenation became increasingly permanent through the Mesozoic, Cd, Cu, Mo, Zn, and nitrate availability increased, which allowed the red lineage to expand (Anbar & Knoll 2002, Falkowski et al. 2004a,b). Temporal changes in the availability of redox-sensitive trace metals are consistent with the biological transition to red-lineage dominance of phytoplankton (Falkowski et al. 2004a,b). Thus, long-term changes in ventilation of the world's oceans ultimately appear to have played a significant role in the rise of the red lineage during the Mesozoic.

The Mesozoic oceans occasionally were punctuated by widespread organic carbon burial events associated with the short-lived (<1 myr) Oceanic Anoxic Events (OAEs) (e.g., Arthur & Sageman 1994). These events briefly altered ocean redox conditions, but the effects were relatively short-lived. The biological impact of OAEs is debated (e.g., Leckie et al. 2002 versus Bown et al. 2004) despite the substantial effort dedicated to examining the phytoplankton and zooplankton communities across these carbon-burial events (e.g., Erbacher & Thurow 1997, Leckie et al. 2002, Roth 1987). Once the red lineage garnered a more secure ecological advantage in the Late Triassic, the fossil records suggest that the OAEs only had a minor influence on the evolutionary trajectories of eukaryotic phytoplankton (Bown et al. 2004, Falkowski et al. 2004a).

THE ROLE OF THE WILSON CYCLE

The radioactive decay of elements within Earth's core produces heat that dissipates to the planet's surface through convection and conduction. Near the surface, the thin oceanic crust conducts heat about three times more efficiently than the thicker continental crust. The differential dissipation of radiogenic heat oceanic basalts versus continental crust leads to the buildup of thermal energy below a fully assembled supercontinent, which causes the continental crust to thin and eventually to fracture and rift apart. A new ocean spreading center and ocean basin form between the continental fragments. Heat-driven convection in the mantle below the lithosphere drives the tectonic plates apart with the fragmented continents attached, and new oceanic crust forms at the intervening spreading center. As it ages, the oceanic crust becomes cooler and denser and eventually subsides as it moves away from the spreading center. When this old crust becomes so dense that it subsides below the adjacent, relatively low-density continental crust, a new subduction zone is created and the ocean basin is consumed as the whole process reverses itself and the continents reassemble. Named after its conceptual discoverer, J. Tuzo Wilson (Wilson 1966), this episodic breakup, dispersal, and subsequent reassembly of supercontinents has become known as the Wilson Cycle, and it occurs over approximately 300-myr to 500-myr intervals (e.g., Fischer 1984, Rich et al. 1986, Valentine & Moores 1974, Worsley et al. 1986).

The Wilson Cycle and Evolutionary Trajectories

Several studies have drawn attention to the correlation between evolutionary pulses in the marine realm and the Wilson Cycle (e.g., Damsté et al. 2004, Fischer 1984, Rich et al. 1986, Valentine & Moores 1974, Worsley et al. 1986). Myriad studies since these early publications provide the foundation for speculating on causal rather than casual linkages for the phytoplankton response to the Wilson Cycle.

Global species diversity reflects both the packing of taxa within communities and the distribution of suitable habitat area (MacArthur & Wilson 1967, Rosenweig 1995). A geological proxy for the former is provided by estimates of within-assemblage diversity through time (e.g., Bambach 1977), whereas paleobiogeographic analyses gauge the contribution of the latter. Diversity increases correlate with continental rifting during early iterations of the Wilson Cycle. Acritarchs appear to have radiated as Rodinia rifted in the Late Proterozoic, and diversified again in the Early Paleozoic as Pannotia rifted (Figure 3). Thus, continental separation may have played a role, along with the ecological drivers that govern within-assemblage diversity, in promoting Cambro-Ordovician phytoplankton (and marine invertebrate [Sepkowski 1997; Bambach 1999]) expansion.

Sea-level rise and flooded continental area also are highly correlated with increasing diversity of Mesozoic calcareous nannoplankton (e.g., coccolithophorids) and dinoflagellates (Figure 5), as well as with their declining diversities in the Cenozoic oceans. Although flooded continental area is a small percentage of the total oceanic area suitable for phytoplankton, the shallow seas appear to have contributed proportionally more to niche space because of high nutrient input, high rates of primary production, and habitat heterogeneity. Flooded continental area provides variable, high-nutrient habitat by creating additional upwelling zones and increasing turbulence and nutrient suspension from below the thermocline. In addition, terrestrial nutrient input likely increases because nutrients that were previously sequestered in the large supercontinent interior are more readily transported to the newly opened, nearby oceans. We note that flooded continents increase the availability of fossiliferous sediments and have the potential to impart a taphonomic and sampling component to diversity compilations. This bias may be most pronounced in the older record because most pre-Jurassic ocean crust has been destroyed.

A simple equation captures the empirical relationship between long-term sea level change and diversities in calcareous nannoplankton, dinoflagellates, and diatoms (Figure 5); this equation provides correlation coefficients, but is not a predictor of sea level based on phytoplankton diversities. Although accurate reconstruction of diversities is inevitably biased by differential preservation and the problem of defining species based on morphologic characters, our approach provides an estimate of the relationship between phytoplankton richness (R) and sea level (meters) that solves for coefficients that link the diversity of phytoplankton taxa to change in habitat area caused by sea level change. This relationship is defined as:

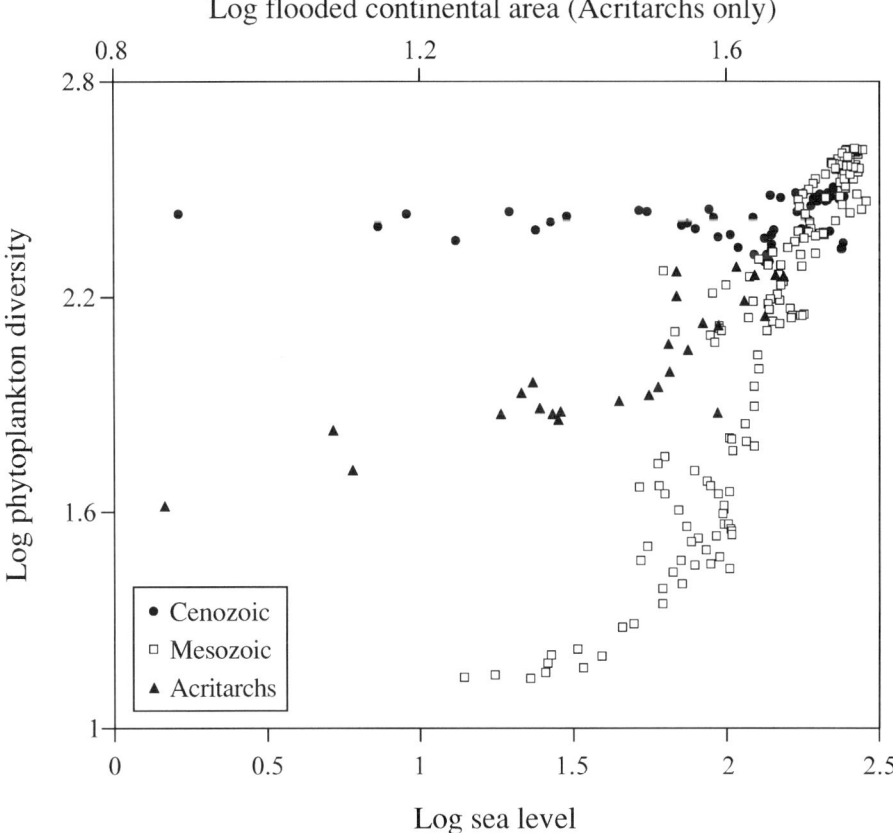

Figure 5 Phytoplankton diversity (sources as in Figure 3) as a function of flooded continental area in the Paleozoic (acritarch genera) and sea level in the Mesozoic (calcareous nannoplankton species + dinoflagellates species). The radiation in diatom species diversity alters the relationship between phytoplankton diversity (calcareous nannoplankton species + dinoflagellate species + diatom species) and sea level change in the Cenozoic. Sea level was translated to obtain positive values. Timescales are as in Figure 3.

$$-0.29\,R_{diatoms} + 0.42\,R_{nannos} + 0.59\,R_{dinos} + 25.4 = \text{sea level } (r^2 = 0.82) \quad (1)$$

Our analysis suggests that whereas calcareous nannoplankton and dinoflagellate diversities are highly correlated to sea level, diatom diversity responds to other environmental factors (see Selection Pressures and Adaptations, above) and has rapidly increased over the past approximately 35 myr.

The tectonic processes that drive the Wilson Cycle not only affect available niche space, but also affect ocean chemistry. As seawater cycles through mid-ocean ridges, magnesium is removed and calcium is added. Total ridge length and seafloor spreading rates can change Mg/Ca ratios in seawater. As a result, high-Mg/Ca ratios tend to occur during times of supercontinent assembly, when ridge length is shortest, characterized by deposition of aragonite and high-Mg calcite (called "aragonite seas"). Low-Mg calcite deposition tends to characterize times of continental breakup (called "calcite seas") (Hardie 1996, Sandberg 1975). Low-Mg/Ca and high-Ca^{+2} concentration in seawater favors calcification in certain groups of marine organisms (including coccolithophores), which results in a correspondence between these organisms and "calcite sea" intervals (Stanley & Hardie 1998). This relationship is illustrated by the expansion of coccolithophores in the Mesozoic calcite seas (Figure 3), along with other low-Mg calcifiers among marine invertebrates. The massive Cretaceous coccolith chalks (e.g., the White Cliffs of Dover) were deposited on continental shelves when margins were flooded, Mg/Ca ratios were low, and dissolved Ca^{+2} concentrations in seawater were high. This correlation further links the expansion of coccolithophores to the opening phase of the current Wilson Cycle. Chalk deposition continued in the Paleocene after the K/T extinctions, but as the Mg/Ca ratio increased through the Cenozoic, the degree of calcification in coccoliths declined, and coccolith size decreased (Bukry 1971, Houghton 1991).

Bottom-Up Control of Marine Invertebrate Fauna

In his classic essay "Seafood Through Time," Bambach (1993) identified a series of changes in late Mesozoic and Cenozoic marine faunas—increases in abundance and mechanical strength of top predators, mean size of marine invertebrates, and mean rates of energy consumption. These changes all require a greater nutritional supply to marine life consistent with observed increases in C to P ratios (Martin 2003). Bambach (1993, 1999) suggested that the necessary nutrients were supplied by angiosperm-facilitated increases in erosional runoff from continents, runoff that would have been further augmented by higher erosion rates associated with increasing continental elevations (see Silica Bioavailability, above). As an additional hypothesis, we propose that the Wilson Cycle and the evolutionary shifts in phytoplankton community composition, in conjunction with increases in primary production and the quantity and quality of export production in the Phanerozoic seas, strongly influenced marine invertebrate faunal evolution.

Marine invertebrate faunal diversity (Bambach 1993) is highly correlated with phytoplankton diversity and community composition (Figure 6), but poorly correlated with flooded continental area ($r^2 = 0.14$) and sea level ($r^2 = 0.05$). An increase in net primary production increases both the availability of organic matter that can be transferred to higher trophic levels in marine invertebrates and the potential to export food to the seafloor (Falkowski et al. 2003). The fossil record indicates that an evolutionary change in phytoplankton diversity and community composition is a predictor of change in the richness in marine invertebrate genera: (*a*) diatom genera richness is highly correlated with Cenozoic invertebrate richness ($r^2 = 0.75$); (*b*) the combined diversity of calcareous nannoplankton and dinoflagellates is highly correlated with Mesozoic invertebrate diversity

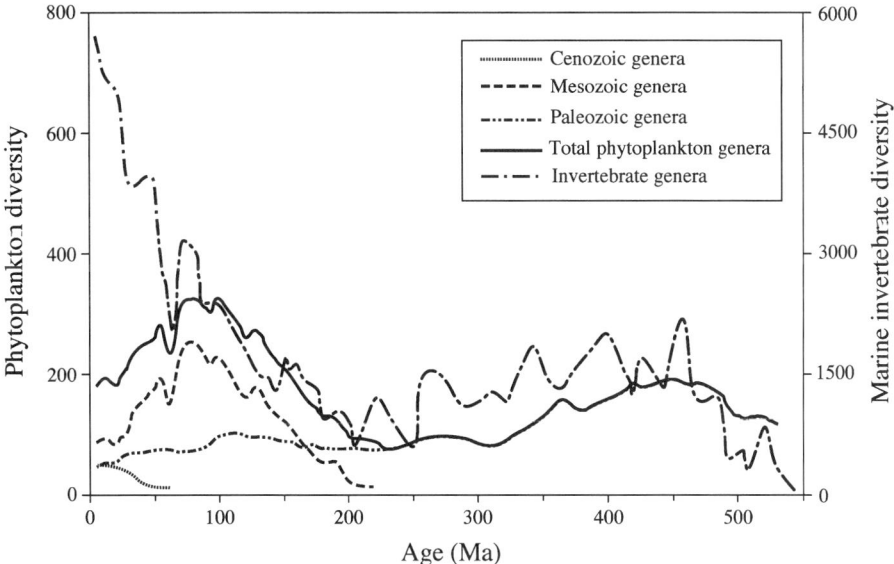

Figure 6 Marine invertebrate diversity (Bambach 1993) and phytoplankton diversity (as in Figure 3) over the Phanerozoic. Timescales are as in Figure 3. Data are available at http://mychronos.chronos.org/~miriamkatz/20040721

($r^2 = 0.89$); and (c) a weak correlation exists between the diversity of Paleozoic marine invertebrates and the diversity of the acritarchs ($r^2 = 0.12$) (Figure 6).

The radiation of diatoms in the Cenozoic demarcates a large change in food-web structure of the Phanerozoic oceans, which altered the relationship of total phytoplankton diversity to flooded continental area and sea level that had persisted through the Paleozoic and Mesozoic (Figure 5). Total phytoplankton diversity is positively correlated with invertebrate diversity in the Paleozoic and Mesozoic ($r^2 = 0.53$), but becomes inversely related to invertebrate diversity in the Cenozoic ($r^2 = 0.61$). This pattern indicates that the transitions from the prasinophyte/acritarch-dominated Paleozoic ocean to the nannoplankton/dinoflagellate-dominated Mesozoic ocean to the diatom-dominated Cenozoic ocean were likely associated with changes in total primary production, nutritional quality, export production efficiency, and food availability for heterotrophic consumers.

The linear relationship between phytoplankton and invertebrate diversities can be used to infer changes in food-web structure and efficiency of trophic transfer of primary production into invertebrate species over time. This comparison highlights an increase in the number of invertebrate species relative to phytoplankton species through the Phanerozoic (Table 2): (a) each Paleozoic phytoplankton genus is associated with five invertebrate genera; (b) each Mesozoic dinoflagellate genus and nannoplankton species is associated with 7 and 14 invertebrate genera, respectively; and (c) each Cenozoic diatom genus is associated with 59 invertebrate

TABLE 2 Invertebrate diversity supported per phytoplankton species or genera (the slope) over different time intervals[a]

Taxa	Time period	Slope	SE[b]	Intercept	SE[b]	r^2
Diatom genera	Cenozoic	59.0	12.1	2308.0	418.6	0.75
Nannoplankton spp.	Mesozoic	14.5	1.3	707.5	100.4	0.85
Dinoflagellate genera	Mesozoic	6.9	0.6	844.1	85.6	0.86
Nannos + Dinos	Mesozoic	4.9	0.4	766.1	79.9	0.89
Acritarch genera	Paleozoic	5.2	2.3	454.1	322.9	0.12
Sum	Phanerozoic	9.0	1.7	187.4	316.0	0.27

[a]Invertebrate diversity (y) is estimated as a linear function of phytoplankton diversity (x) × slope + intercept. Diversity of all groups is represented by number of genera, except calcareous nannoplankton diversity, which is represented by number of species. Data from sources outlined in caption for Figure 3.
[b]SE = standard error.

genera. This large change in the richness of invertebrate genera relative to the phytoplankton species and genera may represent an increase in the relative biomass of each of the phytoplankton species caused by an increase in primary production, an increase in the amount of primary production that is exported, or both. In part, the increasing ratio of preserved invertebrate to phytoplankton taxa also reflects the increasing richness of specialized predators through time; this relationship is another consequence of increasing primary production in the oceans, or increasing biological availability of primary production to consumers, or both (Bambach et al. 2002). In the contemporary oceans, diatoms are the major phytoplankton taxa contributing to export production (Dugdale et al. 1998). The increased export efficiency associated with the ecological success of marine diatoms most likely contributed to the increase in marine invertebrate diversity, size, and rate of energy consumption.

Evolutionary Tempo of Marine Phytoplankton

We compare evolutionary rates among different taxonomic groups based on background rates of extinction and origination normalized by species richness in the fossil record (Table 3). The results of this analysis show a twofold to fourfold decrease in the average longevity of the calcareous nannoplankton, dinoflagellates, and diatom species from the Mesozoic to the Cenozoic. Using a similar approach, Knoll (1994) described an increase in morphological diversity and a decrease in average species longevity in the acritarchs through the Proterozoic into the early Cambrian, and Barron (2003) documented relatively short longevities (~3 myr) in marine diatoms over the past 18 myr from high-resolution records in the Pacific and Southern Oceans. These results highlight the continuing increase in evolutionary tempo of phytoplankton. Extinction rates in dinoflagellates and calcareous nannoplankton have exceeded origination rates through the Cenozoic, which has resulted in declining diversities.

TABLE 3 Estimates of species and genera duration (in Ma ± 1 SE) for different phytoplankton groups

	Cenozoic	Mesozoic
Diatom species	40 (± 6)	
Diatom genera	46 (± 1)	100[a]
Dinoflagellate species	14 (± 1)	22 (± 3)
Dinoflagellate genera	30 (± 4)	73 (± 6)
Nannoplankton species	11 (± 1)	49 (± 6)

	Paleozoic	Proterozoic
Acritarch species cohort[b]	8	102–1960

[a]Species and genera duration was calculated as $1/(E/R\Delta T)$, where E is the number of species extinctions, R is species richness, and ΔT is the duration of the time period, with 1 standard error (SE) in parentheses. To get an estimate of background duration, the middle 50% of the data was used to calculate the average species duration (and 1 SE) for each time period. Data for diatom species are from Spencer-Cervato (1999), diatom genera are from Harwood & Nikolaev (1995), dinoflagellates are from Bujak & Williams (1979), and nannoplankton are from Bown et al. (2004). Because of a small sample size, the duration of Mesozoic diatom genera was determined by a sum of all the extinctions over the Mesozoic and the average R for the whole time interval.

[b]Acritarch species durations were calculated by cohort analysis by Knoll (1994).

THE ROLE OF PLANKTON IN BIOGEOCHEMICAL CYCLES

Geological and biological processes modify atmospheric and seawater chemistry through time and influence global climates. Submarine and subaerial tectonics in combination with erosional processes are the primary suppliers of the major elements in geochemical cycles; the oxidation and reduction processes that alter mobile elemental reservoirs are biologically mediated. Hence, geological and biological processes together form feedback loops in various biogeochemical cycles.

As we have shown in the preceding sections, changing environmental conditions selected for different plankton groups through Earth's history. In turn, phytoplankton have modified various aspects of the environment through time. Perhaps the most notable example of this relationship is the role that plankton played in the oxygenation of Earth's atmosphere and oceans through the evolution of oxygenic photosynthesis. Increasing oxygen levels led to the oxidation of inorganic substrates, particularly iron and sulfur, which altered seawater chemistry. In turn, primary production became iron-limited as the result of iron oxidation in the oceans. In this section, we explore these types of biogeochemical interactions that have left clues preserved in long-term geological records.

Phytoplankton and the Carbon Cycle

The expansion of the red lineage of phytoplankton that began in the Mesozoic had an impact on the long-term carbon cycle, in part by altering the distribution of carbonate and organic carbon buried on the seafloor. Prior to the Mesozoic, most marine calcifying organisms lived in shallow coastal and shelf regions. Carbonate deposition was concentrated in these areas as a result, and deposition of pelagic carbonates was minimal. Two groups of carbonate-secreting plankton successfully competed for limited carbonate resources and expanded in the Mesozoic oceans—coccolithophores, with their calcitic-plated armor, and planktonic foraminifera, with their carbonate tests. As these two groups radiated, the loci of marine carbonate deposition gradually expanded from shallow shelf areas to the deeper ocean (Southam & Hay 1981, Sibley & Vogel 1976) and the carbonate compensation depth (CCD) deepened (e.g., Wilkinson & Algeo 1989). Pelagic carbonate sedimentation has come to dominate as sea level and shelf area has declined since the Late Cretaceous, and the pelagic carbonate reservoir has increased at the expense of the shallow-water carbonate reservoir since the Mesozoic (Wilkinson & Algeo 1989).

At the same time, the expansion of marine phytoplankton in the Mesozoic resulted in greater export production through time (e.g., Bambach 1993) (see above). The newly-emerged eukaryotic phytoplankton efficiently exported organic matter, and the newly formed Atlantic Ocean margins increased the potential storage area for that organic matter. Substantial amounts of organic carbon were sequestered on the passive continental margins of the Atlantic and on flooded continental interiors (e.g., Arthur et al. 1984, Bralower 1999, Claypool et al. 1977, Jenkyns & Clayton 1997) as Pangea broke apart, export production increased, and organic matter was buried before it could be oxidized. At the same time, sedimentary carbon was both recycled at subduction zones and transferred to orogenic metasediments as the Tethys Sea and Pacific Ocean basins shrank. The circum-Atlantic sediments have not yet been recycled through subduction during the current Wilson Cycle, thereby providing long-term storage of large amounts of isotopically light organic carbon. This biologically mediated increase in organic carbon burial may account for as much as half of the long-term increase in $\delta^{13}C$ values recorded in marine carbonates ($\delta^{13}C_{carb}$) (Katz et al. 2004) and organic carbon ($\delta^{13}C_{org}$) (Hayes et al. 1999) from the Jurassic to the Miocene (Figure 7, see color insert). Increased organic carbon burial contributed to a gradual depletion in CO_2 from the ocean-atmosphere system and a simultaneous increase in the oxidation state of Earth's surface (Katz et al. 2004).

The long-term depletion of CO_2 acted as a feedback mechanism that was a key factor that selected β-carboxylation and C_4 photosynthetic pathways in marine and terrestrial photoautotrophs. Diatoms have β-carboxylation pathways (Morris 1980, Reinfelder et al. 2000) and dominate carbon export production in the modern ocean (Smetacek 1999). The rapid radiation of diatoms in the latter half of the Cenozoic enriched the ^{13}C composition of marine organic matter (Figure 7) (Katz et al. 2004).

A decrease in diatom cell size over the Cenozoic may also have affected $\delta^{13}C_{org}$ values (Figure 7) (ZV Finkel, ME Katz, JD Wright, OM Schofield & PG Falkowski, in review). In general, large phytoplankton cells produce high $\delta^{13}C_{org}$ because they tend to have low growth rates and low rates of diffusive flux (Laws et al. 1997, Popp et al. 1998, Rau et al. 1997). Early studies suggested that an increase in diatom size through the Cenozoic might have been responsible for the measured increase in $\delta^{13}C_{org}$ (Hayes et al. 1999). However, the median diatom cell size appears to have decreased through the Cenozoic (ZV Finkel, ME Katz, JD Wright, OM Schofield & PG Falkowski, in review), which contradicts the assertion of Hayes et al. (1999) that a trend toward larger diatoms alone drove the increase in $\delta^{13}C_{org}$. Therefore, an increase in diatom abundance (rather than size) since the mid-Cenozoic likely contributed to the $\delta^{13}C_{org}$ increase (ZV Finkel, ME Katz, JD Wright, OM Schofield & PG Falkowski, in review; Katz et al. 2004).

Terrestrial ecosystems also contributed to the $\delta^{13}C_{org}$ increase. In the late Miocene (6 to 8 Ma), a global expansion of grasslands was coupled with a shift in dominance from C_3 to C_4 grasses. This shift produced ^{13}C-enriched terrestrial biomass (Cerling et al. 1997, Still et al. 2003), some of which was ultimately transferred to and sequestered in the oceans (France-Lanord & Derry 1994, Hodell 1994) at the same that ^{13}C-enriched diatoms continued to expand. These new pathways are responsible for the $\delta^{13}C_{org}$ increase since the mid-Cenozoic, and contributed to the $\delta^{13}C_{carb}$ decrease that began in the mid-Miocene (Figure 7). The abrupt $\delta^{13}C_{org}$ increase occurred without a large change in either the atmospheric oxidation state or an injection of ^{12}C from mantle outgassing, and appears to be a unique event in Earth's history.

Whereas biological fractionation of carbon isotopes by phytoplankton is a major component of the carbon cycle, fractionation of sulfur isotopes occurs primarily through bacterial reduction of sulfate to sulfide. The carbon and sulfur cycles are linked because bacterial sulfate reduction depends on high levels of sedimentary organic matter. Sulfate reduction can result in pyrite burial, which drives $\delta^{34}S$ of marine sulfate higher. For atmospheric oxygen levels to remain stable, the carbon and sulfur cycles must be counterbalanced so that as one reduced reservoir grows, the other shrinks (e.g., more organic carbon and less pyrite). In the simplest scenario, this hypothetical relationship indicates that intervals of high $\delta^{13}C_{carb}$ should coincide with intervals of low $\delta^{34}S_{sulfate}$. In reality, this relationship is complicated by other factors, and redox conditions have fluctuated during times when the two cycles were not counterbalanced, especially on short timescales (e.g., Kump 1993, Payton et al. 1998, Strauss 1999). This situation appears to have been the case not only for some brief intervals, but also for the Jurassic to the Neogene, when the long-term trends in both isotope records show increasing values. These trends indicate that there were increases in the sedimentary reservoirs of reduced carbon (= organic matter) and reduced sulfur (= pyrite), which in turn requires corresponding increases in the oxidized species of both carbon and sulfur. This relationship supports the overall increase in the oxidation state of Earth's surface over this long time period (Katz et al. 2004) (Figure 7).

Primary Productivity and Geochemical Proxy Records

Changes in phytoplankton taxonomic composition can influence the carbon isotope record on shorter timescales as well. Extended intervals of elevated global $\delta^{13}C_{carb}$ values are superimposed on the long-term $\delta^{13}C$ increase from the Jurassic to the mid-Miocene (Figure 7). These intervals typically have been attributed to increases in organic carbon burial relative to carbonate burial that resulted from changes in a combination of surface ocean productivity and/or preservation on the seafloor (e.g., Miller & Fairbanks 1985, Scholle & Arthur 1980, Vincent & Berger 1985). Burial of isotopically light organic carbon leaves the remaining mobile carbon reservoir isotopically heavier, which drives $\delta^{13}C_{carb}$ higher. Comparisons among the records of $\delta^{13}C_{carb}$, $^{87}Sr/^{86}Sr$, and phosphorous flux may provide further insight into these relationships (Figure 7).

Phosphorous is essential for all cells; this element is required for the synthesis of nucleic acids, metabolism of carbohydrates, and formation of membrane lipids. It is supplied only through continental erosion and riverine delivery to the oceans, where it is either used quickly or authigenically precipitated. Phosphorous can be recycled in the upper ocean as organic matter is oxidized and phosphorous is returned to the surface. Because primary production can be limited by phosphorous availability, we compare proxies for productivity ($\delta^{13}C_{carb}$) (Katz et al. 2004) with phosphorous flux (Follmi 1995) (Figure 7). Episodes of elevated $\delta^{13}C_{carb}$ tend to be accompanied by higher phosphorous fluxes prior to the mid-Miocene, although the inverse is not true (this difference may in part reflect the global [$\delta^{13}C_{carb}$] versus regional [phosphorous flux] nature of the records). The phosphorous flux and $\delta^{13}C_{carb}$ curves decouple most notably during two time intervals: (*a*) in the Late Cretaceous, when widespread deposition of chalks occurred in shallow waters, where organic carbon must have been oxidized, and (*b*) in the late Cenozoic, when widespread glaciation increased erosional rates and phosphorous supply to the oceans. These comparisons indicate that phosphorous is not always a limiting nutrient on geological timescales.

Strontium isotopes may provide another clue to link biogeochemical records. In a simple two-source system (Caldeira 1992), strontium is delivered to the oceans through hydrothermal exchange at midocean ridges (low $^{87}Sr/^{86}Sr$ values) and continental erosion (high $^{87}Sr/^{86}Sr$ values); however, the $^{87}Sr/^{86}Sr$ may be complicated by changes in dominant continental source rock type (e.g., Ravizza 1993) and variable riverine fluxes (e.g., Lear et al. 2003). Over the past 200 myr, four out of five decreases in $^{87}Sr/^{86}Sr$ values correspond to major episodes of elevated $\delta^{13}C_{carb}$ values (Figure 7). The fifth decrease occurs in the Turonian across a data gap in our $\delta^{13}C_{carb}$ record, but may correlate to a smaller, shorter interval of elevated $\delta^{13}C_{carb}$ documented in published Tethyan records (Jenkyns et al. 1994, Stoll & Schrag 2000). The causal mechanisms behind these correlations are unclear, but may be related to higher pCO_2 from increased hydrothermal activity (e.g., Berner 1993) that accelerated the geological and biological components of the global carbon cycle.

SUMMARY

Cyclic tectonic changes superimposed on key secular changes in Earth's atmosphere, oceans, and even on land have selected for certain phytoplankton clades through time, with an ever-increasing tempo of phytoplankton evolution. In this paper, we have focused on the evolutionary paths that eventually led to the eukaryotic phytoplankton that dominate the contemporary oceans—coccolithophores, diatoms, and dinoflagellates. The earliest primary producers were prokaryotes. For much of the Archaen and Proterozoic Eons, the oceans were dominated by cyanobacteria, with green and perhaps other algae increasing in importance toward the end of the Precambrian. Planktonic algae radiated in the Early Phanerozoic oceans, cyst-forming dinoflagellates and calcareous nannoplankton dominated in the Mesozoic oceans, and, finally, the diatoms rose to prominence in the latter half of the Cenozoic. The primary producers have always been at the base of the food web; hence, the evolution of organisms at higher trophic levels has depended on the evolutionary trajectories of the phytoplankton. The number of invertebrate species relative to phytoplankton species has increased through the Phanerozoic.

This evolutionary succession of marine phytoplankton was a response to a complex system that cannot be explained by a set of ordinary differential equations. Secular shifts in redox seawater chemistry have influenced phytoplankton evolutionary trajectories, both by altering the trace metal availability in the oceans and by changing the balance of fixed nitrogen between the oxidized form (nitrate) and the reduced form (ammonium). In general, the more reducing conditions of the early oceans favored the green lineage, while the higher oxidation states of the later oceans favored the red lineage. Early eukaroytic phytoplankton were best able to compete in the better-oxygenated coastal regions, while green phytoflagellates thrived in open ocean surface waters, where seawater remained Fe-rich and relatively Zn- and Cd-poor (Anbar & Knoll 2002, Whitfield 2001b). Diversity increases in phytoplankton appear to correlate with continental rifting of Rodinia (Late Proterozoic), Pannotia (Early Paleozoic), and Pangea (Jurassic), which ultimately resulted in the three groups of eukaryotic phytoplankton that dominate the modern ocean: coccolithophores, diatoms, and dinoflagellates.

While changing environmental conditions selected for different plankton groups through Earth history, phytoplankton, in turn, influenced biogeochemical components of the environment, often through the biologically mediated oxidation and reduction processes that alter mobile elemental reservoirs. The best example of this process is the role that plankton played in oxygenating Earth's atmosphere and oceans through the evolution of oxygenic photosynthesis. The subsequent rise of the eukaryotic phytoplankton since the Early Jurassic, coupled with the opening of the Atlantic Ocean basin during the current Wilson Cycle, has increased the efficiency of organic carbon burial and contributed to a gradual depletion of CO_2 from the oceans and atmosphere, with a simultaneous increase in the oxidation state of Earth's surface. Ultimately, this change favors the red lineage.

ACKNOWLEDGMENTS

We thank Ken Miller, Oscar Schofield, and Scott Wing for their comments on this manuscript, Ben Cramer for his assistance with statistical analyses, and Richard Bambach for discussions and for providing his revised version of Jack Sepkoski's marine invertebrate database. This study was supported by NSF OCE 00,84032 Biocomplexity: The Evolution and the Radiation of Eukaryotic Phytoplankton Taxa (EREUPT).

The *Annual Review of Ecology, Evolution, and Systematics* is online at
http://ecolsys.annualreviews.org

LITERATURE CITED

Alexandre A, Meunier J-D, Colin F, Koud J-M. 1997. Plant impact on the biogeochemical cycle of silicon and related weathering processes. *Geochim. Cosmochem. Acta* 61:677–82

Alvarez LW, Alvarez W, Asaro F, Michel HV. 1980. Extraterrestrial cause for the Cretaceous-Tertiary extinction. *Science* 208:1095–108

Anbar AD, Knoll AH. 2002. Proterozoic ocean chemistry and evolution: A bioinorganic bridge? *Science* 297:1137–42

Arthur M, Sageman B. 1994. Marine black shales: depositional mechanisms and environments of ancient deposits. *Annu. Rev. Earth Planet. Sci.* 22:499–51

Arthur MA, Dean WE, Stow DAV. 1984. Models for the deposition of Mesozoic-Cenozoic fine-grained organic-C rich sediment in the deep sea. In *Fine-grained Sediments: Deepwater Processes and Facies*, ed. DAV Stow, DJW Piper, pp. 527–59. London: Geological Society of London Special Publication

Baldauf SL. 2003. The deep roots of eukaryotes. *Science* 300:1703–6

Baldauf SL, Roger AJ, Wenk-Siefert I, Doolittle WF. 2000. A kingdom-level phylogeny of eukaryotes based on combined protein data. *Science* 290:972–77

Bambach RK. 1977. Species richness in marine benthic habitats through the Phanerozoic. *Paleobiology* 3:152–67

Bambach RK. 1993. Seafood through time: changes in biomass, energetics, and productivity in the marine ecosystem. *Paleobiology* 19:372–97

Bambach RK. 1999. Energetics in the global marine fauna: A connection between terrestrial diversification and change in the marine Biosphere. *Geobios* 32:131–44

Bambach RK, Knoll AH, Sepkoski JJ. 2002. Anatomical and ecological constraints on Phanerozoic animal diversity in the marine realm. *Proc. Natl. Acad. Sci. USA* 99:6854–59

Banse K. 1992. Grazing, temporal changes of phytoplankton concentrations, and the microbial loop in the open sea. In *Primary Productivity and Biogeochemical Cycles in the Sea*, ed. PG Falkowski, pp. 409–40. New York: Plenum

Banse K. 1964. On the verticle distribution of zooplankton in the sea. *Prog. Oceanogr.* 2:53–125

Barron JA. 2003. Planktonic marine diatom record of the past 18 m.y.: appearances and extinctions in the Pacific and Southern Oceans. *Diatom Res.* 18:203–24

Bartoli F. 1983. The biogeochemical cycle of silicon in two temperate forest ecosystems. *Ecol. Bull.* 35:469–76

Bekker A, Holland HD, Wang P-L, Rumble III D, Stein HJ, et al. 2004. Dating the rise of atmospheric oxygen. *Nature* 427:117–20

Berggren WA, Kent DV, Swisher CC, Aubry M-P. 1995. A revised Cenozoic geochronology and chronostratigraphy. In *Geochronology, Time Scales and Global Stratigraphic Correlations: A Unified Temporal Framework for an Historical Geology*, Special Publ. No. 54, ed. WA Berggren, DV Kent, J Hardenbol, pp. 129–212. Tulsa, OK: SEPM (Society for Sedimentary Geology)

Berner RA. 1993. Paleozoic atmospheric CO_2: Importance of ocean radiation and plant evolution. *Science* 261:68–70

Bhattacharya D, Medlin L. 1998. Algal phylogeny and the origin of land plants. *Plant Physiol.* 116:9–15

Bidle KD, Azam F. 1999. Accelerated dissolution of diatom silica by marine bacterial assemblages. *Nature* 397:508–12

Bown PR. 1998. *Calcareous Nannofossil Biostratigraphy*. pp. 1–315. London: Kluwer Academic Publishers

Bown PR, Lees JA, Young JR. 2004. Calcareous nannoplankton diversity and evolution through time. In *Coccolithophores— from Molecular Processes to Global Impact*, ed. H Thierstein, JR Young, pp. 481–507, Berlin: Springer-Verlag

Bralower TJ. 1999. The record of global change in mid-Cretaceous (Barremian-Albian) sections from the Sierra Madre, northeast Mexico. *J. Foram. Res.* 29:418–37

Brasier MD, Green OR, Jephcoat AP, Kleppe AK, Van Kranendonk MJ, et al. 2002. Questioning the evidence for Earth's oldest fossils. *Nature* 416:76–81

Brocks JJ, Buick R, Summons RE, Logan GA. 2003. A reconstruction of Archean biological diversity based on molecular fossils from the 2.78 to 2.45 billion-year-old Mount Bruce Supergroup, Hamersley Basin, Western Australia. *Geochim. Cosmochim. Acta* 67:4321–35

Brocks JJ, Logan GA, Buick R, Summons RE. 1999. Archean molecular fossils and the early rise of eukaryotes. *Science* 285:1033–36

Bujak JP, Williams GL. 1979. Dinoflagellate diversity through time. *Mar. Micropal.* 4:1–12

Bukry D. 1971. Discoaster evolutionary trends. *Micropaleontology* 17:43–52

Butterfield NJ. 1997. Plankton ecology and the Proterozoic-Phanerozoic transition. *Paleobiology* 23:247–62

Butterfield NJ. 2000. Bangiomorpha pubescens n. gen., n. sp.: implications for the evolution of sex, multicellularity, and the Mesoproterozoic/Neoproterozoic radiation of eukaryotes. *Paleobiology* 26:386–404

Butterfield NJ. 2002. *A Vaucheria-like fossil from the Neoproterozoic of Spitsbergen*. Presented at Geol. Soc. Am. Abstracts with Programs, Denver, CO

Caldeira K. 1992. Enhanced Cenozoic chemical weathering and the subduction of pelagic carbonate. *Nature* 357:578–81

Canfield DE. 1998. A new model for Proterozoic ocean chemistry. *Nature* 396:450–53

Cavalier-Smith T. 2003. Genomic reduction and evolution of novel genetic membranes and protein-targeting machinery in eukaryote-eukaryote chimaeras (meta-algae). *Philos. Trans. R. Soc. London Ser. B* 358:109–34

Cerling TE, Harris JM, MacFadden BJ, Leakey MG, Quade J, et al. 1997. Global vegetation change through the Miocene/Pliocene boundary. *Nature* 389:153–58

Chacon-Baca E, Beraldi-Campesi H, Cevallos-Ferriz SRS, Knoll AH, Golubic S. 2002. 70 Ma nonmarine diatoms from northern Mexico. *Geology* 30:279–81

Claypool GE, Lubeck CM, Bayeinger JP, Ging TG. 1977. Organic Geochemistry. In *Geological Studies on the COST No. B-2 Well, U.S. Mid-Atlantic Outer Continental Shelf Area*, ed. PA Scholle, pp. 46–59. Reston, VA: United States Geological Survey

Colbath GK, Grenfell HR. 1995. Review of biological affinities of Paleozoic acid-resistant, organic-walled eukaryotic algal microfossils (including "acritarchs"). *Rev. Palaeobot. Palynol.* 86:287–314

Conley DJ. 2002. Terrestrial ecosystems and the global biogeochemical silica cycle. *Glob. Biogeochem. Cycles* 16:1121:doi10.029/2002GB001894

Damsté JSS, Muyzer G, Abbas B, Rampen SW, Massé G, et al. 2004. The rise of the Rhizosolenid diatoms. *Science* 304:584–87

De La Rocha CL, Brzezinski MA, DeNiro MJ. 2000. A first look at the distribution of the stable isotopes of silicon in natural waters. *Geochim. Cosmochim. Acta* 64:2467–77

Delwiche CF. 1999. Tracing the thread of plastid diversity through the tapestry of life. *Am. Nat.* 154:S164–S77

Dugdale R, Wilkerson F, Wilkerson. 1998. Silicate regulation of new production in the equatorial Pacific upwelling. *Nature* 391:270–73

Erbacher J, Thurow J. 1997. Influence of oceanic anoxic events on the evolution of mid-K radiolaria in the North Atlantic and western Tethys. *Mar. Micropaleontol.* 30: 139–58

Falkowski P, Laws EA, Barbar RT, Murray JW. 2003. Phytoplankton and their role in primary, new and export production. In *Ocean Biogeochemistry. The Role of the Ocean Carbon Cycle in Global Change*, ed. MJR Fasham, pp. 99–121. Berlin: Springer-Verlag

Falkowski PG, Katz ME, Knoll A, Quigg A, Raven JA, et al. 2004a. The evolution of modern eukaryotic phytoplankton. *Science* 305:354–60

Falkowski PG, Raven JA. 1997. *Aquatic Photosynthesis*. Malden, MA: Blackwell Sci.. 375 pp.

Falkowski PG, Schofield O, Katz ME, van de Schootbrugge B, Knoll A. 2004b. Why Is the Land Green and the Ocean Red? In *Coccolithophores—from Molecular Processes to Global Impact*, ed. H Thierstein, JR Young, pp. 429–53. Amsterdam: Elsevier

Fast NM, Kissinger JC, Roos DS, Keeling PJ. 2001. Nuclear-encoded, plastid-targeted genes suggest a single common origin for apicomplexan and dinoflagellates plastids. *Mol. Biol. Evol.* 18:418–26

Fensome RA, MacRae RA, Moldowan JM, Taylor FJR, Williams GL. 1996. The early Mesozoic radiation of dinoflagellates. *Paleobiology* 22:329–38

Field C, Behrenfeld M, Randerson J, Falkowski P. 1998. Primary production of the biosphere: integrating terrestrial and oceanic components. *Science* 281:237–40

Fischer AG, ed. 1984. Catastrophes and Earth History. In *The Two Phanerozoic Supercycles*, ed. WA Bergren, JA van Couvering, pp. 129–50 Princeton, NJ: Princeton Univ. Press

Follmi KB. 1995. 160 m.y. record of marine sedimentary phosphorus burial: coupling of climate and continental weathering under greenhouse and icehouse conditions. *Geology* 23:859–62

France-Lanord C, Derry LA. 1994. $\delta^{13}C$ of organic carbon in the Bengal Fan: source evolution and transport of C_3 and C_4 plant carbon to marine sediments. *Geochim. Cosmochim. Acta* 58:4809–14

German TN. 1990. *Organic World One Billion Years Ago*. Leningrad: Nauka.

Gersonde R, Harwood DM. 1990. Lower Cretaceous diatoms from ODP Leg 113 Site 693 (Weddell Sea). Part 1: vegetative cells. In *Proceedings of the Ocean Drilling Program, Scientific Results*, ed. PF Barker, JP Kennett, SB O'Connell, pp. 365–402. College Station, TX: Ocean Drilling Program

Ghil M, Allen MR, Dettinger MD, Ide K, Kondrashov D, et al. 2002. Advanced spectral methods for climatic time series. *Rev. Geophys.* 40:10.1029/2000RG000092

Gradstein FM, Agterberg FP, Ogg JG, Hardenbol H, van Veen P, et al. 1995. A Triassic, Jurassic, and Cretaceous time scale. In *Geochronology, Time Scales and Global Stratigraphic Correlations: A Unified Temporal Framework for an Historical Geology*, Special Publ. No. 54, ed. WA Berggren, DV Kent, J Hardenbol, pp. 95–126. Tulsa, OK: SEPM (Society for Sedimentary Geology)

Grantham PJ, Wakefield LL. 1988. Variations in the sterane carbon number distributions of marine source rock derived crude oils through geological time. *Org. Geochem.* 12: 61–73

Gray MW, Lang BF, Cedergreen R, Golding GB, Lemieux C, et al. 1998. Genome structure and gene content in protist mitochondrial DNAs. *Nucleic Acids Res.* 26:865–78

Grover JP. 1988. Dynamics of competition in a

variable environment: experiments with two diatom species. *Ecology* 69:408–17

Grover JP. 1991. Resource competition in a variable environment: phytoplankton growing according to the variable-internal-stores model. *Am. Nat.* 138:811–35

Grzebyk D, Schofield O, Vetriani C, Falkowski PG. 2003. The Mesozoic radiation of eukaryotic algae: The portable plastid hypothesis. *J. Phycol.* 39:259–67

Hamm CE, Merkel R, Springer O, Jurkojc P, Maier C, et al. 2003. Architecture and material properties of diatom shells provide effective mechanical protection. *Nature* 421:841–43

Han TM, Runnegar B. 1992. Megascopic eukaryotic algae from the 2.1-billion-year-old Negaunee iron-formation, Michigan. *Science* 257:232–35

Haq BU, Hardenbol J, Vail PR. 1987. Chronology of fluctuating sea levels since the Triassic (250 million years ago to present). *Science* 235:1156–67

Hardie LA. 1996. Secular variation in seawater chemistry: an explanation for the coupled secular variation in the mineralogies of marine limestones and potash evaporites over the past 600 m.y. *Geology* 24:279–83

Harper JT, Keeling PJ. 2003. Nucleus-encoded, plastid-targeted glyceraldehyde-3-phosphate dehydrogenase (GAPDH) indicates a single origin for chromalveolate plastids. *Mol. Biol. Evol.* 20:1730–35

Harwood DM, Gersonde R. 1990. Lower Cretaceous diatoms from ODP Leg 113 Site 693 (Weddell Sea). Part 2: Resting spores, Chrysophycean cysts, and endoskeletal dinoflagellate, and notes on the origin of diatoms. In *Proc. Ocean Drilling Program*, ed. PF Barber, JP Kennett, et al., 403–25. College Station, TX: Ocean Drilling Program

Harwood DM, Nikolaev VA. 1995. Cretaceous diatoms: morphology, taxonomy, biostratigraphy. In *Siliceous Microfossils*, ed. CD Blome, PM Whalen, R Katherine, pp. 81–106. Lawrence, KS: Paleontological Society

Hayes JM, Strauss H, Kaufman AJ. 1999. The abundance of ^{13}C in marine organic matter and isotopic fractionation in the global biogeochemical cycle of carbon during the past 800 Ma. *Chem. Geol.* 161:103–25

Hess WR, Rocap G, Ting CS, Larimer F, Stilwagen S, et al. 2001. The photosynthetic apparatus of *Prochlorococcus*: Insights through comparative genomics. *Photosynth. Res.* 70:53–71

Hodell DA. 1994. Magnetostratigraphic, biostratigraphic, and stable isotope stratigraphy of an Upper Miocene drill core from the Sale Briqueterie (northwestern Morocco): a high-resolution chronology for the Messinian stage. *Paleoceanography* 9:835–55

Houghton SD. 1991. Calcareous nannofossils. In *Calcareous Algae and Stromatolites*, ed. R Riding, pp. 217–66. Berlin: Springer-Verlag

Howarth RJ, McArthur JM. 1997. Statistics for strontium isotope stratigraphy: A robust LOWESS fit to the marine Sr-isotope curve for 0 to 206 Ma, with look-up table for drivaiton of numeric age. *J. Geol.* 105:441–56

Isozaki Y. 1997. Permo-Triassic boundary superanoxia and stratified superocean: records from lost deep sea. *Science* 276:235–38

Jacobs BF, Kingston JD, Jacobs LL. 1999. The origin of grass-dominated ecosystems. *Ann. Mo. Bot. Gard.* 86:590–643

Janis CM, Damuth J, 2000. Mammals. In *Evolutionary Trends*, ed. KJ McNamara, pp. 301–45. London: J. Belknap.

Janofske D. 1992. Calcareous Nannofossils of the Alpine Upper Triassic. In *Nannoplankton Research*, ed. B Hamrsmîd, JR Young, 1:87–109. Prague: Knihovnicka ZPZ

Javaux EJ, Knoll AH, Walter MR. 2001. Morphological and ecological complexity in early eukaryotic ecosystems. *Nature* 412:66–69

Jeffrey SW, Mantoura RFC, Wright SW. 1997. *Phytoplankton Pigments in Oceanography: Guidelines to Modern Methods*. Paris: UNESCO Publishing. 661 pp.

Jenkyns HC, Clayton CJ. 1997. Lower Jurassic epicontinental carbonates and mudstones from England and Wales: chemostratigraphic signals and the early Toarcian anoxic event. *Sedimentology* 44:687–706

Jenkyns HC, Gale AS, Corfield RM. 1994. Carbon- & oxygen-isotope stratigraphy of the English chalk & Italian Scaglia & its palaeoclimatic significance. *Geol. Mag.* 131:1–34

Kamykowski D. 1981. Laboratory experiments on the diurnal verticle migration of marine dinoflagellates through temperature gradient. *Mar. Biol.* 62:57–64

Katz ME, Wright JD, Miller KG, Cramer BS, Fennel K, Falkowski PG. 2004. Biological overprint of the geological carbon cycle. *Marine Geol.* In press

Kellogg EA. 2000. The grasses: a case study in macroevolution. *Annu. Rev. Ecol. Syst.* 31:217–38

Knoll AH. 1989. Evolution and extinction in the marine realm: some constraints imposed by phytoplankton. *Phil. Trans. R. Soc. London Ser. B* 325:279–90

Knoll AH. 1992. The early evolution of eukaryotes: A geological perspective. *Science* 256:622–27

Knoll AH. 1994. Proterozoic and Early Cambrian protists: evidence for accelerating evolutionary tempo. *Proc. Natl. Acad. Sci. USA* 91:6743–50

Knoll AH. 2003. *Life on a Young Planet: The First Three Billion Years of Evolution on Earth*. Princeton, NJ: Princeton Univ. Press, 277 pp.

Kump LR, ed. 1993. *The Coupling of the Carbon and Sulfur Biogeochemical Cycles Over Phanerozoic Time, Vol. 14*. Berlin: Springer-Verlag

Laws EA, Bidigare R, Popp BN. 1997. Effect of growth rate and CO_2 concentration on carbon isotopic fractionation by the marine diatom *Phaeodactylum tricornutum* Limnol. *Oceanogr.* 42:1552–60

Lear CH, Elderfield H, Wilson PA. 2003. A Cenozoic seawater Sr/Ca record from benthic foraminiferal calcite and its application in determining global weathering fluxes. *Earth Planet Sci. Lett.* 208:69–84

Leckie RM, Bralower TJ, Cashman R. 2002. Oceanic anoxic events and plankton evolution: Biotic response to tectonic forcing during the mid-Cretaceous. *Paleoceanography* 17:1–29

Li WKW. 2002. Macroecological patterns of phytoplankton in the northwestern North Atlantic Ocean. *Nature* 419:154–57

Lipps JH, ed. 1993. *Fossil Prokaryotes and Protists*. Oxford: Blackwell. 342 pp.

MacArthur RH, Wilson EO. 1967. *The Theory of Island Biogeography*. 203 pp. Princeton, NJ: Princeton University

Maldonado M, Carmona MC, Uriz MJ, Cruzado A. 1999. Decline in Mesozoic reef-building sponges explained by silicon limitation. *Nature* 401:785–88

Margalef R. 1994. Dynamic aspects of diversity. *J. Veg. Sci.* 5:451–56

Martin RE. 2003. The fossil record of biodiversity: nutrients, productivity, habitat area and differential preservation. *Lethaia* 36:179–93

McFadden GI. 1999. Plastids and protein targeting. *J. Eukaryot. Microbiol.* 46:339–46

Medlin LK, Kooistra WCHF, Schmid A-MM. 2000. A review of the evolution of the diatoms—a total approach using molecules, morphology and geology. In *The Origin and Early Evolution of the Diatoms: Fossil, Molecular and Biogeographical Approaches*, ed. A Witkowski, J Sieminska, pp. 13–35. Krakow: Polish Acad. Sci.

Medlin LK, Kooistra WHCF, Potter D, Saunders GW, Andersen RA. 1997. Phylogenetic relationships of the 'golden algae' (haptophytes, heterokont chromophytes) and their plastids. *Plant Syst. Evol.* 11(Suppl.):187–219

Melkonian M, Surek B. 1995. Phylogeny of the Chlorophyta: Congruence between ultrastructural and molecular evidence. *Bull. Soc. Zool. Fr.* 120:191–208

Miller KG, Fairbanks RG. 1985. Oligocene to Miocene carbon isotope cycles and abyssal circulation changes. In *The Carbon Cycle and Atmospheric CO_2: Natural Variations Archean to Present*, ed. ET Sundquist, WS Broecker, pp. 469–86. Washington, DC: American Geophysical Union

Moldowan JM, Dahl J, Jacobson SR, Huizinga

BJ, Fago FJ, et al. 1996. Chemostratigraphic reconstruction of biofacies: molecular evidence linking cyst-forming dinoflagellates with pre-Triassic ancestors. *Geology* 24:159–62

Moldowan JM, Jacobson SR. 2000. Chemical signals for early evolution of major taxa: biosignatures and taxon-specific biomarkers. *Int. Geol. Rev.* 42:805–12

Moldowan JM, Talyzina NM. 1998. Biogeochemical evidence for dinoflagellate ancestors in the Early Cambrian. *Science* 281:1168–70

Molyneux SG, Le Hérissé A, Wicander R. 1996. Paleozoic phytoplankton. In *Palynology: Principles and Applications*, ed. J Jansonius, DC McGregor, pp. 493–530. Amer. Assoc. Strat. Palynol. Found.

Morris I. 1980. Paths of carbon assimilation in marine phytoplankton. In *Primary Productivity in the Sea*, ed. PG Falkowski, pp. 139–59. New York: Plenum

Müller KM, Oliveira MC, Sheath RG, Bhattacharya D. 2001. Ribosomal DNA phylogeny of the Bangiophycidae (Rhodophyta) and the origin of secondary plastids. *Am. J. Bot.* 88:1390–400

Oliveira MC, Bhattacharya D. 2000. Phylogeny of the Bangiophycidae (Rhodophyta) and the secondary endosymbiotic origin of algal plastids. *Am. J. Bot.* 87:482–92

Osyczka A, Moser CC, Daldal F, Dutton PL. 2004. Reversible redox energy coupling in electron transfer chains. *Nature* 427:607–12

Paerl HW, Valdes LM, Pinckney JL, Piehler MF, Dyble J, Moisander PH. 2003. Phytoplankton photopigments as indicators of estuarine and coastal eutrophication. *BioScience* 53:953–64

Palmer JD. 2003. The symbiotic birth and spread of plastids: How many times and whodunit? *J. Phycol.* 39:4–11

Payton A, Kastner M, Campbell D, Thiemens MH. 1998. Sulfur isotopic composition of Cenozoic seawater sulfate. *Science* 282:1459–62

Payton A, Kastner M, Campbell D, Thiemens MH. 2004. Seawater sulfur isotope fluctuations in the Cretaceous. *Science* 304:1663–65

Popp BN, Kenig F, Wakeham SG, Laws EA, Bidigare RR. 1998. Does growth rate affect ketone unsaturation and intracellular carbon isotopic variability in *Emiliania huxleyi*? *Paleoceanography* 13:35–41

Prauss M. 2000. The oceanographic and climatic interpretation of marine palynomorph phytoplankton distribution from Mesozoic, Cenozoic and Recent sections. *Habilitationsschrift, Gött. Arb. Geol. Paläontol.* 76:1–235

Quigg A, Finkel ZV, Irwin AJ, Rosenthal Y, Ho T-Y, et al. 2003. Plastid inheritance of elemental stoichiometry in phytoplankton and its imprint on the geological record. *Nature* 425:291–94

Rapp GJ, Mulholland SC. 1992. *Phytolith Systematics: Emerging Issues*. New York: Plenum

Rau GH, Riebesell U, Wolf-Gladrow DA. 1997. CO_{2aq}-dependent photosynthetic $_{13}C$ fractionation in the ocean: a model versus measurements. *Global Biogeochem. Cycles* 11:267

Raven JA. 1997. The vacuole: A cost-benefit analysis. *Adv. Bot. Res. Adv. Plant Pathol.* 25:59–86

Ravizza G. 1993. Variations of the $^{187}Os/^{186}Os$ ratio of seawater over the past 28 million years as inferred from metalliferous carbonates. *Earth Planet. Sci. Lett.* 118:335–48

Reinfelder JR, Kraepeil AML, Morel FMM. 2000. Unicellular C_4 photosynthesis in a marine diatom. *Nature* 407:996–99

Retallack GJ. 1997. Cenozoic expansion of grasslands and climatic cooling. *J. Geol.* 109:407–26

Retallack GJ. 2001. Neogene expansion of the North American prairie. *Palaios* 12:380–90

Rich JE, Johnson GL, Jones JE, Campsie J. 1986. A significant correlation between fluctuations in seafloor spreading rates and evolutionary pulsations. *Paleoceanography* 1:85–95

Ronov AB. 1994. Phanerozoic transgressions and regressions on the continents: a quantitative approach based on areas flooded by the sea and areas of marine and continental deposition. *Am. J. Sci.* 294:777–801

Rosenweig ML. 1995. *Species Diversity in Space and Time.* Cambridge: Cambridge Univ. Press

Rosing MT. 1999. C-13-depleted carbon microparticles in >3700-Ma sea-floor sedimentary rocks from west Greenland. *Science* 283:674–76

Roth PH. 1987. Mesozoic calcareous nannofossil evolution: relation to paleoceanographic events. *Paleoceanography* 2:601–11

Rothpletz A. 1896. Uber die Flysh-Fucoiden und einzige andere fossile Algen, sowie uber Liasische Diatomeen fuhrende Hornschwamme. *Deutsch. Geol. Ges.* 48:858–914

Sandberg PA. 1975. New interpretations of Great Salt Lake ooids and of ancient nonskeletal carbonate mineralogy. *Sedimentology* 22:497–538

Scholle PA, Arthur MA. 1980. Carbon isotope fluctuations in Cretaceous pelagic limestones: potential stratigraphic and petroleum explorations tool. *Bull. Am. Assoc. Petroleum Geologists:* 64:67–87

Schopf JW. 1993. Microfossils of the Early Archean Apex chert: New evidence of the antiquity of life. *Science* 260:640–46

Schopf JW. 2002. *Life's Origin: The Beginnings of Biological Organization.* Berkeley: Univ. Calif. Press. 208 pp.

Sepkowski J. 1997. Biodiversity: past, present, and future. *J. Paleont.* 71:533–39

Shen Y, Canfield DE, Knoll AH. 2002. The chemistry of mid-Proterozoic oceans: evidence from the McArthur Basin, northern Australia. *Am. J. Sci.* 302:81–109

Shen Y, Knoll AH, Walter MR. 2003. Evidence for low sulphate and anoxia in a mid-Proterozoic marine basin. *Nature* 423:632–35

Sibley DF, Vogel TA. 1976. Chemical Mass Balance of the Earth's crust: the calcium dilemma (?) and the role of Pelagic sediments. *Science* 551–53

Siegel DA. 1998. Resource competition in a discrete environment: Why are plankton distributions paradoxical? *Limnol. Oceanogr.* 43: 1133–46

Smetacek V. 1999. Diatoms and the ocean carbon cycle. *Protist* 150:25–32

Smetacek V. 2001. A watery arms race. *Nature* 411:745

Southam JR, Hay WW. 1981. Global sedimentary mass balance and sea level changes. In *The Sea,* ed. C Emiliani, pp. 1617–84. New York: Wiley-Interscience

Spencer-Cervato C. 1999. The Cenozoic deep sea microfossil record: explorations of the DSDP/ODP sample set using the Neptune database. *Palaeontologia Electronica* 2:1–270

Stanley GD Jr., Hardie LA. 1998. Secular oscillations in the carbonate mineralogy of reef-building and sediment-producing organisms driven by tectonically forced shifts in seawater chemistry. *Palaeogeogr. Palaeoclimatol. Palaeoecol.* 144:3–19

Still CJ, Berry JA, Collatz GJ, DeFries RS. 2003. Global distribution of C_3 and C_4 vegetation: carbon cycle implications. *Glob. Biogeochem. Cycles* 17:doi:1029/2001 GB001807

Stoll HM, Schrag DP. 2000. Coccolith Sr/Ca as a new indicator of coccolithophorid calcification and growth rate. *Geochem. Geophys. Geosystems* http://146.201.54.53/ publicationsfinal/articles/1999GC000015/fs 1999GC.html

Stover LE, Brinkhuis H, Damassa SP, de Verteuil L, Helby RJ, et al. 1996. Mesozoic-Tertiary dinoflagellates, acritarchs & prasinophytes. In *Palynology: Principles and Applications,* ed. J Jansonius, DC McGregor, pp. 641–750: Amer. Assoc. Strat. Palynol Found.

Strauss H. 1999. Geological evolution from isotope proxy signals—sulfur. *Chem. Geol.* 161:89–101

Strelnikova NI. 1991. Evolution of diatoms during the Cretaceous and Paleogene periods. Presented at the 10th Diatom Symposium, 1988, Joensuu, Finland

Summons R, Jahnke L, Hope J, Logan G. 1999. 2-Methylhopanoids as biomarkers for cyanobacterial oxygenic photosynthesis. *Nature* 400:55–57

Summons RE, Thomas J, Maxwell JR, Boreham CJ. 1992. Secular and environmental constraints on the occurrence of dinosterane in sediments. *Geochim. Cosmochim. Acta* 56: 2437–44

Summons RE, Walter MR. 1990. Molecular fossils and microfossils of prokaryotes and protists from Proterozoic sediments. *Am. J. Sci.* 290-A: 212–44

Talyzina NM, Moczydlowska M. 2000. Morphological and ultrastructural studies of some acritarchs from the Lower Cambrian Lukati Formation, Estonia. *Rev. Palaeobot. Palynol.* 112:1–21

Tappan H. 1980. *The Palaeobiology of Plant Protists*. San Francisco: WH Freeman, 1028 pp.

Taylor FJR. 1987. An overview of the status of evolutionary cell symbiosis theories. *Ann. NY Acad. Sci.* 503:1–16

Tilman D. 1977. Resource competition between planktonic algae: an experimental and theoretical approach. *Ecology* 58:338–48

Tomitani A, Okada K, Miyashita H, Matthijs HCP, Ohno T, Tanaka A. 1999. Chlorophyll *b* and phycobilins in the common ancestor of cyanobacteria and chloroplasts. *Nature* 400:159–62

Tozzi S, Schofield O, Falkowski PG. 2004. Historical climate change and ocean turbulence as selective agents for two key phytoplankton functional groups. *Mar. Ecol. Prog. Ser.* 274:123–32

Tréguer P, Nelson DM, van Bennekom AJ, DeMaster DJ, Leynaert A, Quéguiner B. 1995. The silica balance in the world ocean: a reestimate. *Science* 268:375–79

Twitchett R. 1999. Palaeoenvironments and faunal recovery after the end-Permian mass extinction. *Palaeogeogr. Palaeoclimatol. Palaeoecol.* 154:27–37

Vail PR, Mitchum RM, Thompson III S, eds. 1977. Seismic stratigraphy and global changes of sea level, part 4: Global cycles of relative changes of sea level. In *Seismic Stratigraphy and Global Changes of Sea Level*, Vol. 26, ed. CE Payton, pp. 83–97. Tulsa, OK: American Association of Petroleum Geologists Memoirs

Valentine JW, Moores EM. 1974. Plate tectonics and the history of life in the oceans. *Sci. Am.* 230:80–89

Van den Hoek C, Mann DG, Jahns HM. 1995. *Algae: An Introduction to Phycology*. Cambridge: Cambridge Univ. Press. 627 pp.

Villareal TA, Altabet MA, Culver-Rymsza K. 1993. Nitrogen transport by vertically migrating diatoms mats in the North Pacific Ocean. *Nature* 363:709–12

Vincent E, Berger WH. 1985. Carbon dioxide and polar cooling in the Miocene: The Monterey Hypothesis. In *The Carbon Cycle and Atmospheric CO_2: Natural Variations Archean to Present*, ed. ET Sundquist, WS Broecker, pp. 455–68. Washington, DC: American Geophysical Union

Whitfield M. 2001. Interactions between phytoplankton and trace metals in the ocean. *Adv. Mar. Biol.* 41:3–128

Wignall P, Twitchett R. 2002. Permian-Triassic sedimentology of Jameson Land, East Greenland: incised submarine channels in an anoxic basin. *J. Geol. Soc. London* 159:691–703

Wilkinson BH, Algeo TJ. 1989. Sedimentary carbonate record of calcium-magnesium cycling. *Am. J. Sci.* 289:1158–94

Wilson JT. 1966. Did the Atlantic close and then re-open? *Nature* 211:676–81

Worsley TR, Nance RD, Moody JB. 1986. Tectonic cycles and the history of the Earth's biogeochemical and paleoceanographic record. *Paleoceanography* 1:233–63

Xiao S, Knoll AH, Yuan X, Pueschel C. 2004. Phosphatized multicellular algae in the Neoproterozoic Doushantuo Formation, China, and the early evolution of florideophyte red algae. *Am. J. Bot.* 91:214–27

Yoon HS, Hackett JD, Bhattacharya D. 2002a. A single origin of the peridinin- and fucoxanthin-containing plastids in dinoflagellates through tertiary endosymbiosis. *Proc. Natl. Acad. Sci. USA* 99:11724–29

Yoon HS, Hackett JD, Pinto G, Bhattacharya D. 2002b. The single, ancient origin of chromist plastids. *Proc. Natl. Acad. Sci. USA* 99:15507–12

Yoon HS, Hackett JD, Ciniglia C, Pinto G, Bhattacharya D. 2004. A Molecular Timeline for the Origin of Photosynthetic Eukaryotes. *Mol. Biol. Evol* 21:809–18

Zhang Z. 1986. Clastic facies microfossils from the Chuanlingguo Formation near Jixian, north China. *J. Micropaleontol.* 5:9–16

REGIME SHIFTS, RESILIENCE, AND BIODIVERSITY IN ECOSYSTEM MANAGEMENT

Carl Folke,[1,2] Steve Carpenter,[2,3] Brian Walker,[4] Marten Scheffer,[5] Thomas Elmqvist,[1] Lance Gunderson,[6] and C.S. Holling[7]

[1]*Department of Systems Ecology, Stockholm University, SE-106 91 Stockholm, Sweden; email: calle@system.ecology.su.se; thomase@ecology.su.se*
[2]*Beijer International Institute of Ecological Economics, Royal Swedish Academy of Sciences, Stockholm, Sweden*
[3]*Center for Limnology, University of Wisconsin, Madison, Wisconsin 53706; email: srcarpen@wisc.edu*
[4]*Sustainable Ecosystems, CSIRO, Canberra, ACT, 2601, Australia; email: Brian.Walker@csiro.au*
[5]*Aquatic Ecology and Water Quality Management Group, Wageningen Agricultural University, Wageningen, The Netherlands; email: Marten.Scheffer@wur.nl*
[6]*Department of Environmental Studies, Emory University, Atlanta, Georgia 30322; email: lgunder@emory.edu*
[7]*16871 Sturgis Circle, Cedar Key, Florida 32625; email: holling@zoo.ufl.edu*

Key Words alternate states, regime shifts, response diversity, complex adaptive systems, ecosystem services

■ **Abstract** We review the evidence of regime shifts in terrestrial and aquatic environments in relation to resilience of complex adaptive ecosystems and the functional roles of biological diversity in this context. The evidence reveals that the likelihood of regime shifts may increase when humans reduce resilience by such actions as removing response diversity, removing whole functional groups of species, or removing whole trophic levels; impacting on ecosystems via emissions of waste and pollutants and climate change; and altering the magnitude, frequency, and duration of disturbance regimes. The combined and often synergistic effects of those pressures can make ecosystems more vulnerable to changes that previously could be absorbed. As a consequence, ecosystems may suddenly shift from desired to less desired states in their capacity to generate ecosystem services. Active adaptive management and governance of resilience will be required to sustain desired ecosystem states and transform degraded ecosystems into fundamentally new and more desirable configurations.

INTRODUCTION

Humanity strongly influences biogeochemical, hydrological, and ecological processes, from local to global scales. We currently face more variable environments with greater uncertainty about how ecosystems will respond to inevitable increases in levels of human use (Steffen et al. 2004). At the same time, we seem to challenge the capacity of desired ecosystem states to cope with events and disturbances (Jackson et al. 2001, Paine et al. 1998). The combination of these two trends calls for a change from the existing paradigm of command-and-control for stabilized "optimal" production to one based on managing resilience in uncertain environments to secure essential ecosystem services (Holling & Meffe 1996, Ludwig et al. 2001). The old way of thinking implicitly assumes a stable and infinitely resilient environment, a global steady state. The new perspective recognizes that resilience can be and has been eroded and that the self-repairing capacity of ecosystems should no longer be taken for granted (Folke 2003, Gunderson 2000). The challenge in this new situation is to actively strengthen the capacity of ecosystems to support social and economic development. It implies trying to sustain desirable pathways and ecosystem states in the face of continuous change (Folke et al. 2002, Gunderson & Holling 2002).

Holling (1973), in his seminal paper, defined ecosystem resilience as the magnitude of disturbance that a system can experience before it shifts into a different state (stability domain) with different controls on structure and function and distinguished ecosystem resilience from engineering resilience. Engineering resilience is a measure of the rate at which a system approaches steady state after a perturbation, that is, the speed of return to equilibrium, which is also measured as the inverse of return time. Holling (1996) pointed out that engineering resilience is a less appropriate measure in ecosystems that have multiple stable states or are driven toward multiple stable states by human activities (Nyström et al. 2000, Scheffer et al. 2001).

Here, we define resilience as the capacity of a system to absorb disturbance and reorganize while undergoing change so as to retain essentially the same function, structure, identity, and feedbacks (Walker et al. 2004). The ability for reorganization and renewal of a desired ecosystem state after disturbance and change will strongly depend on the influences from states and dynamics at scales above and below (Peterson et al. 1998). Such cross-scale aspects of resilience are captured in the notion of a panarchy, a set of dynamic systems nested across scales (Gunderson & Holling 2002). Hence, resilience reflects the degree to which a complex adaptive system is capable of self-organization (versus lack of organization or organization forced by external factors) and the degree to which the system can build and increase the capacity for learning and adaptation (Carpenter et al. 2001b, Levin 1999).

Several studies have illustrated that ecological systems and the services that they generate can be transformed by human action into less productive or otherwise less desired states. The existence of such regime shifts (or phase shifts) is an area of active research. Regime shifts imply shifts in ecosystem services and consequent impacts on human societies. The theoretical basis for regime shifts has been described by Beisner et al. (2003), Carpenter (2003), Ludwig et al. (1997), Scheffer & Carpenter (2003), and Scheffer et al. (2001).

Here, we review the evidence of regime shifts in terrestrial and aquatic ecosystems in relation to resilience and discuss its implications for the generation of ecosystem services and societal development. Regime shifts in ecosystems are increasingly common as a consequence of human activities that erode resilience, for example, through resource exploitation, pollution, land-use change, possible climatic impact and altered disturbance regimes. We also review the functional role of biological diversity in relation to regime shifts and ecosystem resilience. In particular, we focus on the role of biodiversity in the renewal and reorganization of ecosystems after disturbance—what has been referred to as the back-loop of the adaptive cycle of ecosystem development (Holling 1986). In this context, the insurance value of biodiversity becomes significant. It helps sustain desired states of dynamic ecosystem regimes in the face of uncertainty and surprise (Elmqvist et al. 2003). Strategies for transforming degraded ecosystems into new and improved configurations are also discussed.

REGIME SHIFTS AND DYNAMICS OF RESILIENCE IN ECOSYSTEMS

Ecosystems are complex, adaptive systems that are characterized by historical dependency, nonlinear dynamics, threshold effects, multiple basins of attraction, and limited predictability (Levin 1999). Increasing evidence suggests that ecosystems often do not respond to gradual change in a smooth way (Gunderson & Pritchard 2002). Threshold effects with regime shifts from one basin of attraction to another have been documented for a range of ecosystems (see Thresholds Database on the Web site www.resalliance.org). Passing a threshold marks a sudden change in feedbacks in the ecosystem, such that the trajectory of the system changes direction—toward a different attractor. In some cases, crossing the threshold brings about a sudden, sharp, and dramatic change in the responding state variables, for example, the shift from clear to turbid water in lake systems (Carpenter 2003). In other cases, although the dynamics of the system have "flipped" from one attractor to another, the transition in the state variables is more gradual, such as the change from a grassy to a shrub dominated rangeland (Walker & Meyers 2004). In Table 1, we provide examples of documented shifts between alternate states and expand on some of them in the text.

Temperate Lakes

Lake phosphorus cycles exhibit multiple regimes, each stabilized by a distinctive set of feedbacks. Generally, two regimes have attracted the most interest, although deeper analyses have revealed even greater dynamic complexities (Carpenter 2003,

TABLE 1 Documented shifts between states in different kinds of ecosystems

Ecosystem type	Alternate state 1	Alternate state 2	References
Freshwater systems			
Temperate lakes	Clear water	Turbid eutrophied water	Carpenter 2003
	Game fish abundant	Game fish absent	Post et al. 2002, Walters & Kitchell 2001, Carpenter 2003
Tropical lakes	Submerged vegetation	Floating plants	Scheffer et al. 2003
Shallow lakes	Benthic vegetation	Blue-green algae	Blindow et al. 1993, Scheffer et al. 1993, Scheffer 1997, Jackson 2003
Wetlands	Sawgrass communities	Cattail communities	Davis 1989, Gunderson 2001
	Salt marsh vegetation	Saline soils	Srivastava & Jefferies 1995
Marine systems			
Coral reefs	Hard coral dominance	Macroalgae dominance	Knowlton 1992, Done 1992, Hughes 1994, McCook 1999
	Hard coral dominance	Sea urchin barren	Glynn 1988, Eakin 1996
Kelp forests	Kelp dominance	Sea urchin dominance	Steneck et al. 2002, Konar & Estes 2003
	Sea urchin dominance	Crab dominance	Steneck et al. 2002
Shallow lagoons	Seagrass beds	Phytoplankton blooms	Gunderson 2001, Newman et al. 1998
Coastal seas	Submerged vegetation	Filamentous algae	Jansson & Jansson 2002, Worm et al. 1999
Benthic foodwebs	Rock lobster predation	Whelk predation	Barkai & McQuaid 1988
Ocean foodwebs	Fish stock abundant	Fish stock depleted	Steele 1998, Walters & Kitchell 2001, de Roos & Persson 2002

Forest systems			
Temperate forests	Spruce-fir dominance	Aspen-birch dominance	Holling 1978
	Pine dominance	Hardwood dominance	Peterson 2002
	Hardwood-hemlock	Aspen-birch	Frelich & Reich 1999
	Birch-spruce succession	Pine dominance	Danell et al. 2003
Tropical forests	Rain forest	Grassland	Trenbath et al. 1989
	Woodland	Grassland	Dublin et al. 1990
	Native crab consumers	Invasive ants	O'Dowd et al. 2003
Savanna and grassland			
Grassland	Perennial grasses	Desert	Wang & Eltahir 2000, Foley et al. 2003, van de Koppel et al. 1997
Savanna	Native vegetation	Invasive species	Vitousek et al. 1987
	Tall shrub, perennial grasses	Low shrub, bare soil	Bisigato & Bertiller 1997
	Grass dominated	Shrub dominated	Anderies et al. 2003, Brown et al. 1997
Arctic, sub-Arctic systems			
Steppe/tundra	Grass dominated	Moss dominated	Zimov et al. 1995
	Tundra	Boreal forest	Bonan et al. 1992, Higgins et al. 2002

Scheffer 1997). The two regimes of most concern to people who use the lakes are the clear-water and turbid-water regimes. In the clear-water regime, phosphorus inputs, phytoplankton biomass, and recycling of phosphorus from sediments are relatively low. In the turbid-water regime, these same variables are relatively high. The turbid-water regime provides lower ecosystem services because of abundant toxic cyanobacteria, anoxic events, and fish kills (Smith 1998).

In the clear-water regime of shallow lakes (lakes that do not stratify thermally), extensive beds of higher aquatic plants (macrophytes) stabilize sediments and reduce recycling of phosphorus to phytoplankton (Jeppesen et al. 1998, Scheffer 1997). Macrophyte beds may be lost because of shading by phytoplankton when high phosphorus inputs stimulate phytoplankton growth. Bottom-feeding (benthivorous) fishes that increase in abundance with nutrient enrichment damage the macrophytes and cause turbidity by resuspending sediment. Once the macrophytes are lost, sediments are more easily resuspended by waves, and rapid recycling of phosphorus from sediments maintains the turbid regime. Reversion to the clear-water state requires reduction of phosphorus inputs, but even if phosphorus inputs are reduced, the turbid state may remain resilient. With sufficiently low levels of phosphorus the ecosystem can possibly be perturbed to the clear-water state by temporarily removing fish, which allows macrophytes to recover, stabilizes the sediments, and reduces phosphorus cycling, thereby consolidating the clear-water state.

A different mechanism operates in deep (thermally stratified) lakes, although the clear-water and turbid-water states are similar (Carpenter 2003). Interactions of iron with oxygen are the key (Caraco et al. 1991, Nürnberg 1995). In the clear-water regime, rates of phytoplankton production, sedimentation, and oxygen consumption in deep water are low. Consequently the deep water remains oxygenated most of the time, and iron remains in the oxidized state, which binds phosphorus in insoluble forms. When phosphorus inputs are high, rates of phytoplankton production, sedimentation, and oxygen consumption increase. The deep water is anoxic part if not all of the time, iron is in the reduced state, and phosphorus dissolves into the water. Recycling of phosphorus from sediments makes the turbid-water state resilient. The regime can be shifted back to clear water by extreme reductions of phosphorus input or by various manipulations that decrease recycling of phosphorus (Carpenter et al. 1999, Cooke et al. 1993).

Similar mechanisms may have operated during massive oceanic events in the remote past. Episodes of large-scale phosphorus release in the oceans may represent a regime shift in which high phytoplankton production forms a strong positive feedback with phosphorus recycling from deep waters or sediments (Algeo & Scheckler 1998, Van Capellen & Ingall 1994).

Tropical Lakes

Experiments, field data, and models suggest that a situation with extensive free-floating plant cover and a state characterized by submerged plants tend to be alternate regimes (Scheffer et al. 2003). Dense mats of floating plants have an adverse effect on freshwater ecosystems because they create anoxic conditions, which strongly reduce animal biomass and diversity. Floating plants are superior competitors for light and carbon. Submerged plants are better competitors for nutrients and may prevent expansion of free-floating plants through a reduction of available nutrients in the water column. As a result, over a range of conditions, the lake can exist in either a floating-plant-dominated state or a submerged-plant-dominated state. Both states are resilient, but nutrient enrichment reduces the resilience of the submerged plant state. A single drastic harvest of floating plants can induce a permanent shift to an alternate state dominated by rooted submerged growth forms if the nutrient loading is not too high (Scheffer et al. 2003).

Wetlands, Estuaries, and Coastal Seas

In the Everglades, the freshwater marshes have shifted from wetlands dominated by sawgrass to cattail marshes because of nutrient enrichment. The soil phosphorous content defines the alternate states, and several types of disturbances (fires, drought, or freezes) can trigger a switch between these states (Gunderson 2001).

In Florida Bay, the system has flipped from a clear-water, seagrass-dominated state to one of murky water, algae blooms, and recurrently stirred-up sediments. Hypotheses that have been proposed to explain this shift include change in hurricane frequency, reduced freshwater flow entering the Bay, higher nutrient concentrations, removal of large grazers such as sea turtles and manatees, sea-level rise, and construction activities that restrict circulation in the Bay (Gunderson 2001).

The Baltic Sea is eutrophied and overfished (Elmgren 2001), and a shift has occurred in the coastal subsystem from submerged vegetation dominated by perennial fucoids to filamentous and foliose annual algae with lower levels of diversity (Kautsky et al. 1986). Jansson & Jansson (2002) propose that these conditions represent alternate states. Grazers on turf algae, such as gastropods, contribute to the maintenance of the fucoid-dominated state, but nutrient influx overrides grazing control and shifts the system into a less desired state (Worm et al. 1999). In addition to shading by filamentous algae (Berger et al. 2003), increased sedimentation caused by excessive phytoplankton production hinders recruitment and settlement of fucoids plants (Eriksson & Johansson 2003), which may lock the coastal subsystem into the undesired state.

Coral Reefs

The current shift of coral reefs into dominance by fleshy algae was long preceded by diminishing stocks of fishes and increased nutrient and sediment runoff from land (Jackson et al. 2001). The grazing of algae by fish species and other grazers contributes to the resilience of the hard coral dominated reef by, for example, keeping the substrate open for recolonization of coral larvae after disturbances such as hurricanes (Nyström et al. 2000).

In the Caribbean, overfishing of herbivores (dominated by fishes on reefs) led to expansion of sea urchin populations as the key grazers on invading algae. Thereby, the coral-dominated state was maintained, albeit at low resilience.

The high densities of the sea urchin populations may have contributed to their eventual demise when a disease outbreak spread throughout the Caribbean and reduced their numbers by two orders of magnitude, which precipitated the shift to the algal-dominated state that persists today (Hughes 1994, Knowlton 1992).

In other areas, high densities of grazing echinoids erode the reef matrix and, if unchecked, have the capacity to destroy reefs, as documented in the Galapagos Islands and elsewhere in the East Pacific (Eakin 1996, Glynn 1988). Loss of macrofauna, reduced fish stocks, replacement of herbivorous fishes by a single species of echinoid, overgrazing by food-limited sea urchins, high levels of erosion by echinoids, and reduced coral recruitment make coral reefs vulnerable to change and subject to regime shifts (Bellwood et al. 2004, Done 1992, Hughes et al. 2003). A significant ecological restructuring of reefs towards "weedy" generalist species of low trophic levels that are adapted to variable environments is underway (Knowlton 2001, McClanahan 2002). Bellwood et al. (2004) describe six different reefs transitions towards less desired states as a consequence of human-induced erosion of resilience (Figure 1).

Kelp Forests

Remarkably sharp boundaries are found between kelp forests and neighboring "deforested" areas (Konar & Estes 2003). Also, remarkable switches have occurred

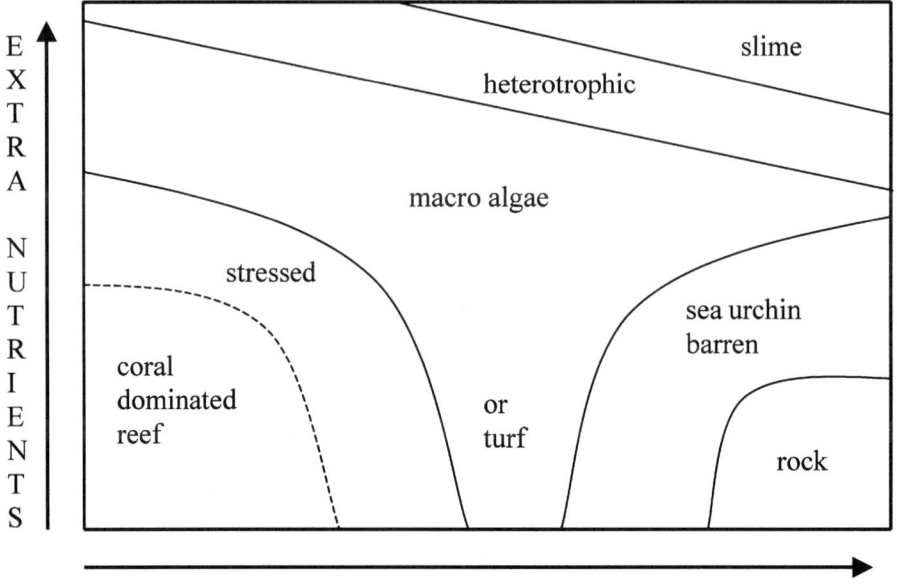

Figure 1 Effects of eutrophication and fishing and observed shifts between states in coral reefs (modified from Bellwood et al. 2004).

between kelp dominance and sea-urchin dominance over time (Steneck et al. 2002). Experiments reveal that sea urchins can control kelp growth in the open areas, but the sweeping of kelp foliage over the rocks in the border region prevents migration of sea urchins into the kelp stands. Thus, kelp stands can withstand herbivory by combining their flexible morphology with the energy of wave-generated surge (Konar & Estes 2003). Kelp forests can recover if the numbers of sea urchins are reduced by an external factor. Proposed mechanisms include increased predation by sea otters, cycles of diseases, and introduction of sea urchin fishery. The fishing down of coastal food webs has in some locations led to the return of kelp forests devoid of vertebrate apex predators. Large predatory crabs have filled this void in areas of the western North Atlantic (Steneck et al. 2002).

Pelagic Marine Fisheries

Pelagic marine fish stocks sometimes exhibit sharp changes consistent with regime shifts (Steele 1996, 1998). Large changes in fish stocks have occurred ever since the introduction of fisheries (Jackson et al. 2001). Similar rapid and massive changes have occurred in freshwater ecosystems subject to sport fishing (Post et al. 2002). Cascading changes are often related to size-structured predation (de Roos & Persson 2002). Larger individuals of one species eat smaller individuals of other species. If larger individuals of one species become rare for some reason, that species' recruitment can be eliminated by predation from the other species and perhaps lead to severe decline of the population of the first species. Walters & Kitchell (2001) call this dynamic "cultivation/depensation." If adults of the former species are abundant, they create favorable conditions for their own offspring by reducing the abundance of the latter species. If adults of the former species are overfished, expansion of latter species may permanently prevent reestablishment of the former species.

Savannas

Marked fluctuations in grass and woody plant biomass are a characteristic feature of savannas, because of their highly variable rainfall, and primary productivity varies up to tenfold from one year to the next (Kelly & Walker 1976). Herbivores cannot respond fast enough to track these fluctuations, and the accumulation of grass during wet periods means periodic accumulation of fuel and, therefore, fires. The net effect of fires has been to maintain savanna rangelands in more open, grassy states than would be achieved without fires (Scholes & Walker 1993).

The interaction of fire, herbivory, and variable rainfall has resulted in a grass-shrub-livestock system that exhibits regime shifts between an open, grassy state and a dense, woody state, particularly where humans have altered the pattern and intensity of grazing (Anderies et al. 2002). Establishment of shrub seedlings occurs in wet periods when the seedlings can get their roots below the grass-rooting zone to survive the first dry season. A vigorous grass layer for the first few years strongly suppresses established seedlings, but once established, grasses have little effect on woody plant growth. Fire has little effect on grasses because it occurs at the end of the dry season when grasses are dormant, but it has a severe effect on woody plants by killing many and reducing others to ground level.

The change from a grassy to a woody state comes about through a combination of sustained grazing pressure and lack of fire. Periods of drought with high stock numbers bring about the death of perennial grasses and lead to reduced grass cover. When followed by good rains this reduced grass cover, in turn, leads to a profusion of woody seedlings. If, at this point, all livestock were removed, enough grass growth would still occur to enable an effective fire and keep the system in a grassy state. However, if grazing pressure is sustained a point is reached in the increasing woody:grass biomass ratio after which, even if all livestock are removed, the competitive effect of the woody plants is such that it prevents the build up of sufficient grass fuel to sustain a fire. The system then stays in the woody state until the shrubs or trees reach full size and, through competition among them, begin to die. The vegetation then opens up for the reintroduction of grass and fire. This process can take 30 or 40 years.

The flip in the rangeland occurs when the resilience of the grassy state has been exceeded—that is, when the amount of change in the ratio of grass:woody vegetation needed to push the system into the woody state falls within the range of the annual fluctuations of this ratio (because of fluctuations in rainfall and grazing pressure). Once this situation is reached, the conditions needed to flip the system (e.g., a period of low rainfall) will inevitably follow.

Forests

The boreal forests of North America experience distinctive outbreaks of the spruce budworm, with 30 to 45 years and occasionally 60 to 100 years between outbreaks. This defoliating insect destroys large areas of mature softwood forests, principally spruce and fir. Once the softwood forest is mature enough to provide adequate food and habitat for the budworm, and if a period of warm dry weather occurs, budworm numbers can increase sufficiently to exceed the predation rate and trigger an outbreak. A local outbreak can spread over thousands of square kilometers and eventually collapse after 7 to 16 years. Programs of spraying insecticide to control spruce budworm outbreaks exacerbated the conditions for outbreaks over even more extensive areas (Holling 1978). After a defoliation event, aspen and birch often dominate the regenerating forest, but over a period of 20 to 40 years, selective browsing by moose can shift this forest back to a state dominated by conifers (Ludwig et al. 1978).

Browsing that causes change in dominance between tree species that have different effects on ecosystem functions can lead to dramatic effects in forest ecosystems. For example, the gradual reduction of willows by ungulates on the Alaskan floodplain makes room for nitrogen-fixing alders that increase soil fertility and cause overall vegetation change. In the mountain range of Scandinavia, birches dominate young stands, followed by Norway spruce in the forest succession. If the birches are heavily browsed by ungulates, spruce does not get shelter and fails. Instead, pines may establish and become dominant, which causes long-term changes in soil fertility (Danell et al. 2003). Forestry and hunting policies affect and shape those trajectories.

Shifts in forest cover, associated with management of fire regimes, in the well-drained soils of the southeastern United States reflect alternate states (Peterson 2002). A pine-dominated savanna, with grasses, palms, or shrubs in the understory, historically covered the region and was the result of frequent ground fires. Hardwood shrubs would invade during fire-free periods but their dominance was inhibited by frequent burning. Because of fire suppression and fragmentation of the landscape, fire frequency decreased and led to either mixed pine-hardwood forests or hardwood forests. Once a canopy of hardwoods is established, the site becomes less flammable and precludes pine regeneration (Peterson 2002).

Regime Shifts and Irreversibility

In some cases, regime shifts may be largely irreversible. Loss of trees in cloud forests is one example. In some areas, the forests were established under a wetter rainfall regime thousands of years previously. Necessary moisture is supplied through condensation of water from clouds intercepted by the canopy. If the trees are cut, this water input stops and the resulting conditions can be too dry for recovery of the forest (Wilson & Agnew 1992).

A continental-scale example of an irreversible shift seems to have occurred in Australia, where overhunting and use of fire by humans some 30,000 to 40,000 years ago removed large marsupial herbivores. Without large herbivores to prevent fire and fragment vegetation, an ecosystem of fire and fire-dominated plants expanded and irreversibly switched the ecosystem from a more productive state, dependent on rapid nutrient cycling, to a less productive state, with slower nutrient cycling (Flannery 1994). Similarly, extinction of megafauna at the end of the Pleistocene in Siberia, possibly through improvement in hunting technology, may have triggered an irreversible shift from steppe grassland to tundra. The resulting increase in mosses led to cooler soils, less decomposition, and greater carbon sequestration in peat (Zimov et al. 1995).

VULNERABILITY THROUGH HUMAN-INDUCED LOSS OF RESILIENCE

As illustrated by the foregoing examples, undesired shifts between ecosystem states are caused by the combination of the magnitudes of external forces and the internal resilience of the system. As resilience declines, the ecosystem becomes vulnerable, and progressively smaller external events can cause shifts. Human actions have increased the likelihood of undesired regime shifts. In Figure 2, we summarize shifts into less desired states as a consequence of human-induced loss of resilience.

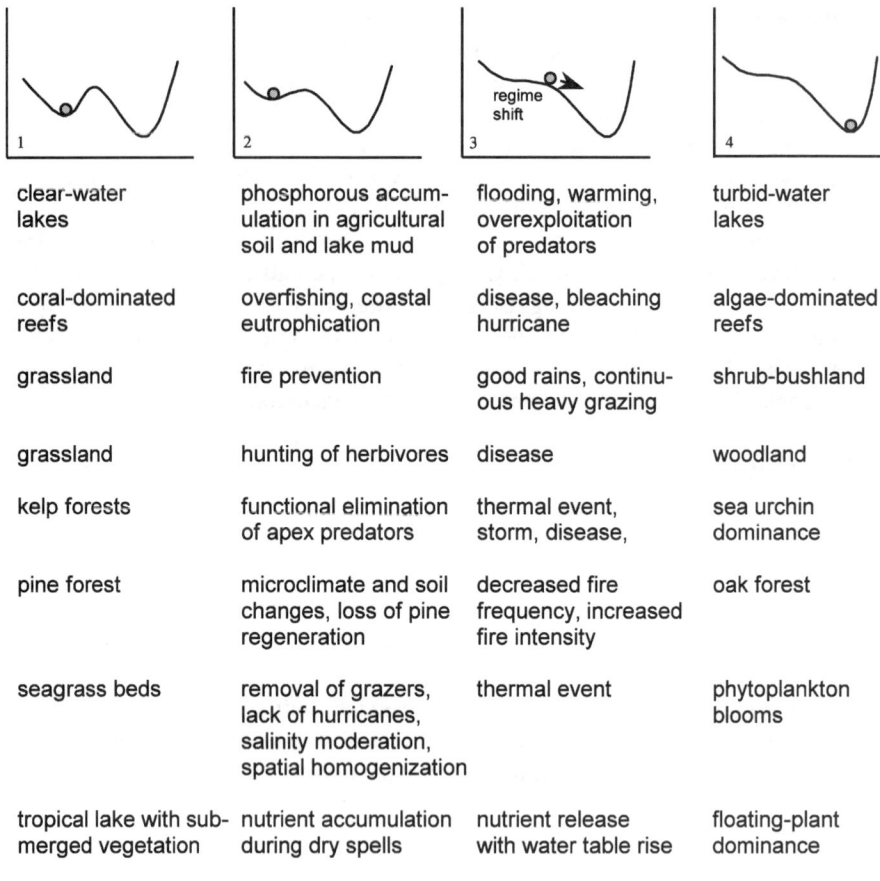

Figure 2 Alternate states in a diversity of ecosystems (1, 4) and the causes (2) and triggers (3) behind loss of resilience and regime shifts. For more examples, see Thresholds Database on the Web site www.resalliance.org.

Humans have, over historical time but with increased intensity after the industrial revolution, reduced the capacity of ecosystems to cope with change through a combination of top-down (e.g., overexploitation of top predators) and bottom-up impacts (e.g., excess nutrient influx), as well as through alterations of disturbance regimes including climatic change (e.g., prevention of fire in grasslands and forest or increased bleaching of coral reefs because of global warming) (Nyström et al. 2000, Paine et al. 1998, Worm et al. 2002). The result of those combined impacts tends to be leaking, simplified, and "weedy" ecosystems characterized by unpredictability and surprise in their capacity to generate ecosystem services.

The likelihood that an ecological system will remain within a desired state is related to slowly changing variables that determine the boundaries beyond which disturbances may push the system into another state (Scheffer & Carpenter 2003).

Consequently, efforts to reduce the risk of undesired shifts between ecosystem states should address the gradual changes that affect resilience rather than focus all effort into trying to control disturbance and fluctuations. The slowly changing variables include such things as land use, nutrient stocks, soil properties, freshwater dynamics, and biomass of long-lived organisms (Gunderson & Pritchard 2002). In the following sections we focus on biological diversity as a slowly changing variable and its significance in ecosystem resilience.

Trophic Cascades

Loss of top predators can increase the vulnerability of aquatic ecosystems to eutrophication by excessive nutrient input (Carpenter 2003). In manipulations of whole-lake ecosystems, removal of top predators allows primary producers to respond much more strongly to experimental inputs of nutrients (Carpenter et al. 2001a). The mechanism is a trophic cascade—in the absence of top predators, planktivorous fishes become abundant, and grazing zooplankton are suppressed. When nutrients are added, phytoplankton grow unconstrained by grazers.

Over human history, removal of top predators from nearshore marine ecosystems through fishing may have increased the vulnerability of the ecosystems to coastal nutrient inputs and paved the way for impacts such as eutrophication, algal blooms, disease outbreaks, and species introductions in coastal areas (Jackson et al. 2001). In the Black Sea, for example, overfishing and eutrophication changed the ecosystem from one dominated by piscivorous fishes (bluefish, bonito) and dolphins to one dominated by jellyfish, small-bodied planktivorous fishes, and phytoplankton (Daskalov 2002, Zaitsev & Mamaev 1997). In the Baltic Sea, removal of top predators such as seals (through pollution and hunting), fishing pressure, and influx of excessive nutrients have caused widespread eutrophication and oxygen deficiency in deeper waters that have wiped out important food chains over nearly 100,000 km^2 of sea bottom (Elmgren 1989).

Trophic cascades occur in a diversity of ecosystem, and examples are known from terrestrial systems (Pace et al. 1999). Trophic cascades are becoming another signature of the vast and growing human footprint (Terborgh et al. 2001). Cascading provides nonlinear and often surprising changes in ecosystem dynamics and may lead to regime shifts (Diaz & Cabido 2001). Trophic cascades seem to be less likely under conditions of high diversity or extensive omnivory in food webs (Pace et al. 1999).

BIODIVERSITY AND RESILIENCE DYNAMICS

The diversity of functional groups in a dynamic ecosystem undergoing change, the diversity within species and populations, and the diversity of species in functional groups appear to be critical for resilience and the generation of ecosystem services (Chapin et al. 1997, Luck et al. 2003). Two aspects of diversity are distinguished: functional-group diversity and functional-response diversity.

Functional-Group Diversity

Functional groups of species in a system refers to groups of organisms that pollinate, graze, predate, fix nitrogen, spread seeds, decompose, generate soils, modify water flows, open up patches for reorganization, and contribute to the colonization of such patches. The persistence of functional groups contributes to the performance of ecosystems and the services that they generate. Loss of a major functional group, such as apex predators, other consumers, or benthic filter-feeders, may, as previously discussed, cause drastic alterations in ecosystem functioning (Chapin et al. 1997, Duffy 2002, Jackson et al. 2001).

However, in systems that lack a specific functional group, the addition of just one species may dramatically change the structure and functioning of ecosystems (Chapin et al. 2000). In Hawaii, the introduced nitrogen-fixing tree *Myrica faya* has dramatically changed the structure and function in ecosystems where no native nitrogen-fixing species had been present. Once established, *M. faya* can increase nitrogen inputs up to five times, thereby facilitating establishment of other exotic species (Vitousek et al. 1987). Studies of coastal environments and reefs suggest that more diverse ecosystems are less sensitive to invasion of exotic species (Knowlton 2001, Stachowicz et al. 1999).

Functional-Response Diversity

Recently, Naeem & Wright (2003) argued that functional-response traits should be considered and distinguished from functional-effect traits when analyzing biodiversity effects on ecosystem functioning. Variability in responses of species within functional groups to environmental change is critical to ecosystem resilience (Chapin et al. 1997, Norberg et al. 2001). Elmqvist et al. (2003) call this property *response diversity*, and it is defined as the diversity of responses to environmental change among species that contribute to the same ecosystem function. For example, in lake systems, animal plankton species with higher tolerance to low pH sustain the grazing function on phytoplankton during acid conditions (Frost et al. 1995). In semiarid rangelands, resilience of production to grazing pressure is achieved by maintaining a high number of apparently less important and less common, or apparently "redundant," species from the perspective of those who want to maximize production, each with different capacities to respond to different combinations of rainfall and grazing pressures. The species replace each other over time, which ensures maintenance of rangeland function over a range of environmental conditions (Walker et al. 1999).

The role of genetic and population diversity for response diversity is illustrated through sockeye salmon production in the rivers and lakes of Bristol Bay, Alaska. Several hundred discrete spawning populations display diverse life-history characteristics and local adaptations to the variation in spawning and rearing habitats. Geographic regions and life-history strategies that were minor producers during a certain climatic regime have been the major producers during other climatic regimes, which allowed the aggregate of the populations to sustain its productivity in fluctuating freshwater and marine environments. The response diversity of the fish stocks has been critical in sustaining their resilience to environmental change. Such management is in stark contrast to the common focus on only the most productive runs at a certain moment in time (Hilborn et al. 2003).

Springer et al. (2003) propose that the decimation of the great whales since the World War II caused their foremost natural predator, killer whales, to begin feeding more intensively on seals, sea lions, and sea otters, which are the major predators on sea urchins, thereby possibly contributing to shifts from kelp-dominated to sea-urchin-dominated coastal areas. In areas where sea-urchin predator diversity is low (e.g., in western North Atlantic and Alaska), the transition between kelp dominance and sea-urchin dominance has been rapid, frequent, widespread, and often long lasting. In southern California, where the diversity of predators, herbivores, and kelps is high, deforestation events have been rare or patchy in space and short in duration, and no single dominant sea-urchin predator exists (Steneck et al. 2002). Functional redundancy and response diversity may contribute to the resilience of kelp forests in California.

An important distinction should be made between real redundancy and apparent redundancy, which involves response diversity within functional groups. Functional redundancy refers to the number of species that perform the same function. Adding more species does not lead to increased system performance where there is real functional redundancy. Furthermore, if this set of functionally redundant species does not exhibit any response diversity, they do not contribute to the insurance value.

We argue that the biodiversity insurance metaphor needs to be revived with a focus on how to sustain ecosystem resilience and the services it generates in the context of multiple-equilibrium systems and human-dominated environments (Folke et al. 1996). Ecosystems with high response diversity increase the likelihood for renewal and reorganization into a desired state after disturbance (Chapin et al. 1997, Elmqvist et al. 2003).

Biodiversity in Ecosystem Renewal and Reorganization

Recovery after disturbance has often been measured as return time to the equilibrium state. Frequently, the sources of ecosystem recovery have been taken for granted and the phases of ecosystem development that prepare the system for succession and recovery largely neglected. Disturbance releases the climax state and is followed by renewal and reorganization. We refer to those phases as the back-loop of ecosystem development (Gunderson & Holling 2002). Functional roles in the back-loop and sources of resilience are critical for sustaining an ecosystem within a desired state in the face of change (Nyström & Folke 2001).

In coral reef systems, three functional groups of herbivores, dominated by fishes, play different and complementary roles in renewing and reorganizing reefs into a coral-dominated state after disturbance. These groups—grazers, scrapers, and bioeroders—prepare the reef for recovery. Bioeroding fishes remove dead corals and other protrusions, which exposes the hard reef matrix for new settlement of coralline algae and corals. Grazers remove seaweed, which reduces coral overgrowth and shading by macroalgae. Scrapers remove algae and sediment by close cropping, which facilitates settlement, growth, and survival of coralline algae and corals. Without bioeroders, recovery may be inhibited by extensive stands of dead staghorn and tabular coral that can remain intact for years before collapsing and taking with them attached coral recruits. Without grazers, algae can proliferate and limit coral settlement and survival of juvenile and adult colonies. Without scrapers, sediment-trapping algal turfs develop that smother coral spat and delay or prevent recovery. The extents to which reefs possess these functional groups are central to their capacity to renew and reorganize into a coral-dominated state in the face of disturbance (Bellwood et al. 2004).

The biological sources of renewal and reorganization for ecosystem resilience consist of functional groups of biological legacies and mobile link species and their support areas in the larger landscape or seascape. For example, large trees serve as biological legacies after fire and storms in forest ecosystems (Elmqvist et al. 2001, Franklin & MacMahon 2000), and seed banks and vegetative propagules play the same role in tundra ecosystems (Vavrek et al. 1999). Mobile link species connect habitats, sometimes widely separated in space and time (Lundberg & Moberg 2003). For example, vertebrates that eat fruit, such as flying foxes, play a key functional role in the regeneration of tropical forests hit by disturbances such as hurricanes and fire by bringing in seeds from surrounding ecosystems that result in renewal and reorganization (Cox et al. 1991, Elmqvist et al. 2001). The functional group of grazers on coral reefs connect a wide range of spatial scales from centimeters, such as amphipods and sea urchins, to thousands of kilometers, such as green turtles (Elmqvist et al. 2003). By operating at different spatial and temporal scales, competition among species within the guild of grazers is minimized, and the robustness over a wider range of environmental conditions is enhanced (Peterson et al. 1998). Such response diversity plays a significant role in the capacity of ecosystems to renew and reorganize into desired states after disturbance.

Metapopulation analyses have largely focused on dispersal, connectivity, recovery, and life history of species, populations, and communities. Great potential lies in redirecting this knowledge into addressing the role of functional groups and response diversity in ecosystem resilience that considers the central role of human actors in this context.

A number of observations suggest that biodiversity at larger spatial scales (i.e., landscapes and regions) ensures that appropriate key species for ecosystem functioning are recruited to local systems after disturbance or when environmental conditions change (Bengtsson et al. 2003, Nyström & Folke 2001, Peterson et al. 1998). The current emphasis on setting aside "hot spot" areas of diversity and protecting species richness in reserves will in itself not be a viable option in human-dominated environments (Folke et al. 1996). Present static reserves should be complemented with dynamic reserves, such as ecological fallows and dynamic successional reserves (Bengtsson et al. 2003), and serve as one important tool that contributes to sustaining the configuration of functional groups and response diversity required for renewing and reorganizing desired ecosystem states after disturbance. In this sense, biological diversity provides insurance, flexibility, and risk spreading across scales in dynamic landscapes and seascapes.

Hence, spatial and temporal relations of functional groups that renew and help reorganize ecosystem development after disturbance, and their response diversity, will influence the ability of ecosystems to remain within desired states.

MANAGING RESILIENCE FOR DEVELOPMENT

Archaeological research indicates that over time human societies have degraded the capacity of ecosystems to sustain societal development (Redman 1999, van der Leeuw 2000). Historical overfishing of coastal waters has created simplified, leaky, and weedy ecosystems that rapidly respond to external influences in an unpredictable fashion (Jackson et al. 2001). A shifting baseline, an incremental lowering of environmental standards over time (Pauly 1995), may occur, and each new human generation may adapt to the new conditions of less diverse and less productive ecosystems.

Our review clearly illustrates that regime shifts in ecosystems are, to a large extent, driven by human actions. A combination of top-down impacts, such as fishing down food webs and losing response diversity and functional groups of species, and bottom-up impacts, such as accumulation of nutrients, soil erosion, or redirection of water flows, as well as altered disturbance regimes, such as suppression of fire and increased frequency and intensity of storms, have shifted several ecosystems into less desired states with diminished capacities to generate ecosystem services.

Shifts from desired to less desired states may often follow gradual loss of ecosystem resilience. Resilience has multiple attributes, but four aspects are critical (Walker et al. 2004):

1. *Latitude* is the maximum amount the system can be changed before losing its ability to reorganize within the same state; basically it is the width of the stability domain or the basin of attraction.

2. *Resistance* is the ease or difficulty of changing the system; deep basins of attraction indicate that greater disturbances are required to change the current state of the system away from the attractor.

3. *Precariousness* is how close the current trajectory of the system is to a threshold that, if breached, makes reorganization difficult or impossible.

4. *Cross-scale relations* (i.e., panarchy) is how the above three attributes are influenced by the states and dynamics of the (sub)systems at scales above and below the scale of interest.

Ecosystem management of resilience, biodiversity, and regime shifts needs to address those attributes. Such an initiative will require adaptability among the actors involved in ecosystem management (Berkes et al. 2003). Adaptability is the capacity of actors in a system to manage resilience in the face of uncertainty and surprise (Gunderson & Holling 2002). Humans are a part of, and not apart from, the trajectory and stability domain of the system and will, to a large extent, determine their own paths through management of the ecosystem. Human actors can (*a*) move thresholds away from or closer to the current state of the system by altering latitude, (*b*) move the current state of the system away from or closer to the threshold by altering precariousness, or (*c*) make the threshold more difficult or easier to reach by altering resistance. Actors can also manage cross-scale interactions to avoid or instigate loss of resilience at larger and more catastrophic scales (Holling et al. 1998).

Human actions have often altered slowly changing ecological variables, such as soils or biodiversity, with disastrous social consequences that did not appear until long after the ecosystems were first affected. A current major problem in this context is the large-scale salinization of land and rivers in Australia. About 5.7 million hectares are currently at risk for dryland salinity, and the amount of land at risk could rise to over 17 million hectares by 2050. Extensive land clearing during the past 200 years has removed native woody vegetation to make way for agricultural crops and pasture grasses that transpire much less water. Thus, more water is infiltrating the soils and causing groundwater tables to rise. The rising water mobilizes salts and causes problems with salinity both in rivers and in the soils, which severely reduces the capacity for plant growth (Gordon et al. 2003). Increased vulnerability, as a consequence of loss of resilience, places a region on a trajectory of greater risk of the panoply of stresses and shocks that occur over time (Kasperson et al. 1995).

Most semiarid ecosystems have suffered from severe overexploitation by excessive grazing and agriculture that resulted in depletion of vegetation biomass and soil erosion. These changes are often difficult to reverse because of positive feedbacks that stabilize the new situation. According to one hypothesis, rainy periods associated with El Niño can be used in combination with grazer control to restore degraded ecosystems (Holmgren & Scheffer 2001). Removing grazers to regenerate vegetation during normal years will not be sufficient, because conditions are too dry. Also, wet years do not allow regeneration if grazers remain present. However, removing grazers during a wet year pulse may tip the balance and allow reorganization into a more desired state, and this pulse management may be sufficient for the state to remain intact, subject to grazing. Clearly, responding to El Niño as an opportunity for shifting an ecosystem back to the desired state demands a highly responsive social system, organized for rapid and flexible adaptation (Berkes et al. 2003).

At times, human societies or groups may find themselves trapped in an undesired basin of attraction that is becoming so wide and so deep that movement to a new basin, or sufficient reconfiguration of the existing basin, becomes extremely difficult. A major challenge in ecosystem management is to develop social and ecological capacity to transform such an undesired basin into a fundamentally new and more desirable configuration, a new stability landscape defined by different state variables or old state variables supplemented by new ones. We call this challenge transformability, that is, the capacity to create untried beginnings from which to evolve a fundamentally new way of living when existing ecological, economic, and social conditions make the existing system untenable (Walker et al. 2004). The new way of living requires social-ecological resilience to cope with future change and unpredictable events (Olsson et al. 2004).

Resilience-building management needs to be flexible and open to learning. It attends to slowly changing, fundamental variables such as experience, memory, and diversity in both social and ecological systems (Folke et al. 2003). The crucial slow variables that determine the underlying dynamic properties of the system and that govern the supply of essential ecosystem services need to be identified and assessed. The processes and drivers that determine the dynamics of this set of crucial variables need to be identified and assessed. The role of biological diversity in ecosystem functioning and response to change should be explicitly accounted for in this context and acknowledged in resilience-building policies.

CONCLUSIONS

Ecosystems can be subject to sharp regime shifts. Such shifts may more easily occur if resilience has been reduced as a consequence of human actions. Human actions may cause loss of resilience through the following methods:

- removal of functional groups of species and their response diversity, such as the loss of whole trophic levels (top-down effects),
- impact on ecosystems via emissions of waste and pollutants (bottom-up effects) and climate change, and
- alteration of the magnitude, frequency, and duration of disturbance regimes to which the biota is adapted.

Loss of resilience through the combined and often synergistic effects of those pressures can make ecosystems more vulnerable to changes that previously could be absorbed. As a consequence, they may suddenly shift from desired to less desired states in their capacity to sustain ecosystem services to society. In some cases, these shifts may be irreversible (or too costly to reverse). Irreversibility is a reflection of changes in variables with long turnover times (e.g., biogeochemical, hydrological, or climatic) and loss of biological sources and interactions for renewal and reorganization into desired states.

In light of these changes and their implication for human well-being, the capacities for self-repair of ecosystems can no longer be taken for granted. Active adaptive management and governance of resilience will be required to help sustain or create desired states of ecosystems. A first step in this direction is to understand better the interactions between regime shifts, biological diversity, and ecosystem resilience.

ACKNOWLEDGMENTS

This work is a product of the Resilience Alliance and has been supported by a grant from the JS McDonnell Foundation. In addition, the work of Carl Folke and Thomas Elmqvist is partly funded by grants from the Swedish Research Council for the Environment, Agricultural Sciences and Spatial Planning (Formas). The Beijer Institute has supported the collaboration with the review article.

The *Annual Review of Ecology, Evolution, and Systematics* is online at http://ecolsys.annualreviews.org

LITERATURE CITED

Algeo TJ, Scheckler SE. 1998. Terrestrial-marine teleconnections in the Devonian: Links between the evolution of land plants, weathering processes, and marine anoxic events. *Philos. Trans. R. Soc. London Ser. B* 353:113–28

Anderies JM, Janssen MA, Walker BH. 2002. Grazing management, resilience, and the dynamics of a fire-driven rangeland system. *Ecosystems* 5:23–44

Barkai A, McQuaid C. 1988. Predator-prey role reversal in a marine benthic ecosystem. *Science* 242:62–64

Beisner BE, Haydon DT, Cuddington K. 2003. Alternative stable states in ecology. *Front. Ecol. Environ.* 1:376–82

Bellwood DR, Hughes TP, Folke C, Nyström M. 2004. Confronting the coral reef crisis. *Nature* 429:827–33

Bengtsson J, Angelstam P, Elmqvist T, Emanuelsson U, Folke C, et al. 2003. Reserves, resilience, and dynamic landscapes. *Ambio* 32:389–96

Berger R, Henriksson E, Kautsky L, Malm T. 2003. Effects of filamentous algae and deposited matter on the survival of *Fucus vesiculosus* L. germlings in the Baltic Sea. *Aquat. Ecol.* 37:1–11

Berkes F, Colding J, Folke C, eds. 2003. *Navigating Social-Ecological Systems: Building Resilience for Complexity and Change.* Cambridge: Cambridge Univ. Press

Bisigato AJ, Bertiller MB. 1997. Grazing effects on patchy dryland vegetation in Northern Patagonia. *J. Arid Environ.* 36:639–53

Blindow I, Anderson G, Hargeby A, Johansson S. 1993. Long-term pattern of alternative stable states in two shallow eutrophic lakes. *Freshw. Biol.* 30:159–67

Bonan GB, Pollard D, Thompson SL. 1992. Effects of boreal forest vegetation on global climate. *Nature* 359:716–18

Brown JH, Valone TJ, Curtin CG. 1997. Reorganization of an arid ecosystem in response to recent climate change. *Proc. Natl. Acad. Sci. USA* 94:9729–33

Caraco NF, Cole JJ, Likens GE. 1991. A cross-system study of phosphorus release from lake sediments. In *Comparative Analysis of Ecosystems,* ed. J Cole, G Lovett, S Findlay, pp. 241–58. New York: Springer-Verlag

Carpenter SR. 2003. *Regime Shifts in Lake Ecosystems: Pattern and Variation.* Excellence in Ecology Series 15. Oldendorf/Luhe, Germany: Ecol. Inst.

Carpenter SR, Cole JJ, Hodgson JR, Kitchell JF, Pace ML, et al. 2001a. Trophic cascades, nutrients, and lake productivity: whole-lake experiments. *Ecol. Monogr.* 71:163–86

Carpenter SR, Ludwig D, Brock WA. 1999. Management of eutrophication for lakes subject to potentially irreversible change. *Ecol. Appl.* 9:751–71

Carpenter SR, Walker B, Anderies JM, Abel N. 2001b. From metaphor to measurement: Resilience of what to what? *Ecosystems* 4:765–81

Chapin FS, Walker BH, Hobbs RJ, Hooper DU, Lawton JH, et al. 1997. Biotic control over the functioning of ecosystems. *Science* 277:500–4

Chapin FS, Zavaleta ES, Eviner VT, Naylor RL, Vitousek PM, et al. 2000. Consequences of changing biodiversity. *Nature* 405:234–42

Cooke GD, Welch EB, Peterson SA, Newroth PR. 1993. *Restoration and Management of Lakes and Reservoirs*. Boca Raton, FL: Lewis

Cox PA, Elmqvist T, Rainey EE, Pierson ED. 1991. Flying foxes as strong interactors in South Pacific Island ecosystems: a conservation hypothesis. *Conserv. Biol.* 5:448–54

Danell K, Bergström R, Edenius L, Ericsson G. 2003. Ungulates as drivers of tree population dynamics at module and genet levels. *Forest Ecol. Manag.* 181:67–76

Daskalov GM. 2002. Overfishing drives a trophic cascade in the Black Sea. *Mar. Ecol. Prog. Ser.* 225:53–63

Davis SM. 1989. Sawgrass and cattail production in relation to nutrient supply in the Everglades. In *Fresh Water Wetlands & Wildlife, 9th Annual Symposium, Savannah River Ecology Laboratory*, 24–27 March 1986, ed. RR Sharitz, JW Gibbons, pp. 325–42. Charleston, SC: US Dep. Energy

de Roos A, Persson L. 2002. Size-dependent life history traits promote catastrophic collapses of top predators. *Proc. Natl. Acad. Sci. USA* 99:12907–12

Diaz S, Cabido M. 2001. Vive la difference: plant functional diversity matters to ecosystem processes. *Trends Ecol. Evol.* 16:646–55

Done TJ. 1992. Phase shifts in coral reef communities and their ecological significance. *Hydrobiologia* 247:121–32

Dublin HT, Sinclair ARE, McGlade J. 1990. Elephants and fire as causes of multiple stable states in the Serengeti-Mara woodlands. *J. Anim. Ecol.* 59:1147–64

Duffy JE. 2002. Biodiversity and ecosystem function: the consumer connection. *Oikos* 99:201–19

Eakin CM. 1996. Where have all the carbonates gone? A model comparison of calcium carbonate budgets before and after the 1982–1983 El Nino at Uva Island in the eastern Pacific. *Coral Reefs* 15:109–19

Elmgren R. 1989. Man's impact on the ecosystems of the Baltic Sea: energy flows today and at the turn of the century. *Ambio* 18:326–32

Elmgren R. 2001. Understanding human impact on the Baltic ecosystem: changing views in recent decades. *Ambio* 30:222–31

Elmqvist T, Folke C, Nyström M, Peterson G, Bengtsson J, et al. 2003. Response diversity and ecosystem resilience. *Front. Ecol. Environ.* 1:488–94

Elmqvist T, Wall M, Berggren AL, Blix L, Fritioff S, Rinman U. 2001. Tropical forest reorganization after cyclone and fire disturbance in Samoa: remnant trees as biological legacies. *Conserv. Ecol.* 5:10. http://www.consecol.org/vol5/iss2/art10

Eriksson BK, Johansson G. 2003. Sedimentation reduces recruitment success of *Fucus vesiculosus* (Phaeophyceae) in the Baltic Sea. *Eur. J. Phycol.* 38:217–22

Flannery T. 1994. *The Future Eaters: An Ecological History of the Australasian Lands and People*. Sydney: Reed New Holland

Foley JA, Coe MT, Scheffer M, Wang G. 2003. Regime shifts in the Sahara and Sahel: interactions between ecological and climatic systems in Northern Africa. *Ecosystems* 6:524–39

Folke C. 2003. Freshwater and resilience: a shift in perspective. *Philos. Trans. R. Soc. London Ser. B* 358:2027–36

Folke C, Carpenter SR, Elmqvist T, Gunderson L, Holling CS, Walker B. 2002. Resilience and sustainable development: building adaptive capacity in a world of transformations. *Ambio* 31:437–40

Folke C, Colding J, Berkes F. 2003. Synthesis: building resilience and adaptive capacity in social-ecological systems. See Berkes et al. 2003, pp. 352–87

Folke C, Holling CS, Perrings C. 1996. Biological diversity, ecosystems and the human scale. *Ecol. Appl.* 6:1018–24

Franklin JF, MacMahon JA. 2000. Enhanced: messages from a mountain. *Science* 288:1183–90

Frelich LE, Reich PB. 1999. Neighborhood effects, disturbance severity and community stability in forests. *Ecosystems* 2:151–66

Frost TM, Carpenter SR, Ives AR, Kratz TK. 1995. Species compensation and complementarity in ecosystem function. In *Linking Species and Ecosystems*, ed. CG Jones, JH Lawton, pp. 224–39. New York: Chapman and Hall

Glynn PW. 1988. El Nino warming, coral mortality and reef framework destruction by echinoid bioerosion in the eastern Pacific. *Galaxea* 7:129–60

Gordon L, Dunlop M, Foran B. 2003. Land cover change and water vapour flows: learning from Australia. *Philos. Trans. R. Soc. London Ser. B* 358:1973–84

Gunderson LH. 2000. Ecological resilience: in theory and application. *Annu. Rev. Ecol. Syst.* 31:425–39

Gunderson LH. 2001. Managing surprising ecosystems in southern Florida. *Ecol. Econ.* 37:371–78

Gunderson LH, Holling CS, eds. 2002. *Panarchy: Understanding Transformations in Human and Natural Systems*. Washington, DC: Island Press

Gunderson LH, Pritchard L, eds. 2002. *Resilience and the Behavior of Large-Scale Ecosystems*. Washington, DC: Island Press

Higgins PAT, Mastrandrea MD, Schneider SH. 2002. Dynamics of climate and ecosystem coupling: abrupt changes and multiple equilibria. *Philos. Trans. R. Soc. London Ser. B* 357:647–55

Hilborn R, Quinn TP, Schindler DE, Rogers DE. 2003. Biocomplexity and fisheries sustainability. *Proc. Natl. Acad. Sci. USA* 100:6564–68

Holling CS. 1973. Resilience and stability of ecological systems. *Annu. Rev. Ecol. Syst.* 4:1–23

Holling CS. 1978. The spruce-budworm/forest-management problem. In *Adaptive Environmental Assessment and Management*. International Series on Applied Systems Analysis, ed. CS Holling, 3:143–82. New York: John Wiley & Sons

Holling CS. 1986. The resilience of terrestrial ecosystems: local surprise and global change. In *Sustainable Development of the Biosphere*, ed. WC Clark, RE Munn, pp. 292–317. Cambridge: Cambridge Univ. Press

Holling CS. 1996. Engineering resilience versus ecological resilience. In *Engineering within Ecological Constraints*, ed. PC Schulze, pp. 31–44. Washington, DC: Natl. Acad. Press

Holling CS, Berkes F, Folke C. 1998. Science, sustainability, and resource management. In *Linking Social and Ecological Systems: Management Practices and Social Mechanisms for Building Resilience*, ed. F Berkes, C Folke, pp. 342–62. Cambridge: Cambridge Univ. Press

Holling CS, Meffe GK. 1996. Command and control and the pathology of natural resource management. *Conserv. Biol.* 10:328–37

Holmgren M, Scheffer M. 2001. El Niño as a window of opportunity for the restoration of degraded arid ecosystems. *Ecosystems* 4:151–59

Hughes TP. 1994. Catastrophes, phase shifts, and large-scale degradation of a Caribbean coral reef. *Science* 265:1547–51

Hughes TP, Baird AH, Bellwood DR, Card M, Connolly SR, et al. 2003. Climate change, human impacts, and the resilience of coral reefs. *Science* 301:929–33

Jackson JBC, Kirb MX, Berher WH, Bjorndal KA, Botsford LW, et al. 2001. Historical overfishing and the recent collapse of coastal ecosystems. *Science* 293:629–38

Jackson LJ. 2003. Macrophyte-dominated and turbid states of shallow lakes: evidence from Alberta lakes. *Ecosystems* 6:213–23

Jansson B-O, Jansson AM. 2002. The Baltic Sea: Reversibly unstable or irreversibly stable? See Gunderson & Pritchard 2002, pp. 71–109

Jeppesen E, Sondergaard M, Sondergaard M, Christofferson K, eds. 1998. *The Structuring*

Role of Submerged Macrophytes in Lakes. Berlin: Springer-Verlag

Kautsky N, Kautsky H, Kautsky U, Waern M. 1986. Decreased depth penetration of *Fucus vesiculosus* (L.) since the 1940s indicates eutrophication of the Baltic Sea. *Mar. Ecol. Prog. Ser.* 28:1–8

Kasperson JX, Kasperson RE, Turner BL. 1995. *Regions at Risk: Comparisons of Threatened Environments.* New York: United Nations University Press

Kelly RD, Walker BH. 1976. The effects of different forms of land use on the ecology of a semi-arid region in south-eastern Rhodesia. *J. Ecol.* 64:553–76

Knowlton N. 1992. Thresholds and multiple stable states in coral reef community dynamics. *Am. Zool.* 32:674–82

Knowlton N. 2001. The future of coral reefs. *Proc. Natl. Acad. Sci. USA* 98:5419–25

Konar B, Estes JA. 2003. The stability of boundary regions between kelp beds and deforested areas. *Ecology* 84:174–85

Levin S. 1999. *Fragile Dominion: Complexity and the Commons.* Reading, MA: Perseus Books

Luck GW, Daily GC, Ehrlich PR. 2003. Population diversity and ecosystem services. *Trends Ecol. Evol.* 18:331–36

Ludwig D, Jones DD, Holling CS. 1978. Qualitative analysis of insect outbreak systems: Spruce-budworm and forest. *J. Anim. Ecol.* 47:315–32

Ludwig D, Mangel M, Haddad B. 2001. Ecology, conservation, and public policy. *Annu. Rev. Ecol. Syst.* 32:481–517

Ludwig D, Walker B, Holling CS. 1997. Sustainability, stability, and resilience. *Conserv. Ecol.* 1:7. http://www.consecol.org/vol1/iss1/art7

Lundberg J, Moberg F. 2003. Mobile link organisms and ecosystem functioning: implications for ecosystem resilience and management. *Ecosystems* 6:87–98

McClanahan TR. 2002. The near future of coral reefs. *Environ. Conserv.* 29:460–83

McCook LJ. 1999. Macroalgae, nutrients and phase shifts on coral reefs: scientific issues and management consequences for the Great Barrier Reef. *Coral Reefs* 18:357–67

Naeem S, Wright JP. 2003. Disentangling biodiversity effects on ecosystem functioning: deriving solutions to a seemingly insurmountable problem. *Ecol. Lett.* 6:567–79

Newman S, Schuette J, Grace JB, Rutchey K, Fontaine T, et al. 1998. Factors influencing cattail abundance in the northern Everglades. *Aquat. Bot.* 60:265–80

Norberg J, Swaney DP, Dushoff J, et al. 2001. Phenotypic diversity and ecosystem functioning in changing environments: a theoretical framework. *Proc. Natl. Acad. Sci. USA* 98:11376–81

Nürnberg GK. 1995. Quantifying anoxia in lakes. *Limnol. Oceanogr.* 40:1100–11

Nyström M, Folke C. 2001. Spatial resilience of coral reefs. *Ecosystems* 4:406–17

Nyström M, Folke C, Moberg F. 2000. Coral-reef disturbance and resilience in a human-dominated environment. *Trends Ecol. Evol.* 15:413–17

O'Dowd DJ, Green PT, Lake PS. 2003. Invasional 'meltdown' on an oceanic island. *Ecol. Lett.* 6:812–17

Olsson P, Folke C, Hahn T. 2004. Social-ecological transformation for ecosystem management: the development of adaptive co-management of a wetland landscape in southern Sweden. *Ecol. Soc.* 9(4):2. http://www.ecologyandsociety.org/vol9/iss4/art2

Pace ML, Cole JJ, Carpenter SR, Kitchell JF. 1999. Trophic cascades revealed in diverse ecosystems. *Trends Ecol. Evol.* 14:483–88

Paine RT, Tegner MJ, Johnson EA. 1998. Compounded perturbations yield ecological surprises. *Ecosystems* 1:535–45

Pauly D. 1995. Anecdotes and the shifting baseline syndrome of fisheries. *Trends Ecol. Evol.* 10:430

Peterson GD, Allen CR, Holling CS. 1998. Ecological resilience, biodiversity, and scale. *Ecosystems* 1:6–18

Peterson GD. 2002. Forest dynamics in the Southeastern United States: managing multiple stable states. See Gunderson & Pritchard 2002, pp. 227–46

Post JR, Sullivan M, Cox S, Lester NP, Walters CJ, et al. 2002. Canada's recreational fisheries: the invisible collapse? *Fisheries* 27:6–17

Redman CL. 1999. *Human Impact on Ancient Environments.* Tucson, AZ: Univ. Arizona Press

Scheffer M. 1997. *The Ecology of Shallow Lakes.* London: Chapman and Hall

Scheffer M, Carpenter SR. 2003. Catastrophic regime shifts in ecosystems: linking theory to observation. *Trends Ecol. Evol.* 18:648–56

Scheffer M, Carpenter SR, Foley J, Folke C, Walker BH. 2001. Catastrophic shifts in ecosystems. *Nature* 413:591–96

Scheffer M, Hosper SH, Meijer ML, Moss B, Jeppesen E. 1993. Alternative equilibria in shallow lakes. *Trends Ecol. Evol.* 8:275–79

Scheffer M, Szabo S, Gragnani A, van Nes EH, Rinaldi S, et al. 2003. Floating plant dominance as a stable state. *Proc. Natl. Acad. Sci. USA* 100:4040–45

Scholes RJ, Walker BH. 1993. *Nylsuley: The Study of an African Savanna.* Cambridge: Cambridge Univ. Press

Smith VH. 1998. Cultural eutrophication of inland, estuarine and coastal waters. In *Successes, Limitations and Frontiers of Ecosystem Science*, eds. ML Pace, PM Groffman, pp. 7–49. New York: Springer-Verlag

Springer AM, Estes JA, van Vliet GB, Williams TM, Doak DF, et al. 2003. Sequential megafaunal collapse in the North Pacific Ocean: an ongoing legacy of industrial whaling? *Proc. Natl. Acad. Sci. USA* 100:12223–28

Srivastava DS, Jefferies RL. 1995. Mosaics of vegetation and soil salinity: a consequence of goose foraging in an arctic salt marsh. *Can. J. Bot.* 73:75–83

Stachowicz JJ, Whitlach RB, Osman RW. 1999. Species diversity and invasion resistance in marine ecosystems. *Science* 286:1577–79

Steele JH. 1998. Regime shifts in marine ecosystems. *Ecol. Appl.* 8:S33–S36

Steele JH. 1996. Regime shifts in fisheries management. *Fish. Res.* 25:19–23

Steffen W, Sanderson A, Jäger J, Tyson PD, Moore B III, et al. 2004. *Global Change and the Earth System: A Planet Under Pressure.* Heidelberg: Springer-Verlag. 336 pp.

Steneck RS, Graham MH, Bourque BJ, Corbett D, Erlandson JM, et al. 2002. Kelp forest ecosystems: biodiversity, stability, resilience and future. *Environ. Conserv.* 29:436–59

Terborgh J, Lopez L, Nunez P, Rao M, Shahabuddin G, et al. 2001. Ecological meltdown in predator-free forest fragments. *Science* 294:1923–26

Trenbath BR, Conway GR, Craig IA. 1989. Threats to sustainability in intensified agricultural systems: analysis and implications for management. In *Agroecology: Researching the Ecological Basis for Sustainable Agriculture*, ed. SR Gliessman, pp. 337–65. Berlin: Springer-Verlag

Van Cappellen P, Ingall ED. 1994. Benthic phosphorus regeneration, net primary production, and ocean anoxia: a model of the coupled marine biogeochemical cycles of carbon and phosphorus. *Paleoceanography* 9:677–92

van de Koppel J, Rietkerk M, Weissing FJ. 1997. Catastrophic vegetation shifts and soil degradation in terrestrial grazing systems. *Trends Ecol. Evol.* 12:352–56

van der Leeuw S. 2000. Land degradation as a socionatural process. In *The Way the Wind Blows: Climate, History, and Human Action*, eds. RJ McIntosh, JA Tainter, SK McIntosh, pp. 357–83, New York: Columbia Univ. Press

Vavrek MC, Fetcher N, McGraw JB, Shaver GR, Chapin FS, Bovard B. 1999. Recovery of productivity and species diversity in Tussock tundra following disturbance. *Arct. Antarct. Alp. Res.* 31:254–58

Vitousek PM, Walker LR, Whiteaker LD, Muellerdombois D, Matson PA. 1987. Biological invasion by *Myrica-Faya* alters ecosystem development in Hawaii. *Science* 238:802–4

Walker BH, Holling CS, Carpenter SR, Kinzig AS. 2004. Resilience, adaptability and transformability. *Ecol. Soc.* In press

Walker BH, Kinzig A, Langridge J. 1999. Plant

attribute diversity, resilience, and ecosystem function: the nature and significance of dominant and minor species. *Ecosystems* 2: 95–113

Walker BH, Meyers JA. 2004. Thresholds in ecological and social-ecological systems: a developing database. *Ecol. Soc.* 9(2):3

Walters CJ, Kitchell JF. 2001. Cultivation/depensation effects on juvenile survival and recruitment: implications for the theory of fishing. *Can. J. Fish. Aquat. Sci.* 58:1–12

Wang GL, Eltahir EAB. 2000. Role of vegetation dynamics in enhancing the low-frequency variability of the Sahel rainfall. *Water Resour. Res.* 36:1013–21

Wilson JB, Agnew ADQ. 1992. Positive-feedback switches in plant communities *Adv. Ecol. Res.* 23:263–336

Worm B, Lotze H, Boström C, Engkvist R, Labanauskas V, Sommer U. 1999. Marine diversity shift linked to interactions among grazers, nutrients and propagule banks. *Mar. Ecol. Prog. Ser.* 185:309–14

Worm B, Lotze H, Hillebrand H, Sommer U. 2002. Consumer versus resource control of species diversity and ecosystem functioning. *Nature* 417:848–51

Zaitsev Y, Mamaev V. 1997. *Marine Biological Diversity in the Black Sea: A Study of Change and Decline*. New York: United Nations Publications

Zimov SA, Chuprynin VI, Oreshko AP, Chapin FS, Reynolds JF, Chapin MC. 1995. Steppe-tundra transition: a herbivore-driven biome shift at the end of the Pleistocene. *Am. Nat.* 146:765–94

ECOLOGY OF WOODLAND HERBS IN TEMPERATE DECIDUOUS FORESTS*

Dennis F. Whigham

Smithsonian Environmental Research Center, Edgewater, Maryland 21037
Harvard Forest, Petersham, Massachusetts 01366; email: whighamd@si.edu

Key Words life history, reproduction, herbivory, mycorrhiza, conservation

■ **Abstract** The diversity of woodland herbs is one of the most striking features of deciduous forests in the temperate zone. Here I review the literature on the ecology of woodland herbs. The review is timely because, since Paulette Bierzychudek's seminal review of the subject in 1982, a number of species have become rare or threatened owing to the conversion of forests to other land uses, competition by alien plant species, and increased abundance of native wildlife that negatively impact woodland herbs (e.g., white-tailed deer). Although the basic biology of woodland herbs is mostly known, few species have been studied in detail, and we are only able to make broad generalities about their ecology. We are especially lacking in information needed to conserve and restore species in altered and threatened habitats.

INTRODUCTION

Woodland herbs account for most of the vascular plant species diversity in deciduous forests in eastern North America (e.g., Ramsey et al. 1993, McCarthy & Bailey 1996, McCarthy 2003), Europe (e.g., Hermy et al. 1999), and Japan (e.g., Kawano 1985). This diverse group of species has long attracted the interest of naturalists, and herbalists have used them since ancient times because many species (e.g., American Ginseng) contain medicinally active and therapeutic chemicals (Lewis & Zenger 1982, Smith et al. 1996). Ecologists have also long been interested in woodland herbs because of the wide array of life history attributes that make them ideal research subjects. In recent years, woodland herbs have attracted attention because landscape alteration and habitat destruction caused many species to become rare or threatened.

Species declines and losses are primarily due to the conversion of forests into nonforest land uses and to competition from invasive species (Jolls 2003, Meekins & McCarthy 2000), leading investigators to discuss conservation and restoration

---

*The U.S. Government has the right to retain a nonexclusive, royalty-free license in and to any copyright covering this paper.

of woodland herbs in deciduous forests (e.g., Drayton & Primack 1996, Robinson et al. 1994, Jolls 2003). Three issues related to the population ecology of woodland herbs have received the most attention in eastern North America: (*a*) Logging (Kochenderfer & Wendel 1983, Meier et al. 1995) and (*b*) deer browsing (Rooney & Dress 1997b, Waller & Alverson 1997, Rooney 2001, Bellemare et al. 2002; Rooney & Gross 2003) both negatively impact herb diversity, and (*c*) clear-cut areas abandoned to secondary succession will slow the recovery of herb diversity (Vellend 2003).

Most studies of the ecology of woodland herbs cite the 1982 review paper by Bierzychudek (1982a). During the intervening 20 years, additional information has been published on a range of topics related to the biology and ecology of woodland herbs (e.g., Gilliam & Roberts 2003). The objective of this contribution is to update and extend the analysis provided by Bierzychudek. My hope is that this synthesis will be useful to individuals who study woodland herbs as well as support efforts to develop more effective research, conservation, and restoration strategies.

The review is organized into topics that contain relevant literature citations and an overview of the current state of knowledge. I consider the topics covered by Bierzychudek (phenology, longevity, modes of reproduction, reproductive effort, seed dispersal and germination, patterns of mortality, population structure and stability) and additional topics (clonality, herbivory, genetic variability, mycorrhizae, nutrient storage and nutrient cycling, responses to disturbance) not included or only briefly considered in the 1982 review. Jolls (2003) has also reviewed topics considered by Bierzychudek and I have attempted to augment that material.

PHENOLOGY

Although most woodland herbs are deciduous, the range of phenological patterns has been more fully examined, and Givnish (1983, 1987) described the diversity of leaf phenology patterns in ecological and evolutionary contexts. Studies in Japan (Kawano 1985, Uemura 1993) provide a detailed comparative analysis of phenological patterns among woodland herbs. Uemura (1993) identified two groups of evergreen and three groups of deciduous species, as well as wintergreen and achlorophyllous species. Uemura's wintergreen groups were based on leaf longevity (one or two years). The deciduous groups, which accounted for most species, were based on whether the species had heteroptic (i.e., plants with both summer green and overwintering leaves), summer green, or spring green leaves. Four wintergreen species had no leaves in the summer, and three achlorophyllous species were in two families (Orobanchaceae and Orchidaceae). Kawano (1985) also recognized a similar range of phenological patterns in Japan and attributed the diversity of phenological patterns to having "differentiated as a result of adaptive response to woodland habitats where conspicuous periodicity in various physical and biotic regimes predominates."

Givnish (1983, 1987) provided rigorous analyses that supported Kawano's comments, and Neufeld & Young (2003) summarized the literature on the ecophysiology of woodland herbs. Excluding winter annuals, Givnish (1987) recognized six guilds on the basis of leaf phenology: spring ephemerals, early summer, late summer, wintergreen, evergreen, and dimorphic. Givnish found clear differences in leaf thickness and width between the spring ephemeral, early summer, and late summer guilds and clear evolutionary patterns within guilds and growth forms. Spring ephemerals had determinate growth and displayed foliage between 5–15 cm above the ground on basal leaves or short umbrellas. Leaves close to the ground have more efficient temperature regulation, permitting efficient use of light during the short period between emergence and development of the tree canopy. Early summer species had determinate growth with leaves from 10–160 cm above the ground, displayed in a variety of umbrella-like structures that minimize shading and maximize light capture with the lowest possible structural costs. Late summer species had indeterminate growth and displayed leaves at a greater height (40–160 cm). Evergreen and wintergreen species and the winter phases of dimorphic species displayed leaves close to the ground, resulting in enhanced winter photosynthesis (Minoletti & Boerner 1993, Tissue et al. 1995).

More recent studies support the earlier conclusions about flowering and fruiting, especially Bierzychudek's comment that caution needs to be applied to information on breeding systems because the degree of compatibility or incompatibility may vary from one location to another and geographic variability in breeding systems should be expected. *Trillium kamtschaticum* in Japan (Ohara et al. 1996) and *Podophyllum peltatum* in North America (Policansky 1983) are two species that have more than one breeding system over their ranges of distribution. Northern populations of *T. kamtschaticum* in Hokkaido, for example, are self-compatible and have a high degree of genetic diversity compared with mostly self-incompatible southern populations.

Several studies (Ashmun et al. 1982, Pitelka et al. 1985a, Collins & Pickett 1988, Givnish 1987) have shown that autumn flowering species respond positively to light gaps and have indeterminate growth and thin, broad leaves displayed along vertically elongating shoots. Most autumn flowering species do not develop preformed buds, whereas most spring flowering species do (Randall 1952, as described in Gerber et al. 1997b). Not described in the 1982 review were species that flower in midsummer, such as *Goodyera pubescens*, an evergreen orchid, and *Tipularia discolor* a wintergreen orchid that is leafless when it flowers (Snow & Whigham 1989, Whigham 1990).

LONGEVITY

Not much was known about the longevity of woodland herbs in 1982, and little has changed since then (Jolls 2003). Most woodland herbs are perennial and clonal. Annuals (Baskin & Baskin 1998) and pseudoannuals (Baskin & Baskin 1988,

Cook 1988) that produce ramets at the ends of short-lived spacers typically live for one year (Wijesinghe & Whigham 1997). Some species (e.g., *Viola* spp.) have individuals that live between 1 and 10 years (Bender et al. 2000, 2002), and others live for as long as 30–50 years. *Clintonia borealis* is an example of a clonal species with long-lived individuals (Pitelka et al. 1985b), and *Panax quinquefolia* (Lewis & Zenger 1982) and species of *Trillium* (Davis 1981) are examples of long-lived nonclonal species.

For some woodland herbs, longevity cannot be easily interpreted. For example, individuals of the nonclonal orchid *Galearis spectabilis* failed to produce seeds in or near permanent plots for more than 10 years, and fewer than 5% of the individuals originally marked are still alive (D. Whigham, unpublished data). Most individuals appear aboveground for 1–2 years, and new plants appear each year. I suspect that the new plants are seedlings or individuals that have persisted underground as heterotrophic plants supported by mycorrhiza (e.g., Gill 1996, Shefferson et al. 2001), and thus establishing ages of individuals is difficult.

PATTERNS OF MORTALITY

Recent studies have provided more information on mortality. Seedling mortality can be negatively (Syrjänen & Lehtilä 1993) or positively (Smith 1983a) related to density. Groups of *Floerkea proserpinacoides* seedlings, for example, have a greater chance of pushing through leaf litter and surviving. Some species have low mortality at all life history stages or mortality may be independent of age (e.g., Solbrig 1981). Mortality rates of most species are, however, higher for juveniles (Meagher & Antonovics 1982b, Inghe & Tamm 1985, Matlack 1987) and smaller plants (Pitelka et al. 1985b, Scheiner 1988, Bloom et al. 2001). In dioecious species, mortality may (Meagher & Antonovics 1982a) or may not (Bawa et al. 1982) be higher for females.

One would expect diseased plants to have higher mortality. Ramets of *Actaea spicata* had higher mortality when infected with the smut *Urocystis carcinodes*, and voles consumed infected plants at a higher rate (Wennström & Ericson 1994). Larger infected plants, however, produced more fruits because large plants may have enough resources to allow the fungus to remain in a latent condition even though mortality would be higher if the smut became virulent.

MODES OF REPRODUCTION

As described in the phenology section, most woodland herbs are perennials, and the majority are clonal. Clonality has arisen separately in many groups of plants, and many of the 21 clonal growth forms identified in the European flora (Klimés et al. 1997) have also been described for North American (Table 1) and Japanese floras (Kawano 1975). However, a complete analysis of the distribution of clonal

TABLE 1 Asexual reproduction (+ or − in Column 2) for woodland herbs. Shown here are ferns and fern allies. For a more complete list, follow the Supplemental Material link from the Annual Reviews home page at http://www.annualreviews.org

Species (Family)	Vegetative propagation	Structure	Growth (cm/yr)	Frequency of branching[a]	Physiological integration	Duration of connection[b]	Number of ramets produced[b]	Source
Ainsliaea apiculata (Asteraceae)	+	Rhizome	NA[c]	NA	NA	NA	NA	Hori & Yokoi 1999
Athyrium filix-femina (Aspidiaceae)	+	Rhizome	<1.0	NA	NA	NA	NA	Sobey & Barkhouse 1977
Dryopteris phegopteris (Aspidiaceae)	+	Rhizome	1.5–3.0	Irregular	NA	NA	NA	Sobey & Barkhouse 1977
D. spinulosa	+	Rhizome	<1–4	NA	NA	NA	NA	Sobey & Barkhouse 1977
Lycopodium lucidulum (Lycopodiaceae)	+	Rhizome	1–3	Irregular	NA	NA	NA	Sobey & Barkhouse 1977
L. annotinum	+	Rhizome	0.1–8	Frequent	Fully integrated	NA	1	Sobey & Barkhouse 1977, Jónsdóttir & Watson 1997
L. clavatum	+	Rhizome	8–74	Frequent	NA	NA	1–4	Sobey & Barkhouse 1977
L. obscurum	+	Rhizome	13–20	Frequent	NA	NA	1	Sobey & Barkhouse 1977

(*Continued*)

TABLE 1 (*Continued*)

Species (Family)	Vegetative propagation	Structure	Growth (cm/yr)	Frequency of branching[a]	Physiological integration	Duration of connection[b]	Number of ramets produced[b]	Source
L. flabelliforme	+	Rhizome	17–31	Frequent	Highly integrated–intermediate	NA	1	Sobey & Barkhouse 1977, Lau & Young 1988, Jónsdóttir & Watson 1997
L. tristrachyum	+	Rhizome	ca. 15	NA	NA	NA	NA	Sobey & Barkhouse 1977
Osmunda claytoniana (Osmundaceae)	+	Rhizome	<0.5	Very infrequent	NA	NA	1–75	Sobey & Barkhouse 1977
Polystichum acrostichoides (Aspidiaceae)	+	Rhizome	<1.0	Infrequent	NA	NA	NA	Sobey & Barkhouse 1977
Pteridium aquilinum (Pteridiaceae)	+	Rhizome	15–30	Frequent	Fully integrated	NA	1	Sobey & Barkhouse 1977, Jónsdóttir & Watson 1997

[a]Sobey & Barkhouse 1977.
[b]per year or as otherwise provided.
[c]NA = Information not provided in reference.

growth forms among herbs in temperate forests of North America, Japan, and other parts of Asia has not yet been done.

Further studies support Bierzychudek's finding that although many woodland herbs reproduce clonally, sexual reproduction is important. Kudoh et al. (1999), for example, found that some patches of *Uvularia perfoliata* contained only one genotype, indicating establishment from a single seed. A patch in a gap habitat, in contrast, had a very high level of genetic diversity, indicating establishment of individuals from many seeds.

In 1982, most woodland herbs were considered hermaphroditic (table 3 in Bierzychudek 1982a). About half were self-compatible and the other half partially or completely self-incompatible. Studies cited in Table 2 indicate that there is an almost equal distribution between the three categories, suggesting that some level of self-incompatibility may be more common than previously thought.

CLONALITY

Many woodland herbs are clonal and have a wide range of growth patterns (Klimés et al. 1997). I now consider aspects of clonality that influence current and future growth and reproduction.

Physiological Integration

Jónsdóttir & Watson (1997) divided clonal species into four categories on the basis of clone size and the degree of physiological integration. Using the same categories to characterize woodland herbs, Table 3 shows that physiological integration has been studied in few woodland herbs and most species examined form large and highly integrated clones (Pitelka & Ashmun 1985). Patterns of physiological integration range from species such as *Aster acuminatus*, which form large patches with short-lived connections and little physiological integration, to species such as *Clintonia borealis*, which form large patches with long-lived connections that are highly integrated (Ashmun et al. 1982). *Clintonia borealis* may be typical of many species of spring flowering herbs, as similar patterns have been shown for *Podophyllum peltatum* (Landa et al. 1992) and *Aralia nudicaulis* (Flanagan & Moser 1985b). Patterns of resource sharing are complex in herbaceous species that have a high degree of physiological integration (e.g., Hutchings & Mogie 1990, Price & Hutchings 1992). Tables 1 and 3 demonstrate, however, that too few woodland herbs have been examined in detail to draw broad generalities regarding physiological integration in woodland herbs.

Foraging for Resources

Foraging has been demonstrated in clonal herbs (Hutchings & de Kroon 1994), but few species have been studied (e.g., de Kroon & Hutchings 1995, Cain et al. 1996, Wijesinghe & Whigham 2001). Wijesinghe & Hutchings (1997) suggested

TABLE 2 Flowering characteristics of woodland herbs

Species	Flowering period	Compatibility	Breeding system	Pollinator limited seed set	Insect visitors	Source
Aralia nudicaulis (Araliaceae)	June	Self-incompatible	Dioecious	Yes/No	Bumblebees, solitary bees, syrphids	Barrett & Helenurm 1987, Flanagan & Moser 1985a
Anemone nemorosa (Ranunculaceae)	NA[a]	Mainly self-compatible	Hermaphrodite	NA	NA	Müller et al. 2000
Carex pedunculata (Cyperaceae)	April	Self-compatible	Monoecious	NA	NA	Handel 1976
Caulophyllum thalictroides (Berberidaceae)	Late April–mid-May	Self-compatible	Hermaphrodite	Yes	Diotera, hemiptera, hymenoptera	Hannan & Prucher 1996
Chamaelirium luteum (Liliaceae)	Mid-May	NA	Dioecious	NA	NA	Meagher & Antonovics 1982a,b
Chimaphila umbellata (Pyrolaceae)	Late July	Partially self-incompatible	Hermaphrodite	Yes	Syrphids	Barrett & Helenurm 1987
Claytonia virginica (Portulacaceae)	Late March–early May	Self-compatible, **Self-incompatible**[b]	Hermaphrodite	Yes	24 sp, Andrenid was primary pollinator	Schemske 1977

Species (Family)	Flowering time	Breeding system	Sexual system	Pollen limitation	Pollinators	References
Clintonia borealis (Liliaceae)	Early June	Partially self-incompatible	Hermaphrodite	Yes	Bumblebees, solitary bees	Barrett & Helenurm 1987, Galen et al. 1985
Cornus canadensis (Cornaceae)	Mid-June	Self-incompatible	Hermaphrodite	Yes	Bumblebees, solitary bees, beeflies, syrphids	Barrett & Helenurm 1987
Corydalis ambigua (Papaveraceae)	Late April–early May	Self-incompatible	Hermaphrodite	Yes	Honeybees, bumblebees	Ohara & Higashi 1994
Cypripedium acaule (Orchidaceae)	Mid-June	Self-incompatible	Hermaphrodite	Yes	Bumblebees	Barrett & Helenurm 1987, Davis 1986
Erythronium albidum (Liliaceae)	Late April	Self-compatible	Hermaphrodite	NA	Bees	Harder et al. 1993
E. americanum	Late April	Self-incompatible	Hermaphrodite	NA	Bees	Harder et al. 1993
Geranium maculatum (Geraniaceae)	May	Partially self-incompatible	Hermaphrodite	No	Halictid bees, flies, bumblebees, beetles, ants	McCall & Primack 1987
Hepatica americana (Ranunculaceae)	March–April	Self-compatible	Hermaphrodite	No	Mostly solitary bees, bee flies, halictid bees	Motten 1982
Jeffersonia diphylla (Berberidaceae)	April	Self-compatible (facultative autogamy)	Hermaphrodite	No	Halictids, honeybees	Smith et al. 1986
Linnaea borealis (Caprifoliaceae)	Early July	Partially self-incompatible	Hermaphrodite	No	Solitary bees, syrphids	Barrett & Helenurm 1987

(*Continued*)

TABLE 2 (*Continued*)

Species	Flowering period	Compatibility	Breeding system	Pollinator limited seed set	Insect visitors	Source
Maianthemum canadense (Liliaceae)	Early June	Self-incompatible	Hermaphrodite	No	Solitary bees, bee flies, syrphids	Barrett & Helenurm 1987, McCall & Primack 1987
Medeola virginiana (Liliaceae)	Late June	Self-incompatible	Hermaphrodite	Yes	Flies?	Barrett & Helenurm 1987, McCall & Primack 1987
Oxalis montana (Oxalidaceae)	Early July	Partially self-incompatible	Hermaphrodite (Chasmogamous, cleistogamous)	Yes	Solitary bees, syrphids, flies, thrips, beetles	Barrett & Helenurm 1987, Jasieniuk & Lechowicz 1987
Orchis (=Galearis) spectabilis (Orchidaceae)	Late April–early June	Partially self-incompatible	Hermaphrodite Polygamodioecious	Yes	Bumblebees	Dieringer 1982
Panax quinquefolium (Araliaceae)	June–August	NA	Hermaphrodite	Yes?	NA	Lewis & Zenger 1982
Podophyllum peltatum (Berberidaceae)	May	Partially self-incompatibility	Polygamodioecious	Yes	Bumblebees	Policansky 1983, Laverty & Plowright 1988
Primula veris (Primulaceae)	May	Self-incompatible	Hermaphrodite (distylous)	NA	NA	Leimu et al. 2002
P. vulgaris	May	Self-incompatible	Hermaphrodite (distylous)	Yes	Bee-flies, bumble bees, butterflies	Boyd et al. 1990

Species	Flowering time	Breeding system	Sex expression	Pollen limited	Pollinators	Reference
Pyrola secunda (Pyrolaceae)	Mid-July	Self-compatible	Hermaphrodite	Yes	Bumblebees, solitary bees	Barrett & Helenurm 1987
Trientalis borealis (Primulaceae)	Early June	Self-incompatible	Hermaphrodite	Yes	Syrphids	Barrett & Helenurm 1987
Trillium kamtschaticum (Liliaceae)	May	Partially self-compatible	Hermaphrodite	No	Bumblebees, butterflies, flies	Ohara et al. 1996
T. undulatum	Late May	Highly autogamous **Obligate apomictic**	Hermaphrodite	Unclear	Solitary bees	Barrett & Helenurm 1987
T. erectum	May	Self-compatible **Highly apomictic**	Hermaphrodite	Yes	Dipterans, Hymenopterans, Coleopterans	Davis 1981, Irwin 2000
T. grandiflorum	May	Self-incompatible and self-compatible	Hermaphrodite	Yes	Dipterans, Hymenopterans, Coleopterans	Kalisz et al. 1999, Irwin 2000
T. nivale	March	Self-compatible	Hermaphrodite	Yes	Honeybees, beetles,	Nesom & LaDuke 1985
Uvularia sessilifolia (Liliaceae)	May	Self-incompatible	Hermaphrodite	No	Halictid bees, beetle, bumblebees	McCall & Primack 1987
Viola mirabilis (Violaceae)	Early May– end of growing season	Self-compatible	Hermaphrodite (Chasmogamous, cleistogamous)	NA	Bumblebees	Mattila & Salonen 1995

[a]Information not provided in original source material.
[b]Bold indicates information also in Bierzychudek (1982a).

TABLE 3 Four categories of ramet systems based on the number of ramets and the degree of physiological integration (Jónsdóttir & Watson 1997). The number of species found in each category is shown in the second column and the number of the species examined that are found in woodland habitats is shown in the third column

Type of integration	Number of species examined	Number of woodland herb species
Restrictive integration in small ramet systems	4	1
Restrictive integration in large ramet systems	25	2
Full integration in small ramet systems	6	0
Full integration in large ramet systems	17	7

that species that have the potential for rapid clonal expansion respond to resource variability differently than species with a limited potential for horizontal expansion. Woodland species that typically produce preformed bulbs, corms, and short rhizomes, for example, may be more responsive to small-scale variations in habitat conditions than species with the potential to spread rapidly by clonal propagation. Wijesinghe & Whigham (2001) found little evidence for nutrient foraging in two clonal species of *Uvularia*, and Cain & Damman (1997) also found that nutrient foraging might not be important in woodland herbs.

Competition

There is evidence for (e.g., Handel 1978, Hughes 1992) and against (Pitelka 1984, Kudoh et al. 1999) intra- and interspecific competition in clonal woodland herbs. Givnish (1982) concluded that the degree of competition depends on habitat characteristics, such as the level of productivity, and that clonal species in productive habitats can minimize interspecific competition through allelopathic interactions or by maintaining a dense leaf cover close to the ground. Pitelka (1984) suggested that clonal species that are highly integrated physiologically regulate shoot densities to minimize intraspecific competition.

REPRODUCTIVE EFFORT

There have been several studies of patterns of biomass allocation in woodland herbs since Bierzychudek (1982a) concluded that the only generalization that could be made was that woodland herbs allocated less biomass to sexual reproduction than species in grasslands or early successional habitats. Biomass of most woodland herbs is allocated between sexual reproductive effort (SRE) and asexual propagation, referred to as vegetative reproductive effort (VRE). Both SRE and VRE have been measured in few species. The mean variation of VRE in *Aster acuminatus* was large (28.2%–35.2%) compared with 5.0%–13.4% for SRE (Pitelka et al. 1980). VRE for *Solidago caesia* (8.3%–33.3%) and *Aster lateriflorus* (11.1%–56.3%) was also highly variable (Gross et al. 1983). Most recent SRE data fall within the range of values reported by Bierzychudek (1982a), but they show that SRE can be highly variable. A highly variable SRE, for example, has been found in the sex-switching genus *Arisaema* (Lovett Doust et al. 1986, Clay 1993). SREs of staminate, carpellate, and monoecious individuals were similar (ca. 15%), but large carpellate plants allocated 44% of the biomass to sexual reproduction, a value much higher than the SRE reported for the species by Bierzychudek (1982a). Factors acting independently or in combination that influence SRE are light (Gross et al. 1983, Jurik 1983, Pitelka et al. 1985a), pollination, and fruit success (Snow & Whigham 1989). Although several authors have demonstrated that fruit and seed production in woodland herbs is often limited (Flanagan & Moser 1985a, Davis 1986, Barrett & Helenurm 1987, Lubbers & Lechowicz 1989, Snow & Whigham 1989, Primack & Hall 1990, Robertson & Wyatt 1990, Bertin & Sholes 1993, Agren & Willson 1992, Primack et al. 1994, Rockwood & Lobstein 1994, Irwin 2000), few studies have examined the impacts of sexual reproduction on SRE or VRE. The most significant impacts of pollinator limitations on fruit and seed set appear to be on future growth and reproduction (Snow & Whigham 1989, Primack & Hall 1990, Gerber et al. 1997b, Matsumura & Washitani 2000).

SEED DISPERSAL AND GERMINATION

Bierzychudek (1982a) found that most forest herbs lack "special dispersal mechanisms," that seeds of most species require a cold treatment, and that dormancy typically lasts less than 6 months (i.e., there is no seed bank). A review by Pickett & McDonnell (1989) supports her conclusion, but several woodland herbs, including weedy species, persist in the seed bank (Baskin & Baskin 1992, Baskin et al. 1993, Leckie et al. 2000).

There have been additional studies of seed germination of woodland herbs (e.g., Lewis & Zenger 1982; Smith 1983b; Matlack 1987; Syrjänen & Lehtilä 1993; Traveset & Willson 1997; Baskin & Baskin 1998, 2002; Kalisz et al. 1999; Bender et al. 2003). Germination characteristics of selected woodland herbs are summarized in Table 4 and a larger listing can be found in table 10.18 in Baskin & Baskin (1998). Seeds of some species are not dormant at maturation (Baskin & Baskin 1998), but most are dormant, and a common dormancy syndrome is morphophysiological (i.e., embryos are both morphologically and physiologically dormant). Embryos of some species mature during a period of warm temperature and germinate in the autumn (Baskin & Baskin 1983, Smith 1983b).

Most seed germination studies have used relatively large seeds. Some woodland herbs (e.g., ferns, parasitic plants, and orchids), however, have very small seeds or spores. "Dust" seeds of terrestrial orchids are dormant when they are dispersed, and germination is spatially and temporally variable (Rasmussen & Whigham

TABLE 4 Seed characteristics of woodland herbs

Species (Family)	Seed weight (mg)	Number seeds per fruit	Seed longevity	Germination	Source
Actaea spicata (Ranunculaceae)	5.9[a]	NA[d]	NA	NA	Ehrlén & Eriksson 2000
Allium ursinum (Liliaceae)	5.4[a]	NA	NA	Winter or spring after cold treatment	Ernst 1979
Arisaema triphyllum (Araceae)	NA	NA	NA	NA	Braun & Brooks 1987
Asarum canadense (Aristolochiaceae)	6.8–14.2[b]	2–30[c]	1 year?	Epicotyle dormancy (radicle growth in autumn, shoot growth following spring)	Baskin & Baskin 1986, Cain & Damman 1997, Heithaus 1981, Smith et al. 1989a
Campanula americana (Campanulaceae)	0.1[a]	20–40	NA	Autumn or spring depending on phenology of mother plant	Galloway 2002
C. latifolia	0.1[a]	NA	NA	NA	Ehrlén & Eriksson 2000
Convallaria majalis (Convallariaceae)	17.0[a]	NA	NA	NA	Ehrlén & Eriksson 2000
Floerkea proserpinacoides (Limnanthaceae)	NA	1	Several years	No dormancy, controlled by environment	Smith 1983b
Hepatica acutiloba (Ranunculaceae)	2.4[a]	9.9[a]	NA	NA	Smith et al. 1989b

Species (Family)					References
Jeffersonia diphylla (Berberidaceae)	3.4–36.8[b] (Embryos undeveloped when seeds dispersed)	~20–40[c]	NA	Spring following dispersal	Baskin & Baskin 1989; Smith et al. 1986, 1989a; Heithaus 1981
Lathyrus vernus (Fabaceae)	12.0[a]	NA	NA	NA	Ehrlén & Eriksson 2000
Paris quadrifolia (Trilliaceae)	4.6[a]	NA	NA	NA	Ehrlén & Eriksson 2000
Panax quinquefolia (Araliaceae)	NA	1.9[a]	2 years?	18–22 months after dispersal	Lewis & Zenger 1982
Podophyllum peltatum (Berberidaceae)	NA	NA	NA	NA	Braun & Brooks 1987
Polygonatum multiflorum (Convallariaceae)	28.6[a]	NA	NA	NA	Ehrlén & Eriksson 2000
Sanguinaria canadensis (Papaveraceae)	15.7[a]	11.2–31.0[b]	NA	NA	Pudlo et al. 1980, Heithaus 1981
Hyacinthoides (= Scilla) nonscripta (Liliaceae)	NA	NA	NA	June–July. Seeds require high temperature conditioning, germinate in autumn	Thompson & Cox 1978
Silene dioica (Caryophyllaceae)	0.7–1.2[b]	NA	Several years	No dormancy (controlled by environment)	Matlack 1987
Trientalis europaea (Primulaceae)	41.1–64.7[c]	0–14[c]	Several years	Autumn or following spring	Hiirsalmi 1969

(Continued)

TABLE 4 (*Continued*)

Species (Family)	Seed weight (mg)	Number seeds per fruit	Seed longevity	Germination	Source
Trillium erectum (Liliaceae)	4.1[a]	9.6–21.6[b]	NA	NA	Davis 1981, Gunther & Lanza 1989
T. grandiflorum	6.5[a]	NA	NA	NA	Gunther & Lanza 1989
T. kamtschaticum	2.9[a]	38–216[c]	NA	NA	Ohara & Higashi 1987, Ohara & Kawano 1986
T. niveale	2.8[a]	31.4[a]	NA	NA	Smith et al. 1989a
T. tschonoskii	3.5[a]	28–168[c]	NA	NA	Ohara & Higashi 1987, Ohara & Kawano 1986
T. undulatum	5.7[a]	NA	NA	NA	Gunther & Lanza 1989
Uvularia perfoliata (Liliaceae)	NA	NA	NA	Double dormancy (radicle in first year, epicotyl in second)	Webb & Willson 1985, Whigham 1974
Viola mirabilis (Violaceae)	2.48–2.56[b]	17.0–19.9[b]	NA	NA	Mattila & Salonen 1995
V. sororia	7.8–9.3[b]	20–54[c]	Unknown but >1 year	Require cold treatment, germinate in spring	Solbrig 1981

[a] Mean
[b] Range of means
[c] Range
[d] Information not provided in original source material.

1993, 1998b; Whigham et al. 2002). Seeds of many woodland orchids will not germinate unless an appropriate mycorrhizal fungus is present (Rasmussen 1995, Whigham et al. 2002), but others germinate without fungi (Baskin & Baskin 1998) and the spatial variation in fungal distribution may determine where successful germination will occur (Rasmussen & Whigham 1998b).

To the list of seed dispersers (Table 5) can be added spiders (Gunther & Lanza 1989), yellow jackets (Zettler et al. 2001), turtles (Braun & Brooks 1987), rabbits (Knight 1964), and both brown and black bear (Traveset & Willson 1997). Long distance dispersal of seeds has also been documented (Willson 1993, Cain et al. 1998, Vellend et al. 2003), and hedgerows that connect forest patches provide intermediate scale corridors for dispersal (Corbit et al. 1999). Although a variety of vertebrates and invertebrates disperse woodland herb seeds (Handel et al. 1981, Vellend et al. 2003), ants are the most common dispersal agent (table 10.21 in Baskin & Baskin 1998). Interactions between ants and other animals can play a critical role in seed dispersal (Webb & Willson 1985; Smith et al. 1989a,b; Valverde & Silvertown 1995). Ground beetles, for example, eat elaiosomes on *Trillium* seeds, resulting in lower rates of ant seed dispersal (Ohara & Higashi 1987). Seed dispersal by ants is also influenced by seed density (Smith et al. 1986), and studies of *Trillium* species have shown that ant-seed interactions are very complex (Gunther & Lanza 1989, Higashi et al. 1989)

Seed dispersal distances vary widely, and data in Table 5 are probably minimum values because no wind-dispersed species (e.g., terrestrial orchids) are included. Mammals (Vellend et al. 2003) and birds can disperse seeds for long distances, but movements of seeds over long distances are difficult to assess. Rodents disperse seeds of woodland herbs (e.g., Valverde & Silvertown 1995), but they also destroy many seeds (Smith et al. 1989a,b).

Models have also been used to test generalities and evaluate the importance of seed dispersal and factors that influence it. Patterns of seed dispersal (Ehrlén & Eriksson 2000) and disturbance to the forest canopy influence woodland herb community diversity (Fröborg & Eriksson 1997, Valverde & Silvertown 1997b), and dispersal of seeds away from parent plants can be beneficial (Cipollini et al. 1993) or harmful (Valverde & Silvertown 1997a).

NUTRIENT STORAGE AND NUTRIENT CYCLING

Herbs account for a relatively small amount of the biomass and nutrient standing stocks in deciduous forests, but nutrient uptake and cycling by herbs can account for significant amounts of total ecosystem nutrient flux (Peterson & Rolfe 1982). The importance of herbs in nutrient cycling processes in forests is, however, undoubtedly related to their patterns of distribution in forests, as the distribution of most species is strongly influenced by nutrient availability and habitat heterogeneity (Crozier & Boerner 1984; Gilliam 1988; Lechowicz et al. 1988; Klinka et al. 1990; Vellend et al. 2000a,b; Miller et al. 2002; Beatty 2003; Small & McCarthy 2003).

TABLE 5 Seed dispersal and predation characteristics of woodland herbs[a]

Species	Dispersal agent	Dispersal distance (m)	Predator	Percent predation	Source
Allium ursinum (Liliaceae)	Gravity and flowering stalks falling over	0.19–0.27	NA[b]	NA	Ernst (1979)
Asarum canadense (Aristolochiaceae)	Ants	Mean = 1.54, Max = 35	Rodent (*Peromyscus leucopus*)	Little to ~50	Heithaus 1981, Smith et al. 1989b, Cain & Damman 1997, Cain et al. 1998
Arabis laevigata var. *laevigata* (Brassicaceae)	Gravity	<0.5	NA	NA	Bloom et al. 2002a
Arisaema triphyllum (Araceae)	Turtle (*Terrapene carolina*)	NA	NA	NA	Braun & Brooks 1987
Carex communis (Cyperaceae)	Ants, mature culms bend toward ground	0.25	NA	NA	Handel 1978
C. platyphylla	May, flower stalks bend when mature	0.01–0.29	NA	NA	Handel 1978
C. umbellata	Ants	0.02–0.16	NA	NA	Handel 1978
Claytonia virginica (Portulaceceae)	Ants, mature inflorescence reflexes toward ground	NA	NA	NA	Handel 1978
Endymion onscriptus (Liliaceae)	Passive expulsion of seeds, rabbits	Mean = 0.40, Max = 0.81	N	NA	Knight 1964
Floerkea poserpinacoides (Limnanthaceae)	Gravity	Up to 0.05	None	NA	Stamp & Lucas 1983
Geranium maculatum (Geraniaceae)	Explosive (e.g., ballistic)	Mean = 3.02, Max = 4.55	NA	NA	Stamp & Lucas 1983
Hepatica acutiloba (Ranunculaceae)	Ants	NA	Rodents	Seed predation is rare	Smith et al. 1989a
Impatiens capensis (Balsaminaceae)	Explosive (e.g., ballistic)	Mean = 0.24, Max = 3.5	NA	NA	Stamp & Lucas 1983, Primack & Miao 1992

Species	Dispersal	Distance	Predators	Predispersal	References
Jeffersonia diphylla (Berberidaceae)	Ants	NA	Rodents	37–54 of capsules prior to dehiscence	Hethaus 1981; Smith et al. 1986, 1989b
Luzula campestris (Juncaceae)	Ants	1	NA	NA	Handel 1978
Panax quinquefolium (Araliaceae)	Gravity	<0.1	Rodents?	NA	Lewis & Zenger 1982
Oxalis acetosella (Oxalidaceae)	Explosive (e.g., ballistic)	Range: 0.6–4.5	NA	NA	Berg 2000
Phytolacca americana (Phytolaccaceae)	Birds	>33	NA	NA	Heppes 1988
Podophyllum peltatum (Berberidaceae)	Turtle (Terrapene carolina)	NA	NA	NA	Braun & Brooks 1987
Primula vulgaris (Primulaceae)	Ants (Myrmica rubra) ca. 20%, rodents ca. 75%	Ants = 0.037–0.099, rodents = unknown	NA	NA	Cahalan & Gliddon 1985; Valverde & Silvertown 1995, 1997a,b
Sanguinaria canadensis (Papaveraceae)	Ants (Aphaenogaster rudis, Formica subsericea, Lasius alienus, Myrmica punctiventris, Stenamma sp.)	0 in disturbed habitats, 17 in undisturbed areas. Mean = 1.38	Rodents	Predispersal consumption of seeds to >65 of post-dehiscence seeds	Pudlo et al. 1980, Heithaus 1981
Silene dioica (Caryophyllaceae)	Gravity	≤0.7	NA	NA	Matlack 1987
Streptopus amplexifolius (Liliaceae)	American robin (Turdus migratorius); varied thrush (Ixoreus naevius); brown bear (Ursus arctos); black bear (Ursus americanus)	NA	NA	NA	Traveset & Willson 1997
Trillium cuneatum, T. erectum, T. grandiflorum, T. undulatum (Liliaceae)	Ants (Myrmica punctiventris) and harvestmen (Arachnida), white-tailed deer, yellow jackets (Vespula sp.)	<1 to >3000	Slugs and harvestmen eat seeds	NA	Gunther & Lanza 1989, Kalisz et al. 1999, Zettler et al. 2001, Vellend et al. 2003
T. kamtschaticum	Ants (Aphaenogaster smythiesi japonica, Myrmica ruginodis)	0.30–3.30	Elaiosomes eaten by ground beetles	85	Chara & Higashi 1987, Higashi et al. 1989

(Continued)

TABLE 5 (*Continued*)

Species	Dispersal agent	Dispersal distance (m)	Predator	Percent predation	Source
T. niveale	Ants	NA	Rodents	Seed predation is rare	Smith et al. 1989a
T. tschonoskii	Ants (*Aphaenogaster smythiesi japonica, Myrmica ruginodis, Lasius niger*)	0.01–2.7	Elaiosome eaten by ground beetles	85	Ohara & Higashi 1987
Uvularia grandiflora (Liliaceae)	Ants (*Myrmica americana, M. aphaenogaster, Camponotus pennsylvanicus, Formica phenolepis*)	<0.1 to >2	*Peromyscus leucopus, Tamius striatus*	NA	Webb & Wilson 1985, Kalisz et al. 1999
U. perfoliata	Ants	≤2	NA	NA	D.F. Whigham, personal observation
Viola sp., *Viola mirabilis* (Violaceae)	Ants	Mean = 0.01–2.1, Max. 0.02–5.4	NA	NA	Beattie & Lyons 1975, Mattila & Salonen 1995
V. pedata	Ants, ballistic	0.25–5.1 Mean = 0.35	NA	NA	Beattie & Lyons 1975, Culver & Beattie 1978
V. pensylvanica	Ants, ballistic	0.72–1.2	NA	NA	Beattie & Lyons 1975, Culver & Beattie 1978
V. rostrata	Ants, ballistic	0.1–4.2	NA	NA	Beattie & Lyons 1975, Culver & Beattie 1978
V. striata	Ants, ballistic	0.2–3.3	NA	NA	Beattie & Lyons 1975, Culver & Beattie 1978, Stamp & Lucas 1983

[a]Dispersal distances are given as ranges or, where indicated, as mean or maximum distances.
[b]Information not provided in source material.

Nutrient cycling is also related to patterns of resource allocation. The majority of woodland herbs are perennial, and most biomass and nutrients are stored in roots, rhizomes, bulbs, and corms (Kawano 1975, Muller 1979, Gross et al. 1983, Piper 1989, Kawano et al. 1992). The seasonal pattern of biomass and nutrient allocation in *Allium tricoccum* seems to be typical of many perennial woodland herbs (e.g., Whigham 1974, Nault & Gagnon 1988). Early in the growing season, carbon and nutrients stored belowground are allocated to aboveground biomass, peaking in leaves and reproductive structures in the spring or early summer. As the growing season progresses, nutrients are resorbed and accumulate in belowground structures, declining in aboveground structures as leaves senesce and fruits and seeds develop. In clonal species, belowground biomass and nutrients are allocated to new ramets, as well as to older structures (Pitelka et al. 1980, Boerner 1986, Benner & Watson 1989, Wijesinghe & Whigham 1997).

Nutrient resorption has been investigated in a few species. DeMars & Boerner (1997) found significant differences in phosphorus and nitrogen resorption and suggested that differences among species were due to phenological characteristics, topographic differences, and nutrient and moisture availability. In *Polystichum acrostichoides*, an evergreen fern, there were no seasonal differences in nutrient resorption, indicating that leaves were maintained during the cold period to allow the species to fix carbon whenever climatic conditions were suitable (Minoletti & Boerner 1993).

Patterns of nutrient uptake and resorption by woodland herbs led to the "vernal dam hypothesis" (Muller 1979, Muller 2003) that described the ability of woodland herbs to immobilize nutrients in biomass that would otherwise be lost from the ecosystem. Others (e.g., Peterson & Rolfe 1982) examined this hypothesis and concluded that plants have the potential to immobilize nitrogen but have little impact on other nutrients. Zak et al. (1990) evaluated the hypothesis in the context of competition for nutrients between the spring geophyte *Allium tricoccum* and microbes. They concluded that microbes immobilized more nitrogen than the spring geophyte, but that microbes and *A. tricoccum* together retained significant amounts of nutrients.

MYCORRHIZAE

Every woodland herb that has been examined has been shown to develop mycorrhizal associations, and the primary benefit to the herb partner is increased nutrient uptake (e.g., Widden 1996, Whitbread et al. 1996, Lapointe & Molard 1997). There appear to be differences in levels of mycorrhizal activity among species, with slow growing species having lower levels of mycorrhizal infection (Brundrett & Kendrick 1990a,b). Differences in mycorrhizal activity also occur across moisture, nutrient, and successional gradients; woodland herbs in wet and nutrient-rich habitats have lower levels of mycorrhizal activity (DeMars & Boerner 1995) and species in late successional habitats have higher levels of vesicular, arbuscular, and ectomycorrhizal activity (Boerner 1992, Boerner et al. 1996).

Terrestrial orchids are an interesting group because they develop mycorrhizal associations that apparently only benefit the orchid partner (Rasmussen 1995). Some woodland orchids are achlorophyllous (e.g., *Corallorhiza* sp.) and obtain all their carbon and other nutrients from mycorrhizae, thus requiring fungal interactions at all life history stages (Zelmer & Currah 1995). Mycorrhizae are also important for woodland orchids that produce leaves but are able to persist belowground for one to many years without forming leaves (Mehrhoff 1989, Gill 1996, Shefferson et al. 2001).

A variety of techniques for studying seed germination and protocorm development in the field portend to expand our understanding of mycorrhizal interactions in terrestrial orchids (Rasmussen & Whigham 1993, 1998a,b, 2002). Much of our knowledge comes from laboratory studies (e.g., Rasmussen 1995), but more recent efforts have included field-based approaches. Seeds of some species, as previously noted, require mycorrhizae for germination, whereas interactions between seeds and mycorrhizal fungi appear to be facultative for other species (Baskin & Baskin 1998; Rasmussen 1995, 2002; Rasmussen & Whigham 1998b; Whigham et al. 2002). Protocorms, the life history stage that follows seed germination, obtain all their resources through mycorrhizal interactions (Rasmussen 1995, 2002). Some woodland orchids (e.g., species of *Corallorhiza*) associated with fungi that form ectomycorrhizal interactions and are indirectly parasitic on trees (Zelmer & Currah 1995).

POPULATION STRUCTURE AND STABILITY

Since 1982, there have been few age structure studies of woodland herbs, and most have employed stage-based models to simulate age-based characteristics (e.g., Bierzychudek 1982b; Meagher & Antonovics 1982a,b; Hara & Wakahara 1994; Bierzychudek 1999; Jolls 2003). Some species have been shown to have stable population structure (Meagher & Antonovics 1982a; Inghe & Tamm 1985; Kawano et al. 1986, 1992), but for most species, population structure is temporally variable (Inghe & Tamm 1985, Matlack 1987, Scheiner 1988, Whitman et al. 1998) due to variation in factors such as weather, disturbance, and levels of nutrient availability (Whigham et al. 1993, Fröborg & Eriksson 1997). Population size may also vary spatially, as has been shown for the rare orchid *Isotria medeoloides* (Mehrhoff 1989). Population trends may also be directional. Diekmann & Dupré (1997), for example, found long-term declines in populations due to regional acidification and eutrophication.

More recent publications suggest that flowering rates may be greater and more variable than the 1%–25% range reported in the 1982 review. Flowering may be size dependent, and once a minimum size is reached, individuals can flower yearly independent of the level of reproductive effort (Lewis & Zenger 1982, Mehrhoff 1989, Cain & Damman 1997). Other species switch back and forth between flowering and nonflowering depending on the level of reproductive effort

(Bierzychudek 1982b, Snow & Whigham 1989, Gerber et al. 1997a). Variability in flowering also appears to be influenced by factors such as genetics (Meagher & Antonovics 1982a), herbivory (Snow & Whigham 1989), weather (Inghe & Tamm 1985), disturbance (Whigham et al. 1993, Primack et al. 1994, Gill 1996), and the costs of producing flowers and fruits (Snow & Whigham 1989, Whigham & O'Neill 1991, Primack & Hall 1990, Falb & Leopold 1993, Gerber et al. 1997b).

GENETIC VARIABILITY

Relatively few studies have examined genetic variability of woodland herbs, and even fewer studies have been based on measurements of genetic identities of plants within or among populations. The most important conclusion thus far is that most populations of clonally reproducing species typically contain more than one genet (e.g., Smith et al. 2002). Patches of some species (e.g., *Clintonia borealis* and *Podophyllum peltatum*) contain few, perhaps one or two, genets (Policansky 1983, Pitelka et al. 1985b). Patches of other woodland herbs (e.g., *Asarum canadense*) contain numerous genotypes (Eriksson 1989, Cain & Damman 1997), and genetic substructuring has been demonstrated for clonal species (Cahalan & Glidden 1985, Kudoh et al. 1999, Ziegenhagen et al. 2003). Some woodland herbs have both high and low genetic diversity, depending on rates of outcrossing and seed set (Ohara et al. 1996).

Uvularia perfoliata (Kudoh et al. 1999) is probably representative of many clonal woodland herbs. An *Uvularia* patch in an area where tree gaps had previously occurred had a high level of genetic variability at small (centimeter) and large (meter) scales. Ramets of different genets intermingled, and there was no evidence of competition between genets. Kudoh et al. (1999) also found genetic substructuring of a few genets with many ramets, and of many genets with few ramets. They attributed this pattern to a few founder genets and subsequent establishment of new genets from seeds that were produced from within-patch outcrossing.

RESPONSES TO DISTURBANCE

Canopy Gaps

Disturbance plays an important role in the dynamics of forests (Pickett & White 1985, Webb 1999), and a close linkage exists between canopy disturbance and dynamics of understory vegetation (Gilliam & Roberts 2003). Collins et al. (1985) reviewed the literature on woodland herb responses to gaps and suggested that herbaceous species can be divided into three guilds (Table 6) and that various physiological, morphological, and ecological responses of species could be predicted for each guild. Woodland herbs clearly respond to changes in light conditions in canopy gaps, but microhabitat heterogeneity may be equally important (Bratton

TABLE 6 Predicted responses (first column) of woodland herbs in gap habitats (Collins et al. 1985). The predicted responses are positive (+), none (0), or negative (−)

Responses	Sun herbs	Light flexible herbs	Shade herbs
Water uptake	0	+	+
Nutrient uptake	+/0	+/0	0
Seedling establishment	+	+	0/−
Leaf duration	+/0	+/0	0/−
Assimilation	+/0	+	−
Pollination	+/0	+	+
Flowering	+/0	+	0/+
Seed set	+/0	+	0
Clonal growth	+/0	+	0/−
Architectural shift	0	+	0
Survivorship	0	+	−

1976). Pits and mounds associated with treefalls, for example, provide a high degree of microhabitat variability and may have a higher species diversity and greater herb cover than adjacent undisturbed areas (Peterson & Campbell 1993). Mounds and logs created during treefalls are sites for colonization by woodland orchids (Rasmussen & Whigham 1998a). Pits and mounds have also been shown to have characteristic species assemblages, depending to a large degree on the dominant tree species (Beatty 1984).

Although microhabitat conditions are important, changes in light quality and quantity associated with gaps generate the greatest responses in understory herbs because most species are light limited (Table 6). Most woodland herbs experience increased growth and reproduction in response to increased light (Collins & Pickett 1988, Neufeld & Young 2003). Positive responses to gaps are related to gap size, but positive responses can be partially offset by negative impacts associated with competition (Hughes 1992).

Individual species have shown a range of responses to increased light in gaps. Gap size has a positive influence on population size (Scheiner 1988), growth rates, and frequency of flowering (Dahlem & Boerner 1987, Whigham et al. 1993, Griffith 1996, Wijesinghe & Whigham 1997). Recently, models have been used to evaluate a range of species responses to gap habitats. Valverde & Silvertown (1997a, 1998) concluded that *Primula vulgaris* formed metapopulations in response to gaps and that local populations formed and became extinct in response to changing habitat conditions associated with light gaps. Cipollini et al. (1993) and Whigham et al. (1993) found similar results for *Cynoglossum virginianum*, a nonclonal woodland herb that persists in a vegetative state for long periods under closed canopy conditions. When gaps form, individuals increase in size and flower until light levels decline, eventually returning to a vegetative state.

Herbivory

Woodland herbs (e.g., ferns) produce chemicals that deter herbivores, but herbivory appears to have a negative impact on most species (Brunet 1993). The impacts of herbivory on growth and reproduction of woodland herbs appears to be, in part, species dependent (Rockwood & Lobstein 1994), but other factors are also important. Plant size (Davis 1981, Rooney & Waller 2001), sex (Delph et al. 1993), and nutritional quality (Ericson & Oksanen 1987) are important. The plant parts attacked by herbivores (Ericson & Oksanen 1987), herbivore-feeding patterns (Ehrlén 1995, Matlack 1987), the phenological status of the plant (de Kroon et al. 1991, Watson 1995, Whigham & Chapa 1999), and the degree of physiological integration in clonal species (Ashmun et al. 1982, Price et al. 1992) are also important.

As might be expected, the greater the amount of tissue removed the greater the impacts on growth and reproduction, but other factors (e.g., the pattern of leaf removal) also influence plant responses to herbivory (Whigham 1990, Price & Hutchings 1992). Perhaps most importantly, most woodland herbs do not produce new aboveground tissues in response to herbivory, which suggests that the timing and amount of herbivory would influence short- and long-term plant responses. Complete or partial leaf removal can result in cessation of or decreased flowering (Whigham 1990, Syrjänen & Lehtilä 1993, Primack et al. 1994). Partial defoliation can, however, also have no effect on growth and reproduction (Agren & Willson 1992).

Most herbivory studies have only considered aboveground plant parts. Herbivores also consume belowground tissues (Ericson & Wennström 1997). Rodent herbivory of underground corms of *Tipularia discolor*, for example, had a greater impact on the long-term dynamics of populations than any other factor (Whigham & O'Neill 1991).

Deer browsing can result in the almost complete elimination of woodland herbs, except for a few unpalatable species (Tilghman 1989). In addition, species that can survive on microsites (e.g., refuges such as mounds, boulders, and logs) that are unavailable to deer can also avoid browsing (Rooney & Dress 1997a). If species can survive browsing, however, there is some evidence that they can recover following elimination of deer browsing (Albert & Barnes 1987).

Waller & Alverson (1997) agreed with McShea & Rappole (1992) in concluding that deer are keystone species with impacts well beyond reducing herb diversity in forests. Deer browsing reduces or eliminates tree regeneration (e.g., Nomiya et al. 2002), and browsing results in the loss of bird species that specialize in the forest understory (Casey & Hein 1983). Deer browsing can also reduce invertebrate diversity (Miller et al. 1992), and increased deer abundance has been linked to Lyme disease (Jones et al. 1998). The evidence seems clear that deer abundance and management will continue to be a major threat to the diversity of woodland herbs in forests of eastern North America.

Urbanization and Forest Conversion

Researchers have examined the fate of woodland herbs in urban settings, concluding that we may not be able to conserve the original species diversity in urban forests (Robinson et al. 1994, Drayton & Primack 1996). Similarly, several studies (Peterken & Game 1984, Falinski et al. 1988, Brunet 1993, Hermy et al. 1993, Hermy 1994, Singleton et al. 2001, Bellemare et al. 2002, Vellend 2003, Verheyen et al. 2003) have shown that it will take long periods of time, perhaps centuries, for herb diversity to recover in forests that developed on lands that had previously been cleared and used for agricultural production. For example, forested areas in the Vosges Mountains that had been cleared and used to support crops, gardens, pastures, or tree plantations had similar species diversity of woodland herbs, but none of them were characteristic of ancient forests (Koerner et al. 1997).

Logging

Woodland herb diversity may be lower in logged forests, and it may never again reach levels found in old-growth forests (Meier et al. 1995). Comparison of different logging activities showed that woodland herbs were negatively impacted by clearcuts (Albert & Barnes 1987, Duffy & Meier 1992) and completely eliminated by clearcutting followed by herbicide application (Kochenderfer & Wendel 1983). Duffy & Meier (1992) concluded that it was unlikely that the herb community would recover in the clearcut stands within normal (40–150 years) cutting cycles. Selective logging had less of an impact on herb diversity in Canada (Reader 1987, 1988; Meier et al. 1995), and Reader & Bricker (1992) suggested that selective logging could be used as a management tool to conserve woodland herbs.

CONCLUSIONS AND RECOMMENDATIONS

Woodland herbs include species of diverse phylogenetic origin that have evolved a wide range of life history adaptations that allow them to persist and flourish in an environment that is often light limited. Light is clearly the most important factor limiting the growth and reproduction of woodland herbs, but some life history strategies (e.g., spring ephemerals, wintergreen species, parasites, etc.) have evolved that allow individuals to complete most of their growth and reproduction prior to full development of the tree canopy and resultant low light conditions.

Although much more has been learned about the ecology of woodland herbs since Bierzychudek's (1982a) review, only a small percentage of the species have been studied in detail, and, at best, we can still offer only broad generalizations about the ecology of woodland herbs. What types of studies will most likely benefit future efforts to understand the ecology of woodland herbs and also support conservation and restoration efforts? Individual species will undoubtedly continue to attract the attention of researchers, and those types of efforts should be encouraged. The most effective studies, however, are likely to be those that use a synthetic approach in which functional groups (e.g., Givnish 1982, Kawano 1985) are identified and used as the basis for testing ideas that further our understanding of woodland herbs as well as support conservation and restoration efforts.

As indicated, only a few broad generalizations about the ecology of woodland herbs are, however, possible because only a limited number of species and have been studied in detail. First, it is clear that most species are adapted to take advantage of canopy disturbances, which result in higher light conditions for a few years or more and higher levels of other resources. Canopy gaps often have higher herb diversity because they have a greater range of microhabitats (e.g., pits and mounds, coarse wood in various stages of decomposition), but light is still the key to success of most species. Most woodland herbs have increased growth, sexual reproduction and asexual propagation in gaps, and seedling recruitment also appears to be more common in gap habitats. It is difficult, however, to evaluate species responses to gap disturbances without either long-term studies or experimental manipulations. Matrix models offer the opportunity to explore broader questions about the range of responses of woodland herbs to gap disturbances. To date, however, matrix models have been applied to few species, and a useful approach would be to use them to address questions related to functional groups. The incorporation of spatially explicit models into studies of woodland herb populations would also be useful because of the high degree of spatial heterogeneity found in forest habitats.

A second broad generalization is that all woodland herbs are mycorrhizal. What are the ecological consequences of mycorrhizal interactions beyond those that have been associated with nutrient uptake? We still know very little about the importance of mycorrhizal interactions in the establishment and growth of seedlings of woodland herbs, especially species that require mycorrhizal associations for germination and survival.

A third generalization is that herbivores and pathogens affects the short- and long-term dynamics of woodland herb populations. Herbivores and pathogens probably influence woodland herbs much more than have been identified to date. Additional studies are needed on the growth and reproductive responses of woodland herbs to pathogens and herbivores, especially the responses of belowground tissues.

Conservation issues have only begun to be addressed, and future research needs to focus on the maintenance and restoration of woodland herbs in landscapes influenced more directly by human activities (e.g., clearcutting and fragmentation of forests) or indirectly (e.g., deer browsing, invasion of alien species). Little research has been conducted on management approaches that might be used to reintroduce species into forests from which they have been eliminated.

Finally, more effort should be placed on long-term population studies of woodland herbs. Inevitably, long-term studies provide valuable information that can be used to design experiments, develop realistic parameters for modeling efforts, and open new lines of investigation. Long-term studies reveal previously unknown facts about woodland herb ecology and the many important interactions that woodland herbs have with other organisms. Although difficult to fund and sustain, long-term studies have low start-up and maintenance costs and are ideal for demonstrating ecological phenomena in teaching environments. Small colleges and universities throughout the eastern United States own or have access to sites that would be ideal for conducting long-term studies of woodland herbs. Datasets that expand from year to year with relatively little effort can be very useful in teaching ecological principles. Hopefully, this review will encourage others to start both short- and long-term studies using this interesting group of plants as models to embellish scientific knowledge and educate the public and future scientists.

In closing, space limitations required me to restrict the number of references cited in the review. A larger list of relevant documents can be found at http://www.serc.si.edu/labs/plant_ecology/biblio.jsp, and additional references are in Gilliam & Roberts (2003).

ACKNOWLEDGMENTS

The review was written while the author was a Bullard Fellow at Harvard Forest. I extend thanks to the staff of the Harvard Forest, especially David Foster, Glen Mozkin, and Ruth Kern for reading early versions and for discussions related to it. I thank Matt Baker, Jerry Baskin, Candy Feller, Anne Innis, Ann Lubbers, Jay O'Neill, Ryan King, Melissa McCormick, Glen Mozkin, Dag-Inge Øien, Mark Vellend, and Dushyantha Wijesinghe for comments on various drafts. The Smithsonian's Environmental Sciences and Scholarly Studies programs have supported the author's research on woodland herbs. Smithsonian Fellowship and Visiting Scientist programs provided funding for post-docs and visitors who work on woodland herbs (Martin Cipollini, Hans de Kroon, Heidrun Huber, Hiroshi Kudoh, Melissa McCormick, Hanne Rasmussen, Allison Snow, Dushyantha Wijesinghe, Jess Zimmerman). The Smithsonian Work-Learn program supported many student interns over a 20-year period. The long-term influence of Shoichi Kawano is much appreciated, as are grants from the Smithsonian Institution and the Japanese government, which have provided resources for long-term interactions between our labs. Finally, special thanks to Jan for her support during the preparation of the manuscript.

The *Annual Review of Ecology, Evolution, and Systematics* is online at
http://ecolsys.annualreviews.org

LITERATURE CITED

Agren J, Willson MF. 1992. Determinants of seed production in *Geranium maculatum*. *Oecologia* 92:177–82

Albert DA, Barnes BV. 1987. Effects of clearcutting on the vegetation and soil of a Sugar Maple–dominated ecosystem, Western Upper Michigan. *For. Ecol. Manag.* 18:283–98

Angevine MW, Handel SN. 1986. Invasion of forest floor space, clonal architecture, and population growth in the perennial herb *Clintonia borealis*. *J. Ecol.* 74:547–60

Antos JA. 1988. Underground morphology and habitat relationships of three pairs of forest herbs. *Am. J. Bot.* 75:106–13

Ashmun JW, Thomas RJ, Pitelka LF. 1982. Translocation of photoassimilates between sister ramets in two rhizomatous forest herbs. *Ann. Bot.* 49:403–15

Barrett SCH, Helenurm K. 1987. The reproductive biology of boreal forest herbs. I. Breeding systems and pollination. *Can. J. Bot.* 65:2036–46

Baskin CC, Baskin JM. 1998. *Seeds: Ecology, Biogeography, and Evolution of Dormancy and Germination.* New York: Academic

Baskin CC, Baskin JM. 2002. Achene germination ecology of the federally threatened floodplain endemic *Boltonia decurrens* (Asteraceae). *Am. Midl. Nat.* 147:16–24

Baskin CC, Baskin JM, Chester EW. 1993. Seed germination ecology of two mesic woodland winter annuals, *Nemophila aphylla* and *Phacelia ranunculaceae* (Hydrophyllaceae). *Bull. Torrey Bot. Club* 120:29–37

Baskin JM, Baskin CC. 1983. Germination ecology of *Collinsia verna*, a winter annual of rich deciduous woodlands. *Bull. Torrey Bot. Club* 110:311–15

Baskin JM, Baskin CC. 1986. Seed germination ecophysiology of the woodland herb *Asarum canadense*. *Am. Midl. Nat.* 116:132–39

Baskin JM, Baskin CC. 1988. The ecological life cycle of *Cryptotaenia canadensis* (L.) DC. (Umbelliferae), a woodland herb with monocarpic ramets. *Am. Midl. Nat.* 119:165–73

Baskin JM, Baskin CC. 1989. Seed germination ecophysiology of *Jeffersonia diphylla*, a perennial herb of mesic deciduous forests. *Am. J. Bot.* 76:1073–80

Baskin JM, Baskin CC. 1992. Role of temperature and light in the germination ecology of buried seeds of weedy species of disturbed forests I. *Lobelia inflata*. *Can. J. Bot.* 70:589–92

Bawa KS, Keegan CR, Voss RH. 1982. Sexual dimorphism in *Aralia nudicaulis* L. (Araliaceae). *Evolution* 36:371–78

Beattie AJ, Lyons N. 1975. Seed dispersal in *Viola* (Violaceae): adaptations and strategies. *Am. J. Bot.* 62:714–22

Beatty SW. 1984. Influence of microtopography and canopy species on spatial patterns of forest understory plants. *Ecology* 65:1406–19

Beatty SW. 2003. Habitat heterogeneity and maintenance of species in understory communities. In *The Herbaceous Layer in Forests of Eastern North America*, ed. FS Gilliam, MR Roberts, pp. 177–97. New York: Oxford Univ. Press

Bell AD. 1974. Rhizome organization in relation to vegetative spread in *Medeola virginiana*. *J. Arnold Arb.* 55:458–68

Bellemare J, Motzkin G, Foster DR. 2002. Legacies of the agricultural past in the forest present: an assessment of historical land-use effects on rich mesic forests. *J. Biogeogr.* 29:1401–20

Bender MH, Baskin JM, Baskin CC. 2000. Ecological life history of *Polymnia canadensis*, a monocarpic species of the North American temperate deciduous forest. *Plant Ecol.* 147:117–36

Bender MH, Baskin JM, Baskin CC. 2002. Phenology and common garden and reciprocal transplant studies of *Polymnia canadensis* (Asteraceae), a monocarpic species of the North American temperate deciduous forest. *Plant Ecol.* 161:15–39

Bender MH, Baskin JM, Baskin CC. 2003. Seed germination ecology of *Polymnia canadensis* (Asteraceae), a monocarpic species of the North American Temperate Deciduous Forest. *Plant Ecol.* 168:221–53

Benner BL, Watson MA. 1989. Developmental ecology of mayapple: seasonal patterns of resource distribution in sexual and vegetative rhizome systems. *Funct. Ecol.* 3:539–47

Berg H. 2000. Differential seed dispersal in *Oxalis acetosella*, a cleistogamous perennial herb. *Acta Oecol.* 21:109–18

Bertin RI, Sholes ODV. 1993. Weather, pollination and phenology of *Geranium maculatum*. *Am. Midl. Nat.* 129:52–66

Bierzychudek P. 1982a. Life histories and demography of shade-tolerant temperate forest herbs: a review. *New Phytol.* 90:757–76

Bierzychudek P. 1982b. The demography of Jack-in-the pulpit, a forest perennial that changes sex. *Ecol. Monogr.* 52:335–51

Bierzychudek P. 1999. Looking backwards: assessing the projections of a transition matrix model. *Ecol. Appl.* 9:1278–87

Bloom TC, Baskin JM, Baskin CC. 2001. Ecological life history of the facultative woodland biennial *Arabis laevigata* variety *laevigata* (Brassicaceae): survivorship. *J. Torrey Bot. Soc.* 128:93–108

Bloom TC, Baskin JM, Baskin CC. 2002. Ecological life history of the facultative woodland biennial *Arabis laevigata* variety *laevigata* (Brassicaceae): seed dispersal. *J. Torrey Bot. Soc.* 129:21–28

Boerner REJ. 1986. Seasonal nutrient dynamics, nutrient resorption, and mycorrhizal infection intensity of two perennial forest herbs. *Am. J. Bot.* 73:1249–57

Boerner REJ. 1992. Plant life span and response to inoculation with vesicular-arbuscular mycorrhizal fungi. III. Responsiveness and residual soil P levels. *Mycorrhiza* 1:169–74

Boerner REJ, DeMars BG, Leicht PN. 1996. Spatial patterns of mycorrhizal infectiveness of soils along a successional chronosequence. *Mycorrhiza* 6:79–90

Boyd M, Silvertown J, Tucker C. 1990. Population ecology of heterostyle and homostyle *Primula vulgaris*: growth, survival and reproduction in field populations. *J. Ecol.* 78:799–813

Bratton SP. 1976. Resource division in an understory herb community: responses to temporal and microtopographic gradients. *Am. Nat.* 110:679–93

Braun J, Brooks GR Jr. 1987. Box turtles (*Terrapene carolina*) as potential agents for seed dispersal. *Am. Midl. Nat.* 117:312–18

Brundrett M, Kendrick B. 1990a. The roots and mycorrhizas of herbaceous woodland plants I. Quantitative aspects of morphology. *New Phytol.* 114:457–67

Brundrett M, Kendrick B. 1990b. The roots and mycorrhizas of herbaceous woodland plants II. Structural aspects of morphology. *New Phytol.* 114:469–79

Brunet J. 1993. Environmental and historical factors limiting the distribution of rare forest grasses in south Sweden. *For. Ecol. Manag.* 61:263–75

Cahalan CM, Gliddon C. 1985. Genetic neighbourhood sizes in *Primula vulgaris*. *Heredity* 54:65–70

Cain ML, Damman H. 1997. Clonal growth and ramet performance in the woodland herb, *Asarum canadense*. *J. Ecol.* 85:883–97

Cain ML, Damman H, Muir A. 1998. Seed dispersal and the Holocene migration of woodland herbs. *Ecol. Monogr.* 68:325–47

Cain ML, Dudle DA, Evans JP. 1996. Spatial models of foraging in clonal plant species. *Am. J. Bot.* 83:76–85

Casey D, Hein D. 1983. Effects of heavy browsing on a bird community in a deciduous forest. *J. Wildl. Manag.* 47:829–36

Cipollini ML, Whigham DF, O'Neill J. 1993. Population growth, structure, and seed dispersal in the understory herb *Cynoglossum virginianum*: a population and patch dynamic model. *Plant Species Biol.* 8:117–29

Clay K. 1993. Size-dependent gender change in green dragon (*Arisaema dracontium*: Araceae). *Am. J. Bot.* 80:769–77

Collins BS, Dunne KP, Pickett STA. 1985. Responses of forest herbs to canopy gaps. See Pickett & White 1985, pp. 218–34

Collins BS, Pickett STA. 1988. Response of herb layer cover to experimental canopy gaps. *Am. Midl. Nat.* 119:282–90

Cook RE. 1983. Clonal plant populations. *Am. Sci.* 71:244–53

Cook RE. 1988. Growth in *Medeola virginiana* clones. I. Field observations. *Am. J. Bot.* 75:725–31

Corbit M, Marks PL, Gardescu S. 1999. Hedgerows as habitat corridors for forest herbs in central New York, USA. *J. Ecol.* 87:220–32

Crozier CR, Boerner REJ. 1984. Correlations of understory herb distribution patterns with microhabitats under different tree species in a mixed mesophytic forest. *Oecologia* 62:337–43

Culver DC, Beattie AJ. 1978. Myrmecochory in *Viola*: dynamics of seed-ant interactions in some West Virginia species. *J. Ecol.* 66:53–72

Dahlem TS, Boerner REJ. 1987. Effects of canopy light gap and early emergence on the growth and reproduction of *Geranium maculatum*. *Can. J. Bot.* 65:242–45

Davis M. 1981. The effect of pollinators, predators, and energy constraints on the floral ecology and evolution of *Trillium erectum*. *Oecologia* 48:400–6

Davis RW. 1986. The pollination biology of *Cypripedium acaule* (Orchidaceae). *Rhodora* 88:445–50

de Kroon H, Hutchings MJ. 1995. Morphological plasticity in clonal plants: the foraging concept reconsidered. *J. Ecol.* 83:143–52

de Kroon H, Whigham DF, Watson MA. 1991. Developmental ecology of mayapple: effects of rhizome severing, fertilization and timing of shoot senescence. *Funct. Ecol.* 5:360–68

Delph LF, Lu Y, Jayne LD. 1993. Patterns of resource allocation in a dioecious *Carex* (Cyperaceae). *Am. J. Bot.* 80:607–15

DeMars BG, Boerner REJ. 1995. Mycorrhizal dynamics of three woodland herbs of contrasting phenology along topographic gradients. *Am. J. Bot.* 82:1426–31

DeMars BG, Boerner REJ. 1997. Foliar phosphorus and nitrogen resorption in three woodland herbs of contrasting phenology. *Castanea* 62:43–54

Diekmann M, Dupré C. 1997. Acidification and eutrophication of deciduous forests in northwestern Germany demonstrated by indicator species analysis. *J. Veg. Sci.* 8:885–64

Dieringer G. 1982. The pollination ecology of *Orchis spectabilis* L. (Orchidaceae). *Ohio J. Sci.* 82:218–24

Drayton B, Primack RB. 1996. Plant species lost in an isolated conservation area in metropolitan Boston from 1894–1993. *Conserv. Biol.* 10:30–39

Duffy DC, Meier AJ. 1992. Do Appalachian herbaceous understories ever recover from clearcutting? *Conserv. Biol.* 6:196–201

Ehrlén J. 1995. Demography of the perennial herb *Lathyrus vernus*. I. Herbivory and individual performance. *J. Ecol.* 83:287–95

Ehrlén J, Eriksson O. 2000. Dispersal limitation and patch occupancy in forest herbs. *Ecology* 81:1667–74

Ericson L, Oksanen L. 1987. The impact of controlled grazing by *Clethrionomys rufocanus* on experimental guilds of boreal forest floor herbs. *OIKOS* 50:403–16

Ericson L, Wennström A. 1997. The effect of herbivory on the interaction between the clonal plant *Trientalis europaea* and its smut fungus *Urocystis trientalis*. *OIKOS* 80:107–11

Eriksson O. 1989. Seedling dynamics and life histories in clonal plants. *OIKOS* 55:231–38

Eriksson O. 1992. Population structure and dynamics of the clonal dwarf-shrub *Linnaea borealis*. *J. Veg. Sci.* 3:61–68

Ernst WHO. 1979. Population biology of *Allium ursinum* in northern Germany. *J. Ecol.* 67:347–62

Falb DL, Leopold DJ. 1993. Population dynamics of *Cypripedium candidum* Muhl. ex Willd., Small White Ladyslipper, in a western New York fen. *Nat. Areas J.* 13:76–86

Falinski JB, Canullo R, Bialy K. 1988. Changes in herb layer, litter fall and soil properties under primary and secondary tree stands in a deciduous forest ecosystem. *Phytocoenosis* 1:1–49

Flanagan LB, Moser W. 1985a. Flowering phenology, floral display and reproductive success in dioecious, *Aralia nudicaulis* L. (Araliaceae). *Oecologia* 68:23–28

Flanagan LB, Moser W. 1985b. Pattern of ^{14}C assimilate distribution in a clonal herb, *Aralia nudicaulis*. *Can. J. Bot.* 63:2111–14

Fröborg H, Eriksson O. 1997. Local colonization and extinction of field layer plants in a deciduous forest and their dependence upon life history features. *J. Veg. Sci.* 8:395–400

Galen C, Plowright RC, Thomson JD. 1985. Floral biology and regulation of seed set and seed size in the lily, *Clintonia borealis*. *Am. J. Bot.* 72:1544–52

Galloway LF. 2002. The effect of maternal phenology on offspring characters in the herbaceous plant *Campanula americana*. *J. Ecol.* 90:851–58

Gerber MA, de Kroon H, Watson MA. 1997a. Organ preformation in mayapple as a mechanism for historical effects on demography. *J. Ecol.* 85:211–23

Gerber MA, Watson MA, de Kroon H. 1997b. Organ preformation, development, and resource allocation in perennials. In *Plant Resource Allocation*, ed. FA Bazzaz, J Grace, pp. 113–41. New York: Academic

Gill DE. 1996. The natural population ecology of temperate terrestrials: Pink Lady's-Slippers, *Cypripedium acaule*. In *North American Native Terrestrial Orchids, Propagation and Production*, ed. C Allen, pp. 91–106. Germantown, MD: N. Am. Native Terr. Orchid Conf.

Gilliam FS. 1988. Interactions of fire with nutrients in the herbaceous layer of a nutrient-poor Coastal Plain forest. *Bull. Torrey Bot. Club* 115:265–71

Gilliam FS, Roberts MR. 2003. *The Herbaceous Layer in Forests of Eastern North America*. New York: Oxford Univ. Press

Givnish TJ. 1982. On the adaptive significance of leaf height in forest herbs. *Am. Nat.* 120:353–81

Givnish TJ. 1983. Biomechanical constraints on crown geometry in forest herbs. In *On the Economy of Plant Form and Function*, ed. TJ Givnish, pp. 525–83. New York: Cambridge Univ. Press

Givnish TJ. 1987. Comparative studies of leaf form: assessing the relative roles of selective pressures and phylogenetic constraints. *New Phytol.* 106:131–60

Griffith C, Jr. 1996. Distribution of *Viola blanda* in relation to within-habitat variation in canopy openness, soil phosphorus, and magnesium. *Bull. Torrey Bot. Club* 123:281–85

Gross KL, Berner T, Marschall E, Tomcko C. 1983. Patterns of resource allocation among five herbaceous perennials. *Bull. Torrey Bot. Club* 110:345–52

Gunther RW, Lanza J. 1989. Variation in attractiveness of *Trillium* diaspores to a seed-dispersing ant. *Am. Midl. Nat.* 122:321–28

Handel SN. 1976. Dispersal ecology of *Carex pedunculata* (Cyperaceae), a new North American myrmecochore. *Am. J. Bot.* 63:1071–79

Handel SN. 1978. New ant-dispersed species in the genera *Carex, Luzula*, and *Claytonia. Can. J. Bot.* 56:2925–27

Handel SN, Fisch SB, Schatz GE. 1981. Ants disperse a majority of herbs in a mesic forest community in New York state. *Bull. Torrey Bot. Club* 108:430–37

Hannan GL, Prucher HA. 1996. Reproductive biology of *Caulophyllum thalictroides* (Berberidaceae), an early flowering perennial of eastern North America. *Am. Midl. Nat.* 136:267–77

Hara T, Wakahara M. 1994. Variation in individual growth and the population structure of a woodland perennial herb, *Paris tetraphylla*. *J. Ecol.* 82:3–12

Harder LD, Cruzan MB, Thomson JD. 1993. Unilateral incompatibility and the effects of interspecific pollination for *Erythronium americanum* and *Erythronium albidum* (Liliaceae). *Can. J. Bot.* 71:353–58

Heithaus ER. 1981. Seed predation by rodents on three ant-dispersed plants. *Ecology* 62:136–45

Hermy M. 1994. Effects of former land use on plant species diversity and pattern in European deciduous woodlands. In *Biodiversity, Temperate Ecosystems, and Global Change*, ed. TJB Boyle, CEB Boyle, pp. 123–44. Berlin: Springer-Verlag

Hermy M, Honnay O, Firbank L, Grashof-Bokdam C-J, Lawesson J-E. 1999. An ecological comparison between ancient and other forest plant species of Europe, and the implications for forest conservation. *Biol. Conserv.* 91:9–22

Hermy M, van den Bremt P, Tack G. 1993.

Effects of site history on woodland vegetation. In *European Forest Reserves*, ed. MEA Broekmeyer, W Vos, H Koop, pp. 219–31. Wageningen, Neth.: Pudoc Sci.

Higashi S, Tsuyuzaki S, Ohara M, Ito F. 1989. Adaptive advantages of ant-dispersed seeds in the myrmecochorous plant *Trillium tschonoskii* (Liliaceae). *OIKOS* 54:389–94

Hiirsalmi H. 1969. *Trientalis europaea* L. A study of the reproductive biology, ecology, and variation in Finland. *Ann. Bot. Fenn.* 6:119–73

Holland PG. 1974. The growth behavior, ecology, and geography of *Erythronium americanum* in northeast North America. *Can. J. Bot.* 52:1765–72

Hoppes WG. 1988. Seedfall pattern of several species of bird-dispersed plants in an Illinois woodland. *Ecology* 69:320–29

Hori Y, Yokoi T. 1999. Population structure and dynamics of an evergreen shade herb, *Ainsliaea apiculata* (Asteraceae), with special reference to herbivore effects. *Ecol. Res.* 14:39–48

Hughes JW. 1992. Effect of removal of co-occurring species on distribution and abundance of *Erythronium americanum* (Liliaceae), a spring ephemeral. *Am. J. Bot.* 79:1329–36

Hutchings MJ, de Kroon H. 1994. Foraging in plants: the role of morphological plasticity in resource acquisition. *Adv. Ecol. Res.* 25:160–238

Hutchings MJ, Mogie M. 1990. The spatial structure of clonal plants: control and consequences. In *Clonal Growth in Plants: Regulation and Function*, ed. J van Groenendael, H de Kroon, pp. 57–76. The Hague: SPB Academic

Inghe O, Tamm CO. 1985. Survival and flowering of perennial herbs. IV. The behaviour of *Hepatica nobilis* and *Sanicula europaea* on permanent plots during 1943–1981. *OIKOS* 45:400–20

Irwin RE. 2000. Morphological variation and female reproductive success in two sympatric *Trillium* species: evidence for phenotypic selection in *Trillium erectum* and *Trillium grandiflorum* (Liliaceae). *Am. J. Bot.* 87:205–14

Jasieniuk M, Lechowicz MJ. 1987. Spatial and temporal variation in chasmogamy and cleistogamy in *Oxalis montana* (Oxalidaceae). *Am. J. Bot.* 74:1672–80

Jolls CL. 2003. Populations and threats to rare plants of the herb layer: more challenges and opportunities for conservation biology. In *The Herbaceous Layer of Forests of Eastern North America*, ed. FS Gilliam, MR Roberts, pp. 105–59. New York: Oxford Univ. Press

Jones CG, Ostfeld RS, Richard MP, Schauber EM, Wolff JO. 1998. Chain reactions linking acorn to gypsy moth outbreaks and Lyme disease risk. *Science* 279:1023–26

Jurik TW. 1983. Reproductive effort and CO_2 dynamics of wild strawberry populations. *Ecology* 64:1329–42

Jónsdóttir IS, Watson MA. 1997. Extensive physiological integration: an adaptive trait in resource-poor environments? In *The Ecology and Evolution of Clonal Plants*, ed. H de Kroon, J van Groenendael, pp. 109–36. Leiden, Neth.: Backhuys

Kalisz S, Hanzawa FM, Tonsor SJ, Thiede DA, Voigt S. 1999. Ant-mediated seed dispersal alters pattern of relatedness in a population of *Trillium grandiflorum*. *Ecology* 80:2620–34

Kawano S. 1975. The productive and reproductive biology of flowering plants. II. The concept of life history strategy in plants. *J. Coll. Lib. Arts Toyama Univ.* 8:51–86

Kawano S. 1985. Life history characteristics of temperate woodland plants in Japan. In *The Population Structure of Vegetation*, ed. J White, pp. 515–49. Dordrecht: Dr. W. Junk

Kawano S, Ohara M, Utech FH. 1986. Life history studies on the genus *Trillium* (Liliaceae) II. Reproductive biology and survivorship of four eastern North American species. *Plant Species Biol.* 1:47–58

Kawano S, Ohara M, Utech FH. 1992. Life history studies on the genus *Trillium* (Liliaceae) VI. Life history characteristics of three western North American species and their evolutionary-ecological implications. *Plant Species Biol.* 7:21–36

Klimés L, Klimesová J, Hendriks R, van Groenendael J. 1997. Clonal plant architecture: a comparative analysis of form and function. In *The Ecology and Evolution of Clonal Plants*, ed. H de Kroon, J van Groenendael, pp. 1–29. Leiden, Neth.: Backhuys

Klinka K, Wang Q, Carter RE. 1990. Relationships among humus forms, forest floor nutrient properties, and understory vegetation. *For. Sci.* 36:564–81

Knight GH. 1964. Some factors affecting the distribution of *Endymion Nonscriptus* (L.) Garcke in Warwickshire Woods. *J. Ecol.* 52:405–56

Kochenderfer JN, Wendel GW. 1983. Plant succession and hydrologic recovery on a deforested and herbicided watershed. *For. Sci.* 29:545–58

Koerner W, Dupouey JL, Dambrine E, Benoît M. 1997. Influence of past land use on the vegetation and soils of present day forest in the Vosges mountains, France. *J. Ecol.* 85:351–58

Kudoh H, Shibaike H, Takasu H, Whigham DF, Kawano S. 1999. Genet structure and determinants of clonal structure in a temperate deciduous woodland herb, *Uvularia perfoliata*. *J. Ecol.* 87:244–57

Landa K, Benner B, Watson M, Gartner J. 1992. Physiological integration for carbon in mayapple (*Podophyllum peltatum*), a clonal perennial herb. *OIKOS* 63:348–56

Lapointe L, Molard J. 1997. Cost and benefits of mycorrhizal infection in a spring ephemeral, *Erythronium americanum*. *New Phytol.* 135:491–500

Lau RR, Young DR. 1988. Influence of physiological integration on survivorship and water relations in a clonal herb. *Ecology* 69:215–19

Laverty TM, Plowright RC. 1988. Fruit and seed set in Mayapple (*Podophyllum peltatum*): influence of intraspecific factors and local enhancement near *Pedicularis canadensis*. *Can. J. Bot.* 66:173–78

Lechowicz MJ, Schoen DJ, Bell G. 1988. Environmental correlates of habitat distribution and fitness components in *Impatiens capensis* and *Impatiens pallida*. *J. Ecol.* 76:1043–54

Leckie S, Vellend M, Bell G, Waterway MJ, Lechowicz MJ. 2000. The seed bank in an old-growth, temperate deciduous forest. *Can. J. Bot.* 78:181–92

Leimu R, Syrjänen K, Ehrlén J. 2002. Pre-dispersal seed predation in *Pirmula veris*: among-population variation in damage intensity and selection on flower number. *Oecologia* 133:510–16

Lewis WJ, Zenger VE. 1982. Population dynamics of the American Ginseng *Panax quinquefolium* (Araliaceae). *Am. J. Bot.* 69:1483–90

Lovett Doust L, Lovett Doust J, Turi K. 1986. Fecundity and size relationships in Jack-in-the-Pulpit *Arisaema triphyllum* (Araceae). *Am. J. Bot.* 73:489–94

Lubbers AE, Lechowicz MJ. 1989. Effects of leaf removal on reproduction vs. belowground storage in *Trillium grandiflorum*. *Ecology* 70:85–96

Marino RC, Eisenberg RM, Cornell HV. 1997. Influence of sunlight and soil nutrients on clonal growth and sexual reproduction of the understory perennial herb *Sanguinaria canadensis* L. *J. Torrey Bot. Soc.* 124:219–27

Martin MC. 1965. An ecological life history of *Geranium maculatum*. *Am. Midl. Nat.* 73:111–49

Matlack GR. 1987. Comparative demographies of four adjacent populations of the perennial herb *Silene dioica* (Caryophyllaceae). *J. Ecol.* 75:113–34

Matsumura C, Washitani I. 2000. Effects of population size and pollinator limitation on seed-set of *Primula sieboldii* populations in a fragmented landscape. *Ecol. Res.* 15:307–22

Mattila T, Salonen V. 1995. Reproduction of *Viola mirabilis* in relation to light and nutrient availability. *Can. J. Bot.* 73:1917–24

McCall C, Primack RB. 1987. Resources limit the fecundity of three woodland herbs. *Oecologia* 71:431–35

McCarthy BC. 2003. The herbaceous layer of

eastern old-growth deciduous forests. See Gilliam & Roberts 2003, pp. 163–76

McCarthy BC, Bailey DR. 1996. Composition, structure, and disturbance history of Crabtree Woods: an old-growth forest of western Maryland. *Bull. Torrey Bot. Club* 123:350–65

McShea WJ, Rappole JH. 1992. White-tailed deer as keystone species within forested habitats of Virginia. *Va. J. Sci.* 43:177–86

Meagher TR, Antonovics JJ. 1982a. Life history variation in dioecious plant populations: a case study of *Chamaelirium luteum*. In *Evolution and Genetics of Life Histories*, ed. H Dingle, JL Hegmann, pp. 139–54. New York: Springer-Verlag

Meagher TR, Antonovics JJ. 1982b. The population biology of *Chamaelirium luteum*, a dioecious member of the lily family: life history studies. *Ecology* 63:1690–700

Meekins JF, McCarthy BC. 2000. Responses of the biennial forest herb *Alliaria petiolata* to variation in population density, nutrient addition and light availability. *J. Ecol.* 88:447–63

Mehrhoff LA. 1989. The dynamics of declining populations of an endangered orchid, *Isotria medeoloides*. *Ecology* 70:783–86

Meier AJ, Bratton SP, Duffy DC. 1995. Possible ecological mechanisms for loss of vernal-herb diversity in logged eastern deciduous forests. *Ecol. Appl.* 5:935–46

Miller SG, Bratton SP, Hadidian J. 1992. Impacts of white-tailed deer on endangered plants. *Nat. Areas J.* 12:67–74

Miller TF, Mladenoff DJ, Clayton MK. 2002. Old-growth northern hardwood forests: spatial autocorrelation and patterns of understory vegetation. *Ecol. Monogr.* 72:487–503

Minoletti ML, Boerner REJ. 1993. Seasonal photosynthesis, nitrogen and phosphorus dynamics, and resorption in the wintergreen fern *Polystichum acrosticoides* (Mich.) Schott. *Bull. Torrey Bot. Club* 120:397–404

Motten AF. 1982. Autogamy and competition for pollinators in *Hepatica americana* (Ranunculaceae). *Am. J. Bot.* 69:1296–305

Mukerji SK. 1936. Contributions to the autecology of *Mercurialis perennis* L. *J. Ecol.* 24:38–81

Müller N, Schneller JJ, Holderegger R. 2000. Variation in breeding system among populations of the common woodland herb *Anemone nemorosa* (Ranunculacae). *Plant Syst. Evol.* 221:69–76

Muller RN. 1979. Biomass accumulation and reproduction in *Erythronium albidum*. *Bull. Torrey Bot. Club* 106:276–83

Muller RN. 2003. Nutrient relations of the herbaceous layer in deciduous forest ecosystems. See Gilliam & Roberts 2003, pp. 15–37

Nault A, Gagnon D. 1988. Seasonal biomass and nutrient allocation patterns in wild leek (*Allium tricoccum* Ait.), a spring geophyte. *Bull. Torrey Bot. Club* 115:45–54

Nesom GL, LaDuke JC. 1985. Biology of *Trillium nivale* (Liliaceae). *Can. J. Bot.* 63:7–14

Neufeld HS, Young DR. 2003. Ecophysiology of the herbaceous layer in temperate deciduous forests. See Gilliam & Roberts 2003, pp. 38–90

Nomiya H, Suzuki W, Kanazashi T, Shibata M, Tanaka H, Nakashizuka T. 2002. The response of forest floor vegetation and tree regeneration to deer exclusion and disturbance in a riparian deciduous forest, central Japan. *Plant Ecol.* 164:263–76

Ohara M, Higashi S. 1987. Interference by ground beetles with the dispersal by ants of seeds of *Trillium* species (Liliaceae). *J. Ecol.* 75:1091–98

Ohara M, Higashi S. 1994. Effects of inflorescence size on visits from pollinators and seed set in *Corydalis ambigua* (Papaveraceae). *Oecologia* 98:25–30

Ohara M, Kawano S. 1986. Life history studies on the genus *Trillium* (Liliaceae) I. Reproductive biology of four Japanese species. *Plant Species Biol.* 1:35–45

Ohara M, Takeda H, Ohno Y, Shimamoto Y. 1996. Variations in the breeding system and the population genetic structure of *Trillium kamtschaticum* (Liliaceae). *Heredity* 76:476–84

Packham JR, Willis AJ. 1982. The influence of

shading and of soil type on the growth of *Galeobdolon luteum*. *J. Ecol.* 70:491–512
Peterken GF, Game M. 1984. Historical factors affecting the number and distribution of vascular plant species in the woodlands of central Lincolnshire. *J. Ecol.* 72:155–82
Peterson CJ, Campbell JE. 1993. Microsite differences and temporal change in plant communities of treefall pits and mounds in an old-growth forest. *Bull. Torrey Bot. Club* 120:451–60
Peterson DL, Rolfe GL. 1982. Nutrient dynamics of herbaceous vegetation in upland and floodplain forest communities. *Am. Midl. Nat.* 107:325–39
Pickett FL. 1915. A contribution to our knowledge of *Arisaema triphyllum*. *Mem. Torrey Bot. Club* 16:1–55
Pickett STA, McDonnell MJ. 1989. Seed bank dynamics in temperate deciduous forest. In *Ecology of Soil Seed Banks*, ed. MA Leck, TV Parker, RL Simpson, pp. 124–48. New York: Academic
Pickett STA, White PS, eds. 1985. *The Ecology of Natural Disturbance and Patch Dynamics*. London: Academic
Piper JK. 1989. Distribution of dry mass between shoot and root in nine understory species. *Am. Midl. Nat.* 122:114–19
Pitelka LF. 1984. Application of the −3/2 power law to clonal herbs. *Am. Nat.* 123:442–49
Pitelka LF, Ashmun JW. 1985. Physiology and integration of ramets in clonal plants. In *Population Biology and Evolution of Clonal Organisms*, ed. JBC Jackson, LW Buss, RE Cook, pp. 399–435. New Haven, CT: Yale Univ. Press
Pitelka LF, Ashmun JW, Brown RL. 1985a. The relationship between seasonal variation in light intensity, ramet size, and sexual reproduction in natural and experimental populations of *Aster acuminatus* (Compositae). *Am. J. Bot.* 72:311–19
Pitelka LF, Hansen SB, Ashmun JW. 1985b. Population biology of *Clintonia borealis*. I. Ramet and patch dynamics. *J. Ecol.* 73:169–83
Pitelka LF, Stanton DS, Peckenham MO. 1980. Effects of light and density on resource allocation in a forest herb, *Aster acuminatus* (Compositae). *Am. J. Bot.* 67:942–48
Policansky D. 1983. Patches, clones and self-fertility of mayapples (*Podophyllum peltatum* L.). *Rhodora* 85:253–56
Price EAC, Hutchings MJ. 1992. Studies of growth in the clonal herb *Glechoma hederacea*. II. The effects of selective defoliation. *J. Ecol.* 80:39–47
Price EAC, Marshall C, Hutchings MJ. 1992. Studies of growth in the clonal herb *Glechoma hederacea*. I. Patterns of physiological integration. *J. Ecol.* 80:25–38
Primack RB, Hall P. 1990. Costs of reproduction in the Pink Lady's Slipper orchid: a four-year experimental study. *Am. Nat.* 136:638–56
Primack RB, Miao SL. 1992. Dispersal can limit local plant distribution. *Conserv. Biol.* 6:513–19
Primack RB, Miao SL, Becker KR. 1994. Costs of reproduction in the Pink Lady's Slipper orchid (*Cypripedium acaule*): defoliation, increased fruit production, and fire. *Am. J. Bot.* 81:1083–90
Pudlo RJ, Beattie AJ, Culver DC. 1980. Population consequences of changes in an ant-seed mutualism in *Sanguinaria canadensis*. *Oecologia* 146:32–37
Ramsey GW, Leys CH, Wright RAS, Coleman DA, Neas AO, et al. 1993. Vascular flora of the James River gorge watersheds in the central Blue Ridge mountains of Virginia. *Castanea* 58:260–300
Rasmussen HN. 1995. *Terrestrial Orchids from Seed to Mycotrophic Plant*. Cambridge, UK: Cambridge Univ. Press
Rasmussen HN. 2002. Recent developments in the study of orchid mycorrhiza. *Plant Soil* 244:149–63
Rasmussen HN, Whigham DF. 1993. Seed ecology of dust seeds in situ: a new study technique and its application in terrestrial orchids. *Am. J. Bot.* 80:1374–78
Rasmussen HN, Whigham DF. 1998a. Importance of woody debris in seed germination

of *Tipularia discolor* (Orchidaceae). *Am. J. Bot.* 85:829–34

Rasmussen HN, Whigham DF. 1998b. The underground phase: a special challenge in studies of terrestrial orchid populations. *Bot. J. Linn. Soc.* 126:49–64

Rasmussen HN, Whigham DF. 2002. Phenology of roots and mycorrhiza in orchid species differing in phototrophic strategy. *New Phytol.* 154:797–807

Reader RJ. 1987. Loss of species from deciduous forest understory immediately following selective tree harvesting. *Biol. Conserv.* 42:231–44

Reader RJ. 1988. Using the guild concept in the assessment of tree harvesting effects on understory herbs: a cautionary note. *Environ. Manag.* 12:803–8

Reader RJ, Bricker BD. 1992. Response of five deciduous forest herbs to partial canopy removal and patch size. *Am. Midl. Nat.* 127:149–57

Robertson JL, Wyatt R. 1990. Reproductive biology of the yellow-fringed orchid, *Platanthera ciliaris*. *Am. J. Bot.* 77:388–98

Robinson GR, Yurlina ME, Handel SN. 1994. A century of change in the Staten Island flora: ecological correlates of species losses and invasions. *Bull. Torrey Bot. Club* 121:119–29

Rockwood LL, Lobstein MB. 1994. The effects of experimental defoliation on reproduction in four species of herbaceous perennials from northern Virginia. *Castanea* 59:41–50

Rooney TP. 2001. Deer impacts on forest ecosystems: a North American perspective. *Forestry* 74:201–8

Rooney TP, Dress WJ. 1997a. Patterns of plant diversity in overbrowsed primary and mature secondary hemlock–northern hardwood forest stands. *J. Torrey Bot. Soc.* 124:43–51

Rooney TP, Dress WJ. 1997b. Species loss over sixty years in the ground-layer vegetation of Heart's Content, an old-growth forest in Pennsylvania, USA. *Nat. Areas J.* 17:297–305

Rooney TP, Gross K. 2003. A demographic study of deer browsing impacts on *Trillium grandiflorum*. *Plant Ecol.* 168:267–77

Rooney TP, Waller DM. 2001. How experimental defoliation and leaf height affect growth and reproduction in *Trillium grandiflorum*. *J. Torrey Bot. Soc.* 128:393–99

Scheiner SM. 1988. Population dynamics of an herbaceous perennial *Danthonia spicata* during secondary forest succession. *Am. Midl. Nat.* 119:268–81

Schemske DW. 1977. Flowering phenology and seed set in *Claytonia virginica* (Portulacaceae). *Bull. Torrey Bot. Club* 104:254–63

Shefferson RP, Sandercock BK, Proper J, Beissinger SR. 2001. Estimating dormancy and survival of a rare herbaceous perennial using mark-recapture models. *Ecology* 82:145–56

Singleton R, Gardescu S, Marks PL, Gerber MA. 2001. Forest herb colonization of postagricultural forests in central New York state, USA. *J. Ecol.* 89:325–38

Slade AJ, Hutchings MJ. 1987. An analysis of the costs and benefits of physiological integration between ramets in the clonal perennial herb *Glechoma hederacea*. *Oecologia* 73:425–31

Small CJ, McCarthy BC. 2003. Spatial and temporal variability of herbaceous vegetation in an eastern deciduous forest. *Plant Ecol.* 164:37–48

Smith BH. 1983a. Demography of *Floerkea proserpinacoides*, a forest-floor annual. I. Density-dependent growth and mortality. *J. Ecol.* 71:391–404

Smith BH. 1983b. Demography of *Floerkea proserpinacoides*, a forest-floor annual. II. Dynamics of seed and seedling populations. *J. Ecol.* 71:413–25

Smith BH, DeRivera CE, Bridgman CL, Woida JJ. 1989a. Frequency-dependent seed dispersal by ants of two deciduous forest herbs. *Ecology* 70:1645–48

Smith BH, Forman PD, Boyd AE. 1989b. Spatial patterns of seed dispersal and predation of two myrmecochorous forest herbs. *Ecology* 70:1649–56

Smith BH, Ronsheim ML, Swartz KR. 1986. Reproductive ecology of *Jeffersonia diphylla* (Berberidaceae). *Am. J. Bot.* 73:1416–26

Smith JL, Hunter KL, Hunter RB. 2002. Genetic variation in the terrestrial orchid *Tipularia discolor*. *Southeast. Nat.* 1:17–26

Smith RG, Caswell D, Carriera A, Zielke B. 1996. Variation in the ginsenoside content of American ginseng, *Panax quinquefolius* L., roots. *Can. J. Bot.* 74:1616–20

Snow A, Whigham DF. 1989. Cost of flower and fruit production in *Tipularia discolor* (Orchidaceae). *Ecology* 70:1286–93

Sobey DG, Barkhouse P. 1977. The structure and rate of growth of the rhizomes of some forest herbs and dwarf shrubs of the New Brunswick–Nova Scotia border region. *Can. Field Nat.* 91:377–83

Solbrig OT. 1981. Studies on the population biology of the genus *Viola*. II. The effect of plant size on fitness in *Viola sororia*. *Evolution* 35:1080–93

Stamp NE, Lucas JR. 1983. Ecological correlates of explosive seed dispersal. *Oecologia* 59:272–78

Syrjänen K, Lehtilä K. 1993. The cost of reproduction in *Primula veris*: differences between two adjacent populations. *OIKOS* 67:465–72

Thompson PA, Cox SA. 1978. Germination of the Bluebell (*Hyacinthoides non-scripta* (L.) Chouard) in relation to its distribution and habitat. *Ann. Bot.* 42:51–62

Tilghman NG. 1989. Impacts of white-tailed deer on forest regeneration in northwestern Pennsylvania. *J. Wildl. Manag.* 53:524–32

Tissue DT, Skillman JB, McDonald EP, Strain BR. 1995. Photosynthesis and carbon allocation in *Tipularia discolor* (Orchidaceae), a wintergreen understory herb. *Am. J. Bot.* 82:1249–56

Traveset A, Willson MF. 1997. Effect of birds and bears on seed germination of fleshy-fruited plants in temperate rainforests of southeast Alaska. *OIKOS* 80:89–95

Uemura S. 1993. Patterns of leaf phenology in forest understory. *Can. J. Bot.* 72:409–14

Valverde T, Silvertown J. 1995. Spatial variation in the seed ecology of a woodland herb (*Primula vulgaris*) in relation to light environment. *Funct. Ecol.* 9:942–50

Valverde T, Silvertown J. 1997a. A metapopulation model for *Primula vulgaris*, a temperate forest understorey herb. *J. Ecol.* 85:193–210

Valverde T, Silvertown J. 1997b. An integrated model of demography, patch dynamics and seed dispersal in a woodland herb, *Primula vulgaris*. *OIKOS* 80:67–77

Valverde T, Silvertown J. 1998. Variation in the demography of a woodland understorey herb (*Primula vulgaris*) along the forest regeneration cycle: projection matrix analysis. *J. Ecol.* 86:545–62

Vellend M. 2003. Habitat loss inhibits recovery of plant diversity as forests regrow. *Ecology* 84:1158–64

Vellend M, Lechowicz MJ, Waterway MJ. 2000a. Environmental distribution of four Carex species (Cyperaceae) in an old-growth forest. *Am. J. Bot.* 87:1507–16

Vellend M, Lechowicz MJ, Waterway MJ. 2000b. Germination and establishment of forest sedges (Carex: Cyperaceae): tests for home-site advantage and effects of leaf litter. *Am. J. Bot.* 87:1517–25

Vellend M, Myers JA, Gardescu S, Marks PL. 2003. Dispersal of *Trillium* seeds by deer: implications for long-distance migration of forest herbs. *Ecology* 84:1067–72

Verheyen K, Guntenspergen GR, Biesbrouck B, Hermy M. 2003. An integrated analysis of the effects of past land use on forest herb colonization at the landscape scale. *J. Ecol.* 91:731–42

Waller DM, Alverson WS. 1997. The white-tailed deer: a keystone herbivore. *Wildl. Soc. Bull.* 25:217–26

Watson MA. 1995. Sexual differences in plant developmental phenology affect plant-herbivore interactions. *Trees* 10:180–82

Webb SL. 1999. Disturbance by wind in temperate-zone forests. In *Ecosystems of Disturbed Ground*, ed. LR Walker, pp. 187–222. Amsterdam, Neth.: Elsevier

Webb SL, Willson MF. 1985. Spatial heterogeneity in post-dispersal predation on *Prunus* and *Uvularia* seeds. *Oecologia* 67:150–53

Wennström A, Ericson L. 1994. The effect of the systemic smut *Urocystis carcinodes* on the long-lived herb *Actaea spicata*. *OIKOS* 71:111–18

Whigham DF. 1974. An ecological life history study of *Uvularia perfoliata* L. *Am. Midl. Nat.* 91:343–59

Whigham DF. 1984. Biomass and nutrient allocation patterns of *Tipularia discolor* (Orchidaceae). *OIKOS* 42:303–13

Whigham DF. 1990. The effect of experimental defoliation on the growth and reproduction of a woodland orchid, *Tipularia discolor*. *Can. J. Bot.* 68:1812–16

Whigham DF, Chapa AS. 1999. Timing and intensity of herbivory: its influence on the performance of clonal woodland herbs. *Plant Species Biol.* 14:29–37

Whigham DF, O'Neill J. 1991. The dynamics of flowering and fruit production in two eastern North American terrestrial orchids, *Tipularia discolor* and *Liparis lilifolia*. In *Population Ecology of Terrestrial Orchids*, ed. TCE Wells, JH Willems, pp. 89–101. The Hague: SPB Academic

Whigham DF, O'Neill J, Cipollini M. 1993. The role of tree gaps in maintaining the population structure of a woodland herb: *Cynoglossum virginianum* L. *Plant Species Biol.* 8:107–15

Whigham DF, O'Neill J, McCormick M, Smith C, Rasmussen H, et al. 2002. Interactions between decomposing wood, mycorrhizas, and terrestrial orchid seeds and protocorms. In *Trends and Fluctuations and Underlying Mechanisms in Terrestrial Orchid Populations*, ed. P Kindlmann, JH Willems, DF Whigham, pp. 117–31. Leiden, Neth.: Backhuys

Whitbread F, McGonigle TP, Peterson RL. 1996. Vesicular-arbuscular mycorrhizal associations of American ginseng (*Panax quinquefolius*) in commercial production. *Can. J. Bot.* 74:1104–12

Whitman AA, Hunter ML Jr., Witham JW. 1998. Age distribution of ramets of a forest herb, wild sarsaparilla, *Aralia nudicaulis* (Araliaceae). *Can. Field Nat.* 112:37–44

Widden P. 1996. The morphology of vesicular-arbuscular mycorrhizae in *Clintonia borealis* and *Medeola virginiana*. *Can. J. Bot.* 74:679–85

Wijesinghe D, Hutchings MJ. 1997. The effects of spatial scale of environmental heterogeneity on the growth of a clonal plant: an experimental study with *Glechoma hederacea*. *J. Ecol.* 85:17–28

Wijesinghe DK, Whigham DF. 1997. Costs of producing clonal offspring and the effects of plant size on population dynamics of the woodland herb *Uvularia perfoliata* (Liliaceae). *J. Ecol.* 85:907–19

Wijesinghe DK, Whigham DF. 2001. Nutrient foraging in woodland herbs: a comparison of three species of *Uvularia* (Liliaceae) with contrasting belowground morphologies. *Am. J. Bot.* 88:1071–79

Willson MF. 1993. Dispersal mode, seed shadows, and colonization patterns. *Vegetatio* 107/108:261–80

Wilson JY. 1959. Vegetative reproduction in the bluebell, *Endymion nonscriptus* (L.) Garcke. *New Phytol.* 58:155–63

Zak DR, Groffman PM, Pregitzer KS, Christensen S, Tiedje JM. 1990. The vernal dam: plant-microbe competition for nitrogen in northern hardwood forests. *Ecology* 71:651–56

Zelmer CD, Currah RS. 1995. Evidence for a fungal liaison between *Corallorhiza trifida* (Orchidaceae) and *Pinus contorta* (Pinaceae). *Can. J. Bot.* 73:862–66

Zettler JA, Spira TP, Allen CR. 2001. Yellow jackets (*Vespula* spp.) disperse *Trillium* (spp.) seeds in eastern North America. *Am. Midl. Nat.* 146:444–46

Ziegenhagen B, Bialozyt R, Kuhlenkamp V, Schulze I, Ulrich A, Wulf M. 2003. Spatial patterns of maternal lineages and clones of *Galium odoratum* in a large ancient woodland: inferences about seedling recruitment. *J. Ecol.* 91:578–86

THE SOUTHWEST AUSTRALIAN FLORISTIC REGION: Evolution and Conservation of a Global Hot Spot of Biodiversity

Stephen D. Hopper[1] and Paul Gioia[2]

[1]*Botanic Gardens and Parks Authority, Kings Park and Botanic Garden, West Perth, Western Australia 6005; Australia and Plant Biology, Faculty of Natural and Agricultural Sciences, The University of Western Australia, Crawley Western Australia, 6009; email: steve.hopper@cyllene.uwa.edu.au*
[2]*Western Australian Herbarium, Department of Conservation and Land Management, Bentley Delivery Center, Western Australia 6983; email: paulg@calm.wa.gov.au*

Key Words phylogeny, fossils, biogeography, speciation, threatened species

■ **Abstract** Like South Africa's Greater Cape Floristic Region, the Southwest Australian Floristic Region (SWAFR) is species rich, with a Mediterranean climate and old, weathered, nutrient-deficient landscapes. This region has 7380 native vascular plants (species/subspecies): one third described since 1970, 49% endemic, and 2500 of conservation concern. Origins are complex. Molecular phylogenies suggest multiple dispersal events into, out of, and within the SWAFR throughout the Cretaceous and Cenozoic; in many phylogenetically unrelated clades; and from many directions. Either explosive speciation or steady cladogenesis occurred among some woody sclerophyll and herbaceous families from the mid-Tertiary in response to progressive aridity. Genomic coalescence was sometimes involved. Rainforest taxa went extinct by the Pleistocene. Old lineages nevertheless persist as one endemic order (Dasypogonales) and 6–11 endemic families. Such a rich flora on old landscapes that have been exposed to European land-use practices is highly threatened. Conservation programs must minimize soil removal and use local germplasm in restoration programs.

INTRODUCTION

The Southwest Australian Floristic Region (SWAFR) occupies 302,627 km^2 on a temperate margin of the world's most arid and insular populated continent. The region is island-like (Carlquist 1974, Hopper 1979), a relatively wet continental refuge, bordered on two sides by ocean, and isolated by arid lands to the north, northeast, and east.

During the past few decades, scientists have made extraordinary leaps forward in the discovery, collection, and description of new flowering plant species in this region, without parallel among the world's temperate floras (Figure 1). These

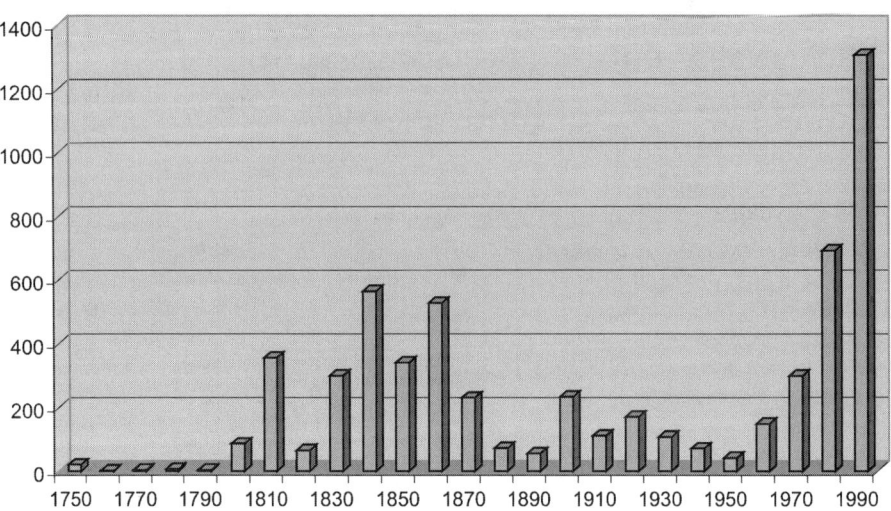

Figure 1 The number of currently recognized native SWAFR vascular plant species described per decade since the 1750s, derived from a total of 6759 described species listed on the Western Australian Herbarium's FLORABASE in February 2004. Intraspecific taxa, unpublished species with a manuscript name, and phrase-name species are not included.

advances highlight an accentuated degree of local speciation, eclipsed in temperate regions only by that of the Greater Cape flora of South Africa (Cowling et al. 1996, Goldblatt & Manning 2002, Linder 2003).

The SWAFR is topographically unique among the world's five regions of Mediterranean climate (Hooker 1860; Hopper 1979, 2004; Cowling et al. 1996, 2004; Dallman 1998): It is a flat, stable, highly weathered low plateau with granite occasionally emergent as domed inselbergs (Anand & Paine 2002). There are a few small quartzitic mountainous areas; the Stirling Range is the tallest, reaching only 1109 m (Figure 2c, see color insert). The SWAFR has enjoyed a maritime climate since the Jurassic, and it has been unglaciated since the Permian. This region is dominated by old landscapes with nutrient-deficient soils (Hopper et al. 1996a,b).

Like other Mediterranean regions, the SWAFR has a fossil record rich in closed rainforest taxa dating back to the Cretaceous (McLoughlin & Hill 1996, McLoughlin & McNamara 2001). However, it is unique in that all such taxa were extinct by the start of the Quaternary, following the onset and persistence of aridity in Australia that began in the Oligocene/Miocene (Frakes 1999). In SWAFR vegetation, only sclerophyllous shrubs, trees, and herbs have adapted to the nutrient-deficient, highly weathered soils that now dominate as either eucalypt forests, woodlands, mallee (lignotuberous, multistemmed eucalypts), or kwongan (shrublands and herbfields) (Figure 2a) (Beard 1990).

Recently the SWAFR was listed among 25 global biodiversity hot spots—those regions on Earth richest in endemic species under threat (Myers et al. 2000). Such plant richness is surprising given the subdued topography and simple rainfall and vegetation gradients evident in the region (Figure 2). This hot-spot status raises two fundamental questions that have challenged many authors since Hooker (1860) first summarized austral phytogeographic data: (a) How has such an extraordinarily rich endemic flora evolved, especially given the region's subdued, nutrient-poor terrain with few mountains? and (b) Why is the flora so threatened and how might its conservation be secured? Since Hopper (1979) addressed aspects of the first question in this series 25 years ago, many important discoveries and insights have been published (references in Pate & Beard 1984; Hobbs 1992; Pate & Hopper 1993; Cowling et al. 1996; Hopper et al. 1996b; Hopper 1997; Crisp et al. 1999; Coates et al. 2000; James 1992, 2000; Beard et al. 2000; Coates & Atkins 2001; Lambers et al. 2003; Lamont 2003). It is timely, therefore, to highlight recent advances in the understanding of the evolution of the SWAFR flora and to explore briefly applications to the challenging task of conservation. A more detailed elaboration and reference list is available elsewhere (Hopper 2004).

EVOLUTIONARY HYPOTHESES

Evidence reviewed by Hopper (1979) led to the formulation of the following hypotheses and conclusions accounting for the extraordinary species richness and high endemism of flora on southwest Australia's subdued landscapes:

1. The vascular flora comprises at least 3600 species; 68%, but possibly 75–80%, of which are endemic based on ongoing taxonomic description of local endemics.

2. The flora is richest in the coastal and inland kwongan of the Transitional Rainfall Zone (300–800 mm rainfall per annum, modified to 300–600 mm herein to embrace the Transitional Rainfall Province (TRP) and Southeast Coastal Province (SCP). The flora is relatively species-poor in the high rainfall forests [800–1500 mm, modified as the High Rainfall Province (HRP) to 600–1500 mm herein] and in arid zone communities (<300 mm) (Figures 2b, 5).

3. There is a subtle trend for recently evolved taxa to occur more frequently than expected in the TRP/SCP compared with the HRP, whereas the reverse trend applies with the persistence of mesothermic relictual sclerophyllous taxa in stable, high-rainfall areas or locally wet habitats elsewhere.

4. These trends are correlated in the SWAFR with patterns of landscape evolution and climatic history. Topography, soils, and climate are considerably more diverse in the TRP/SCP near the west and south coasts and inland down its western margins than in the HRP and arid zone.

5. Today's predominantly sclerophyllous SWAFR flora evolved from old lineages in isolated pockets on nutrient-deficient soils in landscapes dominated by rainforests that are now locally extinct.
6. Explosive recent speciation of sclerophyllous taxa, especially in the TRP/SCP, was caused by greater population dissection and lability than seen in the HRP or arid zone, arising from more diverse topography and the erosional dynamism and climatic stresses of the late Tertiary and Quaternary.
7. Biological, ecological, and genetic correlates and causes of explosive speciation and endemism were poorly known, but they differed from those then recorded in other Mediterranean regions; the southwest flora included few rapidly evolving annuals and few species-specific hymenopteran pollinators.

Thus, like parts of California (Stebbins & Major 1965), the SWAFR's TRP/SCP was hypothesized to be a semiarid speciation hot spot of late-Tertiary antiquity, whereas the HRP was an area more conducive to evolutionary stability and persistence over longer periods. The arid zone is a predictably harsh region favoring relatively few recently evolved taxa able to survive punishing conditions.

Hopper (1979) emphasized that much remained to be done to test these hypotheses. Furthermore, exciting work on the contribution to speciation of chromosome repatterning in small populations had only just begun. This work had led to the development of two further hypotheses (James 1981):

1. Two bursts of speciation had occurred. The first involved widespread dysploid and polyploid speciation in the Cretaceous to early Tertiary, resulting in the evolution of endemic Australian families, tribes, and genera. This was followed by a period of relative cytoevolutionary stability as genera, subgenera, and sections radiated. A more recent, late-Tertiary to Quaternary second phase of intense cytoevolution ensued, giving rise to modern dysploid and polyploid species and races.
2. According to the genomic coalescence hypothesis, such cytoevolutionary change arose under intense selection to conserve heterozygosity in the face of inbreeding imposed by small disjunct population structures (James 1992, 2000).

REVISED RICHNESS AND ENDEMISM ANALYSES

An Unexpected Rate of Recent Descriptions of New Species

For a century following the pioneering botanical work on the SWAFR flora, the estimated number of species remained at 3600 (Hooker 1860, Beard 1970). Many authors regarded the SWAFR as the species-poor cousin of the world's five Mediterranean regions. Few realized that the taxonomic impediment for understanding and conserving the SWAFR flora was as great as that applying to many tropical rainforest regions. Since then, however, the number of described SWAFR species has increased by one third and is still rising significantly (Figure 1), primarily owing to the establishment of systematic and evolutionary research and training programs at universities and herbaria in the 1960s, and subsequently fueled by burgeoning interests in environmental conservation and, most recently, in molecular systematics. The long-term effects of a recent decline in university teaching programs in systematics have yet to impact these trends.

Although increasingly recognized by many authors (Carlquist 1974; Hopper 1979, 1992, 1997; Marchant 1991; Cowling et al. 1996; Ornduff 1996; Goldblatt & Manning 2002), this taxonomic impediment of many recently discovered but yet to be described species is not considered by others (e.g., Crisp et al. 1999, Beard et al. 2000), leading to widely divergent published estimates of the number of species in the SWAFR. However, counting described and manuscript species and infraspecific taxa listed on Western Australian Herbarium databases (Paczkowska & Chapman 2000; S.D. Hopper & P. Gioia, unpublished data) yields a total of 7380 native species/subspecies. We count both species and subspecies because most of the latter ($\sim$8% of the 7380 taxa) have proven to be species when subjected to modern critical study (e.g., Hopper & Brown 2001; Hopper 2004). Given present trends in discovery and description (Figure 1), it seems probable that the SWAFR will have at least 8000 native species when taxonomic survey is close to completion some decades from now (Hopper 1992, Hopper et al. 1996b). More than a decade ago, Marchant (1991) independently estimated 9000 species, a figure that encompasses both native (8000) and naturalized (1000) species/subspecies.

Major Taxa

As established two centuries ago by Robert Brown and colleagues on Flinders' *Investigator* expedition (Hopper 2003, 2004), the rich diversity of the SWAFR flora is primarily among its angiosperms, especially woody families: Myrtaceae (1283 species/subspecies), Proteaceae (859), Fabaceae (540), Mimosaceae (503), Orchidaceae (374), Ericaceae (including Epacridaceae, 297), Asteraceae (280), Goodeniaceae (207), Cyperaceae (199) and Stylidiaceae (178) (Figure 3, see color insert) (Paczkowska & Chapman 2000; Beard et al. 2000; S.D. Hopper & P. Gioia, unpublished data).

The importance of woody taxa is evident also in the ten largest genera. Listed in order of number of species/subspecies, they are *Acacia* (Mimosaceae; 502 species/infraspecies), *Eucalyptus* (Myrtaceae; 362), *Grevillea* (Proteaceae; 229), *Melaleuca* (Myrtaceae; 185), *Stylidium* (Stylidiaceae; 170), *Leucopogon* (Ericaceae; 165), *Caladenia* (Orchidaceae; 162), *Verticordia* (Myrtaceae; 138), *Dryandra* (Proteaceae; 136) and *Hakea* (Proteaceae; 105). Of these, only the herbaceous Triggerplants (*Stylidium*) and the geophytic orchid genus *Caladenia* constitute nonwoody plants.

Endemics on an Old Landscape

Phylogenetically significant endemics confirmed in DNA sequence studies include one monocot order (Dasypogonales) and at least six families (the monocot

Dasypogonaceae, Ecdeiocoleaceae, and Anarthriaceae, and eudicot Cephalotaceae, Emblingiaceae, and Eremosynaceae (APG II 2003) (Figure 4, see color insert). Other high-level SWAFR endemic monocot clades regarded by some as families include Haemodoraceae subfamily Conostylidoideae, *Baxteria* and *Calectasia* of Dasypogonaceae, and *Hopkinsia* and *Lyginia* of Anarthriaceae.

Based on premolecular intuitive classifications of most families and genera, 711 genera with 13% (92) endemic to the region are known (Beard et al. 2000). As a result of molecular phylogenetics, substantial changes in delimitation for many of these genera are proposed (e.g., *Banksia/Dryandra*) (Mast & Givnish 2002).

A surprising finding is that estimates of species-level endemism as high as 80–90% (Hooker 1860, Hopper 1979, Marchant 1991) have required significant downward modification as collecting has intensified. The figure of 53% (Beard et al. 2000) is now superseded by an even lower 49% (S. Hopper and P. Gioia, manuscript in preparation). Even though recently described species/subspecies are often narrow-range endemics, the ranges of many SWAFR species/subspecies have been extended, at least marginally, beyond the region's borders.

The presence of endemic families and the endemic order Dasypogonales signal prolonged conditions for the persistence of relict taxa in the SWAFR. It is indeed one of the oldest landscapes on Earth. Its essential flatness is due to the absence of mountain building since the Carboniferous-Permian glaciation, which lasted from 320 to 290 Mya (Playford 1999). Moreover, some granite inselbergs have had their summits exposed since the mid-Cretaceous (Watchman & Twidale 2002). Combined with the absence of inundation or glaciation since the Permian, this great antiquity of landforms renders the SWAFR among the oldest unglaciated regions on Earth. Younger landscape elements, such as extensive Tertiary laterites and Quaternary coastal limestones, dunes, and wetlands, are also found in the region. However, in comparison with postglacial landscapes common in the northern hemisphere and parts of southeast Australia, most of the SWAFR is extraordinarily old (Anand & Paine 2002, Wyrwoll 1988).

Not surprisingly, with such exceptional opportunities for continuous terrestrial evolution, local endemism is prominent in components of both the flora and the less vagile fauna (Hopper et al. 1990, 1996b; Cowling et al. 1994; Brown et al. 1998; Harvey 2002). Hopper and colleagues (Hopper et al. 1996b; Hopper 1997, 2000) hypothesized that natural selection has resulted in mechanisms promoting local persistence rather than wide dispersal and colonizing. This persistence may be a result of the prolonged absence of major geomorphological agents of soil disturbance and rejuvenation, as seen during glaciation, mountain building, and vulcanism. Only along coastlines and rivers, in other wetlands, and around rock outcrops has wholesale and regular soil disturbance been prominent, with the biota consequently exhibiting better mechanisms for long-distance dispersal and colonization. Indeed, local endemism and attendant landscape continuity seen in the SWAFR may well be a model for what many other regions on Earth were like prior to Quaternary glacial cycles.

Biological Correlates of Narrow Endemism

Cowling et al. (1994) found that local and regional endemics constituted 6% and 22%, respectively, of the 1422 species from communities of the Fitzgerald District (Figure 5, see color insert). Endemics were both overrepresented in species-rich families such as Proteaceae, Myrtaceae, Fabaceae, and Ericaceae and underrepresented in Asteraceae and Orchidaceae. Endemics were also virtually absent from less nutrient-impoverished coastal calcareous sands, but as expected, they constituted up to 30% of kwongan communities on highly infertile quartzites and siliceous sands. Biologically, the narrow-range endemics were equally likely to be shrubs or graminoids. These shrubs were of medium height; with soil or canopy-stored seed dispersed either by wind, vertebrates, ants, or ballistically in roughly equal proportion; and with estimated medium (10–100 m) to short (<10 m) seed-dispersal distances. No local endemics were recorded among tall shrubs (>2 m) or among woody shrubs lacking seed storage [such as *Billardiera* (*Sollya*) *heterophylla*, Pittosporaceae].

Threatened endemic SWAFR taxa similarly are mostly woody perennials, one third of which are short-lived disturbance opportunists and obligate seeders after fire (e.g., species of *Acacia* and *Grevillea*) (Hopper et al. 1990, Bell 2001). Perennial herbs feature prominently among the remaining threatened taxa. More than half of these are orchids. Spring flowering occurs in two thirds of the threatened taxa, and 40% have flowers likely to be pollinated by birds and/or mammals. This is almost three times the proportion (15%) of the SWAFR flora at large that is vertebrate pollinated (Keighery 1982, Hopper & Burbidge 1986, Brown et al. 1997), a striking figure in itself.

TOWARD A NEW PHYTOGEOGRAPHIC UNDERSTANDING

Detailed vegetation mapping of the SWAFR (Figure 2*d*) (Diels 1906, Beard 1990) has been followed up by recent floristic analyses of all 7380 southwest species/subspecies represented in collections of the Western Australian Herbarium (standardized for collection effort; S.D. Hopper & P. Gioia, unpublished data) as well as by plant community studies at the regional and local level. It is now possible to explore phytogeographic patterns in the whole flora. Modern floristic data support the inland boundary of the SWAFR approximating the 300-mm rainfall isohyet (Figure 2*b*) as originally proposed by Diels (1906).

Although the TRP/SCP kwongan districts are richer in species than are the HRP forests (Hopper 1979, 1992), intensified collection of the flora has revealed that there are many centers of species richness throughout the whole SWAFR (Figure 5*a*) (Gioia & Pigott 2000). These centers are most evident close to the coast in both the TRP/SCP and the HRP, but they can also be found inland, tapering off toward the arid margins of the region. They largely coincide with centers of endemism (Figure 5*b*).

Classification of the richness centers according to their composite species revealed the presence of three major phytogeographical provinces and 11 floristic districts in the SWAFR (Figure 5). This new classification closely approximates the scheme proposed by Hopper (1979, 1992), except that, as mentioned above, the HRP extends out to the 600-mm isohyet rather than the 800-mm isohyet used to define the earlier High Rainfall Zone, and the Transitional Rainfall Zone is split into two provinces, the TRP and the SCP (Figure 5). The latter is novel, and it bears limited resemblance to vegetation-based bioregions currently enjoying wide use (Beard 1990; cf. Figure 2). Although some support for the new floristic classification exists, especially in relation to upland refugial centers (Hopper 1979, Hopkins et al. 1983, Lamont et al. 1984, Hopper et al. 1997, Mast 2000), further exploration of the classification is warranted to test its usefulness as a predictive model of species richness and endemism.

Many SWAFR floristic studies have established moderate to high diversity within habitats and among adjacent habitats, with extraordinarily rapid turnover of species across southwest landscapes (Cowling et al. 1994, 1996; Hopper 2004). At the extreme level in kwongan heathlands at Mt. Lesueur, quadrats on the same lateritic upland landform just 1 km apart may display up to a 60% difference in species composition (Hopkins & Griffin 1984).

Richardson et al. (1995) explored correlates of coexistence of *Banksia* species in the SWAFR, finding coexistence to be complexly linked to local and regional processes that are generally unrelated to growth form or regeneration class of sympatric species, as niche theory would predict. Rather, coexistence appears to be mediated by edaphic specialization and marked spatial and temporal variation in recruitment opportunities associated with interactions of soil type, fire, drought, flooding, and disease. Lottery recruitment models are likely to apply, and simplistic explanations should be questioned (Grubb 1992). This insight is not unique to banksias among SWAFR taxa (e.g., Yates et al. 2003). Much more experimental demographic and ecophysiological research is needed to advance understanding of how species-rich communities of SWAFR flora are assembled and ecologically maintained.

NEW INSIGHTS INTO EVOLUTIONARY ORIGINS

Simple ecological hypotheses such as soil mosaics being a primary cause of speciation (Beard et al. 2000) are insufficient to account for the origins of the rich SWAFR flora. Data from ecology, biogeography, genetics, phylogenetics, paleontology, geomorphology, and paleoclimatology are all needed to unravel this challenging question.

The Fossil Record, Geomorphology, and Paleoclimates

Significant advances in paleontology, plate tectonics, and geomorphology during the past two decades have generated fresh insights regarding origins of the SWAFR's flora (Powell et al. 1988; McLoughlin & Hill 1996; Hopper et al.

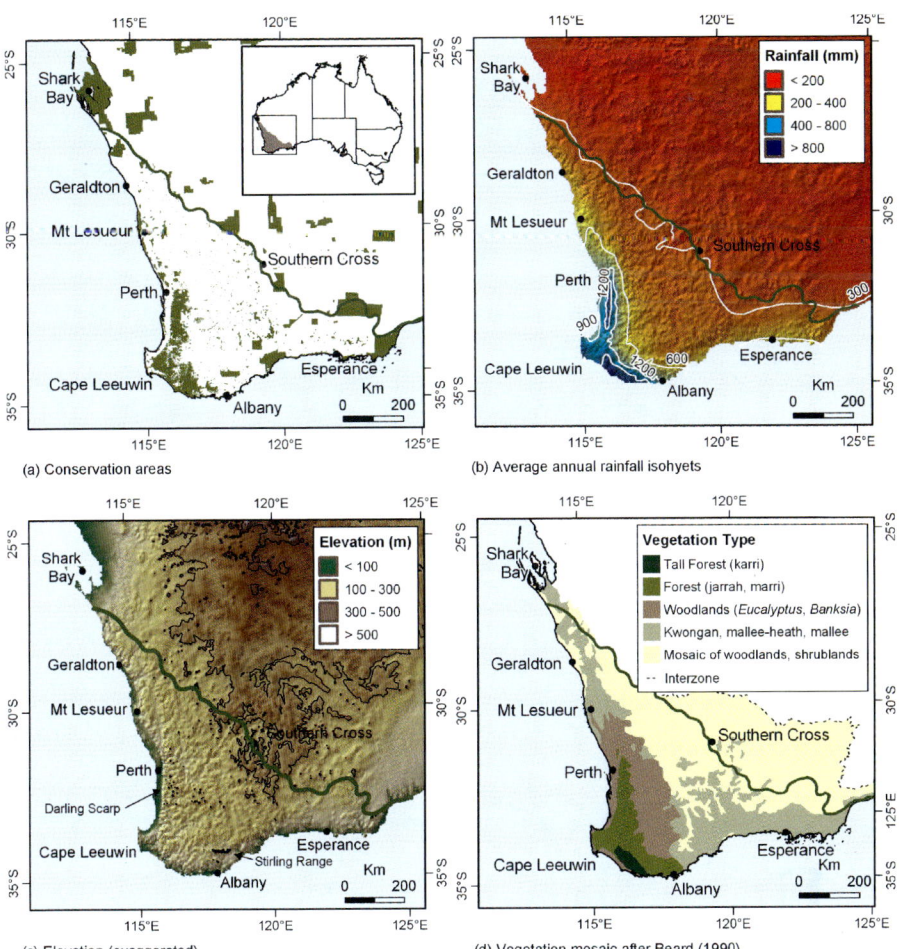

Figure 2 The SWAFR illustrating (*a*) protected conservation areas, (*b*) rainfall, (*c*) topography, and (*d*) vegetation.

Figure 3 Representatives of the most species-rich families of flowering plants in the SWAFR. (*a*) Proteaceae (*Banksia coccinea*) on which is perched a nectar-feeding SWAFR endemic honey possum *Tarsipes rostratus*, (*b*) Ericaceae (*Andersonia axilliflora*), (*c*) Myrtaceae (*Eucalyptus rhodantha*), (*d*) Stylidiaceae (*Stylidium scandens*), (*e*) Mimosaceae (*Acacia glaucoptera*), (*f*) Fabaceae (*Gastrolobium bilobum*), (*g*) Sterculiaceae (*Thomasia montana*), (*h*) Asteraceae (*Rhodanthe manglesii*), (*i*) Cyperaceae (*Lepidosperma costale*), (*j*) Goodeniaceae (*Lechenaultia superba*), (*k*) Droseraceae (*Drosera fimbriata*), and (*l*) Orchidaceae (*Caladenia elegans*). Illustrated and printed with permission by Ellen Hickman.

Figure 4 Representatives of the order, families, and subfamilies endemic to the SWAFR. (*a*) Eremosynaceae (*Eremosyne pectinata*), (*b*) Emblingiaceae (*Emblingia calceoliflora*), (*c*) Dasypogonales (Dasypogonaceae *sens. str.*; *Kingia australis*), (*d*) Dasypogonales (Dasypogonaceae/Calectasiaceae *Calectasia grandiflora*), (*e*) Anarthriaceae/Lyginiaceae (*Lyginia barbata*), (*f*) Haemodoraceae subfamily Conostylidoideae (*Anigozanthos flavidus*), (*g*) Anarthriaceae/Hopkinsiaceae (*Hopkinsia anaectocolea*), (*h*) Ecdeiocoleaceae—sister to the grasses (*Ecdeiocolea monostachya*), (*i*) Anarthriaceae (*Anarthria scabra*), (*j*) Cephalotaceae (*Cephalotus follicularis*), and (*k*) Dasypogonales (Dasypogonaceae/Baxteriaceae *Baxteria australis*). Illustrated and printed with permission by Ellen Hickman.

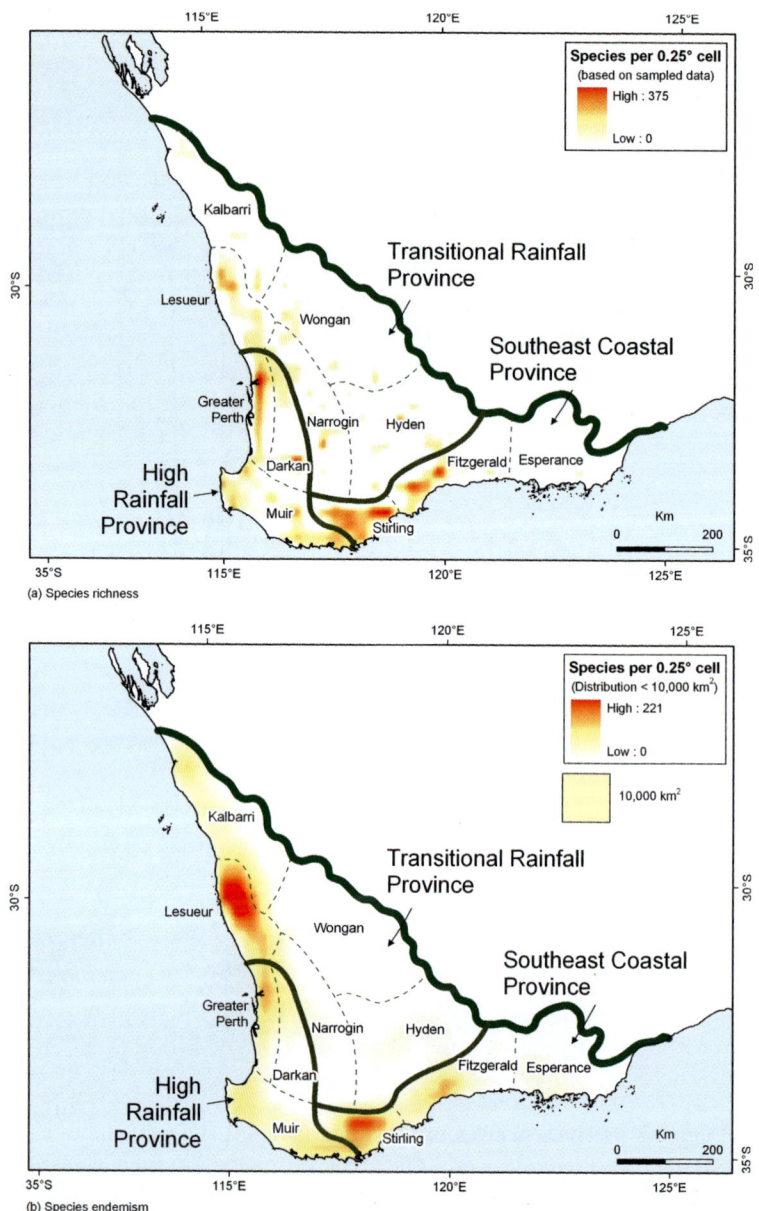

Figure 5 Phytogeographic maps of the SWAFR showing Botanical Provinces and Districts. Shown are (*a*) the species richness in 0.25° latitude by 0.25° longitude grids and (*b*) the richness in 0.25° latitude by 0.25° longitude grids of local endemics with a geographical distribution occupying <10,000 km^2, as determined from specimen labels at the Western Australian Herbarium and adjusted for collection intensity (S.D. Hopper & P. Gioia, unpublished data).

1996a,b; Anderson et al. 1999; McLoughlin & McNamara 2001; McLoughlin 2001; Hopper 2003; Atahn et al. 2004). For example, the discovery of fossils of extant genera such as *Banksia/Dryandra* (Proteaceae) and *Agonis* (Myrtaceae) as old as early- to mid-Tertiary indicate great antiquity and stability of some contemporary plant lineages.

The SWAFR's high-latitude Eocene floras were conspicuously rich in families such as the Myrtaceae and Proteaceae, suggesting that a long history of speciation underpins the species richness of the flora (McLoughlin & Hill 1996). Similar to contemporary SWAFR vegetation, the proportional representation of taxa in these Eocene communities varied over relatively short distances (McLoughlin & Hill 1996, Itzstein-Davey 2003). Perhaps low dispersal capabilities of many components of the flora also have significant antiquity.

The rifting of Antarctica from southern Australia resulted in an uplift of the Stirling Range in the mid-Cretaceous (Powell et al. 1988, McLoughlin 2001). This has remained an important center of floristic richness and endemism in the SWAFR (Figure 5). Central parts of the Australian plate were also downwarped and inundated during the Cretaceous, resulting in many islands along the east, south, and west margins of the SWAFR. These islands may have played a significant part in the early evolution of endemism of the extant flora. In light of new evidence regarding the unexpected antiquity of some flowering plant lineages, subsequent periods of high sea level and island formation, especially in the Eocene, are certain to have played such a part.

The early stages of a shift toward drier conditions across Australia associated with the final rifting of the continent from Antarctica are evident in Oligocene (45–25 Mya) fossil floras. These are less diverse than their Eocene predecessors and have significant proportions of scleromorphous elements, such as *Banksia/ Dryandra*. For example, the West Dale and Tambellup floras have a "striking increase in the representation of myrtacean remains and a corresponding decline in the proportions of *Nothofagus* [Nothofagaceae] and *Gymnostoma* [Casuarinaceae]" (McLoughlin & Hill 1996, p. 76). Extensive silcretes, laterites, and calcretes also developed at this time, possibly suggesting stronger seasonality of rainfall. If so, this would imply a significant increase in fire frequency.

Regional habitat diversity and opportunities for speciation were enhanced by moderate post-Eocene uplift of the west and southern margins of the Yilgarn Craton (Anand & Paine 2002) along the Darling Range and Ravensthorpe Ramp, respectively. Coastal rivers were rejuvenated, dissecting the Cretaceous–early Tertiary plateau duricrusts into numerous insular/peninsula remnants bordered by erosional and depositional soils (see map in Hopper 1979). The plateau margins thereafter had relatively rugged and more diverse topographical features than did inland areas (Figure 2c), albeit at subdued elevations from a global perspective. Evidence is mounting that the flora responded spectacularly to this regional variation in habitat diversity following the mid-Tertiary onset of aridity.

Final separation of Australia from Antarctica in the Oligocene initiated climatic changes that established the Australian deserts, arching inland from the Great

Australian Bight across the Nullarbor Plain to eventually enclose the SWAFR as early as 30 Mya (Frakes 1999). This emplaced a belt of semiarid transitional rainfall on the area of greatest fragmentation of lateritic duircrusts flanking the more mesic, high-rainfall forested areas of the SWAFR, forming one of the world's major plant speciation hot spots (Hopper 1979, 1992).

The downwarping of central Australia in the Cretaceous was terminated when Miocene seas completed formation of the Nullarbor limestones commenced in the Eocene. The Nullarbor Plain formed an edaphic barrier to east-west migration for terrestrial species adapted to acidic soils derived from the granitoid bedrock of much of the SWAFR. Barriers to migration and emigration were thus enhanced for the SWAFR, conserving the flora's long-standing regional endemism.

Not all plant lineages in the SWAFR are so limited in dispersal that they were contained within these borders throughout their phylogeny. There is increasing evidence of multiple dispersal events into, out of, and within the SWAFR throughout the Cretaceous and Cenozoic, in many phylogenetically unrelated clades, and from many directions (Figure 6) (McLoughlin 2001, Crisp et al. 1999).

Essentially modern floras, with the addition of small amounts of closed forest *Nothofagus*, araucarian, and podocarp conifer species, are seen in late-Tertiary and Quaternary fossil deposits. Inland areas had increased abundance and diversity of pollen of Myrtaceae, Poaceae, *Acacia*, and Asteraceae, suggesting seasonal rainfall. A slight rise in temperatures in the early Miocene (25–20 Mya) was accompanied by renewed formation of laterite, silcrete, and calcrete as well as lake sedimentation. This period was followed by sharp declines to cold and substantially fluctuating conditions from the mid- to late Miocene onward.

It seems probable that a mildly seasonal Mediterranean climate was present in southwest Australia from this period 20 Mya to the present. Closure of the Indonesian seaway in the Pliocene, an event of global significance, led to more rapid drying out of Australia from $\sim$3–5 Mya (J. Dodson, personal communication) and possibly signaled the onset of conditions similar to today's.

The similarity of Pliocene floras (Atahan et al. 2004, Dodson & Macphail 2004) to contemporary vegetation suggests only moderate impacts of Quaternary ice age

---

Figure 6 Historical biogeographic tracks along which congruent patterns of speciation have occurred within and from outside the SWAFR, as inferred from molecular phylogenetic studies (from Crisp et al. 1999, Hopper 2004). (*A*) Track names within southwest Australia are (*1*) Transitional Rainfall NW-SE, (*2*) South Coast HRP-SCP, (*3*) West Coast HRP-TRP, (*4*) SW-Inland HRP-TRP, and (*5*) High Rainfall N-S. (*B*) Track names from the SWAFR elsewhere are (*1*) SW-Arid Australia, (*2*) SW-SE Australia, (*3*) SW-Monsoon Australia, (*4*) SW-NE Australia, (*5*) SW-E Australia, (*6*) SW-Tasmania, (*7*) Trans-Indian Ocean, (*8*) Pan-Temperate, and (*9*) South Pacific. Major areas of the southwest identified (*a, b, c, d*) allow for more precise track descriptors. For example, the SWc-Arid Australia track is for speciation events involving SWAFR taxa from only the Southeast Coastal Province (*c*) and sister taxa in Arid Australia.

SOUTHWEST AUSTRALIAN FLORISTIC REGION   633

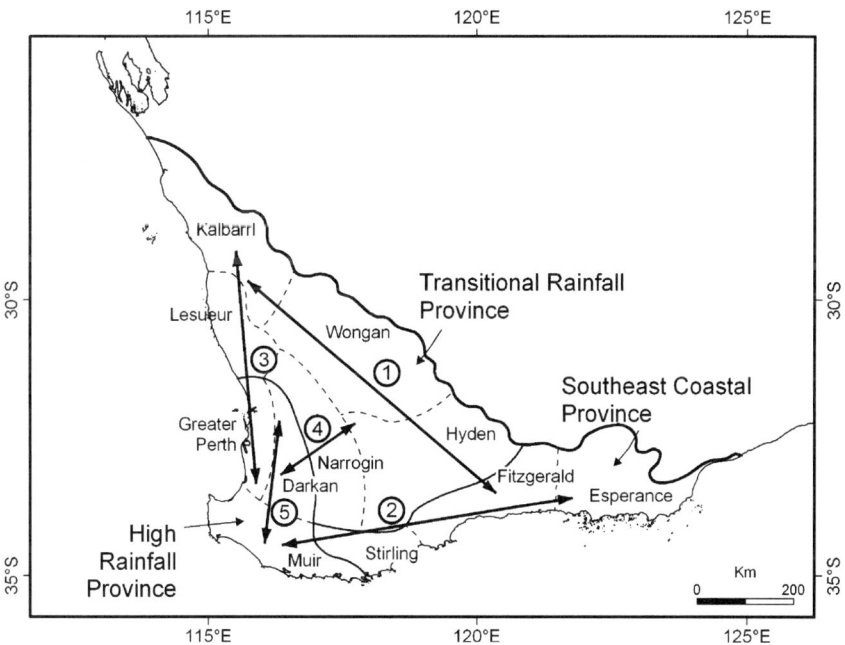

(a) Southwestern Australia

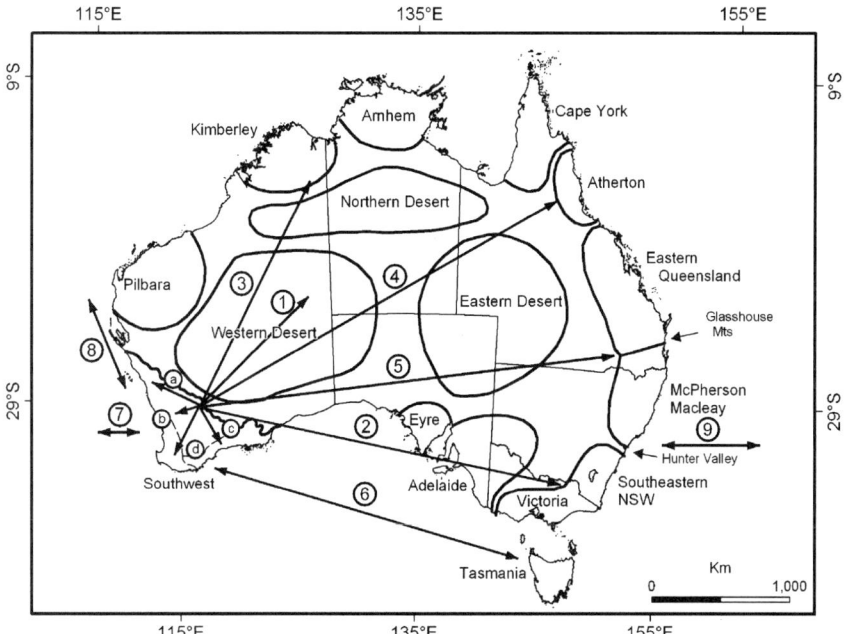

(b) Australian continent

climatic oscillations in the SWAFR, probably most pronounced in the TRP/SCP as hypothesized by several authors (Hopper 1979, 1992; Hopper et al. 1996b). Firm evidence of fire is seen in the common occurrence of charcoal in these Pliocene deposits. Also noteworthy is the persistence of rare *Nothofagus* and podocarp and araucarian rainforest conifers alongside such obvious postfire opportunists as species of Gyrostemonaceae (Pate & Hopper 1993, Bell 2001). The closed rainforest taxa were extinct by 2.6 Mya. Early concepts of a wet mid-Holocene in the SWAFR are not supported by a repeat palynological survey of key lake deposits (Newsome & Pickett 1993).

Molecular Phylogenetic Tests of Evolutionary Hypotheses

Molecular phylogenetics provides powerful tools to test the above hypotheses from independent and rigorously analyzed biological data sets. Although much remains to be done, significant recent progress allows for first glimpses of likely congruence or noncongruence of phylogenetic pattern among SWAFR lineages.

SPECIATION AND EXTINCTION RATES   If, as predicted (Hopper 1979, 1992), a period of explosive speciation in SWAFR taxa occurred in the late Tertiary and Quaternary after a slower period of radiation, unresolved polytomies arising from fewer steadily branching clades or grades and linking many crown taxa should be evident in the molecular phylogenies. Such a pattern was evident for Aizoaceae in the Greater Cape floristic region of South Africa, providing support for the hypothesis that explosive speciation occurred especially among the subfamily Ruschioideae of the succulent karoo (Klak et al. 2004).

Species-level molecular phylogenies are steadily accumulating for SWAFR genera and often contain large late-branching polytomies, e.g., in *Daviesia* (Fabaceae, Crisp & Cook 2003), *Eucalyptus* (Steane et al. 2002), *Acacia* (Miller & Bayer 2001), *Leucopogon* (Ericaceae) (Taffe et al. 2001, Quinn et al. 2003), and *Stylidium* (Coates et al. 2003). Coates et al. (2003) proposed a Pliocene origin for radiation in the *S. caricifolium* complex, and a late-Miocene to early-Pliocene date is also likely for *Conostylis* (Haemodoraceae) (S. Hopper, M. Chase & M. Fay, unpublished data). This is consistent with evidence that climatic variation was significant in the Pliocene as well as the Quaternary (Dodson & Macphail 2004). Large unresolved late-branching polytomies are also seen in molecular cladograms for two major arid zone families: Myoporaceae and Chenopodiaceae (Kelchner 2003, Shepherd et al. 2004).

There are exceptions to this inferred explosive speciation pattern, however, with a more even and steady accumulation of species in late-branching nodes. Examples include *Gastrolobium* (Fabaceae) (Chandler et al. 2001) and Restionaceae (Linder et al. 2003). *Banksia* provides a transitional model with some clades steadily branching and others showing moderately sized terminal polytomies (Mast & Givnish 2002). Further work is needed for a conclusive statement on speciation rates in the SWAFR, but greater complexity than just a major Neogene explosive speciation episode is already evident.

An earlier phase of explosive speciation at tribal/generic level, hypothesized originally from chromosome number patterns (James 1981), is suggested by large polytomies at or near the spine of family or tribal cladistic trees in Rhamnaceae (Richardson et al. 2000), Myrtaceae (Wilson et al. 2001), and Fabaceae (Doyle et al. 2000). Conversely, such a pattern is not evident for Restionaceae (Linder et al. 2003) or Haemodoraceae (Hopper et al. 1999), where a steady accumulation of early branches has been documented. A more complex situation than previously envisioned seems to apply here as well.

Discriminating between patterns due to speciation and those due to extinction processes is a major challenge in interpreting molecular phylogenies. How should an early-branching clade with a few species that is sister to a much richer clade be interpreted—as an evolutionarily quiescent lineage or as the remnants of a richer clade with most members now extinct? In the absence of a fossil record, resolution of this question is problematic. New evidence and approaches are needed.

Although regional extinction of closed rainforest taxa in the Quaternary occurred, it is instructive to enquire where early-branching extant clades of SWAFR taxa persist. Is there evidence of more buffering from extinction in the HRP than in the TRP/SCP (Hopper 1979), or do old relictual clades persist throughout the SWAFR?

OLD RELICTUAL CLADES   Of the 54 species of the psilotid, lycopsid, fern, cycad, cupressid conifer, and podocarp lineages found in the SWAFR, 43 occur in the HRP, 32 in the SCP, and 28 in the TRP. Given that the TRP is three times larger than the HRP and more than twice as large as the SCP, a disproportionate number of these old vascular plant lineages occur in the HRP and SCP. Moreover, 17 of the 54 species extend from the HRP into the TRP, whereas 26 extend from the HRP into the SCP. These statistics support the hypothesis that the more mesic southern Provinces are ancestral to the drier northwest and northern TRP in the phylogenetic radiation of older components of the flora. This pattern is evident in the distributions of *Selaginella gracillima* (Selaginaceae), *Podocarpus drouynianus* (Podocarpaceae), and some but not all *Actinostrobus* and *Callitris* (Cupressaceae).

Members of the endemic order Dasypogonales (Figure 4c,d,k) occur mainly in moist habitats of the HRP and SCP, with some taxa extending into well-drained soils of the TRP. A similar situation prevails for the endemic Cephalotaceae, Eremosynaceae, and oldest members of Colchicaceae (*Burchardia*) (Vinnersten & Reeves 2003) and Haemodoraceae subfamily Conostylidoideae (*Tribonanthes*) (Hopper et al. 1999). Their current distributions support the hypothesis that refugial habitats for old lineages are prevalent in the HRP and SCP.

The Albany Pitcher Plant *Cephalotus follicularis* (Figure 4j), the classic SWAFR endemic discovered in 1802 (Hopper 2004), belongs to Oxalidales (Bradford & Barnes 2001). Its closest relatives are either the monogeneric South American Brunelliaceae or *Brunellia* together with the family Cunoniaceae. In turn, these three families are sister to Elaeocarpaceae. *Cephalotus* is a highly divergent relict of a rainforest lineage that has persisted in the SWAFR by inhabiting swamp margins that are moist year round. However, other endemic SWAFR families, including Anarthriaceae, Ecdeicoleaceae, and Emblingiaceae (Figure 4), are most at home in the moisture-retaining soils of the semiarid TRP, SCP, and HRP.

The divergence time of the two genera of Ecdeiocoleaceae (*Ecdeiocolea* and *Georgeantha*) (Figure 4*h*) was estimated as 75–50 Mya, whereas Ecdeiocoleaceae as a family likely diverged from Poaceae 85–60 Mya (Bremer 2002). The divergence of *Anarthria* from the common ancestor of *Lyginia* and *Hopkinsia* (Figure 4) was estimated at between 70 and 35 Mya (Bremer 2002). Anarthriaceae likely diverged from Restionaceae 100 to 80 Mya. These estimates, derived from molecular phylogenetic data, corroborate the hypothesis based on fossil studies that the TRP kwongan taxa are of considerable antiquity (Hopper et al. 1996b, McLoughlin & McNamara 2001).

It appears that waterlogged or moisture-retaining soils throughout the SWAFR provide habitat for the phylogenetically relictual taxa of the region, with a tendency for these to be concentrated in the southern HRP and SCP rather than the more northerly and drier TRP. These are likely to be the taxa with relatively low extinction rates, best suited to studies aimed at assessing speciation rates and patterns.

RECENTLY SPECIATED CLADES   The patterns emerging from molecular phylogenetic studies are more complex and individualistic than early hypotheses regarding the TRP/SCP speciation hot spot suggested (Hopper 1979, 2004). This complexity is exemplified by an analysis of the distributions of SWAFR taxa in species-level molecular phylogenies of the largely SWAFR endemic poison pea genus *Gastrolobium* (Fabaceae) (Figure 3*f*) and related genera (Chandler et al. 2001).

Within this lineage, the earliest branching clade is the *G. spinosum* group. This group mainly grows on lateritic and granite outcrops in the SCP and TRP, but it also has satellite taxa evolving from independent TRP/SCP groups into the HRP (*G. cuneatum* and *G. bilobum*) (Figure 3*f*). The group also has taxa from the SCP's Esperance District (Figure 5), giving rise to the only two species in the genus extending widely into the Australian arid zone and its margins (bird-pollinated *G. grandiflorum* and *G. brevipes*).

Essentially, these patterns within the SWAFR are repeated in subsequent clades of *Gastrolobium*. There are complex allopatric replacement series across the semi-arid TRP and SCP, with multiple origins of offshoot species to the HRP, local radiations of taxa in the SCP's Stirling Range (Figure 5) (especially of taxa previously placed in *Nemcia*), and a terminal radiation of mainly bird-pollinated species previously placed in *Brachysema* and *Jansonia* that reinvaded and speciated within the HRP.

Thus the hypothesis that recent speciation is accentuated in the TRP/SCP relative to the HRP (Hopper 1979) is supported by the *Gastrolobium* study, but the degree of involvement of HRP taxa is reasonably high and present at all levels in the cladogram except the earliest branches. With the arid zone taxa *G. grandiflorum* and *G. brevipes*, the genus also exemplifies how a predominantly southwestern endemic has spawned widespread congeners out into Australia's deserts.

Cytogeography and Infraspecific Phylogeography

Historical biogeographic patterns revealed by molecular phylogenetic studies of many taxa above the species level are mirrored by patterns of genetic differentiation within species in the SWAFR. For example, a perennial herb of rock outcrops, *Isotoma petraea* (Campanulaceae–Lobeliaceae), ranges across arid southern Australia as a series of uniformly diploid outbreeding, chromosomally invariant populations until it reaches the margins of the SWAFR. There, formation of chromosomal rings in translocation heterozygotes is associated with a shift to inbreeding. The number of chromosomes in the ring increases toward the southwest until all seven pairs are bound up in one super ring in populations at the extreme southwest margin of the Wongan District of the TRP (James 1965, 1992, 2000). Recent DNA phylogeographic studies affirm this cytogeographic hypothesis (Bussell et al. 2002), thus providing an example of directional chromosomal divergence from the arid zone into the SWAFR. A similar phylogeographic pattern has been hypothesized recently for sandalwood (*Santalum spicatum*, Santalaceae) (Byrne et al. 2003).

Evidence of complex intraspecific phylogeographic divergence, some likely to date back to the Pliocene, is emerging from studies of a range of TRP/SCP taxa, including species of *Acacia*, *Eucalyptus*, *Banksia*, *Geleznowia* (Rutaceae), and *Stylidium* (Coates 2000, Byrne 2003). *Lambertia orbifolia* (Proteaceae), *Acacia anomala*, and *Laxmannia sessiflora* (Lomandraceae) provide similar HRP examples (James et al. 1999, Coates 2000). As case after case accumulates, the evidence that this is a flora with anciently and complexly fragmented population systems becomes compelling.

Biogeographical Tracks Within and Outside the SWAFR

The overriding emergent hypothesis from a review of molecular phylogenetic studies (Hopper 2004) is that the SWAFR has a vascular flora of multiple origins. The flora is made up of lineages that originated within and from outside the area and at a range of times from the Carboniferous (*Selaginella*) to the late Cenozoic. As a way of summarizing and synthesizing these multiple origins, Figure 6 depicts a predictive hypothesis on general patterns for the SWAFR that is evident from the literature using track analysis (Hopper 2004). This hypothesis identifies relationships between areas of endemism through lines along which congruent speciation in independent lineages has occurred.

Formal analysis of these new tracks is possible using approaches such as cladistic biogeography (Crisp et al. 1999). An elegant start was made by Mast (2000, Mast & Givnish 2002) on Proteaceae tribe Banksieae (*Banksia/Dryandra*, *Musgravea*, and *Austromuellera*). Dispersal-vicariance analysis proposed an ancestral taxon for the tribe distributed across the Southwest-Northeast Australian Track followed by a vicariance event, possibly triggered by island formation during the late-Cretaceous downwarping and marine inundation of parts of the Australian plate.

A SWAFR origin for the subtribe Banksiinae (*Banksia/Dryandra*) was resolved, with two later expansions east followed by vicariance events along the Southwest-Southeast or Southwest-East Australia Tracks associated with Eocene or Miocene flooding of the Nullarbor Plain and the Oligocene onset of aridity in the same region. Within the SWAFR, two major centers of endemism were identified on areas down the northwest coast and southern coast. Although not elaborated in detail by Mast (2000, Mast & Givnish 2002), an examination of his phylogeny reveals congruent examples of speciation in *Banksia* within the SWAFR along all five tracks illustrated in Figure 6.

Interesting examples highlighting other tracks in Figure 6 include a biogeographic subtree analysis of the *Melaleuca* group (Brown et al. 2001), the papilionoid legume genistoid subtribe Brongniartieae with six Australian and four tropical American genera (Thompson et al. 2001), and the prostrate monotypic shrub *Emblingia calceolifolia* (Emblingiaceae) (Figure 4), which is a highly distinctive member of a widely dispersed group of brassicalean families (Chandler & Bayer 2000). An African connection to the SWAFR is evident in studies of Proteaceae (Barker et al. 2002), *Pelargonium* (Geraniaceae, Bakker et al. 1998), and Aizoaceae (Klak et al. 2003).

Adaptive Radiation

Most work on adaptive radiation in SWAFR plants has been descriptive and narrative, sometimes backed up by experimental studies of function (Carlquist 1974, Pate et al. 1984, Lamont & Enright 2000, Lambers et al. 2003, Lamont 2003). The complexity of growth form and scope for exploration of adaptive radiation in the SWAFR is substantial. Pate et al. (1984) noted the exceptional diversity of form and function within families and genera of what appeared to be a "highly random nature," indicating complex and perhaps ancient but ongoing evolutionary pathways. Only reproductive attributes such as fruit dispersal patterns in kwongan plants aligned with and displayed homogeneity within generic and family-level boundaries. Several books and reviews on the SWAFR richly illustrate this theme (e.g., Pate & McComb 1981, Pate & Beard 1984, Hobbs 1992, Pate & Hopper 1993, Hopper et al. 1996a, Abbott & Burrows 2003).

Narrative accounts of adaptation are fraught with difficulties, however. There is a fundamental difference between a functional evolutionary adaptation and an effect or correlation, and this difference requires sophisticated experimentation and phylogenetic knowledge to resolve (Williams 1966, Brandon 1990).

Additional penetrating recent work addressing adaptive radiation in a phylogenetic context has emerged for thickened roots in SWAFR *Daviesia* (Crisp & Cook 2003) and for sclerophylly and leaf xeromorphy in *Banksia/Dryandra* (Mast & Givnish 2002). However, hypotheses of adaptive radiation are difficult to test unequivocally, even within a well-resolved phylogeny. Much more experimental research at the population level within well-resolved phylogenies is needed to advance understanding of the evolution of functional traits before hypotheses of adaptive radiation can be convincingly invoked.

Genetic Systems and Speciation

Chromosome repatterning is frequently observed among related species or populations, sometimes without obvious ecological or morphological change. This correlation suggests a possible role for chromosome repatterning in speciation. Yet how chromosome changes might become fixed in populations remains a complex and controversial subject that is poorly documented (Levin 2002). The dilemma is that all structural mutations must arise at low frequencies and occur as heterozygotes, where the meiotic irregularities they cause should confer low fitness and high genetic load on their carriers, which should in turn cause natural selection to purge them.

However, under the genomic coalescence hypothesis, James (1992) proposed that chromosomal mutations that link supergenes heterozygous for recessive lethals will be elevated to high frequencies by natural selection in inbreeding populations. This process was demonstrated in a series of papers exploring the evolution of complex translocation heterozygosity in the granite outcrop herb *Isotoma petraea*, as discussed above (James 1965, 1981, 2000; Bussell et al. 2002). In this species, seed abortion reflecting high genetic load is evident in the southwest populations, yet the populations persist through intense natural selection favoring chromosome restructuring to conserve heterozygosity in the face of inbreeding.

Aspects of the genomic coalescence hypothesis have been elaborated in other groups of SWAFR plants, including *Stylidium*, *Drosera*, *Dampiera* (Goodeniaceae), *Boronia* (Rutaceae), and Myrtaceae (Coates & James 1996, James 2000, Coates et al. 2003, Shan et al. 2004). Perhaps the most penetrating recent exploration was the study of chromosome number reduction, self-pollination, and lethal polymorphisms in a population of the paper lily *Laxmannia sessiliflora* (Lomandraceae) (James et al. 1999). This line of research might be profitably applied to other species-rich floras where chromosome repatterning is evident.

Indeed, many fundamental questions regarding plant speciation have been barely addressed in the SWAFR or elsewhere, but new approaches show considerable promise in helping elucidate the processes involved (Coates et al. 2003, Lamont & Weins 2003, Hopper 2004, Rieseberg & Wendel 2004).

Natural Hybridization

In SWAFR *Banksia* and eucalypts, rates of detected natural hybridization are several times less than those found in eastern Australia (Hopper 1994a). This pattern has been attributed to the above-mentioned evolutionary patterns in the SWAFR, resulting in narrower geographical ranges, less sympatry, and long periods of time to evolve divergent genetic systems incapable of coalescing when sympatry is achieved. Nevertheless, some interesting case studies of natural hybridization have emerged in the SWAFR. These include evidence for selective pollinator behavior in *Anigozanthos* (Hopper & Burbidge 1986; Hopper 1994a), for ancient allopolyploid speciation (Krauss & Hopper 2001), and for anthropogenic disturbance that creates habitat that breaks down phenological barriers in *Banksia* species (Lamont et al. 2003).

CONSERVATION BIOLOGY

Changes wrought on the SWAFR following European colonization have profoundly altered most vegetation. The region now has more species of threatened plants (2500) than other Australian states and most countries of the world (Figure 7) (Hopper et al. 1990; Hopper 1997, 2004; Brown et al. 1998; Coates & Atkins 2001). Conservation issues include massive habitat loss and fragmentation, root-rot disease (*Phytophthora*) impacting 2500 species, displacement by 900 invasive weeds, and rising saline groundwater tables threatening 470 taxa. Moreover, a significant taxonomic impediment is patently evident in a comparison of Figures 1 and 7. Until plants are recognized as distinct and named, their specific conservation needs will not be addressed and their fate is reliant on other non-targeted strategies such as the accidental inclusion of populations into protected areas (Hopper 1994b, 1997). This is a significant issue in a flora with 14% of species/subspecies still without names (Paczkowska & Chapman 2000, Hopper 2004).

Overall, the challenge of conserving the SWAFR flora is daunting. Managers need all the help scientists can provide to deal with such biodiversity hot spots. Otherwise diversity becomes tyranny, and conservation outcomes languish as managers ignore the problems or move on to other jobs with such rapidity that there is no hope of understanding or effectively dealing with the hot spot. Such an approach

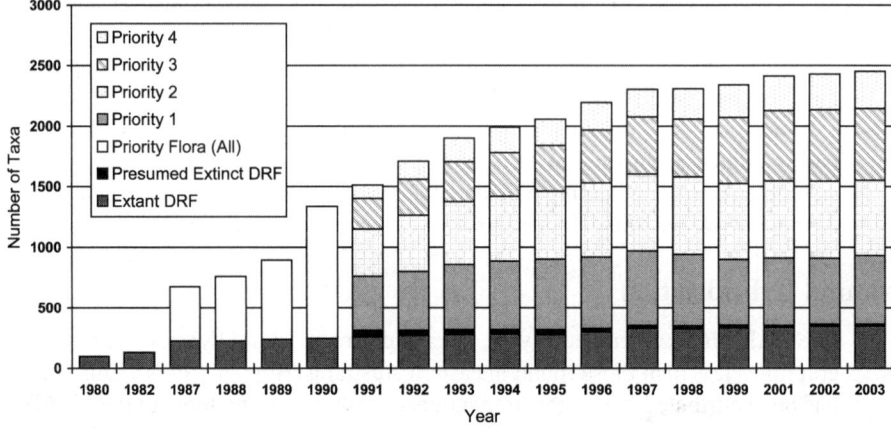

Figure 7 Number of threatened Western Australian plant species/subspecies from 1980–2003 declared as *Rare Flora* (DRF) under the Wildlife Conservation Act, including presumed extinct DRF. Also included are priority taxa in need of urgent survey to determine their conservation status (priority 1–3) and taxa adequately surveyed but in need of monitoring (priority 4). Figure courtesy of Ken Atkins, Department of Conservation and Land Management, Perth, Australia.

is in stark contrast to the world view of indigenous people, who lived sustainably with most biodiversity and passed it on to future generations.

Modern conservation of the SWAFR's flora has relied traditionally on a protected-areas approach (Figure 2a) that has achieved variable outcomes, from comprehensive coverage in forested districts of the HRP to almost no representation in heavily cleared parts of agricultural districts in the TRP. More recently, the emergent disciplines of conservation biology and restoration ecology have been applied as Western Australians grapple with the extraordinary conservation challenge they face (Hobbs 1992, Dixon 1994, Hopper et al. 1996a, Hopper 1997, Coates & Atkins 2001, Hobbs & Yates 2003, Cochrane 2004).

Beginning with the early listing of rare and poorly known flora in 1979 by a handful of botanists, plant conservation biology in the SWAFR has moved on to sophisticated mapping, monitoring, and process-focused research aimed at repair, recovery, and restoration by a diverse cadre of workers. It is much too early to know whether effective and sustainable conservation will be achieved.

Conservation genetics remains a strength of work in the SWAFR (Coates et al. 2000). Perhaps of greatest interest in terms of future research is James's (1992, 2000) proposal that high genetic diversity in some circumstances can be a signal for a species in trouble with impeded evolutionary capability and, conversely, that "allelically depauperate lineages of the Gondwana-like wetter regions are probably not inbred and debilitated, rather they are probably highly adapted with little genetic load and unimpeded recombination systems" (James 2000, p. 346). This hypothesis calls for a significant re-evaluation of how patterns of genetic architecture are used to define population-level conservation priorities.

From an ecological perspective, the SWAFR has a remarkably high level of turnover of plant species across the landscape, requiring caution in the application of broadscale conservation management approaches. Regeneration and recruitment work has highlighted the importance of understanding fire- and smoke-stimulated germination (Flematti et al. 2004); the ephemeral nature of some seed banks; as well as the impact of fragmentation on mating systems, pollination, and seed set. Removal of topsoil makes the landscape especially vulnerable to loss of seed and helpful symbionts of native plants, as well as massive weed invasion. The minimization of bulldozing is therefore a vital conservation strategy. Risk assessment, ex situ strategies, species recovery, and landscape restoration are crucial emergent activities for integrated conservation outcomes, but care for protected areas of remnant vegetation (Figure 2a), especially programs involving local communities as active stewards, remains an important and effective priority.

Barely perceived conservation consequences are arising from altered disturbance regimes such as the loss of digging by locally extinct marsupials (Garkaklis et al. 2003). Moreover, complex ecological interactions between insects and paleoendemic flora such as Ecdeiocoleaceae (Figure 4h) highlight the need for increasingly sophisticated conservation management (Main 1996). There is continued debate on what the appropriate scale and diversity of management regimes for biodiversity conservation, particularly in forests, should be (Abbott & Burrows 1999, 2003; Wardell-Johnson & Horwitz 2000).

Given the geohistorical context of the SWAFR and known high levels of genetic divergence among populations of species, it is reasonable to hypothesize that many rare and threatened plants have evolved over long periods in disjunct small populations. Consequently, they would be more resilient to the effects of contemporary fragmentation than would be plants from elsewhere that have naturally large continuous population structures. Predictions of accelerated extinction rates due to habitat loss may not hold in such circumstances, or at least, these rates may be slower than elsewhere (Hopper et al. 1990, Brooks et al. 2002). Active research is under way to further test this hypothesis (Hobbs & Yates 2003). Potential global climate change also poses significant challenges to the SWAFR (Mooney et al. 2001). Although the prognosis is daunting, given the ongoing interaction of destructive processes in the SWAFR, local commitment to conserve this globally significant floral heritage remains strong.

COMPARISONS WITH SOUTH AFRICA'S GREATER CAPE FLORISTIC REGION

In many respects, the region on Earth most similar to the SWAFR is the Greater Cape Floristic Region in South Africa, including the Cape and Succulent Karoo global biodiversity hot spots (Cowling et al. 1996, Myers et al. 2000, Goldblatt & Manning 2002, Linder 2003). Both the SWAFR and Greater Cape Mediterranean regions are floristically rich; have a climate that has been oceanically moderated since the Jurassic; have been unglaciated since the Permian; and are dominated by old, weathered, nutrient-deficient landscapes. These regions also share rainfall regimes that are more reliable than those seen in California and Spain (Cowling et al. 2004).

The Greater Cape Floristic Region is much richer in species and genera than the SWAFR, but it has approximately the same number of families, including a comparable five to six endemics (Goldblatt & Manning 2002). The exceptional richness of the Greater Cape flora, unparalleled for a temperate region, may be due to the greater (Cretaceous) age of the Namib Desert compared with the Australian deserts of Oligocene age or younger. Climatic transitional areas between the Namib and mesic southwest Cape mountains have existed for a much longer period, facilitating explosive speciation in succulent karoo and adjacent areas (e.g., Klak et al. 2003, 2004). Also, the more rugged topography of the Greater Cape Floristic Region within a similar stable, oceanically moderated Mediterranean region has offered greater opportunities for genetic divergence and speciation (Goldblatt & Manning 2002, Linder 2003). Comparative phylogenetic studies, only just beginning (Bakker et al. 1998, Barker et al. 2002, Linder et al. 2003), need to be undertaken to test such hypotheses and search for general patterns across unrelated lineages.

Some illuminating ecological comparisons between climatically matched sites in kwongan and fynbos (Cowling & Witkowski 1994, Cowling et al. 1994) as well as comparative studies of species flocks in Proteaceae (Cowling et al. 1996, Cowling & Lamont 1998) have been completed. These have demonstrated, for example, that Proteaceae in the SWAFR are richer in species overall and locally, are more common in transitional rather than high rainfall districts, and have higher gamma (geographical) diversity. Proteaceae from the SWAFR also more commonly resprout after fire, produce serotinous fruits, are vertebrate pollinated, and grow taller than their Cape congeners. Harsher climatic conditions in the SWAFR had been hypothesized to favor resprouters over nonsprouters (Cowling & Lamont 1998).

A study of matched sites on south coastal districts on both continents (Fitzgerald River District, SCP, in the SWAFR) showed that woody plants with leaf spines and canopy-stored seed were more common in the SWAFR on nutrient-poor soils, whereas the Cape's less impoverished calcareous and limestone sands had far more species with bird-dispersed fruits and interfire germination (Cowling & Witkowski 1994). Nevertheless, strong convergence was documented between SWAFR and Cape shrublands in a wide range of other growth form and function traits. Narrow-range endemics in the SWAFR are more likely to be graminoids, or shrubs with either canopy-stored or soil-stored seeds of various dispersal agents (wind, vertebrates, ants, and ballistic), compared with the predominant low shrubs with soil-stored, ant-dispersed seeds in the Cape (Cowling et al. 1994). Differences may be due to phylogenetic contingency and regional processes (Herrera 1992), such as the availability of less impoverished coastal habitat in the Cape for nearby rainforest elements to occupy compared with the SWAFR's complete isolation by desert from rainforest taxa. Further work along these lines, especially that addressing phylogenetics and chromosomal aspects of speciation, would be valuable.

Exciting work on the impact of smoke on seed germination in the SWAFR was stimulated by discoveries in South Africa (Dixon et al. 1995, Roche et al. 1997, Tieu et al. 2001, Flematti et al. 2004). This exemplifies the ongoing benefit of collaborative comparative studies.

CONCLUSIONS

Significant advances in understanding the SWAFR flora have occurred since 1979. The SWAFR has proven to be a botanical frontier, as poorly inventoried three decades ago as many tropical rainforest regions. The SWAFR is now recognized as one of the world's 25 global biodiversity hot spots, the only region in Australia accorded this status.

A sharper focus on landscape evolution and the fossil record have highlighted the remarkable antiquity and stability of the SWAFR, which has been free of glaciers and marine inundation since the Permian, with little orogeny since. The species richness, high endemism, and rapid turnover of species over short distances across the landscape are ancient, extending at least as far back as the Eocene. Such antiquity has resulted in remarkably sophisticated evolutionary responses to living on a flat, stable, highly weathered, nutrient-deficient landscape with subtle soil mosaics. Examples of coping with rarity in naturally fragmented populations are common in the flora, and complex interplays between chromosomal systems and ecological adaptation have occurred, with explosive speciation as an incidental by-product in some groups.

The region thus exemplifies plant evolution in temperate environments at its most sophisticated and durable. However, such a flora, exposed to European land-use practices, is highly threatened. Fundamental changes in attitudes toward land use and the intrinsic value of plant life are needed to go hand in hand with a commitment to protect, repair, and restore native vegetation in the face of uncertainty. These changes will require the inspiration and training of new cohorts of plant conservation biologists as well as continuing strong local support, which thankfully is forthcoming.

ACKNOWLEDGMENTS

To all those colleagues who accompanied S.D.H. in fieldwork, assisted our learning, and generously shared their ideas and insights, thanks. We are indebted to the director and staff at the Western Australian Herbarium for their support and data access, and to collectors, past and present, who have contributed specimens. We are grateful to Mark Burgman, Margaret Byrne, David Coates, Richard Cowling, Mike Crisp, John Dodson, Peggy Fiedler, Peter Goldblatt, Bob Hill, Siegy Krauss, Hans Lambers, Byron Lamont, Austin Mast, Vincent Savolainen, and Colin Yates for comments on the manuscript and to Ellen Hickman for her exquisite artwork in Figures 3 and 4.

The *Annual Review of Ecology, Evolution, and Systematics* is online at
http://ecolsys.annualreviews.org

LITERATURE CITED

Abbott I, Burrows N. 1999. Biodiversity conservation in the forests and associated vegetation types of southwest Western Australia. *Aust. For.* 62:27–32

Abbott I, Burrows N, eds. 2003. *Fire in Ecosystems of South-west Western Australia: Impacts and Management*. Leiden: Backhuys Publishers

Anand RR, Paine M. 2002. Regolith geology of the Yilgarn Craton, Western Australia: implications for exploration. *Aust. J. Earth Sci.* 49:3–162

Anderson JM, Anderson HM, Archangelsky S, Bamford M, Chandra S, et al. 1999. Patterns of Gondwana plant colonisation and diversification. *J. Afr. Earth Sci.* 28:145–67

APG II 2003. An update of the Angiosperm Phylogeny Group classification for the orders and families of flowering plants: APG II. *Bot. J. Linn. Soc.* 141:399–436

Atahan P, Dodson JR, Itzstein-Davey F. 2004. A fine-resolution Pliocene pollen and charcoal record from Yallalie, south-western Australia. *J. Biogeogr.* 31:199–205

Bakker FT, Helbrugge D, Culham A, Gibby M. 1998. Phylogenetic relationships within *Pelargonium* sect. *Peristera* (Geraniaceae)

inferred from nrDNA and cpDNA comparisons. *Plant. Syst. Evol.* 211:273–87
Barker NP, Weston PH, Rourke JP, Reeves G. 2002. The relationships of the southern African Proteaceae as elucidated by internal transcribed spacer (ITS) DNA sequence data. *Kew Bull.* 57:867–83
Beard JS. 1970. *A Descriptive Catalogue of West Australian Plants.* Sydney: Society for Growing Australian Plants. 2nd ed.
Beard JS. 1990. *Plant Life of Western Australia.* Sydney: Kangaroo Press
Beard JS, Chapman AR, Gioia P. 2000. Species richness and endemism in the Western Australian flora. *J. Biogeogr.* 27:1257–68
Bell DT. 2001. Ecological response syndromes in the flora of southwestern Australia: fire resprouters versus reseeders. *Bot. Rev.* 67:417–40
Bradford JC, Barnes RW. 2001. Phylogenetics and classification of Cunoniaceae (Oxalidales) using chloroplast DNA sequences and morphology. *Syst. Bot.* 26:354–85
Brandon RN. 1990. *Adaptation and Environment.* Princeton, NJ: Princeton Univ. Press
Bremer K. 2002. Gondwanan evolution of the grass alliance of families (Poales). *Evolution* 56:1374–87
Brooks TM, Mittermeier RA, Mittermeier CG, da Fonseca AB, Rylands AB, et al. 2002. Habitat loss and extinction in the hot spots of biodiversity. *Cons. Biol.* 16:909–23
Brown EM, Burbidge AH, Dell J, Edinger D, Hopper SD, Wills RT. 1997. *Pollination in Western Australia: A Database of Animals Visiting Flowers.* Handb. 15. Perth: West. Aust. Nat. Club
Brown AP, Thomson-Dans C, Marchant N, eds. 1998. *Western Australia's Threatened Flora.* Perth: Dept. Conservation & Land Management
Brown GK, Udovicic F, Ladiges PY. 2001. Molecular phylogeny and biogeography of *Melaleuca, Callistemon* and related genera (Myrtaceae). *Aust. Syst. Bot.* 14:565–85
Bussell JD, Waycott M, Chappill JA, James SH. 2002. Molecular phylogenetic analysis of the evolution of complex hybridity in *Isotoma petraea*. *Evolution* 56:1296–302
Byrne M. 2003. Phylogenetics and the conservation of a diverse and ancient flora. *Comptes Rendus Biol.* 326:73–79
Byrne M, Macdonald B, Brand J. 2003. Phylogeography and divergence in the chloroplast genome of Western Australian Sandalwood (*Santalum spicatum*). *Heredity* 91:389–95
Carlquist S. 1974. *Island Biology.* New York: Colombia Univ. Press
Chandler GT, Bayer RJ. 2000. Phylogenetic placement of the enigmatic Western Australian genus *Emblingia* based on *rbcL* sequences. *Plant. Species Biol.* 15:67–72
Chandler GT, Bayer RJ, Crisp MD. 2001. A molecular phylogeny of the endemic Australian genus *Gastrolobium* (Fabaceae: Mirbelieae) and allied genera using chloroplast and nuclear markers. *Am. J. Bot.* 88:1675–87
Coates DJ. 2000. Defining conservation units in a rich and fragmented flora, implications for the management of genetic resources and evolutionary processes in south-west Australian plants. *Aust. J. Bot.* 48:329–39
Coates DJ, Atkins KA. 2001. Priority setting and the conservation of Western Australia's diverse and highly endemic flora. *Biol. Conserv.* 97:251–63
Coates DJ, James SH. 1996. Chromosome repatterning, population genetic structure and local speciation in southwestern Australian triggerplants (*Stylidium*). See Hopper et al. 1996a, pp. 276–86
Coates DJ, Carstairs S, Hamley VL. 2003. Evolutionary patterns and genetic structure in localized and widespread species in the *Stylidium caricifolium* complex (Stylidiaceae). *Am. J. Bot.* 90:997–1008
Coates DJ, Hopper SD, Farrer SL, eds. 2000. *Genetics and Conservation of Australian Flora. Aust. J. Bot.,* Spec. Issue 48:287–416
Cochrane A. 2004. Western Australia's *ex situ* program for threatened species: a model integrated strategy for conservation. In *Ex Situ Plant Conservation Supporting Species*

Survival in the Wild, ed. EO Guerrant, K Havens, M Maunder, pp. 40–66. Washington, DC: Island Press. 504 pp.

Cowling RM, Lamont BB. 1998. On the origin of Gondwanan species flocks: diversity of Proteaceae in mediterranean south-western Australia and South Africa. *Aust. J. Bot.* 46:335–55

Cowling RM, Witkowski ETF. 1994. Convergence and non-convergence of plant traits in climatically and edaphically matched sites in Mediterranean Australia and South Africa. *Aust. J. Ecol.* 19:220–32

Cowling RM, Witkowski ETF, Milewski AV, Newbey KR. 1994. Taxonomic, edaphic and biological aspects of narrow plant endemism on matched sites in mediterranean South Africa and Australia. *J. Biogeogr.* 22:651–64

Cowling RM, Rundel PW, Lamont BB, Arroyo MK, Arianoutsou M. 1996. Plant diversity in Mediterranean-climate regions. *Trends Ecol. Evol.* 11:362–66

Cowling RM, Ojeda F, Lamont BB, Rundel PW, Lechmere-Oertel R. 2004. Rainfall reliability, plant reproductive traits and regional-scale diversity in fire-prone mediterranean-climate ecosystems. *Oikos.* In press

Crisp MD, Cook LG. 2003. Phylogeny and evolution of anomalous roots in *Daviesia* (Fabaceae: Mirbelieae). *Int. J. Plant Sci.* 164:603–612

Crisp MD, West JG, Linder HP. 1999. Biogeography of the terrestrial flora. In *Flora of Australia*, 1:321–67. Canberra: Aust. Biol. Res. Study. 2nd ed.

Dallman PR. 1998. *Plant Life in the World's Mediterranean Climates.* Berkeley: Calif. Native Plant Soc., Univ. Calif. Press

Diels L. 1906. *The Plant Life of Western Australia South of the Tropics. The Vegetation of the World. VII*, ed. A Engler, O Drude. Leipzig: Engelmann

Dixon KW. 1994. Towards integrated conservation of Australian endangered plants—the Western Australian model. *Biodivers. Conserv.* 3:148–59

Dixon KW, Roche S, Pate JS. 1995. The promotive effect of smoke derived from burnt native vegetation on seed germination of Western Australian plants. *Oecologia* 101:185–92

Dodson JR, Macphail MK. 2004. Palynological evidence for aridity events and vegetation change during the Middle Pliocene, a warm period in Southwestern Australia. *Glob. Planet. Chang.* In press

Doyle JJ, Chappill JA, Bailey DC, Kajita T. 2000. Towards a comprehensive phylogeny of legumes: evidence from *rbcL* sequences and non-molecular data. In *Advances in Legume Systematics 9*, ed. PS Herendeen, A Bruneau, pp. 1–20. Kew: Royal Botanic Gardens

Flematti GR, Ghisalberti EL, Dixon KW, Trengove RD. 2004. A compund from smoke that promotes seed germination. *Science* In press

Frakes LA. 1999. Evolution of Australian environments. *Flora of Australia* 1:163–203. 2nd ed.

Garkaklis MJ, Bradley JS, Wooller RD. 2003. The relationship between animal foraging and nutrient patchiness in south-west Australian woodland soils. *Aust. J. Soil. Res.* 41:665–73

Gioia P, Pigott JP. 2000. Biodiversity assessment: a case study in predicting richness from the potential distributions of plant species in the forests of south-western Australia. *J. Biogeogr.* 27:1065–78

Goldblatt P, Manning JC. 2002. Plant diversity of the Cape Region of Southern Africa. *Ann. Missouri Bot. Gard.* 89:281–302

Grubb PJ. 1992. A positive distrust in simplicity—lessons from plant defences and from competition among plants and among animals. *J. Ecol.* 80:585–610

Harvey MS. 2002. Short-range endemism among the Australian fauna: some examples from non-marine environments. *Invert. Syst.* 16:555–70

Herrera CM. 1992. Historical effects and sorting processes as explanations for contemporary ecological patterns, character syndromes in Mediterranean woody plants. *Am. Nat.* 140:421–46

Hobbs RJ, ed. 1992. *Biodiversity of Mediterranean Ecosystems in Australia.* Chipping Norton, NSW: Surrey Beatty & Sons

Hobbs RJ, Yates CJ. 2003. Turner Review No. 7. Impacts of ecosystem fragmentation on plant populations: generalising the idiosyncratic. *Aust. J. Bot.* 51:471–88

Hooker JD. 1860. *The Botany of the Antarctic Voyage of H. M. Discovery Ships Erebus and Terror in the years 1839–1843. Part III. Flora Tasmaniae.* Vol. I. Dicotyledones. London: Lovell Reeve. 359 pp.

Hopkins AJM, Griffin EA. 1984. Floristic patterns. See Pate & Beard 1984, pp. 69–83

Hopkins AJM, Keighery GJ, Marchant NG. 1983. Species-rich uplands of south-western Australia. *Proc. Ecol. Soc. Aust.* 12:15–26

Hopper SD. 1979. Biogeographical aspects of speciation in the south west Australian flora. *Annu. Rev. Ecol. Syst.* 10:399–422

Hopper SD. 1992. Patterns of diversity at the population and species levels in south-west Australian Mediterranean ecosystems. See Hobbs 1992, pp. 27–46

Hopper SD. 1994a. Evolutionary networks: natural hybridization and its conservation significance. In *Nature Conservation 4: The Role of Networks*, ed. DA Saunders, JL Craig, EM Mattiske, pp. 51–66. Chipping Norton, NSW: Surrey Beatty & Sons

Hopper SD. 1994b. Plant taxonomy and genetic resources: foundations for conservation. In *Conservation Biology in Australia and Oceania*, ed. C Moritz, J Kikkawa, pp. 269–285. Chipping Norton, NSW: Surrey Beatty & Sons

Hopper SD. 1997. An Australian perspective on plant conservation biology in practice. In *Conversation Biology for the Coming Decade*, ed. PL Fiedler, PM Kareiva, pp. 255–78. New York: Chapman and Hall

Hopper SD. 2000. How well do phylogenetic studies inform the conservation of Australian plants? *Aust. J. Bot.* 48:321–28

Hopper SD. 2003. South-western Australia—Cinderella of the world's temperate floristic regions. 1. *Curtis's Bot. Mag.* 20:101–26

Hopper SD. 2004. *A Cinderella Flora—botanical discovery and description in the southwest Australian Global Biodiversity Hotspot.* Nedlands: Univ. West. Aust. Press. In press

Hopper SD, Brown AP. 2001. Contributions to Western Australian Orchidology: 2. New taxa and circumscriptions in *Caladenia* (Spider, Fairy and Dragon Orchids of Western Australia). *Nuytsia* 14:27–314

Hopper SD, Burbidge AH. 1986. Speciation of bird-pollinated plants in south-western Australia. In *The Dynamic Partnership: Birds and Plants in Southern Australia*, ed. HA Ford, DC Paton, pp. 20–31. Adelaide: Govt. Printer

Hopper SD, van Leeuwen S, Brown AP, Patrick SJ. 1990. *Western Australia's Endangered Flora.* Perth: Department of Conservation and Land Management

Hopper SD, Brown AP, Marchant NG. 1997. Plants of Western Australian granite outcrops. In *Granite Outcrops Symposium*, ed. PC Withers, SD Hopper. *J. R. Soc. West. Aust.* 80:141–58

Hopper SD, Chappill JA, Harvey MS, George AS, eds. 1996a. *Gondwanan Heritage: Past, Present and Future of the Western Australian Biota.* Chipping Norton, NSW: Surrey Beatty & Sons

Hopper SD, Harvey MS, Chappill JA, Main AR, Main BY. 1996b. The Western Australian biota as Gondwanan Heritage—a review. See Hopper et al. 1996a, pp. 1–46

Hopper SD, Fay MF, Rossetto M, Chase MW. 1999. A molecular phylogenetic analysis of the bloodroot and kangaroo paw family Haemodoraceae: taxonomic, biogeographic and conservation implications. *Bot. J. Linn. Soc.* 131:285–99

Itzstein-Davey F. 2003. *Changes in the abundance and diversity of the Proteaceae over the Cainozoic in south-western Australia.* PhD thesis. Univ. West. Aust., Nedlands

James SH. 1965. Complex hybridity in *Isotoma petraea* I. The occurrence of interchange heterozygosity, autogamy and a balanced lethal system. *Heredity* 20:341–53

James SH. 1981. Cytoevolutionary patterns, genetic systems and the phytogeography of

Australia. In *Ecological Biogeography of Australia*, ed. A Keast, pp 763–782. The Hague/Boston/London: Dr. W. Junk

James SH. 1992. Inbreeding, self-fertilization, lethal genes and genomic coalescence. *Heredity* 68:449–56

James SH. 2000. Genetic systems and conservation strategies for Australian plant species. *Aust. J. Bot.* 48:341–47

James SH, Keighery GJ, Moorrees A, Waycott M. 1999. Genomic coalescence in a population of *Laxmannia sessiliflora* (Angiospermae, Anthericaceae), an association of lethal polymorphism, self-pollination and chromosome number reduction. *Heredity* 82:364–72

Keighery GJ. 1982. Bird-pollinated plants in Western Australia. In *Pollination and Evolution*, ed. JA Armstrong, JM Powell, AJ Richards, pp. 77–90. Sydney: Royal Botanic Gardens

Kelchner SA. 2003. *Phylogenetic structure, biogeography, and evolution of Myoporaceae*. PhD thesis. The Australian National Univ., Canberra. 222 pp.

Klak C, Khunou A, Reeves G, Hedderson T. 2003. A phylogenetic hypothesis for the Aizoaceae (Caryophyllales) based on four plastid DNA regions. *Am. J. Bot.* 90:1433–45

Klak C, Reeves G, Hedderson T. 2004. Unmatched tempo of evolution in Southern African semi-desert ice plants. *Nature* 427:63–65

Krauss SL, Hopper SD. 2001. From Dampier to DNA: the 300-year-old mystery of the identity and proposed allopolyploid origin of *Conostylis stylidioides* (Haemodoraceae). *Aust. J. Bot.* 49:1–8

Lambers H, Cramer MD, Shane MW, Wouterlood M, Poot P, Veneklass EJ. 2003. Introduction: structure and functioning of cluster roots and plant responses to phosphate deficiency. *Plant Soil.* 248:ix-xix

Lamont BB. 2003. Structure, ecology and physiology of root clusters—a review. *Plant Soil* 248:1–19

Lamont BB, Enright NJ. 2000. Adaptive advantages of aerial seed banks. *Plant. Species Biol.* 15:157–166

Lamont BB, Wiens D. 2003. Are seed set and speciation rates always low among species that resprout after fire, and why? *Evol. Ecol.* 17:277–92

Lamont BB, Hopkins AJM, Hnatiuk RJ. 1984. The flora—composition, diversity and origins. See Pate & Beard 1984, pp. 27–50

Lamont BB, He T, Enright NJ, Krauss SL, Miller BP. 2003. Anthropogenic disturbance promotes hybridisation between *Banksia* species by altering their biology. *J. Evol. Biol.* 16:551–57

Linder HP. 2003. The radiation of the Cape flora, southern Africa. *Biol. Rev.* 78:597–638

Linder HP, Eldenas P, Briggs BG. 2003. Contrasting patterns of radiation in African and Australian Restionaceae. *Evolution* 57:2688–702

Levin DA. 2002. *The Role of Chromosomal Change in Evolution*. Oxford: Oxford Univ. Press. 230 pp.

Main AR. 1996. Case history studies of the effects of vegetation succession and fire on the moth *Fraus simulans* (Lepidoptera, Hepialidae) and its food plant, the sedge *Ecdeiocolea monostachya* (Ecdeiocoleaceae) in the Western Australian wheatbelt: implications for retention of biodiversity. *Pacific Cons. Biol.* 7:93–100

Marchant N. 1991. The vascular flora of south western Australia. *Proc. ASGAP Bienn. Conf., 16th, Perth*, pp. 16–18. Perth: Assoc. Soc. Grow. Aust. Plants

Mast A. 2000. *Molecular systematics of the Subtribe Banksiinae (Banksia and Dryandra: Proteaceae), with insights into the historical biogeography of Australia and the origin of xeromorphic leaf traits*. PhD thesis. Univ. Wisconsin, Madison

Mast AR, Givnish TJ. 2002. Historical biogeography and the origin of stomatal distributions in *Banksia* and *Dryandra* (Proteaceae) based on their cpDNA phylogeny. *Am. J. Bot.* 89:1311–23

McLoughlin S. 2001. The breakup history of Gondwana and its impact on pre-Cenozoic floristic provincialism. *Aust. J. Bot.* 49:271–300

McLouglin S, Hill RS. 1996. The succession of Western Australian Phanerozoic floras. See Hopper et al. 1996a, pp. 61–80

McLoughlin S, McNamara K. 2001. *Ancient Floras of Western Australia*. Perth: West. Aust. Mus.

Miller JT, Bayer RJ. 2001. Molecular phylogenetics of *Acacia* (Fabaceae: Mimosoideae) based on the chloroplast *matK* coding sequence and flanking *trnK* intron spacer regions. *Am. J. Bot.* 88:697–705

Mooney HA, Arroyo MTK, Bond WJ, Canadell J, Hobbs RJ, et al. 2001. Mediterranean-climate ecosystems. In *Global Biodiversity in a Changing Environment*, ed. F Stuart Chapin III, OE Sala, E Huber-Sannwald, pp. 157–99. Heidelberg: Springer

Myers N, Mittermeier RA, Mittermeier CG, da Fonseca GAB, Kent J. 2000. Biodiversity hot spots for conservation priorities. *Nature* 403:803–8

Newsome JC, Pickett EJ. 1993. Palynology and palaeoclimatic implications of two Holocene sequences from southwestern Australia. *Palaeogeogr. Palaeoclim. Palaeoecol.* 101:245–61

Ornduff R. 1996. A Californian's commentary on plant life in Mediterranean climates. See Hopper et al. 1986a, pp. 81–89

Paczkowska G, Chapman AR. 2000. *The Western Australian Flora: A Descriptive Catalogue*. Perth: Wildflower Soc. West. Aust., West. Aust. Herbarium CALM, Bot. Gard. Parks Auth.

Pate JS, Beard JS, eds. 1984. *Kwongan—Plant Life of the Sandplain*. Nedlands: Univ. West. Aust. Press

Pate JS, Hopper SD. 1993. Rare and common plants in ecosystems, with special reference to the south-west Australian flora. In *Biodiversity and Ecosystem Function*, ed. E-D Schultze, HA Mooney, pp. 293–325. Heidelberg: Springer

Pate JS, McComb AJ, eds. 1981. *The Biology of Australian Plants*. Perth: Univ. West. Aust. Press

Pate JS, Weber G, Dixon KW. 1984. Growth and life form of kwongan species. See Pate & Beard 1984, pp. 84–100

Playford PE. 1999. The Permo-Carboniferous glaciation of Gondwana: its legacy in Western Australia. *Geol. Surv. Western Australia Extended Abstr.* 6:15–16

Powell CMcA, Roots SR, Veevers JJ. 1988. Pre-breakup continental extension in East Gondwanaland and the early opening of the eastern Indian Ocean. *Tectonophysics* 155:261–83

Quinn CJ, Crayn DM, Heslewood MM, Brown EA, Gadek PA. 2003. A molecular estimate of the phylogeny of the Styphelieae. *Aust. Syst. Bot.* 16:581–94

Richardson DM, Cowling RM, Lamont BB, van Hensbergen HJ. 1995. Coexistence of *Banksia* species in southwestern Australia: the role of regional and local processes. *J. Veg. Sci.* 6:329–42

Richardson JE, Fay MF, Cronk QCB, Bowman D, Chase MW. 2000. A phylogenetic analysis of Rhamnaceae using *rbcL* and *trnL-F* plastid DNA sequences. *Am. J. Bot.* 87:1309–24

Rieseberg LH, Wendel J. 2004. Plant speciation—rise of the poor cousins. *New Phytol.* 161:3–7

Roche S, Dixon KW, Pate JS. 1997. For everything a season—smoke-induced seed germination and seedling recruitment in a Western Australian Banksia woodland. *Aust. J. Ecol.* 23:111–20

Shan F, Yan G, Plummer JA. 2004. Phylogenetic and cytoevolutionary analysis of the genus *Boronia* (Rutaceae). *Ann. Bot.* In press

Shepherd KA, Waycott M, Calladine A. 2004. Radiation of the Australian Salicornioideae (Chenopodiaceae)—based on evidence from nuclear and chloroplast DNA sequences. *Am. J. Bot.* 91:1387–97

Stebbins GL, Major J. 1965. Endemism and speciation in the California flora. *Ecol. Monogr.* 35:1–35

Steane DA, Nicolle D, MacKinnon GE, Vaillancourt RE, Potts BM. 2002. Higher-level relationships among the eucalypts are resolved by ITS-sequence data. *Aust. Syst. Bot.* 15:49–62

Taffe G, Brown EA, Crayn DM, Gadek PA, Quinn CJ. 2001. Generic concepts in Styphelieae: resolving the limits of *Leucopogon. Aust. J. Bot.* 49:107–20

Thompson IR, Ladiges PY, Ross JH. 2001. Phylogenetic studies of the Tribe Brogniartieae (Fabaceae) using nuclear DNA (ITS-1) and morphological data. *Syst. Bot.* 26:557–70

Tieu A, Dixon KW, Meney KA, Sivasithamparam K. 2001. Interaction of soil burial and smoke on germination patterns in seeds of selected Australian native plants. *Seed Sci. Res.* 11:69–76

Vinnersten A, Reeves G. 2003. Phylogenetic relationships within Colchicaceae. *Am. J. Bot.* 90:1455–62

Wardell-Johnson G, Horwitz P. 2000. The recognition of heterogeneity and restricted endemism in the management of forested ecosystems in south-western Australia. *Aust. For.* 63:218–25

Watchman AL, Twidale CR. 2002. Relative and 'absolute' dating of land surfaces. *Earth Sci. Rev.* 58:1–49

Williams GC. 1966. *Adaptation and Natural Selection. A Critique of some Current Evolutionary Thought.* Princeton, NJ: Princeton Univ. Press

Wilson PG, O'Brien MM, Gadek PA, Quinn CJ. 2001. Myrtaceae revisited: a reassessment of infrafamilial groups. *Am. J. Bot.* 88:2013–25

Wyrwoll K-H. 1988. Time in the geomorphology of Western Australia. *Progr. Phys. Geogr.* 12:237–63

Yates CJ, Hopper SD, Brown A, van Leeuwen S. 2003. Impact of two wildfires on endemic granite outcrop vegetation in Western Australia. *J. Veg. Sci.* 14:185–94

PREDATOR-INDUCED PHENOTYPIC PLASTICITY IN ORGANISMS WITH COMPLEX LIFE HISTORIES

Michael F. Benard

Section of Evolution and Ecology, Center for Population Biology, University of California, Davis, California 95616; email: mfbenard@ucdavis.edu

Key Words predator, life history, induced defense, metamorphosis

■ **Abstract** Predator-induced phenotypic plasticity is widespread in nature and includes variation in life history, morphology, and behavior. In organisms with complex life histories, predator-induced phenotypic plasticity in the larval period has been widely documented. Several models predict how organisms should alter their size at and time to metamorphosis in response to an increased risk of predation. A survey of empirical studies finds that these theoretical predictions are frequently met. However, no one model performs the best. Additionally, there are several results not predicted by any model. Predator-induced plasticity in metamorphic traits may be related to predator-induced changes in larval morphology and behavior. Predictions of predator effects on larval traits are generally met, except for direct costs of predator-induced morphological phenotypes. Future work should incorporate more detailed studies of growth rate, morphology, and behavior during the larval period, as well as studies of size-specific mortality rates in the presence and absence of predators.

INTRODUCTION

The risk of predation causes many organisms to alter foraging behavior, induce morphological defenses, and even alter the time and size at which they undergo life history shifts (e.g., Skelly & Werner 1990, Crowl & Covich 1990, Peckarsky et al. 1993, Tollrian & Dodson 1999, Relyea 2001b, Van Buskirk 2002). For organisms that undergo metamorphosis from a larval stage in one environment to an adult stage in a different environment, metamorphosis represents an opportunity to escape predation risk in the larval environment. Many theoretical studies have predicted how individuals should alter the size and time at which they metamorphose in response to changes in predation risk in the larval environment (Werner 1986, Ludwig & Rowe 1990, Rowe & Ludwig 1991, Abrams & Rowe 1996, Crowley & Johansson 2002). These models frequently differ in assumptions and predictions. Many recent empirical studies have used nonlethal predator cues experimentally to examine plastic responses of age and size at metamorphosis in a wide range of organisms (e.g., Skelly & Werner 1990, Peckarsky et al. 1993). The recent increase in the number of such studies allows us to examine the predictions of theory in light of empirical results. By comparing empirical results with theoretical predictions, we can identify aspects of empirical work not addressed by theory. Similarly, future empirical work can be designed to distinguish better between different theoretical models of life history shifts.

Substantial empirical work has demonstrated that many organisms change foraging behavior and induce defensive morphological changes in response to the risk of predation. Predator-induced behavioral and morphological plasticity can provide the benefit of reduced predation risk, but it can also incur costs, such as reduced growth rates (Skelly & Werner 1990, Peckarsky et al. 1993, Van Buskirk et al. 1997). Because models of optimal age and size at metamorphosis incorporate predation risk and growth rate as variables, it is valuable to consider how induced behavioral and morphological responses can influence age and size at life history shifts. Models of induced defense generally hold that for a defense, such as spines, to evolve to be inducible, the defense must carry a benefit in the presence of a predator and the predator-induced form must carry a cost in the absence of a predator (Harvell 1990). Models of foraging under predation risk generally predict that risky foraging declines as the risk of predation increases (Werner & Anholt 1999).

This review has three main goals. The first goal is to compare the various models of predator effects on metamorphosis and to examine if the empirical results support the model predictions. The second goal is to examine the role of predator-induced behavioral and morphological change on prey growth rates and predation risk. The third goal is to examine how larval morphological and behavioral plasticity can alter growth and mortality values and thus influence the predictions of models of age and size at metamorphosis.

EFFECTS OF PREDATION RISK ON METAMORPHIC TIMING

Theoretical Predictions

Wilbur & Collins (1973) produced the earliest model to predict the direction of plasticity in size at metamorphosis and duration of the larval period in response to different environments. The first models that predicted the direction of plasticity in age and timing of life history shifts in response to changes in predation risk were those of Werner & Gilliam (1984) and Werner (1986). They proposed that an organism should optimize the size and age of a life history shift based on its size-specific mortality (μ) and growth (g) rates in the larval environment and postmetamorphic environments. An organism should metamorphose at the size that minimizes the μ/g ratio of the larval and postmetamorphic environments (Werner 1986). Determining the postmetamorphic mortality and growth rates in many animals is empirically more difficult than determining larval mortality and growth rates. However, one need not know the postmetamorphic μ/g curve to make

TABLE 1 Predictions of the various models of predator-induced plasticity in metamorphosis[a]

Predator effect on age at metamorphosis	Predator effect on size at metamorphosis		
	Larger	Same size	Smaller
Earlier		fixed size	μ/g, RL, flexible, fixed growth effort
Same time		fixed time, fixed size	flexible, fixed time
Later		fixed size	flexible

[a]μ/g: Werner (1986); RL, Rowe & Ludwig (1991); flexible, fixed growth effort, size, and time at metamorphosis (Abrams & Rowe 1996).

predictions on the relative change in age and size at metamorphosis expected in response to a change in predation risk during the larval period. If larval μ increases owing to increased risk of predation while growth rate does not change, metamorphosis will occur at a smaller size and earlier age (Table 1).

Ludwig & Rowe (1990) and Rowe & Ludwig (1991) investigated the optimal size and age at metamorphosis using dynamic optimization models. As in Werner (1986), they considered size and age at metamorphosis to be influenced by larval growth and mortality rates. However, they included time constraints for reproduction in their model, which was not included in Werner (1986). By including time constraints, they were able to model more accurately different life histories of insects and amphibians. For instance, many aquatic insects experience no postmetamorphic growth and must breed in the same season in which they metamorphose. In contrast, many amphibians experience substantial postmetamorphic growth and will breed for multiple seasons after metamorphosis. Unlike Werner (1986), their inclusion of time constraints caused substantial variation in size and age at metamorphosis, depending on the initial size and time of season at which an animal started. However, all other factors being equal, their model still predicted smaller size at metamorphosis in environments with predators than in environments without predators (Table 1).

Abrams & Rowe (1996) presented a broad analysis of how the size at and time of life history thresholds should vary in response to predation risk. Their model assumed that fitness was positively correlated with size at metamorphosis and negatively correlated with age at metamorphosis. Additionally, their model included growth effort as a variable. Growth effort could represent any trait, such as time spent foraging, that increases growth rate but that also increases mortality risk. They examined four possible life histories. In three life histories, either growth effort, size at metamorphosis, or age at metamorphosis were fixed. In the fourth life history, growth effort, size at metamorphosis, and age at metamorphosis were flexible. These different life histories gave different predictions for the effects of predation risk on the optimal age and size at metamorphosis (Table 1). Additionally, Abrams & Rowe (1996) considered two broad ecological contexts in which these four life history strategies could occur. The "direct effect" ecological context had only direct mortality effects of predators. In these cases, there was no indirect effect of predators on the underlying availability of resources in the environment. The direct context was the most appropriate in which to consider the empirical tests of predator-induced plasticity discussed later in this review, so only metamorphic predictions for the direct effects are described. The "indirect effect" ecological context included density-mediated effects of predators on the resource availability for the prey. In these cases, predators consume prey, reducing the number of prey and therefore increasing the per capita resource availability for the prey. The most striking difference between direct and indirect predictions was that in the direct context, predators caused a decrease in size at metamorphosis, whereas in the indirect context predators caused an increase in size at metamorphosis.

Some of the different models predict the same effect of predators on age and size at metamorphosis (Table 1). For instance, smaller and earlier metamorphosis in response to predation risk is predicted by four models: μ/g, Rowe & Ludwig (1991), Abrams & Rowe's (1996) fixed growth rate model, and Abrams & Rowe's (1996) flexible model. Distinguishing among most of these models is empirically difficult. However, distinguishing between the flexible model and the remaining models could be achieved by varying predation risk and resource availability to determine if the organisms varied their growth effort in different environments.

Several other studies have investigated the role of predation in metamorphic timing. However, these models are not included here because they either involved changes in prey density in response to predation (i.e., Crowley & Johansson 2002) or did not provide explicit predictions on how changes in predation risk during the larval period influenced metamorphosis (i.e., Bouskila et al. 1998).

Empirical Results

Numerous studies have tested for predator-induced plasticity in the age and size at metamorphosis of prey. These studies can be classified as two types: experiments using a treatment with free-ranging predators and experiments that use nonlethal predator cues to simulate the risk of predation without predators killing focal prey animals. To assess the models of optimal life history transitions, the experiments with nonlethal predator cues are the most appropriate because they generally hold resource availability between predator-cue treatments and nonpredator treatments constant. Therefore, only perceived predation is varied. In experiments in which predators are free to kill prey, the per capita resource level of the prey increases as the predators remove prey.

I searched the literature for empirical studies that tested for the effect of nonlethal predator cues on age and size at metamorphosis in organisms that transformed from a larval form in one environment to a postmetamorphic form in another environment. Thus, many studies, such as those on life history shifts in freshwater cladocera (Tollrian & Dodson 1999), were not included. Additionally, studies had to use a nonlethal predator cue and test for a change in age and size at metamorphosis to be included. Studies that tested for a change in growth rate but did not follow the animals to metamorphosis were not included in this summary. For all the studies included, I recorded if there was a significant predator effect on size at metamorphosis and age at metamorphosis. Additionally, I recorded if the study tested for predator effects on growth rate to metamorphosis, foraging behavior, or morphology. There were 16 studies on amphibians and 11 studies on insects with aquatic larvae. Several different studies looked at the same species, and several studies examined more than one amphibian. The amphibians in the study included 2 species of salamander in 2 different families (Ambystomatidae and Salamandridae), and 16 species of anurans from 4 families (Bufonidae, Hylidae, Pelobatidae, Ranidae). The insect studies included 10 species of insects from 3 orders (Diptera, Ephemeroptera, and Odonata).

Several taxa were not included in this study because they did not meet one of the requirements to test the model. For instance, predator cues may delay settlement and metamorphosis in some marine invertebrates, but the effect of the cues on mass at settlement is not reported (Forward et al. 2001). Similarly, herbivorous terrestrial insects exhibit a trade-off between growth rate and predation or parasitism risk (e.g., Clancy & Price 1987, Benrey & Denno 1997), but no studies have examined how predation risk influences metamorphosis in these insects. Studies examining the effect of predation risk on the age and size at metamorphosis in these groups would be valuable.

Most models predicted that larvae should metamorphose smaller and earlier under the risk of predation (Table 1). However, this prediction was only met in 2 out of the 40 experiments (Table 2, Lardner 2000, Dahl & Peckarsky 2003).

TABLE 2 Summary of results of studies testing for effects of predator cues on metamorphosis[a]

Predator effect on age at metamorphosis	Predator effect on size at metamorphosis		
	Larger	Same size	Smaller
Amphibians			
Earlier		2	1
Same time		8	3
Later	6	4	1
Insects			
Earlier			1
Same time		5	3
Later		3	3

[a]Each cell has the number of experimental results that fall into each category of size at metamorphosis and age at metamorphosis. A particular study may be represented more than once if there was an interaction between sex and predator-cue treatment, or food treatment and predator-cue treatment. Similarly, several different studies used the same predator and prey species, and those are counted multiple times.

The results of many of the remaining experiments agreed with predictions of one of the various models. Results consistent with Abrams & Rowe's fixed size model occurred in 22 experiments, and 19 experiments were consistent with their fixed time model. Their flexible development model's predictions were met in 12 experiments. Six experiments, all from studies performed on amphibians, had a result that no model predicted. In these six studies, amphibians metamorphosed later and larger in the presence of nonlethal predator cues than in the absence of predator cues. Three of these studies also tested for an effect of predators on growth rate but found no effect. Thus, in at least these three studies, larvae remained longer in the riskier (predator cue) habitat, even though they gained no detectable growth advantage. Thirteen experiments failed to detect predator effects on age and size at metamorphosis. A lack of power to detect an effect may explain the lack of a predator effect in some studies, but it does not explain it for all (e.g., Caudill & Peckarsky 2003).

The results from 22 of the 40 experiments explicitly tested for an effect of predator cues on growth rate (Table 3). Of these, six experiments detected a significant reduction in growth rate, and the remaining 16 experiments detected no significant effect of predator cues on growth rate to metamorphosis. Two studies found that toad (*Bufo* sp.) larvae metamorphosed at the same size, but earlier in the presence of predators (Laurila et al. 1998, Chivers et al. 1999). Laurila et al. (1998) found no predator effect on growth rate early in the larval period but did not test for a predator effect on final growth rate to metamorphosis. Chivers et al. (1999) did not explicitly test for a predator-cue effect on growth rate. The results of both of these studies indicated that growth rate and development rate were accelerated in the presence of predators.

In 20 experiments, predator-induced behavioral shifts were studied in addition to predator-induced changes in age and size at metamorphosis. In two experiments, predators induced a behavioral effect but no metamorphic response; in one experiment, predators induced a metamorphic response but no larval behavioral response; and in three experiments, predators induced no behavioral and no metamorphic responses. In the remaining 14 experiments, predators induced a behavioral and a metamorphic response.

Fifteen experiments tested for predator-induced plasticity in morphology in addition to predator-induced plasticity in age and size at metamorphosis. In five cases, predators induced a presumed morphological defense in the larvae, but did not induce a change in age or size at metamorphosis. There were six cases in which predators induced a morphological response and a metamorphic response, one case in which predators induced a metamorphic response but no morphological response, and three cases in which predators induced neither a morphological nor metamorphic response. Only one study tested for predator effects on larval behavior, larval morphology, and the age and size at metamorphosis. In this case, cues from predatory dragonflies caused larval newts to increase refuge use, increase tail size, and metamorphose larger and later than newts raised in the absence of predators (Van Buskirk & Schmidt 2000).

TABLE 3 Summary of studies testing for predator-induced changes in age and size at metamorphosis

			Predator effect on				
Prey taxa	Predator	Size at met.	Time to met.	Growth rate	Morph.	Behavior	Study
Aedes triseriatus, Mosquito	*Toxorhynchites rutilis*, mosquito	None	Males none, females later	None	Not tested	No effect	Hechtel & Juliano 1997
Baetis bicaudatus, mayfly	*Salvelinus fontinalis*, brook trout	Smaller	None	Not tested	Not tested	Increased nocturnal drift, activity	Peckarsky & McIntosh 1998
Baetis bicaudatus, mayfly	*Megarcys signata*, stonefly	Smaller	Males none, females later	Not tested	Not tested	Increased nocturnal drift, activity	Peckarsky & McIntosh 1998
Baetis bicaudatus, mayfly	*Megarcys signata*, stonefly	Smaller	None	Not tested	Not tested	Reduced foraging	Peckarsky et al. 1993
Baetis bicaudatus, mayfly	stonefly and fish	Smaller	None	Not tested	Not tested	Increased nocturnal drift, activity	Peckarsky & McIntosh 1998
Baetis tricaudatus, mayfly	*Rhinicthys. cataractae*, longnose dace	Smaller	Later	Lower	Not tested	Not tested	Scrimgeour & Culp 1994
Callibaetis ferrugineus hageni, mayfly	*Salvelinus fontinalis*, brook trout	None	None	Not tested	Not tested	No effect	Caudill & Peckarsky 2003
Chironomus tentans, dipteran	*Lepomis gibbosus*, sunfish	None	Males later, females none	Lower	Not tested	Not tested	Ball & Baker 1996
Drunella coloradensis, mayfly	*Salvelinus fontinalis*, brook trout	None	None	No difference	Heavier exoskeleton, longer caudal filaments	Not tested	Dahl & Peckarsky 2002
Ephemerella invaria, mayfly	*Rhinicthys cataractae*, longnose dace, and *Etheostoma flabellare*, fantail darters	Smaller	Earlier	No difference	No effect	Not tested	Dahl & Peckarsky 2003
Ephemerella subvaria, mayfly	*Luxilus cornutus*, commno shiner	None	Later	Not tested	Not tested	Not tested	Tseng 2003
Lestes sponsa, damselfly	*Perca fluviatilis*, perch	Smaller	Later	Lower	Not tested	Reduced foraging	Johansson et al. 2001

(*Continued*)

TABLE 3 (Continued)

Prey taxa	Predator	Size at met.	Time to met.	Growth rate	Morph.	Behavior	Study
Leucorrhinia dubia, dragonfly	Perca fluviatilis, perch	None	None	None	Longer dorsal spines	Not tested	Johansson 2002
Ambystoma macrodatylum, long toed salamander	Ambystoma Savannamacrodactylum	None	Later	Lower	Not tested	Not tested	Wildy et al. 1999
Bufo americanus, American toad	Anax junius, dragonfly	Smaller	None	Not tested	Not tested	Reduced activity	Skelly & Werner 1990
Bufo boreas, western toad	Notonecta sp., backswimmer	None	Earlier	Not tested	Not tested	Not tested	Chivers et al. 1999
Bufo boreas, western toad	simulated cues	None	None	Not tested	No difference	Not tested	Benard & Fordyce 2003
Bufo bufo, common toad	Aeshna juncea, dragonfly	None	Earlier	No effect halfway into larval period	Not tested	Reduced activity	Laurila et al. 1998
Bufo bufo, common toad	Dytiscus marginalis, beetle	Smaller	Earlier	None	Narrower tail fin	Not tested	Lardner 2000
Bufo calamita, natterjack toad	Dytiscus marginalis, beetle	None	None	None	No difference	Not tested	Lardner 2000
Hyla arborea, common tree frog	Dytiscus marginalis, beetle	None	None	None	Significantly deeper tail	Not tested	Lardner 2000
Hyla versicolor, gray treefrog	Anax sp., dragonfly	None	None	None	Deeper tail fin depth, body morphology	Not tested	Relyea & Hoverman 2003
Pelobates fuscus, garlic toad	Dytiscus marginalis, beetle	None	None	None	No difference	Not tested	Lardner 2000
Rana arvalis, moor frog	Dytiscus marginalis, beetle	None	None	None	Tail relatively deeper	Not tested	Lardner 2000
Rana aurora, red legged frog	Aeshna palmata, dragonfly	Larger	Later	Not tested	Not tested	Reduced activity, spatial avoidance	Barnett & Richardson 2002

Species	Predator	Size	Timing		Morphology	Behavior	Reference
Rana aurora, red legged frog	Taricha granulosa, roughskinned newt and simulated cues	Smaller	None	Not tested	Not tested	Not tested	Kiesecker et al. 2002
Rana dalmatina, agile frog	Dytiscus marginalis, beetle	Smaller	Later	Lower	Relatively deeper tail	Not tested	Lardner 2000
Rana lessonae, pool frog	Anax imperator, dragonfly	Larger	Later	Not tested	Not tested	Reduced activity	Altwegg 2002
Rana pretiosa, Oregon spotted frog	Aeshna palmata, dragonfly	None	None	None	Not tested	Reduced activity spatial avoidance	Barnett & Richardson 2002
Rana ridibunda, water frog	Aeshna cyanea, dragonfly	None	Later	Not tested	Deeper tail fin depth	Not tested	Van Buskirk & Saxer 2001
Rana sphenocephala, Southern Leopard frog	Anax junius, dragonfly	Low food = larger; high food = no difference	Later	None	Not tested	Reduced activity	Babbitt 2001
Rana sylvatica, wood frog	Anax junius and Anax longipes, dragonflies	None	Later	Not tested	Induced morphology	Not tested	Relyea 2001a
Rana temporaria, common frog	Aeshna juncea, dragonfly	None	None	Not tested	Not tested	Reduced activity, spatial avoidance	Laurila & Kujasalo 1999
Rana temporaria, common frog	Aeshna juncea, dragonfly	Larger	Later	No effect halfway into larval period	Not tested	Reduced activity	Laurila et al. 1998
Rana temporaria, common frog	Dytiscus marginalis, beetle	Smaller	None	Lower	Tail relatively deeper	Not tested	Lardner 2000
Rana temporaria, common frog	Salmo salar, Atlantic salmon	Larger	Later	None	Not tested	Reduced activity	Nicieza 2000
Triturus alpestris	Aeshna cyanea, dragonfly	Larger	Later	None	Induced morphology and darker tail color	Increased refuge use	Van Buskirk & Schmidt 2000

A few species were used in multiple studies. For instance, for the mayfly, *Baetis bicaudatus*, multiple studies found almost identical results (Table 2). In contrast, the common frog, *Rana temporaria*, had a range of very different metamorphic responses across four different studies (Table 2). These differences within *R. temporaria* could be due to geographic variation between populations or to an interaction between the environments of the studies (e.g., different resource levels or temperatures in different experiments) and predator cues. Additionally, most of the studies examined had only a predator/no predator treatment and did not vary food levels as an additional treatment. The few that did vary resource availability found cases in which there were both additive and interactive effects between resource availability and predator cues (Skelly & Werner 1990, Nicieza 2000, Babbitt 2001).

There are also several notable patterns within the predator/prey combinations. Of the 16 cases that examined predator-induced plasticity in insect metamorphosis, five used other insects as predators, ten used fish as predators, and one combined fish and insects as predators. There were no substantial differences between fish predation cues and invertebrate predation cues in the effect on metamorphosis in freshwater invertebrates. Most of the cases that examined amphibian metamorphic responses used insects as predators. Only two studies that investigated amphibian metamorphic plasticity used another amphibian as a predator, and only one used a fish as a predator.

Other Types of Life History Shifts

Metamorphosis is a particularly interesting life history shift because organisms have to balance the estimate of growth rate and mortality in their current environment with the unknown mortality risk and growth rate in the postmetamorphic environment. There are other life history shifts in which organisms can assess predation risk in their current life stage and the predation risk in a later life history stage. Timing of hatching is one example of this phenomenon. Several amphibians have been shown to detect cues indicating the risk of predation while in their eggs. In some cases, the predators feed on the developing embryos while in the egg, and in accordance with predictions based on minimizing risk, the embryos hatch smaller, but earlier (Chivers et al. 2001). Similarly, predators that feed on hatchlings but not eggs may produce cues that cause larvae to hatch later, but at a larger size (Sih & Moore 1993, Moore et al. 1996, Jones et al. 2003). The larger hatchlings are less likely to be killed by the posthatching predators. In other experiments, embryos have also been shown to hatch earlier in response to egg predators when they cannot assess the relative predation risk of the posthatching environment (Warkentin 1995, 2000; Warkentin et al. 2001).

Implications of the Empirical Results

As more studies are published on predator effects on metamorphosis, researchers will be able to draw broader conclusions about the evolution of plasticity in life history characteristics. Do different types of life history responses represent adaptations to environmental variation? For instance, why do some species appear to metamorphose with constrained size at metamorphosis, others to metamorphose with constrained age at metamorphosis, and some to have no plasticity in metamorphosis in response to predation? Species-level comparisons of metamorphic responses to predators would provide insight into this question. Such comparisons have been made between larval defensive plasticity and palatability, and between predator type and amount of environmental heterogeneity (Kats et al. 1988, Van Buskirk 2002), but they have yet to be made with metamorphic characters.

An additional question is prompted by the wide range of metamorphic responses to predation risk during the larval period: What are the postmetamorphic consequences of predator-induced changes in age and size at metamorphosis? Individuals that metamorphose earlier and larger are more likely to survive to reproduce, are larger at reproduction, and/or reproduce earlier than those that metamorphose smaller and later (Berven & Gill 1983, Banks & Thompson 1987, Smith 1987, Semlitsch et al. 1988, Scott 1994, Taylor et al. 1998, Sokolovska et al. 2000, Altwegg & Reyer 2003). However, there is little empirical evidence to evaluate the effect of the interaction between size and age at metamorphosis on fitness. The one comprehensive selection study to test for a correlation between survival and both age and size at metamorphosis found that for *Rana lessonae*, individuals that metamorphosed small and early could have no statistical difference in probability of survival than individuals that emerged sufficiently large but later (Altwegg & Reyer 2003). Thus, larvae that metamorphose smaller and earlier in response to the risk of predation may have equal fitness to those metamorphosing later and larger.

EFFECT OF PREDATION RISK ON LARVAL MORPHOLOGY AND BEHAVIOR

In addition to inducing changes in the size and age at metamorphosis, predation risk frequently induces behavioral or morphological plasticity in prey. Predator-induced behavioral and morphological plasticity can influence both mortality and growth rates. Predictions from models of predator effects on metamorphosis contain larval mortality and growth rates as parameters (Werner 1986, Ludwig & Rowe 1990, Rowe & Ludwig 1991, Abrams & Rowe 1996). Because larval mortality and growth rates may be influenced by predator-induced morphological and behavioral responses, a comparison of larval morphological and behavioral plasticity with metamorphic plasticity may provide insight on the empirical results of predator-induced metamorphic plasticity. The body of theory motivating empirical studies of predator-induced changes in behavior and morphology is separate from theory motivating studies of predator-induced changes in metamorphic traits.

Animals clearly alter their behavior in response to the risk of predation (Sih 1987, Lima & Dill 1990). Theory that predicts how animals should alter foraging consider foraging in several different ways. Some models examine movement between two patch types that differ in risk and resource availability (e.g., Gilliam & Fraser 1987). Other models examine the amount of time that prey spend foraging in response to resource availability and risk of mortality (e.g., Werner & Anholt 1993). Increasing the amount of time spent foraging is assumed to increase the amount of resources acquired, which translates into increased growth and thus higher reproductive potential. However, increasing the amount of time spent foraging carries the cost of increasing the risk of predation. As the risk of predation increases, foraging should also decrease.

In addition to predator-induced plasticity in behavior, predator-induced changes in morphology are common in larval amphibians and insects as well. Studies of induced morphological defenses in organisms with complex life histories were generally designed to test models of the evolution of induced defenses. There are four conditions under which we predict the evolution of plasticity in a defense: (*a*) the induced defense must provide a benefit in the presence of a predator, (*b*) the induced defense must bear a cost in the absence of a predator, (*c*) development in a predator-inhabited environment is unpredictable, and (*d*) reliable cues signal the risk of predation (Harvell 1990, Lively 1999). Most empirical work in organisms with complex life histories has tested the first two conditions. Only one has tested the third condition (Van Buskirk 2002). Most studies examining the fourth condition use nonlethal cues and demonstrate that they are reliable signals.

Behavior, Predation Risk, and Growth Rate

Many studies have demonstrated that organisms in predator-cue treatments spend less time foraging and have reduced growth rates compared with those in control treatments. Although it is reasonable to assume that the reduced foraging rates are the cause of the reduced growth rates, few studies directly test this assumption with correlations between foraging and growth. The empirical establishment of a correlation between foraging rate and growth rate was only directly done in one of the 27 studies testing the effect of predator cues on metamorphosis (Table 3; Skelly & Werner 1990). However, in anurans, short-term studies over a portion of the larval period have found a positive relationship between foraging and growth rate (e.g., Skelly 1992). Also in accordance with model predictions, many studies have found that increasing foraging directly increases the risk of being killed by a predator (Skelly 1994; Anholt & Werner 1995, 1998; Eklov & Werner 2000; Sih et al. 2000). Stream-dwelling mayflies show different behavioral responses to different types of predators (i.e., fish or stoneflies) (Wooster & Sih 1995). Peckarsky et al. (1993) demonstrated that the mayfly *Baetis bicaudatus* drifted off the stream substrate more often in the presence of predatory stoneflies than in their absence. *Baetis bicaudatus* had less full guts and lower growth rates in the presence of predatory stoneflies, indicating a relationship between increased predator avoidance behavior and reduced growth.

Morphology, Predation Risk, and Growth Rate

Predator-induced changes in morphology are nearly ubiquitous among amphibian larvae. One of the most common amphibian responses to predation risk is an increase in tail depth. Specific changes in morphology vary with prey and predator species and environmental conditions (McCollum & Van Buskirk 1996, Lardner 2000, Relyea 2001b, Van Buskirk & Arioli 2002). Tadpoles with induced tail shape are more likely to survive than tadpoles without the induced shape in short-term predation trials (Van Buskirk et al. 1997, Van Buskirk & Relyea 1998, Van Buskirk & McCollum 2000a). Morphological plasticity in amphibians is not induced solely by cues from predators; competitors can also induce morphological changes in amphibian larvae (Relyea 2002b). The functional mechanism underlying increased survival of tadpoles with larger tails has not been determined, but the tail may serve as a target to distract strikes from the body (Hoff & Wassersug 2000, Van Buskirk & McCollum 2000b, Van Buskirk et al. 2003). Alternatively, relatively larger tails may confer higher propulsion in some larval amphibians (Fitzpatrick et al. 2003, but see Van Buskirk & McCollum 2000b). Although the function of the induced changes has not been confirmed, the induced morphology clearly confers a survival advantage.

Larval invertebrates also induce morphological defenses in response to predation risk. Dahl & Peckarsky (2003) did not find induced morphological defenses in the mayfly *Ephemerella invaria*, despite finding predator-induced plasticity in metamorphic traits. However, the mayfly *Drunella coloradensis* induced heavier exoskeletons and longer caudal filaments when exposed to chemical cues from brook trout (Dahl & Peckarsky 2002). *Drunella coloradensis* with artificially shortened spines were eaten by brook trout at a higher rate than *D. coloradensis* with longer spines (Dahl & Peckarsky 2002). Mayflies also use caudal filaments to avoid predation by stoneflies (Peckarsky 1987). The dragonfly *Leucorrhinia dubia* induced longer caudal spines in the presence of cues from perch (Johansson 2002). Larger spines appear to be more effective than smaller spines in allowing dragonfly nymphs to escape fish predation (Johansson & Samuelsson 1994). Thus, predator-induced morphological changes aid larval insects by reducing the risk of mortality from predators.

Although tests of a relationship between induced morphology and predation risk have generally found such a relationship, tests for costs to induced morphology have been less successful. In many studies, a cost of an induced phenotype is inferred if a nonlethal predator cue causes a larva to induce both a defensive phenotype and a reduced growth rate. Many of these studies assume that the reduced growth rate is a cost of the induced phenotype without an explicit test. However, life history shifts and morphological defenses may be separate responses, rather than the life history shift being a cost of the morphological defense (Tollrian & Dodson 1999). A method to avoid confounding these responses would be to test for phenotypic or genetic correlations between morphology and growth rate within experimental treatments as well as between treatments.

Several studies have found a negative phenotypic correlation between the degree of a predator-induced morphology and growth rate. Van Buskirk et al. (1997) found that for chorus frog tadpoles, larval dragonflies (*Anax*) induced larger tadpole tail muscle depth, muscle width, and depth. They found a negative relationship between tail muscle depth and growth rate, but not between tail muscle width or tail depth and growth rate. The negative correlation that they found suggests there was a direct growth cost of inducing larger tail muscles. In contrast, several studies have found no evidence for a cost of an induced defense in amphibians. Van Buskirk & Schmidt (2000) found that in the newt *Triturus helviticus*, individuals induced larger tails in the presence of dragonflies, and that individuals with larger tails were more likely to survive in the presence of dragonflies. But there was no correlation between *T. helviticus* growth rate and morphology. Van Buskirk & Relyea (1998) found no significant relationship between specific predator-induced morphological traits and growth rate in wood frogs. Similarly, Relyea (2001a) found no correlation between larval morphology and mass/age at metamorphosis. Johansson (2002) found that in the presence of predatory perch chemical cues, the dragonfly *Leucorrhinia dubia* induced longer dorsal spines, yet there was no correlation between spine length and growth rate. Although the mayfly *Drunella coloradensis* grew longer caudal filaments in the presence of predator cues, there was no predator-cue effect on growth or metamorphosis in *D. coloradensis*.

Overall, most studies did not find the expected growth cost of induced morphological phenotypes in amphibians or insects. In many cases, the tests for growth costs are based on correlations of morphology and growth across multiple tanks containing groups of larvae. However, no studies have followed individual morphology and growth within a single experimental population. Growth costs may also be rare. Instead, costs of induced defense may come in a different form, such as increased postmetamorphic susceptibility to predation (Benard & Fordyce 2003).

PREDATOR-INDUCED METAMORPHIC PLASTICITY IN LIGHT OF LARVAL PLASTICITY

Correlations Between Larval and Metamorphic Traits

Few studies of predator-induced phenotypic plasticity examine more than two types of trait at once (Relyea 2004). However, complex interactions among different traits are likely (Relyea 2004). For instance, Van Buskirk & McCollum (2000a) described an interaction between the effects of larval morphology and behavior on survival. Dewitt et al. (1999) proposed that sets of traits could be viewed as being codependent, complementary, cospecialized, or compensatory. If growth costs of an induced morphological defense are rare, induced morphological defenses may work in compensation with life history shifts. Individuals that are more defended, such as anurans with deeper tails, may spend more time in the larval environment exposed to predators (trait compensation), whereas less-defended individuals metamorphose earlier. Of 12 studies in Table 3 that detected a predator-induced morphological shift, 5 found no predator-induced shift in age and size at metamorphosis. In contrast, of 16 studies that found predator-induced changes in behavior, 14 found a corresponding change in age or size at metamorphosis. The apparently stronger relationship between behavioral plasticity and metamorphic plasticity may be because changes in behavior are more closely tied to changes in growth rate than are changes in morphology. These correlations could be examined on a within-species level, as Relyea (2001b) tested for within several anurans. Alternatively, larger species-level patterns could be tested for as well. Across-species comparisons have been done for several larval characteristics in anurans (Richardson 2001a,b, 2002; Van Buskirk 2002).

Predator-induced changes in larval morphology and behavior may also help to explain cases in which larvae exposed to predator cues metamorphosed larger and later than larvae not exposed to predator cues. There are several possible explanations: (*a*) indirect effects, (*b*) reaching a predator-free size threshold, or (*c*) behaviorally or morphologically mediated mortality and growth rates.

Indirect Effects Explanation

Predation risk can indirectly increase the growth rate of prey by suppressing prey foraging, which releases resources to grow rapidly (Peacor 2002). This scenario was considered in the indirect effect models of Abrams & Rowe (1996). Altwegg (2002) suggested that such indirect effects explained why, in his experiment, *Rana lessonae* metamorphosed larger and later in the presence of predation cues than in the absence of predation cues. Indirect predator effects may also explain Van Buskirk & Schmidt's (2000) finding for the newt *Triturus alpestris*. The newt larvae were raised in outdoor tanks, and reduced newt activity in predator-cue treatments may have allowed invertebrate prey numbers to increase more rapidly than in the control-cue treatments. However, indirect supplementation is not likely to explain larger and later size at metamorphosis in the presence of predator cues in the other experiments because these studies were conducted in laboratory environments with controlled food rations (Laurila et al. 1998, Nicieza 2000, Babbitt 2001, Barnett & Richardson 2002).

To explain why larval leopard frogs metamorphosed larger and later in the presence of predator cues, Babbitt (2001) proposed that at low food conditions, the no-predator-treatment animals were often foraging when no additional food was available, whereas the predator-cue animals were not foraging. Therefore, the no-predator-treatment animals were using energy when no food was available to replace it, but the predator-cue treatment animals conserved their energy stores because they reduced activity to avoid mortality. However, this argument does not explain why the leopard frog larvae did not metamorphose earlier to escape the low-quality predation environment.

Size Threshold Explanation

An alternative explanation for larger and later metamorphosis relies on empirical evidence that susceptibility to predation decreases as size increases (Crump 1984,

Travis 1983, Travis et al. 1985, Formanowicz 1986, Semlitsch & Gibbons 1988, Kehr & Schnack 1991). A life history strategy not considered in models of life history transitions under the threat of predation is that a larva may have depressed foraging rates early in ontogeny, which reduced growth but also reduced the risk of predation. Once the larva reaches a size threshold at which it is safe from the risk of predation, it increases its foraging rate. To make up for the reduced growth early in ontogeny, the organism remains as a larva for a longer period, but with an increased growth rate now that it is at a safe size. Thus, it metamorphoses later but at a larger size than individuals developing in an environment without predators. This was the case for the newt *Triturus alpestris* (Van Buskirk & Schmidt 2000). Evaluating this hypothesis should involve studies of the fitness functions for age and size at metamorphosis (Altwegg & Reyer 2003).

Plasticity Explanation

Variation in larval behavior and morphology clearly can effect larval growth rate and predation risk. Can these predator-induced changes in growth and mortality rates explain cases of larger and later metamorphosis, while remaining consistent with models of metamorphosis based on mortality risk and growth rates? For these changes to occur in the "minimize μ/g" model (Werner 1986), the predator-induced plasticity would have to decrease the overall mortality rate to below the mortality rate for a tadpole in the absence of a predator. At the same time, growth rate would have to remain high enough so that the ratio of μ/g actually increases (Figure 1).

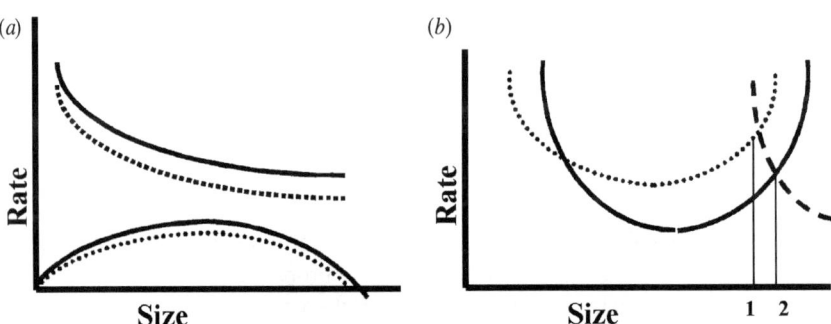

Figure 1 Modification of Werner's (1986) model for optimal age and size at metamorphosis that could result in larger, later metamorphosis. (*a*) In this unlikely scenario, reduced foraging lowers mortality to below mortality rates for the nonpredator environment; although growth is also reduced, it is not reduced as much. Solid lines denote mortality or growth rates in the absence of predators, and dashed lines represent mortality or growth rates in the presence of predators. (*b*) The μ/g curves for predator absence (*dashed line*) and presence (*solid line*). The 1 represents size at metamorphosis for no-predator environment, and 2 represents size at metamorphosis for the predator environment.

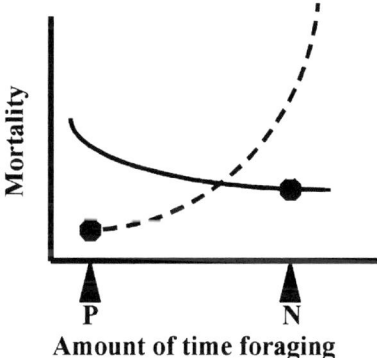

Figure 2 Hypothetical relationship between mortality and amount of time spent foraging for an environment without predators (*solid line*) and an environment with predators (*dashed line*). If resources are limited, in the absence of predators individuals that move little are more likely to starve because more actively foraging individuals consume all the resources. In the presence of predators, individuals that move frequently are consumed. The result is that overall mortality (*black circles*) is actually lower in the presence of predators than their absence.

Under these conditions, the larval μ/g curve will intercept the postmetamorphic μ/g curve at a higher body size. Are there conditions under which this outcome can occur?

Empirical studies have demonstrated that at least some amphibian larvae can assess changes in the risk of predation stemming from differences in conspecific death rates and density of predators (e.g., Anholt et al. 2000, Van Buskirk & Ariolio 2002). However, no studies have determined the relationship between the level of a nonlethal cue and mortality risk in the wild. Most mesocosm and enclosure experiments have found greater individual mortality risk in the presence of predators than in the absence of predators (Morin 1981; Wilbur 1972, 1987; Wilbur & Fauth 1990; Wilbur et al. 1983; Morin 1986; Van Buskirk 1988; Werner & McPeek 1994; Skelly 1995). However, in some studies, per capita survivorship was higher in environments with predators than environments without predators (Morin 1981, Wilbur 1987, Van Buskirk 1988, Gascon & Travis 1992, Skelly 1995). A lower mortality risk in the presence of predators could be mediated through predator-induced changes in prey foraging or defensive morphology (Figure 2). These empirical examples of lower mortality in the presence of predators indicate that the models of predator effects on metamorphosis may not be incorrect in the case of the six species metamorphosing larger and later in the presence of predators. Instead, more detailed empirical studies that combine estimates of per capita mortality risk across a range of resource levels should be coupled with changes in metamorphosis in response to nonlethal predator cues.

CONCLUSIONS

The broad range of metamorphic responses to predation risk (Table 2) presents an exciting challenge to the study of predator effects on life history shifts. No model predicted that larva would metamorphose later and at a larger size in response to the risk of predation in a constant resource environment. Future theoretical work can provide insights into this common phenomenon. Testing predictions of theory will require more than measuring size and age at metamorphosis. Several models predict similar metamorphic outcomes. Distinguishing between models that give the same prediction of metamorphic traits owing to different biological reasons requires study of aspects of the larval period. For instance, multiple measurements of growth rate throughout the larval period can demonstrate if larval growth in the presence of predators is slower than growth in the absence of predators until some safe size threshold is reached. Similarly, to provide valuable insights into the evolutionary reason for predator-induced shifts in age and size at metamorphosis, future experiments will need to include multiple resource levels, prey densities, or levels of predation risk. Additionally, consideration of the interaction among behavioral, morphological, and life history plasticity will provide further insight into variation in metamorphic responses. For instance, why are predator-induced metamorphic responses less common when there are predator-induced larval morphological defenses than when there are predator-induced larval behavioral responses? Finally, studies should include a broader range of both prey and predator taxa.

ACKNOWLEDGMENTS

I thank Beverly Aije, Jim Fordyce, C. Darrin Hulsey, Jake Kerby, Barney Luttbeg, Brad Shaffer, Andy Sih, Anita Stone, Josh Van Buskirk, Earl Werner, and an anonymous reviewer for helpful comments that improved this review. I also thank Jim Fordyce, Brad Shaffer, and Josh Van Buskirk for the critical early encouragement that got me started on this review. I was supported by the Center for Population Biology at UC Davis and an Achievements Rewards for College Scholars (ARCS) Foundation fellowship.

The *Annual Review of Ecology, Evolution, and Systematics* is online at
http://ecolsys.annualreviews.org

LITERATURE CITED

Abrams PA, Rowe L. 1996. The effects of predation on the age and size of maturity of prey. *Evolution* 50:1052–61

Altwegg R. 2002. Predator-induced life-history plasticity under time constraints in pool frogs. *Ecology* 83:2542–51

Altwegg R, Reyer HU. 2003. Patterns of natural selection on size at metamorphosis in water frogs. *Evolution* 57:872–82

Anholt BR, Werner E, Skelly DK. 2000. Effect of food and predators on the activity of four larval ranid frogs. *Ecology* 81:3509–21

Anholt BR, Werner EE. 1995. Interaction between food availability and predation mortality mediated by adaptive behavior. *Ecology* 76:2230–34

Anholt BR, Werner EE. 1998. Predictable changes in predation mortality as a consequence of changes in food availability and predation risk. *Evol. Ecol.* 12:729–38

Babbitt KJ. 2001. Behaviour and growth of southern leopard frog (*Rana sphenocephala*) tadpoles: effects of food and predation risk. *Can. J. Zool.* 79:809–14

Ball SL, Baker RL. 1996. Predator-induced life history changes: antipredator behavior costs or facultative life history shifts. *Ecology* 77:1116–24

Banks MJ, Thompson DJ. 1987. Lifetime reproductive success of females of the damselfly *Coenagrion puella*. *J. Anim. Ecol.* 56:815–32

Barnett HK, Richardson JS. 2002. Predation risk and competition effects on the life-history characteristics of larval Oregon spotted frog and larval red-legged frog. *Oecologia* 132:436–44

Benard MF, Fordyce JA. 2003. Are induced defenses costly? Consequences of predator-induced defenses in western toads, *Bufo boreas*. *Ecology* 84:68–78

Benrey B, Denno RF. 1997. The slow-growth–high-mortality hypothesis: a test using the cabbage butterfly. *Ecology* 78:987–99

Berven KA, Gill DE. 1983. Interpreting geographic variation in life history traits. *Am. Zool.* 23:85–97

Bouskila A, Robinson ME, Roitberg BD, Tenhumberg B. 1998. Life-history decisions under predation risk: importance of a game perspective. *Evol. Ecol.* 12:701–15

Caudill CC, Peckarsky BL. 2003. Lack of appropriate behavioral or developmental responses by mayfly larvae to trout predators. *Ecology* 84:2133–44

Chivers DP, Kiesecker JM, Maro A, DeVito J, Blaustein AR. 2001. Predator-induced life history changes in amphibians: egg predation induced hatching. *Oikos* 92:135–42

Chivers DP, Kiesecker JM, Marco A, Wildy EL, Blaustein AR. 1999. Shifts in life history as a response to predation in western toads (*Bufo boreas*). *J. Chem. Ecol.* 25:2455–63

Clancy KM, Price PW. 1987. Rapid herbivore growth enhances enemy attack: sublethal plant defenses remain a paradox. *Ecology* 68:733–37

Crowl TA, Covich AP. 1990. Predator-induced life history shifts in a freshwater snail. *Science* 247:949–51

Crowley PH, Johansson F. 2002. Sexual dimorphism in Odonata: age, size, and sex ratio at emergence. *Oikos* 96:364–78

Crump ML. 1984. Ontogenetic changes in vulnerability to predation in tadpoles of *Hyla pseudopuma*. *Herpetologica* 40:265–71

Dahl J, Peckarsky BL. 2002. Induced morphological defenses in the wild: predator effects on a mayfly, *Drunella coloradensis*. *Ecology* 83:1620–34

Dahl J, Peckarsky BL. 2003. Developmental responses to predation risk in morphologically defended mayflies. *Oecologia* 137:188–94

Dewitt TJ, Sih A, Hucko JA. 1999. Trait compensation and cospecialization in a freshwater snail: size, shape and antipredator behaviour. *Anim. Behav.* 58:397–407

Eklov P, Werner EE. 2000. Multiple predator effects on size-dependent behavior and mortality of two species of anuran larvae. *Oikos* 88:250–58

Fitzpatrick BM, Benard MF, Fordyce JA. 2003. Morphology and escape performance of tiger salamander larvae (*Ambystoma tigrinum mavortium*). *J. Exp. Zool.* 297:147–59

Formanowicz DR Jr. 1986. Anuran tadpole/aquatic insect predator-prey interactions: tadpole size and predator capture success. *Herpetologica* 42:367–73

Forward RBJ, Tankersley RA, Rittschof D. 2001. Cues for metamorphosis of Brachyuran crabs: an overview. *Am. Zool.* 41:1108–22

Gascon C, Travis J. 1992. Does the spatial scale of experimentation matter? A test with tadpoles and dragonflies. *Ecology* 73:2237–43

Gilliam JF, Fraser DF. 1987. Habitat selection under predation hazard: test of a model with stream-dwelling minnows. *Ecology* 68:1856–62

Harvell CD. 1990. The ecology and evolution of inducible defenses. *Q. Rev. Biol.* 65:323–40

Hechtel LJ, Juliano SA. 1997. Effects of a predator on prey metamorphosis: plastic responses of prey or selective mortality? *Ecology* 78:838–51

Hoff KV, Wassersug RJ. 2000. Tadpole locomotion: axial movement and tail functions in a largely vertebraeless vertebrate. *Am. Zool.* 40:62–76

Johansson F. 2002. Reaction norms and production costs of predator-induced morphological defences in a larval dragonfly (*Leucorrhinia dubia*: Odonata). *Can. J. Zool.* 80:944–50

Johansson F, Samuelsson L. 1994. Fish-induced variation in abdominal spine length of *Leucorrhinia dubia* (Odonata) larvae? *Oecologia* 100:74–79

Johansson F, Stoks R, Rowe L, De Block M. 2001. Life history plasticity in a damselfly: effects of combined time and biotic constraints. *Ecology* 82:1857–69

Jones M, Laurila A, Peuhkuri N, Piironen J, Seppa T. 2003. Timing an ontogenetic niche shift: responses of emerging salmon alevins to chemical cues from predators and competitors. *Oikos* 102:155–63

Kats LB, Petranka JW, Sih A. 1988. Antipredator defenses and the persistence of amphibian larvae with fishes. *Ecology* 69:1865–70

Kehr AI, Schnack JA. 1991. Predator-prey relationship between giant water bugs (*Belostoma oxyurum*) and larval anurans (*Bufo arenarum*). *Alytes* 9:61–69

Kiesecker JM, Chivers DP, Anderson M, Blaustein AR. 2002. Effect of predator diet on life history shifts of red-legged frogs, *Rana aurora*. *J. Chem. Ecol.* 28:1007–15

Lardner B. 2000. Morphological and life history responses to predators in larvae of seven anurans. *Oikos* 88:169–80

Laurila A, Kujasalo J. 1999. Habitat duration, predation risk and phenotypic plasticity in common frog (*Rana temporaria*) tadpoles. *J. Anim. Ecol.* 68:1123–32

Laurila A, Kujasalo J, Ranta E. 1998. Predator-induced changes in life history in two anuran tadpoles: effects of predator diet. *Oikos* 83:307–17

Lima SL, Dill LM. 1990. Behavioral decisions made under the risk of predation a review and prospectus. *Can. J. Zool.* 68:619–40

Lively CM. 1999. Developmental strategies in spatially variable environments: Barnacle shell dimorphism and strategic models of selection. See Tollrian & Harvell 1999, pp. 245–58

Ludwig D, Rowe L. 1990. Life-history strategies for energy gain and predator avoidance under time constraints. *Am. Nat.* 135:686–707

McCollum SA, Van Buskirk J. 1996. Costs and benefits of a predator-induced polyphenism in the gray treefrog *Hyla chrysoscelis*. *Evolution* 50:583–93

Moore RD, Newton B, Sih A. 1996. Delayed hatching as a response of streamside salamander eggs to chemical cues from predatory sunfish. *Oikos* 77:331–35

Morin PJ. 1981. Predatory salamanders reverse the outcome of competition among three species of anuran tadpoles. *Science* 212:1284–86

Morin PJ. 1986. Interactions between intraspecific competition and predation in an amphibian predator-prey system. *Ecology* 67:713–20

Nicieza AG. 2000. Interacting effects of predation risk and food availability on larval anuran behaviour and development. *Oecologia* 123:497–505

Peacor SD. 2002. Positive effect of predators on prey growth rate through induced modifications of prey behaviour. *Ecol. Lett.* 5:77–85

Peckarsky BL. 1987. Mayfly cerci as defense against stonefly predation: deflection and detection. *Oikos* 48:161–70

Peckarsky BL, Cowan CA, Penton MA, Anderson C. 1993. Sublethal consequences of stream-dwelling predatory stoneflies on mayfly growth and fecundity. *Ecology* 74:1836–46

Peckarsky BL, McIntosh AR. 1998. Fitness and

community consequences of avoiding multiple predators. *Oecologia* 113:565–76
Relyea RA. 2001a. The lasting effects of adaptive plasticity: predator-induced tadpoles become long-legged frogs. *Ecology* 82:1947–55
Relyea RA. 2001b. Morphological and behavioral plasticity of larval anurans in response to different predators. *Ecology* 82:523–40
Relyea RA. 2002. Competitor-induced plasticity in tadpoles: consequences, cues, and connections to predator-induced plasticity. *Ecol. Monogr.* 72:523–40
Relyea RA. 2004. Integrating phenotypic plasticity when death is on the line: insights from predator-prey systems. In *The Evolutionary Biology of Complex Phenotypes*, ed. M Pigliucci, K Preston. 1:176–94. Oxford: Oxford Univ. Press. 464 pp.
Relyea RA, Hoverman JT. 2003. The impact of larval predators and competitors on the morphology and fitness of juvenile treefrogs. *Oecologia* 134:596–604
Richardson JML. 2001a. A comparative study of activity levels in larval anurans and response to the presence of different predators. *Behav. Ecol.* 12:51–58
Richardson JML. 2001b. The relative roles of adaptation and phylogeny in determination of larval traits in diversifying anuran lineages. *Am. Nat.* 157:282–99
Richardson JML. 2002. A comparative study of phenotypic traits related to resource utilization in anuran communities. *Evol. Ecol.* 16:101–22
Rowe L, Ludwig D. 1991. Size and timing of metamorphosis in complex life cycles time constraints and variation. *Ecology* 72:413–27
Scott DE. 1994. The effect of larval density on adult demographic traits in *Ambystoma opacum*. *Ecology* 75:1383–96
Scrimgeour GJ, Culp JM. 1994. Foraging and evading predators: the effect of predator species on a behavioural trade-off by a lotic mayfly. *Oikos* 69:71–79
Semlitsch RD, Gibbons JW. 1988. Fish predation in size-structured populations of treefrog tadpoles. *Oecologia* 75:321–26
Semlitsch RD, Scott DE, Pechmann JHK. 1988. Time and size at metamorphosis related to adult fitness in *Ambystoma talpoideum*. *Ecology* 69:184–92
Sih A. 1987. Predators and prey lifestyles: an evolutionary and ecological overview. In *Predation: Direct and Indirect Impacts on Aquatic Communities*, ed. WC Kerfoot, A Sih, pp. 203–24. London: Univ. Press New Engl. 386 pp.
Sih A, Kats LB, Maurer EF. 2000. Does phylogenetic inertia explain the evolution of ineffective antipredator behavior in a sunfish-salamander system? *Behav. Ecol. Sociobiol.* 49:48–56
Sih A, Moore RD. 1993. Delayed hatching of salamander eggs in response to enhanced larval predation risk. *Am. Nat.* 142:947–60
Skelly DK. 1992. Field evidence for a cost of behavioral antipredator response in a larval amphibian. *Ecology* 73:704–8
Skelly DK. 1994. Activity level and the susceptibility of anuran larvae to predation. *Anim. Behav.* 47:465–68
Skelly DK. 1995. A behavioral trade-off and its consequences for the distribution of *Pseudacris* treefrog larvae. *Ecology* 76:150–64
Skelly DK, Werner EE. 1990. Behavioral and life-historical responses of larval American toads to an odonate predator. *Ecology* 71:2313–22
Smith DC. 1987. Adult recruitment in chorus frogs: effects of size and date at metamorphosis. *Ecology* 68:344–50
Sokolovska N, Rowe L, Johansson F. 2000. Fitness and body size in mature odonates. *Ecol. Entomol.* 25:239–48
Taylor BW, Anderson CR, Peckarsky BL. 1998. Effects of size at metamorphosis on stonefly fecundity, longevity, and reproductive success. *Oecologia* 114:494–502
Tollrian R, Dodson SI. 1999. Inducible defenses in Cladocera: constraints, costs, and multipredator environments. See Tollrian & Harvell 1999, pp. 177–202

Tollrian R, Harvell CD, eds. 1999. *The Ecology and Evolution of Inducible Defenses.* Princeton, NJ: Princeton Univ. Press. 383 pp.

Travis J. 1983. Variation in growth and survival of *Hyla gratiosa* larvae in experimental enclosures. *Copeia* 1983:232–37

Travis J, Keen WH, Juilianna J. 1985. The role of relative body size in a predator-prey relationship between dragonfly naiads and larval anurans. *Oikos* 45:59–65

Tseng M. 2003. Life-history responses of a mayfly to seasonal constraints and predation risk. *Ecol. Entomol.* 28:119–23

Van Buskirk J. 1988. Interactive effects of dragonfly predation in experimental pond communities. *Ecology* 69:857–67

Van Buskirk J. 2002. A comparative test of the adaptive plasticity hypothesis: relationships between habitat and phenotype in anuran larvae. *Am. Nat.* 160:87–102

Van Buskirk J, Anderwald P, Lupold S, Reinhardt L, Schuler H. 2003. The lure effect, tadpole tail shape, and the target of dragonfly strikes. *J. Herpetol.* 37:420–24

Van Buskirk J, Arioli M. 2002. Dosage response of an induced defense: How sensitive are tadpoles to predation risk? *Ecology* 83:1580–85

Van Buskirk J, McCollum SA. 2000a. Functional mechanisms of an inducible defence in tadpoles: morphology and behaviour influence mortality risk from predation. *J. Evol. Ecol.* 3:336–47

Van Buskirk J, McCollum SA. 2000b. Influence of tail shape on tadpole swimming performance. *J. Exp. Biol.* 203:2149–58

Van Buskirk J, McCollum SA, Werner EE. 1997. Natural selection for environmentally induced phenotypes in tadpoles. *Evolution* 51:1983–92

Van Buskirk J, Relyea RA. 1998. Selection for phenotypic plasticity in *Rana sylvatica* tadpoles. *Biol. J. Linn. Soc.* 65:301–28

Van Buskirk J, Saxer G. 2001. Delayed costs of an induced defense in tadpoles? Morphology, hopping, and development rate at metamorphosis. *Evolution* 55:821–29

Van Buskirk J, Schmidt BR. 2000. Predator-induced phenotypic plasticity in larval newts: trade-offs, selection, and variation in nature. *Ecology* 81:3009–28

Warkentin KM. 1995. Adaptive plasticity in hatching age: a response to predation risk trade-offs. *Proc. Natl. Acad. Sci. USA* 92:3507–10

Warkentin KM. 2000. Wasp predation and wasp-induced hatching of red-eyed treefrog eggs. *Anim. Behav.* 60:503–10

Warkentin KM, Currie CR, Rehner SA. 2001. Egg-killing fungus induces early hatching of red-eyed treefrog eggs. *Ecology* 82:2860–69

Werner EE. 1986. Amphibian metamorphosis: growth rate predation risk and the optimal size at transformation. *Am. Nat.* 128:319–41

Werner EE, Anholt BR. 1993. Ecological consequences of the trade-off between growth and mortality rates mediated by foraging activity. *Am. Nat.* 142:242–72

Werner EE, Anholt BR. 1999. Density-dependent consequences of induced behavior. See Tollrian & Harvell 1999, pp. 218–30

Werner EE, Gilliam JF. 1984. The ontogenetic niche and species interactions in size-structured populations. *Annu. Rev. Ecol. Syst.* 15:393–26

Werner EE, McPeek MA. 1994. Direct and indirect effects of predators on two anuran species along an environmental gradient. *Ecology* 75:1368–82

Wilbur HM. 1972. Competition, predation and the structure of the *Ambystoma-Rana sylvatica* community. *Ecology* 53:3–21

Wilbur HM. 1987. Regulation of structure in complex systems experimental temporary pond communities. *Ecology* 68:1437–52

Wilbur HM, Collins JP. 1973. Ecological aspects of amphibian metamorphosis. *Science* 182:1305–14

Wilbur HM, Fauth JE. 1990. Experimental aquatic food webs interactions between two predators and two prey. *Am. Nat.* 135:176–204

Wilbur HM, Morin PJ, Harris RN. 1983. Salamander predation and the structure of experimental communities: anuran responses. *Ecology* 64:1423–29

Wildy EL, Chivers DP, Blaustein AR. 1999. Shifts in life-history traits as a response to cannibalism in larval long-toed salamanders (*Ambystoma macrodactylum*). *J. Chem. Ecol.* 25:2337–46

Wooster D, Sih A. 1995. A review of the drift and activity responses of stream prey to predator presence. *Oikos* 73:3–8

THE EVOLUTIONARY ECOLOGY OF NOVEL PLANT-PATHOGEN INTERACTIONS

Ingrid M. Parker[1] and Gregory S. Gilbert[2]

[1]*Ecology and Evolutionary Biology,* [2]*Environmental Studies, University of California, Santa Cruz, California 95064; email: parker@biology.ucsc.edu, ggilbert@ucsc.edu*

Key Words emergent diseases, biological invasions, plant disease epidemics, biological control, serial passage experiments

■ **Abstract** Novel plant-pathogen combinations occur whenever pathogen or plant species are introduced to regions outside their native range. Whether a pathogen is able to acquire a new host depends on the genetic compatibility between the two, through either preadaptation of the pathogen or subsequent evolutionary change. The ecological outcome of the novel interaction—for example, a spreading disease epidemic or the extinction of an incipient plant invasion—depends on the life history of the pathogen, opportunities for rapid evolution of virulence or resistance, and the presence of a suitable environment. We review recent work on the biology of pathogen virulence and host resistance, their mechanisms, and their costs. We then explore factors influencing the ecological and evolutionary dynamics of novel plant-pathogen interactions, using that evolutionary ecology framework to provide insight into three important practical applications: emerging diseases, biological invasions, and biological control.

INTRODUCTION

Advances in transportation technology and expanding global trade have greatly accelerated the rate of biological invasions (Perrings et al. 2002), which has in turn driven an increase in the number of novel encounters between plants and pathogens. In the past century, plant pathogens introduced from distant continents have caused the large-scale transformation of native ecosystems around the world, either by attacking a broad range of host species and altering the landscape through diffuse impacts on many plant species [e.g., *Phytophthora ramorum* in California oak woodlands (Rizzo & Garbelotto 2003)] or by dramatically reducing the populations of single species that played crucial roles in ecosystem structure or function [e.g., *Cryphonectria parastica* in North America and Europe (Anagnostakis 1987)]. At the same time, invasions by weedy plant species are ever more common, and plant pathogens native to their new ranges may play a role in regulating their spread (Mack 1996, Duncan & Williams 2002, Beckstead & Parker 2003). Conversely, under the assumption that escape from native pathogens may help explain invasion success (Blaney & Kotanen 2001, Mitchell & Power 2003), pathogens are sometimes introduced to control weedy invaders (Charudattan & Dinoor 2000). Whether a novel pathogen-plant combination will lead to disease development and major outbreaks is of key concern for understanding the consequences of novel epidemics in conservation biology, for the control of invasive weeds and the safe introduction of biological control agents, and for the development of robust policy for international trade that protects plant and ecosystem health. Outcomes and dynamics of the interactions between hosts and pathogens are shaped by a complex set of ecological and evolutionary influences (Alexander et al. 1996, Simms 1996). Here we review the biological context, theory, empirical studies, and applications of the evolutionary ecology of novel plant-pathogen interactions following from the introduction of plants and/or pathogens into new environments.

We examine the initial interactions and potential for rapid evolutionary changes in novel plant-pathogen interactions for four scenarios, depending on whether the plant, pathogen, or both are novel to the environment (Table 1). Native plant-pathogen interactions (Table 1A) provide a reference for "normal" ecological and coevolutionary dynamics, with fluctuating selection on both hosts and pathogens. Costs of resistance genes should maintain variation in wild populations of the host (Roy & Kirchner 2000), and costs of virulence genes, as well as tradeoffs between transmission and damage to the host, should maintain variation for virulence in populations of the pathogen (Jarosz & Davelos 1995, Jenner et al. 2002). Because coevolutionary interactions are temporally dynamic, pathogens and hosts at a site may or may not be closely coadapted (Kaltz & Shykoff 1998).

Newly introduced plant pathogens (Table 1B) face both a novel environment and novel host species. The first hurdle for these microbes is to arrive in the new range by exploiting introduction pathways (Goodell et al. 2000, Brown & Hovmøller 2002), but environmental conditions and host availability also place strong limits on survival and spread (Aylor 2003). Pathogens must first be able to survive the abiotic conditions of the new location, including extremes of temperature, moisture, and UV irradiation (Gilbert 2002). Those that survive must encounter and infect suitable susceptible hosts, then reproduce and disperse. Each of these factors presents a strong selection filter, which may lead to rapid adaptation to new abiotic conditions and rapid evolution of novel host use. Particulars of the pathogen's life

TABLE 1 Four scenarios for the emergence of novel plant-pathogen interactions

Plant	Pathogen	
	Stay	Move
Stay	Native [A]	Novel Epidemic [B]
Move	Invasive Plant [C]	Biological control [D]

history and degree of host specialization are important determinants of the invasion process.

Newly introduced plant species (Table 1C) face a similar set of challenges, including surviving, growing, and reproducing in the new habitat. Tolerance of local abiotic conditions is most likely the dominant factor influencing initial survival of most species. However, if the naive host plant is highly susceptible to locally abundant plant pathogens, these pathogens could play an important role in limiting incipient invasions (Mack 1996) or preventing cultivation of agronomic species (Coutinho et al. 1998, Wingfield et al. 2001). Local pathogens have the potential to be particularly damaging if they are generalists or preadapted to the new host because they would already be well suited to local environmental conditions and would have a ready source of inoculum on their native host (Watson 1970). However, until an invading plant becomes dominant in the new habitat, there should be only weak selection on local pathogens to increase their ability to infect the new species. Therefore, rapid evolution probably plays only a small role for the pathogen. In contrast, introduced plants, if susceptible to local generalist pathogens, should experience strong selection pressure to develop resistance mechanisms. If not susceptible to local generalist pathogens, introduced plants may experience a demographic release associated with escape from their pathogens from the home range (Blaney & Kotanen 2001, Beckstead & Parker 2003, Mitchell & Power 2003).

Finally, both host and pathogen may find themselves in a novel environment (Table 1D). The two players may be introduced independently to a new area, or the pathogen may hitchhike on the introduced host, or the pathogen may be a biological control agent brought from the native range to control an invasive plant. Because both the pathogen and host are experiencing new conditions, there may be selection on both to adapt to the local environment. Although conditions in the new range may be less optimal for disease development, in some cases the reverse may also occur; a pathogen considered insignificant in its native range has sometimes caused severe outbreaks on the same host under new climatic conditions (Zwolinski et al. 1990, Wingfield et al. 2001). If resistance is costly, there may be initial selection on the invading plant against resistance to pathogens left behind, but then strong selection for resistance after the interaction is re-established.

Our goal is to present a synthetic context for explaining and predicting patterns in novel plant-pathogen interactions. Therefore, we begin by reviewing the biology of plant-pathogen interactions relevant to understanding ecological and evolutionary host shifts and dynamics. We then explore in greater depth predictions for the initial interactions, subsequent changes, and final outcomes of the four scenarios presented in Table 1. Throughout, we consider both numerical (ecological) dynamics and rapid evolutionary dynamics. Finally, we consider fundamental questions underlying three practical applications of scenarios B, C, and D of Table 1: unintended novel epidemics, introduced and invasive plants, and pathogens introduced as classical biological control agents for invasive plants.

CONCEPTS AND MECHANISMS OF DISEASE INTERACTIONS

Definitions

Terminology used in the plant pathology literature has a conflicted history, and meanings may differ from those in studies of animal diseases. In this review we follow the conventions adopted by the American Phytopathological Society (D'Arcy et al. 2001). Pathogenicity is the ability of a pathogen to cause disease on a particular host—it is a qualitative term. Virulence is then a quantitative measure of pathogenicity denoting the degree of damage caused on that host, usually assumed to correlate negatively with host fitness. Aggressiveness (as used by Vanderplank 1968) is then a synonym of virulence. Virulence is not necessarily correlated with factors that determine pathogen fitness, such as the efficiency with which a pathogen uses the host plant as a substrate or its ability to colonize new plants (Statler & Jones 1981, Welz & Leonard 1994). Mechanistically, the ability of many plant pathogens to cause disease on a particular host is regulated at least in part by gene-for-gene interactions between resistance (R) genes in the host and specific avirulence (Avr) genes in the pathogen (Flor 1956). In that context, Avr genes code for products (effectors) recognized by the corresponding R genes and lead to an incompatible interaction (no disease development); pathogens lacking those Avr genes (i.e., genotype avr) are termed virulent on that host. The combination of Avr genes in a pathogen strain determines the race of the pathogen. When we use the terms resistance and virulence in the context of gene-for-gene interactions, we hope to avoid confusion by indicating the special use.

We use the term host shift to refer to the acquisition of a new host, which may or may not involve evolutionary change in the pathogen. When their implications differ, we distinguish between evolutionary host shifts and ecological host shifts, which are often referred to in the biological control literature as host switches.

The Disease Triangle

A central principle of plant pathology is the disease triangle. That is, development of plant disease requires the junction of three equally important components: (*a*) a susceptible host, (*b*) a virulent pathogen, and (*c*) suitable environmental conditions. Even if a virulent pathogen is introduced into an environment with a genetically susceptible host, suboptimal moisture, temperature, or soil conditions may prevent disease development (e.g., Weste & Marks 1987). The environment can also affect the rate of reproduction of a pathogen (Garbelotto et al. 2003), or determine whether it can reproduce sexually (Adams & Line 1984). For novel plant-pathogen interactions, changes in pathogen virulence, host resistance, and pathogen tolerance of environmental conditions can dramatically affect disease impacts and evolutionary dynamics.

Host-Pathogen Recognition and Infection

Agricultural disease management relies primarily on deploying cultivars with *R* genes that provide resistance to pathogens with corresponding *Avr* genes. Such gene-for-gene interactions determine specificity for particular cultivars (genotypes) or species of crop plants and are important in many wild plant-pathogen interactions (Burdon 1991, Holub 2001, Takabayashi et al. 2002, Thrall & Burdon 2003). Characterization of numerous *R* and *Avr* genes shows that most *R* genes encode proteins thought to act as receptors that recognize specific *Avr* gene products in the pathogen (reviews in Leach & White 1996, Laugé & De Wit 1998, Martin et al. 2003, Nimchuk et al. 2003). Such recognition genes are diverse and common in plants—as much as 1% of the total genome (Ellis et al. 2000, Mondragón-Palomino et al. 2002, Nimchuk et al. 2003). Unlike plant *R* genes, the primary function of avirulence effectors in pathogens is not (needless to say) to trigger defense responses in plants. *Avr* gene products include a diversity of extracellular proteins, as well as viral coat proteins (Laugé & De Wit 1998). The recognition of an *Avr* gene product by the corresponding *R* gene product elicits plant defense mechanisms that produce an incompatible reaction (Jia et al. 2000), often through production of programmed host cell death called a hypersensitive response. Modification of an *Avr* (avirulent) allele to an *avr* (virulent) allele allows the pathogen to defeat host resistance by preventing recognition by the host.

Cost of Overcoming *R* Genes

Of particular interest for novel plant-pathogen interactions is genetic variation for virulence and resistance. One mechanism thought to maintain polymorphisms in *Avr* genes and *R* genes in natural populations is the cost of virulence (in the gene-for-gene sense) and resistance. Because recognition by the host is detrimental to the pathogen, we would expect avirulent *Avr* alleles to be quickly lost through selection, unless they have important functions that contribute to other aspects of pathogen fitness (Simms 1996). In the absence of a cost, pathogens would be expected to accumulate and retain the ability to circumvent many different host resistance genotypes. However, as demonstrated in early survey work, pathogen populations are often dominated by races that carry no *avr* alleles beyond those for the corresponding *R* genes present in the crop cultivar from which they were isolated (termed "unnecessary *avr* alleles") (reviewed in Vanderplank 1968, but see Parlevliet 1981 for counter examples). In addition, time series show a loss of *avr* (virulent) alleles in pathogens 6–10 years after a crop line with the corresponding *R* genes is removed from cultivation (e.g., Grant & Archer 1983). Similar loss might be expected in introduced pathogens after they leave their prior hosts behind. Vanderplank (1968) argued that such observations and *R* gene removal studies implied a fitness cost to the pathogen of carrying unnecessary *avr* alleles. However, such studies cannot rule out the possibility that the original *avr* mutation occurred in a pathogen lineage that was otherwise less fit; under this scenario, without the benefit associated with the presence of the corresponding *R* gene, this lineage would then decline (Parlevliet 1981). As a more direct test of the cost of carrying extra *avr* alleles, known mixtures of pathogen spores with different virulence phenotypes have been passed through susceptible hosts. In some studies, races with more *avr* diversity were rapidly selected against in susceptible hosts (Watson & Singh 1952, Leonard 1969, Chin & Wolfe 1984, Thrall & Burdon 2003), whereas others showed no effect or the opposite pattern (review in Vanderplank 1968, Parlevliet 1981). Recent molecular genetic studies on pathogens have begun to elucidate some important functions performed by *Avr* gene products (review in Laugé & De Wit 1998). For instance, all *Phytophthora* species produce extracellular 98–amino acid proteins (elicitins) that are important in sterol scavenging, and they also induce a range of defense responses in plants (reviewed in Tyler 2002). Other *Avr* gene products play key roles in inhibiting the hypersensitive response (Abramovitch et al. 2003), promoting growth or reproduction within the host plant (Leach & White 1996), acting as plant toxins (Wevelsiep et al. 1993), or masking other *Avr* genes (Tsiamis et al. 2000). In plant breeding, crop resistance is more durable when the targeted *R* genes recognize pathogen effectors with particularly important housekeeping functions because changes to these genes inflict a large fitness cost on the pathogen (Vera Cruz et al. 2000).

Cost of Resistance

Polymorphism in resistance within and among host populations affects the dynamics of novel interactions. Pathogens can impose strong selection for particular *R* alleles in a host population. Unless there is a cost to maintaining resistance alleles in the absence of corresponding pathogen races, plant lineages should accumulate *R* alleles and become universally resistant. Although the many examples of polymorphisms for resistance in plant populations imply a significant cost to resistance (e.g., Parker 1988, Burdon 1991), the effects of genetic background, environment, and genetic linkage have made measuring a cost of resistance to plant pathogens difficult and controversial (Parker 1990, Schmid 1994, Brown 2002). In a review of the literature, Bergelson & Purrington (1996) found that 56% of 55 comparisons of resistant and susceptible host genotypes showed significant costs of resistance. For 11 pathogen species and 7 host species, they found the mean cost to fitness to be about 4% in the absence of disease. Tian et al. (2003), using isogenic pairs of transgenic host lines, estimated a 9% cost of specific *R* genes.

The reasons for such a cost of resistance are unclear because there is little evidence that *R* genes have important pleiotropic functions outside of recognizing corresponding pathogens. No doubt there is a significant energetic and fitness cost to mounting a defense response (Smedegaard-Petersen & Stølen 1981), but in the absence of elicitors from the specific pathogens, why should *R* genes be costly? Basal levels of expression or accidental induction by environmental factors or conserved *Avr* gene products from nonpathogenic microbes may be responsible for the costs (Tyler 2002, Brown 2003). There are numerous examples of *R* gene products recognizing multiple elicitors (Parker et al. 1991, Tyler 2002). Individual *R* genes can recognize different *Avr* gene products in a single pathogen species

[*RPM1*, (Grant et al. 1995)], recognize *Avr* gene products from different, closely related species [*RPW8*, (Xiao et al. 2001)], or even provide resistance to two pathogens as different as a bacterium and an Oomycete [*NPR1*, (Cao et al. 1998)]. Plants must face tradeoffs between being too selective—and potentially incurring costs from disease when pathogens are not detected—and the costs of unnecessarily expressing defense responses when induced by nonpathogens. These tradeoffs may be crucial in determining preadaptation in novel plant-pathogen encounters.

Interestingly, costs of resistance (as well as virulence) may decline over time, as modifier genes are selected to mask the disadvantageous traits that are genetically correlated with the virulence or resistance factor. In bacteria, this compensatory evolution has been shown to influence the maintenance of antibiotic resistance after removal of the antibiotic (Levin et al. 2000) and maintenance of resistance to viral infection (Lenski 1988).

Quantitative Resistance and Virulence/Aggressiveness

Whereas *R* genes and *Avr* genes determine the compatibility of a plant-pathogen interaction and regulate host shifts, numerous quantitative resistance and virulence factors determine the amount of damage a pathogen causes on a compatible host. Such genes are also likely to be extremely important in the evolutionary ecology of novel interactions. Pathogens produce toxins against host cells (Wolpert et al. 2002) and enzymes to detoxify host defenses (George & VanEtten 2001) as well as many factors important in colonization (e.g., Saile et al. 1997). Such factors have quantitative effects on disease development, disease severity, and pathogen fitness, and although they are not involved in gene-for-gene associations between plant and pathogen, they are under similar selection for rapid coevolutionary adaptation (Bishop et al. 2000).

Similarly, quantitative traits in plants are involved in many defense-related responses (Holub 1997), including the production of chitinases and endoglucanases that degrade fungal cell walls (Bishop et al. 2000), phytoalexins involved in the hypersensitive response (Bennett et al. 1994), enzymes that detoxify pathogen toxins (George & VanEtten 2001), and morphological traits conferring resistance (Bradley et al. 2003). Although all plant-pathogen interactions are likely to involve defense strategies that combine qualitative and quantitative resistance mechanisms (e.g., Bevan et al. 1993), the specific genetic basis of resistance can vary among populations of a single host species (Parker 1988, 1991a). Plants may also respond to selection from a pathogen by evolving tolerance, that is, the ability to maintain fitness at high levels of pathogen infection (Roy & Kirchner 2000). Tolerance evolves differently from resistance and may influence selection on resistance (Roy & Kirchner 2000, Mauricio 2001).

Transmission and the Evolution of Virulence

Until the emergence of Darwinian medicine in the early 1990s (Ewald 1993), there was a widely held view that host-pathogen interactions should evolve toward a lower level of virulence because causing harm to the host should ultimately prove destructive to the pathogen as well (e.g., Alexander 1981). Therefore, novel, emergent epidemics were thought to be caused by maladapted pathogens whose impacts should decrease over time. Models of the evolution of virulence in animal pathogens, such as classic work by Anderson & May (1982), relied on the assumption that the evolution of increased virulence would be limited by opportunities for transmission to a new susceptible host. Although virulence should decrease if that enhances the probability of vertical transmission (Kover & Clay 1998), evolutionary theory suggests that complex tradeoffs between transmission modes, reproductive rates, and other life history traits determine whether natural selection should lead to increased or decreased virulence (Williams & Hesse 1991, Ewald 1993, Bull 1994, Lenski & May 1994). Caution should also be used in applying the predictions of the classic, animal-based models too strictly in the case of novel plant pathogens, as several assumptions of the models may not hold for plant diseases (Jarosz & Davelos 1995). In fact, in a review of the evidence using a series of different pathogen types (local versus systemic, soil-borne, foliar, floral, etc.), Jarosz & Davelos (1995) found little support for trends toward decreased virulence in natural plant-pathogen systems.

One key to predicting the trajectory of virulence in novel host-pathogen interactions is the relationship between virulence and pathogen fitness. Within a host, high virulence may be a necessary consequence of having higher pathogen titer (Chang et al. 1995), or it may directly increase pathogen fecundity (Fox & Williams 1984). However, several studies indicate that the fecundity of pathogens is sometimes uncorrelated, or even negatively correlated, with virulence (Johnson 1947, Imhoff et al. 1982, Robert et al. 2002, Zhan et al. 2002). Within-host fitness must in turn be linked to transmission, or between-host fitness (Bull 1994). Greater virulence may increase horizontal transmission if higher fecundity also means greater spore dispersal (Fox & Williams 1984). In contrast, serial passage studies of RNA viruses have found strong tradeoffs between virulence within the host (here tightly linked with pathogen fitness) and transmission by either insect or fungal vectors (Tamada & Kusume 1991, Hernandez et al. 1996). In one of the few empirical demonstrations of reduction of virulence in field populations, Escriu et al. (2000, 2003) investigated cucumber mosaic virus (CMV) and the replacement of virulent (necrosis causing) satellite RNA (satRNA) by nonvirulent satRNA. The more virulent satRNA replaced nonvirulent satRNA in mixtures, showing higher within-host fitness. However, virulent satRNA led to a depression in the accumulation of CMV, which led to a reduction in aphid transmission of the virus. This case demonstrates a tradeoff between virulence and transmission, suggesting that a pathogen may experience fitness tradeoffs associated with being highly virulent on a host. On the other hand, Zhan et al. (2002) used DNA fingerprinting to track the relative fitness of 10 strains of the fungus *Mycosphaerella graminicola* in the field and found no significant correlation between fitness and virulence. The relationship between fitness, virulence, and transmission appears to be complex.

PREDICTIONS FOR NOVEL PLANT-PATHOGEN INTERACTIONS

Conservation biologists, land managers, trade officials, and policy makers would like to be able to predict the trajectory of novel plant-pathogen interactions. The course of novel interactions is determined primarily by (*a*) the likelihood of an initial host shift, (*b*) the expected numerical dynamics of both the host and the pathogen, (*c*) the probability of evolution of resistance in the host, and (*d*) the probability of evolution of virulence in the pathogen. Here we examine how the factors described above may influence each of these steps.

What Is the Likelihood of an Initial Host Shift?

Estimating the probability that a pathogen will successfully shift to infect a novel host is a major challenge. Because most plants do not act as hosts for most plant pathogens, we expect that most encounters between novel combinations of plant and pathogen species never result in a compatible, disease-causing interaction. But host shifts do happen, and the likelihood of a shift occurring depends on the particular pathogen, host, and environmental conditions. Host shifts may be purely ecological, when a pathogen is preadapted to attack a newly encountered host species (Anagnostakis 1987). Alternatively, host shifts may involve an evolutionary change to permit infection of a host. For example, plant pathogens have acquired new hosts through hybridization (Brasier 2001) or by adapting to environmental conditions that allow access to new hosts (McDonald & Hoff 2001). Some have argued that ecological host shifts play the dominant role in novel host-pathogen interactions (Schrag & Wiener 1995, Altizer et al. 2003), although this area needs further study (Secord & Kareiva 1996).

Four factors are particularly important in determining the chance that a host shift occurs: (*a*) the degree of dependence of the pathogen on live hosts (i.e., pathogen survival and saprotrophic abilities), (*b*) the degree of specialization of the pathogen, (*c*) the phylogenetic distance between the novel potential host and hosts with which the pathogen is familiar, and (*d*) the degree of ecological association between the pathogen and the potential host.

SURVIVAL/SAPROTROPHIC ABILITIES OF PATHOGEN The life history of the pathogen is an important consideration in predicting which pathogens are most likely to infect invading plants or invade themselves. Some pathogens are obligate biotrophs, which require a living host to complete their life cycle (e.g., rusts, smuts, and powdery mildews). However, many pathogens are facultative saprobes and can persist, grow, and sporulate on dead tissue. Most published work on plant-pathogen interactions in natural systems excludes this important factor because researchers have chosen to focus on tightly linked, two-species interactions that are exclusively biotrophic (e.g., Burdon 1991, Alexander et al. 1996, Roy 2001). However, simply because modes of long-distance transportation should give an advantage to pathogens that survive well without a living host (as resting spores or as saprotrophs), one would expect that many successful invasive microbes are facultative saprobes. Long-lived resting stages or saprotrophic abilities in the new habitat should increase the opportunities for and likelihood of host shifts.

DEGREE OF SPECIALIZATION OF PATHOGEN Some pathogen species are highly specialized on one or a few closely related host species, and individual genotypes (races) may be even more specialized on particular host genotypes. When such highly specialized pathogens are introduced to new regions (Table 1B), they are unlikely to survive the initial introduction unless their host is broadly distributed across both the native and introduced range. Biological control pathogens (Table 1D) are highly specialized by design, implying a low probability of host shift. Interestingly, some have argued that high host specificity may be correlated with evolutionary lability (Brooks & McLennan 1993), an idea that raises concerns about the long-term environmental safety of biological control agents (Secord & Kareiva 1996). For a newly introduced plant (Table 1C), host-generalist pathogens should be more likely to colonize and have the largest negative effect. For instance, the idea of biotic resistance to invasion requires that invaders are repelled by aggressive local pathogens (Elton 1958, Mack 1996, Blaney & Kotanen 2001); this idea relies on the ability of pathogens with broad host ranges to attack novel, naive hosts. In a comparison of sympatric suites of 18 native and non-native clovers, we found no difference in pathogen diversity, infection, leaf damage, or fitness effects of foliar and damping-off fungi (I.M. Parker & G.S. Gilbert, unpublished data). Fungicide experiments revealed significant effects of pathogens on plants in the field, but no difference between native and non-native species. In this system, host-generalist fungi dominate the relationship between plants and their pathogens, leveling the playing field between native and non-native hosts.

The life history of the pathogen plays an important role in its host range. We examined published records of host distribution for necrotrophs (pathogens that kill host tissue and live off the dead material) and biotrophic rusts and smuts (Farr et al. 2004) and found that species in all three groups can attack dozens or hundreds of plant species. However, rusts and smuts are generally limited to hosts from just one order or family (two, in the case of macrocyclic heteroecious rusts), whereas necrotrophs like *Alternaria alternata* or *Verticillium dahliae* have been found attacking plants from 29 or more plant orders (G.S. Gilbert & I.M. Parker, unpublished data). These broad patterns are intriguing, but a more detailed understanding of patterns of association between fungi and hosts is currently limited by unequal effort in different taxonomic groups and a lack of experimental cross-inoculations.

The degree of specialization affects not only whether the pathogen and host interact at all, but also how tightly the numerical dynamics of the pathogen are linked to the host. Invading pathogens that have a broad host range are less dependent on stochastic events, such as dispersing to the right host at the right time. They also have the ability to build up inoculum on a common host and then disperse onto a second host in high numbers, increasing the chances that a pathogen genotype adapted to that second host may successfully colonize it.

PHYLOGENETIC DISTANCE AMONG HOSTS Along the gradient from highly specialized to highly generalized pathogens, host ranges are not a random selection of taxa. Host ranges often have predictable phylogenetic structure, with closest relatives being more likely to share pathogens. For example, of the 70 recorded fungal species from the common California coastal woodland tree *Quercus agrifolia* (Farr et al. 2004), 51% were recorded only on *Q. agrifolia*, and an additional 13% were restricted to the genus *Quercus*. Eighty percent were restricted just to the Fagales. This suggests that the success of an invading pathogen depends in part on the phylogenetic distance between its host(s) in the native range and available potential hosts in the new range. Similarly, whether local pathogens are capable of attacking a new introduced host depends in part on how closely related this plant species is to the resident native species. Mack (1996) tested this idea by looking at patterns of invasiveness in several floras, finding that invaders were more likely to be in families or genera that contain no native species, and attributed this to pressure from native pests and pathogens. However, taking a similar approach but controlling for introduction opportunities, Duncan & Williams (2002) found that introduced species in genera that already had resident natives were more likely, not less likely, to naturalize successfully in New Zealand. To understand how phylogeny may influence patterns of host specialization and host shift, we need detailed studies using experimental cross-inoculations among hosts. Such linkages of ecological and phylogenetic information would also help in assessing risk in phytosanitary policy decisions and biological control cases.

ECOLOGICAL ASSOCIATION Upon careful inspection of the phylogenetic patterns of host use in pathogens, one notices that despite marked structure, seemingly unpredictable host shifts also occur onto widely divergent taxa (Weste & Marks 1987, Eckenwalder & Heath 2001). Recently, researchers have emphasized the importance of ecological association as a driver of host shifts. For instance, Roy (2001) constructed phylogenies for flower-mimic rusts (genus *Puccinia*) and their hosts in the Brassicaceae. Major jumps occurred between distant clades, and overall there was no tight phylogenetic congruence between pathogen and host. Instead, patterns of host use showed strong geographic clustering, such that physical proximity appeared to play an important role in host shifts. Similarly, published records (Farr et al. 2004) demonstrate that despite the strong phylogenetic signal in host range for the fungi that attack *Quercus agrifolia* (Fagaceae), a number of fungi are shared with the phylogenetically distant, but commonly co-occurring Ericaceae.

Factors that increase the likelihood of contact between current and potential hosts should also increase the probability of a host shift onto an introduced plant. In addition to geographic range overlap, other factors include ecological requirements, phenology, and environmental drivers such as fog-drip or the seasonality of rainfall, which could make disease development synchronous within regions and asynchronous among regions (or habitats). Insect vectors that use multiple host species increase the ecological association of these species beyond that caused by physical distribution alone, thereby increasing the opportunity for a host shift. In the case of introduced pathogens (Table 1B), the geographic range of potential hosts relative to the main ports of entry for importation of microbes should have a large influence on whether host shift and invasion will occur.

In addition to factors that increase the probability that new hosts will be in spatial and temporal proximity to current hosts, density of the new host also plays an important role. As an invading plant (Table 1C) increases in density and becomes locally dominant, both ecological opportunity and selection pressure for host switching should also increase. The subsequent dynamics of the plant-pathogen interaction also depend critically on this increase in host density.

What Are the Expected Numerical Dynamics of Pathogen and Plant?

After the initial shift of a pathogen to utilize a novel host, a number of factors influence the epidemiology of the pathogen over time. Here we consider the potential roles of (*a*) pathogen response to host density and (*b*) genetic diversity in the host population.

DENSITY DEPENDENCE   In most plant pathosystems that have been studied, disease development is dependent on host density (reviews in Burdon & Chilvers 1982, Gilbert 2002). Density-dependent disease development may arise from increased transmission rates mediated by decreasing distance between hosts, or indirectly through intraspecific competition effects on host vigor (reviewed in Gilbert 2002). The importance of density dependence is strongly influenced by the degree of host specialization of the pathogen; the dynamics of most strict host specialists are strongly tied to the density of their hosts, whereas populations of generalist pathogens are decoupled from the density of any single host. In more complex systems, pathogens with multiple hosts may respond to the joint population densities of several host species (Garbelotto et al. 2003), they may be influenced by competition between alternative host species (reviews in Alexander & Holt 1998, Gilbert 2002), or, for heteroecious rusts with two obligate alternate and competing hosts, they could show negative density-dependent patterns (Burdon & Chilvers 1982). For the sake of argument, however, here we follow the case of a single host and a single pathogen.

The case of density-dependent transmission has special implications for novel host-pathogen interactions because of the interplay between invasion dynamics and disease dynamics. For an invading pathogen (Table 1B) with density-dependent transmission, the relative density of different prospective native hosts influences both the probability that an epidemic will be initiated and the rate at which it spreads. For an invading plant (Table 1C), the rate of pathogen transmission should be low at the early stages of invasion, making epidemics unlikely for density-dependent pathogens. As the invasion proceeds, the plant reaches higher densities; in the case of weedy, high-impact invaders, plant densities may be very high relative to the co-occurring native species. At these high host densities, a pathogen that gains the ability to infect the host is able to reach maximum transmission rates, and epidemics should proceed extremely rapidly. The same is true for introduced biological control agents (Table 1D), which by definition are usually released onto invasive plants that have already reached high density. In the case of intentionally introduced plant species, native pathogens have produced dramatic effects on plants important in agriculture or forestry (Coutinho et al. 1998, McDonald & Hoff 2001). We could find no explicit examples of widespread invasive weeds experiencing substantial attack by a native pathogen, which perhaps supports the contention that delayed, evolutionary host shifts are rare. However, some invaders have been known to grow exponentially to high densities and then crash mysteriously, and pathogens have often been suggested (although with little empirical support) as a possible mechanism (Simberloff & Gibbons 2003).

Clearly, we need more studies on the timing of host shifts relative to the density and spread of novel invasive hosts. If pathogens play a role in the precipitous crash of certain established introduced species, then investigators should make a concerted effort to identify these pathogens and understand their origin. That is, are these pathogens persisting at low numbers when hosts are rare and becoming more prevalent in a host-density-dependent fashion, or do they undergo host shifts only when plants reach high density? These questions are interesting not only where pathogens have caused population crashes, but for all introduced species. Studies should be initiated to track the accumulation of pathogens in relation to the timing of host arrival and host density. Note that highly virulent, generalist pathogens may *not* show density-dependent responses, and as discussed above, they may prevent introduced plants from spreading. Such events are extremely difficult to observe.

EFFECTS OF HOST VARIATION ON EPIDEMIOLOGY   One of the most common generalizations in the field of plant-pathogen interactions is that disease should be lower in more genetically diverse host populations (Adams et al. 1971, Harlan 1976, Barrett 1988, Mundt 2002). Highly inbred agricultural species planted in large monocultures have been vulnerable to devastating disease epidemics; one response has been to plant multilines of different resistance genotypes together (Mundt 2002). Outside of agriculture, however, the evidence for a direct link between genetic variation and disease pressure has been mixed (Kranz 1990, Roy 1993, Thrall & Burdon 2000). Roy (1993) observed that correlations have been found primarily in systems with very low host genetic diversity—that is, just a handful of clones or selfed lines—whereas studies with slightly higher diversity tend not to show a relationship. In her study of three populations with a range of 6–27 genotypes, the population with lowest disease incidence was unexpectedly the one with lowest clone diversity and highest density. However, most such studies focus on snapshots in time, and temporal fluctuations in disease pressure could obscure the relationship between disease and host genetic diversity.

Introduced plants could provide an interesting context in which to test ideas about how genetic variation should influence disease incidence and dynamics. The introduction bottleneck should lead to reduced genetic variation, at least for Mendelian traits such as molecular markers (Barrett & Husband 1990, Amsellem et al. 2000, Lee 2002); R genes also fall into this category and should be genetically depauperate in invading plants. The analogous low diversity of R genes in crop plants leads breeders to return to wild crop progenitors in the habitat of origin to access a diversity of R genes (Hoisington et al. 1999). As an introduced species spreads, the colonization process results in different subsets of the original diversity in different sites. Pathogens, either native (Table 1C) or introduced (Table 1D), on these introduced hosts could provide a simplified system for looking at the ability of a pathogen to colonize and spread in populations of differing genetic diversity. Similarly, the impact of successful pathogens should be greater in populations that are less diverse. This idea has important implications for strategies in biological control. For example, researchers have long thought that sexually reproducing weeds are harder to control because their higher levels of genetic variation confer greater resistance (Burdon & Marshall 1981, but see Chaboudez & Sheppard 1995).

The best-studied case of classical biological control by a pathogen is that of the rust *Puccinia chondrillina* on *Chondrilla juncea* L. (rush skeletonweed, Asteraceae), a perennial weed of cereal crops native in Europe and introduced to Australia and North America (Burdon et al. 1981, Panetta & Dodd 1995). An interesting aspect of the *Chondrilla/Puccinia* system is the extreme level of host specificity found in the pathogen (Hasan 1972). *Chondrilla* is a triploid apomict in Australia, where only three clones are known and where the pathogen is also exclusively asexual. The initial control program introduced a pathogen isolate highly virulent to the dominant and most widespread clone (McVean 1966), which resulted in replacement by a different clone rather than elimination of the weed (Burdon et al. 1981, Chaboudez & Sheppard 1995). In light of the way host genetic diversity foiled the success of *Chondrilla* biological control in Australia, it is interesting that the level of disease pressure experienced by populations of *Chondrilla* is not strongly influenced by the degree of clonal diversity in its native range (Chaboudez & Sheppard 1995).

What Is the Probability of Evolution of Resistance in the Host?

In addition to host shifts and numerical dynamics, coevolutionary dynamics of the host and pathogen also influence the trajectories of novel plant-pathogen interactions. These coevolutionary dynamics involve both resistance of the host and virulence of the pathogen. For both players, the rate of evolution of a trait is proportional to (a) the strength of selection and (b) the amount of genetic variation for the trait (Fisher 1930, Crow 2002).

STRENGTH OF SELECTION If selection on the host by the pathogen is very weak, evolution of resistance will be imperceptibly slow or swamped by other factors. Different types of plant pathogens provide different strengths of selection (reviewed in Gilbert 2002). Pathogens that cause damping-off of seedlings, and many root rots, wilts, and canker diseases of mature plants can cause a high rate of mortality in host populations. Many smuts and some systemic pathogens can castrate hosts. In contrast, although most foliar diseases reduce host survival and reproduction, their effect may vary from insignificant to very strong.

For a native pathogen on a relative of an invading plant (Table 1C) that does overcome resistance in the invader, it is likely to sweep quickly through the susceptible host population and could provide strong selection pressure on the host to evolve resistance. For a native plant battling an introduced pathogen (Table 1B), evolving resistance to the new pathogen may come at a cost, either in terms of energy allocated to defense, or by modifying R genes that might otherwise be committed to conferring resistance to local native pathogens. Thus, plants may also experience fluctuating or stabilizing selection on resistance traits.

GENETIC VARIANCE FOR RESISTANCE Even if selection is strong, evolution will occur only if there is genetic variation in resistance for selection to act upon. The many examples of rapid evolution of herbicide tolerance in crop weeds (Heap 1997) suggest that there is often enough genetic variation to respond to strong selection. However, in contrast to herbicides, pathogens and herbivores themselves evolve in response to the host, which may exhaust genetic variation for resistance more rapidly. In several cases of novel forest epidemics, local variation for resistance has been quickly exhausted, and forest pathologists have resorted to bringing in resistance genes from distant regions or related species (Anagnostakis 1992, Smalley & Guries 1993). Mutation will, over long periods, be expected to generate new resistant forms; however, the great difference in generation times and population size between the pathogen and host puts the plant at a disadvantage (Schafer & Roelfs 1985, Hafner et al. 1994). In fact, if the impact of the pathogen on the host is extreme, then resistance may never evolve because the host is eliminated first. For example, the rapid sweep of the chestnut blight *Cryphonectria parasitica* through the distribution of the American chestnut eliminated every large adult and, in effect, keeps the species from reproducing (Anagnostakis 1987); therefore it is not clear if resistance will ever evolve in that system. In novel epidemics (Table 1B), we may expect to see resistance evolving in the host most readily when the pathogen's effect on the host is strong, but not exceedingly strong.

For an invading plant (Table 1C), we would predict that little specific resistance would be present in the population at the start, unless it happened to express an R gene receptor that recognized a corresponding Avr gene product in a local pathogen. Because individual R genes may code for resistance to multiple pathogens, the responses to different pathogens can be genetically correlated. If the same allele recognizes multiple pathogens, a positive correlation will result, whereas if alternate alleles at the same locus recognize different pathogens, a negative correlation will result. Novel hosts therefore may be preadapted to resist a pathogen, or may rapidly lose resistance to previous pathogens when faced with selection by a new suite of pathogens. Such tradeoffs could make invaders more susceptible to introduced biological control pathogens (Table 1D).

What Is the Probability of Evolution of Virulence in the Pathogen?

As in the host, evolution of the pathogen will depend on genetic variance and the strength of selection. However, predicting how selection should act is not so straightforward in the case of pathogen virulence. As discussed above, early conventional wisdom suggested that virulence should start high on a naive host and evolve to a lower, intermediate level (reviewed in Bull 1994, Lenski & May 1994). The classic case of the myxamatosis virus introduced to control invasive rabbits in Australia (Fenner & Fantini 1999) suggests that invading host species might be the best place to look for examples of loss of virulence over time. However, currently we have few examples, or even indirect evidence, of reduction in virulence in plant pathogens through natural selection (Jarosz & Davelos 1995).

Key factors that influence the evolution of virulence in plant pathogens are (*a*) linkage between virulence and pathogen fitness, (*b*) availability of alternative hosts, and (c) ability to reproduce on dead plant material or produce long-term survival structures. Evolution of increased virulence will be limited if it decreases the probability of transmission to a new host either by reducing pathogen fecundity or by reducing access to suitable hosts. However, if a pathogen can reproduce abundantly on an alternative host on which it has low virulence, the pathogen may reach very high virulence on some hosts while maintaining population size on the sympatric host. This may be the case in the Sudden Oak Death epidemic, where *Phytophthora ramorum* reproduces prolifically on the reasonably tolerant *Arbutus menziesii* but is often lethal on *Lithocarpus* and *Quercus*, where it has low fecundity (Garbelotto et al. 2003, Rizzo & Garbelotto 2003). Similarly, if the pathogen is able to grow and reproduce as a facultative saprobe, or if it produces durable resting structures (e.g., oospores and chlamydospores produced by many species of *Phytophthora*), the pathogen may be able to wait for years to encounter a suitable host and favorable environmental conditions. Such saprotrophic/long-term survival strategies could allow evolution of very high virulence without a fitness cost to the pathogen.

What Constrains Evolution?

There are constraints on the degree of local adaptation that occurs in plant-pathogen interactions, which can affect pathogen virulence or host resistance and complicate associations between *R* genes and recessive alleles at *Avr* loci. Gene flow may swamp local adaptation, and gene flow in fungal pathogens has the potential to be very large, as many of these organisms travel long distances by wind dispersal (Fitt et al. 1987, Brown & Hovmøller 2002, Aylor 2003). Studies that have looked for evidence of local adaptation in pathogens have provided a mixed picture. Using experimental inoculations of hosts collected over a range of scales from meters to kilometers apart, some studies have found that incidence or virulence of pathogens is higher on their original hosts (Parker 1985, 1991a), whereas others have found no significant "home-host advantage" (Parker 1989, Davelos et al. 1996, Zhan et al. 2002). This lack of consistent local adaptation may obscure our ability to validate predictions about the evolution of virulence (Zhan et al. 2002).

Because genotypes of hosts and pathogens should fluctuate asynchronously, host-pathogen coevolution is difficult to infer from a single sample in time (Dybdahl & Lively 1995, Kaltz & Shykoff 1998). In only a few cases do we have data on changing genetic structure of host and pathogen in natural systems, making it difficult to assess how often these change in a way consistent with local adaptation. So far, studies that have followed the dynamics of evolutionary changes in resistance have not provided clear evidence that these changes are easily predictable (Burdon & Jarosz 1992, Burdon & Thompson 1995, Parker 1991b). As populations of both pathogen and host are linked by gene flow, a metapopulation perspective may be necessary to make sense of long-term dynamics (Thompson 1994). What this implies for novel host-pathogen interactions is that, even if these interactions are governed by predictable evolutionary processes, it may be difficult to discern how selection operates from invasion studies unless they are conceived in a broad geographic framework.

The lack of observed local adaptation in pathogens also has important implications for the practice of classical biological control. Biocontrol strategies are increasingly employing genetic analysis in the careful matching of agent genotypes with the population of origin for the weed (e.g., Hasan et al. 1996, Holden & Mahlberg 1996). This practice is based on the assumption that pathogens from the population of origin are better adapted to exploit the invasive host, which may not always be true. For example, Yugoslavia is the putative site of origin for invasive *Chondrilla juncea* in the western United States (Hasan et al. 1996). *Puccinia chondrillina* genotypes collected from Yugoslavia showed high virulence on some U.S. genotypes, but other genotypes were little affected. In addition, some *P. chondrillina* isolates from other regions also showed high virulence. Our review of the evidence for local adaptation in natural plant pathosystems suggests that close genetic matching of hosts between the native and introduced range may not be an efficient biological control strategy.

CONCLUSIONS: UNRESOLVED APPLIED QUESTIONS

Ecological and coevolutionary interactions between pathogens and their plant hosts help shape the structure and dynamics of natural plant populations and communities. Novel plant-pathogen interactions are a consequence of long-distance movement of pathogens, plants, or both into regions outside of their historical distributions. Most often, these movements are driven by global trade, transportation technology, and changing land-use patterns. In this review we have attempted to create an evolutionary ecology framework for thinking about novel plant-pathogen interactions that will be useful for understanding, predicting, and managing novel disease epidemics and invasions by introduced plants.

We have outlined the key factors affecting novel plant-pathogen interactions and offered predictions for how they should influence the three forms of novel interaction: novel epidemics, invasive hosts, and biological control (Table 1B,C,D). However, these predictions are conjecture based largely on application of principles learned from studies in agriculture or a few native wild systems, with few studies from wild novel interactions themselves. Currently underutilized, studies of novel epidemics and novel host introductions should be used to illuminate these unresolved questions about the evolutionary ecology of plant-pathogen interactions.

In this final section we focus on the practical side of the three forms of novel interactions, outlining some ways in which an evolutionary ecology perspective is necessary for improving policy and management.

Novel Epidemics

Novel disease epidemics are best controlled through prevention. Effective trade regulations, quarantine policy, and land-use planning are essential tools in the prevention of novel epidemics. An increasingly globalized economy ensures a continued increase in the movement of plants and other materials between otherwise biologically isolated regions. Changing land-use patterns, including increasingly fragmented landscapes, often with close juxtaposition of remaining wildlands with low-diversity agricultural and forestry systems, may increase opportunities for the development of novel epidemics. We suggest that the answers to several questions are key to effectively mitigating the effects of novel epidemics. (*a*) Can we design more effective quarantine procedures by understanding which pathogen life history traits are most likely to lead to epidemic development? (*b*) Are considerations of ecological adaptations to novel environmental conditions as important as pathological attributes? (*c*) In creating lists of pathogens for quarantine exclusion, what is the relative importance of preadaptation to particular hosts versus the likelihood of evolutionary host shifts with ecological or economic consequence? (*d*) Could an understanding of the phylogenetic structure of plant communities and the importance of host phylogeny to host shifts be incorporated into land-use planning, in order to minimize the probability of novel epidemics arising at the borders between agricultural and wildlands?

Invasive Hosts

Although most introduced plants do not invade, the few that do represent one of our most challenging environmental problems (D'Antonio & Vitousek 1992, Office of Technology Assessment 1993, Vitousek et al. 1996, Parker et al. 1999, Mack et al.

2000). The central questions for plant invasion biologists are why most plants fail to establish viable populations when introduced into a new habitat, and why a few become noxious invasive weeds. The Biotic Resistance Hypothesis suggests that in natural habitats, native pests colonize the naive exotic hosts and eliminate them before they can become established (Elton 1958, Simberloff 1986, Mack 1996, Duncan & Williams 2002). In contrast, the Natural Enemies Hypothesis proposes that successful invaders leave behind their regulating insect pests and pathogens (Darwin 1859, Crawley 1987, Blossey & Nötzold 1995, Maron & Vila 2001, Siemann & Rogers 2001, Keane & Crawley 2002, Beckstead & Parker 2003). Because empirical studies testing both of these central ideas are surprisingly limited, there are significant unresolved issues for those trying to understand, manage, and prevent plant invasions. In particular: (*a*) How often do native pathogens contribute to biotic resistance in reducing the success of potential invaders, and what does this imply for debates about the value of biodiversity for ecosystem invasibility? (*b*) Is genetic variation for pathogen resistance particularly low for invasive plants relative to native species? (*c*) If so, why do we not see more dramatic epidemics emerging on high-density, invasive plants? (*d*) Is "escape from natural enemies" a common phenomenon in introduced plants, and does it contribute significantly to invasiveness? (*e*) Can we expect aggressive plant invader populations to eventually accumulate enough pathogens to reduce their ecological impacts on the invaded ecosystem?

Biological Control

Biological control of invasive weeds is unusual in that disease epidemics are the desired outcome. The short-term and long-term success of particular biological control introductions depends on host and pathogen numerical dynamics and the way these dynamics are influenced by density, frequency of diseased hosts, and genetic variation. In particular, (*a*) What is the effect of novel environmental conditions on the dynamics of the host-pathogen interaction? (*b*) Is control more successful in genetically depauperate weeds? (*c*) Do transmission rate and demographic impact of the pathogen decline as the host population declines? (*d*) Do pathogen and host reach a stable equilibrium or do they depend on metapopulation dynamics to persist in the landscape?

When the host is genetically depauperate and the pathogen has been chosen for its virulence, the host should be at a relative disadvantage. It is then somewhat surprising that we do not see more cases of spectacularly successful biological control with pathogens. Evolutionary change after introduction may help explain the varied effectiveness of biological controls, but such changes are almost unexplored; we could find no studies that have tracked changes in pathogen virulence or host resistance for a biological control system outside of *Chondrilla juncea* (Chaboudez & Sheppard 1995, Hanley & Groves 2002). Modeling efforts suggest that virulence of the pathogen, and whether it influences host survival or fecundity, will affect the evolution of resistance in the weed population and the long-term effectiveness of biological control agents (Thrall & Burdon 2004).

Because of the implications for policy and risk assessment, we also need to understand the ecological and evolutionary factors influencing shifts to nontarget hosts. Specifically, we need to ask (*a*) Does virulence of biological control pathogens change over time, and has this increased or decreased the success of control? (*b*) Has the host developed resistance over time? And finally, (*c*) Has host specificity changed over time, and are the shifts predictable based on phylogenetic distance from known hosts?

Novel plant-pathogen interactions pose significant threats to natural and managed ecosystems. A better understanding of the evolutionary ecology of such interactions in different contexts may improve management options. Simultaneously, novel epidemics, invasive hosts, and biological control efforts provide numerous underexploited opportunities to increase our understanding of the basic biology of novel plant-pathogen interactions.

ACKNOWLEDGMENTS

We thank D. Schemske, D. Futuyma, B. Ayala, K. Dlugosch, J. Hagen, C. Hays, B. Hardcastle, J. Hein, A. Herre, J. Hoeksema, R. Hufft, S. Lambrecht, S. Langridge, M. Los Huertos, D. Plante, W. Satterthwaite, B. Smarr, Y. Springer, and an anonymous reviewer for helpful discussions and critical comments on drafts of this manuscript. Preparation was supported in part by NSF grants DEB-9808501 to I.M.P., DEB-0096298 and DEB-0096398 to G.S.G., and USDA-NRI #2000-00891 to I.M.P.

The *Annual Review of Ecology, Evolution, and Systematics* is online at http://ecolsys.annualreviews.org

LITERATURE CITED

Abramovitch RB, Kim Y-J, Chen S, Dickman MB, Martin GB. 2003. *Pseudomonas* type III effector AvrPtoB induces plant disease susceptibility by inhibition of host programmed cell death. *EMBO J.* 22:60–69

Adams EB, Line RF. 1984. Epidemiology and host morphology in the parasitism of rush skeletonweed *Chondrilla juncea* by *Puccinia chondrillina*. *Phytopathology* 74:745–48

Adams MW, Ellingboe AH, Rossman EC. 1971. Biological uniformity and disease epidemics. *BioScience* 21:1067–70

Alexander HM, Holt RD. 1998. The interaction between plant competition and disease. *Perspect. Plant Ecol. Evol. Syst.* 1:206–20

Alexander HM, Thrall PH, Antonovics J, Jarosz AM, Oudemans PV. 1996. Population dynamics and genetics of plant disease: a case study of anther-smut disease. *Ecology* 77:990–96

Alexander M. 1981. Why microbial predators and parasites do not eliminate their prey and hosts. *Annu. Rev. Microbiol.* 35:113–33

Altizer S, Harvell D, Friedle E. 2003. Rapid evolutionary dynamics and disease threats to biodiversity. *Trends Ecol. Evol.* 18:549–604

Amsellem L, Noyer JL, Le Bourgeois T, Hossaert-McKey M. 2000. Comparison of genetic diversity of the invasive weed *Rubus alceifolius* Poir. (Rosaceae) in its native range and in areas of introduction, using amplified fragment length polymorphism (AFLP) markers. *Mol. Ecol.* 9:443–55

Anagnostakis SL. 1987. Chestnut blight: the

classical problem of an introduced pathogen. *Mycologia* 79:23–37
Anagnostakis SL. 1992. Measuring resistance of chestnut trees to chestnut blight. *Can. J. For. Res.* 22:568–71
Anderson RM, May RM. 1982. Coevolution of hosts and parasites. *Parasitology* 85:411–26
Aylor DE. 2003. Spread of plant disease on a continental scale: role of aerial dispersal of pathogens. *Ecology* 84:1989–97
Barrett JA. 1988. Frequency-dependent selection in plant-fungal interactions. *Philos. Trans. R. Soc. London Ser. B* 319:473–84
Barrett SCH, Husband BC. 1990. The genetics of plant migration and colonization. In *Plant Population Genetics, Breeding, and Genetic Resources*, ed. AHD Brown, MT Clegg, AL Kahler, BS Weir, pp. 254–78. Sunderland, MA: Sinauer
Beckstead J, Parker IM. 2003. Invasiveness of *Ammophila arenaria*: Release from soilborne pathogens? *Ecology* 84:2824–31
Bennett MH, Gallagher MDS, Bestwick CS, Rossiter JT, Mansfield JW. 1994. The phytoalexin response of lettuce to challenge by *Botrytis cinerea*, *Bremia lactucae* and *Pseudomonas syringae* pv. *phaseolicola*. *Physiol. Mol. Plant Pathol.* 44:321–33
Bergelson J, Purrington CB. 1996. Surveying patterns in the cost of resistance in plants. *Am. Nat.* 148:536–58
Bevan JR, Crute IR, Clarke DD. 1993. Diversity and variation in expression of resistance to *Erysiphe fischeri* in *Senecio vulgaris*. *Plant Pathol.* 42:647–53
Bishop JG, Dean AM, Mitchell-Olds T. 2000. Rapid evolution in plant chitinases: molecular targets of selection in plant-pathogen coevolution. *Proc. Natl. Acad. Sci. USA* 97:5322–27
Blaney CS, Kotanen PM. 2001. Effects of fungal pathogens on seeds of native and exotic plants: a test using congeneric pairs. *J. Appl. Ecol.* 38:1104–13
Blossey B, Nötzold R. 1995. Evolution of increased competition ability in invasive nonindigenous plants: a hypothesis. *J. Ecol.* 83:887–89

Bradley DJ, Gilbert GS, Parker IM. 2003. Susceptibility of clover species to fungal infection: the interaction of leaf surface traits and environment. *Am. J. Bot.* 90:857–64
Brasier CM. 2001. Rapid evolution of introduced plant pathogens via interspecific hybridization. *BioScience* 51:123–33
Brooks DR, McLennan DA. 1993. *Parascript: Parasites and the Language of Evolution*. Washington, DC: Smithson. Inst. Press
Brown JKM. 2002. Yield penalties of disease resistance in crops. *Curr. Opin. Plant Biol.* 5:339–44
Brown JKM. 2003. A cost of disease resistance: Paradigm or peculiarity? *Trends. Genet.* 19:667–71
Brown JKM, Hovmøller MS. 2002. Aerial dispersal of pathogens on the global and continental scales and its impact on plant disease. *Science* 297:537–41
Bull JJ. 1994. Perspective: virulence. *Evolution* 48:1423–37
Burdon JJ. 1991. Host-pathogen interactions in natural populations of *Linum marginale* and *Melampsora lini*: I. Patterns of resistance and racial variation in a large host population. *Evolution* 45:205–17
Burdon JJ, Chilvers GA. 1982. Host density as a factor in plant disease ecology. *Annu. Rev. Phytopathol.* 20:143–66
Burdon JJ, Groves RH, Cullen JM. 1981. The impact of biological control on the distribution and abundance of *Chondrilla juncea* in south-eastern Australia. *J. Appl. Ecol.* 18:957–66
Burdon JJ, Jarosz AM. 1992. Temporal variation in the racial structure of flax rust (*Melampsora lini*) populations growing on natural stands of wild flax (*Linum marginale*): local versus metapopulation dynamics. *Plant Pathol.* 41:165–79
Burdon JJ, Marshall DR. 1981. Biological control and the reproductive mode of weeds. *J. Appl. Ecol.* 18:649–58
Burdon JJ, Thompson JN. 1995. Changed patterns of resistance in a population of *Linum marginale* attacked by the rust pathogen *Melampsora lini*. *J. Ecol.* 83:199–206

Cao H, Li X, Dong XN. 1998. Generation of broad-spectrum disease resistance by overexpression of an essential regulatory gene in systemic acquired resistance. *Proc. Natl. Acad. Sci. USA* 95:6531–36

Chaboudez P, Sheppard AW. 1995. Are particular weeds more amenable to biological control? A re-analysis of mode of reproduction and life history. In *Biological Control of Weeds: Proc. 7th Int. Symp. Biol. Control Weeds*, ed. ES Delfosse, RR Scott, pp. 95–102. Melbourne, Aust.: CSIRO

Chang YC, Borja M, Scholthof HB, Jackson AO, Morris TJ. 1995. Host effects and sequences essential for accumulation of defective interfering RNAs of cucumber necrosis and tomato bushy stunt tombusviruses. *Virology* 210:41–53

Charudattan R, Dinoor A. 2000. Biological control of weeds using plant pathogens: accomplishments and limitations. *Crop Prot.* 19:691–95

Chin KM, Wolfe MS. 1984. Selection on *Erysiphe graminis hordei* in pure and mixed stands of barley. *Plant Pathol.* 33:89–100

Coutinho TA, Wingfield MJ, Alfenas AC, Crous PW. 1998. Eucalyptus rust: a disease with the potential for serious international implications. *Plant Dis.* 82:819–25

Crawley MJ. 1987. What makes a community invasible? In *Colonization, Succession and Stability: 26th Symp. Br. Ecol. Soc.*, ed. AJ Gray, MJ Crawley, PJ Edwards, pp. 429–54. Oxford, UK: Blackwell Sci.

Crow JF. 2002. Perspective: Here's to Fisher, additive genetic variance, and the fundamental theorem of natural selection. *Evolution* 56:1313–16

D'Antonio CM, Vitousek PM. 1992. Biological invasions by exotic grasses, the grass/fire cycle and global change. *Annu. Rev. Ecol. Syst.* 23:63–87

D'Arcy CJ, Eastburn DM, Schumann GL. 2001. Illustrated glossary of plant pathology. In *The Plant Health Instructor, DOI: 10.1094/PHI-1-2001-0219-01*

Darwin C. 1859. *On the Origin of Species by Means of Natural Selection, or, The Preservation of Favoured Races in the Struggle for Life*. London: Murray

Davelos AL, Alexander HM, Slade NA. 1996. Ecological genetic interactions between a clonal host plant (*Spartina pectinata*) and associated rust fungi (*Puccinia seymouriana* and *Puccinia sparganioides*). *Oecologia* 105:205–13

Delfosse ES, Scott RR, eds. 1996. *Biological Control of Weeds*. Melbourne, Aust.: CSIRO

Duncan RP, Williams PA. 2002. Darwin's naturalization hypothesis challenged. *Nature* 417:608–9

Dybdahl MF, Lively CM. 1995. Host-parasite interactions: infection of common clones in natural populations of a freshwater snail (*Potamopyrgus antipodarum*). *Proc. R. Soc. London Ser. B* 260:99–103

Eckenwalder JE, Heath MC. 2001. The evolutionary significance of variation in infection behavior in two species of rust fungi on their hosts and related nonhost plant species. *Can. J. Bot.* 79:570–77

Ellis J, Dodds P, Pryor T. 2000. Structure, function, and evolution of plant resistance genes. *Curr. Opin. Plant Biol.* 3:278–84

Elton CS. 1958. *The Ecology of Invasions by Animals and Plants*. London: Methuen

Escriu F, Fraile A, García-Arenal F. 2000. Evolution of virulence in natural populations of satellite RNA of *Cucumber mosaic virus*. *Phytopathology* 90:480–85

Escriu F, Fraile A, García-Arenal F. 2003. The evolution of virulence in a plant virus. *Evolution* 57:755–65

Ewald PW. 1993. *Evolution of Infectious Disease*. Oxford: Oxford Univ. Press

Farr DF, Rossman AY, Palm ME, McCray EB. 2004. *Fungus-Host Distributions, Fungal Databases, Systematic Botany and Mycology Lab., ARS/USDA*. http://nt.ars-grin.gov/fungaldatabases/

Fenner F, Fantini B. 1999. *Biological Control of Vertebrate Pests: the History of Myxomatosis—An Experiment in Evolution*. Wallingford, UK: CABI

Fisher RA. 1930. *The Genetical Theory of Natural Selection*. Oxford: Clarendon

Fitt BDL, Gregory PH, Todd AD, McCartney HA, MacDonald OC. 1987. Spore dispersal and plant disease gradients; a comparison between two empirical models. *J. Phytopathol.* 118:227–42

Flor HH. 1956. The complementary genic systems in flax and flax rust. *Adv. Genet.* 8:29–54

Fox DT, Williams PH. 1984. Correlation of spore production by *Albugo candida* on *Brassica campestris* and a visual white rust rating-scale. *Can. J. Plant Pathol.* 6:175–78

Garbelotto M, Davidson JM, Ivors K, Maloney PE, Hüberli D, et al. 2003. Non-oak native plants are main hosts for sudden oak death pathogen in California. *Calif. Agric.* 57:18–23

George HL, VanEtten HD. 2001. Characterization of pisatin-inducible cytochrome P450s in fungal pathogens of pea that detoxify the pea phytoalexin pisatin. *Fungal Gen. Biol.* 33:37–48

Gilbert GS. 2002. Evolutionary ecology of plant diseases in natural ecosystems. *Annu. Rev. Phytopathol.* 40:13–43

Goodell K, Parker IM, Gilbert GS. 2000. Biological impacts of species invasions: implications for policy makers. In *Incorporating Biological, Natural, and Social Sciences in Sanitary and Phytosanitary Standards in International Trade*, ed. J Caswell, pp. 87–117. Washington, DC: Natl. Acad. Press

Grant MR, Godiar L, Straube E, Ashfield T, Lewald J, et al. 1995. Structure of the *Arabidopsis RPM1* gene enabling dual specificity disease resistance. *Science* 269:843–46

Grant MW, Archer SA. 1983. Calculation of selection coefficients against unnecessary genes for virulence from field data. *Phytopathology* 73:547–51

Hafner MS, Sudman PD, Villablanca FX, Spradling TA, Demastes JW, Nadler SA. 1994. Disparate rates of molecular evolution in cospeciating hosts and parasites. *Science* 265:1087–90

Hanley ME, Groves RH. 2002. Effect of the rust fungus *Puccinia chondrillina* TU 788 on plant size and plant size variability in *Chondrilla juncea*. *Weed Res.* 42:370–76

Harlan JR. 1976. Diseases as a factor in plant evolution. *Annu. Rev. Phytopathol.* 14:31–51

Hasan S. 1972. Specificity and host specialization of *Puccinia chondrillina*. *Ann. Appl. Biol.* 72:257–63

Hasan S, Chaboudez P, Espiau C. 1996. Isozyme patterns and susceptibility of North American forms of *Chondrilla juncea* to European strains of the rust fungus *Puccinia chondrillina*. See Delfosse & Scott 1996, pp. 367–73

Heap IM. 1997. The occurrence of herbicide-resistant weeds worldwide. *Pestic. Sci.* 51:235–43

Hernandez C, Carette JE, Brown DJF, Bol JF. 1996. Serial passage of tobacco rattle virus under different selection conditions results in deletion of structural and nonstructural genes in RNA 2. *J. Virol.* 70:4933–40

Hoisington D, Khairallah M, Reeves T, Ribaut J-M, Skovmand B, et al. 1999. Plant genetic resources: What can they contribute toward increased crop productivity? *Proc. Natl. Acad. Sci. USA* 96:5937–43

Holden ANG, Mahlberg PG. 1996. Rusts for the biological control of leafy spurge (*Euphorbia esula*) in North America. See Delfosse & Scott 1996, pp. 419–24

Holub EB. 1997. Organization of resistance genes in *Arabidopsis*. In *The Gene-for-Gene Relationship in Plant-Parasite Interactions*, ed. IR Crute, EB Holub, JJ Burdon, pp. 5–26. Wallingford, UK: CABI

Holub EB. 2001. The arms race is ancient history in *Arabidopsis*, the wildflower. *Nat. Rev. Genet.* 2:516–27

Imhoff MW, Leonard KJ, Main CE. 1982. Patterns of bean rust lesion size increase and spore production. *Phytopathology* 72:441–46

Jarosz AM, Davelos AL. 1995. Effects of disease in wild plant populations and the evolution of pathogen aggressiveness. *New Phytol.* 129:371–87

Jenner CE, Wang X, Ponz F, Walsh JA. 2002. A fitness cost for *Turnip mosaic virus* to overcome host resistance. *Virus Res.* 86:1–6

Jia Y, McAdams SA, Bryan GT, Hershey HP,

Valent B. 2000. Direct interaction of resistance gene and avirulence gene products confers rice blast resistance. *EMBO J.* 19:4004–14

Johnson J. 1947. Virus attenuation and the separation of strains by specific hosts. *Phytopathology* 37:822–37

Kaltz O, Shykoff JA. 1998. Local adaptation in host-parasite systems. *Heredity* 81:361–70

Keane RM, Crawley MJ. 2002. Exotic plant invasions and the enemy release hypothesis. *Trends Ecol. Evol.* 17:164–70

Kover PX, Clay K. 1998. Trade-off between virulence and vertical transmission and the maintenance of a virulent plant pathogen. *Am. Nat.* 152:165–75

Kranz J. 1990. Tansley Review No. 28: Fungal diseases in multispecies plant communities. *New Phytol.* 116:383–406

Laugé R, De Wit PJGM. 1998. Fungal avirulence genes: structure and possible functions. *Fungal Gen. Biol.* 24:285–97

Leach JE, White FF. 1996. Bacterial avirulence genes. *Annu. Rev. Phytopathol.* 34:153–79

Lee CE. 2002. Evolutionary genetics of invasive species. *Trends Ecol. Evol.* 17:386–91

Lenski RE. 1988. Experimental studies of pleiotropy and epistasis in *Escherichia coli* II. Compensation for maladaptive effects associated with resistance to virus T4. *Evolution* 42:433–40

Lenski RE, May RM. 1994. The evolution of virulence in parasites and pathogens: reconciliation between two competing hypotheses. *J. Theor. Biol.* 169:253–65

Leonard KJ. 1969. Selection in heterogeneous populations of *Puccinia graminis* f. sp. avenae. *Phytopathology* 59:1851–57

Levin BR, Perrot V, Walker N. 2000. Compensatory mutations, antibiotic resistance and the population genetics of adaptive evolution in bacteria. *Genetics* 154:985–97

Mack RN. 1996. Biotic barriers to plant naturalization. In *Proc. 9th Int. Symp. Biol. Control Weeds*, ed. VC Moran, JH Hoffman, pp. 39–46. Stellenbosch, S. Afr.: Univ. Cape Town

Mack RN, Simberloff D, Lonsdale WM, Evans H, Clout M, Bazzaz FA. 2000. Biotic invasions: causes, epidemiology, global consequences, and control. *Ecol. Appl.* 10:689–710

Maron JL, Vila M. 2001. When do herbivores affect plant invasion? Evidence for the natural enemies and biotic resistance hypotheses. *Oikos* 95:361–73

Martin GB, Bogdanove AJ, Sessa G. 2003. Understanding the functions of plant disease resistance proteins. *Annu. Rev. Plant Biol.* 54:23–61

Mauricio R. 2001. Natural selection and the joint evolution of tolerance and resistance as plant defenses. *Evol. Ecol.* 14:491–507

McDonald GI, Hoff RJ. 2001. Blister rust: an introduced plague. In *Whitebark Pine Communities. Ecology and Restoration*, ed. DF Tomback, SF Arno, RE Keane, pp. 193–220. Washington, DC: Island Press

McVean DN. 1966. Ecology of *Chondrilla juncea* L. in southeastern Australia. *J. Ecol.* 54:345–65

Mitchell CE, Power AG. 2003. Release of invasive plants from fungal and viral pathogens. *Nature* 421:625–27

Mundt CC. 2002. Use of multiline cultivars and cultivar mixtures for disease management. *Annu. Rev. Phytopathol.* 40:381–410

Nimchuk Z, Eulgem T, Holt BF III, Dangl JL. 2003. Recognition and response in the plant immune system. *Annu. Rev. Genet.* 37:579–609

Off. Technol. Assess. 1993. *Harmful Non-Indigenous Species in the United States. Rep. OTA-F-565*, US Congr.

Mondragón-Palomino M, Meyers BC, Michelmore RW, Gaut BS. 2002. Patterns of positive selection in the complete *NBS-LRR* gene family of *Arabidopsis thaliana. Genome Res.* 12:1305–15

Panetta FD, Dodd J. 1995. *Chondrilla juncea.* In *The Biology of Australian Weeds*, ed. RH Groves, RCH Shepherd, RG Richardson, 1:67–84. Melbourne, Aust.: Richardson

Parker IM, Simberloff D, Lonsdale WM, Goodell K, Wonham M, et al. 1999. Impact: toward a framework for understanding the ecological effects of invaders. *Biol. Invasions* 1:3–19

Parker JE, Schulte W, Hahlbrock K, Scheel D. 1991. An extracellular glycoprotein from *Phytophthora megasperma* f. sp. *glycinea* elicits phytoalexin synthesis in cultured parsley cells and protoplasts. *Mol. Plant Microbe. Int.* 4:19–27

Parker MA. 1985. Local population differentiation for compatibility in an annual legume *Amphicarpaea bracteata* and its host-specific fungal pathogen *Synchytrium decipiens*. *Evolution* 39:713–23

Parker MA. 1988. Polymorphism for disease resistance in the annual legume *Amphicarpaea bracteata*. *Heredity* 60:27–31

Parker MA. 1989. Disease impact and local genetic diversity in the clonal plant *Podophyllum peltatum*. *Evolution* 43:540–47

Parker MA. 1990. The pleiotropy theory for polymorphism of disease resistance genes in plants. *Evolution* 44:1872–75

Parker MA. 1991a. Local genetic differentiation for disease resistance in a selfing annual. *Biol. J. Linn. Soc.* 42:337–50

Parker MA. 1991b. Nonadaptive evolution of disease resistance in an annual legume. *Evolution* 45:1209–17

Parlevliet JE. 1981. Stabilizing selection in crop pathosystems: An empty concept or a reality? *Euphytica* 30:259–69

Perrings C, Williamson M, Barbier EB, Delfino D, Dalmazzone S, et al. 2002. Biological invasion risks and the public good: an economic perspective. *Conserv. Ecol.* 6:Article 1 (online). http://www.consecol.org/vol6/iss1/art1

Rizzo DM, Garbelotto M. 2003. Sudden oak death: endangering California and Oregon forest ecosystems. *Front. Ecol. Environ.* 1:197–204

Robert C, Bancal M-O, Lannou C. 2002. Sheat leaf rust uredospore production and carbon and nitrogen export in relation to lesion size and density. *Phytopathology* 92:762–68

Roy BA. 1993. Patterns of rust infection as a function of host genetic diversity and host density in natural populations of the apomictic crucifer, *Arabis holboellii*. *Evolution* 47:111–24

Roy BA. 2001. Patterns of association between crucifers and their flower-mimic pathogens: Host jumps are more common than coevolution or cospeciation. *Evolution* 55:41–53

Roy BA, Kirchner JW. 2000. Evolutionary dynamics of pathogen resistance and tolerance. *Evolution* 54:51–63

Saile E, McGarvey JA, Schell MA, Denny TP. 1997. Role of extracellular polysaccharide and endoglucanase in root invasion and colonization of tomato plants by *Ralstonia solanacearum*. *Phytopathology* 87:1264–71

Schafer JF, Roelfs AP. 1985. Estimated relation between numbers of urediniospores of *Puccinia graminis* f. sp. *tritici* and rates of occurrence of virulence. *Phytopathology* 75:749–50

Schmid B. 1994. Effects of genetic diversity in experimental stands of *Solidago altissima*— evidence for the potential role of pathogens as selective agents in plant populations. *J. Ecol.* 82:165–75

Schrag SJ, Wiener P. 1995. Emerging infectious disease: What are the relative roles of ecology and evolution? *Trends Ecol. Evol.* 10:319–24

Secord D, Kareiva P. 1996. Perils and pitfalls in the host specificity paradigm. *BioScience* 46:448–53

Siemann E, Rogers WE. 2001. Genetic differences in growth of an invasive tree species. *Ecol. Lett.* 4:514–18

Simberloff D. 1986. Introduced insects: a biogeographic and systematic perspective. In *Ecology of Biological Invasions of North American and Hawaii*, ed. HA Mooney, JA Drake, pp. 3–26. New York: Springer-Verlag

Simberloff D, Gibbons L. 2003. Now you see them, now you don't!—Population crashes of established introduced species. *Biol. Invasions* 6:161–72

Simms EL. 1996. The evolutionary genetics of plant-pathogen systems. *BioScience* 46:136–45

Smalley EB, Guries RP. 1993. Breeding elms for resistance to Dutch elm disease. *Annu. Rev. Phytopathol.* 31:325–52

Smedegaard-Petersen V, Stølen O. 1981. Effect of energy-requiring defense reactions on yield and grain quality in a powdery mildew-resistant barley cultivar. *Phytopathology* 71:396–99

Statler GD, Jones DA. 1981. Inheritance of virulence and uredial color and size in *Puccinia recondita tritici*. *Phytopathology* 71:652–55

Takabayashi N, Tosa Y, Oh HS, Mayama S. 2002. A gene-for-gene relationship underlying the species-specific parasitism of *Avena/Triticum* isolates of *Magnaporthe grisea* on wheat cultivars. *Phytopathology* 92:1182–88

Tamada T, Kusume T. 1991. Evidence that the 75k readthrough protein of beet necrotic yellow vein virus RNA-2 is essential for transmission by the fungus *Polymyxa betae*. *J. Gen. Virol.* 72:1497–504

Thompson JN. 1994. *The Coevolutionary Process*. Chicago: Univ. Chicago Press. 387 pp.

Thrall PH, Burdon JJ. 2000. Effect of resistance variation in a natural plant host-pathogen metapopulation on disease dynamics. *Plant Pathol.* 49:767–73

Thrall PH, Burdon JJ. 2003. Evolution of virulence in a plant host-pathogen metapopulation. *Science* 299:1735–37

Thrall PH, Burdon JJ. 2004. Host-pathogen life-history interactions affect the success of biological control. *Weed Technol.* In press

Tian D, Traw MB, Chen JQ, Kreitman M, Bergleson J. 2003. Fitness costs of R-gene-mediated resistance in *Arabidopsis thaliana*. *Nature* 423:74–77

Tsiamis G, Mansfield JW, Hockenhull R, Jackson RW, Sesma A, et al. 2000. Cultivar-specific avirulence and virulence functions assigned to avrPphF in *Pseudomonas syringae* pv. *phaseolicola*, the cause of bean halo-blight disease. *EMBO J.* 19:3204–14

Tyler BM. 2002. Molecular basis of recognition between *Phytophthora* pathogens and their hosts. *Annu. Rev. Phytopathol.* 40:137–67

Vanderplank JE. 1968. *Disease Resistance in Plants*. New York: Academic

Vera Cruz CM, Bai J, Oña I, Leung H, Nelson RJ, Mew T-W. 2000. Predicting durability of a disease resistance gene based on an assessment of the fitness loss and epidemiological consequences of avirulence gene mutation. *Proc. Natl. Acad. Sci. USA* 97:13500–5

Vitousek PM, D'Antonio CM, Loope LL, Westbrooks R. 1996. Biological invasions as global environmental change. *Am. Sci.* 84:468–78

Watson IA. 1970. Changes in virulence and population shifts in plant pathogens. *Annu. Rev. Phytopathol.* 8:209–30

Watson IA, Singh D. 1952. The future for rust resistant wheat in Australia. *J. Aust. Inst. Agric. Sci.* 18:190–97

Welz HG, Leonard KJ. 1994. Genetic analysis of two race 0 × race 2 crosses in *Cochliobolus carbonum*. *Phytopathology* 84:83–91

Weste G, Marks GC. 1987. The biology of *Phytophthora cinnamomi* in Australasian forests. *Annu. Rev. Phytopathol.* 25:207–29

Wevelsiep L, Rüpping E, Knogge W. 1993. Stimulation of barley plasmalemma H^+-ATPase by phytotoxic peptides from the fungal pathogen *Rhynchosporium secalis*. *Plant Physiol.* 101:297–301

Williams GC, Hesse RM. 1991. The dawn of Darwinian medicine. *Q. Rev. Biol.* 66:1–22

Wingfield MJ, Slippers B, Roux J, Wingfield BD. 2001. Worldwide movement of exotic forest fungi, especially in the tropics and the southern hemisphere. *BioScience* 51:134–40

Wolpert TJ, Dunkle LD, Cuiffetti LM. 2002. Host-selective toxins and avirulence determinants: What's in a name? *Annu. Rev. Phytopathol.* 40:251–85

Xiao SY, Ellwood S, Calis O, Patrick E, Li TX, et al. 2001. Broad-spectrum mildew resistance in Arabidopsis thaliana mediated by *RPW8*. *Science* 291:118–20

Zhan J, Mundt CC, Hoffer ME, McDonald BA. 2002. Local adaptation and effect of host genotype on the rate of pathogen evolution: an experimental test in a plant pathosystem. *J. Evol. Biol.* 15:634–47

Zwolinski JB, Swart MJ, Wingfield MJ. 1990. Economic impact of post-hail outbreak of die-back induced by *Sphaeropsis sapinea*. *Eur. J. For. Pathol.* 20:405–11

Subject Index

A

Adaptive radiation
 in Southwest Australian
  Floristic Region
  (SWAFR) plants, 638
Africa
 Greater Cape Floristic
  Region in South Africa,
  642–43
 Pliocene mammal record,
  303–4
Agricultural land
 rivers and, 263–66, 277
 stream condition and,
  272–73
 See also Crops; Farmland
  diversity
Alzheimer's disease, 351
Ambulacraria, 236–37
Amphibians
 widespread decline of, 92
 See also Salamanders
Amphipod
 genetically based variation
  in, 350, 355
Angiosperm(s)
 Cretaceous, 16–20
 Southwest Australian
  Floristic Region
  (SWAFR), 627
 vertebrate dispersal and,
  1–2, 6–9, 11–13, 15, 16,
  18–20
Animal(s)
 Cenozoic era
  vertebrate dispersal and,
   14–15
 cognition
  evolutionary biology of,
   347–67
 introduced, 43

Mesozoic era
 vertebrate dispersal and,
  9–14
multispecies plant-animal
 interactions, 435–59
Paleozoic era
 vertebrate dispersal and,
  7–8
phylogeny, 229–48
species
 cascading effects of deer
  on, 113, 126–30
 vertebrate dispersal and, 5
Annelida, 239–40
Ant(s)
 dispersal and, 3, 6
 nectar-thieving, 440
 seed dispersal and, 444,
  599
Anthropogenic change
 ocean disease and, 38–44,
  47
 See also Anthropogenic
  land use; Anthropogenic
  stress(es); Human actions;
  Human activities;
  Humans
Anthropogenic land use
 stream ecosystems and,
  268–69
 See also Anthropogenic
  change; Anthropogenic
  stress(es); Human actions;
  Human activities;
  Humans
Anthropogenic stress(es)
 ecological indicators and,
  91
 ecological responses to, 94
 marine ecosystems and,
  37–38

organism response to,
 94–95
See also Anthropogenic
 change; Anthropogenic
 land use; Human actions;
 Human activities;
 Humans
Antipredator behavior, 364
 evolution of, 361
 fitness and, 357
Aquatic communities
 mutualisms in, 175–91
Aquatic habitats
 salamanders and, 410–12,
  415, 417, 419–20, 426
Archaen
 oceans, 547
Aridity
 in Australia, 623, 624
Armor
 eukaryotic phytoplankton
  and, 535
Articulata
 hypothesis, 245
Assemblages
 persistence of, 285–314
Atmospheric circulation
 patterns, 313
Attention
 in animals, 348
 genetic variation in, 352
Attention deficit hyperactivity
 disorder (ADHD), 352
Australia
 irreversible regime shifts
  in, 567
 reporting on ecological
  indicators, 90
 salinization of land and
  rivers in, 574
 Southwest Australian

701

702 SUBJECT INDEX

Floristic Region (SWAFR), 623–44
Australian plate, 631
Avian extinctions, 323–39
Avifauna
  forest, 336
  lowland, 326–27
  tropical, 338–39
Avirulence
  plant diseases and, 679

B

Bacteria
  mutualistic, 183
Baltic Sea
  regime shifts and, 563
Barro Colorado Island, Panama
  bird species and, 325, 329, 330, 335
Bats
  dispersal and, 15
Bee(s)
  Africanized honey bees, 443–44
  pollination, 440
  transgenic crops and, 152
  See also Bumblebees
Behavioral characteristics
  extinction proneness and, 331–33
Benthic invertebrate faunas, 304–9
Best management practices (BMPs)
  stream ecosystems and, 275–76
Bilateria, 229, 234–35
Biodiversity
  conservation, 642
  deep impacts on, 133
  deer management and, 134
  ecosystem renewal and restoration and, 571–72
  ecosystem resilience dynamics and, 569–73

hot spots, 625, 640, 642, 643
  indigenous people and, 641
  neutral theory of biogeography and, 294–95
  river systems and, 260
  tropics and subtropics and, 324
Biofacies, 289–91, 300
Biogeochemical cycles
  plankton and, 543–47
Biogeographic patterns, 314
Biogeographical tracks
  Southwest Australian Floristic Region (SWAFR), 638–39
Biogeography
  extinction proneness and, 329–30
  neutral theory of biodiversity and, 294–95
Bioindicators, 94
Bioinformatics, 103
Biological communities
  ecological indicators and, 100–1
Biological control
  agent(s), 676, 677, 688, 693
  genetic analysis and, 691
  of invasive weeds, 693–94
  pathogens, 688, 690
Biological uniformitarianism
  vertebrate dispersal and, 2, 4, 16–17
Biology
  conservation, 640–42
Biomarkers
  pollution stress and, 102
  stressors and, 95
Biome(s)
  climatic changes and, 312
  Late Paleozoic tropics, 301
Bird(s)
  bird community index (BCI), 97

Breeding Bird Survey (BBS), 99–100
  deer browsing and, 607
  extinctions from tropical forests, 323–39
  guilds of, 97
  indicator species and, 98
  learning ability, 359
  population
    transgenic crops and, 154–55
  predictive edge response model and, 503–6
  vertebrate dispersal and, 11–12, 14
Bird song(s), 55–77
  causes of evolution, 63–69
  cultural drift, 64
  cultural selection, 65
  evolution
    intrasexual selection, 56, 68
    natural selection, 65–67, 76
    sexual selection, 56, 67–70, 76, 77
  genetic drift, 64–65
  geographic variation in, 59
  mechanisms, 57, 59–66, 76–77
  memes, 59–60, 63, 76, 77
  pure-tonal structure, 75–76
  repertoire size, 69–72
  song rate, 72–73, 76
  song trait evolution, 69–76
  song traits, 56, 59, 76
  substrates of, 57–63
  trill performance, 73–75
Bivalves, 177–78
Blowflies
  learning and, 351
Body size
  extinction and, 330–31
Bodyguards
  mutualistic, 184–85
Brachiopoda, 229, 241, 247
Brain parts

SUBJECT INDEX 703

volume of, 360
Bryozoa, 229, 241
Budworm
  forest regime shifts and, 566
Bulldozing
  minimization of, 641
Bumblebees
  alterations in host use, 444
  transgenic crops and, 152
Burrows
  salamanders and, 421–22
Butterflies
  predictive edge response model and, 503–6
  umbrella species and, 98

C

Calcium (Ca)
  ocean chemistry and, 539–40
Cambrian
  acritarchs, 542
  eukaryotic phytoplankton, 530
  explosion in the fossil record, 245–46
Canada
  reporting on ecological indicators, 90
Canopy gaps
  woodland herbs and, 605–7, 609
Carbon (C)
  coral reefs and, 178
  cycle
    phytoplankton and, 544–46
  cycling
    deer overabundance and, 113, 126
  isotope
    phytoplankton and, 544–46
  mycorrhizal fungi and, 450
  salamanders and, 418, 419
  spawning salmon

and, 190–91
Carbon dioxide (CO_2)
  marine phytoplankton and, 544, 546, 547
  salamanders and, 419
Carbonate compensation depth (CCD), 544
Carboniferous
  Southwest Australian Floristic Region (SWAFR) flora, 637
  wetland assemblages, 299–301
Caribbean
  sea urchin in, 563–64
Carnivores
  vertebrate dispersal and, 8
Cats
  ground-nesting birds and, 328
Cenozoic
  angiosperms, 9
  diatom(s) and, 532–34, 541, 542, 544–45
  Early Cenozoic Thermal Maximum (ECTM), 302
  ecological dynamics, 20
  era
    vertebrate dispersal and, 14–15
  marine faunas
    phytoplankton and, 540–42
  Southwest Australian Floristic Region (SWAFR), 623, 632, 637
Chiton
  seaweed and, 185–86
Chlorohydrocarbons
  salamanders and, 420
Chordates, 229, 236–37
Chromosome repatterning
  in Southwest Australian Floristic Region (SWAFR) plants, 639
Chronic-wasting disease (CWD), 118–19

Chronofauna, 293
Cladistic haplotype aggregation (CHA)
  species delimitation and, 210–11, 213, 217–19
Clean Water Act of 1972, 90
Climate(s)
  deer overabundance and, 116
  modes of dispersal and, 18
  tracking, 312
Climate change(s)
  coral disease model and, 38
  environmental tracking, 312
  global, 642
  ocean disease and, 35
  oysters and, 45
  vegetational persistence and, 300
Climate warming
  ocean disease and, 31, 39–40, 47
  See also Global warming
Climatic forcing
  interspecific synchrony and, 479
Cnidarians, 229, 233–34
Coal beds
  fossils in, 299–300
Coastal development
  ocean disease and, 42
Coccolithophores, 523, 525, 530–33, 535, 538–40, 547
Coevolution
  diffuse, 436
  plant-pathogen interactions and, 690–91
  vertebrate dispersal and, 3, 5, 16–17, 20
Coevolutionary dynamics
  plant-pathogen, 688–91
Cognitive traits
  effects on evolution, 365
  evolution of, 348, 357–61, 366

704 SUBJECT INDEX

fitness and, 347–48, 354–57, 366
genetic variation in, 347–54, 364–66
phenotypic plasticity of, 362–64, 366
Color vision
  evolution of, 358
  individual variation, 348–49
Communities
  ecological indicators and, 94
  See also Community
Community
  assemblage persistence, 292–95
  composition
    effects on plants, 455
  drift, 295
  dynamics
    modeling, 294–95
  organization
    organismal and individualistic, 286
  paleocommunity, 290–92
  stratigraphic persistence, 293–94
  structure
    edge responses and, 514
    mutualisms and, 175–91
  See also Communities
Competition
  mutualisms and, 188–89, 191
  woodland herbs and, 594
Complexity
  ecological, 288, 289
Composite species concept (CSC), 208–10, 218
Conservation
  Southwest Australian Floristic Region (SWAFR) and, 640–42
Coprolites
  vertebrate dispersal and, 4, 12, 13

Coral(s)
  bleaching, 32, 37, 40, 47
  disease(s), 32, 34–38, 40, 44, 47
  mutualism with crabs, 184–85
  See also Coral reef(s)
Coral reef(s)
  assemblage persistence and, 313, 314
  Coral Reef Monitoring Project, 34
  decline in, 32
  of the Florida Keys, 32, 34, 38, 44
  functional groups and, 571–72
  mutualisms and, 178–80, 188–89
  Quaternary, 309–10
  regime shifts and, 563–64
  See also Coral(s); Reefs
Correlated distance matrices
  species delimitation and, 205–6, 218
Crabs
  mutualism with corals, 184–85
Cretaceous
  angiosperms, 9, 13, 16–20
  birds, 12
Cretaceous-Paleogene boundary
  insect herbivore damage, 302
Cretaceous-Tertiary (K/T) boundary
  biotic dispersal and, 19–20
  mass extinction event, 532
  extinction event, 302
  fleshy fruits, 13–14
  Southwest Australian Floristic Region (SWAFR), 623, 624, 626, 631, 632

tropics, 18
Crops
  deer damage to, 117–18
  transgenic, 149–64
Ctenophores, 229, 233–34
Cultural drift
  in bird song memes, 64
Cytogeography
  Southwest Australian Floristic Region (SWAFR), 637–38

D

Damselfishes
  seaweeds and, 186
Dasypogonales
  Southwest Australian Floristic Region (SWAFR), 623, 627–28
DDT
  species decline and, 92
Decision making
  animals and, 348
  evolution of, 360–61
  fitness and, 356–57
  genetic variation in, 352–54
  phenotypic plasticity and, 364, 366
Deer, 607–8
  abundance
    woodland herbs and, 607–8
  keystone species, 607
  limiting impacts of, 134–36
  management, 113, 132–34
  overabundance
    assessing ecological effects of, 119–21
    causes of, 115–16
    defined, 114
    ecological consequences of, 121–32
    ecological impacts of, 113–36
    research needs, 132–34
    social and economic

SUBJECT INDEX 705

consequences of,
  117–19
Deforestation
  avian extinctions and,
  323–33, 335, 338, 339
Dentition
  diet and, 15–16
  vertebrate dispersal and,
  4–5
Detrius-litter food webs
  salamanders and, 417–19
Deuterostomia, 229, 235–38
Devonian
  phytoplankton and, 530
Dialects
  bird song, 59–60, 76
Diatom(s), 523, 525, 530,
  531, 533, 538–39, 542,
  547
  Cenozoic and, 532–34,
  541, 542, 544–45
  silica bioavailability and,
  534–35
Dinoflagellates, 523, 525,
  530–33, 535, 538–39,
  542, 547
Dinosaurs
  vertebrate dispersal and, 11
Disease(s)
  coral, 32, 34–37, 40, 44, 47
  extinctions and, 324
  ocean, 31–48
  wildlife, 118–19
Dispersal
  consumers as agents of,
  187–88
  events, 454, 632
  persistence and, 628
  population synchrony and,
  469, 478–81
  vertebrate, 1–20
  coevolution and, 16–17
  origins of, 5–6
Distemper virus (CDV)
  in dogs and marine
  mammals, 43
Disturbances

stream ecosystems and,
  261–63
Diversity
  Southwest Australian
  Floristic Region
  (SWAFR), 627
Dogs
  avian extinctions and, 328
Dynamical systems theory
  phase locking in, 476

E

Early Cenozoic Thermal
  Maximum (ECTM), 302
Ecdysozoa, 229, 243–45, 248
Echinoderms, 229, 236–37
Ecological benefits
  of transgenic crops, 164
Ecological concepts and data
  versus paleological
  concepts and data, 287–92
Ecological consequences
  of transgene escape into
  wild populations, 161–63,
  165
Ecological flows
  habitat edges and, 491,
  497–98, 500–1, 515
Ecological functions
  of salamanders, 405–26
Ecological impacts
  of deer overabundance,
  113–36
  of extinctions, 335–36
Ecological indicators, 89–104
  applications of, 99–100
  biological communities
  and, 100–1
  criticisms of, 101–2
  defined, 91
  economic indicators and,
  103–4
  future of, 102–4
  monitoring program, 95
  species, 95–99
  use of, 90, 92–95
Ecological integrity

indicators of, 274
Ecological locking, 294
Ecological monitoring
  programs, 95, 99
Ecological persistence,
  285–314
Ecological pleiotropy, 439,
  458
Ecological response
  to habitat edges, 491–515
Ecological risks
  of transgene escape into
  wild populations, 149,
  156–61, 165
  transgenic crops and,
  149–51, 164
Ecological specialization,
  383–84
  evolutionary specialization
  and, 380–81
Ecological stasis, 285–314
Ecologic-Evolutionary Units
  (EEUs), 292, 293
Ecologists
  paleontologists and,
  286–87
Ecology
  basic terms in, 288–89
  paleoecology and, 310–11
  restoration, 641
  wildlife, 113
  woodland herbs and,
  583–610
Ecomorphs, 289, 291
Economic indicators
  ecological indicators and,
  103–4
  environmental indicators
  and, 92
Ecosystem(s)
  effects of deer population
  on, 125–26
  functions
  salamanders and, 414,
  424
  human actions and, 558,
  559, 564–67, 573–75

management, 134, 573–75
regime shifts in, 557–75
salamanders and, 423
structure
  deer impacts on, 114–15
  trophic cascades in, 569
Ectomycorrhizal fungi, 450
Edge responses
  conceptual framework,
    495–97
  depth of edge influence
    (DEI) and, 512–13
  edge contrast and, 509–10,
    515
  edge orientation and,
    507–8, 515
  extrapolating to larger
    scales, 494, 511–15
  interactions affecting, 494,
    506–11
  mechanistic model of, 491,
    493–501, 515
  multiple edge effects and,
    513–14
  population and community
    consequences, 514
  predictive model of, 491,
    494, 501–6, 515
Ehrlichiosis, 118
El Niño
  ecosystem management
    and, 575
Endangered Species Act, 406
Endangered Species List, 32
Endimism
  Southwest Australian
    Floristic Region
    (SWAFR), 623, 626–29
Endozoochory
  vertebrate dispersal and, 2,
    5
Endyphyte
  herbivore interactions with,
    449–50
Environment
  ecological condition of,
    104

ecological indicators and,
  90, 92
effects on cognitive traits,
  362–66
indicator species and, 96
reporting on ecological
  condition of, 94
Environment Canada
  reporting on ecological
    indicators, 90
Environmental change
  ecological indicators and,
    89, 92
  indicator species and,
    96–97
Environmental conditions
  ecological indicators and,
    90–91
  indicator species and, 95
Environmental conservation,
  627
Environmental control, 313
Environmental decay
  harbingers of, 424
Environmental degradation
  ocean disease and, 47
Environmental health
  human health and, 103
Environmental indicators, 91
  economic indicators and,
    92
Environmental integrity, 101
Environmental stressors
  stream health and, 274
Environmental sustainability
  index (ESI)
  five major categories, 91
Environmental tracking,
  312–13
Eocene
Eocene/Oligocene
  boundary
  phytoplankton and, 532
  Southwest Australian
    Floristic Region
    (SWAFR) vegetation,
    631, 643

Estuaries
  regime shifts and, 563
Eukaryotic phytoplankton,
  523–47
  armor and, 535
  carbon cycle and, 544–46
  fossil evidence, 529–30
  photosynthesis and,
    524–25, 528–29, 547
  red lineage of, 530–32,
    537, 547
Europe
  reporting on ecological
    indicators, 90
European colonization
  Southwest Australian
    Floristic Region
    (SWAFR) and, 640
European land use practices,
  644
Eutrophication, 42, 47
  marine ecosystems and,
    569
Everglades
  regime shifts and, 563
Evolution
  bird song, 55–77
  substrates of, 57–63, 76
  cognitive traits and, 365
  ecology and, 286–87
  pairwise, 436–37
Evolutionary biology
  of animal cognition,
    347–67
Evolutionary ecology
  of novel plant-pathogen
    interactions, 675–94
Evolutionary specialization
  ecological specialization
    and, 380–81
Exozoochory
  vertebrate dispersal and, 2
Extinction(s)
  avian, 323–39
  lag-time in, 325–26, 336
  mechanisms, 324,
    326–29

patterns, 323–24
proneness, 329–35
ecological impacts of,
    335–36
event
    Cretaceous, 302
    global, 482
    local, 482
    mass
        Cretaceous/Tertiary
            boundary (K/T), 532
        Middle-Late Pennsylvanian
            boundary, 300
    rates
        habitat loss and, 642
        Southwest Australian
            Floristic Region
            (SWAFR), 634–35
        regional, 313
Extirpations
    avian, 324–26, 328–29

F

Farm Scale Evaluation (FSE)
    transgenic crops and,
    154–55
Farmland biodiversity
    transgenic crops and,
    154–55
    See also Agricultural land
Feces
    fossil
        vertebrate dispersal and,
        4, 11, 13
Field for recombination
    (FFR)
    species delimitation and,
    207–8, 217, 218
Fire(s)
    regimes
        management of, 567
        savannas and, 565–66
Fish(es)
    color vision and, 349
    coral reefs and, 563,
        571–72
    corals and, 177, 179–80

dispersal and, 6, 8
flood-plain forest and, 188
index for biotic integrity
    (IBI) and, 100
pollution and, 40
predatory
    mutualism and, 189
reports of disease in, 36
transgenic, 161
Fishing
    marine ecosystems and, 37
    ocean disease and, 42
    parasites and, 41–42
    regime shifts and, 565, 573
Fitness
    and cognitive traits, 347,
        348, 354–57, 366
    decision making and,
        356–57
    learning and, 355–56
    perception and, 355
Flora
    Southwest Australian
        Floristic Region
        (SWAFR), 623–44
Floral diversification,
    375–77, 379, 391
Floral diversity, 377, 381, 394
Floral evolution, 375–95
Floral specialization
    pollination syndromes and,
    375–95
Floral visitation
    pollinators and, 443
Florida Bay
    regime shifts and, 563
Florida Keys
    coral reefs of, 32, 34, 38,
    44
Florida Keys National Marine
    Sanctuary, 34
Florivory, 444–45
Flowers
    pollinators and, 375–93
    See also Plant(s)
Food safety
    transgenic crops and, 150

Forage
    deer overabundance and,
    116
Foraging
    genetic variation in,
    353–54
Forest(s)
    and agricultural land use,
    264–65
    cloud, 567
    deer impacts on, 115, 117,
    125, 126, 130
    ecosystems
        deer and, 114
        endangered, 406
        establishment of
            closed-canopy, 15
        food-plain
            fishes and, 188
        health, 424
        Holocene, 312, 314
        maturation
            extinctions and, 324
        regime shifts and, 566–67
        reserves
            effective, 337
        rules of assembly of, 424
        salamanders and, 407–8,
            412–14, 417–19, 421,
            424–25
        shifts in cover, 567
        spawning salmon and,
            190–91
    succession
        effects of deer
            population on, 113,
            125, 130
        salamanders and,
            412–14
    tropical
        avian extinctions from,
        323–39
        woodland herbs and,
        583–610
    See also Forested land;
        Tree(s)
Forested land

SUBJECT INDEX

biological metrics, 277
See also Forest(s)
Forestry practices
  in the tropics, 338
Fossil
  of eukaryotic
    phytoplankton, 529–30
  insects
    Quaternary, 312
  record(s)
    Cambrian explosion in,
      245–46
    ecological persistence
      and, 285–314
    Southwest Australian
      Floristic Region
      (SWAFR), 630–34
    vertebrate dispersal and,
      1–20
Frugivores
  avian extinction and, 331,
    336
Fruit flies
  foraging gene of, 353–54
  habitat choice, 352–53
  learning ability in, 351
  sexual behavior in, 354
  smell and, 349–50
Fruits
  vertebrate dispersal and,
    1–20 passim
Functional group(s)
  coral reefs and, 571–72
  diversity, 570, 571
  ecosystem resilience and,
    569, 571–73
  of pollinators
    plant evolution and,
      375–81, 383–91,
      394–95
Functional-response diversity,
  557, 570–71
Fungal farmers
  mutualism, 187
Fungal infections
  plant performance
    and, 450

G

Game laws
  deer overabundance and,
    115, 116
Gene(s)
  color vision and, 349
  Hox, 229, 233
  memory and, 351–52
  plant diseases and, 676,
    678, 680, 688–90
  taste and, 349
  transfer
    horizontal, 156
  Trojan gene hypothesis,
    161
Genealogical exclusivity
  species delimitation and,
    211–15, 217–19
Genetic analysis
  biological control and, 691
Genetic distance
  species delimitation and,
    204–5, 217–18
Genetic divergence
  in Greater Cape Floristic
    Region, 642
Genetic diversity, 641
Genetic drift
  bird song evolution and,
    64–65
Genetic variation
  in attention, 352
  cognitive traits and,
    347–54, 366
  in decision making, 352–54
  in learning and memory,
    350–52
  in mating behavior, 354
  in perception, 348–50
  for reaction norms of
    cognitive traits, 364–66
Genetics
  avian extinction proneness
    and, 333
  conservation, 641
Genomic coalescence
  hypothesis

Southwest Australian
  Floristic Region
  (SWAFR), 626, 639
Genuine Progress Indicator,
  92
Geographical information
  systems (GIS)
  ecological indicators and,
    103
Geomorphology
  Southwest Australian
    Floristic Region
    (SWAFR), 630–34
Glaciation
  Southwest Australian
    Floristic Region
    (SWAFR) and, 624, 628,
    642, 643
Global Magnetic Polarity
  Time Scale, 299
Global positioning systems
  (GPS)
  ecological indicators and,
    103
Global warming
  ocean disease and, 40
  salamanders and, 419
  See also Climate warming
Gnatcatcher, 98
Grasses
  savannas and, 565–66
  silica cycle and, 534–35
  See also Grassland(s)
Grassland(s)
  ecosystems, 312–13
  habitats
    salamanders and, 409
  mutualisms, 313
  See also Grasses
Grazers
  degraded ecosystems and,
    574
  regime shifts and, 563
  See also Grazing
Grazing
  grasslands, 312–13
  regime shifts and, 565

SUBJECT INDEX 709

See also Grazers
Great Lakes
  ecological indicators and, 91
Greater Cape Floristic Region in South Africa, 642–43
Guilds
  as indicators, 97
Gut cavity fossils
  vertebrate dispersal and, 4, 11, 13
Gut endosymbionts, 183
Gymnosperm
  vertebrate dispersal and, 1–2, 8–9, 11–13, 17, 19, 20

H

Habitat(s)
  and agricultural land use, 264
  choice
    decision making and, 352–53
  degradation
    human disturbance and, 41
  destruction
    extinctions and, 324
  diversity
    Southwest Australian Floristic Region (SWAFR), 631
  edges
    access to resources and, 491, 498, 500–1, 515
    ecological flows and, 491, 497–98, 500–1, 515
    ecological response to, 491–515
    resource mapping and, 491, 498–501, 515
    species interactions and, 491, 499–501, 515
  fragmentation
    edge responses
      and, 509, 515
      links, 337
    loss
      avian extinctions and, 325–27, 330, 333, 336
      conservation biology and, 640
      extinction rates and, 642
      metapopulation persistence and, 482–83
      preservation, 338
      river systems and, 258–60
      transgenic crops and, 153–55
      transgenic-wild plants and, 162–63, 165
      urban land use and, 267
Headwater habitats
  salamanders and, 410–11
Hearing
  genetic variation in, 350
Hemichordates, 229, 236–37
Herbicide
  stream biota and, 265
  tolerance, 150
  use, 164
Herbivore(s)
  as biological switches, 130
  coral reefs and, 571–72
  corals and, 179, 185
  defense theory, 302
  as dispersal agents, 188
  effects on plant fitness, 454, 455
  effects on plants, 439
  endophyte interactions with, 449–50
  floral traits and, 440–41
  interactions, 438
  interactions with pollinators, 444–45
  keystone, 114
  marine communities and, 183–84
  mycorrhizal interactions with, 450–51
  overfishing of, 563
  pathogen interactions, 443
  plant defense against, 440–41
  plant fitness and, 446–48
  plant tolerance and resistance to, 121–24
  seaweed-herbivore mutualisms, 185–86
  seed predator/disperser interactions with, 445
  transgenic crops and, 150–53
  vertebrate dispersal and, 4–5, 7–8, 11, 13, 17
  woodland herbs and, 607–9
Herbivory
  effects on plant fitness, 457
  insect, 6
  woodland herbs and, 607–8
Herbs
  Southwest Australian Floristic Region (SWAFR), 629
  woodland, 583–610
High Rainfall Province (HRP) in Australia, 625, 626, 629, 630, 635, 636
Hippocampus
  neurogenesis, 363–64
Holocene
  forest composition, 314
  pollen, 312
Honeybees
  transgenic crops and, 152
Human actions
  ecosystems and, 558, 559, 564, 565, 567, 573–75
  See also Anthropogenic change; Anthropogenic land use; Anthropogenic stress(es); Human activities; Humans
Human activities
  deer overabundance and, 117–18
  ecosystem resilience and, 553, 557, 567–69

stream ecosystems and,
258, 263–68, 273–74,
276–77
See also Anthropogenic
change; Anthropogenic
land use; Anthropogenic
stress(es); Human actions;
Humans
Human health
environmental health and,
103
Humans
decline in forest species
and, 328
stress marine ecosystems,
47
See also Anthropogenic
change; Anthropogenic
land use; Anthropogenic
stress(es); Human actions;
Human activities; Human
health
Hunting
of birds for international
trade, 328
deer overabundance and,
115, 116, 135–36
ocean disease and, 42
See also Overhunting
Hybrid zone barrier
species delimitation and,
203–4, 217
Hybridization
in Southwest Australian
Floristic Region
(SWAFR) plants, 639
Hydrocarbon-seep
communities
mutualisms and,
177–78

I
Impervious area
stream condition and, 272
Impervious surface
urban land use and, 266–67
Index of Biotic Integrity

(IBI), 98, 100, 260, 274,
277
fish, 270–72
Index of Sustainable
Economic Welfare, 92
Indexes
of biological processes,
102
Indigenous people
biodiversity and, 641
Infectious disease(s)
environmental change and,
37–38, 47
marine environments and,
38–39
marine organisms and, 32
in the ocean, 40
toxic environments and, 41
See also Disease(s)
Insect(s)
crop resistance to, 150
dispersal and, 6
effects on plant fitness, 438
fossil
Quaternary, 312
fossil patterns, 301–3
fungal symbionts and, 451
as indicator species, 98
phytophagous, 360–61
plants more susceptible to,
443
Southwest Australian
Floristic Region
(SWAFR) and, 641
transgenic-wild plants and,
163
Insecticide
stream biota and, 265
Insectivores
avian extinction and,
331–32
International Society for
Ecological Economics,
103
Intrasexual selection
bird song evolution and,
56, 68

Introgression
transgenes and, 149,
157–59, 165
Invasions
biological, 675–77
Invertebrate
diversities
phytoplankton and,
541–42
estuarine, 98
Iron (Fe)
temperate lakes and, 562

J
Jurassic
eukaryotic phytoplankton
and, 530
Southwest Australian
Floristic Region
(SWAFR), 624

K
Kelp
forests, 571
mutualisms and, 181–82
regime shifts and,
564–65
sea urchin and, 571

L
Lakes
temperate
regime shifts and,
559–60, 562
tropical
regime shifts and,
562–63
Land use
attitudes toward, 644
legacy effects, 273–74
mechanisms that influence
stream ecosystems, 262,
274
on rivers, 263–67
stream ecosystems and,
257–77
Landscape(s)

SUBJECT INDEX   711

ecology, 257–58
extrapolating edge
  responses to, 494, 511–15
focal species and, 97
oldest, 628
Larval behavior
  predator risks and, 651,
    661–62
Larval morphology
  predation risk and, 661–64
Larval plasticity
  metamorphic plasticity
    and, 664–68
Law of large numbers, 285,
  292, 310, 313, 314
Learning
  in animals, 347
  bird song, 59–60, 76
  evolution of, 358–60
  fitness and, 355–56
  genetic variation in,
    350–52
  phenotypic plasticity and,
    363–64, 366
Life history traits
  avian extinction proneness
    and, 333
Light
  woodland herbs and, 605–7
Limbo Hunt Club
  bird species and, 329, 330
Lizards
  vertebrate dispersal and, 7,
    12, 13
Logging
  woodland herbs and, 584,
    608
Longevity
  of woodland herbs, 585–86
Lophotrochozoa, 229,
  238–43, 247, 248
Lyme disease, 118

M

Macrofauna
  transgenic crops and, 153
Magnesium (Mg)

ocean chemistry and,
  539–40
Mammal record
  Pliocene
    East African, 303–4
  Mammalian faunas
    Miocene
      of Pakistan, 303
    persistence and, 293
Mammals
  predictive edge response
    model and, 503–6
Management
  best management practices
    (BMPs), 275–76
  See also Fishing; Forestry
    practices; Game laws;
    Hunting; Marine mammal
    protection regulations;
    Overfishing; Overhunting
Mangroves
  mutualisms and, 180
Marine invertebrate
  diversity
    phytoplankton and,
      540–42, 547
  fauna, 540–42
Marine mammal protection
  regulations
    ocean disease and, 42
Marine organisms
  anthropogenic change and,
    38–44
Mast seeding
  plants
    and seed predators, 471
    spatial synchrony and, 479
Mate choice
  fitness and, 357
Mathematical models
  population synchrony and,
    476–78
  See also Modeling; Models
Mathematical patterns
  of taxa in a community, 424
Mating behavior
  genetic variation in, 354

Memes
  bird song, 59–60, 63, 76,
    77
  cultural drift, 64
  cultural selection, 65
  natural selection, 66
Memory
  in animals, 347–48
  evolution of, 358–60
  genetic variation in,
    350–52
  phenotypic plasticity and,
    363–64, 366
Mercury
  salamanders and, 420
Mesozoic
  ecological dynamics, 20
  era
    vertebrate dispersal and,
      8–14
  invertebrate and
    phytoplankton diversity,
      541–42
  marine faunas
    phytoplankton and,
      540–42
  oceans, 547
  red eukaryotic
    phytoplankton and,
      530–32, 537
Metallothioneins
  biomarker, 95
Metamorphic plasticity
  larval plasticity and,
    664–68
Metamorphosis
  predation risk and timing
    of, 652–56, 660
  predator-induced plasticity
    and, 651–68
Metapopulations
  synchrony and, 482
Metazoan
  basal clades, 231–34
  phylogeny, 229–48
Microbes
  mutualistic, 183

Microbial communities
 transgenic crops and, 153
Microorganisms
 transgenic crops and, 153
Miocene
 mammalian faunas of
  Pakistan, 303
 phytoplankton and, 532
Mississippian
 fleshy seeds of, 17
 phytoplankton and, 530
Modeling
 of community dynamics,
  294–95
 See also Mathematical
  models; Models;
  Structural-equation
  modeling
Models
 for "scaling up" edge
  responses, 511–12
 See also Mathematical
  models; Modeling
Molecular phylogenetics
 Southwest Australian
  Floristic Region
  (SWAFR) evolutionary
  hypothesis and, 634–36
Molecular systematics, 627
Montréal Process, 90
Moose, 115, 117
 overpopulation, 130
Moran effect
 population dynamics and,
  467, 469, 478–80
Morphological plasticity
 predation risk and, 661–64
Morphology
 species delimitation and,
  206–7, 218
Mountain lions
 deer overabundance and,
  116
Multimetric indexes
 biological communities
  and, 100
Multispecies interactions

plant-animal, 435–59
 traits involved in, 438–39
Multivariate indexes
 biological communities
  and, 100–1
Mutualisms, 175–91
 bodyguards and, 184–85
 cleaner-client, 183
 community context and,
  185–90
 competitors and, 188–89,
  191
 consumer-prey, 185–88
 coral reefs and, 178–80
 deep-sea, 177–78
 defense, 449
 diffuse, 190–91
 foundation species and,
  177–82, 191
 host-microbe, 178, 183–84
 kelp forests and, 181–82
 mangroves and, 180
 multispecies, 444
 nitrogen fixation and,
  182–83, 191
 parasite-host, 189–91
 phytoplankton-
  zooplankton,
  186–87
 saltmarsh communities
  and, 180–81
 seaweed-herbivore, 185–86
Mutualists
 interactions, 443–44
Mycorrhizae
 herbivore interactions with,
  450–51
 woodland herbs and,
  603–4, 609

N
Nannoplankton, 538–39,
 542–43
National Forest Management
 Act of 1976, 90
Nectar thieves, 440
Nematoida, 244

Neoecological studies, 290
Neoecologists, 314
Neoecology, 285, 292, 313
Neogene
 Southwest Australian
  Floristic Region
  (SWAFR), 634
Nested clade analysis (NCA)
 species delimitation and,
  215–16
Nitrogen (N)
 cycling
  deer overabundance and,
   126
  ectomycorrhizal fungi and,
   450
 fertilizer, 425
 fixation
  mutualisms and, 182–83,
   191
 nitrogen-fixing species,
  570
 salamanders and, 418
 spawning salmon and,
  190–91
Nutrient(s)
 and agricultural land use,
  264–65
 cycling
  deer overabundance and,
   113, 126
  marine ecosystems and,
   569
  spawning salmon and,
   190–91
  storage and cycling
  woodland herbs and,
   599, 603
 stream ecosystems and,
  270

O
Ocean(s)
 chemistry
  phytoplankton
   and, 536–37, 539–40,
   547

disease(s), 31–48
  anthropogenic change
   and, 38–44, 47
  case study of oysters,
   44–47
  climate warming and,
   31, 39–40, 47
  direct observations of,
   33–34
  evidence for change,
   33–37
  pathogen pollution,
   43–44, 47
  patterns in the literature
   on, 35–37
  pollution and, 31, 40–41
  large-scale phosphorus
   release in, 562
  See also Oceanic Anoxic
   Events (OAEs); Oceanic
   circulation patterns;
   Sea(s)
Oceanic Anoxic Events
  (OAEs)
  eukaryotic phytoplankton
   and, 537
Oceanic circulation patterns,
  313
Olfaction
  evolution of, 358
Orchids
  woodland, 604
Ordovician
  eukaryotic phytoplankton,
   530
Oscillations
  dynamical systems theory
   and, 476, 478
Oscillators
  dynamical systems theory
   and, 476, 478
Overfishing
  in coastal waters, 573
  See also Fishing
Overhunting
  extinctions and, 324
  See also Hunting

Oxygen (O)
  phytoplankton and, 536–37
  temperate lakes and, 562
Oysters
  case study for diseases,
   44–47

P
Pakistan
  Miocene mammalian
   faunas, 303
Paleobiology
  evolutionary, 292
Paleoclimates
  Southwest Australian
   Floristic Region
   (SWAFR), 630–34
Paleoclimatology
  World Data Center for
   Paleoclimatology
   (WDCP), 295, 299
Paleocommunities, 296–98
  of marine invertebrates,
   305–7
Paleocommunity, 290–92
Paleoecologists, 289–90
Paleoecology, 285, 295
  ecology and, 310–11
  time-averaging, 290
Paleogene
  community composition,
   299
Paleological concepts and
   data
  versus ecological concepts
   and data, 287–92
Paleontologists
  and ecologists, 286–87
Paleozoic
  invertebrate and
   phytoplankton diversity,
   541–42
  Late
   tropics, 301
  phytoplankton, 530
  vertebrate dispersal, 7–8
Palms

Cretaceous, 18
Palynofloras, 295–99
Panarthropoda, 244–45
Pangea
  eukaryotic phytoplankton
   and, 531
Parasites
  climate warming and,
   39–40
  environmental change and,
   37–38
  healthy ecosystems and, 41
  of introduced species, 43
  ocean disease and, 38–39
  oysters and, 45
  pollution and, 41
  See also Parasitism
Parasitism
  extinctions and, 336
  See also Parasites
Path analysis
  in plant-animal interaction
   studies, 456–58
Pathogen(s)
  degree of specialization of,
   684–85
  environmental change and,
   37–38
  oysters and, 45
  plant-pathogen
   interactions, 675–94
  pollution
   ocean disease and,
    43–44, 47
  survival abilities of,
   683–84
  woodland herbs and,
   607–9
Pathogenicity
  defined, 678
PCBs
  See Polychlorinated
   biphenyls (PCBs)
Pennsylvanian
  plant morphology diversity,
   1, 7
  seed morphologies, 8

vertebrate dispersal, 17
Perception
  defined, 347
  evolution of, 358
  fitness and, 355
  genetic variation in,
    348–50
  phenotypic plasticity and,
    363
Peregrine falcon
  decline of, 92
Permian
  Early
    floras, 301
  Southwest Australian
    Floristic Region
    (SWAFR), 624, 628, 642,
    643
  vertebrate dispersal, 17
  vertebrate herbivores and,
    1, 8
Persistence
  of community
    composition, 292
  dispersal and, 628
  ecological, 285–314
  law of large numbers and,
    314
  species assemblages and,
    285–314
  stratigraphic, 293–94
Pesticide(s)
  and agricultural land use,
    264, 265
  forest birds and, 328
  reduction, 164
  salamanders and, 420
Pet trade, 330
Phase analysis
  spatial dynamics and, 468
Phase locking
  dynamical systems theory
    and, 476, 478
Phase synchrony, 471–72,
    475–76
Phenology
  woodland herbs, 584–85

Phenotypic plasticity
  of cognitive traits, 362–64,
    366
  predator-induced, 651–68
Phoronida, 229, 241, 247
Phosphorus (P)
  biogeochemical records
    and, 546
  lake cycles, 559, 562
  mycorrhizal fungi and, 450
  salamanders and, 418
  spawning salmon and,
    190–91
Photoautotrophs
  oxygenic, 523–27
Photosynthesis, 535
  zooxanthellae and, 178
Phylogenetic niche
  conservatism, 312
Phylogenetic species concept
  (PSC)
  species delimitation and,
    208–10, 218
Phylogeny
  animal, 229–48
  avian extinction and,
    333–35
  Metazoan, 229–48
  plant-pathogen interactions
    and, 685, 692
Phylogeography
  Southwest Australian
    Floristic Region
    (SWAFR), 637–38
Phytoplankton
  biogeochemical cycles and,
    543–47
  carbon cycle and, 544–46
  eukaryotic, 523–47
  evolutionary tempo of,
    542–43
  marine ecosystems and,
    569
  marine invertebrate
    diversity and, 540–42,
    547
  marine invertebrate fauna

and, 540–42
nitrogen and, 182–83
ocean chemistry and,
  536–37, 539–40, 547
phytoplankton-
  zooplankton mutualisms,
  186–87
regime shifts and, 562, 563
Wilson Cycle and, 538–42
Placozoa, 229, 234, 247
Plankton
  biogeochemical cycles and,
    543–47
  nitrogen-fixing, 182–83
  See also Nannoplankton;
    Phytoplankton;
    Zooplankton
Plant(s)
  attraction and defense
    traits, 439
  Cenozoic era
    vertebrate dispersal and,
      14
  deer browsing pressure on,
    131
  deer impacts on
    communities, 114–15
  deer overabundance and,
    120–21
  defense, 440
  disease triangle, 678
  effects of deer population
    on, 124–25, 130
  effects of multiple
    interactors on, 455
  evolution
    pollinator functional
      groups and, 375–81,
      383–91, 394–95
  fitness
    multispecies effects on,
      446–48
  fossil, 299–306
  intrinsic value of, 644
  invasions, 675–77, 684,
    686, 687, 689, 692
  invasive, 692–94

SUBJECT INDEX 715

measuring multispecies
 effects on, 452–58
Mesozoic era
 vertebrate dispersal and,
 1, 8–14
multispecies plant-animal
 interactions, 435–59
mycorrhizal relationships
 with, 450–51
Paleozoic era
 vertebrate dispersal and,
 7–8
pathology, 675–94
plant-herbivore functional
 response model, 121
plant-pathogen
 interactions, 675–94
 predictions for, 683–91
predictive edge response
 model and, 503–6
reactions to damage, 436
species
 Southwest Australian
 Floristic Region
 (SWAFR), 623–44
tolerance and resistance to
 herbivory, 121–24
transgenic-wild plants,
 161–63
 insects and, 163
 population dynamics and
 habitat use, 161–63
tropical lakes and, 562–63
vertebrate dispersal and,
 1–20
See also Crops; Flowers;
 Plant diseases; Transgenic
 crops; Weed(s);
 Woodland herbs
Plant diseases, 675–94
 avirulence and resistance
 genes and, 678–80
 genetic diversity and,
 687–88
 host density and, 686–87
 host-pathogen recognition
 and infection, 679

initial host shift and,
 683–86
novel epidemics, 682, 692,
 694
resistance and, 676–81,
 688–91, 693, 694
virulence and, 676,
 678–82, 690–91, 693, 694
See also Plant(s)
Plastids
 green line of, 523–25,
 528–29, 536, 547
 red line of, 523, 525,
 528–29, 536, 537
Plato, 89
Platyzoa, 240–41, 247
Pliocene
 Southwest Australian
 Floristic Region
 (SWAFR), 633, 634
Policy makers
 policy-relevant indicators
 and, 93
Pollen
 data
 fossil, 295, 299
 Holocene, 312
 Quaternary, 286
Pollination
 specialization, 375–95
 comparative data on,
 387–91, 394
 syndrome(s), 375–95
 defined, 376
 floral specialization and,
 375–95
Pollinator(s)
 effects on plants, 439
 floral traits and, 440–41
 floral visitation and, 443
 flowers and, 375–93
 functional groups of
 plant evolution and,
 375–81, 383–91,
 394–95
 interactions with
 herbivores, 444–45

plant fitness and, 438,
 446–48, 454
relationships with plants,
 436, 455
seed predator interactions
 with, 445, 449
transgenic crops and, 152
Pollutants
 agricultural land use and,
 264
 urban land use and, 266,
 267
See also Pollution;
 Pollutogens
Pollution
 coral disease model and, 38
 ocean disease and, 31, 35,
 40–41
 stress
 biomarkers and, 102
See also Pollutants;
 Pollutogens
Pollutogens
 ocean disease and, 44, 47
See also Pollutants;
 Pollution
Polychlorinated biphenyls
 (PCBs)
 salamanders and, 420
Population(s)
 dynamics
 edge responses and, 514
 spatial synchrony and,
 467–84
 ecological indicators and,
 94
 population aggregation
 analysis (PAA)
 species delimitation and,
 207, 210–11, 213,
 216–18
 synchrony, 468, 476–81
 mathematical models of,
 476–78
Predators
 deer overabundance and,
 116

extinctions and, 324
forest reserves and, 337
keystone, 417, 423, 424
loss of, 569
phenotype plasticity and,
  651–68
population synchrony and,
  478–79
role of, 130, 132
salamanders as, 405, 407,
  414–17, 423, 424
seed, 471
top, 42, 47
transgenic crops and, 152
Prey
  consumer-prey
    mutualisms, 185–88
  salamanders as, 421
Proterozoic
  acritarchs, 542
  oceans, 547
Protostomia, 238
Pterosaurs
  vertebrate dispersal and,
    12–13
Public
  ecological indicators and,
    93

Q
Quaternary
  community composition,
    299
  coral reef, 309–10
  fossil insects, 312
  nonanalog floras, 312
  pollen, 286
  Southwest Australian
    Floristic Region
    (SWAFR), 624, 626, 628,
    632, 634, 635

R
Rats
  avian extirpations and,
    328–29
  learning ability, 350–51

See also Rodents
Redundancy
  ecosystem, 289
Reefs
  consumer-prey mutualisms
    on, 185–86
  See also Coral reef(s)
Regime shifts
  and ecosystem resilience,
    559, 562–67
  in ecosystems, 557–75
  irreversibility, 567, 575
Relocation
  deer overabundance and,
    135
Remote-sensing technology
  ecological indicators and,
    103
Reptiles
  dispersal and, 8, 9, 12, 20
Rescue effect
  metapopulations and, 482
Resilience
  ecological meaning, 289
  ecosystem, 557–75
  biodiversity and, 569–73
  human activities and,
    553, 557, 567–69
  managing for
    development, 573–75
  tropical bird species and,
    336–37
Resistance
  ecological meaning, 289
  plant diseases and, 676–81,
    688–91, 693, 694
Resource mapping
  habitat edges and, 491,
    498–501, 515
Rifting
  of Australia from
    Antarctica, 631
Riparian ecotone
  salamanders and, 420
Riparian habitats
  salamanders and, 409–10
River(s)

assessment of health of,
  260–63
land use and, 257–77
salinization of, 574
See also Riverscapes
Riverscapes, 258
See also River(s)
Rodents
  as seed dispersers, 444
  See also Rats
Rodinia
  rifting, 538, 547

S
Salamanders, 405–26
  aquatic habitats and,
    410–12, 415, 417,
    419–20, 426
  coupling aquatic and
    terrestrial habitats,
    419–20
  detrius-litter food webs
    and, 417–19
  forest succession and,
    412–14
  grassland habitats and, 409
  patterns of functional
    dominance, 414–23
  patterns of structural
    dominance, 406–14
  predatory effects of,
    414–17
  as prey, 421
  regulation of diversity,
    420–21
  riparian habitats and,
    409–10
  terrestrial habitats and,
    406–9, 415, 417, 419–20,
    426
Saline groundwater tables,
  640
Salinization
  of land and rivers, 574
Salmon
  trees and, 190–91
Saltmarsh communities

mutualisms and, 180–81
Savannas
 regime shifts and, 565–66
Scalidophora, 244
Sea(s)
 coastal
  regime shifts and, 563
 sea surface temperatures
  increased, 39
 See also Ocean(s); Sea
  level
Sea level
 phytoplankton and, 538–39
Sea otters
 kelp forests and, 182
Sea urchin
 Caribbean, 563–64
 kelp and, 571
Seagrasses
 mutualisms and, 180–81
Seaweed(s)
 corals and, 179
 seaweed-herbivore
  mutualisms, 185–86
Sediments
 agricultural land use and,
  264, 265
 stream ecosystems and,
  270
Seed(s)
 dispersal
  bird frugivores and, 336
  Southwest Australian
   Floristic Region
   (SWAFR), 629
  by waterfowl, 187–88
  woodland herbs and,
   595, 599–602
 germination
  woodland herbs and,
   595–99, 604
 predator(s)
  herbivore interactions
   with, 445
  plant defense against,
   440
  plant fitness and, 438

pollinator interactions
 with, 445, 449
 vertebrate dispersal of,
  1–20
Sewage runoff
 pollutogens and, 44
Sexual dimorphism
 avian, 331
Sexual reproductive effort
 (SRE)
 woodland herbs and,
  594–95
Siberia
 irreversible regime shifts
  in, 567
Silica
 bioavailability
  diatoms and, 534–35
Sitka spruce, 191
Smell
 genetic variation in,
  349–50
Snails
 mutualism with fungi, 187
Software tools
 for "scaling up" edge
  responses, 511–12
Soil bacteria
 transgenes and, 156
Soil community
 transgenic crops and, 153
Songbirds
 See Bird song
South Africa
 Greater Cape Floristic
  Region in, 642–43
 Southeast Coastal Province
  (SCP)
  in Australia, 625, 626, 629,
   630, 634–35, 643
 Southwest Australian
  Floristic Region
  (SWAFR), 623–44
  compared to Greater Cape
   Floristic Region, 642–43
  conservation biology,
   640–42

evolutionary hypothesis,
  625–26
 molecular phylogenetics
  and, 634–36
 evolutionary origins of,
  630–39
Spatial dynamics
 in natural systems, 482–84
Spatial synchrony, 467–84
 in population dynamics,
  467–84
Speciation
 Greater Cape Floristic
  Region, 642
 Southwest Australian
  Floristic Region
  (SWAFR), 626, 634–35,
  639, 644
Species
 assemblages
  persistence of, 285–314
 associations
  mutualisms and, 176
 composite species concept
  (CSC), 208–10, 218
 concepts and criteria, 201
 delimitation, 199–221
  empirical, 200–1
  nontree-based methods,
   201–8
  tree-based methods,
   208–16, 218
 distribution, 313
 ecological indicator, 95–99
  flagship, 97–99
  focal, 97–99, 101
  foundation
   mutualisms and, 177–82,
    191
  indicator, 101–2
  indicator criteria, 424
  individuality, 312
  interactions
   habitat edges and, 491,
    499–501, 515
  introduced, 47
  parasites of, 43

SUBJECT INDEX

keystone, 99
  deer as, 607
  neutral modeling of
    community dynamics
    and, 294–95
  persistent community
    assemblages and, 292–95
  phylogenetic species
    concept (PSC)
    species delimitation and,
    208–10, 218
  rare, 101
  Southwest Australian
    Floristic Region
    (SWAFR), 626–27
    in trouble, 641
  umbrella, 97–99, 101
Sponge(s), 229, 247
  coral reef
    mutualism and, 188–89
  mangroves and, 180
  paraphyly, 231–33
Stability
  ecological meanings of,
    288–89
Stasis
  coordinated, 293–94
  temporal scale and, 295–99
State of the Lakes Ecosystem
  Conference (SOLEC),
  91–92
Stochasticity
  spatial synchrony and,
    480–81
Stratigraphic persistence
  community, 293–94
Stream ecosystems
  land use and, 257–77
  management and, 274–77
Strontium (Sr) isotopes
  biogeochemical records
    and, 546
Structural-equation modeling
  in ecological and
    evolutionary studies, 456
Sulfur isotopes
  fractionation, 545

Supercontinents
  Wilson Cycle and, 537
  See also Pangea; Rodinia
Swallows
  nesting tree, 99–100
Symbionts
  nitrogen-fixing, 182–83
Synapsids
  vertebrate dispersal and, 6,
    8, 13
Synchrony
  causes of, 476–81
  measuring, 471–76
  metapopulation persistence
    and, 482
  spatial, 467–84

T

Taste
  genetic variation in, 349
Taxonomy
  avian extinction and,
    333–35
Templeton's tests of cohesion
  species delimitation and,
    215–16, 218, 219
Temporal effects
  edge responses and, 508–9,
    515
Temporal scale
  stasis and, 295–99
Terrestrial habitats
  salamanders and, 406–9,
    415, 417, 419–20, 426
Tertiary
  angiosperms, 18–19
  biotic dispersal in, 14–15
  coevolutionary
    relationships in, 20
  Cretaceous-Tertiary (K/T)
    boundary
      biotic dispersal and,
      19–20
      mass extinction event,
      532
  dispersal modes, 18
  Southwest Australian

Floristic Region
  (SWAFR) speciation, 623,
  626, 628
Tick-borne diseases, 118
Time
  Global Magnetic Polarity
    Time Scale, 299
  scales
    ecological versus
      paleontologists, 287–88
    time-averaging in
      paleoecology, 290
  See also Temporal scale
Topsoil
  removal of, 641
Trace elements
  phytoplankton and, 536–37
Transgenes
  current and future, 163–64
  escape into wild
    populations, 149, 156–61
  wild populations and,
    161–63, 165
Transgenic crops, 149–64
  ecological risks and,
    149–51
  feral, 153–54
  habitat and, 153–55
  nontarget effects, 151–55
  potential benefits, 164
  potential risks, 150
  species effects of, 151–53
  wild populations and, 149,
    156–61
Transitional Rainfall Province
  (TRP)
  in Australia, 625, 626, 629,
    630, 634–36, 641
Transitional Rainfall Zone
  in Australia, 625
Tree(s)
  deer overabundance and,
    124–25
  herbivores and, 122
  herbivory and, 123–24
  patterns
    mutualisms, 313

salmon and, 190–91
shifts in dominance
species, 124
See also Forest(s); Wood
Triassic
eukaryotic phytoplankton
and, 530
Trinitrotoluene (TNT)
salamanders and, 420
Trojan gene hypothesis, 161
Trophic cascades
ecosystems and, 569
Tropical forests
avian extinctions from,
323–39
Tropics
Late Paleozoic, 301
vertebrate dispersal and,
17–18
Trout
whirling disease, 189–90
Tuberculosis
bovine, 118
Tubeworm, 177–78
Tunicata, 237
Turnover
in paleoecology and
ecology, 291
soft-bottom benthic
invertebrate faunas, 304–9
Turtles
dispersal and, 12

U

Underground retreats
salamanders and, 421–23
United States
reporting on ecological
indicators, 90
U.S. Environmental
Protection Agency (US
EPA), 90
U.S. Geological Society
(USGS)
Breeding Bird Survey
(BBS), 99–100
Universities

woodland herb studies and,
610
University teaching programs
in systematics, 627
Uplift
in Australia, 631
Urban land
rivers and, 263, 266 67, 277
stream condition and, 272
Urbanization
woodland herbs and, 608

V

Variability
ecological meaning, 289
Vegetative reproductive effort
(VRE)
woodland herbs and,
594–95
Vehicle collisions
with deer, 118
Vernal dam hypothesis, 603
Vertebrate(s)
dispersal, 1–20
origins of, 5–6
as umbrella species, 98
Virulence
plant diseases and, 676,
678–82, 690–91, 693, 694
Viruses
transgenic crops and, 156
Vision
color
evolution of, 358
trichromatic and
dichromatic, 355

W

Waters
report every two years on,
90
See also Lakes; River(s);
Stream ecosystems
Weather
population synchrony and,
478
synchrony, 481

Weed(s)
control, 688
invasion, 641
invasive, 640, 675–76
biological control,
693–94
transgenic crops and,
154–55
See also Weediness
Weediness
transgenes and, 149, 161,
165
See also Weed(s)
Western Australian
Herbarium, 629
Western hemlock, 191
Wetland assemblages
Carboniferous, 299–301
Wetlands
regime shifts and, 563
Whales, 571
Whirling disease, 189–90
Wiens-Penkrot phylogenetic
models
species delimitation and,
215–19
Wild populations
transgenes escape into,
149, 156–63, 165
Wildlife managers, 134
Wilson Cycle
explained, 537
phytoplankton and, 538–42
Wolves, 130
deer overabundance and,
116
Wood
in streams, 266
Woodland herbs, 583–610
canopy gaps and, 605–7,
609
clonality of, 589–94
competition and, 594
flowering characteristics
of, 590–93
genetic variability of, 605
herbivory and, 607–8

logging and, 584, 608
longevity of, 585–86
long-term studies of, 610
mortality of, 586
mycorrhizae and, 603–4, 609
nutrient foraging and, 589, 594
nutrient storage and cycling, 599, 603
physiological integration, 589, 594
population structure and stability, 604–5
reproduction of, 586–89, 594–95
responses to disturbance, 605–9
seed characteristics of, 596–98
seed dispersal and, 595, 599–602
seed germination and, 595–99, 604
urbanization and, 608
See also Plant(s)
World Data Center for Paleoclimatology (WDCP), 295, 299

X

Xenoturbella, 229, 237–38

Z

Zoonoses
 tick-borne, 118
Zooplankton
 eukaryotic phytoplankton and, 535
 marine ecosystems and, 569
 phytoplankton-zooplankton mutualisms, 186–87
 populations
  ecological indicators and, 94
Zooxanthellae
 corals and, 178–80

Cumulative Indexes

CONTRIBUTING AUTHORS, VOLUMES 31–35

A
Abrams PA, 31:79–105
Ackerly DD, 33:475–505
Adams E, 32:277–303
Agrawal AA, 33:641–64
Allan JD, 35:257–84
Allendorf FW, 32:277–303
Altizer S, 34:517–47
Anderson S, 31:61–77
Antonovics J, 34:517–47
Arbogast BS, 33:707–40
Armbruster WS, 35:375–403

B
Badyaev AV, 34:27–49
Baker AC, 34:661–89
Barton NH, 34:99–125
Battin J, 35:491–522
Baughman S, 32:305–32
Bazzaz FA, 35:323–45
Beerli P, 33:707–40
Behrensmeyer AK, 35:285–322
Benard MF, 35:651–73
Bennett AF, 31:315–41
Bilton DT, 32:159–81
Bininda-Emonds OR, 33:265–89
Bjørnstad ON, 35:467–90
Blackburn TM, 34:71–98
Bobe R, 35:285–322
Brakefield PM, 34:633–60
Brawn JD, 32:251–76
Bull J, 32:183–217
Burkepile DE, 35:175–97

C
Cabin RJ, 32:305–32
Cadenasso ML, 32:127–57

Carlton JT, 31:481–531
Carpenter S, 35:557–81
Case TJ, 33:181–233
Caudill CC, 35:175–97
Chapin FS III, 34:455–85
Charlesworth B, 34:99–125
Charlesworth D, 34:99–125
Chave J, 34:575–604
Chequer AD, 35:175–97
Chesson P, 31:343–66
Chiappe LM, 33:91–124
Clout M, 31:61–77
Cohen JE, 32:305–32
Cole CT, 34:213–37
Cory JS, 34:239–72
Costanza R, 32:127–57
Côté SD, 35:113–47
Cowan JH Jr, 34:127–51
Craig J, 31:61–77
Creese B, 31:61–77
Crist TO, 31:265–91
Cunnningham AA, 34:517–47

D
Daehler C, 34:183–211
Davic RD, 35:405–34
Dawson TE, 33:507–59
Day TA, 33:371–96
Dayan T, 34:153–81
Dayton PK, 33:449–73
Dicke M, 32:1–23
DiMichele WA, 35:285–322
Dobson AP, 34:517–47
Donoghue MJ, 33:475–505
Dudash MR, 35:375–403
Dudgeon D, 31:239–63
Dukas R, 35:347–74
Duncan RP, 34:71–98

Dussault C, 35:113–47
Dyke GJ, 33:91–124
Dynesius M, 33:741–78

E
Edwards SV, 33:707–40
Ellstrand NC, 32:305–32
Elmqvist T, 35:557–81
Etter RJ, 32:51–93
Eviner VT, 34:455–85
Ezenwa V, 34:517–47

F
Fahrig L, 34:487–515
Falkowski PG, 35:523–56
Falster DS, 33:125–59
Farnsworth E, 31:107–38
Feder ME, 31:315–41
Festa-Bianchet M, 31:367–93
Fialho RF, 32:481–508
Finkel ZV, 35:523–56
Fletcher RJ Jr, 35:491–522
Fofonoff PW, 31:481–531
Folke C, 35:557–81
Ford SE, 35:31–54
Freeland JR, 32:159–81
French V, 34:633–60
Funk DJ, 34:397–423

G
Gaillard J-M, 31:367–93
Garland T Jr, 32:367–96
Gilbert GS, 35:675–700
Giordano R, 32:481–508
Gioia P, 35:623–50
Gittleman JL, 33:265–89; 34:517–47
Gooday AJ, 32:51–93
Goulson D, 34:1–26

721

Grostal P, 32:1–23
Grove JM, 32:127–57
Grove SJ, 33:1–23
Grzebyk D, 35:523–56
Gunderson L, 35:557–81
Gunderson LH, 31:425–39

H
Haddad BM, 32:481–517
Haig D, 31:9–32
Halanych KM, 35:229–56
Hallinan ZP, 35:175–97
Harshman LG, 32:95–126
Hay ME, 35:175–97
Hedrick PW, 31:139–62
Heil M, 34:425–53
Hessler RR, 32:51–93
Hill GE, 34:27–49
Hines AH, 31:481–531
Hochwender CG, 31:565–95
Holland EA, 32:547–76
Holland SM, 33:561–88
Holling CS, 35:557–81
Holt JS, 32:305–32
Holway D, 33:181–233
Hopper SD, 35:623–50
Huber SK, 35:55–87
Huey RB, 31:315–41

I
Irschick DJ, 32:367–96
Irwin RE, 35:435–66

J
Jaenike J, 32:25–49
Jansson R, 33:741–78
Johannes RE, 33:317–40
Johnson JB, 33:665–706
Johnston RF, 31:1–7
Jones KE, 34:517–47

K
Kalinowski ST, 31:139–62
Karban R, 33:641–64
Kareiva PM, 33:665–706
Katz ME, 35:523–56
Kaufman DM, 34:273–309

Kellogg EA, 31:217–38
Kelly D, 33:427–47
Kershaw P, 32:397–414
Kidwell SM, 33:561–88
Knoll AH, 35:523–56
Koenig WD, 35:467–90
Kornfield I, 31:163–96
Kronfeld-Schor N, 34:153–81

L
Labandeira CC, 35:285–322
Lach L, 33:181–233
Lafferty KD, 35:31–54
Levin LA, 32:51–93
Levin P, 33:665–706
Levin SA, 34:575–604
Levine JM, 34:549–74
Lieberman BS, 34:51–69
Liebhold A, 35:467–90
Liow LH, 35:323–45
Lodge DM, 32:305–32
Loison A, 31:367–93
Ludwig D, 32:441–79

M
MacFadden BJ, 31:33–59
MacMahon JA, 31:265–91
Mambelli S, 33:507–59
Mangel M, 32:441–79
Marquis RJ, 31:565–95
Marshall JC, 35:199–227
Marshall VG, 31:395–423
McCauley DE, 32:305–32
McDonald ME, 35:89–111
McKey D, 34:425–53
McPeek MA, 33:475–505
Meyer A, 34:311–38
Meyer JL, 32:333–65
Milsom CV, 31:293–313
Mitchell N, 31:61–77
Moles AT, 33:125–59
Molofsky J, 32:305–32
Monson RK, 32:509–32
Moritz C, 31:533–63
Mull JF, 31:265–91
Muller-Landau HC, 34:575–604

Murrell DJ, 34:549–74
Myers JH, 34:239–72

N
Nathan R, 34:575–604
Neale PJ, 33:371–96
Niemi GJ, 35:89–111
Nilon CH, 32:127–57
Noor MAF, 34:339–64
Nunn CL, 34:517–47

O
Ogden J, 31:61–77
Okamura B, 32:159–81
Olivera B, 33:25–47
Olszewski TD, 35:285–322
Omland KE, 34:397–423
O'Neil P, 32:305–32

P
Palmer AR, 31:441–80
Pandolfi JM, 35:285–322
Pannell JR, 33:397–425
Parker IM, 32:305–32;
 35:675–700
Parker JD, 35:175–97
Patton JL, 31:533–63
Paul MJ, 32:333–65
Pawson D, 32:51–93
Pedersen AB, 34:517–47
Pickett STA, 32:127–57
Pilson D, 35:149–74
Pineda J, 32:51–93
Plamboeck AH, 33:507–59
Podos J, 35:55–87
Porter JW, 35:31–54
Poss M, 34:517–47
Pouyat RV, 32:127–57
Prendeville HR, 35:149–74
Pulliam JRC, 34:517–47

R
Rabalais NN, 33:235–63
Rahel FJ, 33:291–315
Ramsey J, 33:589–639
Rand DM, 32:415–48
Rex MA, 32:51–93

Ries L, 35:491–522
Rigby S, 31:293–313
Roberts M, 31:61–77
Robinson SK, 32:251–76
Rooney TP, 35:113–47
Roopnarine PD, 34:605–32
Rose KA, 34:127–51
Rubidge BS, 32:417–40
Ruckelshaus M,
 33:665–706
Ruiz GM, 31:481–531
Rusek J, 31:395–423

S
Sakai AK, 32:305–32
Sanderson MJ, 33:49–72
Scheffer M, 35:557–81
Schemske DW, 33:589–639
Schneider CJ, 31:533–63
Servedio MR, 34:339–64
Shaffer H, 33:49–72
Sidor CA, 32:417–40
Simms EL, 31:565–95
Sisk TD, 35:491–522
Sites JW Jr, 35:199–227
Slowinski JB, 33:707–40
Smith CR, 32:51–93
Smith PF, 31:163–96
Smith TB, 31:533–63
Sodhi NS, 35:323–45
Sol D, 34:71–98
Sork VL, 33:427–47
Steel MA, 33:265–89

Stevens L, 32:481–508
Stevens RD, 34:273–309
Stewart IRK, 34:365–96
Stowe KA, 31:565–95
Strauss SY, 35:435–66
Stuart CT, 32:51–93
Suarez AV, 33:181–233
Swanson WJ, 33:161–79

T
Taft B, 35:55–87
Templer PH, 33:507–59
Thewissen J, 33:73–90
Thompson FR III, 32:251–76
Thompson JN, 32:305–32
Thomson JD, 35:375–403
Thrall PH, 34:517–47
Thrush S, 33:449–73
Tiffney BH, 35:1–29
Toïgo C, 31:367–93
Tremblay J-P, 35:113–47
Tsutsui ND, 33:181–233
Tu KP, 33:507–59
Turner RE, 33:235–63

U
Ussher G, 31:61–77

V
Vacquier VD, 33:161–79
Van Auken OW, 31:197–215
Vanni MJ, 33:341–70
Vesk PA, 33:125–59

W
Wagstaff B, 32:397–414
Wakeley J, 33:707–40
Walker B, 35:557–81
Waller DM, 35:113–47
Watson DM, 32:219–49
Webb CO, 33:475–505
Weller SG, 32:305–32
Welsh HH Jr, 35:405–34
Westneat DF, 34:365–96
Westoby M, 33:125–59
Whigham DF, 35:583–621
Wichman HA, 32:183–217
Williams EM, 33:73–90
Willig MR, 34:273–309
Wilson AE, 35:175–97
Wilson P, 35:375–403
Wing SL, 35:285–322
Wiseman WJ Jr,
 33:235–63
With KA, 32:305–32
Wonham MJ, 31:481–531
Wright IJ, 33:125–59

Y
Yoccoz NG, 31:367–93

Z
Zardoya R, 34:311–38
Zera AJ, 32:95–126
Zipperer WC,
 32:127–57
Zwaan BJ, 34:633–60

CHAPTER TITLES, VOLUMES 31–35

Volume 31 (2000)

Preface: A Millennial View of Ecology and Systematics, and *ARES* at Age 30	RF Johnston	31:1–7
The Kinship Theory of Genomic Imprinting	D Haig	31:9–32
Cenozoic Mammalian Herbivores from the Americas: Reconstructing Ancient Diets and Terrestrial Communities	BJ MacFadden	31:33–59
Conservation Issues in New Zealand	J Craig, S Anderson, M Clout, B Creese, N Mitchell, J Ogden, M Roberts, G Ussher	31:61–77
The Evolution of Predator-Prey Interactions: Theory and Evidence	PA Abrams	31:79–105
The Ecology and Physiology of Viviparous and Recalcitrant Seeds	E Farnsworth	31:107–38
Inbreeding Depression in Conservation Biology	PW Hedrick, ST Kalinowski	31:139–62
African Cichlid Fishes: Model Systems for Evolutionary Biology	I Kornfield, PF Smith	31:163–96
Shrub Invasions of North American Semiarid Grasslands	OW Van Auken	31:197–215
The Grasses: A Case Study in Macroevolution	EA Kellogg	31:217–38
The Ecology of Tropical Asian Rivers and Streams in Relation to Biodiversity Conservation	D Dudgeon	31:239–63
Harvester Ants (*Pogonomyrmex* spp.): Their Community and Ecosystem Influences	JA MacMahon, JF Mull, TO Crist	31:265–91
Origins, Evolution, and Diversification of Zooplankton	S Rigby, CV Milsom	31:293–313
Evolutionary Physiology	ME Feder, AF Bennett, RB Huey	31:315–41
Mechanisms of Maintenance of Species Diversity	P Chesson	31:343–66

Temporal Variation in Fitness Components and Population Dynamics of Large Herbivores	J-M Gaillard, M Festa-Bianchet, NG Yoccoz, A Loison, C Toïgo	31:367–93
Impacts of Airborne Pollutants on Soil Fauna	J Rusek, VG Marshall	31:395–423
Ecological Resilience—In Theory and Application	LH Gunderson	31:425–39
Quasi-Replication and the Contract of Error: Lessons from Sex Ratios, Heritabilities and Fluctuating Asymmetry	AR Palmer	31:441–80
Invasion of Coastal Marine Communities in North America: Apparent Patterns, Processes, and Biases	GM Ruiz, PW Fofonoff, JT Carlton, MJ Wonham, AH Hines	31:481–531
Diversification of Rainforest Faunas: An Integrated Molecular Approach	C Moritz, JL Patton, CJ Schneider, TB Smith	31:533–63
The Evolutionary Ecology of Tolerance to Consumer Damage	KA Stowe, RJ Marquis, CG Hochwender, EL Simms	31:565–95

Volume 32 (2001)

Chemical Detection of Natural Enemies by Arthropods: An Ecological Perspective	M Dicke, P Grostal	32:1–23
Sex Chromosome Meiotic Drive	J Jaenike	32:25–49
Environmental Influences on Regional Deep-Sea Species Diversity	LA Levin, RJ Etter, MA Rex, AJ Gooday, CR Smith, J Pineda, CT Stuart, RR Hessler, D Pawson	32:51–93
The Physiology of Life History Trade-Offs in Animals	AJ Zera, LG Harshman	32:95–126

Urban Ecological Systems: Linking Terrestrial Ecological, Physical, and Socioeconomic Components of Metropolitan Areas	STA Pickett, ML Cadenasso, JM Grove, CH Nilon, RV Pouyat, WC Zipperer, R Costanza	32:127–57
Dispersal in Freshwater Invertebrates	DT Bilton, JR Freeland, B Okamura	32:159–81
Applied Evolution	JJ Bull, HA Wichman	32:183–217
Mistletoe—A Keystone Resource in Forests and Woodlands Worldwide	DM Watson	32:219–49
The Role of Disturbance in the Ecology and Conservation of Birds	JD Brawn, SK Robinson, FR Thompson III	32:251–76
Approaches to the Study of Territory Size and Shape	ES Adams	32:277–303
The Population Biology of Invasive Species	AK Sakai, FW Allendorf, JS Holt, DM Lodge, J Molofsky, KA With, S Baughman, RJ Cabin, JE Cohen, NC Ellstrand, DE McCauley, P O'Neil, IM Parker, JN Thompson, SG Weller	32:305–32
Streams in the Urban Landscape	MJ Paul, JL Meyer	32:333–65
Integrating Function and Ecology in Studies of Adaptation: Investigations of Locomotor Capacity as a Model System	DJ Irschick, T Garland Jr.	32:367–96
The Southern Conifer Family Araucariaceae: History, Status, and Value for Palaeoenvironmental Reconstruction	P Kershaw, B Wagstaff	32:397–414

The Units of Selection on Mitochondrial DNA	DM Rand	32:415–48
Evolutionary Patterns Among Permo-Triassic Therapsids	BS Rubidge, CA Sidor	32:449–80
Ecology, Conservation, and Public Policy	D Ludwig, M Mangel, BM Haddad	32:481–517
Male-Killing, Nematode Infections, Bacteriophase Infection, and Virulence of Cytoplasmic Bacteria in the Genus Wolbachia	L Stevens, R Giordano, RF Fialho	32:519–45
Biospheric Trace Gas Fluxes and Their Control Over Tropospheric Chemistry	RK Monson, EA Holland	32:547–76

Volume 33 (2002)

Saproxylic Insect Ecology and the Sustainable Management of Forests	SJ Grove	33:1–23
Conus Venom Peptides: Reflections from the Biology of Clades and Species	BM Olivera	33:25–47
Troubleshooting Molecular Phylogenetic Analyses	MJ Sanderson, HB Shaffer	33:49–72
The Early Radiations of Cetacea (Mammalia): Evolutionary Pattern and Developmental Correlations	JGM Thewissen, EM Williams	33:73–90
The Mesozoic Radiation of Birds	LM Chiappe, GJ Dyke	33:91–124
Plant Ecological Strategies: Some Leading Dimensions of Variation Between Species	M Westoby, DS Falster, AT Moles, PA Vesk, IJ Wright	33:125–59
Reproductive Protein Evolution	WJ Swanson, VD Vacquier	33:161–79
The Causes and Consequences of Ant Invasions	DA Holway, L Lach, AV Suarez, ND Tsutsui, TJ Case	33:181–233

Gulf of Mexico Hypoxia, a.k.a. "The Dead Zone"	NN Rabalais, RE Turner, WJ Wiseman Jr	33:235–63
The (Super)Tree of Life: Procedures, Problems, and Prospects	ORP Bininda-Emonds, JL Gittleman, MA Steel	33:265–89
Homogenization of Freshwater Faunas	FJ Rahel	33:291–315
The Renaissance of Community-Based Marine Resource Management in Oceania	RE Johannes	33:317–40
The Role of Animals in Nutrient Cycling in Freshwater Ecosystems	MJ Vanni	33:341–70
Effects of UV-B Radiation on Terrestrial and Aquatic Primary Producers	TA Day, PJ Neale	33:371–96
The Evolution and Maintenance of Androdioecy	JR Pannell	33:397–425
Mast Seeding in Perennial Plants: Why, How, Where?	D Kelly, VL Sork	33:427–47
Disturbance to Marine Benthic Habitats by Trawling and Dredging–Implications for Marine Biodiversity	Thrush, PK Dayton	33:449–73
Phylogenies and Community Ecology	CO Webb, DD Ackerly, MA McPeek, MJ Donoghue	33:475–505
Stable Isotopes in Plant Ecology	TE Dawson, S Mambelli, AH Plamboeck, PH Templer, KP Tu	33:507–59
The Quality of the Fossil Record: Implications for Evolutionary Analyses	SM Kidwell, SM Holland	33:561–88
Neopolyploidy in Flowering Plants	DW Schemske, J Ramsey	33:589–639
Herbivore Offense	R Karban, AA Agrawal	33:641–64
The Pacific Salmon Wars: What Science Brings to the Challenge of Recovering Species	M Ruckelshaus, P Levin, JB Johnson, PM Kareiva	33:665–706

Estimating Divergence Times from Molecular Data on Phylogenetic and Population Genetic Timescales	BS Arbogast, SV Edwards, J Wakeley, P Beerli, JB Slowinski	33:707–40
The Fate of Clades in a World of Recurrent Climatic Change—Milankovitch Oscillations and Evolution	R Jansson, M Dynesius	33:741–78

Volume 34 (2003)

Effects of Introduced Bees on Native Ecosystems	D Goulson	34:1–26
Avian Sexual Dichromatism in Relation to Phylogeny and Ecology	AV Badyaev, GE Hill	34:27–49
Paleobiogeography: The Relevance of Fossils to Biogeography	BS Lieberman	34:51–69
The Ecology of Bird Introductions	RP Duncan, TM Blackburn, D Sol	34:71–98
The Effects of Genetic and Geographic Structure on Neutral Variation	B Charlesworth, D Charlesworth, NH Barton	34:99–125
Data, Models, and Decisions in U.S. Marine Fisheries Management: Lessons for Ecologists	KA Rose, JH Cowan Jr.	34:127–51
Partitioning of Time as an Ecological Resource	N Kronfeld-Schor, T Dayan	34:153–81
Performance Comparisons of Co-Occuring Native and Alien Invasive Plants: Implications for Conservation and Restoration	CC Daehler	34:183–211
Genetic Variation in Rare and Common Plants	CT Cole	34:213–37
The Ecology and Evolution of Insect Baculoviruses	JS Cory, JH Myers	34:239–72
Latitudinal Gradients of Biodiversity: Pattern, Process, Scale, and Synthesis	MR Willig, DM Kaufman, RD Stevens	34:273–309
Recent Advances in the (Molecular) Phylogeny of Vertebrates	A Meyer, R Zardoya	34:311–38

The Role of Reinforcement in Speciation: Theory and Data	MR Servedio, MAF Noor	34:339–64
Extra-Pair Paternity in Birds: Causes, Correlates, and Conflict	DF Westneat, IRK Stewart	34:365–96
Species-Level Paraphyly and Polyphyly: Frequency, Causes, and Consequences, with Insights from Animal Mitochondrial DNA	DJ Funk, KE Omland	34:397–423
Protective Ant-Plant Interactions as Model Systems in Ecological and Evolutionary Research	M Heil, D McKey	34:425–53
Functional Matrix: A Conceptual Framework for Predicting Plant Effects on Ecosystem Processes	VT Evinerm, FS Chapin III	34:455–85
Effects of Habitat Fragmentation on Biodiversity	L Fahrig	34:487–515
Social Organization and Disease Risk in Mammals: Integrating Theory and Empirical Studies	S Altizer, CL Nunn, PH Thrall, JL Gittleman, J Antonovics, AA Cunningham, AP Dodson, V Ezenwa, KE Jones, AB Pedersen, M Poss, JRC Pulliam	34:517–47
The Community-Level Consequences of Seed Dispersal Patterns	JM Levine, DJ Murrell	34:549–74
The Ecology and Evolution of Seed Dispersal: A Theoretical Perspective	SA Levin, HC Muller-Landau, R Nathan, J Chave	34:575–604
Analysis of Rates of Morphologic Evolution	PD Roopnarine	34:605–32
Development and the Genetics of Evolutionary Change Within Insect Species	PM Brakefield, V French, BJ Zwaan	34:633–60

Flexibility and Specificity in Coral-Algal Symbiosis: Diversity, Ecology, and Biogeography of *Symbiodinium*	AC Baker	34:661–89

Volume 35 (2004)

Vertebrate Dispersal of Seed Plants Through Time	BH Tiffney	35:1–29
Are Diseases Increasing in the Ocean?	KD Lafferty, JW Porter, SE Ford	35:31–54
Bird Song: The Interface of Evolution and Mechanism	J Podos, SK Huber, B Taft	35:55–87
Application of Ecological Indicators	GJ Niemi, ME McDonald	35:89–111
Ecological Impacts of Deer Overabundance	SD Côté, TP Rooney, J-P Tremblay, C Dussault, DM Waller	35:113–47
Ecological Effects of Transgenic Crops and the Escape of Transgenes into Wild Populations	D Pilson, HR Prendeville	35:149–74
Mutualisms and Aquatic Community Structure: The Enemy of My Enemy Is My Friend	ME Hay, JD Parker, DE Burkepile, CC Caudill, AE Wilson, ZP Hallinan, AD Chequer	35:175–97
Operational Criteria for Delimiting Species	JW Sites Jr, JC Marshall	35:199–227
The New View of Animal Phylogeny	KM Halanych	35:229–56
Landscapes and Riverscapes: The Influence of Land Use on Stream Ecosystems	JD Allan	35:257–84
Long-Term Stasis in Ecological Assemblages: Evidence from the Fossil Record	WA DiMichele, AK Behrensmeyer, TD Olszewski, CC Labandeira, JM Pandolfi, SL Wing, R Bobe	35:285–322

Avian Extinctions from Tropical and Subtropical Forests	NS Sodhi, LH Liow, FA Bazzaz	35:323–45
Evolutionary Biology of Animal Cognition	R Dukas	35:347–74
Pollination Syndromes and Floral Specialization	CB Fenster, WS Armbruster, P Wilson, MR Dudash, JD Thomson	35:375–403
On the Ecological Roles of Salamanders	RD Davic, HH Welsh Jr	35:405–34
Ecological and Evolutionary Consequences of Multispecies Plant-Animal Interactions	SY Strauss, RE Irwin	35:435–66
Spatial Synchrony in Population Dynamics	A Liebhold, WD Koenig, ON Bjørnstad	35:467–90
Ecological Responses to Habitat Edges: Mechanisms, Models, and Variability Explained	L Ries, RJ Fletcher Jr, J Battin, TD Sisk	35:491–522
Evolutionary Trajectories and Biogeochemical Impacts of Marine Eukaryotic Phytoplankton	ME Katz, ZV Finkel, D Grzebyk, AH Knoll, PG Falkowski	35:523–56
Regime Shifts, Resilience, and Biodiversity in Ecosystem Management	C Folke, S Carpenter, B Walker, M Scheffer, T Elmqvist, L Gunderson, CS Holling	35:557–81
Ecology of Woodland Herbs in Temperate Deciduous Forests	DF Whigham	35:583–621
The Southwest Australian Floristic Region: Evolution and Conservation of a Global Hot Spot of Biodiversity	SD Hopper, P Gioia	35:623–50
Predator-Induced Phenotypic Plasticity in Organisms with Complex Life Histories	MF Benard	35:651–73
The Evolutionary Ecology of Novel Plant-Pathogen Interactions	IM Parker, GS Gilbert	35:675–700

ANNUAL REVIEWS
Intelligent Synthesis of the Scientific Literature

Annual Reviews – Your Starting Point for Research Online
http://arjournals.annualreviews.org

- Over 900 Annual Reviews volumes—more than 25,000 critical, authoritative review articles in 31 disciplines spanning the Biomedical, Physical, and Social sciences—available online, including all Annual Reviews back volumes, dating to 1932
- Current individual subscriptions include seamless online access to full-text articles, PDFs, Reviews in Advance (as much as 6 months ahead of print publication), bibliographies, and other supplementary material in the current volume and the prior 4 years' volumes
- All articles are fully supplemented, searchable, and downloadable—see http://ecolsys.annualreviews.org
- Access links to the reviewed references (when available online)
- Site features include customized alerting services, citation tracking, and saved searches

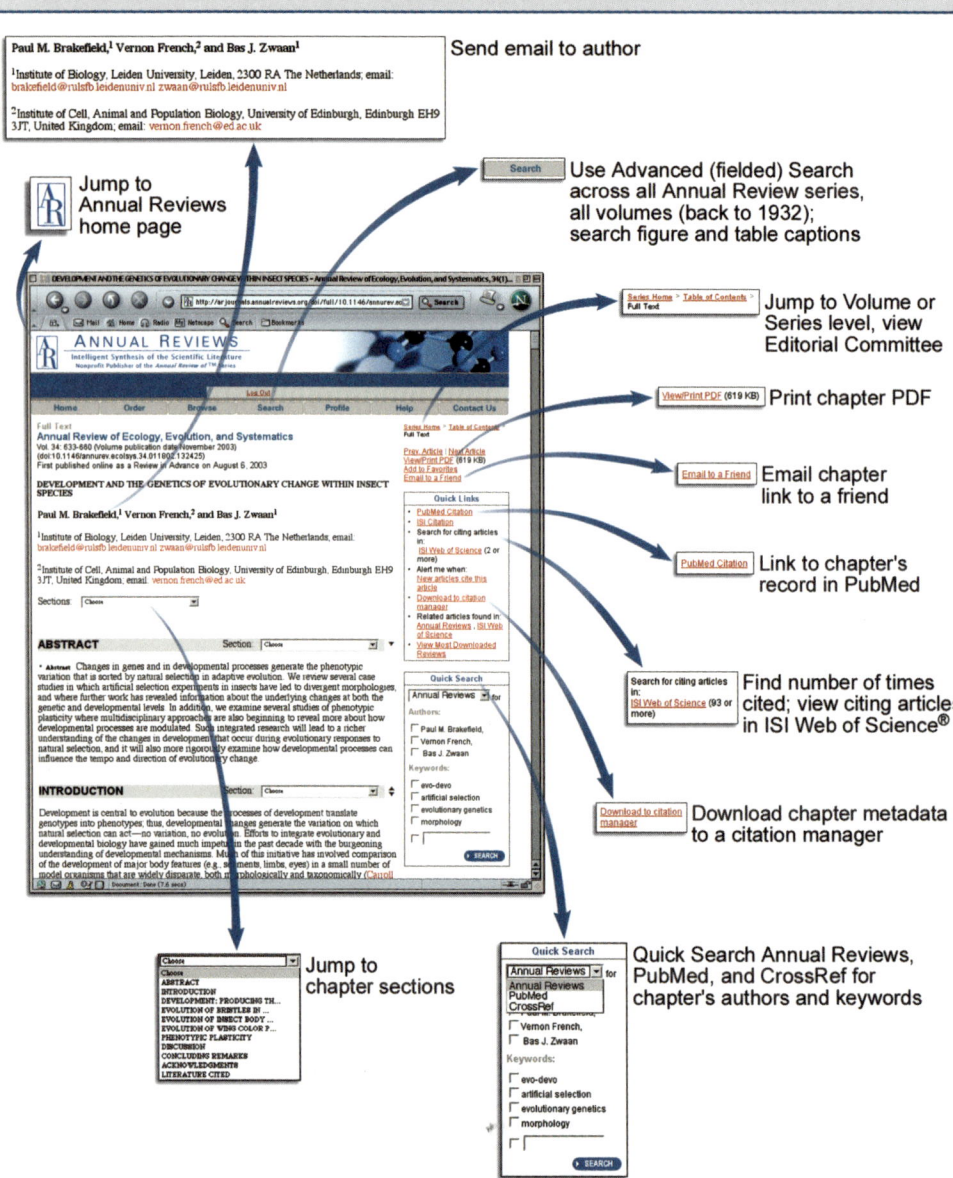

Copyright © 2004 Annual Reviews, Nonprofit Publisher of the *Annual Review of* ™ Series